AVID PERUGIM

INTRODUCTION TO ELECTRIC CIRCUITS

SIXTH EDITION

INTRODUCTION TO ELECTRIC CIRCUITS

Herbert W. Jackson

Prentice-Hall, Englewood Cliffs, New Jersey 07632

Library of Congress Cataloging-in-Publication Data

JACKSON, HERBERT W.
    Introduction to electric circuits.

    Includes index.
    1. Electric circuits.   I. Title.
TK454.J28 1986   621.319'2   85–24422
ISBN   0–13–481425–8

Editorial/production supervision: Tom Aloisi
Interior design: Anne T. Bonanno
Cover design: Anne T. Bonanno
Manufacturing buyer: Gordon Osbourne
Page layout: Meg Van Arsdale
Cover photo courtesy Computervision Corporation

About the cover: The computer-generated cover photo of a
multilayer circuit design is printed by courtesy of Computervision Corporation. Note that each layer is color coded,
which enables designers to follow the circuit logic more
easily.

CONTENTS

PREFACE   xiii

PART ONE THE BASIC ELECTRIC CIRCUIT   1

1 INTRODUCTION   2

1-1 Circuit Diagrams 2
1-2 The International System of Units 3
1-3 Electronic Calculators 6
1-4 Numerical Accuracy 7

1-5 Scientific Notation 7
1-6 SI Unit Prefixes 9
1-7 Conversion of Units 12
1-8 Personal Computers 13

2 CURRENT AND VOLTAGE   19

2-1 The Nature of Electricity 19
2-2 The Atom 22
2-3 Combination of Atoms 25
2-4 Electric Current 26
2-5 The Coulomb 29
2-6 The Ampere 30

2-7 Potential Difference 32
2-8 Generating a Potential Difference 34
2-9 The Volt 36
2-10 EMF, Potential Difference, and Voltage 37
2-11 Conventional Current Direction 39

3 CONDUCTORS AND INSULATORS, SEMICONDUCTORS, AND VOLTAGE SOURCES   43

3-1 Conductors 43
3-2 Insulators 45
3-3 Energy-Band Diagrams 46
3-4 Insulator Breakdown 48
3-5 Electrolytic Conduction 49

3-6 Conduction in Vacuum and Gases 51
3-7 Semiconductors 52
3-8 pn-Junctions 57
3-9 Batteries 58
3-10 Other Voltage Sources 64

4 RESISTANCE   68

4-1 Ohm's Law of Constant Proportionality 68
4-2 The Nature of Resistance 71

4-3 Factors Governing the Resistance of Metallic Conductors 72
4-4 Resistivity 73

4-5 Effect of Temperature on Resistance 76

4-6 Temperature Coefficient of Resistance 79

4-7 Linear Resistors 83

4-8 Nonlinear Resistors 85

4-9 Volt-Ampere Characteristics 88

4-10 Ohm's Law Applied 89

5 WORK AND POWER 95

5-1 Energy and Work 95

5-2 Power 97

5-3 Efficiency 100

5-4 The Kilowatthour 101

5-5 Interrelationship of Basic Electrical Units 103

PART TWO RESISTANCE NETWORKS 109

6 SERIES AND PARALLEL CIRCUITS 110

6-1 Resistors in Series 110

6-2 Polarity of Voltage Drop in a Series Circuit 112

6-3 Double-Subscript Notation 114

6-4 Kirchhoff's Voltage Law 115

6-5 Characteristics of Series Circuits 116

6-6 Internal Resistance 118

6-7 Maximum Power Transfer 120

6-8 Resistors in Parallel 124

6-9 Kirchhoff's Current Law 125

6-10 Conductance 126

6-11 Conductivity 129

6-12 Characteristics of Parallel Circuits 129

7 SERIES-PARALLEL CIRCUITS 135

7-1 Series-Parallel Resistors 135

7-2 Equivalent-Circuit Method 136

7-3 Kirchhoff's-Laws Method 140

7-4 Voltage-Divider Principle 142

7-5 Voltage Dividers 144

7-6 Current-Divider Principle 149

8 RESISTANCE NETWORKS 159

8-1 Network Equations from Kirchhoff's Laws 159

8-2 Constant-Voltage Sources 160

8-3 Constant-Current Sources 162

8-4 Source Conversion 164

8-5 Kirchhoff's Voltage-Law Equations: Loop Procedure 166

8-6 Networks with More Than One Voltage Source 173

8-7 Loop Equations in Multisource Networks 174

8-8 Mesh Equations 181
8-9 Kirchhoff's Current-Law Equations 185
8-10 Nodal Analysis 188
8-11 Superposition Theorem 194

9 EQUIVALENT-CIRCUIT THEOREMS 206

9-1 Equivalent Circuits 206
9-2 Thévenin's Theorem 207
9-3 Norton's Theorem 215
9-4 Thévenin's Theorem in Multisource Networks (Millman's Theorem) 218
9-5 Dependent Sources 223
9-6 Delta-Wye Transformation 230
9-7 Substitution Theorem 234
9-8 Reciprocity Theorem 235

10 ELECTRICAL MEASUREMENT 242

10-1 Mechanical Torque from Electric Current 242
10-2 The Ammeter 246
10-3 The Voltmeter 250
10-4 Voltmeter Loading Effect 252
10-5 The Potentiometer 254
10-6 Resistance Measurement 256
10-7 Moving-Coil Movement in AC Circuits 261
10-8 Moving-Iron Movement 263
10-9 Electro-Dynamometer Movement 264
10-10 Digital Meters 266

PART THREE CAPACITANCE AND INDUCTANCE 273

11 CAPACITANCE 274

11-1 Static Electricity 274
11-2 The Nature of an Electric Field 275
11-3 Electrostatic Induction 279
11-4 Dielectrics 280
11-5 Capacitors 282
11-6 Capacitance 284
11-7 Factors Governing Capacitance 286
11-8 Dielectric Constant 289
11-9 Capacitors in Parallel 290
11-10 Capacitors in Series 291

12 CAPACITANCE IN DC CIRCUITS 298

12-1 Charging a Capacitor 298
12-2 Rate of Change of Voltage 300
12-3 Time Constant 303
12-4 Graphical Solution for Instantaneous Potential Difference 304
12-5 Discharging a Capacitor 306

12-6 *CR* Waveshaping Circuits 309
12-7 Calculator Solution for Instantaneous Potential Difference 313

12-8 Transient Response 318
12-9 Energy Stored by a Capacitor 322
12-10 Characteristics of Capacitive DC Circuits 324

13 MAGNETISM 333

13-1 Electricity and Magnetism 333
13-2 The Nature of a Magnetic Field 334
13-3 Characteristics of Magnetic Lines of Force 336
13-4 Magnetic Field Around a Current-Carrying Conductor 339
13-5 Magnetic Flux 343
13-6 Magnetomotive Force 345
13-7 Reluctance 345
13-8 Permeability 346
13-9 Flux Density 347

13-10 Magnetic Field Instensity 348
13-11 Diamagnetic and Paramagnetic Materials 349
13-12 Ferromagnetic Materials 351
13-13 Permanent Magnets 352
13-14 Magnetization Curves 353
13-15 Permeability from the *BH* Curve 355
13-16 Hysteresis 357
13-17 Eddy Current 359
13-18 Magnetic Shielding 360

14 MAGNETIC CIRCUITS 363

14-1 Practical Magnetic Circuits 363
14-2 Long Air-Core Coils 364
14-3 Simple Magnetic Circuit 366
14-4 Linear Magnetic Circuits 367
14-5 Nonlinear Magnetic Circuits 368
14-6 Leakage Flux 370

14-7 Series Magnetic Circuits 371
14-8 Air Gaps 374
14-9 Parallel Magnetic Circuits 377
14-10 Tractive Force of an Electromagnet 379

15 INDUCTANCE 383

15-1 Electromagnetic Induction 383
15-2 Faraday's Law 385
15-3 Lenz's Law 386
15-4 Self-Induction 388

15-5 Self-Inductance 389
15-6 Factors Governing Inductance 391
15-7 Inductors in Series 392
15-8 Inductors in Parallel 393

16 INDUCTANCE IN DC CIRCUITS 396

16-1 Current in an Ideal Inductor 396
16-2 Rise of Current in a Practical Inductor 398
16-3 Time Constant 401

16-4 Graphical Solution for Instantaneous Current 403
16-5 Calculator Solution for Instantaneous Current 406

16-6 Energy Stored by an Inductor 409
16-7 Fall of Current in an Inductive Circuit 412
16-8 Calculator Solution for Instantaneous Discharge Current 417
16-9 Transient Response 417
16-10 Characteristics of Inductive DC Circuits 420

PART FOUR ALTERNATING CURRENT 425

17 ALTERNATING CURRENT 426

17-1 A Simple Rotating Generator 426
17-2 The Nature of the Induced Voltage 427
17-3 The Sine Wave 429
17-4 The Peak Value of a Sine Wave 432
17-5 The Instantaneous Value of a Sine Wave 433
17-6 The Radian 435
17-7 Instantaneous Current in a Resistor 437
17-8 Instantaneous Power in a Resistor 439
17-9 Periodic Waves 441
17-10 The Average Value of a Periodic Wave 442
17-11 The RMS Value of a Sine Wave 443

18 REACTANCE 449

18-1 The Nature of the Instantaneous Current in an Ideal Inductor 450
18-2 Inductive Reactance 452
18-3 Factors Governing Inductive Reactance 453
18-4 The Nature of the Instantaneous Current in a Capacitor 454
18-5 Capacitive Reactance 456
18-6 Factors Governing Capacitive Reactance 457
18-7 Resistance, Inductive Reactance, and Capacitive Reactance 459

19 PHASORS 462

19-1 Addition of Sine Waves 462
19-2 Addition of Instantaneous Values on a Linear Graph 464
19-3 Representing a Sine Wave by a Phasor Diagram 466
19-4 Letter Symbols for Phasor Quantities 467
19-5 Phasor Addition by Geometrical Construction 468
19-6 Addition of Phasors Which are at Right Angles 470
19-7 Expressing a Phasor Quantity in Rectangular Coordinates 472

19-8 Phasor Addition by Rectangular Coordinates 477

19-9 Subtraction of Phasor Quantities 479

19-10 Multiplication and Division of Phasor Quantities 481

20 IMPEDANCE 486

20-1 Resistance and Inductance in Series 486

20-2 Impedance 488

20-3 Practical Inductors 491

20-4 Resistance and Capacitance in Series 494

20-5 Resistance, Inductance, and Capacitance in Series 495

20-6 Resistance, Inductance, and Capacitance in Parallel 498

20-7 Conductance, Susceptance, and Admittance 500

20-8 Impedance and Admittance 503

21 POWER IN ALTERNATING-CURRENT CIRCUITS 511

21-1 Power in a Resistor 511

21-2 Power in an Ideal Inductor 513

21-3 Power in a Capacitor 516

21-4 Power in a Circuit Containing Resistance and Reactance 517

21-5 The Power Triangle 519

21-6 Power Factor 523

21-7 Power Factor Correction 526

21-8 Measuring Power in AC Circuits 532

21-9 Effective Resistance 534

21-10 The Decibel 536

PART FIVE IMPEDANCE NETWORKS 543

22 SERIES AND PARALLEL IMPEDANCES 544

22-1 Resistance and Impedance 544

22-2 Impedances in Series: Kirchhoff's Voltage Law 545

22-3 Impedances in Parallel: Kirchhoff's Current Law 548

22-4 Series-Parallel Impedances 555

22-5 Equivalent Circuits 557

22-6 Source Conversion 562

22-7 Maximum Power Transfer 564

22-8 Circle Diagrams 565

23 IMPEDANCE NETWORKS 573

23-1 Loop Equations 574

23-2 Mesh Equations 580

23-3 Superposition Theorem 582

23-4 Thévenin's Theorem 585

23-5 Norton's Theorem 590
23-6 Nodal Analysis 593

23-7 Delta-Wye Transformation 598
23-8 Alternating-Current Bridges 601

24 RESONANCE 613

24-1 Effect of Varying Frequency in a Series *RLC* Circiut 613
24-2 Series Resonance 617
24-3 *Q* of Resonant Circuits 619
24-4 Resonant Rise of Voltage 620
24-5 Selectivity 623
24-6 Theoretical Parallel-Resonant Circuit 625

24-7 Practical Parallel-Resonant Circuits 629
24-8 Selectivity of Parallel-Resonant Circuits 632
24-9 Resonant Filter Networks 634

25 TRANSFORMERS 640

25-1 Mutual Induction 640
25-2 Transformer Action 640
25-3 Transformation Ratio 643
25-4 Impedance Transformation 645
25-5 Leakage Reactance 647
25-6 Open-Circuit and Short-Circuit Tests 648

25-7 Efficiency 650
25-8 Effect of Loading a Transformer 651
25-9 Autotransformers 654
25-10 Audio Transformers 656

26 COUPLED CIRCUITS 660

26-1 Determining Coupling Network Parameters 660
26-2 Open-Circuit Impedance Parameters 662
26-3 Short-Circuit Admittance Parameters 668
26-4 Hybrid Parameters 671

26-5 Air-Core Transformers 676
26-6 Mutual Inductance 677
26-7 Coupled Impedance 680
26-8 Tuned Transformers 683

27 THREE-PHASE SYSTEMS 691

27-1 Advantages of Polyphase Systems 691
27-2 Generation of Three-Phase Voltages 695
27-3 Double-Subscript Notation 697
27-4 Four-Wire Wye-Connected System 699

27-5 Delta-Connected System 703
27-6 Wye-Delta System 708
27-7 Power in a Balanced Three-Phase System 711
27-8 Measuring Power in a Three-Phase System 712

27-9 Two-Wattmeter Measurement of Three-Phase Power 715
27-10 Phase Sequence 719
27-11 Unbalanced Three-Wire Wye Loads 723

28 HARMONICS 732

28-1 Nonsinusoidal Waves 732
28-2 Fourier Series 734
28-3 Addition of Harmonically Related Sine Waves 735
28-4 Generation of Harmonics 738
28-5 Generation of Harmonics in an Electronic Amplifier 740
28-6 Generation of Harmonics in an Iron-Core Transformer 742
28-7 RMS Value of a Nonsinusoidal Wave 743
28-8 Square Waves and Sawtooth Waves 744
28-9 Nonsinusoidal Waves in Linear Impedance Networks 746

APPENDICES 752

1 DETERMINANTS 752

2 CALCULUS SOLUTIONS 756

2-1 Maximum Power-Transfer Theorem 756
2-2 Instantaneous PD in a CR Circuit 756
2-3 Energy Stored by a Capacitor 758
2-4 Instantaneous Current in an LR Circuit 759
2-5 Energy Stored by an Inductor 760
2-6 Rms and Average Values of a Sine Wave 761
2-7 Inductive Reactance 762
2-8 Capacitive Reactance 763
2-9 General Transformer Equation 764
2-10 Maximum Transformer Efficiency 764

3 AMERICAN WIRE GAUGE TABLE 765

4 RESISTOR COLOR CODE 766

ANSWERS TO PROBLEMS 768

INDEX 773

PREFACE

As the title suggests, this book is intended as a textbook for the all-important first- and second- semester college-level introductory electric circuits course for electrical/electronics tecnician/technologist students. It assumes that the student has a working knowledge of algebra and elementary trigonometry. (For students with some calculus background, alternative calculus solutions for some examples are included in the Appendices.) Unlike some recent circuit-analysis textbooks, *Introduction to Electric Circuits* does not assume that the student already has a satisfactory understanding of the principles of electricity and magnetism through a separate physics course.

When the first edition was published a quarter of a century ago, emphasis in most technical institute and community college introductory electric circuits courses was on *understanding* circuit *operation*. Over the intervening years, the increasing sophistication of electronic devices and systems has required the emphasis to shift gradually toward circuit-*analysis* techniques. Since increasing course content is seldom accompanied by additional classroom time, it is my belief that an introductory circuits textbook must compensate for the diminishing classroom time available for reviewing fundamental principles of electric circuit behavior. I am convinced that a successful career as an engineering technician/technologist depends on a sound *understanding* of these fundamentals which provide the foundation for later applications courses. Consequently, over the six editions while material on circuit analysis has been increasing, the original concept of understanding fundamental principles has also been expanded and updated in such a manner that these sections can be assigned for self-study and review.

In preparing the manuscript for this sixth edition, I have attempted to respond to three significant factors in the recent evolution of the typical college-level introductory circuits course—increasing emphasis on circuit-analysis techniques, the pressure from concurrent systems courses to get to circuit analysis in the first semester, and increasing accessibility of personal computers. I have expanded the chapters on circuit-analysis techniques to include more examples and problems, and additional topics which are appearing in the introductory course such as dependent sources. Chapters 8 and 9 dealing with analysis of dc resistance networks have been rewritten so that they can directly follow series-parallel resistors fairly early in the first semester. At the same time, I have tried to package the course material into chapters in such a way that Chapters 8 and 9 can be readily postponed in course outlines which wish to proceed to basic alternating current before considering circuit-analysis techniques.

As the number of topics to be included in a typical first-year electric circuits course expands (and the number of pages in an accompanying textbook also increases), most contemporary textbooks will contain more sections than required for a specific course. So that students may know what is expected of them, I suggest that instructors provide a course outline keyed to the text showing what sections will be covered in what order, and which sections are the students' responsibility as review or supplementary material.

Development of personal computers is a mixed blessing for an introductory electric circuits course. First-year college students now have fairly ready access to computing facilities. Indeed, some students own their own personal computer and many know how to use computers from secondary school experience. Personally, I prefer an electronic calculator for the introductory course where the emphasis should be on learning *electric circuit* behavior. Care must be taken that the old "plugging-numbers-into-formulas" syndrome doesn't get a new lease on life as an exercise in feeding numbers into a computer in unthinking response to requests generated by software packages.

Nevertheless, graduates will eventually use computers as circuit design tools and need to become accustomed to their methods. Since both hardware and software are still developing rapidly at the moment, it is not feasible for a circuits textbook to venture into specific computer instruction. Instead, I have tried in this edition to illustrate at appropriate points in the course just how a personal computer can be an asset. For students who wish to experiment with a computer's role in the solution of circuit analysis examples, a computer assignment is included (marked by an asterisk) in the Review Questions at the end of most chapters.

To ensure that beginning students are not burdened with obsolescent terminology and jargon, all units, symbols, practical examples, and references in this edition have been updated to represent "state of the art" in the study of electric circuits. I have been guided in the selection of content and the order of presentation for this edition by the many suggestions from users of the preceding edition. My sincere appreciation goes to all who have taken the time to respond to requests for such comments. Once again, I hope this new edition will make life a little more enjoyable for both students and instructors.

H. W. JACKSON
Toronto, Canada

PART I

THE BASIC ELECTRIC CIRCUIT

ELECTRIC CIRCUIT THEORY consists of a few basic laws and a series of mathematical equations which state quite clearly the relationships that must hold true in all electric and electronic circuits. Consequently, much of an introductory electric circuits course deals with the solution of numerical examples. However, developing an understanding of why electric circuits must behave as they do involves much more than entering numbers into a computer program. To be able to apply basic electric circuit principles to more sophisticated electrical and electronic systems, we also need to form clear concepts of the physical basis for electric circuit theory.

The chapters of Part I are devoted to developing these basic concepts in a detailed, step-by-step manner. This material has been included primarily for students who do not have a sound knowledge of the principles of electricity from a prior course in physics.

1

INTRODUCTION

Our standard of living in today's technological society is very much dependent on an abundant, convenient, and economical supply of **energy** in its many forms and on the relative ease with which we can *convert* energy from one form to another. We need energy to heat (or cool) and light our homes, to drive our cars, to run the machinery of industry, and to operate the many household appliances which we take for granted.

Electricity is a form of energy which was no more than a scientific curiosity in the nineteenth century. But once human beings found a means of *converting* vast amounts of natural energy into an electrical form and then *converting* it at will into heat, light, mechanical, or other useful forms of energy, electricity became the very foundation of modern technology.

Even more significant is the ease and precision with which we can *control* electric energy. As a result, electric and electronic control systems have invaded every branch of engineering technology. To be able to work effectively with these systems, we must first investigate the basic principles which govern the behavior of electric circuits. Once we understand why electric circuits must behave as they do, we can then proceed to develop techniques for analyzing the performance of practical electric and electronic circuits as we may encounter them.

1-1 CIRCUIT DIAGRAMS

Two energy conversions take place in the basic **electric circuit** illustrated in Figure 1-1. Chemical energy stored in the battery is converted into electric energy. This energy is conveyed by a pair of copper electric **conductors** to the lamp, which, in turn,

2

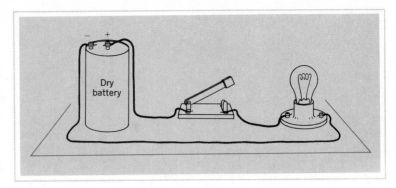

FIGURE 1-1
Pictorial Representation of a Basic Electric Circuit.

converts the electric energy into light (and some heat) energy. An electric **switch** in this circuit allows us to start and stop the energy conversion at will by interrupting the transmission of electric energy from the energy **source** (battery) to the **load**[†] (lamp).

Since we can show the exact interconnection of the various components of an electric circuit much more clearly with a sketch than with a description in words, we shall follow the customary practice of drawing a **circuit diagram** for each electric circuit we discuss. As we progress to more elaborate circuits, the pictorial representation of Figure 1-1 would have serious limitations. To keep circuit diagrams as simple and as clear as possible, we represent the various circuit elements by standard graphic symbols rather than by pictorial sketches. Although there have been some variations in graphic symbols in the past, those shown in Table 1-1 are the accepted standard in North America at present. As we encounter a symbol for the first time in a circuit diagram, we can refer back to Table 1-1 for its interpretation. The circuit diagram for the basic electric circuit of Figure 1-1 is shown in Figure 1-2.

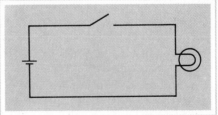

FIGURE 1-2
Circuit Diagram for a Basic Electric Circuit.

1-2 THE INTERNATIONAL SYSTEM OF UNITS

In 1883, Lord Kelvin observed, "I often say that when you can measure what you are speaking about, and express it in numbers, you know something about it; but when you cannot measure it, when you cannot express it in numbers, your knowledge is of a meagre and unsatisfactory kind." We can appreciate the significance of this remark

---

[†]In talking about electric circuits, we use the general term **load** for any device that receives electric energy and converts it to some other form of energy. In digital electronics terminology, we encounter the term **sink** to describe a device that *drains* energy from an electric circuit.

TABLE 1-1

GRAPHIC SYMBOLS USED IN THIS TEXT (IEEE STD 315-1975)

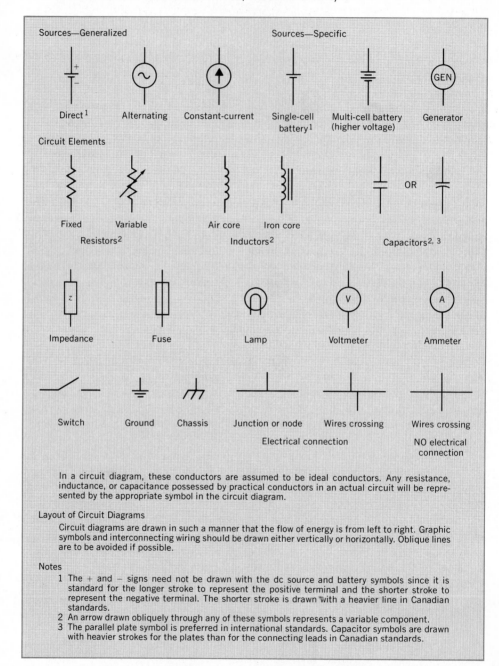

Sources—Generalized

Direct[1]   Alternating   Constant-current

Sources—Specific

Single-cell battery[1]   Multi-cell battery (higher voltage)   Generator

Circuit Elements

Fixed   Variable

Resistors[2]

Air core   Iron core

Inductors[2]

OR

Capacitors[2, 3]

Impedance   Fuse   Lamp   Voltmeter   Ammeter

Switch   Ground   Chassis   Junction or node   Wires crossing   Wires crossing

Electrical connection   NO electrical connection

In a circuit diagram, these conductors are assumed to be ideal conductors. Any resistance, inductance, or capacitance possessed by practical conductors in an actual circuit will be represented by the appropriate symbol in the circuit diagram.

Layout of Circuit Diagrams

Circuit diagrams are drawn in such a manner that the flow of energy is from left to right. Graphic symbols and interconnecting wiring should be drawn either vertically or horizontally. Oblique lines are to be avoided if possible.

Notes

1 The + and − signs need not be drawn with the dc source and battery symbols since it is standard for the longer stroke to represent the positive terminal and the shorter stroke to represent the negative terminal. The shorter stroke is drawn with a heavier line in Canadian standards.

2 An arrow drawn obliquely through any of these symbols represents a variable component.

3 The parallel plate symbol is preferred in international standards. Capacitor symbols are drawn with heavier strokes for the plates than for the connecting leads in Canadian standards.

when we realize that the scientists of Lord Kelvin's time had no knowledge of atomic structure and the free electron. To apply his philosophy to our study of electric circuits, we must have at our disposal a universally recognized system of measuring units.

The English-speaking world in Lord Kelvin's time still used an awkward and haphazard set of measuring units which included inches, feet, yards, rods, chains, fathoms, pints, bushels, teaspoons, cups, etc., ad infinitum. Most of the rest of the world had already switched to a simple decimal system of units which originated in France—the **metric** system. Almost a hundred years would elapse before North America finally began to adopt an updated version of the metric system for everyday measurement. However, in October 1965, the Institute of Electrical and Electronics Engineers adopted the **International System of Units** proposed at the eleventh General Conference on Weights and Measures held in France in 1960. The SI units[†] represent a rationalized selection of the MKS system of units which has been universally used by scientists for many years. Once the three basic MKS units—meter, kilogram, and second—are defined, we can *derive* all additional units that we might need in mechanics from these base units. By defining one additional base unit—the ampere—we can derive all the units required for electrical measurement as we proceed through this book.

The base unit of **length** or **distance** in the International System of Units is the **meter.** Originally calculated as one ten-millionth (1×10^{-7}) of the distance at sea level from the earth's equator to the pole, the meter is now more accurately defined in terms of the wavelength of the atomic radiation of krypton-86.

The original metric unit of mass—the *gram*—was related to the meter by defining the gram as the mass of one cubic centimeter of pure water at 4°C. The gram was too small a unit for practical purposes. Hence the SI base unit for **mass** is 1000 grams—the **kilogram.** The international standard for the kilogram is a platinum-iridium cylinder preserved by the International Bureau of Weights and Measures at Sévres in France.

The SI unit of **time** is the **second.** Originally defined as $1/86\,400$ of a mean solar day, we now define the second more accurately in terms of an "atomic clock" based on the atomic radiation of cesium. The second is not a true decimal unit, since the SI system still recognizes *minutes* and *hours* for longer periods of time.

The fourth base SI unit which will extend the system of units to electrical measurement is the **ampere.** In order to relate electrical measurement to mechanical measurement, the ampere is defined in terms of the electro-magnetic force between two current-carrying electric conductors. We shall define the ampere when we first encounter it in Section 2-6, but we leave details of the international standard ampere until Section 13-4.

As we derive additional units from these base SI units, we shall want to show the relationship among the units by mathematical equations. Hence we need letter symbols to represent both the quantities being measured and the SI units of measurement. Table 1-2 shows the letter symbols for those base SI units that we use in electricity and electronics.

[†]The abbreviation SI comes from the original French—Système Internationale d'Unités.

TABLE 1-2

LETTER SYMBOLS FOR SI BASE UNITS			
Quantity	Quantity Symbol	SI Base Unit	Unit Symbol
Length or distance	$\left.\begin{matrix} \ell \\ s \end{matrix}\right\}$	Meter	m
Mass	m	Kilogram	kg
Time	t	Second	s
Electric current	I	Ampere	A

Note the use of *italic* letters for quantity symbols and roman letters for unit symbols.

Occasionally, we may encounter an *abbreviation* of a unit name. For example, electricians often abbreviate *ampere* to *amp*. Such abbreviations are not used as substitutes for the designated **unit symbol** in equations.

EXAMPLE 1-1

When the switch is closed in the basic electric circuit of Figure 1-1, the electric current through the lamp is 0.25 ampere. Express this statement as a mathematical equation.

SOLUTION

$$I = 0.25 \text{ A}$$

1-3 ELECTRONIC CALCULATORS

Following Lord Kelvin's advice, we find that numerical examples using SI units of measurement provide the best means of understanding the behavior of electric and electronic circuits. Although the mathematics required can be kept reasonably simple, the calculations themselves can often be tedious. The examples throughout this book can all be solved by "longhand" techniques. However, the speed and accuracy of a hand-held electronic calculator are well worth the modest investment. For most examples in an introductory electric circuits course, the simplicity and portability of a hand-held calculator are preferable to the sophistication of a personal computer. Hence, solutions of numerical examples in this book assume that students are using a suitable pocket calculator. From time to time, the solutions include hints on calculator techniques.

You should choose a "scientific notation" calculator, preferably with an Eng key which will give answers in powers of 10 corresponding to SI unit prefixes. It does not have to have expensive "programmable" features; but make sure that it can solve exponentials e^x and convert rectangular to polar coordinates $\rightarrow$ **P,** and vice versa $\rightarrow$ **R.** Some so-called "scientific" calculators do not have these functions.

1-4 NUMERICAL ACCURACY

At first glance, it might seem that an electric current of 12 amperes is exactly the same as a current of 12.0 amperes. However, there is a subtle but significant difference between 12.0 and 12. A current stated as 12 amperes means that the exact value is closer to 12 than it is to either 11 or 13. Possibly the ammeter is calibrated in 1-ampere intervals so that it can be read only to the closest whole number of amperes. A current of 12.0 amperes is closer to 12.0 than it is to 11.9 or to 12.1; consequently, it must have been possible to read the ammeter in this case to the nearest tenth of an ampere. We say that the figure 12 has two significant digits, whereas the figure 12.0 has three significant digits.

An electronic calculator assumes that any numerical quantity we enter is precise, no matter how many digits it contains. Hence, if we enter 34 ÷ 2.3, the calculator will show 14.782608. But if the original quantities are accurate to only two digits, the string of digits displayed by the calculator can only tell us that the answer is closer to 15 than it is to 14. In performing chain calculations with a calculator, we can leave the strings of digits contained in the calculator until the final answer. However, in writing down the *final* answer, we should avoid trying to show more significant digits in the answer than the least accurate of any figures representing measured quantities used in the calculations. If we have to write down the calculator results for an intermediate step in the examples in this book, it is satisfactory to record *four* significant digits.

1-5 SCIENTIFIC NOTATION

Using the SI *base* units, we will often encounter some very large quantities—such as a radio frequency of 455 000 hertz—and some very small quantities—such as an inductance of 0.000 75 henry. Scientists use powers of 10 to avoid having to write down long strings of zeros.

Given the frequency of a radio signal as 455 000 hertz, if we shift the decimal point three places to the left, it is equivalent to *dividing* the original quantity by 1000. We can restore the original value by then *multiplying* by 1000 or 10^3. Hence,

$$f = 455\,000 \text{ Hz} = 455 \times 1000 \text{ Hz} = 455 \times 10^3 \text{ Hz}$$

Similarly, if we shift the decimal point *five* places to the left, multiplying by 10^5 restores the original magnitude,

$$f = 455\,000 \text{ Hz} = \mathbf{4.55 \times 10^5 \text{ Hz}}$$

Given the inductance of a radio tuning circuit as 0.000 75 henry, if we shift the decimal point four places to the right, it is equivalent to *multiplying* by 10 000. Consequently, to restore the original value, we must *divide* 7.5 by 10 000 or *multiply* by 10^{-4}. Hence,

$$L = 0.000\,75 \text{ H} = \frac{7.5}{10\,000} \text{H} = 7.5 \times 10^{-4} \text{ H}$$

To express a quantity in **scientific notation,** we shift the decimal point until there is *one* significant digit to the left of the decimal point, and then we multiply the result by the appropriate power of 10 to restore the quantity to its original value.

EXAMPLE 1-2

Express 839 000 meters in scientific notation.

SOLUTION

To obtain one digit to the left of the decimal point, we must shift the decimal point *five* places to the left. To restore the quantity to its original value, we must then multiply by 10^5.

$$\therefore 839\,000 \text{ m} = \mathbf{8.39 \times 10^5 \text{ m}}$$

EXAMPLE 1-3

Express 1/200 second in scientific notation.

SOLUTION

In a longhand solution, the first step is to divide the fraction out into decimal form.

$$\frac{1}{200}\text{s} = 0.005 \text{ s}$$

We now shift the decimal point three places to the right and multiply the result by 10^{-3}, obtaining

$$\frac{1}{200}\text{s} = \mathbf{5.0 \times 10^{-3} \text{ s}}$$

Most scientific-notation calculators automatically combine these two steps.

We can illustrate the advantage of scientific notation for calculator computation by the following example based on the equation for inductive reactance (which we shall derive in a later chapter).

EXAMPLE 1-4

Given $X_L = 2\pi f L$, $f = 455\,000$ Hz, and $L = 0.000\,75$ H, calculate X_L.

SOLUTION

Substituting the numerical values in the equation gives

$$X_L = 2 \times 3.1416 \times 455\,000 \times 0.000\,75 \; \Omega \; (\text{ohms}) \qquad (1\text{-}1)$$

If we enter the quantities in this form into a calculator, we may have trouble keeping track of the number of zeros. However, if we express f and L in scientific notation, we have

$$X_L = 2 \times \pi \times 4.55 \times 10^5 \times 7.5 \times 10^{-4} \ \Omega \qquad (1\text{-}2)$$

Using the scientific-notation feature of the calculator, we enter this equation using the $\boxed{\text{EE}}$ key[†] as follows:

$$X_L = 2 \ \boxed{\times} \ \boxed{\pi} \ \boxed{\times} \ 4.55 \ \boxed{\text{EE}} \ 5 \ \boxed{\times} \ 7.5 \ \boxed{\text{EE}} \ \boxed{+/-} \ 4 \ \boxed{=} \ 2144 \ \Omega \qquad (1\text{-}3)$$

Even in a longhand solution, scientific notation speeds up the computation since, to *multiply* powers of 10, we simply *add* their indices algebraically. Hence, Equation (1-2) becomes

$$X_L = 6.2832 \times 4.55 \times 7.5 \times 10 = 2144 \ \Omega \qquad (1\text{-}4)$$

1-6 SI UNIT PREFIXES

Engineering personnel prefer to use a variation of scientific notation to express numerical quantities clearly. We prefer to tack an appropriate *prefix* onto the *root* unit[‡] of measurement to represent the equivalent power of 10. To create decimal multiples or submultiples of root units, SI provides the standard unit prefixes listed in Table 1-3.

TABLE 1-3

SI UNIT PREFIXES		
Prefix	Factor by Which the Root Unit Is Multiplied	Letter Symbol
exa	$1\,000\,000\,000\,000\,000\,000 = 10^{18}$	E
peta	$1\,000\,000\,000\,000\,000 = 10^{15}$	P
tera	$1\,000\,000\,000\,000 = 10^{12}$	T
giga	$1\,000\,000\,000 = 10^{9}$	G
mega	$1\,000\,000 = 10^{6}$	M
kilo	$1\,000 = 10^{3}$	k
hecto	$100 = 10^{2}$	h
deka	10	da
deci	$0.1 = 10^{-1}$	d
centi	$0.01 = 10^{-2}$	c
milli	$0.001 = 10^{-3}$	m
micro	$0.000\,001 = 10^{-6}$	μ*
nano	$0.000\,000\,001 = 10^{-9}$	n
pico	$0.000\,000\,000\,001 = 10^{-12}$	p
femto	$0.000\,000\,000\,000\,001 = 10^{-15}$	f
atto	$0.000\,000\,000\,000\,000\,001 = 10^{-18}$	a

*The symbol for micro is the Greek letter μ (mu).

[†] EE key = Enter Exponent. Note that the calculator's permanent memory includes the numerical value for π so that the operator need not remember that $\pi = 3.1416$, as in the longhand solution (or with most personal computer programs).

[‡]*Root* units have no prefixes. In most cases in SI, root units are also base units. The exception is the base unit for mass—the kilogram—which historically already has the prefix *kilo* and is, therefore, not a root unit.

For engineering purposes, we prefer to shift the decimal point in multiples of *three*. Hence, the prefixes *hecto, deka, deci,* and *centi* are seldom used in working with electric circuits. The other unit prefixes shown in lightface type in Table 1-3 are used with quantities much larger or much smaller than we shall encounter in this book. We sometimes refer to quantities expressed with SI unit prefixes as **engineering notation.**

Shifting the decimal point three places at a time means that some numerical values will have more than one digit to the left of the decimal point. Consequently, when using prefixes with root units, SI recommends that we select prefixes that will permit the numerical values of the quantities to be between 0.1 and 1000.

EXAMPLE 1-2A

Express 839 000 meters in a more convenient size of unit.

SOLUTION

Shifting the decimal point *three* places to the left gives a figure of 839, which we must now multiply by 1000 or 10^3 to restore its proper magnitude. Hence,

$$839\,000 \text{ m} = 839 \times 10^3 \text{ m}$$

From Table 1-3, we can replace 10^3 by tacking the prefix *kilo* onto the root unit.

$$\therefore 839\,000 \text{ m} = 839 \times 10^3 \text{ m} = \textbf{839 km} \text{ (kilometers)}$$

The kilometer is 1000 times as large as the base unit, the meter.

EXAMPLE 1-3A

Express 1/200 second in a more appropriate form using SI unit prefixes.

SOLUTION

Again the first step is to divide the fraction out into decimal form.

$$\frac{1}{200} \text{ s} = 0.005 \text{ s}$$

If we shift the decimal point *three* places to the right to obtain the figure 5.0, we must multiply by 0.001 or 10^{-3} to restore the original magnitude. From Table 1-3,

$$\frac{1}{200} \text{ s} = 5.0 \times 10^{-3} \text{ s} = \textbf{5.0 ms} \text{ (milliseconds)}$$

When we make calculations involving quantities expressed in units with prefixes, we must remember that the prefixes represent powers of 10. These powers of 10 *must* be included in the numerical computation for the proper magnitude to be maintained in the answer. We can demonstrate this point with the following example. Using unit prefixes, Equation (1-2) in Example 1-4 becomes

$$X_L = 2\pi \times 455 \text{ kHz (kilohertz)} \times 750 \text{ } \mu\text{H (microhenrys)} \qquad (1\text{-}5)$$

EXAMPLE 1-4A
Given $X_L = 2\pi f L$, $f = 455$ kHz and $L = 750$ μH, calculate X_L.

INCORRECT SOLUTION
If we forget the prefixes and just carry out the arithmetic in Equation (1-5), we get

$$X_L = 2 \times \pi \times 455 \times 750 = 2\,144\,000$$

Comparing with the answer to Example 1-4, we note that the answer is *not* in ohms. (To be numerically correct, we must write the symbol for milliohms with the answer above.)

CORRECT SOLUTION
To make sure that we know that the numerical answer is in base units, we include the unit prefixes in the calculation by pressing the calculator's keys as follows:

$$X_L = 2 \times \boxed{\pi} \times 455 \,\boxed{\text{EE}}\, 3 \times 750 \,\boxed{\text{EE}}\,\boxed{+/-}\, 6 = 2144 \,\Omega$$

$$\qquad\qquad \underset{\text{kilo}}{\uparrow} \qquad\qquad \underset{\text{micro}}{\underbrace{\qquad}} \qquad \underset{\text{root unit}}{\uparrow}$$

Since it is not good form in SI to leave a numerical value greater than 1000, we now make the conversion

$$2144 \,\Omega = 2.144 \times 10^3 \,\Omega = \mathbf{2.144 \; k\Omega \; (kilohms)}$$

EXAMPLE 1-5
The calculator display shows a readout of 7.5×10^{-4} henry. Express this answer in a more suitable dimension.

SOLUTION
Table 1-3 has no unit prefix for 10^{-4}. Hence, we must start by simultaneously shifting the decimal point and changing the power of 10 (exponent) until we reach an exponent which is a + or − multiple of 3 and a numerical quantity (mantissa) between 0.1 and 1000. In this particular example, we can shift in either direction. We can *lower* the exponent from −4 to −6 and compensate by shifting the decimal point two places to the *right*, or we can *raise* the exponent from −4 to −3 and shift the decimal point of the mantissa one place to the *left*. Hence,

$$7.5 \times 10^{-4} \text{ H} = 750 \times 10^{-6} \text{ H} = 0.75 \times 10^{-3} \text{ H}$$

from which

$$7.5 \times 10^{-4} \text{ H} = \mathbf{750 \; \mu H \; or \; 0.75 \; mH}$$

Engineering calculators have an $\boxed{\text{Eng}}$ key which automatically converts a scientific notation answer to display an exponent which is a + or − multiple of 3 and a mantissa between 0.1 and 1000. Other scientific calculators have $\boxed{\text{EE}\uparrow}$ and $\boxed{\text{EE}\downarrow}$

keys so that we can shift both the decimal point and the exponent simultaneously in the proper direction one place at a time. In this example we can press the $\boxed{\text{EE}\downarrow}$ key *twice* to lower the exponent from -4 to -6, or we can press the $\boxed{\text{EE}\uparrow}$ key once to raise the exponent from -4 to -3.

1-7 CONVERSION OF UNITS

When only a single conversion from a root unit to a prefixed unit is involved, as in the last step of Example 1-4A, we can usually convert the calculator readout mentally from root units to more appropriate multiple or submultiple units. But some chain calculations may require several conversions before the final numerical answer is displayed. To assist us in determining the correct unit for the final answer, we can resort to an orderly procedure based on **dimensional analysis.** With this technique, we write the units into an algebraic equation along with the numerical quantities. We cancel out the same units in the numerator and denominator and then collect the remaining units to establish the proper units (or *dimension*) of the answer.

EXAMPLE 1-6
Convert $\frac{1}{4}$ hour to seconds.

SOLUTION
We require two conversion factors—hours to minutes and minutes to seconds. We can write the first as

$$60 \text{ minutes} = 1 \text{ hour}$$

We can rearrange the form of this conversion factor by dividing both sides of the equation by 1 hour:

$$\frac{60 \text{ min}}{1 \text{ h}} = \frac{1\cancel{h}}{1\cancel{h}} = 1$$

Thus when we express a conversion factor as a ratio, we obtain a dimensionless number equal to unity. We can, therefore, multiply any quantity by 60 min/1 h, or its reciprocal, without altering the value of the original quantity. Hence,

$$\frac{1}{4} \text{ h} = \frac{1}{4} \cancel{h} \times \frac{60 \cancel{\text{min}}}{1 \cancel{h}} \times \frac{60 \text{ s}}{1 \cancel{\text{min}}} = 900 \text{ s}$$

We can also use the dimensional analysis technique to convert form one *system* of units into another. Since electrical engineering in North America has already adopted the International System of Units (SI), all calculations in this book will be made in SI units. But for some time to come, we shall encounter quantities in the now obsolescent foot-pound system of units. To make the necessary conversion into SI units for examples in this book, we can probably get by with the two conversion

TABLE 1-4

CONVERSION FACTORS

1 inch = 2.54 centimeters
1 pound = 0.4536 kilogram

factors of Table 1-4. The conversion from inches to centimeters is exact. The conversion from pounds to kilograms is an approximation accurate to four digits.

EXAMPLE 1-7

An electric conductor is 25 ft long. Express its length in SI units.

SOLUTION

$$25 \text{ ft} = 25 \text{ ft} \times \frac{12 \text{ in}}{1 \text{ ft}} \times \frac{2.54 \text{ cm}}{1 \text{ in}} \times \frac{1 \text{ m}}{10^2 \text{ cm}} = 7.62 \text{ m}$$

1-8 PERSONAL COMPUTERS

At the beginning of the 1970s, educational computers were generally limited to costly "mainframe" computers located in college computing centers. A few students in advanced electrical/electronics technology courses had access to such terminals, but computers were not usually available to students in introductory electric circuits courses. By the beginning of the 1980s, rapid advances in VLSI[†] technology made it possible to place thousands of electronic circuits on a silicon chip no larger than a postage stamp. Mass production of **microprocessor** chips, in turn, made it possible to manufacture **microcomputers** with considerable memory and data processing capacity in a package the size of a portable typewriter and at a cost of only a few hundred dollars.

The relatively low cost of the current generation of microcomputers makes it possible for many students to own their own **personal computers.** For the same reason, electrical/electronics technology departments can now provide all students with access to personal computers to assist in electric circuit computations. Since many college students have learned to use such computers in the public school system, we must now consider what place personal computers have in the study of basic electric circuit theory.

The main difference between an electronic calculator and a personal computer is that the computer has a built-in *main memory* that can store many thousands of bits of data and processing instructions. Since each bit of data goes to a specific *address* in the computer's *random-access memory* (RAM), the *control unit* of the computer can locate and process any particular data in microseconds. Most calculators have only a limited memory for temporarily storing intermediate steps in a long chain calcu-

[†]Very-large-scale integration.

lation. Scientific calculators have considerable mathematical data built into a *read-only memory* (ROM). Such data are retained even when the calculator is switched off. They include such data as a numerical value for the constant π, trigonometry and exponential tables, and instructions to make the calculator carry out the appropriate computation when function keys such as $+$, $\div$, and x^2 are pressed.

When we use an electronic calculator, the actual sequence of numerical computations takes place directly as we press each mathematical function key in the proper order, and we see the result of each calculation almost instantly on the display. This direct involvement in the calculation chain is desirable in studying basic electric circuits, since it requires us to think our way through the steps in setting up each numerical example.

Personal computers, on the other hand, use their main memories to store temporarily a *program* for carrying out a series of computations automatically in a predetermined sequence. If the calculation is fairly simple, we usually type the program directly into the computer's input keyboard. For a complex computation or for a special display format, we *load* a prepared program into the computer's main memory from a *software* package consisting of a *permanent memory* file recorded on a magnetic tape cassette or flexible magnetic diskette. The advantage of the personal computer in electric circuit computation is that once a program has been prepared properly and checked, errors are limited to *inputting* incorrect data in response to the computer's requests. The disadvantage is that our attention is focused on the computer's requests for data input, rather than on the electric circuit principles being illustrated by the numerical example. Hence, while we are learning basic electric circuit principles, we should stick to hand-held calculators for simple numerical examples and save the personal computer for more tedious, repetitive calculations.

When a computer is switched off, it loses all the data temporarily stored in its main memory. When it is next switched on, the computer has to reload its main memory with sufficient data and instructions to recognize what we enter via the keyboard and how we want the data to be processed and displayed. For the calculations we encounter in an electric circuits course, personal computers are loaded with one of the variations of BASIC[†] computer language. In spite of the many similarities in the increasing selection of personal computers, each model has its own peculiarities and variations in program format. Hence, it is not feasible for an electric circuits textbook to provide specific computer programming instruction. To duplicate the computer examples on subsequent pages on your computer, you need to be quite familiar with BASIC programming for your particular model. You may also have to modify some of the *keywords* and program formats in the examples.

The computer examples in this book are intended to illustrate the differences between calculator and computer routines for solving various electric circuit examples. In this way, we not only gain some familiarity with personal computers, but also develop some appreciation of where the computer can be more effective than an electronic calculator, and vice versa. We start by considering how we might use a personal computer to solve Example 1-4.

[†]Beginner's All-Purpose Symbolic Instruction Code.

EXAMPLE 1-4B

Given $X_L = 2\pi fL$, $f = 455$ kHz, and $L = 750$ μH, calculate X_L.

SOLUTION 1

With BASIC loaded into the computer, we can use the computer as a simple electronic calculator by omitting the *line numbers* that appear at the beginning of each *statement* on the display. We type into the keyboard

```
PRINT 2*3.1416*455E3*750E-6
```

This statement covers all the numerical data and mathematical operators required to solve the example. We now press the ENTER key and the computer promptly displays **2144.142**

Each statement starts with a *keyword* which the computer recognizes from a selected BASIC vocabulary. Since some computer keyboards do not have an $\times$ key (like calculators), BASIC uses an asterisk $*$ to represent multiplication. For the same reason, BASIC uses a solidus slash / to represent division. Note that BASIC does not provide a numerical value for the constant π.

SOLUTION 2

The *direct-mode* operation of the computer in Solution 1 is an inefficient use of its capabilities. To use the computer's memory, we need a *program* to tell the computer what to do, step by step. The minimum program requires three steps (plus a statement to indicate to the computer that the program is complete). The three steps are (1) input of data, (2) processing the data, and (3) output of results. This minimum program is displayed as follows:

```
10 INPUT F,L
20 LET X=2*3.1416*F*L
30 PRINT X
40 END
```

To start this program, we type RUN and press the ENTER key. The computer displays a ? asking us to enter values for F and L. Once we type and enter the given values, the computer will promptly print the answer **2144.142**

SOLUTION 3

In practice, we would not write such a "bare-bones" program as Solution 2. With nothing but ? showing on the display, we may not remember what we are supposed to input. A more typical program might take the following format:

```
10 REM CALCULATING INDUCTIVE REACTANCE
20 INPUT "ENTER F,L ";F,L
30 LET X=2*3.1416*F*L
40 PRINT "INDUCTIVE REACTANCE= ";X;" OHMS"
50 END
```

When we RUN this program, it starts off by displaying the identifying caption for the program—CALCULATING INDUCTIVE REACTANCE. The next line displays ENTER F,L followed by a ? symbol. The operator responds by typing in the variables in the order requested, 455E3, 750E-6. The computer will display the answer for X, but it is now part of the statement in quotation marks in line 40:

```
INDUCTIVE REACTANCE=2144.142 OHMS
```

PROBLEMS

1-1. Express each of the following quantities in scientific notation.
   (a) 455   (b) 59 000
   (c) 10 000   (d) 0.0765
   (e) 0.000 37   (f) 0.547

1-2. Express each of the following scientific-notation quantities in conventional form.
   (a) 1.2×10^6 (b) 2.37×10^2
   (c) 6.28×10 (d) 5.45×10^{-3}
   (e) 7.7×10^{-6} (f) 8.432×10^{-1}

1-3. Express each of the following quantities in the preferred engineering notation.
   (a) 47 000 ohms   (b) 45 000 000 hertz
   (c) 1500 watts   (d) 0.0505 second
   (e) 0.0005 volt   (f) 0.000 000 000 39 farad

1-4. Convert the following quantities into base units, first in scientific notation and then in conventional notation.
   (a) 3.3 megohms   (b) 35 millimeters
   (c) 0.746 kilowatt   (d) 56 microvolts
   (e) 56 kilometers   (f) 56 kilograms

1-5. Perform the following calculations using an electronic calculator, first in the form given, then in scientific notation, and finally in engineering notation.
   (a) $0.005 \times 47\,000 - 958 =$
   (b) $0.005(47\,000 - 958) =$
   (c) $54\,000 - 75 \times 634 =$
   (d) $0.0005^4 =$
   (e) $\dfrac{423\,000 + 17\,400}{560 - 0.04} =$
   (f) $\sqrt{125^2 + 4 \times 5600} =$

1-6. Assuming that all numerical quantities represent measured quantities, express the solutions to the following computations with the appropriate number of significant digits.
   (a) $2.34 + 110.7 + 4.569 =$
   (b) $180 - 51.74 + 9.3 =$
   (c) $456 \times 0.0050 =$
   (d) $1225 \div 33 =$
   (e) $59.4 + 17 \times 6.28 =$
   (f) $\dfrac{59.4 - 17}{6.28} =$

1-7. The distance between two towns is 25 km (kilometers). Express this in meters, using scientific notation.

1-8. A coulomb is an electrical unit of quantity representing

$$6\,240\,000\,000\,000\,000\,000 \text{ electrons}$$

Express this in scientific notation.

1-9. The mass of an electron is

$$0.000\,000\,000\,000\,000\,000\,000\,000\,000\,899\,9 \text{ g (grams)}$$

Express this in scientific notation.

1-10. The effective diameter of an electron is

$$0.000\,000\,000\,000\,2 \text{ cm (centimeters)}$$

Express this in scientific notation.

1-11. Express 1650 g in kilograms.

1-12. Express 850 mm (millimeters) in meters.

1-13. Express 1/60 s in milliseconds.

1-14. Express 0.000 067 km in millimeters.

1-15. Express 2.2×10^{-2} kg in grams.

1-16. Express 0.04 m^2 (square meters) in square centimeters.

1-17. Express the cross-sectional area of an electric conductor whose diameter is 4.0 cm in square millimeters.

1-18. Express 4.08×10^7 cm^3 (cubic centimeters) in cubic meters.

1-19. Solve Example 1-4A using dimensional-analysis techniques.

1-20. Derive the conversion factor for miles into kilometers.

1-21. Derive the conversion factor for square feet into square meters.

1-22. Derive the conversion factor for ounces into grams.

1-23. Express a velocity of 50 mi/h (miles per hour) in meters per second.

1-24. Express a velocity of 50 ft/s in kilometers per hour.

1-25. The No. 14 AWG (American wire gauge) conductors used in house wiring have a diameter of 64.08 mils. (A mil is one-thousandth of an inch.) What is the cross-sectional area of a No. 14 AWG conductor in square millimeters?

REVIEW QUESTIONS

1-1. What is meant by the term **energy?**

1-2. What is the significance of the **law of conservation of energy?**

1-3. Trace all the energy conversions as we use electricity from a coal-fired generating station to operate a home air conditioner.

1-4. A flashlight is one of the simplest complete electric circuits. It consists of a battery, copper conductors, a switch, and a small lamp. Using standard graphic symbols, draw a circuit diagram for a flashlight.

1-5. Using the flashlight as an example, explain the **law of conservation of energy.**

1-6. Describe a device for converting heat energy into mechanical energy.

1-7. What do the letters **MKS** represent?

1-8. What advantage has a system of units, most of which can be derived from a few basic units?

1-9. What is meant by **scientific notation** in expressing a numerical quantity? What is the advantage of expressing numerical quantities in this manner?

1-10. How does the scientific method of expressing numerical quantities differ from using SI unit prefixes?

1-11. Why is it important that we write down the proper units when we record a calculator solution to a numerical example?

1-12. In the MKS system of units, the wavelength of light was expressed in **millimicrons.** What would be the preferred terminology for the equivalent SI unit?

1-13. What is the meaning of the term **dimensional analysis?**

1-14. How would you apply dimensional analysis when you are doing a chain calculation with your electronic calculator?

1-15. One butcher shop advertises sirloin steak at $3.59 a pound; another sells sirloin steak at $7.59 a kilogram. Assuming equal quality, which is the better price?

1-16. In SI, automobile gasoline consumption is expressed in liters per hundred kilometers. How would you convert miles per gallon to liters per 100 km?

*1-17. Write a BASIC program to accomplish the conversion in Question 1-16.

*1-18. Write a BASIC program to convert Fahrenheit temperature readings into degrees Celsius.

*For students who have access to a personal computer.

2

CURRENT AND VOLTAGE

2-1 THE NATURE OF ELECTRICITY

To understand the behavior of electric circuits, we must know something of the nature of the electrical property of matter. We are probably more familiar with a different property of matter known as its **mass.** This mass property of matter leads to the branch of physical science called **mechanics** and is the basis for mechanical energy in both static and kinetic forms. Although we may not know just why it should exist, we are certainly aware of the force of attraction between two masses. It is this **gravitational force** between the mass of the earth and the mass of objects at the surface of the earth which is responsible for what we call the **weight** of these objects.

To push open a door, we must apply sufficient mechanical force to overcome the mechanical force tending to keep the door closed. This opposing force might be the tension of the spring in the door-closer, or it might be just the friction of the door hinges. The force we apply with our hand in this example is a *contact* force, since it does not act on the door until our hand touches the door. Gravitational force, on the other hand, is not a contact force. Gravitational force can act at a considerable distance, as for example the gravitational force between the mass of the moon and the mass of the earth, which governs the orbit of the moon around the earth. A force, such as gravitational force, which can act at a considerable distance without contact is called a **field force.**

If we carry a heavy parcel against the force of gravity up several flights of stairs, we soon become aware that we are expending a considerable amount of energy. This conversion of the chemical energy stored in the tissue of our body to mechanical energy, which, in turn, moves a mass against the force of gravity is called **work.**

Work is the accomplishment of motion against the action of a force which tends to oppose the motion.

Because the energy needed to perform this work was present before we climbed the stairs,

Energy is the capacity to do work.

One further basic concept which we must note before we investigate the electrical property of matter is the principle of conservation of energy, which states that

In any closed system, the total energy remains constant.

Energy can be transformed from one form into another, but the total energy after the transformation must exactly equal the total energy going into the transformation. Consequently, **energy** and **work** are numerically the same and we may use the same unit of measurement for both.

In searching for the electrical property of matter which will permit useful work to be accomplished in the basic electric circuit of Figure 1-1, we must look for an electric force and for motion of a particle under the influence of this force. Atoms of matter in their normal state are electrically neutral. Consequently, we do not observe the effect of electric force as readily as we observe the effect of gravitational force.

Some 2500 years ago, Thales of Miletus was one of the first to discover a means of disturbing the normal electrical balance of matter. He noted that when he rubbed a piece of amber,[†] it acquired an ability to attract light pieces of straw and dust. However, it was not until the eighteenth century that serious experiments were conducted in an effort to learn the nature of this force which attracted small particles of straw and dust.

We can duplicate Thales' experiment by rubbing a glass rod with a silk cloth, or an ebonite rod with a piece of cat's fur. Since these rods can then attract bits of paper against the force of gravity, work is being accomplished and energy is being expended. Rubbing the rods must have provided them with some form of energy which they did not reveal in their normal state. We say that we "placed" an electric **charge** on the rod, using the term *charge* to imply potential energy in much the same sense that we speak of a charge of dynamite placed in a drill hole for a blasting operation.

We can investigate the effect of the force produced by rubbing glass or ebonite rods by means of lightweight balls suspended by silk threads, as shown in Figure 2-1. We discover that we can transfer some of the charge from the glass or ebonite rods to the balls by touching them with a charged rod. If we touch a charged glass rod to both balls, we discover that they tend to *repel* one another, as in Figure 2-1(a). However, if we touch a charged glass rod to the left ball and a charged ebonite rod to the right ball, we find that they tend to *attract* one another, as in Figure 2-1(b).

From these observations, we can state that the force between the charged balls is a *field* force, since the balls are not in contact. But the force is *not* gravitational force, since gravitational force is always a force of attraction, never repulsion. Sec-

[†]Our word *electron* comes from *elektron*, the Greek word for amber.

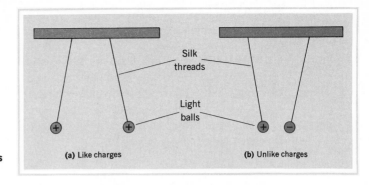

FIGURE 2-1
Like Charges Repel, Unlike Charges Attract.

(a) Like charges   (b) Unlike charges

Silk threads

Light balls

ond, the charge on the ebonite rod must be different from that on the glass rod, since in one example we produce a force of repulsion, and in the other a force of attraction. Since both balls in Figure 2-1(a) were charged from the same glass rod, the two balls must possess **like charges.** With no knowledge of the exact nature of these charges, the eighteenth-century scientists could only label the charges in Figure 2-1(b) as **unlike charges,** and state that

Like charges repel, unlike charges attract.

We now distinguish the two different types of electric charge by the terms positive charge and negative charge, a designation first introduced by Benjamin Franklin.

Using a more precise version of the apparatus in Figure 2-1, a French physicist, Charles Augustin Coulomb, was able to show that the force between two electrically-charged bodies is directly proportional to the product of the magnitudes of the two charges and inversely proportional to the square of the distance between them. We can state this relationship more clearly by writing a mathematical equation in which we represent the various quantities in the statement above by standard letter symbols. Hence, Coulomb's "law" becomes

$$F = k \frac{Q_1 Q_2}{s^2} \qquad (2\text{-}1)$$

where F is the force between the two electrically charged bodies, Q_1 and Q_2 are the respective electric charges, s is the distance between the charged bodies, and k is a numerical constant which makes the equation balance with the units used for the other quantities. We shall leave consideration of suitable units until Chapter 11.

The property of matter which allows objects to become electrically charged remained a mystery for almost two centuries until Lord Rutherford's experiments with radiation led him to the now familiar concept of the structure of an atom consisting of a **nucleus** around which **electrons** revolve in a manner similar to that in which the planets of our solar system follow orbits around the sun.

From a study of chemistry, we know that all substances can be formed by chemical combination from about one hundred known **elements,** each possessing its own distinctive properties. More than 2000 years ago, Democritus of Thrace proposed that, if we keep dividing a sample of an element into smaller and smaller pieces, we reach a limit beyond which it is impossible for us to continue to divide without losing the characteristic properties of the element. Democritus called this indivisible sample of an element an **atom.**

Rutherford's model of an atom consists of subatomic particles which are held together by *electric* force. These particles are the same for all elements. The atoms of one element differ from those of another only in terms of the number and arrangement of subatomic particles in the model. If we can determine the nature of these particles and the forces acting on them within an atom, we shall then know something about the electrical property of matter.

2-2 THE ATOM

We shall soon discover that the behavior of electric circuits is quite dependent on the behavior of **electrons,** the subatomic particles which travel in planetary orbits around the **nucleus** of an atom. From the orbital motion of electrons within an atom, we can deduce several characteristics of an electron: (1) it must possess energy; (2) it appears to possess a mass which is a small fraction of the mass of the nucleus around which it travels; and (3) there must be a force of attraction between the electron and the nucleus for it to travel in a curved orbit.

Unlike the force which governs the orbits of the planets of the solar system, the force between electron and nucleus is not a gravitational attraction based on mass, since we can show that electrons *repel* one another. The force of attraction between orbital electrons and the nucleus of an atom is an electric force, similar to that illustrated in Figure 2-1(b). Since atomic theory will not permit subdivision of an electron, it follows that every electron must possess as part of its nature a unit of electric charge. The charge on every electron is identical.

For an electric force of attraction to exist between electrons and nucleus, the nucleus must contain particles which possess an *opposite* or *unlike* charge. These particles are called **protons.** Using Franklin's nomenclature to identify the opposite charges of electrons and protons, we find that electrons possess a *negative* electric charge and protons have a *positive* electric charge. The charge on a proton is equal in magnitude to that of an electron. The nucleus of an atom contains as many protons as there are orbital electrons in the atom in its normal state. Consequently, the net charge of a complete atom in its normal state is zero and the atom is electrically neutral. However, protons have a mass that is 1837 times the apparent mass of an electron.

The electrons of every atom are identical. The atoms of one element differ from those of another in terms of the number of orbital electrons and the number of protons in the nucleus. From experimental evidence, it appears that the lightest element, hydrogen, has one electron, as shown in Figure 2-2(a). The next lightest is helium which has two electrons, as shown in Figure 2-2(b). If we arrange all the elements in order of increasing mass, each element has one more electron than the one immediately preceding it in the list. For example, uranium is the ninety-second element in the list and the uranium atom has 92 electrons and 92 protons. To account for the atomic weights of the various elements in this list, we find that all nuclei except hydrogen must possess particles that are slightly heavier than protons. These particles are neither attracted nor repelled by electrons or protons. They are electrically neutral and are called **neutrons.**

22

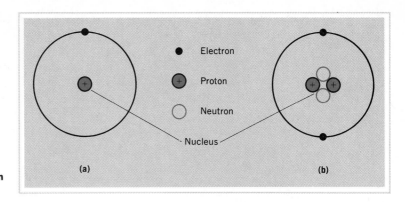

FIGURE 2-2
Atomic Models: (a) Hydrogen Atom; (b) Helium Atom.

Since a force of attraction exists between an orbital electron and a nucleus, in order for the electron to maintain a given orbit, it must move at a velocity around the orbit which will develop the centrifugal force required to exactly balance this attraction. And since an electron has a certain apparent mass, the moving electron must possess a kinetic energy proportional to its mass and the square of its average velocity ($W = \frac{1}{2}mv^2$). Therefore, to occupy a certain orbit, an electron must possess a definite amount of energy. Hence, we can discuss electron orbits in terms of the **energy levels** of the electrons in them. To increase the radius of an electron orbit, the electron must acquire additional *potential* energy to move it against the force of attraction between the electron and the nucleus. It follows that the lowest energy level in an atom is the electron orbit closest to the nucleus.

Electrons sometimes behave as if they were electromagnetic waves rather than particles. To behave in this manner, the orbits along which the electrons move must be an integral number of wavelengths. For a given wavelength, only certain radii will permit circumferences that fulfill this condition. Therefore, the electrons of an atom can appear only at certain definite energy levels.[†] The spacing between energy levels is such that, when we catalog the chemical properties of the various elements, it is convenient to group several closely spaced *permissible* energy levels together into **electron shells** and **subshells,** as indicated in Figure 2-3. The gap in energy level between one shell and the next is much greater than the gaps between energy levels within a shell.

The maximum number of electrons which each subshell and shell or energy level can contain is the same for all atoms. The first shell can contain only two electrons, the second eight, the third 18, and so on to a seventh shell for the heaviest atoms.[‡] Since the innermost shell is the lowest energy level, the shells begin to fill up from the nucleus outward as the atomic number of the element increases. Accordingly, no electrons can appear in the second shell until the first shell has its maximum complement of two electrons. And no electrons will appear in the third shell until the second has its full eight electrons. However, after the inner two subshells of third shell are filled, either one or two electrons will appear in the fourth shell before the outer

---

[†]Niels Bohr proposed this concept of the atom in 1913.
[‡]The maximum number of electrons in a shell is $2n^2$, where n is the shell number, counting from the nucleus.

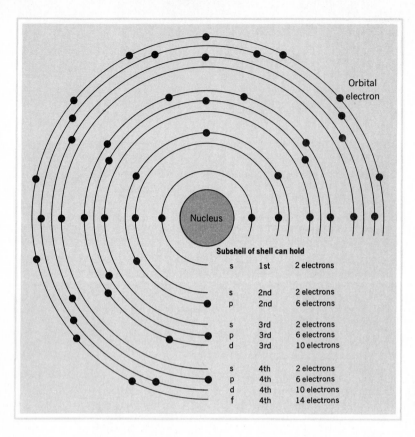

Orbital electron

Nucleus

Subshell of shell can hold

s	1st	2 electrons
s	2nd	2 electrons
p	2nd	6 electrons
s	3rd	2 electrons
p	3rd	6 electrons
d	3rd	10 electrons
s	4th	2 electrons
p	4th	6 electrons
d	4th	10 electrons
f	4th	14 electrons

FIGURE 2-3
Possible Location of Electrons in the First Four Energy Levels of an Atom.

subshell of the third shell can accept any electrons. Variations of this type become more involved in the fourth, fifth, and sixth shells; nevertheless, a definite pattern is followed.

For our investigation of the electrical property of matter, we need consider only those electrons in the highest occupied energy level of the shell structure of a particular atom. These electrons are called **valence** electrons. By adding a small amount of energy to the valence electrons of an atom, it is possible to increase the radius of their orbits so that they can escape from the attraction of the nucleus. Whenever valence electrons are thus removed from an atom, the atom now contains more protons than electrons. Consequently, it is no longer electrically neutral but displays a net *positive* charge.

It is also possible to *add* electrons to the valence subshell of an atom. Such atoms display a net *negative* charge. In Section 2-3, we shall discover how the necessary energy to redistribute valence electrons can be provided by chemical reaction. We can also supply the necessary energy in some cases by friction. Hence, the left-hand ball in Figure 2-1(b) is given a positive charge by removing some electrons from its atoms. The right-hand ball is given a negative charge by adding electrons to its normally neutral atoms.

2-3 COMBINATION OF ATOMS

The number of valence electrons possessed by an atom of a particular element governs the manner in which atoms will combine chemically to form stable **molecules.** An atom is in its most stable state when the number of electrons associated with each nucleus in a molecule exactly fills a subshell in the orbital structure. Elements such as helium with its two orbital electrons, and neon with 10 electrons, have no tendency to combine chemically with other elements since they possess completely filled shells. Helium and neon are therefore **inert** gases. However, there are many elements whose atoms have either a few electrons short of a complete subshell or a few electrons in excess of a complete subshell. These atoms tend to combine into chemical compounds in a manner that will permit all atoms in a molecule of a compound to have completely filled subshells. There are three types of atomic **bonding** that can achieve this requirement.

Sodium chloride (common salt) is an example of one form of atomic bonding. A sodium atom has 11 electrons. The first two shells of this atom are filled and there is one electron in the third shell. Chlorine atoms have 17 electrons, one short of a complete subshell. On combining, the sodium atom is quite eager to lend its valence electron to the chlorine atom so that both atoms may have completely filled subshells. Since the sodium atom has lost an electron, it is no longer electrically neutral. An atom that has either lost or gained electrons does not have the same chemical properties as the neutral atom and is called an **ion.** When the sodium atom loses an electron, the positive charge of its nucleus predominates, thus forming a *positive* sodium ion. Similarly, the chlorine has become a *negative* chlorine ion. The attraction between the unlike charges of the sodium and chlorine ions binds them together to form a stable molecule of sodium chloride. This type of bonding is called **ionic bonding.**

Hydrogen, with only one electron, is a very active element. If it cannot acquire an electron to provide two electrons for its first shell, it would just as soon give up the one electron it does possess. As a result, many ionic compounds contain positive hydrogen ions. But to form a stable molecule of pure hydrogen gas, two hydrogen atoms can combine in such a manner that they *share* each other's one electron, as suggested by Figure 2-4. Consequently, each hydrogen atom in the molecule appears to possess two electrons (a completely filled first shell); yet, the molecule as a whole is electrically neutral. This electron sharing is called **covalent bonding.**

Atoms of a group of elements called **metals** are held together by a third form of bonding. The metals with which we are concerned each have a single valence electron. For example, copper atoms have 29 electrons in orbit around each nucleus. After the first three shells are filled ($2 + 8 + 18 = 28$), there is one lone electron in the fourth shell. Since the copper atom would be more stable without this valence electron, only a very weak attraction holds it in orbit around the nucleus (unlike the 28 tightly bound electrons which fill the first three shells). As a result, the valence electron often finds itself midway between two nuclei, as shown in Figure 2-5. In this position of balanced attraction, the valence electron is then *free* to leave its parent nucleus and to move into the fourth shell of the neighboring atom.

The net effect of the valence electrons of a metal being free to move from one

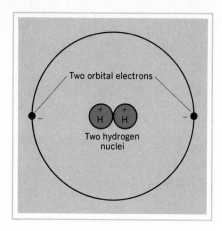

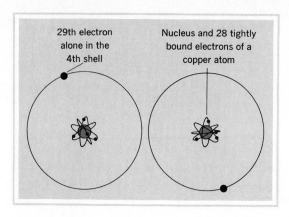

FIGURE 2-4
Atomic Model of a Hydrogen Molecule.

FIGURE 2-5
Atomic Model of Two Adjacent Copper Atoms.

atom to another is that a seemingly solid piece of copper resembles the steel skeleton of a skyscraper with the wind passing through it. The copper atoms donate their valence electrons to form a vast cloud of **free electrons,** thus becoming stable positive copper ions with their 28 tightly bound orbital electrons. Since the positive metal ions repel one another, they will take up fixed geometric positions within the electron cloud, thus giving the metal object its shape. The atoms are held together by the attraction between the positive metal ions and the common cloud of free electrons. This is called **metallic bonding.** Knowing the mass of a single copper atom, we can estimate that there are approximately 85 000 million, million, million (8.5×10^{22}) free electrons in a cubic centimeter of copper.

The manner in which various materials behave in electric circuits depends on which of the three types of atomic bonding holds the atoms of each molecule together. We shall investigate some of these distinctions more thoroughly in Chapter 3. However, we can now specify the nature of the particles which flow through the copper conductors of the simple electric circuit of Figure 1-1. Just as a wind will make air flow through the steel skeleton of a building, we can force **free electrons** to move through the lattice of positive metal ions in the seemingly solid copper conductors. This flow of electrically charged particles under the influence of an electric force constitutes the necessary requirement for the transmission of energy in an electrical form. This flow of free electrons in an electric conductor represents an **electric current.**

2-4 ELECTRIC CURRENT

In Section 2-3, we noted that atoms of copper arrange themselves into a crystalline structure so that the positive copper ions consisting of a nucleus and 28 tightly bound electrons are located in a pattern which will allow the twenty-ninth, or valence electron to move freely from atom to atom. The lattice pattern for metals such as copper is in the form of a cube with a metal ion in each of the eight corners of the cube and also

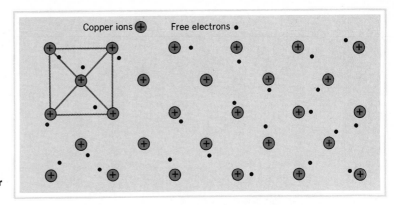

FIGURE 2-6
Cross Section of a Copper Conductor.

in the center of each of the six faces of the cube. Figure 2-6 shows a cross section of such a lattice structure with the outline of one crystal unit sketched in.

Although there is a free electron for each copper ion, Figure 2-7 shows only one of these free electrons so that we may trace its motion through the lattice structure. In a section of a copper conductor lying on a table, this free electron moves from atom to atom, its direction at any time depending on random bits of energy it picks up in its travels from repulsion by other electrons, attraction by copper ions, or from the thermal energy which causes the ions to vibrate in their fixed lattice positions. These random motions average out over a period of time so that there is no *net* motion in any one direction in Figure 2-7(a). Figure 2-7(b) shows the same conductor connected in an electric circuit which is conveying energy. The battery in the basic electric circuit of Figure 1-1 provides an external source of energy for the free electrons in the conductor. The energy picked up by the electrons from this additional source superimposes a net **electron drift** to the right on the random motion of the free electrons.

To produce this electron drift, the battery in Figure 2-8 adds surplus electrons to the left-hand end of the conductor and, at the same time, removes an equivalent number of electrons from the right-hand end of the conductor. At the expense of chemical energy the battery is able to maintain a surplus of electrons (hence, a net negative charge) at its negative terminal, and a deficiency of electrons (net positive charge) at its positive terminal. As we noted in Section 2-1, an electric force of repulsion exists between like electric charges, and an electric force of attraction exists

FIGURE 2-7
Random Motion of a Free Electron in a Copper Conductor.

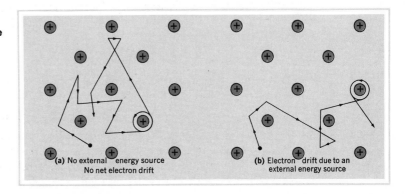

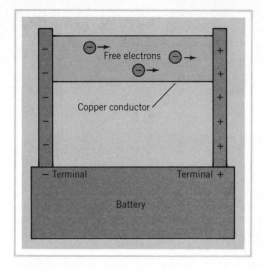

FIGURE 2-8
Motion of Free Electrons due to Electric Energy Supplied by a Battery.

between unlike charges. Since the free electrons in the conductor possess elemental negative charges, they are repelled by the negative charge at the negative terminal of the battery and are attracted by the positive charge at the positive terminal of the battery. Since these electrons are free to move from atom to atom in the lattice of copper ions, the result is a net electron drift from left to right in the conductor in Figure 2-8. As we shall discover in Section 2-8, chemical action within the battery will replenish the supply of surplus electrons at the negative terminal as fast as the free electrons in the conductor drift to the positive terminal of the battery.

To determine the *net* movement of free electrons in an electric conductor, we can consider an imaginary plane cutting across the conductor at right angles to its length, as shown in Figure 2-9. If we were able to observe the free electrons in their random wandering, we would discover billions of electrons crossing the plane in both directions in a given time interval. Figure 2-9(a) represents the conductor lying on the table with no external source of energy applied to it. In this case, the number of electrons crossing the plane in one direction in a given period of time equals the number crossing in the opposite direction in the same time interval. There is no net electron drift, no energy is conveyed along the length of the conductor. Hence, by definition, there is no electric current. In Figure 2-9(b), an external energy source causes more electrons

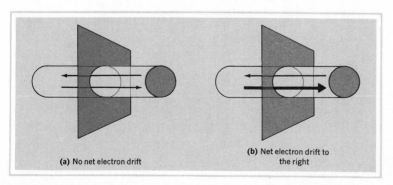

(a) No net electron drift

(b) Net electron drift to the right

FIGURE 2-9
Cumulative Effect of Motion of Billions of Free Electrons.

to cross the imaginary plane from left to right than from right to left. Consequently, there is now a net electron drift to the right along the length of the conductor. It is this *net* drift of charge carriers which constitutes the **electric current** which we must now be able to express in numerical terms.

Electric current is the net *flow of electric charge carriers past a certain point in an electric circuit in a given period of time.*

Since electrons can be neither created nor destroyed, if there is a flow of charge carriers from the energy source to the load in one electric conductor in the basic electric circuit of Figure 1-1, we must provide a second conductor so that an equivalent number of electrons can circulate from the load back to the energy source.[†] Consequently, electric lamp and appliance cords have two current-carrying conductors.

2-5 THE COULOMB

Since each electron possesses the same elemental quantity of electric charge, we may express the net electron drift across the imaginary plane in Figure 2-9 either in terms of the number of electrons or in terms of the total electric charge possessed by this number of electrons. The first step in expressing electric current in numerical terms is to be able to express the net **quantity** of electric charge crossing the imaginary plane in Figure 2-9.

The elemental quantity of electric charge carried by a single electron is too small for practical purposes. A practical unit for expressing quantity of electric charge must represent the charge carried by many billions of electrons.

The coulomb is the SI unit of quantity of electric charge.[‡]

Physicists have determined that

A coulomb represents the quantity of electric charge carried by 6.24×10^{18} electrons.

In technical language, we use algebraic equations to state definitions and data in concise and precise form. Therefore, we must have standard *letter symbols* to represent all variable terms in these equations and *unit symbols* for the units of measurement.

The letter symbol for quantity of electric charge is Q.
The unit symbol for coulomb is C.

EXAMPLE 2-1
If there are 2.40×10^{19} free electrons in a certain piece of copper wire, how many coulombs of electric charge does this represent?

[†]Check the meaning of *circuit* in a dictionary.
[‡]Named in honor of Charles Augustin Coulomb.

SOLUTION

$$Q = \frac{2.40 \times 10^{19}\text{e}}{6.24 \times 10^{18}\text{e/C}} = \mathbf{3.85\ C}$$

To enter this equation into a calculator, it becomes

$$Q = 2.4\,\boxed{\text{EE}}\ 19 \div 6.24\,\boxed{\text{EE}}\ 18 = \mathbf{3.846\ C}$$

2-6 THE AMPERE

From our general definition of electric current in Section 2-4, we require units for *two* variables: (1) quantity of electric charge passing a certain point in an electric circuit (coulombs) and (2) a given period of time. Since the SI unit of time is the *second,* it is quite in order for us to express the magnitude of an electric current in **coulombs per second.**

The International System of Units prefers not to use compound units such as coulombs per second as basic units. As a result, a special unit of electric current has been established to perpetuate the name of one of the pioneer electrical physicists, André Marie Ampère. Thus,

The ampere is the SI unit of electric current.
One ampere is the rate of flow of electric charge (electric current) when one coulomb of electric charge carriers passes a certain point in an electric circuit in one second.

Amperes and **coulombs per second** are synonymous, and in using the ampere as a unit of electric current, we must keep reminding ourselves that current is *rate of charge flow* in coulombs per second.

Any unit which involves *time* is called a *rate* unit. However, there are several types of rate units. *Distance* per unit time is called **velocity**. *Volume* or *quantity* per unit time is called **rate of flow,** or simply **flow.** For example, natural gas *flow* in a pipeline is measured in cubic meters per second. This flow is not the same as the speed or *velocity* at which the gas molecules are moving along the pipe. Similarly, in an electric circuit, we must be careful not to confuse the *rate of flow* of electric charge with the *velocity* of the charge carriers. The ampere is a rate unit in terms of *quantity* per second, not *distance* per second. A traffic engineer measures the rate of flow of cars on a highway by using a traffic counter, which tells him how many cars pass a certain point on the highway in a given period of time. The velocity of the individual cars (in kilometers per hour) may vary considerably and bears no fixed relationship to the number of cars using the highway per day. Similarly, when the electric current in a typical conductor is one ampere, the net *velocity* of electron drift along the length of the conductor is only about one-tenth of a millimeter per second.

The letter symbol for electric current is I. [†]
The unit symbol for ampere is A.

[†]I is from the French, *Intensité* of electron flow. The letter symbol C is reserved to represent capacitance.

We can express the equality between amperes and coulombs per second in equation form as

$$I = \frac{Q}{t} \tag{2-2}$$

where I is current in amperes, Q is quantity of electric charge in coulombs, and t is time in seconds.

EXAMPLE 2-2

What is the current in an electric circuit when 75 C of electric charge pass a certain point in the circuit in half a minute?

SOLUTION

$$I = \frac{Q}{t} = \frac{75 \text{ C}}{30 \text{ s}} = 2.5 \text{ C/s} = \mathbf{2.5 \text{ A}}$$

$$\underbrace{\phantom{I=\frac{Q}{t}}}_{I} \qquad \underset{II}{\uparrow} \qquad \underset{III}{\uparrow} \qquad \underset{IV}{\uparrow}$$

As we become more familiar with our calculators, there is a tendency to simply enter the data and come up with a numerical answer all in one step. However, in these early stages, a systematic solution will greatly assist in understanding the behavior of electric circuits. Therefore, we shall take time to identify the sequence of steps involved in solving Example 2-2.

Step I. Note that the problem states two pieces of information and asks for one. We express this information in equation form. *Write down* the symbol for the unknown quantity on the left of the *equals* sign and the symbols for the given data (in their proper relationship) on the right of the *equals* sign.

Step II. Substitute the given data into the equation making sure that powers of ten or unit prefixes are included to preserve the proper magnitude.

Step III. Perform the numerical computation.

Step IV. The magnitude of the calculator display will be in *root* units. Express the answer in appropriate units, using unit prefixes, if necessary, to conform to SI recommended format.

EXAMPLE 2-3

How long will it take 4.0 mC of electric charge to pass through a fuse in an electric circuit if the current is 50 A?

SOLUTION
Since

$$I = \frac{Q}{t} \qquad \therefore t = \frac{Q}{I}$$

$$t = \frac{4 \text{ mC}}{50 \text{ C/s}} = 4\,\boxed{\text{EE}}\ +/-\ 3\ \boxed{\div}\ 50\ \boxed{=}\ 8 \times 10^{-5}\text{ s} = 80 \times 10^{-6}\text{ s} = 80\ \mu\text{s}$$

2-7 POTENTIAL DIFFERENCE

In Section 2-4, we noted that there can be no electric current in the conductors of our basic electric circuit until we add a device, such as a battery, which can impart energy to the free electrons in the conductors in such a manner that they flow along the length of the conductor. We say that the battery is the source of an **electron-moving force** or **electromotive force,** usually abbreviated to **emf** (ee-em-eff).

Since we are more familiar with the effects of gravitational force than we are with the effects of electric force, we can consider first of all the operation of the hydroelectric generating station shown in simplified cross section in Figure 2-10. Water from above a waterfall is diverted so that it flows through a "waterwheel" or turbine and is then discharged into the river below the waterfall. In falling 100 meters, the water loses some of its potential energy. In accordance with the law of conservation of energy, the potential energy lost by the water is converted into mechanical energy by the turbine and is then converted into electric energy by the electric generator or **dynamo.**

Since objects at the surface of the earth are 6400 kilometers away from the center of gravity of the earth, the difference in gravitational force acting on a cubic meter of

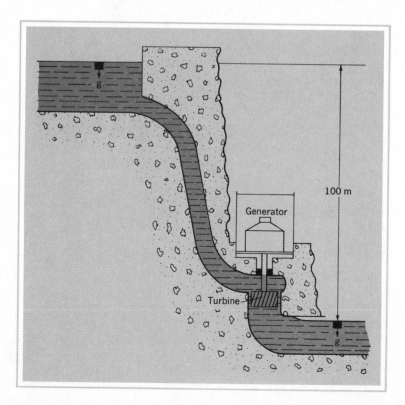

FIGURE 2-10
Simplified Cross Section of a Hydroelectric Generating Station.

32   PART I   THE BASIC ELECTRIC CIRCUIT

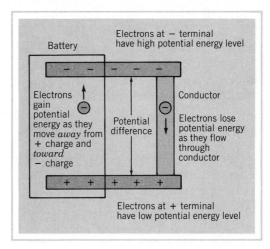

Electrons at − terminal
have high potential energy level

Battery

Electrons gain potential energy as they move *away* from + charge and *toward* − charge

Potential difference

Conductor

Electrons lose potential energy as they flow through conductor

Electrons at + terminal
have low potential energy level

FIGURE 2-11
Electric Potential Difference.

water above and below the generating station shown in Figure 2-10 is negligible. But there is an appreciable difference in the potential energy of a cubic meter of water above and below the station, as we would discover if we had to carry or pump a cubic meter of water from the river below the station 100 meters uphill against the force of gravity to the top of the escarpment. The "force" which operates the hydroelectric generating station is not simply gravitational force. Rather, it is the potential energy difference between a unit quantity of water above and below the generating station. The law of conservation of energy requires the potential energy difference, or simply **potential difference,** between a unit quantity of water above and below the generating station to be equal to the energy expended in raising the unit quantity of water the 100 meters against the force of gravity.

We can now return to the electric circuit of Figure 2-8, in which we connected an electric conductor between the terminals of a battery. This circuit is shown again in Figure 2-11. In this case, we have turned the battery on its side so that the terminology we use in the electric circuit matches the terminology of the hydraulic system of Figure 2-10. Since the battery is the source of the equivalent electromotive "force" in Figure 2-11, we must now find out what happens inside the battery.

We noted in Section 2-4 that free electrons in the conductor flow away from the negative terminal of the battery and toward the positive terminal. To maintain the negative and positive charges at the two battery terminals, *inside* the battery an equivalent number of electrons must be removed from the positive terminal and added to the negative terminal. Thus, inside the battery, electrons must move *away* from the positive terminal and *toward* the negative terminal. In other words, electrons must move *against* the electric forces acting on them, just as water moves against gravitational force when it is pumped uphill. To accomplish this, the electrons must acquire potential energy at the expense of the chemical energy stored by the battery. Consequently, an electron arriving at the − terminal is at a higher potential energy level than it was when it left the + terminal of the battery. There is an electric **potential** energy **difference** between free electrons at the negative and positive terminals of the battery, with electrons at the negative terminal being at a higher potential than those at the positive terminal.

Under the influence of gravitational force, the water in the hydroelectric generating system of Figure 2-10 always tends to fall to a lower potential energy level. Similarly, electrons at the negative terminal of a source tend to "fall" to a lower potential energy level. They can do this via the external conductor connected between the battery terminals. In flowing from the − terminal to the + terminal in the external circuit, electrons lose as much potential energy as they gained in being moved from the + terminal to the − terminal inside the battery. This energy "lost" in the external circuit is converted into light, heat, or some other useful form of energy, depending on the nature of the load component in the electric circuit. In traveling the complete circuit around the closed loop of Figure 2-11, electric charges experience a **potential rise** within the battery and an equivalent **potential fall** or **potential drop** in the external circuit.

2-8 GENERATING A POTENTIAL DIFFERENCE

Before we consider a suitable unit for expressing electric potential difference, we shall consider briefly how a battery converts chemical energy into an electromotive force which, in turn, generates a potential difference between the battery's terminals. Electromotive force is a property of the battery in the electric circuit of Figure 2-11 which distinguishes it from the external load. Emf is a measure of the amount of energy per unit of electric charge which the battery *converts* from chemical to electrical form in the process of *separating electric charges* in order to create a potential difference between its terminals. Electromotive *force* is, therefore, not a true force which we would measure in newtons. Emf is measured in joules per coulomb. Consequently, although we sometimes refer to a battery as a *source of emf*, the term *electromotive force* is disappearing from electrical terminology as we pay more attention to the **potential difference** generated by the battery.

As the term implies, a **battery** originally meant a *group* of chemical cells, but common usage allows us to refer to the simple **voltaic cell** of Figure 2-12 as a single-cell battery. When hydrogen with its one valence electron combines with chlorine, which has 17 orbital electrons, to form the ionic compound hydrochloric acid, the chlorine borrows the one electron from the hydrogen to complete a full subshell in its atomic structure. (Check this by referring to Figure 2-3.) In doing so, the hydrogen atom loses an electron, as indicated by the + sign in Figure 2-12, and the chlorine acquires a surplus electron.

The chemical makeup of hydrogen chloride and zinc chloride is such that energy must be expended to transfer a chlorine atom from a zinc chloride molecule to a hydrogen chloride molecule. Therefore, energy must be released when the chlorine atom moves from a hydrogen chloride molecule to a zinc chloride molecule. This energy is transferred to the surplus electron of the negative chloride ion in the acid as the chloride ion moves toward the zinc rod to combine with atoms of zinc to form zinc chloride. During this chemical process, the surplus electron that the chlorine borrowed from the hydrogen is deposited on the zinc rod, thus building up a surplus of electrons or a negative charge on the zinc rod. As each chlorine ion combines with the zinc, the hydrogen ion which lost the electron to the chlorine ion can now go in search of a

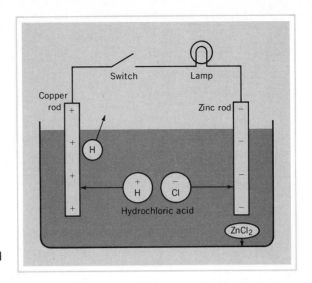

FIGURE 2-12
Producing an Electric Potential Difference with a Simple Chemical Cell.

source of free electrons to regain its normal state as an atom of hydrogen. The copper rod represents such a source, and as the hydrogen ions rob the copper of free electrons, a positive charge builds up on the copper rod. The number of electrons taken from the copper rod by hydrogen ions is equal to the number of electrons deposited on the zinc rod by chlorine ions.

Thus the electromotive-force property of the battery disturbs the normal ion balance in the hydrochloric acid by *separating* the electrically-charged ions. Negative charges are forced to move to the zinc electrode while positive charges go to the copper rod. Hence, we may think of the battery (or any electric energy source) as an electric **charge-separating device.** Or, if we wish to think in terms of our hydroelectric analogy, we may think of an electric energy source (source of emf) as a pump building up a potential difference between its terminals by seemingly pumping electrons from the positive to negative terminal against the electric force opposing such motion.

If we leave the switch open, the chemical action can proceed until sufficient potential difference is established between the copper and the zinc that the negatively charged chlorine ions do not have sufficient chemical attraction to reach the zinc against this electric force. The formation of zinc chloride will then cease. When we close the switch there will be a flow of free electrons through the conductors and the lamp as the electrons in the conductors are repelled by the negative charge on the zinc rod and attracted by the positive charge on the copper rod. This flow is attempting to return the potential difference in the circuit to zero. But as soon as the potential difference drops the least bit, the chemical action recommences, depositing electrons on the zinc rod and removing them from the copper rod. The result will be a steady flow of free electrons (electric current) as the tendency for the electric load (lamp) to lower the potential difference is offset by the action of the cell to raise the potential difference to a level determined by the two particular metals used in the cell. We call this steady flow of electric charge in one direction around the loop a **direct current** (dc).

2-9 THE VOLT

The electric potential difference (PD) between any two points in an electric circuit is the rise (or fall) in potential energy involved in moving a unit quantity of electric charge from one point to the other.

In Section 2-1, we noted that, as a consequence of the law of conservation of energy, work and energy are numerically equal. And we defined work as the accomplishment of motion against the action of a force which tends to oppose the motion.

The letter symbol for work and energy is W.

From the definition of work

$$W = Fs \tag{2-3}$$

where F is force in newtons and s is distance in meters.

We could express work and energy in *newton meters*. But the International System of Units prefers to avoid the compound unit. Hence,

The joule is the SI unit of work and energy.[†]
*The unit symbol for joule is **J**.*

From the definitions above, we can express electric potential difference in terms of **joules per coulomb.** Again, the International System of Units provides a special *derived* unit for joules per coulomb.

The volt is the SI unit of potential difference.[‡]

From the definition of potential difference,

One volt is the potential difference between two points in an electric circuit when the energy involved in moving one coulomb of electric charge from one point to the other is one joule.
The letter symbol for potential difference is E or V.[§]
*The unit symbol for volt is **V**.*

We may express this relationship in equation form as

$$E(\text{or } V) = \frac{W}{Q} \tag{2-4}$$

where E(or V) is the potential difference in volts, W is energy in joules, and Q is quantity of electric charge in coulombs.

[†]Named in honor of the English physicist James Joule.
[‡]Named in honor of the Italian physicist Alessandro Volta.
[§]We shall consider the distinction between E and V in the next section.

EXAMPLE 2-4

During the chemical action in the wet cell of Figure 2-12, 50 C of electrons is effectively transferred from the copper to the zinc by the release of 55 J of energy. What is the potential difference between the two rods?

SOLUTION

$$E = \frac{W}{Q} = \frac{55 \text{ J}}{50 \text{ C}} = 1.1 \text{ V}$$

EXAMPLE 2-5

A current of 0.3 A flowing through the filament of a TV picture tube releases 9.45 J of heat energy in 5 s. What is the PD across the filament of the tube?

SOLUTION
Since $I = Q/t$,

$$\therefore Q = It = 0.3 \text{ A} \times 5 \text{ s} = 1.5 \text{ C}$$

and

$$V = \frac{W}{Q} = \frac{9.45 \text{ J}}{1.5 \text{ C}} = 6.3 \text{ V}$$

2-10 EMF, POTENTIAL DIFFERENCE, AND VOLTAGE

At the beginning of Section 2-7, we used the term **electromotive force** to represent the concept that a generating device must have some property which will *force* electric charge carriers to flow in a closed electric circuit. Emf is, however, not simply the electric force (in newtons) that acts on the electrically charged particles in the circuit to cause a net drift in one direction. By definition, the electromotive force of a generating device is numerically equal to the energy *converted* per unit quantity of electric charge moved from one terminal to the other inside the device. Hence, emf can be expressed in joules per coulomb or volts. But emf and potential rise are numerically equal only when the electric circuit is open ($I = 0$). When there is any current in the electric circuit, inefficiencies in the energy-conversion process mean that the potential rise between the terminals of the generating device will be numerically smaller than the internal emf.

Electromotive force is a means to an end. The energy conversion is, in turn, responsible for generating a potential difference between the terminals of the generating device. It is this potential difference (often abbreviated to PD) which causes an electric current when we close the switch in the electric circuit. In the electric circuit of Figure 2-13, the voltmeter cannot get at the emf of the battery directly. It can only measure the potential difference between the battery terminals. We can *indirectly* determine the magnitude of the emf if the switch is open and if the voltmeter draws no current from the battery.

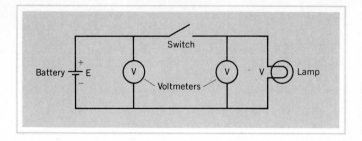

FIGURE 2-13
Illustrating the Distinction between Source Voltage and Voltage Drop in a Basic Electric Circuit.

In studying electric circuits, it has often been customary to assume that the generating device is perfectly efficient—in which case the potential difference between its terminals and the internal emf would be numerically equal. Such an assumption has allowed us to become careless in our terminology by using the terms *emf* and *potential rise* interchangeably. In this book, we shall follow the more modern practice of avoiding the term emf when we really mean the potential rise (in volts) between the terminals of a generating device.

Many years ago, someone coined the term **voltage** as a simpler expression than *potential difference*. Since "voltage" basically means something that is measured in volts, it could be used in place of potential difference, potential rise, potential drop (or even emf). Over the years, increasing usage has led to voltage being defined as being the same as potential difference. Hence, **voltage** is now the preferred term in much of the technical literature.

In a way, it is unfortunate that the term *potential difference* or PD is not more popular. It is a very descriptive term in that its wording reminds us that *two* particular points in an electric circuit must be stated before we can state the potential *difference*. In the electric circuits of Figures 2-11 and 2-12, we noted that there are two kinds of potential difference. There is the potential *rise* that takes place in the battery and the potential *fall* or *drop* that takes place in the electric conductors and the lamp. We can maintain the distinction between rise and fall in potential and still adopt the more popular terminology by distinguishing between the **source voltage** or **applied voltage** appearing between the terminals of a source as a result of its energy-converting ability and the **voltage drop** that occurs when electric current flows through the external circuit.

With the popularity of the more general term *voltage,* much of the distinction between source voltage or applied voltage and a voltage drop has disappeared from technical literature. Some writers use the same letter symbol (either E or V) for both applied voltage and voltage drop. Our objective in this book is to *understand* why electric circuits must behave as they do. Hence, we shall maintain a distinction between source voltage and voltage drop by stating that

The letter symbol for source voltage or applied voltage is E.
The letter symbol for voltage drop is V.

That there is indeed a difference between a source voltage and a voltage drop is illustrated in Figure 2-13. When the switch is closed, both voltmeters show the same reading as the current through the lamp creates a voltage drop across the lamp equal to the applied voltage or source voltage between the terminals of the battery. When the switch is open, the voltmeter across the battery terminals still registers the source voltage resulting from the chemical action within the battery. But the voltmeter across the lamp terminals reads zero. A voltage drop can appear across the lamp only when electrons are flowing through it in an attempt to return the potential difference in the circuit to zero.

It is also worth noting at this point the prepositions which we can use with the terms current and voltage. Remembering that current is the rate of flow of electrons, we speak of current *in* or *through* a conductor or other circuit component. to talk about "current *across* the lamp" would be meaningless. Since we are not accustomed to using the term *potential difference,* we must remind ourselves that there is no such thing as a "voltage *at* a certain point." The applied voltage from a **voltage source** and the voltage drop of a load must be measured *between* two points, or *from* one point *with respect to* another, or *across* a certain circuit element. Talking about "voltage *through* a component" would also be meaningless.

We can also invent the term **amperage,** which means something that is measured in amperes. That something is, of course, electric current. The term *amperage* is sometimes used by electricians in referring to the electric current normally carried by an electrical device. The *National Electrical Code*® also uses the term **ampacity** to state the maximum allowable current for a given size of electric conductor in amperes.

2-11 CONVENTIONAL CURRENT DIRECTION

The most common form of electric conductor is the metallic conductor described in Section 2-4. The charge carriers in metallic conductors are free electrons which flow from the negative terminal of the voltage source toward the positive terminal. However, as we shall discover in Chapter 3, there are several other forms of electrical conduction. Although electric current always consists of a flow of electrically charged particles, these particles are not necessarily free electrons. For example, inside the battery in Figure 2-12, the charge carriers are *not* free electrons. When the switch is closed, *two* types of charge carriers move simultaneously in opposite directions inside the battery. Negatively charged chlorine ions move toward the negative terminal of the battery and positive hydrogen ions move toward the positive terminal.

However, when we are working with pencil-and-paper circuit diagrams, it would be very helpful if we could agree on a standard **conventional direction** for electric current. This is essential if we are to be able to calculate the *algebraic* sum of the currents at a junction point in an electric circuit. For such a chore, the *physical* nature of the actual *flow* of the charge carriers is not important.

Since most electric circuit work deals with metallic conductors, it would have been a very happy arrangement to select the direction of electron flow in such conductors as the mathematically conventional current direction. But experimenters in the days of Benjamin Franklin and Michael Faraday had no knowledge of the atomic structure of matter. They selected a conventional direction for electric current on the basis of observable effects. The clearly observable result of electric current through Faraday's electrolytic cell (described in Chapter 3) suggested that electric current consists of a flow of positively charged particles. As we have noted, positive charge carriers flow in the *opposite* direction to negative charge carriers such as free electrons and negative ions.

Therefore, all the laws and rules for electric circuits were based on a direction *from the positive terminal of the source, through the external circuit, and back to the negative terminal of the source*. As long as we keep in mind the purpose of a mathematical convention for current direction, the fact that the *physical* direction of electron flow is just opposite to the *mathematical* convention due to bad luck should not present any problem. In this text, we associate the term **current** with conventional direction and mark our circuit diagrams accordingly. If we find occasion to refer to the physical process of electrical conduction, we refer to specific charge carriers by using such terms as **electron flow.**

PROBLEMS

2-1. What is the current through a switch in an electric circuit if it takes 6 s for 30 C of electric charge to pass through it?

2.2. What is the current through a transistor if it takes 1 min for 1 C of charge carriers to arrive at the collector?

2-3. If the net drift of electrons across the imaginary cross section in Figure 2-9(b) is 10^{15} electrons per second, what is rate of electric charge flow in amperes?

2-4. Express an electric current of 4.0 mC/min in appropriate SI units.

2-5. How many coulombs of electric charge pass through the lamp in Figure 2-12 in 1 min if the current is a steady 300 mA?

2-6. How many coulombs of electric charge are added to each of the terminals of the battery by chemical action in Figure 2-12 in $\frac{1}{60}$ s if the current drain is a steady $\frac{1}{2}$ A?

2-7. How long will it take for 8 C of electric charge to flow through the switch in Figure 2-12 if the current is 250 mA?

2-8. How long will it take for 10^{20} chlorine ions to combine with the zinc if the current through the lamp in Figure 2-12 is a steady 300 mA?

2-9. If it takes 5 J of chemical energy to move 20 C of electric charge between the positive and negative terminals of a battery, what is the PD between its terminals?

2-10. What PD must be developed across the terminals of a lamp in order for a flow of 0.05 C from one terminal to the other to release 6 J of energy?

2-11. How much electric charge moves from one terminal of a load in an electric circuit to the other terminal if 50 J of energy is released at a potential drop of 120 V?

2-12. What quantity of electrons will move from the positive terminal to the negative terminal (effectively) of a battery whose terminal voltage is 1.5 V during the conversion of $\frac{1}{4}$ J of chemical energy?

2-13. What energy is required to move 5 mC of electric charge through a potential rise of 60 V?

2-14. How much electric energy is converted into light and heat energy by an electric lamp when there is a potential drop of 3.0 V as 600 mC of charge carriers move from one terminal to the other?

2-15. How much energy is involved in maintaining a 0.5-A current through a battery for 0.5 min if the PD between the battery terminals is 1.1 V?

2-16. How long will a steady 250-mA current have to flow through a flashlight lamp with a 3-V drop between its terminals for 6 J of energy to be transferred to the lamp?

2-17. What is the applied voltage of a battery in a transistor radio if 13 J of chemical energy is used up in 1 min as the battery maintains a steady current of 24 mA while the radio is operating?

2-18. An automobile storage battery is designed to maintain a constant terminal voltage of 12V as 3.5×10^6 J of chemical energy are converted into electric energy. How long can this battery maintain a constant current of 20 A in expending the stored chemical energy?

2-19. What current must be flowing through a load in an electric circuit if energy is being delivered to the load at the rate of 18 J/s with a 120-V drop across its terminals?

2-20. Assuming perfect efficiency, at what rate is chemical energy being converted in a battery whose voltage rise is 3 V if a lamp connected to the battery draws a steady 150-mA current?

REVIEW QUESTIONS

2-1. The SI unit of energy is the **joule.** The joule is a derived unit equal to one newton meter. Why is it possible to express energy in terms of force and distance?

2-2. How can we demonstrate that electric force is a *field* force rather than a *contact* force?

2-3. What effect would be observed if we touched a charged *ebonite* rod to both balls in Figure 2-1?

2-4. Compare Coulomb's law, which states the factors governing the magnitude of an electric force, with Sir Isaac Newton's law of universal gravitation.

2-5. What is the distinction between the **valence** electrons of an atom and the remainder of the orbital electrons?

2-6. What is the distinction among an **atom,** a **positive ion,** and a **negative ion?**

2-7. How does a **molecule** differ from an **atom?**

2-8. What is the significance of the **shell** concept of atomic structure in determining whether an element is likely to be a good electric conductor?

2-9. Argon has an atomic number of 18 (18 orbital electrons). With reference to the shell structure of an atom, discuss the probable chemical activity of argon.

2-10. Silver has an atomic number of 47. Show how the shell structure of an atom supports the observation that silver is an excellent electric conductor.

2-11. What is the significance of the term **free electron?**

2-12. What is the relationship between free electrons and a metal's ability to act as an electric conductor?

2-13. Distinguish between the behavior of valence electrons in **ionic bonding** and **covalent bonding.**

2-14. From our present knowledge of the structure of atoms, explain why rubbing a glass rod with a silk cloth places a positive charge on the glass rod.

2-15. Although the current in the electric circuit of Figure 2-12 is zero when the switch is open, free electrons in the conductors are not stationary. Describe their motion.

2-16. How is the random motion of free electrons shown in Figure 2-7(a) accounted for?

2-17. What change in electron motion occurs in the circuit of Figure 2-12 when the switch is closed?

2-18. Figure 2-9(b) indicates electrons simultaneously moving in opposite directions in a current-carrying conductor. How is this possible?

2-19. Distinguish between electric **current** and the effective **velocity** at which electrons move along the length of an electric conductor.

2-20. Why is the term **drift** appropriate for describing electric current in a metallic conductor?

2-21. What is the magnitude (in coulombs) of the negative charge of a single electron?

2-22. Although not an approved SI unit, the **ampere-second** is a unit of electrical measurement. In the measurement of what electrical property would it be used?

2-23. Explain the forces acting on a free electron at the center of the conductor shown in Figure 2-8.

2-24. The water in Figure 2-10 loses potential energy as it flows through the turbine of the generating station. How is the energy replenished to maintain the operation of the station over the years?

2-25. In the hydroelectric generating system of Figure 2-10, a potential difference is produced by moving a unit quantity of water against a gravitational force. In an electric circuit, what is the equivalent force which makes a potential difference between two terminals possible?

2-26. Both gravitational force and electric force are expressed in newtons. Why is electromotive "force" not a true force?

2-27. Explain how the copper rod in the simple battery in Figure 2-12 becomes positively charged.

2-28. Inside the battery in Figure 2-12, positively charged hydrogen ions are acting as electric charge carriers. Insofar as these positive charge carriers are concerned, which terminal of the battery is considered to be at the higher potential? Why?

2-29. When the switch in the electric circuit in Figure 2-12 is closed, light energy is produced by the lamp. Where does this energy originally come from?

2-30. Although charge carriers flow from the load back to the source in one of the conductors of an electric circuit, energy is not transferred from the load back to the source. Explain this in terms of potential rise and potential fall.

2-31. Why does chemical action cease in a battery when the switch in the electric circuit is opened?

2-32. Discuss, using numerical examples, how the particular arrangement of letter symbols in the equation $V = W/Q$ satisfies the definition of the **volt.**

2-33. Why is the letter symbol E used in Example 2-4 and the letter symbol V in Example 2-5?

2-34. If we connect a voltmeter across the *open* switch terminals in the circuit of Figure 2-13, it will show the same "voltage" as the voltmeter connected across the battery. Explain.

3

CONDUCTORS AND INSULATORS, SEMICONDUCTORS, AND VOLTAGE SOURCES

As we noted in Section 2-11, we can proceed with a numerical analysis of pencil-and-paper electric circuit diagrams without reference to the physical process of electrical conduction. Chapter 3 is not part of a typical "electric circuits" course. No new numerical units are introduced. Hence, if time is limited, any or all sections of this chapter may be skipped or assigned for self-study. Chapter 3 is included for students who have not had an opportunity to study the physics of electric circuits separately. So that we may develop a better understanding of the behavior of some of the components we shall encounter in electric circuits, we shall pause in our numerical analysis to consider the electrical properties of several important nonmetals on a qualitative basis.

3-1 CONDUCTORS

In Section 2-4, we discovered that electric current consists of a net drift or *flow* of electrically charged particles. To be suitable as an electric conductor, a material must possess a large number of such charge carriers per unit volume. It also follows that the greater the number of charge carriers per unit volume, the better that material will be as an electric conductor. As we noted in Section 2-4, metals possess a very large number of charge carriers per unit volume in the form of free electrons. All metals may be used as electric conductors, although some metals possess more free electrons per unit volume than others.

Copper is the most popular electric conductor material. It is reasonably inexpensive, can readily be drawn into wire, and rates second only to silver (which is much more costly) in its abundance of free electrons. We have estimated that a cubic centimeter of copper contains approximately 8.5×10^{22} free electrons. Silver has about 5% more free electrons per unit volume than copper. Aluminum is used commercially as an electric conductor material to some extent because, although it has only 60% as many free electrons per unit volume as copper, it is much lighter and less expensive. We can compensate for aluminum's decreased conductivity by increasing the diameter of aluminum wires. Some of the metal alloys we use in electric heater elements have less than 1% of the number of free electrons that copper has. This still represents something like 6×10^{20} free electrons per cubic centimeter. Such alloys are, therefore, reasonably good electric conductors.

The most familiar form of metallic electric conductor is the copper wire, usually with an insulating sheath. The diameter of a copper conductor depends on the magnitude of the electric current it is required to conduct. SI provides for standard diameters for electric conductors with the SI wire gauge number being 10 times the conductor diameter in millimeters. However, in North America, the standard diameters of conductors are still specified in **mils** (thousandths of an inch) using the American wire gauge table in Appendix 3. To maintain flexibility, the cross-sectional area of larger conductors is broken up into several strands twisted together to form a *stranded* conductor.

The high degree of miniaturization of electronic components resulting from a manufacturing technique called **large-scale integration** (LSI) has led to an alternative form of metallic conductors. Instead of interconnecting electronic circuit components with copper wires, the required circuit layout is printed on a copper-coated insulating board. The unwanted copper is then etched away to leave a printed-wiring configuration which is usually much more elaborate than that shown in Figure 3-1. The

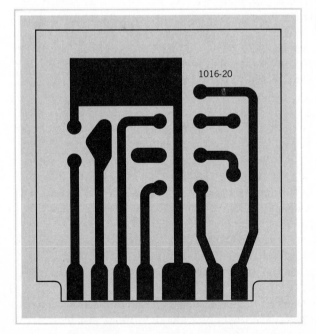

FIGURE 3-1
Small Printed-Wiring Board.

electronic components are connected to the **printed-circuit board** (PCB) by poking their leads through the proper holes in the board. Electrical connection is then assured by running the PCB over a bath of molten solder.

3-2 INSULATORS

Fortunately, the air surrounding the conductors in the basic electric circuit of Figure 1-1 has relatively few electric charge carriers per unit volume. If air were a reasonably good conductor of electricity, we would not be able to contain electric current within the desired circuit. Also, electric current would continue to flow when we *open* the switch to the position shown in Figure 1-1. Dry air consists mainly of oxygen molecules and nitrogen molecules in which the orbital electrons are all firmly bound to their respective nuclei. Consequently, it is quite difficult for the charge carriers (free electrons) in the metal conductors to cross the boundary and to move from atom to atom through the air surrounding the conductors. Therefore, we rate dry air as a nonconductor of electricity or an electric **insulator.**

In some electric circuits where it is unlikely that anything other than air will come into contact with the conductors, no other insulation is required. In the electric circuit illustrated in Figure 1-1, we can safely use bare copper conductors. Similarly, the conductors of the high-tension transmission lines crossing the countryside suspended from steel towers can be bare metal. However, they must be suspended from the towers by porcelain insulators to prevent leakage of current via the metal of the towers.

When there is a possibility that the conductors of an electric circuit may touch one another (for example, the two conductors of a lamp cord), we must use conductors that are surrounded by a suitable solid insulating material. In ionic-bonded molecules, each atom achieves a filled subshell in its orbital electron structure by one atom lending its valance electron(s) to the other. As a result, all the valence electrons are bound to the negative ions. Hence, ionic solids are (theoretically) electric insulators since none of the valence electrons are free to act as energy carriers. In practice, we find that it is fairly easy to break the ionic bond between the atoms of an ionic solid by adding energy to the molecule. Consequently, we seldom use ionic compounds in manufacturing practical insulating materials.

Polyethylene molecules are formed by the covalent bonding of carbon and hydrogen atoms, as illustrated in Figure 3-2. Carbon has six electrons, two in the first shell and four in the second shell. Four more electrons are needed to completely fill the second shell. In the polyethylene molecule, each carbon atom shares all of its second-shell electrons with neighboring carbon and hydrogen atoms, as shown. One electron is shared with the carbon atom on its left, one with the hydrogen atom above it, one with the carbon atom on its right, and one with the hydrogen atom below it. This particular form of covalent bonding creates a long, filament-like molecule in which every atom now has a completely filled outermost shell. The binding force in such a molecule is so great that it is very difficult for an electron to break away from the pattern. Thus, covalent-bonded materials of this type, called **polymers,** are excellent electric insulators. Most modern insulating materials such as rubber, plastics, varnish, enamel, silk, cotton, paper, oil, mica, and asbestos are polymers.

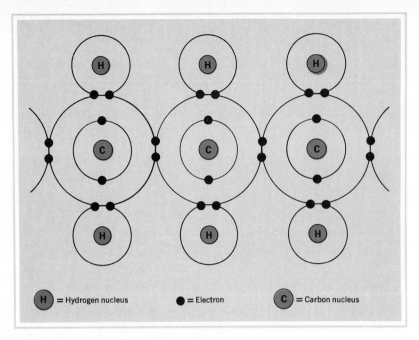

H = Hydrogen nucleus   ● = Electron   C = Carbon nucleus

FIGURE 3-2
Structure of a Polyethylene Molecule.

If we could manufacture a 100% pure polymer, it would possess no free electrons and would, therefore, be an ideal insulator. But practical insulating materials contain some impurities which do have a few free valence electrons that are not included in the pattern of covalent bonds. Oxidation of polymer molecules makes additional free electrons available for conduction. (Rubber deteriorates rapidly when exposed to ozone, which is created from oxygen by an electric spark.) Moisture absorbed by an insulator introduces ions which, being in a liquid state, can migrate slowly through the insulator. Consequently, polystyrene, which we consider a very good electric insulator, has over 6×10^{10} free electrons per cubic centimeter. Although this seems like a large number, compared with copper it is extremely small. Therefore, for practical purposes, the leakage of the charge carriers through a layer of polystyrene or other polymer is negligible.

3-3 ENERGY-BAND DIAGRAMS

Energy-band diagrams provide us with a quick means of determining whether a solid material will behave as a conductor or an insulator in electric circuits. In considering a model of an atom in Chapter 2, we noted that the greater the radius of an electron orbit, the greater must be the energy level of the electrons in that orbit. We also noted that the wave-like motion of electrons allowed them to appear only at definite *permissible* energy levels. Instead of showing permissible energy levels as complete orbits, as in Figure 2-3, we can draw a diagram showing permissible energy levels for a single atom on a thermometer-like scale, as in Figure 3-3.

Above the permissible energy levels on the scale in Figure 3-3, we can mark an

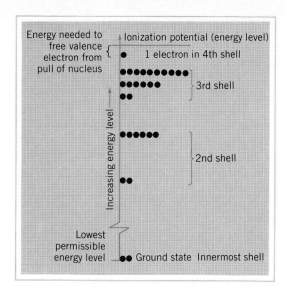

FIGURE 3-3
Energy-Level Diagram for a Single Copper Atom.

energy level which must be acquired by an electron to escape completely from the attraction of the nucleus. This energy level is called the **ionization potential.** It is appearent from Figure 3-3 that only the electrons farthest from the nucleus and in the highest-occupied energy levels will have an opportunity to acquire sufficient energy to reach the ionization potential. Hence, we can discard the lower portion of this diagram and show only the outer-most occupied or **valence** energy level and above.

As many atoms of a material are bonded together to form a solid, the energy levels of every atom are affected by the proximity of adjacent atoms. As a result, there are now many more permissible energy levels available to each atom. As shown in the energy diagrams of Figure 3-4, the *unoccupied* permissible energy levels now form a *band* of closely spaced energy levels reaching all the way to the ionization potential. Any electron which has sufficient energy to fall within this band can readily acquire

FIGURE 3-4
Energy-Band Diagrams for Solids.

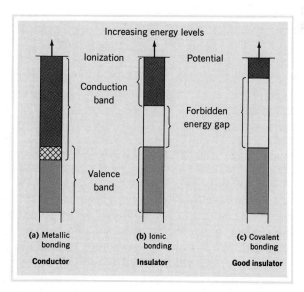

enough energy to move up through the unoccupied energy levels of this band until it has sufficient energy to be *free* from the attraction of the parent nucleus. These electrons become the *free electrons* which, in turn, become electric charge carriers. Consequently, this band is called the **conduction band.** Similarly, the single energy level occupied by valence electrons of a single atom now becomes a band of permissible energy levels called the **valence band.** The width of the **forbidden-energy-**level gap between the valence band and the conduction band depends on the exact manner in which adjacent atoms interact as they bond themselves together to form a solid.

In metallic bonding, the valence electrons do not fill all permissible energy levels in the valence band. There are unoccupied levels in the conduction band which are at the same energy level as some of the valence electrons. Consequently, in the energy-band diagram for metals, the conduction and valence bands overlap. This allows the valence electrons of metals to move into the conduction band and to acquire sufficient energy to become free from the parent nucleus, as was illustrated in Figure 2-5. Any solid in which there is no forbidden energy gap in the energy-band diagram has valence electrons which can readily move up through the conduction band to become charge carriers. Hence, any solid with an energy-band diagram similar to that shown in Figure 3-4(a) is classified as an electric conductor.

In Section 2-3, we discovered that atoms of an ionic-bonded solid arrange themselves so that every atom will fill a particular permissible energy level by either *lending* one or more valence electrons to or *borrowing* valence electrons from an adjacent atom. Consequently, all permissible energy levels in the valence band of an ionic-bonded solid are filled, and there is an appreciable forbidden energy gap between the highest permissible level in the valence band and the lowest permissible level in the unoccupied conduction band. The only way an electron in an ionic solid can acquire sufficient energy to reach an unoccupied level in the conduction band is for it to receive enough energy to suddenly jump across the forbidden energy gap. Such a move requires the breaking of an ionic bond. The presence of a forbidden energy gap in the energy-band diagram of Figure 3-4(b) represents an electric insulator. The only electrons present in the conduction band to act as electric charge carriers are those which reached that level as a result of a breaking of ionic bonds by heat energy possessed by the solid, by impurities in the solid, or by strong external electric forces.

Covalent-bonded atoms achieve full valence bands by *sharing* their valence electrons. Again, no electrons appear in the conduction band of the energy diagram unless covalent bonds are broken. Since the energy required to break the covalent bonds which hold polymers together is comparatively high, the forbidden energy gap in the energy-band diagram of Figure 3-4(c) is quite wide.

3-4 INSULATOR BREAKDOWN

In practice, no insulator is as perfect as energy-band diagrams might lead us to believe. There is always some energy present which causes some of the valence bonds to be broken. Even an excellent insulator such as polystyrene has a considerable number of electrons in the conduction band at room temperature. In general, as the temperature increases, we can expect to find even more electrons occupying energy levels in the conduction band. These charge carriers permit a small leakage current through practical insulators.

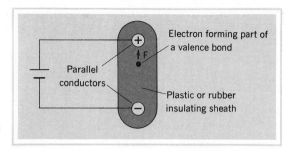

FIGURE 3-5
Electric Stress on Valence Bonds in an Insulator.

Figure 3-5 shows a voltage source connected between the two conductors of a lamp cord. As we noted in Chapter 2, electric force is a *field* force. Hence, the potential difference between the two conductors gives rise to an electric force which acts on the electrons in the valence bonds of the insulating material between the conductors. As the potential difference of the source is increased, a critical level of stress is reached at which billions of valence bonds are broken and billions of electrons jump the forbidden energy gap into the conduction band, resulting in a *rupture* or *breakdown* of the insulator. The potential difference required to rupture a particular insulator will depend on the spacing between the two conductors and the strength of the valence bonds of the insulating material. Some insulating materials can withstand a much greater electric stress for a given thickness than others.[†] Because of the smaller number of atoms per unit volume, gases break down more readily than solids. Consequently, air rates rather poorly in terms of dielectric strength, but it is cheap enough so that we can compensate for this by using a thicker sheath of air around electric conductors such as those used in overhead transmission lines.

3-5 ELECTROLYTIC CONDUCTION

In developing our concept of the **coulomb** and the **ampere,** we considered only free electrons as electric charge carriers. Although by far the most significant type of charge carrier, free electrons are not the only electrically charged particles which can act as charge carriers in an electric circuit. In the chemical battery in Figure 2-12, we found that electric current consists of positive and negative ions moving simultaneously in opposite directions. Michael Faraday made a fairly thorough study of the observable effects of **electrolytic conduction** during the 1830s.

Figure 3-6 shows an **electrolytic cell** or electroplating bath consisting of a silver (Ag) **electrode** and a steel (Fe) **electrode** in the form of a spoon dipped into a solution of silver nitrate ($AgNO_3$). This current-carrying solution is called an **electrolyte.** The silver electrode is connected to the positive terminal of the battery and is called the **anode.** The spoon is connected to the negative terminal of the source and is called the **cathode.**[‡] Michael Faraday reasoned that, since he could measure an electric current in the copper conductors connecting the source to the electrodes, electric current must

---

[†]See Table 11-1 on page 281.

[‡]Faraday is credited with originating the terminology, including electrode, electrolyte, and electrolysis, which is now universally used with reference to electrolytic conduction of electric current.

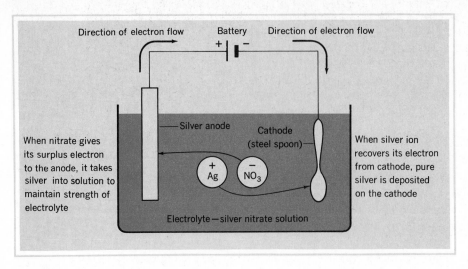

Direction of electron flow — Battery — Direction of electron flow

Silver anode

Cathode (steel spoon)

+ Ag

− NO$_3$

When nitrate gives its surplus electron to the anode, it takes silver into solution to maintain strength of electrolyte

When silver ion recovers its electron from cathode, pure silver is deposited on the cathode

Electrolyte—silver nitrate solution

FIGURE 3-6
Simple Electrolytic Cell.

be flowing through the silver nitrate solution to complete the circuit. Indeed, he noticed that, as current flowed through the cell, silver disappeared from the anode and the same amount of silver was added to the cathode. With no knowledge of the electron (circa 1835), Faraday was able to state his first law of electrolysis from his observations.

The mass of a substance liberated by the anode and deposited on the cathode varies directly with the quantity (Q) of electricity passing through the cell.

At one time, the principle of electrolysis formed the basis for the international standard for the ampere, which was the current that would deposit silver at the rate of 0.001 118 gram per second with an apparatus similar to that of Figure 3-6.

The basis for our present-day concept of the nature of the charge carriers in an electrolyte was suggested in 1887 by Svante August Arrhenius. As a result, we can explain both the apparent transportation of the silver from the anode to the cathode and the process by which electric current can pass through the silver nitrate solution. As shown in Figure 3-6, silver nitrate is an ionic compound in which the silver lends its valence electron to the nitrate radical to form a molecule consisting of a positive silver ion and a negative nitrate ion. Arrhenius suggested that, when ionic compounds such as acids, bases, and salts are dissolved in water, the ions tend to **dissociate;** that is, the silver ions and nitrate ions can roam around independently in the solution.

If we connect a voltage source to the electrodes, we create a potential difference between them. And since unlike charges have a force of attraction between them, the dissociated positive silver ions are attracted to the negatively charged cathode. As the silver ion touches the metallic cathode and regains its missing electron, it is deposited out of solution on the cathode as pure silver. At the same time an equal number of negative nitrate ions are attracted to the positively charged anode. As the NO_3^- gives up its surplus electron to the positively charged anode, it becomes chemically active and takes an atom of silver from the anode into solution with it, thus maintaining the strength of the electrolyte. No *chemical* energy is used up in this cell, since identical amounts of silver go into solution at the anode and are deposited out at the cathode. The work done in transporting the silver from the anode to the cathode is done at the expense of electric energy taken from the source.

We can now make an important observation which was completely hidden from Faraday. At the same time that electric current in the copper conductors connecting the source to the electrolytic cell consists of a unidirectional flow of free electrons, **two different types of charge carriers—positive silver ions and negative nitrate ions—are flowing simultaneously in opposite directions through the electrolyte.** But the observable effect some 100 years ago was that, when electric current passed through the electrolytic cell of Figure 3-6, the anode lost weight and the cathode gained weight. Hence, it was assumed that the direction of electric current was from anode to cathode to convey the silver with it. As we noted in Section 2-11, this became the **conventional** direction for electric current.

3-6 CONDUCTION IN VACUUM AND GASES

If we connect a voltage source to a pair of electrodes in a vacuum, as in Figure 3-7(a), no electric current will flow since there are no charge carriers in a vacuum. The free electrons in the electrodes do not possess sufficient energy at room temperature to escape from the surface of the metal. However, if we supply additional energy to the cathode, usually by heating it to red heat with an electric heater, as in Figure 3-7(b), some of the free electrons in the cathode attain sufficient velocity to escape from the surface of the cathode (just as a rocket must attain an escape velocity of 11.2 kilometers per second to escape the earth's gravitational attraction). Once **emitted** from the cathode, these electrons are attracted by the positively charged anode (or plate) and electric current flows around the circuit. Hence, under proper conditions, the charge carriers in a vacuum are electrons, just as in metallic conductors.

Suppose we reverse the terminals of the battery in Figure 3-7(b) so that the anode

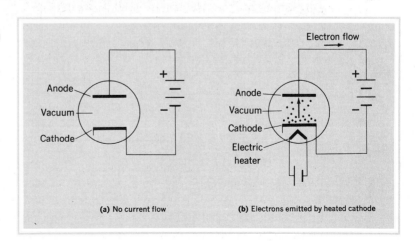

FIGURE 3-7
Electric Current in a Vacuum.

(a) No current flow

(b) Electrons emitted by heated cathode

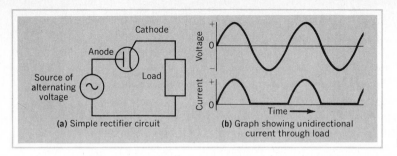

(a) Simple rectifier circuit (b) Graph showing unidirectional current through load

FIGURE 3-8
Vacuum-Tube Rectifier Circuit.

of the vacuum tube is now connected to the negative terminal of the voltage source and the cathode is connected to the positive terminal. The voltage source will attempt to make electrons flow in the reverse direction through the tube. Since the anode is not heated, it cannot emit electrons as does the cathode. Consequently, just as in Figure 3-7(a), no electric current can flow through the tube, even though the cathode is still red hot. Thermionic vacuum tubes, therefore, allow electric current only in one direction. By connecting a vacuum **diode** (two-element tube) in series with a source of alternating voltage[†] and a load, as shown in Figure 3-8(a), current can flow through the load only on the half-cycle when the anode of the diode is positive with respect to the cathode, as shown in Figure 3-8(b). The vacuum tube in this circuit performs the function of a **rectifier.**

If we now introduce a small amount of a gas such as neon or mercury vapor into our vacuum tube, as is the case in neon signs and fluorescent lamps, still another type of electric conduction takes place. The electrons emitted by the cathode pick up kinetic energy as they are attracted toward the anode. As these high-speed electrons collide with the gas atoms between the cathode and the anode, they possess enough energy to knock some of the electrons of the gas atoms out of orbit, leaving the gas as positively charged ions. There will now be an increase in the number of electrons arriving at the anode. There will also be a simultaneous flow of positive gas ions to the cathode to recover their missing electrons.

Since the vacuum tube was the heart of almost all electronic circuits until some 25 years ago, many electronic circuits textbooks chose electron flow as a reference direction for electric current. It was rather awkward to talk about the conventional direction of current from anode to cathode in view of the actual physical operation of vacuum tubes. However, as transistors have taken over most vacuum-tube functions in electronic circuits and free electrons are not the only charge carriers involved, it is once more desirable to maintain a clear distinction between the actual motion of charge carriers and the conventional direction used in analyzing pencil-and-paper circuit diagrams.

3-7 SEMICONDUCTORS

Column IV of the periodic table of elements includes carbon (six electrons), silicon (14 electrons), and germanium (32 electrons). These elements are grouped because each has four electrons which can act as valence electrons, and each is short four electrons to fill a subshell or energy level. (Check with Figure 2-3.) These three

---

[†]See Section 17-2 for a description of an alternating voltage.

52 PART I THE BASIC ELECTRIC CIRCUIT

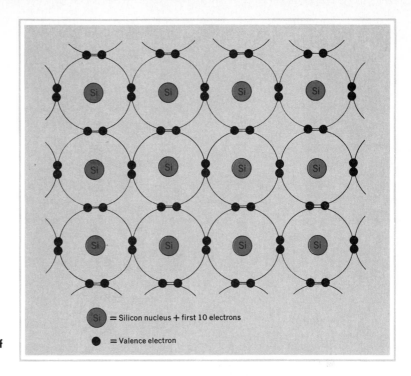

FIGURE 3-9
Covalent Bonds in a Crystal of
Pure Silicon.

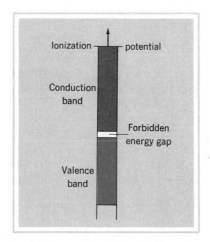

FIGURE 3-10
Energy-Band Diagram for a Crystal of
Pure Silicon.

elements tend to form a crystal structure by a covalent-bonding pattern. Pure crystals
are rare in nature, but we can "grow" pure or **intrinsic** crystals of germanium and
silicon from the purified molten metal. Although the crystal is a three-dimensional
structure, we can represent the covalent bonds by a two-dimensional sketch, as in
Figure 3-9. Each silicon (or carbon or germanium) atom shares one of its four valence
electrons with four adjacent atoms.

Drawn on the same scale as the energy-band diagrams of Figure 3-4, the
energy-band diagram of intrinsic silicon in Figure 3-10 has a forbidden energy gap
between the completely filled valence band and the unoccupied conduction band,
which is much smaller than that of good insulators. This forbidden energy gap is small

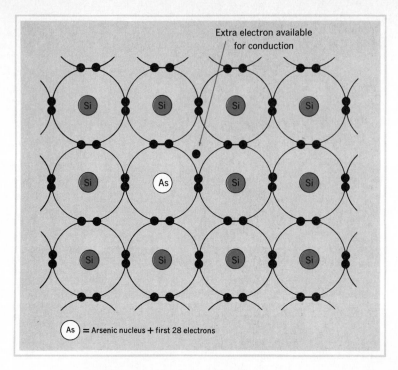

Extra electron available for conduction

$\left(\begin{array}{c}As\end{array}\right)$ = Arsenic nucleus + first 28 electrons

FIGURE 3-11
Donor Atom in an *n*-Type Semiconductor Crystal.

enough that, at normal temperatures, many millions of valence electrons can acquire sufficient energy from the heat energy possessed by the crystal to break from their valence bonds and jump into an energy level in the conduction band. As fast as these electrons give up their newfound energy and reform covalent bonds, they are replaced by more conduction electrons from freshly broken bonds. Hence, at room temperature, there are always sufficient charge carriers in these crystals that we would have to rate them as rather poor insulators. In spite of the forbidden energy gap in the energy-band diagram, we can make use of the equally poor electrical *conductivity* of germanium and silicon crystals in developing circuit elements with rather special electrical characteristics. We, therefore, designate such materials as **semiconductors.**

With no forbidden energy gap in the energy-band diagram for metals such as copper and aluminum, there is a conduction electron for every atom. But, in semiconductors, the number of charge carriers can vary considerably, depending on the energy available for breaking valence bonds. Hence, we must expect the electrical conductivity of semiconductors to vary considerably under the influence of various forms of energy. We can also control the number and type of charge carriers in semiconductors by controlling the impurities introduced into a semiconductor crystal.

In growing a silicon crystal, suppose we *dope* the pure silicon by adding one part in 10 million of a *pentavalent* element such as arsenic from column V of the periodic table. Arsenic atoms take up positions in the lattice structure of the crystal, as shown in Figure 3-11. But only four of the five valence electrons can enter into valence bonds. The fifth electron is only loosely held by the attraction of the arsenic nucleus. If we examine the energy-band diagram for this doped crystal on the expanded scale of Figure 3-12(a), we find a new permissible energy level just below the conduction band of the pure crystal. Very little energy is required to raise these electrons to the conduction band, that is, to break them *free* of the attraction of the arsenic nucleus.

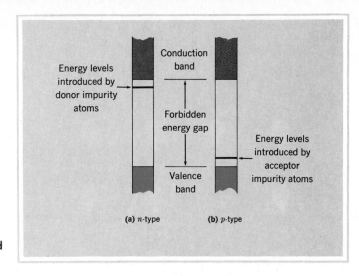

Conduction band

Energy levels introduced by donor impurity atoms

Forbidden energy gap

Energy levels introduced by acceptor impurity atoms

Valence band

(a) n-type   **(b)** p-type

FIGURE 3-12
Energy-Band Diagrams for n- and p-Type Semiconductor Crystals.

Adding a pentavalent impurity to a semiconductor crystal thus makes a controlled number of electrons available as electric charge carriers. Hence, the arsenic atoms are called electron-**donor** atoms. The doped semiconductor is called an **n-type** semiconductor, since negatively charged electrons are available as charge carriers.

Suppose we dope the pure silicon with a very small amount of a *trivalent* element such as indium from column III of the periodic table. As indium atoms take up positions in the lattice structure of the semiconductor crystal, one valence bond cannot be completed, since the indium atom has only three valence electrons. As shown in Figure 3-13, there is a **hole** in the covalent bonding where an electron is missing. Since indium has only three valence electrons as compared with four for the pure semicon-

FIGURE 3-13
Acceptor Atom in a p-Type Semiconductor Crystal.

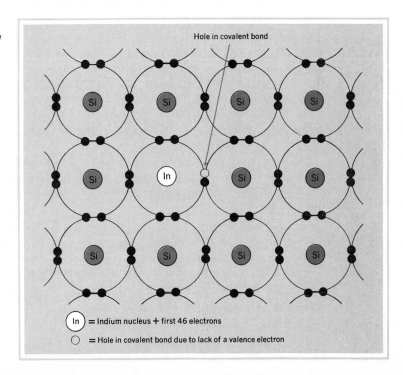

Hole in covalent bond

In = Indium nucleus + first 46 electrons
◯ = Hole in covalent bond due to lack of a valence electron

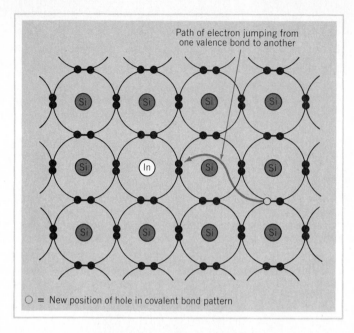

Path of electron jumping from one valence bond to another

○ = New position of hole in covalent bond pattern

FIGURE 3-14
Hole Flow in a *p*-Type Semiconductor.

ductor atoms, the normal energy level for the three valence electrons of indium comes slightly above the upper limit of the valence band in the energy-band diagram for the pure crystal, as shown in Figure 3-12(b). As a result, very little energy is required for an electron to break a valence bond and move into this new electron-**acceptor** energy level. Relating this to a sketch of a semiconductor with an **acceptor** impurity, we note that very little energy is required for an electron to break out of an existing covalent bond to move into the hole in one of the valence bonds of the indium acceptor atoms, as shown in Figure 3-14. If we were not quick enough to see this electron move, the observable difference between Figure 3-13 and Figure 3-14 would be that the **hole** has apparently moved to a new location. Another nearby valence bond would break and once again the hole would appear to have moved.

It takes a while to become used to the concept of a moving hole where an electron should be but is not. Suppose we place a string of 12 colored lights on a Christmas tree only to discover that we have only 11 lamps. We move lamps from one socket to another as we try to leave the empty socket in the most inconspicuous place. What attracts our attention with each move is not the lamps which have been moved; rather, it is the empty socket, which appears to move every time we switch a lamp to another socket.

In dealing with semiconductors with trivalent impurities, we find it more convenient to consider the **hole flow** than to keep track of electrons jumping from one valence bond to another. Holes move in the opposite direction to electrons, and since a hole is the absence of a negative electron, the electric charge where a hole is located is no longer electrically neutral. Hence, holes are considered to have a *positive* charge equal in magnitude to the negative charge of an electron. Semiconductors doped with acceptor impurities are, therefore, called *p*-**type** semiconductors. These semiconductors allow for electric conduction via positive holes which flow in the valence band of the energy-band diagram.

3-8 *pn*-JUNCTIONS

We can illustrate the justification for the cost of developing two types of semiconductor crystals having different types of charge carriers by examining an electron device consisting of a layer of a *p*-type semiconductor diffused onto a layer of *n*-type semiconductor, as shown in Figure 3-15. In considering the mechanics of electrical conduction in semiconductor crystals, we have overlooked the equal number of electrons in the conduction band and holes in the valence band created as a result of covalent bonds broken by the heat energy possessed by the crystals. The quantity of such charge carriers which give rise to **leakage current** in semiconductors is negligible in good semiconductors. The **majority carriers** are those specifically created as described in Section 3-7. In the sketch in Figure 3-15, only the majority carriers are shown.

When *p*-type and *n*-type semiconductors are diffused together, as shown in Figure 3-15, some of the conduction electrons near the junction in the *n*-type semiconductor cross into the *p*-type semiconductor to fill the holes in the valence bonds near the junction. This leaves a **depletion layer** on both sides of the junction where the concentration of charge carriers is greatly reduced. The flow of electrons from left to right across the junction also creates a small potential difference between the two types of semiconductors, with the *p*-type semiconductor now becoming slightly more negative than the *n*-type, which has lost some electrons. Just as the chemical action in the battery in Figure 2-12 ceases when the potential difference builds up to a certain magnitude, electrons will cease flowing across the junction in search of holes in valence bonds when this depletion-layer potential or **barrier potential** reaches a certain magnitude.[†]

We can increase the potential difference between the two semiconductors by connecting them to an external voltage source. If the polarity of the external voltage

FIGURE 3-15
Cross Section of a *pn*-Junction.

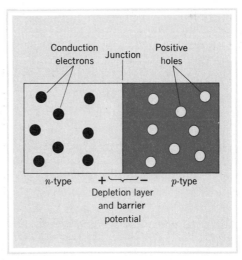

[†]$V_B = 0.3$ V for a germanium *pn*-junction and 0.7 V for a silicon *pn*-junction.

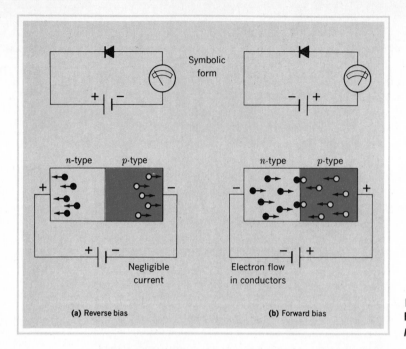

Symbolic form

n-type p-type

Negligible current

n-type p-type

Electron flow in conductors

(a) Reverse bias   (b) Forward bias

FIGURE 3-16
Rectifier Action of a *pn*-Junction.

source *increases* the natural contact potential, as shown in Figure 3-16(a), the electrons in the *n*-type material are attracted *away from* the junction. Similarly, positive holes in the *p*-type semiconductor are attracted away from the junction. As a result, electric current in the circuit is very small. This polarity of external source voltage **reverse biases** the *pn*-junction.

If, on the other hand, we apply the external source voltage with the polarity shown in Figure 3-16(b), the charge carriers in both types of semiconductor are driven *toward* the junction where the holes and electrons recombine into valence bonds. Replacement charge carriers are introduced into the semiconductors by the voltage source and a considerable electric current is maintained. This polarity of external source voltage **forward biases** the *pn*-junction. Hence, the *pn*-junction performs as a more compact and more efficient **rectifier** than the vacuum diode described in Section 3-6.

3-9 BATTERIES

For an electric circuit to convey energy from one location to another, we require a device that can generate an electric pressure or potential rise which, in turn, will force free electrons to flow around the circuit. To accomplish this, a practical **voltage source** (such as the simple chemical battery of Figure 2-12) seemingly pumps electrons from one of its terminals to the other. Since this involves moving electrons within the voltage source away from the positive terminal (which attracts electrons) and toward the negative terminal (which repels electrons), some form of energy must be expended. Therefore, as the current in an electric circuit conveys energy to a load, energy must be expended in the voltage source in order to maintain a constant potential rise between its terminals. Actually then, a practical voltage source is an energy *convertor,* which converts some readily available form of energy into electric energy in the form of charge carriers flowing under pressure around the circuit. Similarly, a load in an electric circuit is a device that can *convert* electric energy into heat, light, mechanical, or other useful forms of energy.

We have already encountered one simple voltage source in the form of the simple wet-cell battery of Section 2-8. As long as the switch is closed in Figure 2-12, energy is taken from a chemical form and conveyed by the electric current to the lamp where it is converted into light energy and heat energy. While this energy transfer is taking place, the combining of chlorine ions with the zinc dilutes the acid and eats away the zinc rod. In time then, this battery can no longer maintain a potential difference. We have noted that when we open the switch, chemical action will cease when the potential difference builds up to a level at which the chlorine ion is no longer able to reach the zinc atoms. If we raise the potential difference above this level by connecting some other electric generator to the battery, it is possible to separate the chlorine ions from the zinc chloride, thus **recharging** the battery at the expense of electric energy from the external generator. However, recharging this type of battery in this manner is not economically practical. It is cheaper to replace the used up acid and zinc rod. Cells that are discarded when their chemical energy is exhausted are called **primary** cells.

The practical limitations of the simple wet-cell battery restrict its use to laboratory experiments. The liquid nature of the **electrolyte** means that the battery is not readily portable. Another disadvantage results from the release of hydrogen gas at the copper rod as the hydrogen ions recover their missing electrons from the copper. After the switch in Figure 2-12 has been closed for a few minutes, the copper is blanketed with hydrogen bubbles. Since hydrogen gas is an electric insulator (see Figure 2-4), this layer of hydrogen bubbles prevents hydrogen ions from getting to the copper rod. This action, which is called **polarization,** appreciably limits the current in the circuit.

The familiar **dry-cell** battery of Figure 3-17 overcomes these two disadvantages

FIGURE 3-17
Simplified Cross Sections: (a) Zinc-Carbon Dry Cell; and (b) Manganese-Alkaline Dry Cell.

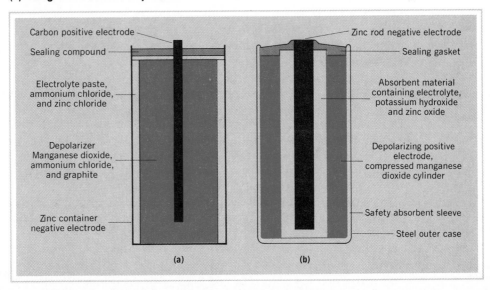

by virtue of its construction. The electrolyte must still be *wet* so that positive and negative ions can migrate to the respective electrodes. But in the so-called dry battery, the electrolyte either is mixed with a suitable binder to form a paste or saturates an absorbent spacer. Polarization is prevented by surrounding the positive electrode with a chemical **depolarizer** which will absorb the hydrogen as fast as it is formed.

Several combinations of electrodes and electrolytes are currently used in the manufacture of commercial dry-cell batteries. The original dry-cell construction illustrated in Figure 3-17(a) uses a zinc can as the negative electrode, a carbon rod as the positive electrode, and an electrolyte paste consisting of ammonium chloride and zinc chloride. As shown in Figure 3-17(a), most of the space in the **zinc-carbon cell** is taken up with a depolarizer, the active ingredient of which is manganese dioxide.

Although the chemical action is not quite so simple as that of the metal-acid wet cell, the net results are the same. Negative ions in the electrolyte attempt to corrode the zinc and, in doing so, add surplus electrons to the zinc electrode. Hydrogen ions then rob the carbon rod of electrons, thus building up a positive charge on the carbon electrode. Again, hydrogen bubbles try to form on the positive electrode, but the manganese dioxide depolarizer combines with the hydrogen to prevent polarization from taking place. Some of the electrolyte material and graphite (which is an electric conductor) are mixed with the depolarizer thus effectively increasing the surface area of the positive electrode. Since hydrogen gas is not allowed to form, the cell can be sealed to make the dry battery completely portable. The voltage generated by a standard zinc-carbon dry battery is 1.5 volts, as compared to 1.1 volts for the zinc-copper wet cell.

But the zinc-carbon battery is not without its drawbacks. Impurities within the cell allow *local* chemical *action* to continue, even when the external circuit is open. Hence, the *shelf life* of zinc-carbon batteries is limited. An improvement on the zinc-carbon battery is the **manganese-alkaline cell,** shown in Figure 3-17(b). As the name indicates, the electrolyte is an alkaline solution of potassium hydroxide and zinc oxide. The negative electrode is still zinc, but in a highly purified form to minimize self-discharge through local action. The positive electrode is made by compressing pure manganese dioxide into cylinders. As a result, the positive electrode provides its own depolarizing action. As shown in Figure 3-17(b), the manganese-alkaline cell is built "inside-out" from the zinc-carbon battery. This provides the outer electrode with maximum surface area to take care of polarization. The complete cell is sealed in a leakproof steel case. The manganese-alkaline cell also generates a voltage of 1.5 volts. Due to the compact, "inside-out" nature of the self-depolarizing positive electrode, the manganese-alkaline battery can maintain its terminal voltage at a given current drain for more than twice as long as a zinc-carbon battery of the same physical size. This increased operating life more than offsets the higher cost associated with the more precise construction of manganese-alkaline batteries.

A third type of dry battery also has a zinc negative electrode and a potassium hydroxide electrolyte. The positive electrode of the **mercury cell** is a self-depolarizing cylinder of mercuric oxide. In addition to the familiar cylindrical form, the mercury cell adapts well to the miniature button form used in hearing aids and electric watches. Such a cell consists of small wafers of amalgamated zinc and mercuric oxide separated by a wafer of absorbent material saturated with potassium hydroxide electrolyte solution. Again the higher cost of the mercury dry battery is offset by its greater output. The nominal terminal voltage (1.3 V) of the mercury cell is slightly lower than that of the zinc-carbon and manganese-alkaline batteries. Dry cells are usually constructed to be used as primary cells. Figure 3-18 shows the manner in which the three types of dry batteries attempt to maintain their terminal voltage as three cells of the same physical size are discharged continuously into identical loads.

One particular wet cell worthy of note at this point is the lead-acid storage battery shown in Figures 3-19 and 3-20. The chemical action is such that no gas is liberated as it discharges its chemical energy, and therefore polarization does not occur. Absence of polarization, along with the ability to convert chemical energy into electric energy very rapidly, allows this battery to maintain an appreciable potential difference between its terminals even when the current through the circuit approaches 100 amperes. The high rate of energy conversion possible with the **lead-acid** battery makes it particularly useful for industrial applications. In this case, recharging the battery by converting surplus electric energy back into energy stored in chemical form is economically feasible. A cell which is recharged in this manner is called a **secondary cell.** The ability of the lead-acid battery to store energy in chemical form accounts for its common classification as a **storage battery.** The electric system of an automobile is a good example of the charge and discharge cycle of the lead-acid storage battery. When we start the car, energy stored in chemical form is converted to electric energy to operate the starter motor. Then, when the car is being driven down the highway, surplus electric energy from the generator (obtained at the expense of mechanical energy from the engine, which in turn came from the chemical energy of

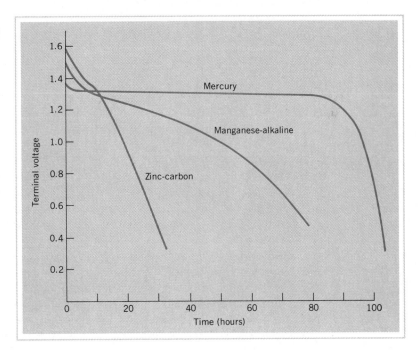

FIGURE 3-18
Discharge Characteristics of Dry Batteries.

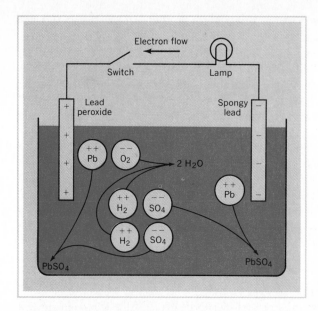

FIGURE 3-19
Chemical Action of a Lead-Acid Storage Cell When Discharging.

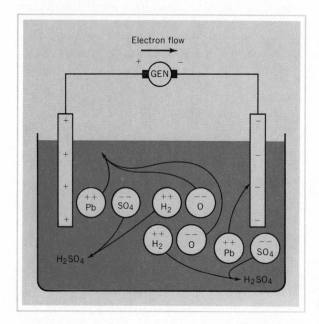

FIGURE 3-20
Chemical Action of a Lead-Acid Storage Cell When Charging.

the fuel, etc.) reverses the chemical action in the battery to store up energy for the next time the car has to be started.

The electrolyte of the lead-acid cell consists of sulfuric acid diluted with water. The positive electrode consists of lead peroxide supported by a grid of lead-antimony alloy. The negative plate consists of a similar grid packed with pure lead in a spongy form. In solution, the sulfuric acid forms sulfate (SO_4) ions, which borrow an electron from each of the two hydrogen atoms in the acid molecule. These sulfate ions are chemically very active and combine with the lead in *both* plates to form lead sulfate. Since lead sulfate is insoluble, each of the electrodes will receive two electrons. However, the combining of two molecules of acid to form one molecule of lead sulfate at each plate leaves four hydrogen ions on their own to look for their missing electrons. They can obtain these electrons by combining with the oxygen that is made available when the lead in the lead peroxide of the positive plate forms lead sulfate. In the overall action, for two molecules of acid, two electrons were deposited on the negative plate, and although two electrons were deposited at the positive plate, *four* electrons were taken from it as the hydrogen combined with the oxygen to form additional water. Therefore, the net effect at the positive plate is the loss of two electrons. As the battery discharges through the lamp in Figure 3-19, the chemical action continues to reduce both plates to lead sulfate, and the acid becomes weaker and weaker as hydrogen sulfate disappears and water is produced. As the battery discharges, electrons will flow from the negative terminal through the lamp and the switch to the positive terminal.

To recharge the battery, the chemical action must be reversed. First, the hydrogen and oxygen ions in the water must be separated. This is done by connecting an external dc generator, which can take electrons from the positive plate of the battery to the negative terminals of the battery via the generator. This electron flow outside the battery is in the *opposite* direction from the discharge current in Figure 3-19. As a result of this removal of electrons from the positive terminal and adding of electrons to the negative terminal, the potential difference between the two plates in Figure 3-20 is raised above that normally created by the chemical action of the battery itself. This additional potential difference attracts the oxygen ions to the positive plate and takes their surplus electrons away from them. Robbed of these surplus electrons, the oxygen becomes very active and unites with the lead of the lead sulfate to reform the lead peroxide on the positive plate. The external generator must now use up energy to move the two electrons it took from the oxygen ion around to the negative plate.

The recombination of the oxygen from two molecules of water and one molecule of lead sulfate frees four hydrogen ions and one sulfate ion. Two of the hydrogen ions will combine with the one sulfate ion to form a molecule of sulfuric acid. The other two are attracted to the negative plate where they will regain the two missing electrons. This allows the hydrogen to become active and combine with sulfate from nonionized lead sulfate at the negative plate, thus forming another molecule of sulfuric acid and depositing pure spongy lead on the negative plate.

Another form of secondary battery is the **nickel-cadmium cell.** This cell is much more rugged than the lead-acid battery, and it can be completely sealed, thus making it suitable for portable applications. For many years, the high cost of this battery limited its use to industrial applications such as railway signaling systems. But recently, the demand for a small secondary cell which can be used for electronic photoflash units, cordless electric shavers, etc., has increased the popularity of the nickel-cadmium battery. The positive plate of the nickel-cadmium cell is packed with nickel hydroxide and graphite. The negative plate consists of a mixture of cadmium and iron oxides. The electrolyte is potassium hydroxide. The voltage generated by the nickel-cadmium cell is only 1.2 volts compared to 2.0 volts for a lead-acid cell.

3-10 OTHER VOLTAGE SOURCES

Batteries are a convenient and not too costly voltage source for such portable applications as flashlights, hearing aids, and portable radios, where the amount of energy converted is relatively small. Batteries are also used for emergency power sources for telephone exchanges, and for supplying energy to the starter motor in an automobile. However, batteries are quite impractical when it comes to supplying the huge amounts of electric energy required by the industries and residences of a modern city.[†]

Most of the world's electric energy is generated at the expense of mechanical energy by rotating machines called **dynamos** (or, through common usage, simply **generators).** In some parts of the world, this mechanical energy is obtained from falling water, as discussed in Section 2-7 when we considered the operation of the hydroelectric generating station shown in Figure 2-10. Much of the required mechanical energy comes from the conversion of the chemical energy of coal and oil into heat energy and then the conversion of this heat energy into mechanical energy by the use of some form of heat engine such as a steam or gas turbine or a diesel engine. The mechanical output of the heat engine, in turn, drives the rotating shaft of a suitable dynamo. In the more recent generating stations, nuclear energy supplies the heat energy to operate the turbines. Since there has been some public resistance in recent years to building more nuclear generating stations, several alternative systems are being investigated.

Until recently, we have not had an efficient means of converting large amounts of chemical energy directly into electric energy. Such a device would be very desirable since it would greatly reduce the initial cost and the maintenance problems associated with rotating machinery. Recent developments indicate that **fuel cells** could become quite important as practical voltage sources in the not too distant future. Consolidated Edison is now operating a demonstration 4.8-megawatt fuel cell generating station in New York City. It is expected that present fuel cell technology can at least match the overall efficiency of coal-fired plants.

The chemical reaction accompanying the conversion of energy in a battery includes either one or both of the electrodes. Consequently, every so often we must restore the electrodes by either replacing the battery or electrically recharging it. But the chemical action in a fuel cell does not affect the electrodes; only the chemical fuel is consumed. Therefore, a fuel cell can maintain a terminal voltage as long as we supply a suitable fuel whose chemical energy is converted directly into electric energy by the cell.

The basic fuel cell uses the energy released as hydrogen and oxygen combine to form water to separate electric charges in the form of a + charge at one electrode and a − charge at the other. Water is an ionic compound in which two hydrogen atoms each lend their single valence electron to an oxygen atom. To create a fuel cell, this combination must take place in two steps. First, the hydrogen and oxygen must be ionized at *different* electrodes in the fuel cell, and then the ions combine to form water. Figure 3-21 is a simplified representation of how this may be achieved. Hydrogen and

---

[†]Some electric utility companies are experimenting with huge "storage batteries" in the form of large tanks. The purpose of such devices is to store surplus electric energy generated in off-peak hours (during the night). High-efficiency semiconductor rectifiers and converters allow this stored energy to be fed back into the system to supplement generating capacity during peak-load periods.

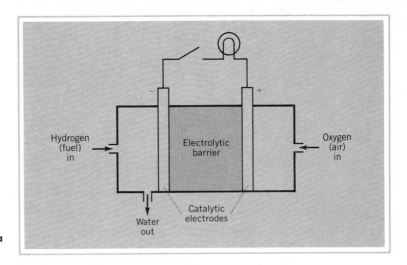

FIGURE 3-21
Simplified Cross Section of a Fuel Cell.

oxygen (the oxygen may come from air) are fed into separate chambers, each with a catalytic electrode to encourage ionization. The chambers are separated by an electrolytic barrier which permits the flow of ions but which will not allow the passage of gas molecules. Consequently, for the hydrogen and oxygen to combine, hydrogen must give its valence electrons to the electrode in the left chamber and the oxygen must acquire electrons from the electrode in the right chamber.

By keeping the oxygen and the hydrogen "fuel" separated during the ionizing process, the fuel cell develops an electric potential difference between its two electrodes. When the switch is closed, this potential difference maintains an electric current through the load. Like the battery, when the switch is open, the potential difference builds up to a certain magnitude where it exceeds the chemical attraction for hydrogen and oxygen ions to combine. Ionization then ceases, and, since gas molecules cannot pass through the electrolytic barrier, consumption of the fuel gas also ceases. Many thousands of individual fuel cells are connected in a series-parallel arrangement to generate sufficient voltage and current for a practical generating unit.

We can also construct practical devices for generating a potential difference directly from other sources of energy such as heat, light, and sound energy. The efficiency of such energy conversions is not great enough at present to allow us to use these voltage sources with household or industrial loads. However, they do perform very useful functions in measuring instruments and in automatic control systems. We can generate a potential difference by applying heat energy directly to the welded junction of a pair of certain dissimilar metals. By connecting the leads of this **thermocouple** to a sensitive meter, we can use it to measure temperatures beyond the range of conventional thermometers. Similarly, the light meters used by photographers operate on the electric current which flows through a meter movement as a result of a potential difference generated by light energy falling on a **photovoltaic cell.**

An increased knowledge of semiconductor behavior has permitted development of more efficient photovoltaic devices. Figure 3-22 shows a cross section of a *pn*-junction created by diffusing trivalent boron atoms into one surface of a slab of *n*-type silicon. The *p*-type layer thus formed is so thin that light energy can penetrate through

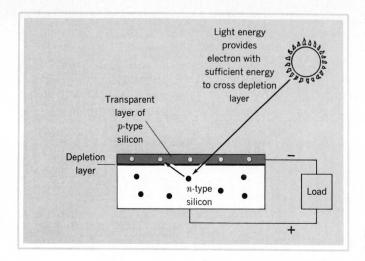

FIGURE 3-22
Cross Section of a Solar Cell.

it to the n-type base. As we noted in Figure 3-15, electrons near the junction will combine with the holes in the p-type layer, forming a depletion layer which is almost devoid of charge carriers. Light energy can penetrate this depletion layer in the solar cell of Figure 3-22 and provide the conduction electrons in the n-type layer with sufficient energy to drive them through the depletion layer into the thin p-type layer. Since these electrons cannot return through the depletion layer, they must eventually return through the external load, thus constituting an electric current. Banks of these cells form the **solar batteries** which can convert sufficient light energy from the sun to operate the radio receivers and transmitters of communications satellites for an almost unlimited period of time. Solar cell technology has developed to the point where it is now pratical in some sunny localities to use arrays of solar cells to feed considerable electric energy into local electric power systems.

We can even use sound energy to generate a potential difference. In one popular type of microphone, sound waves striking a diaphragm cause it to place a mechanical stress on a Rochelle salt crystal. Stressing a crystal in this manner causes a displacement of valence electrons within the crystal so that a surplus of electrons appears at one side of the crystal and an equivalent deficiency at the other. Generation of a potential difference in this manner is called **piezoelectric effect.**

REVIEW QUESTIONS

3-1. What is the charge (coulombs) of the available charge carriers in a cubic centimeter of copper?

3-2. Both copper and sodium chloride (common table salt) are crystalline solids; yet, copper is a good conductor of electricity and NaCl is an insulator. What distinction in the bonding process accounts for this difference?

3-3. Given energy-band diagrams for copper and sodium chloride, how do you identify which is an electric conductor and which is an insulator?

3-4. Why does the energy-band diagram for an electric conductor show energy levels in the conduction band dipping down to the same energy level as some of the energy levels in the valence band?

3-5. In the energy-band diagram for a theoretical insulator, all energy levels in the valence band are occupied and all energy levels in the conduction band are unoccupied. Explain this in terms of a covalent-bonded material.

3-6. The energy-band diagram for polystyrene shows no electrons in the conduction band, yet a cubic centimeter of polystyrene can have some 6×10^{10} conduction electrons. Where do these conduction electrons come from?

3-7. Why is the term **ionization potential** appropriate in describing the energy level which an electron must possess to escape the attraction of its parent nucleus?

3-8. What relationship is there between the width of the forbidden energy gap in an energy-band diagram and the performance of the material as an electric insulator? Explain this relationship.

3-9. Valence bonds in an insulator are broken by heat energy possessed by the insulator. They can also be broken by applying a large electric potential difference across a layer of insulating material. What is the observable difference in the behavior of the material under these two conditions?

3-10. Explain how the silver anode in the electrolytic cell shown in Figure 3-6 obtains the positive charge required to attract nitrate ions.

3-11. Why could electric current not pass through the solution of an ionic compound if dissociation of the ions did not occur?

3-12. What is the difference between the process of electric conduction in a liquid and in a metallic conductor?

3-13. As in incandescent lamps, the heater of a vacuum tube gradually becomes thinner due to vaporization of the metal and finally burns out. What effect does the burning out of the heater have on the operation of a vacuum-tube rectifier?

3-14. At room temperature, there are as many electrons in the conduction band of a *pure* semiconductor crystal as there are positive holes in the valence band. Explain the source of these charge carriers.

3-15. Explain the mechanism of hole flow in a semiconductor.

3-16. When electric current flows in an intrinsic semiconductor or in an insulator, equal numbers of electrons and positive holes flow in opposite directions. How does this compare with electric charge flow in an electrolyte?

3-17. In a crystal of *n*-type germanium at room temperature, electrons are the **majority** carriers and positive holes are the **minority** carriers. Why are there more electrons than holes?

3-18. The magnitude of the leakage current when a *pn*-junction is **reverse biased** depends on the concentration of *minority* carriers. Explain this with reference to Figure 3-16(a).

3-19. As a mercury cell is discharged, the zinc electrode is oxidized into zinc oxide by combination with negative oxygen ions from the electrolyte. The positive potassium ions reduce the mercuric oxide electrode to mercury, thus maintaining the concentration of the solution. Explain, with a sketch, how this chemical action generates a voltage between the terminals of the cell.

3-20. Describe the various energy conversions that take place from the time some of the chemical energy of an automobile storage battery is used to start the engine of the car until this energy is restored to the battery by the car's electric generator.

4

RESISTANCE

In the basic electric circuit diagram of Figure 4-1, we have used the generalized form of the graphic symbol to represent a voltage source (Table 1-1). Although the symbol is the same as that for a single-cell battery, we shall use it in the following pages to represent any direct-current voltage source when we are concerned only with the magnitude of the potential rise produced and are not particularly interested in the details of the manner in which the voltage is generated. All that we usually "see" of the voltage source provided by an electrical utility company, for example, are the two terminals of the outlet into which we plug a lamp, toaster, etc. We can show that these two terminals do represent a voltage source by connecting a voltmeter *across* them (see Figure 2-13). The magnitude of the applied voltage is a property of the voltage source and is essentially independent of the electric circuit connected to the source.

We now want to find out just what governs the numerical magnitude of the electric current that will flow when we close the switch in the basic electric circuit of Figure 4-1.

4-1 OHM'S LAW OF CONSTANT PROPORTIONALITY

Over a hundred years ago, Georg Simon Ohm discovered that every time he closed the switch in a circuit such as that of Figure 4-1, the current became the same constant value. He also discovered that, providing the temperature of the conductor did not change, doubling the applied voltage doubled the current and tripling the applied voltage tripled the current. In other words,

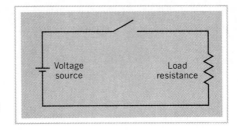

FIGURE 4-1
Schematic Representation of a Basic Electric Circuit.

For a given circuit, the ratio of the applied voltage to the current is a constant.

This statement became known as **Ohm's law of constant proportionality.** We can express Ohm's law in equation form as

$$\frac{E}{I} = k \tag{4-1}$$

where E is the applied voltage in volts, I is the resulting current in amperes, and k is a numerical constant.

Carrying the discovery a step further, Ohm found that changing the conductors or the load to some other size resulted in a different magnitude of electric current, hence a different value for the numerical constant. If the current is one ampere when the applied voltage is 10 volts, the constant becomes 10/1, or simply 10; and if the current is 0.5 ampere when the applied voltage is 10 volts, the constant becomes 10/0.5, or 20. From these results, Ohm concluded that this constant E/I ratio for a given circuit is, therefore, a property of that circuit. Since, for a given applied voltage, the current must *decrease* if the numerical value of the constant *increases,* we can think of this constant as representing the *opposition* of the circuit to flow of charge carriers. This property should then be given a name which suggests *opposition,* such as **resistance.** Therefore,

Resistance is the opposition of an electric circuit to electric current through that circuit.
The letter symbol for resistance is R.

We can substitute the symbol R for the constant k in Equation (4-1). There is another change that we should make in Equation (4-1). As we noted in Figure 2-13, if we open the switch in the circuit of Figure 4-1, the current becomes zero. According to Equation (4-1), the resistance of the circuit is now infinitely high. This is quite true when we remember that the switch is part of the complete circuit. The resistance of an open switch should be infinitely high. But we are usually more interested in determining the resistance of the *load*. As we noted in Section 2-10, when the switch is open and the current is zero, there is also zero voltage drop across the load. Hence, for Equation (4-1) to represent the resistance of the load only, we must think in terms of the voltage drop across the load, rather than of the voltage applied to the total circuit.

Although it would be quite proper to express resistance in terms of **volts per ampere,** it was decided to honor Ohm's original discovery by stating that

The ohm is the SI unit of electric resistance.
The unit symbol for ohm is the Greek letter Ω (omega).

Since Equation (4-1) has now become

$$R = \frac{V}{I} \tag{4-2}$$

where R is the resistance in ohms, V is the voltage drop across the resistance in volts, and I is the current through the resistance in amperes, the size of the ohm is automatically established. Therefore,

An electric circuit has a resistance of one ohm when a current of one ampere through the circuit causes a voltage drop across the circuit of one volt.

Few people nowadays recall the original statement of Ohm's law of constant proportionality which allowed us to define resistance and to establish a unit of resistance. As a result, Ohm's law is usually stated simply by the equation

$$R = \frac{V}{I} \tag{4-2}$$

EXAMPLE 4-1
What is the resistance of a lamp if a current of 150 mA flows through the lamp when a voltage of 6.0 V is applied to its terminals?

SOLUTION

$$R = \frac{V}{I} = \frac{6.0 \text{ V}}{150 \text{ mA}} = 40 \ \Omega$$

EXAMPLE 4-2
What value of resistance is required to limit the current through the resistance to 20 μA if the voltage drop across the resistance is to be 480 mV?

SOLUTION

$$R = \frac{V}{I} = \frac{480 \text{ mV}}{20 \ \mu A} = 24\,000 \ \Omega = 24 \text{ k}\Omega$$

$$(R = 480 \text{ EE} +/- 3 \ \div \ 20 \text{ EE} +/- 6 \ = \ 24\,000 = 24 \times 10^3)$$

4-2 THE NATURE OF RESISTANCE

In Section 2-4, we noted that free electrons in metallic conductors move around in a random manner among the metal ions in the crystal structure of the conductor, alternately giving up or adding to their kinetic energy. Most of this energy exchange is due to the vibration of the metal ions in their lattice positions as a consequence of the heat energy possessed by the conductor. When we apply a voltage to the conductor, the random motion continues, but superimposed on this random motion is a net drift along the length of the conductor, as shown in Figure 2-7(b). Since an electron represents a certain mass, as it accelerates along the length of the conductor, potential energy provided by the source is converted into kinetic energy associated with the velocity of the electron.

Before the accelerating free electron gets very far in its journey along the length of the conductor, it collides with one of the metal atoms. This collision results in a considerable reduction in the speed of the electron. As a result, some of the kinetic energy gained by the electron is transferred to the atom with which it collided in the form of heat energy (just as vigorous hand clapping results in a transfer of mechanical energy to heat energy). After the collision, the free electron is again urged by the applied voltage to accelerate and acquire more kinetic energy. Again it will collide with one of the metal atoms and give up some of its kinetic energy in the form of heat. Therefore, the progress of the electrons which constitute the charge carriers in a metallic conductor is characterized by alternate acceleration (during which interval potential energy from the source is transferred to the electron in kinetic form) and sudden deceleration (at which time some of the kinetic energy of the electron is transferred to the atoms of the conductor material in the form of heat).

These many collisions, which constitute an *opposition to electric current,* account for the property of electric conductors known as **resistance.** The consequence of resistance in an electric circuit is the transfer of energy taken from the source into heat energy whenever we force charge carriers to flow in conductors possessing this property.

Resistance is a desirable property in such circuit components as lamps and electric stove elements in which we are interested in producing heat energy. In circuit elements such as the conductors connecting the load to the source in Figure 4-1, resistance is an undesirable property. Not only is some of the energy in the system wasted in the form of heat in the conductors, but if heat is developed at a greater rate than the conductors can dissipate it or pass it on to the surrounding air, the temperature of the conductors can rise to the point where they become a fire hazard.

We often insert resistance into an electric circuit as a means of limiting current. For example, if we wish to dim the lamp in the circuit of Figure 4-2, we can connect

FIGURE 4-2
Using a Resistor to Limit Current.

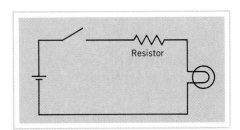

71

a circuit element possessing appreciable resistance in *series* with the lamp and the battery. Now the free electrons have to flow through both the lamp and the added resistance in completing their journey from the negative terminal of the battery to the positive terminal. Since this additional resistance produces heat when current flows through it, rather than connecting the lamp to the battery with conductors possessing higher resistance, we prefer to lump the extra resistance into a single electric component termed a **resistor,** which has been designed to dissipate the heat adequately without damage to either itself or nearby objects.

4-3 FACTORS GOVERNING THE RESISTANCE OF METALLIC CONDUCTORS

In stating his law of constant proportionality, Ohm had to add the provision that the temperature must be kept constant. Therefore, temperature must have some effect on the resistance of an electric conductor. We can readily check this in the laboratory by measuring the V/I ratio (resistance) of a lamp operating at the rated value of applied voltage (white hot) and then measuring the V/I ratio when the applied voltage is reduced to 10% of the rated value and the filament gives off practically no light. We shall investigate the exact manner in which temperature affects the resistance of a metallic conductor a little later in this chapter. For the moment, we assume temperature to remain constant at normal room temperature (20° C).

If we cut two lengths of wire from the same reel, one being twice the length of the other, in order for electrons to make the complete trip through the longer wire, they will have twice as many opportunities to collide with atoms of the conductor material. Consequently, the opposition of the longer wire to electric current is twice as great as that of the shorter wire. Therefore,

The resistance of a metallic electric conductor is directly proportional to its length.

If we select a wire size which has double the cross-sectional area of another wire size and cut equal lengths of each, the wire with the larger cross section would have the same cross-sectional area as two pieces of the wire with the smaller cross section connected in *parallel*. Since each of the two smaller diameter wires in parallel will pass the same current when connected to a single voltage source, the two parallel-connected wires pass twice the current of one wire. Since $R = V/I$, double current with no change in the applied voltage represents half as much resistance. Since doubling the cross-sectional area of a single conductor is the same as using two of the smaller diameter wires in parallel, doubling the cross-sectional area cuts the resistance in half. Therefore,

The resistance of a metallic electric conductor is inversely proportional to its cross-sectional area.

We noted in Section 3-1 that some materials possess more free electrons per unit volume than others and that the greater the number of free electrons per unit volume, the better conductor that material becomes. The greater the number of charge carriers in a conductor of certain dimensions, the smaller the percentage of the total energy that is carried by any one electron. Hence, the amount of energy that is converted into heat by collision with atoms of conductor material is proportionately reduced. Thus, a silver wire has a lower resistance than a copper wire with the same dimensions, and the copper wire has a lower resistance than an aluminum wire with the same dimensions. Therefore,

The resistance of a metallic electric conductor is dependent on the type of conductor material.

4-4 RESISTIVITY

Since we know that the resistance of a conductor is directly proportional to its length, if we are given the resistance of unit length of wire, we can readily calculate the resistance of any length of wire of that particular material having the same diameter. Similarly, since we know that the resistance of a conductor is inversely proportional to its cross-sectional area, if we are given the resistance of a length of wire with unit cross-sectional area, we can calculate the resistance of a similar length of wire of the same material with any cross-sectional area. Combining both of these statements, we find that if we know the resistance of a given conductor, we can calculate the resistance for *any* dimensions of a conductor of the same material at the same temperature.

EXAMPLE 4-3

An electric conductor 1 m long with a cross-sectional area of 1 mm^2 has a resistance of 0.017 Ω. What is the resistance of 50 m of wire of the same material with a cross-sectional area of 0.25 mm^2?

SOLUTION
From the relationships stated above, we can write the equation

$$\frac{R_2}{R_1} = \frac{l_2}{l_1} \times \frac{A_1}{A_2} \tag{4-3}$$

$$\therefore R_2 = 0.017 \ \Omega \times \frac{50 \ m}{1 \ m} \times \frac{1 \ mm^2}{0.25 \ mm^2} = 3.4 \ \Omega$$

As we become more familiar with our calculators, we should resist the tendency to enter data directly from the statement of the example into the calculator. We should always follow the steps outlined in Example 2-2 (as we have done above), so that we have a better chance of spotting an error in the way we made the calculation. To enter the exponent for *square* millimeters in this example, we should note that the prefix *milli* is also squared. (One square meter = 10^6 square millimeters.) However, in this particular example, we might also notice that mm^2 in the denominator of the equation above cancels out mm^2 in the numerator. Hence, we do not need to enter the exponents if we so wish.

As Example 4-3 suggests, a convenient way of showing the effect of the type of material on the resistance of a given conductor is to state the resistance of a standard conductor with unit length and unit cross-sectional area made from that material.

The resistance of a unit length and cross-section conductor of a certain material is called the resistivity of that material.[†]
The letter symbol for resistivity is the Greek letter ρ (rho).

Since the SI base unit for both length and cross section is the meter, the SI standard conductor must take the form of a cube with each of its edges 1 meter in length. The resistivity of a certain electric conducting material will therefore be the resistance between opposite faces of the 1-meter cube. To determine the SI unit of resistivity, we can apply dimensional analysis techniques. Since $R = V/I$, ohms = volts ÷ amperes. But voltage is measured between the ends or across the *length* of a conductor, and current is measured through the *area* of the cross section. Hence, from the definition of resistivity,

$$\rho = \text{volts per unit length} \div \text{amperes per unit cross-sectional area}$$

$$= \frac{\text{volts}}{\text{meter}} \times \frac{\text{meter}^2}{\text{amperes}} = \frac{\text{volts}}{\text{amperes}} \times \text{meters}$$

$$= \textbf{ohm meters}$$

Therefore,

The ohm meter is the SI unit of resistivity.
The unit symbol for ohm meter is $\Omega \cdot \text{m}$.

Since temperature has an effect on the resistance of electric conductors, values of resistivity for various conducting materials hold only for a certain specified temperature, usually 20 degrees Celsius.[‡] Table 4-1 gives the resistivity of the more common metallic conductor materials at 20°C.

In Equation (4-3), if we set $l_1 = 1$ m and $A_1 = 1$ m^2, then R_1 becomes ρ and has a value as given in Table 4-1. We can now rewrite Equation (4-3) to calculate the resistance of any electric conductor at normal room termperature:

$$R = \rho \times \frac{l}{A} \tag{4-4}$$

where R is the resistance of the conductor in ohms, l is the length of the conductor

[†]Sometimes called **specific resistance.**

[‡]The Celsius temperature scale, which places the freezing point of water at 0°C and the boiling point of water at +100°C, is used with SI units even though °C is not a true SI unit. (The SI base unit of temperature is the *kelvin*.)

in meters, A is the cross-sectional area in square meters, and ρ is the resistivity of the conductor material in ohm meters at 20°C.

TABLE 4-1

RESISTIVITY OF SOME COMMON ELECTRIC CONDUCTOR MATERIALS

Conductor Material	Resistivity (ohm meters at 20°C)
Silver	1.64×10^{-8}
Copper (annealed)	1.72×10^{-8}
Aluminum	2.83×10^{-8}
Tungsten	5.5×10^{-8}
Nickel	7.8×10^{-8}
Iron (pure)	12.0×10^{-8}
Constantan	49×10^{-8}
Nichrome II	110×10^{-8}

EXAMPLE 4-4

What is the resistance (at 20°C) of 200 m of an aluminum electric conductor whose cross-sectional area is 4 mm²?

SOLUTION

Using the value of ρ for aluminum from Table 4-1,

$$R = \rho \times \frac{l}{A} = 2.83 \times 10^{-8} \ \Omega \cdot m \times \frac{200 \ m}{4 \times 10^{-6} \ m^2} = 1.415 \ \Omega$$

In entering the cross-sectional area of the conductor in the equation above, the conversion from milli squared to an exponent is shown in order to point out how the unit *meters* cancels out in Equation (4-4).

Most electric conductors have a circular cross section. If we know the diameter of the conductor, we must then express its cross-sectional area in square meters.

EXAMPLE 4-5

What is the resistance at normal room temperature of 60 m of copper wire having a diameter of 0.64 mm?

SOLUTION

$$A = \pi r^2 = \frac{\pi}{4} d^2 = \frac{\pi}{4} \times 0.64^2 \ mm^2 = 3.217 \times 10^{-7} \ m^2$$

$$R = \rho \times \frac{l}{A} = 1.72 \times 10^{-8} \ \Omega \cdot m \times \frac{60 \ m}{3.217 \times 10^{-7} \ m^2} = 3.2 \ \Omega$$

4-5 EFFECT OF TEMPERATURE ON RESISTANCE

We have already noted that temperature will have some effect on the resistance of an electric conductor, but so far we have not had to consider it since we have assumed that the temperature remained constant at 20°C. By checking the effect of temperature on the V/I ratio (resistance) of a conductor as previously suggested, we find that for most conducting materials, the resistance increases linearly with an increase in temperature over normal temperature ranges. In Figure 4-3, if R_1 is the resistance of a certain conductor at temperature T_1, then R_2 is its resistance at temperature T_2. The straight line CF represents the manner in which the resistance varies for temperatures between T_1 and T_2. The *slope* of the line CF is dependent on the type of material. Some alloys have been developed (e.g., **constantan**) for which the line CF is almost horizontal. Temperature has very little effect on the resistance of such materials. There are a few materials (e.g., **carbon** and other semiconductors) which have a *negative* temperature effect; that is, the resistance *decreases* as the temperature *increases*.

Since CF in Figure 4-3 is a straight line, we can project it to the left to cut the baseline (zero ohms) at point A, thus producing two similar triangles ABC and ADF. Since these triangles are similar, we can show that

$$\frac{DF}{BC} = \frac{AD}{AB} = \frac{AO + OD}{AO + OB}$$

Substituting the electrical properties that these geometric dimensions represent gives

$$\frac{R_2}{R_1} = \frac{x + T_2}{x + T_1} \tag{4-5}$$

where R_1 is the resistance of the conductor at temperature T_1, R_2 is the resistance of the conductor at temperature T_2, T_1 and T_2 are the temperatures of the conductor in degrees Celsius, and x is a constant which depends on the type of material used in the conductor.

Table 4-2 gives the numerical value for x for some common electric conductor materials. The value of x is a constant for any given material. From the value of x for

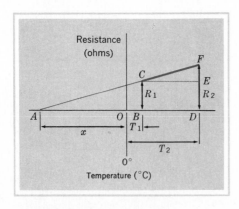

FIGURE   4-3
Effect of Temperature on Resistance.

TABLE 4-2

VALUES OF x FOR SOME COMMON ELECTRIC CONDUCTOR MATERIALS (SEE FIGURE 4-3)

Conductor Material	$x(°C)$
Silver	243
Copper	234.5
Aluminum	236
Tungsten	202
Nickel	147
Iron	180
Nichrome II	6250
Constantan (55% Cu, 45% Ni)	125 000

such materials as Nichrome II and constantan, it is apparent that x does not represent the actual temperature at which resistance becomes zero ohms. The resistance of metallic conductor materials approaches zero at absolute zero temperature, $-273°C$.[†] The thin line AC in Figure 4-3 is a geometric projection of the linear portion of the resistance-temperature characteristic of various metallic conductor materials so that we can solve for the effect of temperature on resistance by the linear Equation (4-5).

EXAMPLE 4-6

A copper conductor has a resistance of $12\,\Omega$ at $20°C$. What is its resistance at $100°C$?

SOLUTION

$$\frac{R_2}{R_1} = \frac{x + T_2}{x + T_1} \quad \text{or} \quad R_2 = R_1 \frac{x + T_2}{x + T_1}$$

$$\therefore R_2 = 12\,\Omega \times \frac{(234.5 + 100)}{(234.5 + 20)} = \frac{12 \times 334.5}{254.5} = 15.77\,\Omega$$

We use parentheses when entering the data in the equation above to remind us to use the *parenthesis* keys of our calculator. Parentheses are used to ensure that the mathematical operations are performed by the calculator in the proper sequence. The chain calculation becomes

$$R_2 = 12 \times (234.5 + 100) \div (234.5 + 20) = 15.77\,\Omega$$

Note that the units $°C$ cancel out in the equation above. The second last step is included for those solving the example by longhand. These figures will also appear on the calculator display as the chain calculation proceeds.

[†]By cooling metallic conductors to near absolute zero with liquid helium, it is possible to reduce their resistance to almost zero. Such a condition is called **superconductivity,** and it is practical for such devices as the electromagnets of nuclear particle accelerators where extremely large currents are required.

EXAMPLE 4-7

A precision resistor made of constantan wire has a resistance of $10\,000\ \Omega$ at 20°C. What is its resistance when its temperature rises 20°C?

SOLUTION

$$R_2 = R_1 \frac{x + T_2}{x + T_1} = 10\,000 \times \frac{(125\,000 + 40)}{(125\,000 + 20)}$$

$$R_2 = \mathbf{10\,001.6\ \Omega}$$

In addition to being able to calculate the effect that temperature has on resistance, we can use the change in resistance of a certain material to calculate the temperature in locations where it is difficult to place a thermometer.

EXAMPLE 4-8

The copper winding of an electric motor which has been standing for several hours in a room at 20°C has a resistance of $0.20\ \Omega$. When the motor has been in use for some hours the resistance of the winding is found to be $0.22\ \Omega$. Calculate the temperature rise of the winding.

SOLUTION

$$\frac{R_2}{R_1} = \frac{x + T_2}{x + T_1}$$

$$\frac{0.22}{0.20} = \frac{234.5 + T_2}{234.5 + 20}$$

To enter the data into a calculator, we rearrange this equation to place T_2 by itself on the left-hand side of the equals sign.

$$T_2 = \frac{0.22}{0.20} \times (234.5 + 20) - 234.5 = 45.45°C$$

$$\therefore \text{ temperature rise} = T_2 - T_1 = \mathbf{25.45\,°C}$$

If we now use Equation (4-4) to substitute for R_1 in Equation (4-5), we obtain

$$R = \rho \frac{l}{A} \left(\frac{x + T}{x + 20} \right) \tag{4-6}$$

With the values of ρ and x given in Tables 4-1 and 4-2, we can now determine the resistance of a given metallic conductor at any temperature.

EXAMPLE 4-9

What is the resistance of 300 m of copper wire with a cross-sectional area of 1.5 mm^2 at 40°C?

SOLUTION
From Table 4-1,

$$\rho = 1.72 \times 10^{-8} \; \Omega \cdot m$$

From Table 4-2,

$$x = 234.5°C$$

$$R = \rho \frac{l}{A}\left(\frac{x + T}{x + 20}\right)$$

$$= 1.72 \times 10^{-8} \; \Omega \cdot m \times \frac{300 \; m}{1.5 \; mm^2} \times \frac{(234.5 + 40)°C}{(234.5 + 20)°C}$$

$$= 3.44 \times \frac{274.5}{254.5} = \textbf{3.71} \; \boldsymbol{\Omega}$$

The calculator chain is

$$R = 1.72 \; \boxed{EE} \; \boxed{+/-} \; 8 \; \times \; 300 \; \div \; 1.5 \; \boxed{EE} \; \boxed{+/-} \; 6 \; \times \; (\; 234.5 \; + \; 40\;)$$
$$\div (\; 234.5 \; + \; 20\;) \; = \; 3.71 \; \Omega$$

4-6 TEMPERATURE COEFFICIENT OF RESISTANCE

Although the method we have just derived is quite adequate for calculating the effect of temperature on resistance, the constant x is based on a geometric projection, rather than on a property of the conductor material. So let us return to geometric construction on Figure 4-3 to derive an alternative method of showing the effect of temperature on the resistance of an electric conductor. The line CE is drawn parallel to the baseline. Since resistivity (ρ) is usually stated for a temperature of 20°C, we shall let T_1 represent 20°C.

Since triangle ABC is similar to triangle CEF, we can show that

$$\frac{FE}{BC} = \frac{CE}{AB} \qquad \text{or} \qquad \frac{FE}{R_1} = \frac{\Delta T}{x + 20}$$

where ΔT is the difference in temperature (degrees Celsius) between T_2 and 20°C.
Therefore,

$$FE = R_1\left(\frac{\Delta T}{x + 20}\right)$$

Since CE and BD are parallel,

$$R_2 = R_1 + FE$$

$$\therefore R_2 = R_1 + R_1\left(\frac{\Delta T}{x + 20}\right) = R_1\left(1 + \frac{\Delta T}{x + 20}\right)$$

If we now call $1/(x + 20)$ the **temperature coefficient of resistance** (at 20°C) and represent it by the Greek letter α (alpha),

$$R_2 = R_1(1 + \alpha \Delta T) \tag{4-7}$$

where R_2 is the resistance of a conductor at any specified temperature, R_1 is its resistance at 20°C, α is the temperature coefficient of resistance at 20°C $[\alpha = 1/(x + 20)]$, and ΔT is the difference (degrees Celsius) between the specified temperature and 20°C.

Since R_1 represents the resistance of a given conductor at 20°C,

$$R_1 = \rho \frac{l}{A} \tag{4-4}$$

Therefore, the general equation for the resistance of a metallic conductor, including the effect of temperature, becomes

$$R = \rho \frac{l}{A}(1 + \alpha \Delta T) \tag{4-8}$$

where R is the resistance of the conductor in ohms, ρ is the resistivity of the material in ohm meters at 20°C, l is the length in meters, A is the cross-sectional area in square meters, α is the temperature coefficient of resistance of the material (ohmic change per degree per ohm at 20°C), and ΔT is the difference between the operating temperature and 20°C.

Since the temperature coefficient $(\alpha_{20°}) = 1/(x + 20)$ and x is dependent on the type of material, the temperature coefficient of resistance represents the manner in which each ohm of resistance of that material changes for each degree change in temperature from the reference temperature of 20°C. We can compute a table of temperature coefficients from the relationship we have already stated between α and x (Table 4-3).

TABLE 4-3

TEMPERATURE COEFFICIENT OF RESISTANCE OF SOME COMMON ELECTRIC CONDUCTOR MATERIALS AT 20°C

Conductor Material	$\alpha_{20°}$
Silver	0.0038
Copper	0.003 93
Aluminum	0.0039
Tungsten	0.0045
Nickel	0.006
Iron	0.0055
Nichrome II	0.000 16
Constantan	0.000 008
Carbon	−0.0005

EXAMPLE 4-9A

What is the resistance of 300 m of copper wire with a cross-sectional area of 1.5 mm^2 at 40°C?

SOLUTION

From Table 4-1,

$$\rho = 1.72 \times 10^{-8}\,\Omega \cdot m$$

From Table 4-3,

$$\alpha = 0.003\,93$$

$$R = \rho\frac{l}{A}(1 + \alpha\Delta T)$$

$$= 1.72 \times 10^{-8}\,\Omega \cdot m \times \frac{300\ m}{1.5\ mm^2} \times (0.003\,93 \times 20 + 1)$$

$$= 3.44 \times 1.079 = \mathbf{3.71\ \Omega}$$

[Note that in entering the data into the equation, we changed $(1 + \alpha\Delta T)$ to $(\alpha\Delta T + 1)$ so that we could accomplish the calculation on a calculator having only a single level of parenthesis.]

EXAMPLE 4-10

An electric heater made from Nichrome II wire has a resistance of 16 Ω at 1500°C. What is its resistance at normal room temperature?

SOLUTION

In this example, R_2 in Equation (4-7) is given and R_1 is unknown. Hence,

$$R_1 = \frac{R_2}{(1 + \alpha\Delta T)} = \frac{16}{(0.000\,16 \times 1480 + 1)}$$

$$= \frac{16}{1.237} = \mathbf{12.9\ \Omega}$$

If the required temperature is below 20°C, ΔT will be a *negative* quantity and Equation (4-8) becomes effectively:

$$R = \rho\frac{l}{A}(1 - \alpha\Delta T)$$

Solving Equation (4-8) on a calculator involves a fairly long chain of operations, as we discovered in Example 4-9A. If we are only required to solve a few examples to become familiar with Equation (4-8), the calculator is the simplest method. But if we are given an assignment to determine the resistance of a longer list of electric conductors, it may be worth while preparing a personal computer program. Once we have written the program in Example 4-11(a), we can RUN it as many times as we like by inserting values for the variables, as in Example 4-11(b).

EXAMPLE 4-11
(a) Write a BASIC computer program to calculate the resistance of any circular-cross-section metallic conductor.
(b) Use this program to determine the resistance of 75 m of aluminum wire with a diameter of 2 mm at a temperature of 36°C.

SOLUTION
(a) Since there are five independent variables in Equation (4-8), we can spread the INPUT statements out into a separate line for each variable. A typical program might consist of

```
100 REM  RESISTANCE OF A METALLIC CONDUCTOR
110 INPUT "ENTER RESISTIVITY FOR TYPE OF METAL ";P
120 INPUT "ENTER LENGTH OF CONDUCTOR IN METERS ";L
130 INPUT "ENTER DIAMETER OF CONDUCTOR IN MILLIMETERS ";D
140 INPUT "ENTER TEMP COEFF FOR TYPE OF METAL ";A
150 INPUT "ENTER TEMP OF CONDUCTOR IN DEGREES C ";T
```

Before we write the equation for resistance into the processing statement of the program, we must modify Equation (4-8) to take into account the way the program requests data on conductor *diameter* and temperature. Whereas Equation (4-8) requires cross-sectional area in square meters, our program allows us to enter the conductor's diameter in millimeters. To do this,

$$\text{cross-sectional area} = \frac{\pi D^2}{4} \times 10^{-6}\ m^2$$

$$= 7.854E-7*D^2$$

($^$ is the BASIC symbol for an exponent.) Also,

$$\Delta T = T - 20$$

Although BASIC can handle two levels of parentheses, it may be easier if we have to troubleshoot or debug our program to solve Equation (4-8) by splitting it into two functions. Hence, our computer program might continue as

```
160 LET X = 1+A*(T-20)
170 LET R = P*L/(7.854E-7*D^2)*X
180 PRINT "RESISTANCE OF CONDUCTOR = ";R;" OHMS"
190 END
```

(b)

```
RUN
ENTER RESISTIVITY FOR TYPE OF METAL ? 2.83E-8
ENTER LENGTH OF CONDUCTOR IN METERS ? 75
ENTER DIAMETER OF CONDUCTOR IN MILLIMETERS ? 2
ENTER TEMP COEFF FOR TYPE OF METAL ? 0.0039
ENTER TEMP OF CONDUCTOR IN DEGREES C ? 36
RESISTANCE OF CONDUCTOR = 0.717769 OHMS
```

4-7 LINEAR RESISTORS

Most electric conductor materials behave in the manner discovered by Georg Ohm in establishing his law of constant proportionality; that is, doubling the applied voltage causes the current to double, tripling the applied voltage triples the current, and so on. We can show this relationship graphically, as in Figure 4-4. The straight-line graph represents a constant value of resistance; the smaller the resistance, the steeper the slope of the linear graph. A resistor that maintains this constant V/I ratio is known as a **linear** resistor. As current through a resistor is increased, electric energy is converted into heat at an increased rate, resulting in a rise in the temperature of the resistor. As indicated in Table 4-3, this increase in temperature causes a slight *increase* in the resistance of most conductor materials. But the percentage change in resistance over the usual temperature ranges encountered in practice is small enough in the case of the common electric conductor materials that the deviation of their resistance graphs from the straight lines of Figure 4-4 is hardly perceptible. Therefore, copper and aluminum are classed as linear resistors. Special alloys such as *constantan* with a temperature coefficient of almost zero have been developed for the production of resistance wire.

As the name suggests, we construct **wire-wound resistors** by winding a wire of one of the metal alloys such as constantan on a hollow porcelain tube and then sealing the wire in position with a porcelain coating. Since temperature has an effect on the resistance of metallic conductors, and since the temperature of a resistor must rise when we increase the current passing through it, the alloy chosen for constructing wire-wound resistors must have almost zero resistance change over normal temperature ranges. The larger the physical dimensions of a resistor, the more readily it can dissipate heat to the surrounding air. Hence, the power rating of a resistor is proportional to its physical size.

FIGURE 4-4
Volt-Ampere Graphs of Linear Resistors.

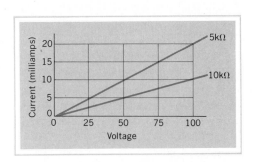

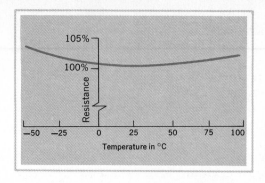

FIGURE 4-5
Resistance-Temperature Character-istic of Carbon Composition Resistors.

Resistors used with the electrical measuring instruments of Chapter 10 must be accurate to within 1% of the specified resistance if the measuring devices are to maintain a similar accuracy. The older form of **precision resistor** consists of a long, thin, insulated wire of an alloy with a very small temperature coefficient. In manufacturing this wire, care is taken to maintain a uniform cross section so that the resistance is a linear function of the length of wire used. The selected length of wire is then spooled on a small bobbin and connected to the terminals on the ends of the bobbin. A new manufacturing technique consists of depositing a thin film of the appropriate metal on a small ceramic cylinder, after which leads are attached. The metal film may be etched if necessary to adjust the resistance to the specified value. Finally, the metal film precision resistor is given a protective coating of an insulating material.

Resistors used in electronic devices usually have a resistance of thousands of ohms (kilohms) and pass currents of only a few milliamperes. If, as is usually the case, the conversion of electric energy into heat is at the rate of less than one or two joules per second, a resistance element consisting of a mixture of finely ground carbon and an insulating composition pressed into a tiny rod less than a centimeter long is much more economical than wire-wound resistance elements. The resistance of **carbon-composition resistors** is governed by the ratio of carbon to insulating composition in the resistance element. Electrical connection to the resistance element is made through tinned copper "pigtail" leads imbedded in the ends of the element. The resistance element is sealed in a plastic jacket for protection.[†]

Carbon-composition resistors do not have the linear resistance/temperature characteristic of metal wire and film resistors shown in Figure 4-3. The resistance of a carbon-composition resistor increases if the temperature varies appreciably in *either* direction from normal room temperature, as shown by the expanded scale in Figure 4-5. Again, the physical dimensions of the resistor govern the rate at which it can dissipate heat. If the resistor is operated within its normal rating limits, the temperature variation is small enough for us to think of carbon-composition resistors as linear resistors.

With the development of small transistors, designers of electronic equipment were able to discard *hand-wiring* techniques, which used plastic-insulated copper wire to interconnect components. The next step was the printed-circuit board shown in Figure 3-1. Carbon-composition resistors are used with both wiring techniques. A more recent step in *solid-state* circuit design is to do away with interconnecting wiring

[†]See Appendix 4 for resistor color code.

altogether by combining many circuit components into tiny modules known as **integrated circuits** (ICs). The transistors at the heart of many ICs are created by depositing various combinations of p-type and n-type semiconductor material in microminiature patterns. The same technique is used to add many almost microscopic resistor elements to the circuit configuration of the ICs.

As noted in Section 3-6, the covalent bonding of germanium or silicon atoms in an intrinsic semiconductor crystal produces a theoretical insulator which, in practice, becomes a poor insulator due to the number of charge carriers made available through broken valence bonds. At room temperature, semiconductors are electric conductors. They are not good conductors like copper because of the limited number of charge carriers. They are, therefore, semiconductors or good resistance material. Since the resistance of a strip of semiconductor material depends on the number of available charge carriers, any desired resistance within certain limits can be manufactured by regulating the number of electron-donor impurity atoms introduced into the pure semiconductor crystal. We can expect to see an increase in the use of semiconductor material as resistance elements in electronic circuits.

4-8 NONLINEAR RESISTORS

The tungsten filament of an incandescent lamp and the heater of a cathode-ray tube are examples of resistors where the temperature range from room temperature to the operating temperature is such that there is a considerable change in resistance. A 60-watt lamp operating from a 120-volt source has a white hot resistance of 240 ohms to limit the current through it to $I = V/R = 120/240 = 0.5$ ampere. But when the lamp is turned off, its cold resistance is only about 18 ohms. Since its resistance is not reasonably constant over its normal temperature range, we can think of an incandescent lamp as a **nonlinear resistor.**

In the case of the incandescent lamp and tube heater, this nonlinear characteristic is a disadvantage. With a cold resistance of 18 ohms, the *inrush* current at the instant when the lamp is turned on (before the filament has had a chance to heat up) is $I = V/R = 120/18 = 6.6$ amperes. Fortunately, the mass of the lamp filament is small enough that it can reach white heat in less than a millisecond. Therefore, the current surge is of very short duration, as shown in Figure 4-6. Nevertheless, electric

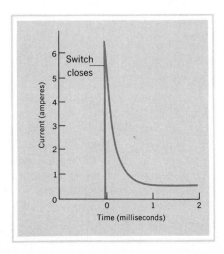

FIGURE 4-6
Inrush Current of an Incandescent Lamp.

85

switches used with incandescent lamps have to be designed to withstand the inrush current. If the heating elements of electric stoves had the same temperature coefficient as tungsten, the surge would be more serious since stove elements take several seconds to reach red heat. Consequently, electric stove elements are manufactured from an alloy such as Nichrome II, which has a very small temperature coefficient.

A nonlinear resistance characteristic is not always a disadvantage. In the circuit of Figure 4-2, maybe we can arrange for the series resistor to have a large *negative* temperature coefficient so that its resistance is quite high at the instant the switch is closed. As the lamp warms up, its resistance increases. But, as the negative-temperature-coefficient resistor heats up, its resistance *decreases*. The series combination of these two nonlinear resistors tends to provide a more uniform current when the switch is first closed, rather than the surge of Figure 4-6. Such negative-temperature-coefficient resistors are available under various trade names for use in limiting the initial surge which occurs when electronic devices are first switched on. These nonlinear resistors have a resistance of over 100 ohms at room temperature, but with a current of approximately 1 ampere through them, their resistance drops to less than 1 ohm after 10 to 15 seconds.

We again look to the mechanics of electrical conduction in semiconductors to account for this remarkable change in resistance. Certain metallic oxides form semiconductor crystals which are mixed with a ceramic binder to form resistors in which the breaking and reforming of covalent bonds make only a limited number of electric charge carriers available at normal room temperature. When electric current flows through these resistors, their temperature rises. With the increase in temperature, many more covalent bonds are broken, many more charge carriers are created, and, consequently, the resistance drops to a fraction of its value at room temperature.

In addition to their application in inrush surge-limiting resistors, tiny beads of these metallic oxide semiconductors can be calibrated in terms of resistance versus temperature to become sensitive temperature-measuring devices. These tiny nonlinear resistors are called **thermistors.** A decrease in temperature of less than 20°C will more than double their resistance.

Semiconductor conduction also accounts for the nonlinear resistance characteristic of a device called a **varistor.** Varistors depend on the nonlinear resistance characteristic of zinc oxide or silicon carbide crystals, which are formed into wafers with a clay binder. Zinc oxide varistors are used to protect low-voltage power supplies

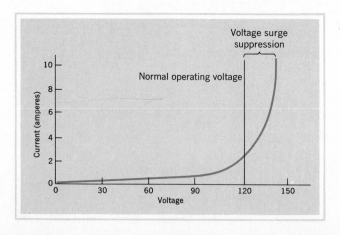

FIGURE 4-7
Typical Varistor Characteristic.

TABLE 4-4

GRAPHIC SYMBOLS FOR NON-LINEAR RESISTORS

Nonlinear Resistor	Graphic Symbol
General	(resistor symbol)
Thermistor	(resistor symbol with t°)
Varistor OR Varistor	(resistor symbol with V) (diode-like symbol)
Photoresistor	(resistor symbol with light arrows)

from voltage surges. Silicon carbide varistors (commonly known by the trade name **thyrite**) are designed to protect higher voltage systems—lightning arrestors on power transmission lines.

Actually, temperature has little effect on the resistance of a varistor. The rapid increase in the number of charge carriers depends on the potential difference across the varistor wafer. When the applied voltage is increased beyond a certain magnitude, there is a marked *increase* in the breaking of valence bonds and a consequent sharp *decrease* in resistance. As Figure 4-7 indicates, the decrease in resistance is such that many times the normal current can flow through the varistor without appreciably increasing the voltage drop across it. Therefore, connecting a varistor in parallel with an electrical device provides that portion of the system with a means of coping with a sudden surge without significantly affecting the voltage across (and the current through) the protected device.

A more recent nonlinear resistor is the **photoresistor.** As we noted in Figure 3-22, light falling on certain semiconductor materials breaks down valence bonds and thus creates additional charge carriers. Unlike the *photovoltaic* cell with its *pn*-junction, the photoresistor consists of a thin zigzag strip of cadmium sulphide behind a window only a few millimeters in diameter. However, its resistance can range from hundreds of kilohms in the dark to less than 100 ohms in bright daylight. Photo-resistors have all but replaced photovoltaic cells in photographic exposure meters (auto-exposure cameras) and simple light-operated switches for turning lights on at dusk and off at dawn.

Semiconductor technology has led to the development of three quite different types of nonlinear resistors:

Thermistor: resistance decreases as *temperature* increases.
Varistor: resistance decreases as *applied voltage* increases.
Photoresistor: resistance decreases as *incident light* increases.

To identify these nonlinear resistors in circuit diagrams, we modify the basic resistor symbol as shown in Table 4-4.

4-9 VOLT-AMPERE CHARACTERISTICS

Ohm's law of constant proportionality [Equation (4-1)] applies only to linear resistors, where the V/I ratio is a reasonably constant value expressed in ohms. However, in Section 4-8 we have used the term **resistance** with nonlinear resistors as well. The difference is that the resistance is no longer a constant, but is a variable depending on such factors as temperature, applied voltage, and incident light. By retaining the concept of resistance, the basic relationship derived from Ohm's law still holds true:

$$R = \frac{V}{I} \tag{4-2}$$

Since $R = V/I$, it is often convenient to show variations in resistance graphically by plotting this volts-per-ampere ratio for the resistor in question. In Figure 4-4, we discovered that the volt-ampere graph for a linear resistor is a straight line whose slope is inversely proportional to the resistance. Figure 4-7 shows a volt-ampere graph for a nonlinear varistor.

In drawing a volt-ampere characteristic, it is customary to plot voltage drop as an independent variable on the x-axis. Current then becomes the dependent variable on the y-axis, as shown in Figure 4-8. Consequently, a low resistance, which by definition has a low volts-per-ampere ratio, appears as a very steep graph. Similarly, a high resistance with its high volts-per-ampere ratio is represented by a graph with a very shallow slope.

Both graphs in Figure 4-8 represent linear resistors. For a linear resistor, we can read the resistance from the graph by simply taking the ratio of voltage to current at any point on the graph. All points on a linear characteristic have the same ratio. If the volt-ampere characteristic is not perfectly linear, we often consider the *dynamic* value

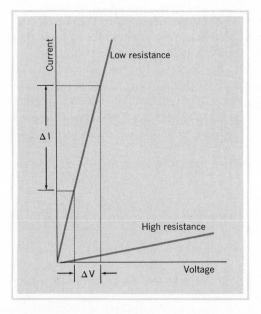

FIGURE 4-8
Volt-Ampere Characteristics of Low and High Linear Resistances.

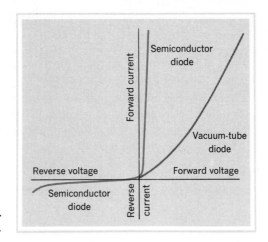

FIGURE 4-9
Volt-Ampere Characteristics of Semi-conductor and Vacuum-Tube Diodes.

of resistance in terms of the *change* in voltage drop versus the *change* in current associated with this change in voltage drop, as indicated in Figure 4-8. Hence, for a given ΔV, the ΔI is much greater for a small resistance than it is for a high resistance.

Since linear resistors conform to Ohm's law of constant proportionality within normal operating limits, we seldom bother to plot their straight-line volt-ampere characteristics. We can simply print the resistance in ohms on these resistors either in numerical or color code form.[†] However, the most convenient way of discussing the behavior of a nonlinear resistor such as the varistor of Figure 4-7 is to draw a graph of its particular volt-ampere characteristic.

Both vacuum tubes and transistors behave as nonlinear resistances. Hence, any study of electronic circuits will frequently resort to volt-ampere characteristics. The graphs in Figure 4-9 compare the volt-ampere characteristics of the vacuum-tube diode[‡] and the semiconductor diode described in Chapter 3. Since we want to show what happens when we *reverse* the polarity of the applied voltage, we draw the negative x and y axes as well as the usual positive axes. The volt-ampere characteristics show us that both diodes permit electric current when they are forward biased but that the internal resistance of the vacuum-tube diode is much higher than that of the semiconductor diode. When we reverse the polarity of the applied voltage, no current flows through a vacuum-tube diode, whereas the semiconductor diode does permit a small reverse current. Under these conditions, the internal resistance of the semiconductor is very high and the internal resistance of the vacuum-tube diode is infinitely high.

4-10 OHM'S LAW APPLIED

At the beginning of the chapter, we set out to determine the factors that govern the current in an electric circuit. By transposing Equation (4-2), which defines Ohm's law, we get $I = V/R$. If we are considering the resistance of the total circuit connected

---

[†]See Appendix 4 for resistor color code.
[‡]*Diode* means a two-element device.

between the terminals of the voltage source, the total voltage drop V will be equal to the applied voltage E of the source (as we noted in Figure 2-13). Hence, the magnitude of the current is dependent on the applied voltage and the total resistance of the circuit. In the electric circuits we shall be considering in this book, applied voltage and resistance are *independent* variables. This means that the potential difference of the batteries of Chapter 3 depends on the two particular metals used and is essentially *independent* of the resistance and the current. Similarly, the resistance of an electric circuit depends on such controllable physical factors as length, diameter, type of material, and temperature. Current, on the other hand, is the *dependent* variable. Whenever the switch is closed in the circuits we have discussed, the current *must* automatically take on a value which will satisfy the relationship

$$I = \frac{E}{R} \tag{4-9}$$

where I is the current that must flow in the circuit (in amperes), E is the applied voltage (in volts), and R is the total resistance of the circuit (in ohms).

In the circuit of Figure 2-13, we noted that a voltage drop can appear across a resistor *only* when current flows through it. Transposing Ohm's law, Equation (4-2) once more gives

$$V = IR \tag{4-10}$$

where I is the current through a resistor (in amperes), R is the resistance of the resistor (in ohms), and V is the resulting voltage drop across the resistor (in volts). Because of this relationship, it is quite common to hear *voltage drop* across a resistance referred to as an **IR drop.**

EXAMPLE 4-12
Compute the circuit current when a 4.7-kΩ resistor is connected across a 9-V source.

SOLUTION

$$I = \frac{E}{R} = \frac{9 \text{ V}}{4.7 \text{ k}\Omega} = 1.9 \times 10^{-3} \text{ A} = \textbf{1.9 mA}$$

EXAMPLE 4-13
Calculate the voltage drop that will be developed across a 560-Ω resistor by a 15-mA current through the resistor.

SOLUTION

$$V = IR = 15 \text{ mA} \times 560 \text{ } \Omega = \textbf{8.4 V}$$

PROBLEMS

4-1. What is the resistance of an electric circuit which draws a 2.5-A current from a 120-V source?

4-2. What resistance is required to limit the current in a circuit to 2 A when the applied voltage is 30 V?

4-3. What is the resistance of an electric stove element when a 5-A current through it causes a 110-V voltage drop across it?

4-4. If a current of 0.15 A through the resistor in Figure 4-2 causes a 1.0-V drop across it, what is its resistance?

4-5. What current will flow through a 20-Ω resistor connected across a 100-V source?

4-6. What current through a 250-Ω resistor will cause a voltage drop of 75 V across the resistor?

4-7. A 12-Ω heating element has a voltage drop of 98 V across it. What current is flowing through the heater?

4-8. What current will flow in a circuit whose total resistance is 17 Ω when it is connected to a 117-V source?

4-9. What voltage must be applied to a 15-Ω resistor to make it pass a 3-A current?

4-10. What is the voltage drop across a 125-Ω resistor when a 0.2-A current flows through it?

4-11. What voltage drop is produced by a 1.6-A current flowing through a 56-Ω resistor?

4-12. What applied voltage is required to cause current at the rate of 0.35 A in a circuit where the total resistance is 124 Ω?

4-13. A 15-μA current through a resistor in a radio receiver causes a 30-mV drop across it. What is its resistance?

4-14. What is the resistance of an ammeter whose full-scale reading is 10 A with a 30-mV drop across the meter?

4-15. What value of bias resistor is required for an amplifier in order to produce a 7-V drop across the resistor when the current through the resistor is 12 mA?

4-16. What value of resistor will draw a 32-mA current when connected to a 2.5-kV source?

4-17. A fuse in the power supply of a transistor amplifier has a resistance of 0.02 Ω. What current through the fuse will develop a 500-μV drop across it?

4-18. What current will flow when a 50-μV source is connected to a 5-MΩ resistor?

4-19. What current will flow through a 2.2-kΩ resistor connected across a 480-mV source?

4-20. What current must flow through a 6.8-kΩ voltage-dropping resistor to produce an 80-V drop across the resistor?

4-21. What voltage must be applied to a 220-kΩ bleeder resistance in a piece of electronic equipment if the bleeder current is to be 20 μA?

4-22. What IR drop is produced by a 6-mA current passing through a 15-kΩ load resistor?

4-23. What voltage drop appears across the 2.2-kΩ collector load resistor of a transistor amplifier when the collector current is 1.8 mA?

4-24. The operating coil of a relay, which is designed to energize with a current of 24 mA through the coil, has a resistance of 3.7 kΩ. What is the minimum applied voltage which will energize the relay?

NOTE Normal room temperature is assumed in Problems 4-25 to 4-40.

4-25. An electric conductor with a cross-sectional area of 10 mm² has a resistance of 1.72 Ω/km of length. What is the resistance of 200 m of wire of the same material with a cross-sectional area of 4 mm²?

4-26. A conductor with a cross-sectional area of 2.5 mm² and a length of 50 m has a resistance of 1.42 Ω. What length of wire of the same material with a cross-sectional area of 0.5 mm² will have a resistance of 5.0 Ω?

4-27. An electric conductor with a diameter of 2.0 mm and a length of 80 m has a resistance of 0.44 Ω. What is the resistance of a conductor of the same material having a length of 50 m and a diameter of 0.8 mm?

4-28. A conductor 40 m in length and having a diameter of 1.2 mm has a resistance of 17.3 Ω. What diameter of wire of the same material will have a resistance of 1.0 Ω per meter of length?

4-29. Using the data of Table 4-1, calculate the resistance of a copper wire 200 m in length with a cross-sectional area of 4 mm².

4-30. What is the resistance of a copper bar with a rectangular cross section of 2 × 4 cm and a length of 1.5 m?

4-31. Calculate the resistance of 160 m of aluminum wire having a diameter of 1.6 mm.

4-32. What is the resistance of a tungsten filament 12 cm in length and 0.1 mm in diameter?

4-33. What length of silver wire 0.5 mm in diameter has a resistance of 0.04 Ω?

4-34. What diameter of copper wire has a resistance of 1 Ω per meter of length at normal room temperature?

4-35. A relay coil requires 144 m of copper wire. If its resistance is to be 50 Ω, what diameter of wire is required?

4-36. A coil of constantan wire 0.4 mm in diameter, wound in a single layer on a cylindrical form so that the mean diameter of each turn is 2 cm, has a resistance of 24 Ω. How many turns of wire are there in this wire-wound resistor?

4-37. A carbon rod 0.5 cm in diameter and 10 cm in length has a resistance of 0.153 Ω. What is the resistivity of carbon at 20°C in ohm meters?

4-38. An electric conductor 40 m in length and 1.8 mm in diameter has a resistance of 17.3 Ω. What material is used for this conductor?

4-39. A stranded cable consists of eight copper conductors, each having a diameter of 1.0 mm. Twisting the wires to form the cable requires the length of each conductor to be 3% longer than the length of the cable in its final form. What is the resistance of 1 km of this cable?

4-40. What diameter of solid copper conductor will have the same resistance per kilometer as a stranded cable consisting of six aluminum conductors each having a diameter of 1.3 mm? Allow an increase of 3% in the length of the individual aluminum conductors for twisting in forming the cable.

4-41. A length of copper telephone line has a resistance of 24 Ω at 20°C. What is its resistance on a hot summer day when its temperature rises to 36°C?

4-42. What is the resistance of an aluminum conductor at −20°C if its resistance at +30°C is 1.25 Ω?

4-43. What is the resistance at 60°C of 25 m of copper wire having a diameter of 0.5 mm?

4-44. What length of Nichrome II wire having a diameter of 0.64 mm has a resistance of 48 Ω at 200°C?

4-45. If the resistance-temperature graph for brass is extended in a straight line until the resistance becomes zero, the corresponding temperature would be $-480°C$. What is the temperature coefficient of brass at $20°C$?

4-46. The temperature coefficient of platinum at $20°C$ is 0.003 per degree per ohm. Find the temperature at which the resistance of platinum would become zero if the resistance-temperature graph were extended as a straight line.

4-47. An incandescent lamp draws a 1.0-A current from a 110-V source to raise the temperature of its tungsten filament to $2800°C$. What is its resistance at normal room temperature?

4-48. By how many degrees must the temperature of a nickel rod be changed to increase its resistance to 105% of its resistance at normal room temperature?

4-49. A certain conductor has a resistance of 10 Ω at $20°C$ and 11.35 Ω at $50°C$. If you consider only those materials listed in Tables 4-2 and 4-3, of what material is this conductor made?

4-50. An electric motor is to be operated at a distance of 250 m from a 120-V source. The current drawn by the motor is 8.0 A. This current raises the temperature of the copper conductors feeding the motor to $40°C$. What is the minimum diameter of wire that can be used without the voltage drop in the conductors exceeding 10% of the applied voltage?

REVIEW QUESTIONS

4-1. Why can Ohm's law be described in terms of **constant proportionability?**

4-2. Why does the V/I ratio of an electric circuit indicate its ability to oppose electric current rather than its ability to permit current?

4-3. If there were no such unit as the **ohm,** how could you express the resistance of an electric circuit?

4-4. If resistance is the *opposition* to electric current, energy must be expended in forcing free electrons to move *against* this opposition. Where does this energy come from? Where does it go?

4-5. What relationship exists between the resistance of a given conductor and the number of free electrons in the conductor?

4-6. An electric fuse consists of a small strip of metal with a low melting temperature. The current in the protected electric circuit flows through this strip of metal. Which will have the greater resistance, a 10-A fuse or a 20-A fuse? Explain.

4-7. The *National Electrical Code* ® limits the current through No. 14 AWG[†] house wiring to 15A. What is the reason for doing so? Is it possible for a current of greater than 15 A to flow through No. 14 wire?

4-8. A resistor made of Nichrome wire wound on a ceramic form has the same resistance and is passing the same current as the filament of an incandescent lamp, yet the temperature of the lamp filament is many times that of the Nichrome resistor. Explain.

4-9. Why does shortening the length of a given conductor decrease its resistance?

4-10. Why is resistance inversely proportional to the *square* of the diameter of a conductor?

[†]American wire gauge; see Appendix 3.

4-11. Why does the term **resistivity** apply to the *material* of an electric conductor, rather than to a particular conductor?

4-12. Given an accurate resistance-measuring device, how would you go about determining the resistivity of a sample of an unknown alloy?

4-13. Draw a graph of the type shown in Figure 4-3 for carbon, using the data given in Table 4-3. What would the temperature x be for carbon?

4-14. In checking the temperature rise of the copper field winding of an electric motor by measuring its resistance, before the motor was started, we found that its resistance (at room temperature) was 50 Ω. After 30 min of operation, its resistance was 53 Ω. A half-hour later its resistance was 54.3 Ω, and after a further 40 min had elapsed, the field coil resistance was 54.7 Ω. It took 3 h in all for the resistance to settle at a steady value of 55 Ω. Plot a graph of temperature rise against time for this heat run. Explain the reasons for the shape of the graph.

4-15. Draw a graph similar to Figure 4-4 for a resistor having a fairly pronounced negative temperature coefficient. Would such a resistor be useful as a current regulator? Explain.

4-16. Find the temperature coefficient of copper at 0°C. From our derivation of temperature coefficient, discuss why the temperature coefficient should be different when 0°C is specified rather than 20°C.

4-17. The volt-ampere characteristic of a carbon-composition resistor is a straight line for all values of current (and voltage) within its normal operating limits. If the graph is extended beyond these limits, the volt-ampere characteristic starts to curve slightly. In which direction will it bend? Why?

4-18. Whereas resistors made from resistance wire or metal films usually have a *positive* temperature coefficient of resistance, resistors made from semiconductor materials generally have a *negative* temperature coefficient. Account for this distinction.

4-19. Whereas a resistor made from a pure semiconductor material is likely to be quite nonlinear, a resistor of *n*-type germanium is reasonably linear over its normal operating range. Account for this difference in behavior.

4-20. In the equation $I = E/R$, why do we refer to current as a *dependent variable* in a simple electric circuit?

*4-21. Referring to the resistor color code in Appendix 4, write a BASIC program which will permit you to enter the nominal value of any color-coded resistor and its tolerance, and print out the maximum and minimum resistance values for that resistor.

*4-22. A nonlinear resistor has a volt-ampere characteristic such that $I = 5.0 \times 10^{-3}V^2$. Write a program from which you can plot a volt-ampere characteristic for values of V from 0 to 20 V in 1-V steps. (If you are familiar with BASIC routines for plotting graphs, write a program to display the resulting volt-ampere graph on fully labeled coordinates.)

5

WORK AND POWER

5-1 ENERGY AND WORK

In Chapter 2, we defined **work** in mechanical terms:

Work is the accomplishment of motion against the action of a force which tends to oppose the motion.

From this definition, the SI unit for work, the **joule,** is defined as the work accomplished in moving (or the energy required to move) an object through a distance of one meter against an opposing force of one newton.

We also discovered that energy must be expended when we move an object against an opposing force, as in the example of carrying a heavy parcel up several flights of stairs against the opposition of the force of gravity. And since energy can neither be created nor destroyed, energy and work are numerically equal. In the electric circuit shown in Figure 5-1, mechanical energy supplied to the shaft of the dynamo is *converted* into electric energy which is transmitted by a pair of electric conductors to the heater unit where the electric energy is, in turn, *converted* into heat energy. From the sequence of events in this example, it follows that **work is performed whenever energy is converted from one form into another.**

Since the law of conservation of energy must apply to the circuit of Figure 5-1, and since we have already defined the unit of energy and work for a mechanical system, we can also express both electric energy and work and heat energy and work in joules. Since the energy conversions illustrated in Figure 5-1 are typical in practical

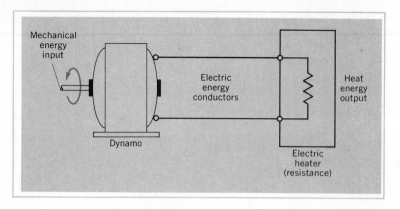

FIGURE 5-1
Energy Conversions in a Simple Electric Circuit.

electric circuits, the advantage of expressing mechanical, electric, and heat energy all in joules is obvious. Hence, we can now state that

The joule is the SI unit of electric energy and work.

From Equation (2-4),

One joule of electric energy is required to raise one coulomb of electric charge through a potential difference of one volt.

To illustrate the equivalence among electric, mechanical, and heat energy, we shall consider the following examples. Then, for the remainder of this book, we shall make our energy and work calculations in the *electrical* part of the system.

EXAMPLE 5-1

How much electric energy must be supplied to an electric motor to raise an elevator having a mass of 4 megagrams[†] a distance of 40 m? Assume that 10% of the electric energy supplied to the motor is lost in the form of heat.

SOLUTION

To raise the elevator car, it is necessary to move it against the force of gravity acting on it. From Newton's second principle, the **weight** of an object is its mass times gravitational acceleration.

$$\therefore F_g = mg = kg \cdot m/s^2 = newtons$$

$$= 4 \; Mg \times 9.8 \; m/s^2 = 4 \times 10^3 \; kg \times 9.8 \; m/s^2 = 39\,200 \; N^{\ddagger}$$

$$\therefore mechanical \; work, \; W = F_g s = 39\,200 \; N \times 40 \; m = 1\,568\,000 \; J$$

[†]In the International System of Units, large masses are expressed in *megagrams*. However, SI recognizes that, for some time to come, the old MKS term *tonne* (metric ton) will appear in commercial usage. One tonne = one megagram.

[‡]In entering the exponent for the unit of mass into the calculator in this example, we encounter the one anomaly in SI base units. The base unit of mass is the *kilo*gram.

Although the electric energy going toward raising the elevator is also 1 568 000 J, this represents only 90% of the electric energy supplied to the motor. Therefore, the total electric energy required is

$$W_T = \frac{1\,568\,000 \text{ J}}{0.9} = 1\,742\,000 \text{ J}$$

EXAMPLE 5-2

A teakettle contains 1.6 kg of water at normal room temperature (20°C). If 20% of the electric energy supplied to the stove element under the kettle is lost by radiation and in heating the kettle itself, how much electric energy must be supplied to the stove element to bring the water to the boiling point? (The heat energy required to raise 1 kg of water through 1°C is 4185 J.)

SOLUTION

Heat energy, $W = 1.6 \text{ kg} \times 4185 \text{ J}/(\text{kg} \times °\text{C}) \times (100 - 20)°\text{C} = 535\,680 \text{ J}$

The electric energy going toward heating the water is also 535 680 J. However, the total electric energy supplied to the stove element must be

$$W_T = \frac{535\,680 \text{ J}}{0.8} = 670\,000 \text{ J}$$

5-2 POWER

By using a suitable train of gears that would multiply mechanical force by dividing down its speed by the same factor, we could use a small electric motor to raise the elevator in Example 5-1. But in doing so, it would take about five minutes for the elevator to move through the 40 meters. This might be satisfactory for a freight elevator but a passenger elevator would have to be supplied with a much larger motor so that the required amount of mechanical work could be accomplished in a much shorter time. Therefore, in energy-conversion systems, the *time rate* of doing work is just as important as the amount of work done and is given the name **power.**

Power is the rate of doing work.
The letter symbol for power is P.

Since the **joule** is the basic unit of work and the **second** is the basic unit of time, it would be quite satisfactory to express power in **joules per second.** Expressing power in joules per second automatically calls our attention to the fact that power is the *rate* of doing work. But it was decided to honor James Watt for his work in converting heat energy into mechanical energy with his steam engine by calling the unit of electric power the **watt.** Therefore,

The watt is the SI unit of electric power.
The unit symbol for watt is **W.**

The magnitude of the watt is automatically established in terms of joules and seconds:

One watt is the rate of doing work when one joule of work is done in one second.

Expressing this in equation form, we have

$$P = \frac{W}{t} \tag{5-1}$$

where P is power in watts, W is work in joules, and t is time in seconds. Since the term **watt** does not suggest the idea of *rate* as does the term **joules per second,** we must keep reminding ourselves when calculating power that **watts** are exactly the same as **joules per second** and that **power** is the *rate* of doing work.

EXAMPLE 5-3
At what rate must electric energy be supplied to the electric motor in Example 5-1 if the elevator is raised through a distance of 40 m in 4 min?

SOLUTION

$$P = \frac{W}{t} = \frac{1\,742\,000 \text{ J}}{(4 \times 60) \text{ s}} = 7258 \text{ W} \quad \text{or} \quad \textbf{7.26 kW}$$

Equation (5-1) is the basic equation which defines power and establishes the magnitude of the watt. But it does not provide the most convenient method of determining power in electric circuits. By means of the following algebraic substitution, we can develop a more useful relationship for calculating electric power:
From Equation (2-4) defining potential difference,

$$V = \frac{W}{Q} \qquad \therefore W = QV$$

and from Equation (2-2) defining current,

$$I = \frac{Q}{t} \qquad \therefore t = \frac{Q}{I}$$

substituting for W and t in Equation (5-1) gives

$$P = \frac{W}{t} = \frac{QV}{Q/I} = QV \times \frac{I}{Q} = V \times I$$

$$\therefore P = VI \tag{5-2}$$

where P is power in watts, V is voltage drop in volts, and I is current in amperes. Also, since $V = IR$ [from Ohm's law, Equation (4-2)], substituting in Equation (5-2) gives

$$P = IR \times I$$

$$\therefore P = I^2R \qquad (5\text{-}3)$$

where P is power in watts, I is current through the resistance in amperes, and R is resistance in ohms. Since $I = V/R$ [from Ohm's law, Equation (4-2)], substituting in Equation (5-2) gives

$$P = \frac{V^2}{R} \qquad (5\text{-}4)$$

where P is power in watts, V is voltage drop across the resistance in volts, and R is resistance in ohms.

EXAMPLE 5-4

A lamp draws a 2-A current when connected to a 120-V source. What is the power rating of the lamp?

SOLUTION

$$P = EI = 120 \text{ V} \times 2 \text{ A} = 240 \text{ W}$$

EXAMPLE 5-5

A 10-kΩ resistor is connected into a circuit where the current through it is 50 mA. What is the minimum safe power rating for a resistor to be used in this circuit?

SOLUTION

$$P = I^2R = (50 \text{ mA})^2 \times 10 \text{ k}\Omega = 25 \text{ W}$$

EXAMPLE 5-6

What is the highest voltage that can be applied to a 5-kΩ 2-W resistor without exceeding its heat-dissipating capability?

SOLUTION
Since

$$P = \frac{V^2}{R}, \quad V^2 = PR \quad \text{and} \quad V = \sqrt{PR}$$

$$\therefore E = V = \sqrt{2 \text{ W} \times 5 \text{ k}\Omega} = 100 \text{ V}$$

5-3 EFFICIENCY

Examples 5-1 and 5-2 suggest that in converting energy from one form into another, some of the input energy is converted to a form of energy that is not useful in performing the work which the equipment was set up to do. This wasted energy is usually in the form of heat. Some of this heat comes from mechanical friction, some of it from current in an electric resistance, and some of it is lost by radiation, as in Example 5-2. Not only does this wasted energy cost money to supply, but the equipment has to be designed to dissipate the wasted heat energy safely.

The ability of an energy-converting device (such as the electric motor of Example 5-1 and the electric stove element of Example 5-2) to convert as much of the input energy as possible into *useful* work is specified in terms of its **efficiency.**

Efficiency is the ratio of useful output energy to total input energy.
The letter symbol for efficiency is the Greek letter η (eta).

$$\eta = \frac{W_{out}}{W_{in}} \tag{5-5}$$

We usually express efficiency as a percentage. For example, the efficiency of the system in Example 5-1 is 90%.

EXAMPLE 5-7

What is the efficiency of an electric hoist if 60 000 J of electric energy has to be supplied to the motor in order to raise a 300-kg mass through 18 m?

SOLUTION

$$\text{Mechanical work, } W = 300 \text{ kg} \times 9.8 \text{ m/s}^2 \times 18 \text{ m} = 52\,920 \text{ J}$$

$$\text{Efficiency, } \eta = \frac{52\,920 \text{ J}}{60\,000 \text{ J}} \times 100\% = \mathbf{88.2\%}$$

From Equation (5-1) defining power, it follows that $W = Pt$. If we substitute this form for W_{out} and W_{in} in Equation (5-5),

$$\text{Efficiency, } \eta = \frac{P_{out} \times t}{P_{in} \times t}$$

from which

$$\eta = \frac{P_{out}}{P_{in}} \tag{5-6}$$

In applying Equation (5-6) to electric motors, P_{out} is *mechanical* power output, and P_{in} is *electric* power input. In the International System of Units, the **watt** applies to both electric and mechanical power. Unfortunately, until conversion to metric units in North America is well established, we shall find that mechanical power output of motors is expressed in terms of a unit invented by James Watt—the **horsepower.** For a few years, therefore, we require a conversion factor to convert mechanical power expressed in *horsepower* into SI units—*watts*:

$$1 \text{ hp} = 746 \text{ W} \qquad (5\text{-}7)$$

EXAMPLE 5-8

What electric power input must be provided to an electric motor which develops mechanical energy at the rate of 24 hp with an 85% efficiency?

SOLUTION

Mechanical power output, $P_{out} = 24 \times 746 = 17\,904$ W or 17.9 kW

$$P_{in} = \frac{P_{out}}{\eta} = \frac{17.9 \text{ kW}}{0.85} = \textbf{21 kW}$$

We must note that the **horsepower** is an *English* unit of *mechanical* power. It is not an approved SI unit. Therefore, the horsepower must not be used to express *electric* power.

5-4 THE KILOWATTHOUR

In the examples in this chapter, we purposely left the answers for *work* in base units to show that a relatively small amount of work results in a large numerical answer when we express the answer in joules. Therefore, the base unit of electric energy, the **joule,** is considered to be too small a unit for practical purposes such as computing a monthly electric power bill. The proper SI unit for such purposes is the **megajoule (MJ).** However, for commercial purposes, another practical unit, the **kilowatthour,** is used. Although the International System of Units prefers the term **megajoule,** it does permit the use of the **kilowatthour** for practical purposes for the time being.
    Since

$$P = \frac{W}{t} \qquad \therefore W = Pt$$

If P is power in **watts** and t is time in **seconds,** then W must be work in **watt-seconds** or **joules.**
    But if P is power in **kilowatts** and t is time in **hours,** then W must be work in **kilowatthours.**

The kilowatthour is the practical unit of electric work or energy.
The unit symbol for kilowatthours is **kWh.**

The kilowatthour is defined by the equation

$$W = Pt \qquad (5\text{-}8)$$

where W is work or energy in kilowatthours, P is power in kilowatts, and t is time in hours. Since 1 kW = 1000 W and 1 h = 3600 s,

$$1 \text{ kWh} = 1000 \text{ W} \times 3600 \text{ s} = 3.6 \times 10^6 \text{ J} = 3.6 \text{ MJ} \qquad (5\text{-}9)$$

Note: Kilowatthours = kilowatts *times* hours (not kilowatts/hour).

EXAMPLE 5-9

At 7¢ per kilowatthour, how much will it cost to leave a 60-W lamp burning for 5 days?

SOLUTION

$$W = Pt = 60 \text{ W} \times (24 \times 5) \text{ h} = 7200 \text{ Wh} = 7.2 \text{ kWh}$$

$$\text{Cost} = 7.2 \text{ kWh} \times 7¢/\text{kWh} = \mathbf{50.4¢}$$

The amount of electric energy consumed by a residence or business is measured by a **kilowatthour meter.** It consists of an aluminum disk which rotates on an induction-motor principle at a speed which is directly proportional to the applied voltage and the current drawn by the consumer ($P = EI$). The disk drives a series of 10-to-1 reduction gears connected to dials. Hence the dial reading corresponds to speed times accumulated time, $P \times t$ = kilowatthours. Since there are some costs involved in providing an electric utility service regardless of the energy used, most utilities charge for electric energy on a sliding scale that decreases as more energy is used during the billing period.

EXAMPLE 5-10

A typical electric utility company rate for residential service is

First 100 kWh:	9.9¢/kWh
Next 400 kWh:	5.3¢/kWh
All additional kWh:	3.6¢/kWh

If the present meter reading is 38 770 kWh and the previous reading was 37 530 kWh, what is the bill for the electric energy used in the period between readings?

SOLUTION

Amount of energy used = 38 770 − 37 530 = 1240 kWh

Cost of first 100 kWh = 100 × 9.9¢ = $ 9.90

Cost of next 400 kWh = 400 × 5.3¢ = $21.20

Cost of last 740 kWh = 740 × 3.6¢ = $26.64

Total bill $57.74

TABLE 5-1

INTERRELATIONSHIP OF BASIC ELECTRICAL UNITS

Function	Defining Equation	Unit and Unit Symbol	Definition of Unit	Useful Derived Equations	Derivation
Voltage	$V = W/Q$	Volt V	Joules per coulomb	$V = IR$ $V = P/I$ $V = \sqrt{PR}$	Transpose equation (4-2) Transpose equation (5-2) Transpose equation (5-4)
Current	$I = Q/t$	Ampere A	Coulombs per second	$I = V/R$ $I = P/V$ $I = \sqrt{P/R}$	Transpose equation (4-2) Transpose equation (5-2) Transpose equation (5-3)
Resistance	$R = V/I$	Ohm Ω	Volts per ampere	$R = V^2/P$ $R = P/I^2$	Transpose equation (5-4) Transpose equation (5-3)
Power	$P = W/t$	Watt W	Joules per second	$P = VI$ $P = I^2R$ $P = V^2/R$	Equation (5-2) Equation (5-3) Equation (5-4)
Work and energy	$W = Fs$	Joule J	Newton meters	$W = Pt$	Transpose equation (5-1)

Note: E is interchangeable with V in this table.

5-5 INTERRELATIONSHIP OF BASIC ELECTRICAL UNITS

If we gather together all the equations which define the basic electrical units which we have discussed so far, we find that if we know three such circuit parameters as voltage, current, and elapsed time, we can determine any of the other parameters such as resistance, work, and power. This is achieved by a process of algebraic substitution in the defining equations as we did to develop a practical equation for power. Table 5-1 shows some of the more useful derived equations. The student should derive each of the equations shown in this table from the basic defining equations as an aid to understanding and remembering the interrelationship among the basic electrical units. Note that not all possible combinations are shown. Only those which are useful in solving and understanding electric circuits have been included. We may substitute the symbol E for V in this table when our attention is directed toward applied voltage rather than voltage drop in a particular circuit.

PROBLEMS

5-1. How many joules of energy is required to carry a 20-kg parcel up four flights of stairs through a vertical distance of 11 m?

5-2. How many joules of energy is required to pump 2000 liters† of water to the surface from a well which is 5 m deep?

† The liter is the common metric unit of volume. One liter is equal to 10^3 cubic centimeters. Hence a liter of water has a mass of 1 kilogram.

5-3. How many joules of energy is required to raise the temperature of 4 liters of water by 35°C?

5-4. How many joules of heat energy must be developed by an electric stove element to heat 1 liter of water from 20°C to 85°C, assuming that the container and radiation account for one-third of the heat energy produced by the stove element?

5-5. What is the power rating of a toaster which draws a 5-A current from a 120-V source?

5-6. What is the power rating of a soldering iron which has a 110-V drop between its terminals when the current through it is 1.2 A?

5-7. At what rate is electric energy converted into heat in the heater of a cathode-ray tube that is rated at 6.3 V 0.6 A?

5-8. A 4-mA current through a resistor causes a voltage drop across the resistor of 120 V. What is the minimum "wattage" rating this resistor must have?

5-9. At what rate must a 75-Ω resistor be able to dissipate heat if the current through it is 2.5 A?

5-10. At what rate must a resistor convert electric energy into heat if 6000 J of heat is produced in 10 min?

5-11. What power rating must a 100-Ω resistor have to pass a current of 20 mA without overheating?

5-12. What "wattage" rating must a 10-kΩ resistor have to pass a current of 250 mA safely?

5-13. What "wattage" rating would you select for a 560-Ω bias resistor which has to pass a current of 36 mA?

5-14. At what rate is electric energy being converted into heat in a 1.5-Ω motor-starting resistor when the starting current through it is 60 A?

5-15. What is the maximum current that a 20-kΩ 10-W resistor can handle without overheating?

5-16. What is the maximum voltage that can be safely applied to a 20-kΩ 10-W resistor?

5-17. What voltage drop will there be across a 1-kW stove element whose resistance when hot is 40 Ω?

5-18. What voltage drop will there be across a 75-Ω 3-W Christmas tree lamp?

5-19. A certain voltmeter has a resistance of 150 kΩ. While it is measuring a voltage source, it draws energy from the source at the rate of 96 mW. What is the PD of the source?

5-20. If the 600-mA heater of a television picture tube must produce heat at the rate of 3 W, what is the voltage drop across the heater?

5-21. What current will an electric motor draw from a 250-V source if it is developing mechanical energy at the rate of 5 hp at 90% efficiency?

5-22. What mechanical work can be done by a 75% efficient electric motor drawing 2 A from a 120-V source for 8 h?

5-23. The armature winding of an electric motor has a resistance of 0.2 Ω. At what rate is electric energy being used up by this particular motor loss when the armature current is 40 A?

5-24. How many joules of heat are produced in 1 h in the winding of a motor which has an I^2R loss of 320 W?

5-25. If 2.5×10^6 J of electric energy are supplied to the motor operating an elevator which raises a total mass of 5 tonnes through 45 m, what is the efficiency of the system?

5-26. With the friction of the pipes and pump bearings, the overall efficiency of a water pump is 86%. What mechanical energy is expended in raising 1600 liters of water 5 m to the surface?

5-27. What is the efficiency of an electric motor that develops a 2.0-hp mechanical output when the electric power input is 1.8 kW?

5-28. What horsepower electric motor is required to operate a pump which can pump water at the rate of 1600 liters per hour through a vertical distance of 5 m if the overall efficiency of the pumping system is 86%?

5-29. How long will it take to raise the temperature of 1 liter of water by 40°C if energy is imparted to the water at the rate of 1 kW?

5-30. If 60% of the electric energy input to an electric stove element is effective in heating the water in the kettle, what power rating must the stove element have if the temperature of a liter of water is to be raised from 20°C to 85°C in 4 min?

5-31. A fuse element whose resistance is 0.02 Ω is designed to "blow" when its electric power dissipation exceeds 5 W. What is the current rating of the fuse?

5-32. A power supply for a transistor amplifier develops an output of 25 V at 2.4 A. If the overall efficiency of the power supply is 80%, what input current will it draw from a 120-V source?

5-33. What is the minimum value of resistance that can be placed across the terminals of a 120-V source if the power drawn from the source is not to exceed 0.48 kW?

5-34. An electric heater produces heat at the rate of 50 000 J/min when the current through it is 10 A. What is its resistance?

5-35. The current drawn by a room air conditioner from a 230-V source is measured as 12.0 A. If the overall efficiency of the unit is 80%, how many joules of heat can the air conditioner remove from the room in 1 hour?

5-36. In an automobile electric system, the following devices are simultaneously drawing energy from the storage battery whose terminal voltage is 6.0 V:
(a) two 6-W taillight lamps;
(b) two parking lamps with a hot resistance of 12 Ω each;
(c) a radio drawing a 6.0-A current; and
(d) the heater fan motor which develops $\frac{1}{20}$ hp at 80% efficiency. What is the total rate at which the battery is supplying electric energy?

5-37. How much electric energy is used by a 750-W heater in 24 h?

5-38. If the overall efficiency of a radio transmitter is 48%, how much electric energy is required to produce a power output of 50 kW from 7:00 A.M. to midnight?

5-39. How much will it cost to operate a 4000-Ω electric clock from a 110-V power line for one year if electric energy costs 7¢/kWh?

5-40. How much electric energy did Mr. Jones use in May? Assume that:
(a) Electric energy costs 6¢/kWh.
(b) The meter reading at the end of May was 74 267 kWh.
(c) He receives a bill from the power company every two months.
(d) The meter reading at the end of February was 73 067 kWh.
(e) His bill for March and April was $48.00.

REVIEW QUESTIONS

5-1. What is the distinction between **work** and **energy**?

5-2. Why is the joule not commonly used in calculating electric energy?

5-3. What is the significance of the unit **watt second** found in some textbooks?

5-4. Why is the amount of energy represented by a joule quite small?

5-5. Why is the joule selected as the basic unit for electric energy even though it is not used in practice?

5-6. Why would a 5-hp motor with a 90% efficiency be physically smaller than a 5-hp motor with a 60% efficiency?

5-7. Why is it possible to express efficiency in terms of the ratio of output **power** to input **power** even though efficiency is defined as the ratio of output **energy** to input **energy**?

5-8. Why must electric kettles be equipped with thermostatic switches to open the circuit when the kettle boils dry?

5-9. Some small heaters consist of an electric heating coil and a small fan. It is noted that the heating element is brighter when the fan is turned off. Explain.

5-10. Is the heater coil in Question 5-9 drawing more current when the fan is turned off? Explain.

5-11. How would you apply the observations of Question 5-9 to the miniaturization of an aircraft radio?

5-12. Two 1000-Ω resistors are wound from Nichrome wire on ceramic tubes. One is rated at 5 W and the other at 50 W. Compare the two resistors for conductor size, overall dimensions, and temperature at rated dissipation.

5-13. A 35-W soldering iron is used to solder miniature radio components, whereas a 150-W iron is needed to solder a lead to the radio chassis. Explain. What would be the effect of using the 150-W iron on the miniature components?

5-14. SI defines the volt as the difference in electric potential between two points of a conductor carrying a constant current of 1 A when the power dissipated between these points is equal to 1 W. Show that this definition is consistent with the one used in Chapter 2.

5-15. Why is it customary to find larger electric motors used on passenger elevators than freight elevators although the freight elevators are the heavier?

5-16. What is wrong with the wording of the question: "How much power is consumed by a toaster drawing 3 A from a 110-V source?"

5-17. What is wrong with the wording of the question: "How many joules are there in a horsepower?"

5-18. What is wrong with the wording of the question: "What is the efficiency of an electric motor which has a power input of 5 hp and a power output of 4 hp?"

REVIEW PROBLEMS

Interrelationship of basic electrical units.

NOTE The purpose of these short drill problems is to help the student to recognize the relationships that must exist among the basic electrical units. Each of these problems should be solved by the method outlined with Example 2-2. These problems should be attempted without constant reference to Table 5-1.

5-41. What voltage drop is produced across a 42-Ω resistor by a 2.4-A current through it?

5-42. At what rate is electric energy converted into heat in the resistor in Problem 5-41?

5-43. How long will it take the resistor in Problem 5-41 to convert 2 kWh of energy?

5-44. What voltage must be applied to a 34-Ω resistor to make it dissipate energy at the rate of 180 W?

5-45. What is the power rating of a 72-Ω heater which passes a 6-A current?

5-46. How long will it take the heater in Problem 5-45 to consume 9400 J of electric energy?

5-47. How many coulombs of electrons pass through the heater in Problem 5-46?

5-48. What current is flowing in a 17-Ω resistor while it is dissipating energy at the rate of 520 W?

5-49. What is the resistance of a load which draws 1.4 A from a 24-V source?

5-50. What current will a 40-W lamp draw from a 117-V source?

5-51. What resistance will dissipate energy at the rate of 50 W when connected to a 110-V source?

5-52. What voltage must be applied to a 555-W heater to make it pass a 12-A current?

5-53. What current will a 45-Ω resistor pass while it is producing heat at the rate of 400 W?

5-54. What voltage must be applied to a 27-Ω resistor to make it convert $1\frac{1}{2}$ kWh in 7 h?

5-55. What is the efficiency of a motor with a 1-kW input and a 1-hp output?

5-56. What voltage drop is created by a 6.8-mA current flowing through a 27-kΩ load resistor?

5-57. What voltage is developed across a 300-Ω resistance by a power input to the resistance of 50 μW?

5-58. What value of resistance is required to obtain a voltage drop of 7 V when the current is 18 mA?

5-59. What power rating must the resistor in Problem 5-58 have?

5-60. What current will flow when a 20-μV signal is applied to a 75-Ω load?

5-61. What power is fed to the load in Problem 5-60?

5-62. What current will a 4-kW load draw from a 220-V source?

5-63. If an automobile starter motor develops $1\frac{1}{4}$ hp at 80% efficiency, what current does it draw from a 12-V battery?

5-64. How much work can a fully charged 12-V storage battery accomplish if it is rated at 80 ampere hours?

5-65. What is the resistance of a voltmeter which reads 120 V when it has a 40-μA current passing through it?

5-66. How long will it take for 2.8 C of electric charge to pass through the meter in Problem 5-65?

5-67. What is the resistance of the copper bus bars feeding an aluminum refining cell if a 4000-A current through them causes a 620-mV drop across them?

5-68. How much electric energy is lost as heat in the bus bars of Problem 5-67 in 8 h?

5-69. What is the resistance of an ammeter shunt if it is designed to have a 5-mV drop across it with a 9.99-A current through it?

5-70. What must be the resistance of a 15-A fuse if heat must be developed at the rate of 4.3 W to melt it?

PART II

RESISTANCE NETWORKS

PRACTICAL ELECTRIC CIRCUITS are never as simple as the basic electric circuit of Figure 1-1. Nevertheless, the basic relationships among current, voltage, resistance, work, and power summarized in Table 5-1 must hold true for all electric circuits, no matter how complex they may be.

In Part 1, we developed concepts of the physical nature of electric current, potential difference or voltage, resistance, work, and power. Such concepts are part of a qualitative consideration of the principles of electricity. We also learned how to assign numerical values to units of current, voltage, resistance, work, and power so that we can now examine the behavior of practical electric circuits from a quantitative point of view.

Part 2 groups together in four chapters all the principles and techniques we require for the analysis of any resistance network from a simple series circuit to a complicated array including more than one voltage source. The topics are developed step by step in the most popular sequence. Chapter 7 logically follows Chapter 6. But Chapters 8 and 9 are self-sufficent packages which can be reversed in order, or postponed to permit further study of the principles of circuit operation in the first term. Chapter 10 is a supplementary chapter which is intended to show how electrical measurement is an integral part of electric circuit analysis techniques.

6

SERIES AND PARALLEL CIRCUITS

6-1 RESISTORS IN SERIES

In the last two chapters, we have been treating electric current as a numerical quantity expressed in amperes. At this point, we should remind ourselves that electric current is defined as the *rate of flow* of electric charge. In metallic conductors, the flow of electric charge consists of free electrons moving through the circuit as a result of an electric potential difference generated by a voltage source. Keeping this in mind as we examine the simple series circuit of Figure 6-1, a second part of Ohm's law becomes apparent:

The current must be the same in all parts of a simple series circuit.[†]

Common sense tells us that there cannot be more electrons entering R_1 per second than leave R_1 in a second. Nor can more electrons enter R_1 in a second than leave R_2 in the same period of time. Similarly, the same number of electrons must pass through the source per second as pass through any other part of the series circuit in the same time interval. Since we express *electrons per second* in *amperes,* the same current flows through all components of the simple series circuit of Figure 6-1. Because the truth of this statement is quite evident once we understand what constitutes an electric current, Ohm seldom receives credit nowadays for having been the first person to state this as a basic law of electric circuits.

[†]According to the dictionary, *same* means *identical in all respects.* If R_1 and R_2 are equal in the circuit of Figure 6-1, there will be *equal but different* voltage drops across them, not the *same* voltage drop.

110

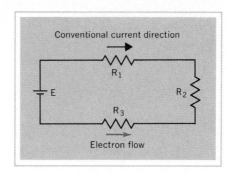

FIGURE 6-1
Simple Series Circuit.

Sometimes series-connected components are described in terms of the manner in which they are physically connected. In the series circuit of Figure 6-1, R_1 and R_2 are connected in series because no other component or branch is connected to the junction of R_1 and R_2. Also from Figure 6-1, we can say that E, R_1, R_2, and R_3 are all in series because there is only *one* path for electrons to follow through them. All characteristics of series circuits are based on the current being common to all components. Hence, we prefer to define a series circuit in terms of this common current:

Two or more electric components are considered to be in series if the same (common) current flows through all these components.

Since the current in a series circuit is common to all components, it is not necessary to use any subscript with the letter symbol I for the circuit of Figure 6-1. Here I represents the current through R_1, R_2, and R_3, and also the current through the source and the wire connecting the components together.

In Chapter 4, we decided that current in simple electric circuits is a *dependent* variable. It depends on what voltage is applied to the circuit and the total resistance of the circuit. Therefore, in solving the simple series circuit of Figure 6-1, we must be able to determine the total resistance of the circuit. Suppose that resistor R_1 is constructed of 2 meters of Nichrome wire, resistor R_2 is made from 1 meter of the same wire, and R_3 consists of 3 meters of the same resistance wire. Therefore, for an electron to travel around the circuit from one terminal of the generator to the other, it must travel through $3 + 1 + 2 = 6$ meters of Nichrome wire. Since the resistance of an electric conductor is directly proportional to its length, and since the total length of resistance wire used in this example is the sum of the individual lengths, it follows that the total resistance of this circuit will be the sum of the individual resistances. Therefore,

The total resistance of a series circuit is

$$R_T = R_1 + R_2 + R_3 + etc. \tag{6-1}$$

Once we determine the total resistance of a series circuit, we can then solve for the common current on which all the characteristics of a series circuit are based by Ohm's law: $I = E/R_T$. Actually, the current must take on a magnitude such that it will produce a total voltage drop V_T equal to the applied voltage E.

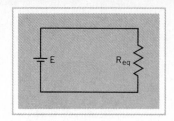

FIGURE 6-2
Equivalent Circuit of Figure 6-1.

EXAMPLE 6-1

What current will flow in a series circuit consisting of 20-Ω, 10-Ω, and 30-Ω resistors connected to a 45-V source?

SOLUTION

$$R_T = R_1 + R_2 + R_3 = 20 + 10 + 30 = 60 \ \Omega$$

$$I = \frac{E}{R_T} = \frac{45 \text{ V}}{60 \ \Omega} = 0.75 \text{ A}$$

As far as the source is concerned, the current drawn from it will be exactly the same for a single 60-ohm resistor connected to its terminals as for the 20-ohm, 10-ohm, and 30-ohm resistors in series. Therefore, if it will aid in analyzing circuit performance, we may think of the total resistance as being replaced by a single *equivalent* resistance R_{eq}. And the circuit of Figure 6-2 becomes the **equivalent circuit** of the original circuit of Figure 6-1. The solution we have used in Example 6-1 consists of reducing the original series circuit to a simple equivalent circuit and then solving for current by Ohm's law. As we start to analyze the behavior of more elaborate electric circuits in later chapters, we often resort to circuit simplification by replacing two or more components in a circuit diagram by a single equivalent circuit component.

EXAMPLE 6-2

In the series circuit of Figure 6-1, R_1 and R_3 are 10 kΩ and 22 kΩ, respectively. If the potential difference produced by the source is 16 V, what resistance is required for R_2 to limit the current drawn from the source to 340 μA?

SOLUTION

$$R_T = \frac{E}{I} = \frac{16 \text{ V}}{340 \ \mu\text{A}} = 47 \text{ k}\Omega$$

$$R_2 = R_T - (R_1 + R_2) = 47 \text{ k}\Omega - (10 \text{ k}\Omega + 22 \text{ k}\Omega) = 15 \text{ k}\Omega$$

6-2 POLARITY OF VOLTAGE DROP IN A SERIES CIRCUIT

In Figure 6-3, we have redrawn the simple series circuit of Figure 6-1 and added a voltmeter and an ammeter. Moving the positions of R_1 and R_3 in the circuit diagram, so that R_1, R_2, and R_3 are all in a vertical row, in no way alters the electrical behavior of the circuit, because we have not changed the sequence of the components through which the common current must flow. Since the ammeter is to show the magnitude of the common current, it must be connected in *series* with the circuit so that the current being measured will flow *through* it. Since the voltmeter shows the potential difference *between* two points in a circuit, it is connected *between* points B and C in this example. In other words, the voltmeter is connected *across* or *in parallel with* R_2. The ideal voltmeter draws negligible current; consequently, it does not alter the series circuit behavior for present discussions.

The analog direct-current meters we shall encounter in Chapter 10 either mark their terminals with $+$ and $-$ signs, as in Figure 6-3, or at least mark one terminal with a $+$ sign. If we reversed the voltmeter or ammeter leads in a practical circuit, the meters would read off-scale to the left of the zero scale marking. To make sure that we connect direct-current meters properly, we must become quite familiar with the **polarity** of applied voltages and voltage drops in dc circuits.

In Section 2-10, we noted that applied voltage E represents a potential rise, and voltage drop V represents a potential drop or fall. In the circuit of Figure 6-3, electrons experience a potential rise *inside* the voltage source as they are moved from the $+$ terminal to the $-$ terminal *against* the electric force which tends to make them flow in the opposite direction. In the *external* circuit, electrons flow *with* the electric field force from the $-$ terminal of the source back to the $+$ terminal. As electrons flow around this external circuit, they gradually lose the potential energy they gained in their trip through the source. Hence, we can say that the electrons at point D are at a higher potential than those at point C, and electrons at point B are at a higher potential than those at point A. Similarly, electrons at point B are at a lower potential than those at point C, and so on.

In Chapter 3 we discovered that there were quite a few different types of electric charge carriers in various electrical devices. We can divide them into two groups. All *negative* charge carriers will behave like the free electrons in the metallic conductors of Figure 6-3. All *positive* charge carriers flow in the *opposite* direction. Hence,

FIGURE 6-3
Polarity of Voltage Drops in a Series Circuit.

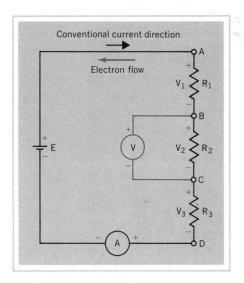

positive charge carriers will receive a rise in potential as they move from the − terminal to the + terminal inside the source and will gradually lose the potential energy thus gained as they then flow through the external circuit. It follows the + charge carriers have a higher potential at point A than at point B, and so on.

We can avoid possible confusion by concentrating on the **polarity of the voltage drop** rather than on potential rise and fall. As we have noted, electrons are moved from the + to the − terminal inside the voltage source. We call a voltage source an **active** circuit element since it generates electric energy (at the expense of some other form of energy). In **passive** circuit elements (which consume electric energy) electrons flow from the − to the + terminal, the direction in which they are urged to move by the potential difference in the circuit. Note that an ammeter is also a passive circuit element. Hence, we mark its terminals in the same manner. Electrons enter the ammeter through its − terminal and leave through its + terminal.

Given the polarity of the voltage source in the circuit of Figure 6-3, electrons will flow in a counterclockwise direction around the circuit. We can, therefore, proceed to mark the polarities of all voltage drops with − and + signs. The end of the resistor at which electrons enter must be marked − and the end of the resistor from which electrons leave must be marked + so that the potential difference will have the proper polarity to move electrons *with* the resulting electric field force. In a series circuit, it follows that the + end of the voltage drop across R_3, is connected to the − end of the voltage drop across R_2, and so on. We can now talk in terms of point C being positive with respect to point D, and point B being even more positive with respect to point D. We also say that point C is negative with respect to point A, and so on.

In our pencil-and-paper circuit diagrams, the nature of the actual charge carriers is not particularly important. But when we come to parallel circuits where current divides into two or more branches, it is important that we be able to keep track of current direction. As noted in Section 2-11, to avoid confusion which would result from observing the direction of flow of the actual charge carriers, the current practice in electrical engineering is to adopt the direction of flow of positive charge carriers as the **conventional direction** for electric current. Although both conventional current direction and electron flow are marked in Figure 6-3, only conventional direction will be marked in the remainder of the text (unless it is necessary to show the physical behavior of actual charge carriers).

When we mark polarities on circuit diagrams using conventional current direction, positive charge carriers must "flow" from + to − through any passive circuit element. As shown in Figure 6-3, we end up with exactly the *same* polarities from either point of view.

6-3 DOUBLE-SUBSCRIPT NOTATION

In the circuit diagram of Figure 6-3, the symbol V_1 identifies the voltage drop across R_1. The **polarity** of V_1 is obvious from Figure 6-3. Terminal A is positive with respect to terminal B.

Double-subscript notation provides us with a technique for including polarity information on the voltage drop across R_1. By identifying this voltage drop as V_{AB}, we are talking about the potential at point A *with respect to* point B. **It is standard procedure that the *second* subscript will designate the reference** (with respect to) **point or *node*.** Hence in Figure 6-3, V_{AB} is a positive quantity, whereas V_{BA} is a negative quantity and $V_{AB} = -V_{BA}$.

In dealing with three-phase circuits in Chapter 27, we shall find that double-subscript notation is essential in keeping track of the polarity of voltages and the directions of currents. In dealing with the dc circuits of the next few chapters, double-subscript notation is optional and we use it only when it is essential for clarity. In most examples, single-subscript notation is adequate.

6-4 KIRCHHOFF'S VOLTAGE LAW

Since the rules of algebra allow us to multiply both sides of an equation by the same quantity without changing the equality of the statement, let us multiply both sides of Equation (6-1) by I.

$$IR_T = IR_1 + IR_2 + IR_3 + \text{etc.}$$

Since
$$I = \frac{V_T}{R_T} \quad \therefore IR_T = E \text{ (the applied voltage)}$$

and
$$IR_1 = V_1 \text{ (the voltage drop across resistor } R_1)$$

$$\therefore E = V_1 + V_2 + V_3 + \text{etc.} \tag{6-2}$$

Therefore,

In a series circuit, the sum of all the voltage drops across the individual resistances must equal the applied voltage.

Gustav Kirchhoff discovered that this applied to any complete electric circuit, whether a simple series circuit as in Figure 6-1 or one loop of an elaborate network. This principle is stated as

Kirchhoff's voltage law: In any complete electric circuit, the algebraic sum of the applied voltages must equal the algebraic sum of the voltage drops.

Kirchhoff's voltage law provides us with an alternate solution to Example 6-1. In this case, the numerical work is essentially the same, but only one step is involved, as compared with the first method in which the first step was to replace the three separate resistances with a single equivalent resistance.

EXAMPLE 6-1A
What current will flow in a series circuit consisting of 20-Ω, 10-Ω, and 30-Ω resistors connected to a 45-V source?

SOLUTION

$$E = IR_1 + IR_2 + IR_3$$

$$45 = 20I + 10I + 30I = 60I$$

$$\therefore I = \frac{45 \text{ V}}{60 \text{ }\Omega} = 0.75 \text{ A}$$

6-5 CHARACTERISTICS OF SERIES CIRCUITS

From Ohm's law, $I = V_1/R_1$, and since the current is the *same* through all components of a series circuit,

$$I = \frac{V_1}{R_1} = \frac{V_2}{R_2} = \frac{V_3}{R_3} = \frac{V_T}{R_T}$$

If we transpose any pair of these equal terms, say $V_1/R_1 = V_2/R_2$, we get $V_1/V_2 = R_1/R_2$. From this, it follows that

In a series circuit, the ratio between any two voltage drops is the same as the ratio of the two resistances across which these voltage drops occur.

We can use this knowledge in solving for some specific information about a circuit without having to complete a detailed solution of the whole circuit.

EXAMPLE 6-3

A 20-kΩ resistor and a 15-kΩ resistor are connected in series to a 140-V source. What is the voltage drop across the 15-kΩ resistor?

Note: There is less chance of making an error in the solution of an electric circuit if, as the first step in any solution, we draw a fully labeled schematic diagram to work from (Figure 6-4).

LONG SOLUTION

$$R_T = R_1 + R_2 = 20 \text{ k}\Omega + 15 \text{ k}\Omega = 35 \text{ k}\Omega$$

$$I = \frac{E}{R_T} = \frac{140 \text{ V}}{35 \text{ k}\Omega} = 4 \times 10^{-3} \text{ A} = 4 \text{ mA}$$

$$V_2 = IR_2 = 4 \text{ mA} \times 15 \text{ k}\Omega = 60 \text{ V}$$

SHORT SOLUTION

$$\frac{V_2}{E} = \frac{R_2}{R_T} \qquad \therefore V_2 = E\frac{R_2}{R_T} = 140 \text{ V} \times \frac{15 \text{ k}\Omega}{35 \text{ k}\Omega} = 60 \text{ V}$$

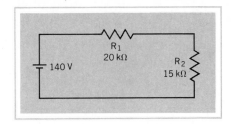

FIGURE 6-4
Circuit Diagram for Example 6-3.

The calculator chain for the short solution is noticeably simpler than that for the long solution which is

$$140 \div (20\,\text{EE}\,3 + 15\,\text{EE}\,3) \times 15\,\text{EE}\,3 = 60.$$

Another characteristic of a series circuit that is based on the *same* current flowing through all components is that any change to *one* component of a series circuit will have an effect on the current through *all* the components. Therefore, we must connect in *series* with a load, such *control* components as switches which turn the current on and off, fuses which open the circuit if the current becomes excessive, and variable resistors (rheostats) which control the magnitude of the current, as in Figure 6-5.

To be able to distinguish readily between series and parallel circuit characteristics, we must be thoroughly familiar with the characteristics of series circuits which we have just developed. We can summarize these characteristics as follows:

1. The *current* is the same in all parts of a series circuit.
2. The *total resistance* is the sum of all the individual resistances.

$$R_T = R_1 + R_2 + R_3 + \text{etc.}$$

3. The *applied voltage* is equal to the sums of all the individual voltage drops.

$$E = V_1 + V_2 + V_3 + \text{etc.}$$

4. The *ratio* between voltages is the same as the resistance ratio.
5. Any change to *any* component of a series will affect the current through *all* components.

FIGURE 6-5
Control Components in Series with a Load.

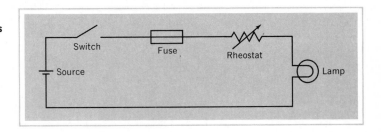

6-6 INTERNAL RESISTANCE

In Section 2-10 we noted that the terminal voltage generated by a source is numerically equal to the emf produced by the energy-converting action of the source only under open-circuit ($I = 0$) conditions. The emf of a given source depends on the nature of the energy-converting action and is independent of the circuit current. However, as shown in the graph of Figure 6-6, the generated potential difference that appears between the two terminals of a practical voltage source *decreases* as the external load is adjusted to draw more current from the source. An example of this change in terminal voltage is the way automobile headlights dim when the starter motor is drawing current from the car battery.

In many cases, the change in terminal voltage with an increase in circuit current is so slight that we can ignore it. But in other cases, particularly in electronic circuits, the decrease in the terminal voltage of a source as the current through it increases must be taken into account. A convenient way of representing this effect is to assume that the *practical* voltage source consists of (1) an *ideal* source which generates a constant potential difference at any current and (2) a resistance connected in series with this ideal source and the actual terminals of the practical source as shown in the schematic representation of Figure 6-6. We call this resistance the **internal resistance** of the source. In addition to causing the terminal voltage to decrease as the current increases, this internal resistance concept can be used to calculate the loss of electric energy through conversion into heat within the practical generator.

If we can measure the voltage between the terminals of the practical source without drawing any current from the source, there will be no voltage drop across the internal resistance (since $V_{int} = IR_{int}$ and $I = $ zero). Therefore, the terminal voltage under these conditions is equal to the constant potential difference generated by the ideal source portion of the practical generator. From this method for determining the potential difference of the ideal voltage source, we call this potential difference the **open-circuit voltage** of the source. In equations we shall represent this open-circuit voltage generated by the ideal portion of the practical source by the letter symbol E. The actual terminal voltage of the source becomes V_T.

The following example illustrates how we can determine the internal resistance of a practical voltage source.

FIGURE 6-6
Schematic and Graphic Representation of a Practical Voltage Source.

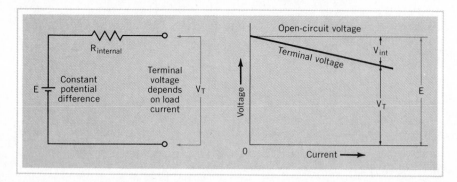

EXAMPLE 6-4

When a voltmeter which draws negligible current is the only circuit connected to the terminals of a battery, it reads 6.0 V. When a 5-Ω resistor is connected to the battery terminals, the voltmeter reads 5.0 V. What is the internal resistance of the battery?

SOLUTION

Since the open-circuit voltage is 6.0 V, the potential difference of the ideal source (Figure 6-7) is 6.0 V. And since the voltage drop across the 5-Ω load is 5.0 V, according to Kirchhoff's voltage law, the voltage drop across the internal resistance is

$$E = V_{int} + V_{load}$$

$$\therefore V_{int} = E - V_{load} = 6.0 - 5.0 = 1.0 \text{ V}$$

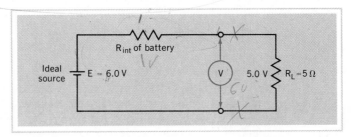

FIGURE 6-7
Circuit Diagram for Example 6-4.

From Ohm's law, the current in the circuit with the 5-Ω resistor connected is

$$I = \frac{V_L}{R_L} = \frac{5.0 \text{ V}}{5 \text{ }\Omega} = 1 \text{ A}$$

$$\therefore R_{int} = \frac{V_{int}}{I} = \frac{1.0 \text{ V}}{1 \text{ A}} = 1 \text{ }\Omega$$

ALTERNATE SOLUTION

$$\frac{R_{int}}{R_L} = \frac{V_{int}}{V_L}$$

$$\therefore R_{int} = R_L \frac{V_{int}}{V_L} = 5 \text{ }\Omega \times \frac{1 \text{ V}}{5 \text{ V}} = 1 \text{ }\Omega$$

From Example 6-4, we note that

1. The *terminal voltage* V_T of a practical voltage source is equal to the open-circuit voltage of the source only if there is no voltage drop across its internal resistance.
2. The terminal voltage V_T is equal to the total voltage drop across the load. ($V_T = V_L$). Hence, we may say that V_T also represents *total* load voltage drop.

3. From Kirchhoff's voltage law, we may write an equation for the terminal voltage of generators as

$$V_T = E - IR_{int} \tag{6-3}$$

where V_T is the terminal voltage of a practical voltage source, E is the open-circuit voltage of the source, I is the current drawn from the source, and R_{int} is its internal resistance.

Equation (6-3) suggests an interesting possibility—suppose that we can arrange to have the emf of the voltage source *increase* as the current drawn from the source *increases*. If we arrange that the increase in generated voltage (E) just matches the increase in voltage drop (IR_{int}), the terminal voltage (V_T) will remain constant over a wide range of load currents. We call such an arrangement **voltage regulation.** Since the generated emf of a battery depends on the chemical reaction in the cell, we cannot apply voltage regulation to battery voltage sources. However, the control room operators in a utility generating station can certainly increase the voltage generated by rotating machines as the current demand increases. In electronic power supplies, we can readily accomplish control of the source voltage by "feedback" signals monitoring the output terminal voltage. Such a voltage source then becomes a **controlled source.** We shall examine controlled sources in more detail in a later chapter.

As we noted at the beginning of this section, the effect of internal resistance for many practical voltage sources is small enough that we can say that the terminal voltage of the source is essentially the same as its open-circuit voltage. This is generally the case when we are using heavy-duty sources, such as a storage battery or the local electric company supply, which are designed to supply a load current of many amperes. But if the voltage source is an electronic device which has a high internal resistance, we must take care that we are not misled by this circuit parameter, which usually is not shown in the circuit diagram. In solving the problems in this text, if no specific value of internal resistance is stated, we shall consider the internal resistance of the source to be zero, in which case $V_T = E$.

6-7 MAXIMUM POWER TRANSFER

One effect of internal resistance in a voltage source is a decrease in generator terminal voltage as the current drawn from the source is increased. To determine what other effects internal resistance may have on electric circuit behavior, we can substitute various values of load resistance in the circuit of Figure 6-8. For every value of load

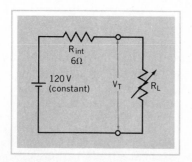

FIGURE 6-8
Circuit Diagram for Table 6-1.

120

resistance, the internal resistance and applied voltage remain constant. ($R_{int} = 6\ \Omega$ and $E = 120$ V.) We need to perform the following calculations for a range of values of load resistance and to display the results in table format.

$$I = \frac{E}{R_L + R_{int}} = \frac{120}{R_L + 6}$$

$$V_L = IR_L$$

$$P_L = I^2 R_L$$

$$P_{int} = I^2 R_{int} = 6\ I^2$$

$$\eta = \frac{P_L}{P_L + P_{int}} \times 100\% = \frac{100 P_L}{P_L + P_{int}}$$

We have now found an example where a computer can outperform a pocket calculator. To produce the data for Table 6-1 on a calculator, we have to perform over 100 separate calculations. We can program the computer to perform the five calculations listed above, entering E and R_{int} as numerical constants and R_L as the only independent variable. Instead of entering a single value for R_L, we can use the computer's "loop" ability to repeat all calculations automatically with R_L increased by a predetermined amount over a predetermined range—for example, from 0 to 40 in 2-ohm steps. (We selected 21 increments for R_L since a typical video screen displays 24 lines with 80 characters per line.) We start the program by instructing the computer how to label and space the columns and how to round off the calculations. The latter chore would be simpler if we had some idea what the final table should look like; hence, our first attempt may produce a ragged table. A typical program might be

```
100 REM    MAXIMUM POWER TRANSFER THEOREM
110 F1$ = "\      \       \       \        \       \        \       \         \        \  \                 \    "
120 F2$ = "  \      \        \       \        \        \        \         \        \    \                 \    "
130 F3$ = "  ##          ##.#            ###           ###           ####            ## "
140 PRINT USING F1$; "R LOAD";"CURRENT";"LOAD";"LOAD";"INTERNAL";"PERCENT"
150 PRINT USING F2$; "OHMS";"AMPERES";"VOLTAGE";"POWER";"POWER";"EFFICIENCY"
160 FOR R = 0 TO 40 STEP 2
170 LET I = 120/(R + 6)
180 LET VL = I*R
190 LET PL = I^2*R
200 LET PI = 6*I^2
210 LET EFF = 100*PL/(PL + PI)
220 PRINT USING F3$; R;I;VL;PL;PI;EFF
230 NEXT R
240 END
```

When we enter RUN, the computer should then print Table 6-1 (p. 122).

To assist us in analyzing the significance of the data displayed in Table 6-1, we can plot graphs of some of the columns of dependent figures against the stepped values of R_L. The personal computer is quite capable of generating such graphs from the data

TABLE 6-1

R LOAD OHMS	CURRENT AMPERES	LOAD VOLTAGE	LOAD POWER	INTERNAL POWER	PERCENT EFFICIENCY
0	20.0	0	0	2400	0
2	15.0	30	450	1350	25
4	12.0	48	576	864	40
6	10.0	60	600	600	50
8	8.6	69	588	441	57
10	7.5	75	563	338	63
12	6.7	80	533	267	67
14	6.0	84	504	216	70
16	5.5	87	476	179	73
18	5.0	90	450	150	75
20	4.6	92	426	128	77
22	4.3	94	404	110	79
24	4.0	96	384	96	80
26	3.8	98	366	84	81
28	3.5	99	349	75	82
30	3.3	100	333	67	83
32	3.2	101	319	60	84
34	3.0	102	306	54	85
36	2.9	103	294	49	86
38	2.7	104	283	45	87
40	2.6	104	272	41	87

it has already calculated in the program above. But to guard against losing sight of the purpose of this exercise to the novelty of the computer, we have plotted the graphs of Figure 6-9 from the data of Table 6-1 without resorting to computer graphics. Note in Table 6-1 that the graph for load power rises to a maximum when R_L reaches six ohms and then tapers off gradually as R_L increases further. If we use a linear x-axis scale for R_L, we will produce a lopsided graph (as the computer will). For the graphs

FIGURE 6-9
Graph Showing Requirements for Maximum Power Transfer.

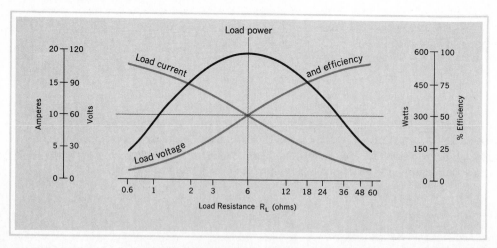

of Figure 6-9, we have used a *logarithmic* scale for the x-axis so that the spacing between 2 and 6 ohms is the same as the spacing between 6 and 18 ohms ($6 = 3 \times 2$ and $18 = 3 \times 6$).

On the y-axis, we first plot load current (column 2) and load voltage (column 3). As we increase the load resistance from 0 ohms (short circuit) to 40 ohms, the current decreases from its short-circuit value of 20 A to 2.6 A when $R_L = 40$ ohms. Voltage across the load behaves in the opposite manner—rising from 0 volts at short circuit to 104 volts when $R_L = 40$ ohms. Note that the percentage efficiency graph (from column 6) is the same as the load voltage graph if we use suitable scales.

When we superimpose the graph of load power (from column 4), we immediately note that

1. *Maximum power output (into the load) occurs when*

$$R_L = R_{int} \tag{6-4}$$

Also, when R_L is selected for maximum load power,

$$I = \tfrac{1}{2} I_{sc} \qquad \text{(short-circuit current)} \tag{6-5}$$

$$V_L = \tfrac{1}{2} E \qquad \text{(open-circuit voltage)} \tag{6-6}$$

$$\eta = 50\% \qquad \text{(one-half maximum efficiency)} \tag{6-7}$$

Note that maximum power output does *not* coincide with maximum efficiency. When a load resistance is selected for maximum power output, there is an equal power dissipation inside the source.

2. If we want to increase the efficiency, a load resistance of from two to three times the internal resistance of the generator results in appreciable reduction in wasted power (as heat in the generator) for only a small reduction in power output.

3. A load resistance less than the internal resistance of the generator not only results in a reduction in power output but also causes a very high dissipation within the generator. In practice, this condition of operation is termed **overload** and must be avoided.

4. If we are interested more in *voltage* output than power output (as in transistor voltage amplifiers), the load resistance should be high in comparison to the internal resistance of the source.

$$R_L \cong 5 \times R_{int} \tag{6-8}$$

From Figure 6-9 we note that the *slope* of the load power graph becomes zero at maximum power. Hence, if we are familiar with elementary differential calculus, we can prove the maximum power transfer theorem of Equation (6-4) with a single calculation, as shown in Appendix 2-1.

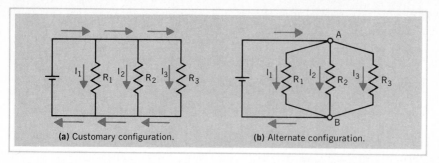

(a) Customary configuration. (b) Alternate configuration.

FIGURE 6-10
Simple Parallel Circuit.

6-8 RESISTORS IN PARALLEL

In dealing with parallel-(or shunt) connected resistors, we find a set of characteristics which, in many respects, are similar but opposite to those of series circuits.[†] Figure 6-10 shows two different ways of drawing a circuit diagram for a simple parallel circuit. Since we usually prefer to draw interconnecting conductors as either horizontal or vertical lines, the diagram of Figure 6-10(a) is the customary configuration. However, to illustrate the definition of a parallel circuit, we can redraw the circuit diagram as in Figure 6-10(b).

We can define a parallel circuit in terms of the physical connection of the components. In Figure 6-10, E, R_1, R_2, and R_3 are all in parallel because they are all connected between the *same two* **junction points** or **nodes** (A and B). Since each of the resistors is connected directly across the generator terminals, $V_1 = V_2 = V_3 = E$. And since this voltage is common to all components in parallel, we may omit the subscript. In keeping with our definition of a series circuit, we prefer to define a parallel circuit in terms of this common voltage:

Two or more electric components are considered to be in parallel if the same (common) voltage appears across all these components.[‡]

In solving the series circuit, we determined the total resistance in order to find the current in the circuit. In the **total current method** of solving parallel circuits, the steps are reversed. If we are given the open-circuit voltage of the source (assuming negligible internal resistance) and the values of each resistance in the circuit of Figure 6-10, we can solve for the current in each branch as an individual circuit by using Ohm's law. If we think of the current in each branch in terms of electrons per second

---

[†]We encounter this "similar-but-opposite" characteristic of series and parallel electric circuits quite often. It is sometimes referred to as the principle of **duality.**

[‡]Once again we must note that *same* does *not* mean *equal but different.*

flowing through the branch, it is apparent that the generator current must be the sum of the branch currents.[†] Therefore,

In a simple parallel circuit, the total current is the sum of all the branch currents.

$$I_T = I_1 + I_2 + I_3 + \text{etc.} \tag{6-9}$$

EXAMPLE 6-5

With reference to Figure 6-10, R_1 is 40 Ω, R_2 is 30 Ω, R_3 is 20 Ω, and E is 120 V. What single resistance would draw the same current from the source?

SOLUTION

$$I_1 = \frac{V_1}{R_1} = \frac{120 \text{ V}}{40 \text{ }\Omega} = 3 \text{ A}$$

$$I_2 = \frac{V_2}{R_2} = \frac{120 \text{ V}}{30 \text{ }\Omega} = 4 \text{ A}$$

$$I_3 = \frac{V_3}{R_3} = \frac{120 \text{ V}}{20 \text{ }\Omega} = 6 \text{ A}$$

$$I_T = I_1 + I_2 + I_3 = 3 + 4 + 6 = 13 \text{ A}$$

$$\therefore R_{eq} = \frac{E}{i_T} = \frac{120 \text{ V}}{13 \text{ A}} = 9.23 \text{ }\Omega$$

From Equation (6-1), we note that the *total* resistance of a series circuit is *greater* than the resistance of any individual resistor. But the answer in Example 6-5 is *less* than the resistance of any individual resistor. Hence, in dealing with parallel circuits, we do not speak of total resistance. Instead, we speak of the **equivalent resistance** (R_{eq}) of two or more resistors in parallel.

6-9 KIRCHHOFF'S CURRENT LAW

Kirchhoff applied the total-current principle of Equation (6-9) to any electric circuit in the form of

Kirchhoff's current law: At any junction point or node in an electric circuit, the algebraic sum of the currents enterinq the point must equal the algebraic sum of the currents leaving the point.

[†]In working with circuits in which there is more than one path for current, it will help us to keep track of the various branch currents if we mark current direction on the schematic diagram with arrows. Note that we prefer to use the *conventional* current direction for this purpose.

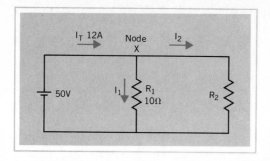

FIGURE 6-11
Circuit Diagram for Example 6-6.

EXAMPLE 6-6

What current is flowing in the R_2 branch of the circuit of Figure 6-11?

SOLUTION

$$I_1 = \frac{V}{R_1} = \frac{50 \text{ V}}{10 \text{ }\Omega} = 5 \text{ A}$$

The current flowing into node X is I_T and the currents flowing away from node X are I_1 and I_2. Therefore,

$$I_T = I_1 + I_2 \quad \text{or} \quad I_2 = I_T - I_1$$
$$\therefore I_2 = 12 - 5 = 7 \text{ A}$$

6-10 CONDUCTANCE

As the results in Examples 6-5 and 6-6 indicate, the total current must always be greater than the current through any branch of a parallel circuit. Therefore, the equivalent resistance must always be *less* than the smallest of the branch resistances. If we continue this line of thought, the *more* resistors we connect in parallel, the *smaller* the equivalent resistance becomes. We can appreciate the behavior of parallel circuits more readily if we turn our attention away from resistance and restate the above remark thus: The more resistors we connect in parallel, the more readily the circuit can pass current since there are more parallel branches for current to flow through. Therefore, since resistance is a measure of the ability of a circuit to *oppose* electric current,

> *Conductance is a measure of the ability of an electric circuit to pass current.*
> *The letter symbol for conductance is G.*
> *The siemens is the SI unit of conductance.* [†]
> *The unit symbol for siemens is S.*

[†]Before the IEEE adopted the SI name **siemens** for the unit of conductance, the common name for the unit of conductance was the **mho** (ohm spelled backward). The unit symbol was (logically) an *inverted* Greek letter "omega." The siemens was named in honor of a British engineer, Sir William Siemens.

Since conductance is simply the reciprocal point of view of resistance, we can define the size of the siemens by the equation

$$G = \frac{1}{R} \tag{6-10}$$

where G is the conductance of a circuit in siemens and R is the resistance of the same circuit in ohms.

When we are given the voltage across a group of parallel branches, the total-current method described in the preceding section is a convenient method of solving parallel circuits. Even if this voltage is not known, we can assume a suitable voltage in order to solve for the equivalent resistance. But, in this case, the concept of conductance offers a more straightforward approach to the solution of parallel circuits. If we divide both sides of Equation (6-9) by E (or V, since they are the same for simple parallel circuits), we get

$$\frac{I_T}{E} = \frac{I_1}{V} + \frac{I_2}{V} + \frac{I_3}{V} + \text{etc.}$$

But $R = V/I$, and since $G = 1/R$, $G = I/V$

$$\therefore G_T = G_1 + G_2 + G_3 + \text{etc.} \tag{6-11}$$

Therefore,

In parallel circuits, the total conductance is equal to the sum of the conductances of all the individual branches.

Having determined the total conductance, the equivalent resistance is simply $R_{eq} = 1/G_T$.

EXAMPLE 6-5A

With reference to Figure 6-10, R_1 is 40 Ω, R_2 is 30 Ω, and R_3 is 20 Ω. What single resistance would draw the same current from the source?

SOLUTION

$$G_T = G_1 + G_2 + G_3 = \frac{1}{40} + \frac{1}{30} + \frac{1}{20} \text{S}$$

$$= 2.5 \times 10^{-2} + 3.33 \times 10^{-2} + 5 \times 10^{-2} = 0.1083 \text{ S}$$

$$\therefore R_{eq} = \frac{1}{G_T} = \frac{1}{0.1083 \text{ S}} = 9.23 \ \Omega$$

Using the reciprocal $1/x$ key, the calculator chain for G_T becomes 40 $1/x$ + 30 $1/x$ + 20 $1/x$ = 0.1083 S. Pressing the reciprocal key once more converts G_T into R_{eq}.

When only *two* resistors in parallel are involved, we can reduce Equation (6-11) to a convenient form for determining the equivalent resistance directly.

$$G_T = G_1 + G_2 = \frac{1}{R_1} + \frac{1}{R_2} = \frac{R_1 + R_2}{R_1 R_2}$$

$$\therefore R_{eq} = \frac{R_1 R_2}{R_1 + R_2} \tag{6-12}$$

Therefore,

For two resistors in parallel, the equivalent resistance equals their product over their sum.

This does *not* apply to more than two resistors, as we shall discover by trying to simplify Equation (6-11) with three or more terms.

To deal with the situation where we want to know what resistance R_2 we must connect in parallel with a given resistance R_1 to obtain a certain equivalent resistance, we can rearrange Equation (6-12) as

$$R_2 = \frac{R_1 R_{eq}}{R_1 - R_{eq}} \tag{6-13}$$

To determine the "wattage" rating needed for resistor R_2, we want to express its current I_2 as a fraction of the total current through the parallel combination of R_1 and R_2. Since $V_2 = V_1$, $I_2 R_2 = I_1 R_1$. And from Kirchhoff's current law, $I_2 R_2 = (I_T - I_2)R_1$ from which,

$$I_2 = I_T \left(\frac{R_1}{R_1 + R_2} \right) \tag{6-14}$$

Another special example of Equation (6-11) is the case of N *equal* resistors in parallel. A moment's reflection will show us that

$$R_{eq} = \frac{R}{N} \tag{6-15}$$

where R is the resistance of each of the parallel resistors, and N is the number of resistors connected parallel.

EXAMPLE 6-7

What is the equivalent resistance of a 1-kΩ and a 4-kΩ resistor in parallel?

SOLUTION

$$R_{eq} = \frac{R_1 R_2}{R_1 + R_2} = \frac{1 \text{ k}\Omega \times 4 \text{ k}\Omega}{(1 \text{ k}\Omega + 4 \text{ k}\Omega)} = \frac{4\,000\,000}{5000} = \mathbf{800 \ \Omega}$$

(For verification, solve this example using the calculator procedure of Example 6-5A.)

6-11 CONDUCTIVITY

We defined resistivity as a property of a conducting *material,* being the resistance of a unit length and cross-section conductor of that material. Similarly, then,

The conductivity of a material is the conductance of a unit length and cross-section conductor of that material.

The letter symbol for conductivity is the Greek letter σ (sigma).

Since conductance is the reciprocal of resistance, conductivity is the reciprocal of resistivity. Therefore,

$$\sigma = \frac{1}{\rho} \tag{6-16}$$

Conductivity is expressed in **siemens per meter.**

6-12 CHARACTERISTICS OF PARALLEL CIRCUITS

From Ohm's law, $V = IR$, and since $R = 1/G$, $V = I/G$, and since the voltage is the *same* across all components of a parallel circuit,

$$V = \frac{I_1}{G_1} = \frac{I_2}{G_2} = \frac{I_3}{G_3} = \frac{I_T}{G_T} \tag{6-17}$$

If we take any pair of these equal terms, say $I_1/G_1 = I_2/G_2$ and transpose it we get $I_1/I_2 = G_1/G_2 = R_2/R_1$. From this, it follows that

In a parallel circuit, the ratio between any two branch currents is the same as the ratio of their conductances or the inverse of their resistance ratio.

EXAMPLE 6-8
The total current drawn by a 12.5-kΩ resistor (R_1) and a 50-kΩ resistor (R_2) in parallel is 15 mA. What is the current through the 50-kΩ resistor?

SOLUTION 1
From the circuit diagram of Figure 6-12,

FIGURE 6-12
Circuit Diagram for Example 6-8.

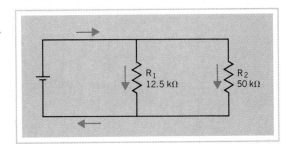

$$I_1 + I_2 = 15 \text{ mA}$$

$$\frac{I_1}{I_2} = \frac{R_2}{R_1} = \frac{50 \text{ k}\Omega}{12.5 \text{ k}\Omega} = 4$$

$$\therefore I_1 = 4I_2$$

Substituting this value of I_1 in the first equation gives

$$4I_2 + I_2 = 15 \text{ mA}$$

$$\therefore I_2 = \frac{15 \text{ mA}}{5} = 3 \text{ mA}$$

SOLUTION 2

$$R_{eq} = \frac{R_1 R_2}{R_1 + R_2} = \frac{12.5 \text{ k}\Omega \times 50 \text{ k}\Omega}{(12.5 \text{ k}\Omega + 50 \text{ k}\Omega)} = 10 \text{ k}\Omega$$

$$V = I_T R_{eq} = 15 \text{ mA} \times 10 \text{ k}\Omega = 150 \text{ V}$$

$$\therefore I_2 = \frac{V}{R_2} = \frac{150 \text{ V}}{50 \text{ k}\Omega} = 3 \text{ mA}$$

Note: There is more than one way of solving most electric circuit problems. Although it is wise to try to understand the various methods, we should select the method by which we can visualize what each step represents. It is much better to achieve accurate results by a longer method than to make an error in a shortcut through not appreciating the significance of each step of the solution.

Another characteristic of parallel circuits is based on the internal resistance of the source being negligible. If such is the case, altering the resistance of one branch will not affect the voltage across and thus the current through the other branches. Therefore, changes in one branch of a parallel circuit have negligible effect on the other branches. In house wiring, each lighting circuit is connected in parallel with the others to the 117-volt source so that switching one circuit on or off does not affect the operation of the other circuits, as shown in Figure 6-13.

We may summarize the characteristics of parallel circuits as follows:

1. The *voltage* is the same across all components in a parallel circuit.
2. The *total conductance* is the sum of all the individual branch conductances.

$$G_T = G_1 + G_2 + \text{etc.}$$

3. The total *current* is the sum of all the individual branch currents.

$$I_T = I_1 + I_2 + I_3 + \text{etc.}$$

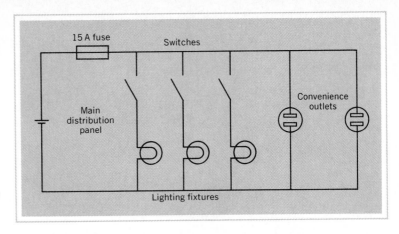

FIGURE 6-13
Parallel Connection of Loads to a Branch Circuit in House Wiring.

4. The *ratio* between branch currents is the *same* as the conductance ratio and the *inverse* of the resistance ratio.
5. Each branch is essentially independent of any changes in the other branches providing the voltage across the parallel circuit is kept constant.

PROBLEMS

Draw a fully labeled schematic diagram for each problem.

6-1. Given $R_1 = 5\ \Omega$, $R_2 = 10\ \Omega$, and $R_3 = 15\ \Omega$, what is the total resistance when they are connected in series?

6-2. What is the total resistance of a series circuit consisting of $R_1 = 47\ k\Omega$, $R_2 = 330\ k\Omega$, and $R_3 = 2.2\ k\Omega$?

6-3. If the series circuit of Problem 6-1 is connected to a 120-V source, what is the voltage drop across each resistor?

6-4. What resistance must be connected in series with a vacuum-tube heater rated at 150 mA with a hot resistance of 330 Ω in order to operate it safely from a 117-V source? What power rating must this resistor have?

6-5. A 6-V car radio draws a current of 6.5 A. What resistance must be connected in series with it to operate it in a car with a 12-V battery? What "wattage" resistor is required?

6-6. An electric stove element is rated at 300 W when connected to a 110-V source. Assuming negligible change in resistance for any change in temperature, determine the total rate of energy conversion when two of these elements are connected in series.

6-7. Three resistors are connected in series to a 120-V generator. The first has a resistance of 50 Ω, the second passes a current of 0.5 A, and the third has a voltage drop across it of 50 V. Calculate the resistance of the second and third resistors.

6-8. Three resistors are connected in series to a 120-V source. The voltage drop across R_1 and R_2 together is 80 V, and the voltage drop across R_2 and R_3 together is 90 V. If the total resistance is 8 kΩ, what is the resistance of each of the three resistors?

6-9. A 10-kΩ 20-W; an 80-kΩ 100-W; and a 20-kΩ 200-W resistor are connected in series. What is the maximum voltage that can be applied to the network without exceeding the power rating of any resistor?

6-10. A Christmas tree light set for use with a 110-V source consists of eight 6-W lamps in series. What is the hot resistance of each lamp?

6-11. What resistance must be connected in series with a 100-Ω resistor for the 100-Ω resistor to dissipate heat at the rate of 30 W when the combination is connected to a 120-V source?

6-12. What resistance must be connected in series with a 100-Ω resistor for the unknown resistance to dissipate heat at the rate of 30 W when the combination is connected to a 120-V source?

6-13. A storage battery has an open-circuit voltage of 6 V and an internal resistance of 0.05 Ω.
(a) What is the terminal voltage of the battery when a 0.2-Ω load is connected to it?
(b) What power will be dissipated by a 0.5-Ω load?
(c) What is the efficiency of the system when a 0.15-Ω load is connected to the battery?
(d) What is the maximum power that a load can draw from this battery?

6-14. The generator of a lighting plant driven by a gasoline engine has an open-circuit voltage of 32 V and an internal resistance of 0.2 Ω. The generator is designed for a constant-duty power output of 300 W.
(a) What value of load resistor will draw energy from the generator at the rate of 300 W?
(b) What is the efficiency of the generator under this condition?

6-15. The high-voltage power supply of a television receiver produces an open-circuit voltage of 15 kV. Its internal resistance is 2 MΩ.
(a) What is the voltage at the high voltage anode of the picture tube when the anode current is 600 μA?
(b) What will the anode voltage be if the anode current is increased by 50%?
(c) What is the short-circuit current of this power supply?

6-16. A phonograph pickup develops an open-circuit voltage of 50 mV and has an internal resistance of 1200 Ω.
(a) What voltage will it develop across a 4-kΩ load resistance?
(b) What value of load resistor must be used to obtain a 75% efficiency.
(c) What is the maximum power output of the pickup?

6-17. A certain generator has a terminal voltage of 110 V when a 5.5-Ω load is connected to its terminals. The terminal voltage becomes 105 V when the load is 3.5 Ω. What is the internal resistance of the generator?

6-18. A 200-Ω resistor dissipates heat at the rate of 8 W when connected to a certain generator. If a 300-Ω resistor is connected in series with the 200-Ω resistor, the dissipation of the 200-Ω resistor becomes 2 W. What is the open-circuit voltage of the generator?

6-19. A generator whose internal resistance is 1 Ω feeds a load at the end of a line, each wire of which has a resistance of 2 Ω. The generator voltage is adjusted to produce 120 V across the load at full load. If the total dissipation of the two wires is not to exceed 10% of the total generated power, what is the maximum power that can be delivered to the load?

6-20. An audio amplifier has an internal resistance at its output terminals of 4 Ω. The loudspeaker is connected at the end of a line, each wire of which has a resistance of 2 Ω. What resistance load must the loudspeaker present to obtain maximum power in the loudspeaker?

6-21. Given R_1 is 5 Ω, R_2 is 10 Ω, and R_3 is 15 Ω, what is the total current when they are connected in parallel to a 120-V source?

6-22. What is the equivalent resistance of a parallel circuit consisting of $R_1 = 47$ kΩ, $R_2 = 330$ kΩ, and $R_2 = 2.2$ kΩ?

6-23. A circuit element having a conductance of 150 μS is connected in parallel with a branch having a conductance of 750 mS. What is the equivalent resistance of the circuit?

6-24. Three lamps operating in parallel on a 110-V circuit are rated at 40 W, 60 W, and 100 W, respectively. What is the equivalent hot resistance of this load?

6-25. Three resistors in parallel pass a total current of 0.6 A. The first resistor has a resistance of 400 Ω, the second passes a current of 60 mA, and the third has a voltage drop across it of 150 V. Calculate the resistance of the second and third resistors.

6-26. If the three resistors in Problem 6-22 are each rated at $\frac{1}{2}$ W, what is the maximum total current that the network can handle without overheating any resistor?

6-27. What resistance must be placed in parallel with a 15-kΩ resistor to reduce the equivalent resistance to 10 kΩ?

6-28. The equivalent resistance of three resistors in parallel is 2.5 kΩ. If R_1 is 10 kΩ and R_2 is 20 kΩ, what is the resistance of R_3?

6-29. The total current passed by a 10-kΩ, a 15-kΩ, and a 20-kΩ resistor in parallel is 20 mA. What is the current through each branch?

6-30. Three resistors connected in parallel have an equivalent resistance of 2.5 kΩ. R_1 has a resistance of 15 kΩ, R_2 has a voltage drop of 25 V, and R_3 dissipates electric energy at the rate of 25 mW. Determine R_2 and R_3.

REVIEW QUESTIONS

6-1. How would you justify the statement that the current is the same in all parts of a simple series circuit?

6-2. How would you justify the statement that the total resistance of a series circuit is the sum of all the individual resistances?

6-3. In the electric circuit of Figure 6-14, one of the "black boxes" represents a voltage source and the other a resistor. The voltmeter and ammeter read correctly when their terminals are connected with the polarity markers as shown. Which "black box" is the source?

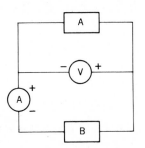

FIGURE 6-14
Polarities in an Electric Circuit.

6-4. Kirchhoff's voltage law may be stated thus: "The voltage between any two points in an electric network is the same via any path between those two points." Prove this statement numerically by considering the two ends of R_2 in Figure 6-1, using the values given in Example 6-1.

6-5. Prove that in a series circuit, the ratio between any two power dissipations is the same as the ratio between the two resistances.

6-6. A lamp, rated at 120 V, 100 W and one rated at 120 V, 25 W are connected in series to a 120-V source. Which one will glow more brightly? Explain.

6-7. Explain the visible effect when the 25-W lamp in Question 6-6 is (a) open-circuit and (b) short-circuit.

6-8. What disadvantage does the "series string" type of Christmas tree lamp possess as compared to the 110-V parallel type of Christmas tree lamp?

6-9. How would you go about determining the internal resistance of a given flashlight battery?

6-10. Explain why the kitchen light becomes a bit dimmer when a toaster is turned on.

6-11. When a storage battery is discharging, its terminal voltage is less than its open-circuit voltage; but when it is being charged, its terminal voltage is greater than its open-circuit voltage. Explain.

6-12. Using the data of Table 6-1, draw a graph of load power vs. load resistance using linear graph paper. Explain the shape of the graph.

6-13. Using the data of Table 6-1, draw a graph of load power vs. load current. Express the condition for maximum power output in relation to short-circuit and open-circuit current.

6-14. There are two numerically correct answers to Problem 6-14(a). But we select only one as the electrically correct answer. Why?

6-15. Generator B develops three times the open-circuit voltage of generator A but has three times as great an internal resistance. Each generator is developing the same power in its load. Which has the greater efficiency?

6-16. A generator is operated in such a manner that its terminal voltage is kept constant by increasing its generated potential difference as the load current is increased. Does this mean that the efficiency is 100%? Explain.

6-17. In operating electronic equipment, it is customary to make the load resistance equal to the internal resistance. Why would we do this for electronic apparatus but not in the case of a power-generating station?

6-18. Why must the equivalent resistance of a parallel circuit always be less than that of the smallest of the branch resistances?

6-19. Show that the statement that the total current in a parallel circuit is the sum of all the branch currents confirms Kirchhoff's current law.

6-20. Why is it preferable to think in terms of **conductance** in working with parallel circuits rather than in terms of resistance?

6-21. If the two lamps of Question 6-6 are connected in parallel, explain the visible effect when the 25-W lamp is (a) open-circuit and (b) short-circuit.

6-22. An electric stove element has a resistance of 50 Ω with a center-tap connection. Draw circuit diagrams to show three means of connecting it to a 110-V source to obtain three different rates of conversion of electric energy to heat.

6-23. Most homes have a lighting fixture that can be switched on or off from two different locations. This is accomplished with two single-pole, double-throw switches (which electricians call "three-way switches"). Draw a circuit diagram.

6-24. Figure 6-13 is not a simple parallel circuit because the fuse is in *series* with the remainder of the circuit. Why must it be located in the position shown?

6-25. Two 50-kΩ resistors are connected in series across a voltage source. If a voltmeter which has a resistance of 100 kΩ is connected first across one resistor and then the other and finally across both, the sum of the first two readings does not equal the third reading. Is this third reading greater or less than the sum of the other two? Explain.

*6-26. Adapt the program shown in Section 6-7 for your computer and then program the computer to draw a graph of load power vs. load resistance.

134

7

SERIES-PARALLEL CIRCUITS

7-1 SERIES-PARALLEL RESISTORS

In Chapter 6, we considered only simple series and simple parallel circuits. In practice, electric networks are seldom this simple. If we consider the circuit of Figure 7-1 as a *whole*, it does not comply with the characteristics of either a series or a parallel circuit. But if we consider only R_2 and R_3, since they are connected between the same two junction points in the circuit, they must both have the *same* voltage drop. Therefore, R_2 and R_3 *do* fill the definition of a parallel circuit. If we are given the values for R_2 and R_3, we can use the rules of parallel circuits to solve for a single equivalent resistance. As far as the generator is concerned, the simplified circuit of Figure 7-2 is the same as the original circuit of Figure 7-1. In Figure 7-2, R_1 is in series with the equivalent resistance of R_2 and R_3 in parallel. Therefore, we can solve the circuit of Figure 7-2 by the rules of series circuits.

Another arrangement of series-parallel resistors is shown in Figure 7-3(a). In this case, R_2 and R_3 have the same current through them and, therefore, are in series. We may replace them, if we wish, by an equivalent resistor, as in Figure 7-3(b), which is equal to $R_2 + R_3$. We can now solve the simplified circuit of Figure 7-3(b) by the rules for simple parallel circuits.

As these examples show, we may define a series-parallel circuit as one in which some portions of the circuit have the characteristics of simple series circuits and other portions have the characteristics of simple parallel circuits. Whenever two or more components of an electric network are in parallel, all the characteristics of parallel circuits must apply to these components. And, whenever two or more components are in series, all the characteristics of series circuits must apply.

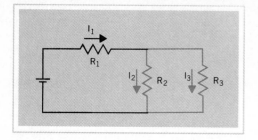

FIGURE 7-1
Simple Series-Parallel Circuit.

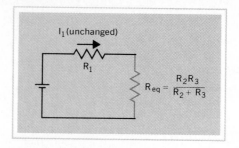

FIGURE 7-2
Equivalent Circuit of Figure 7-1.

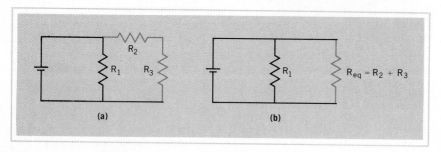

FIGURE 7-3
(a) Parallel-Series Circuit; (b) Simplified Version.

7-2 EQUIVALENT-CIRCUIT METHOD

Figures 7-2 and 7-3(b) suggest a method whereby we may solve some of the simpler series-parallel networks, that is, by substituting the equivalent resistance for various portions of the circuit until the original circuit is reduced to a simple series or parallel circuit.

EXAMPLE 7-1
Complete Table 7-1 with reference to the circuit diagram of Figure 7-1.

TABLE 7-1

Component	Resistance	Voltage	Current	Power
R_1	12 Ω			
R_2	10 Ω			
R_3	40 Ω			
Totals		100 V		

SOLUTION

Step I.   Draw a fully labeled schematic diagram for this particular circuit.

Step II.   Visual inspection of Figure 7-4 shows that R_2 and R_3 are in parallel.

136

$$\therefore R_{eq} = \frac{R_2 R_3}{R_2 + R_3} = \frac{10 \times 40}{(10 + 40)} = 8 \ \Omega$$

With reference to Figure 7-4, the total resistance becomes

$$R_T = R_1 + R_{eq} = 12 + 8 = 20 \ \Omega$$

Enter this answer in the appropriate blank of Table 7-1.

Step III. From Ohm's law,

$$I_T = \frac{E}{R_T} = \frac{100 \ V}{20 \ \Omega} = 5 \ A$$

Step IV. Since R_1 is directly in series with the source,

$$I_1 = I_T = 5 \ A$$

Step V. From Ohm's law,

$$V_1 = I_1 R_1 = 5 \ A \times 12 \ \Omega = 60 \ V$$

Step VI. From Kirchhoff's voltage law,

$$V_{eq} = E - V_1 = 100 - 60 = 40 \ V$$

Returning now to the original circuit,

$$V_2 = V_3 = V_{eq} = 40 \ V$$

Step VII. From Ohm's law,

$$I_2 = \frac{V_2}{R_2} = \frac{40 \ V}{10 \ \Omega} = 4 \ A$$

$$I_3 = \frac{V_3}{R_3} = \frac{40 \ V}{40 \ \Omega} = 1 \ A$$

FIGURE 7-4
Circuit Diagram for Example 7-1.

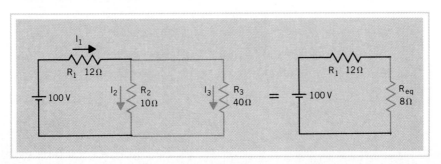

137

As a check on our calculations, we can note that according to Kirchhoff's current law, $I_1 = I_2 + I_3$ for this circuit, which does check with our calculations.

Step VIII. Since $P = VI$,

$$P_1 = V_1 I_1 = 60 \text{ V} \times 5 \text{ A} = 300 \text{ W}$$

$$P_2 = V_2 I_2 = 40 \text{ V} \times 4 \text{ A} = 160 \text{ W}$$

$$P_3 = V_3 I_3 = 40 \text{ V} + 1 \text{ A} = 40 \text{ W}$$

$$P_T = EI_T = 100 \text{ V} \times 5 \text{ A} = 500 \text{ W}$$

Again as a check, P_T should equal $P_1 + P_2 + P_3$, which does check with our calculations.

EXAMPLE 7-2

What is the resistance at the input terminals of the **ladder** network in Figure 7-5(a) with the output terminals open-circuit and (b) with a 600-Ω load resistance connected to the output terminals?

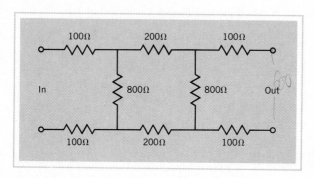

FIGURE 7-5
Circuit Diagram for Example 7-2.

SOLUTION

(a) *Step I.* With the output terminals open-circuit, no current flows in the two right-hand 100-Ω resistors. Thus they are effectively out of the circuit, and it is equivalent to Figure 7-6(a).

Step II. The two 200-Ω resistors and the right-hand 800-Ω resistor have the *same* current through them. Hence they must have the characteristics of a *series* circuit.

$$\therefore R_{eq} = 200 + 800 + 200 = 1200 \ \Omega$$

This gives the equivalent circuit of Figure 7-6(b).

Step III. Since the 800-Ω and 1200-Ω resistors in the equivalent circuit of Figure 7-6(b) are connected between the same junction points, they are in *parallel*.

$$\therefore R_{eq} = \frac{800 \times 1200}{(800 + 1200)} = 480 \ \Omega$$

138

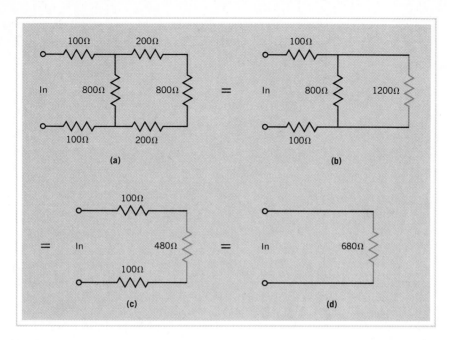

FIGURE 7-6
Equivalent Circuits for Example 7-2(a).

Step IV. The equivalent circuit of Figure 7-6(c) is a simple series circuit.

$$\therefore R_{in} = 100 + 480 + 100 = \mathbf{680 \; \Omega}$$

(b) *Step I.* When a 600-Ω load resistor is connected to the output terminals, it forms a series circuit with the two right-hand 100-Ω resistors. Hence,

$$R_{eq} = 100 + 600 + 100 = 800 \; \Omega$$

Step II. We can replace the two 800-Ω resistors in parallel in the equivalent circuit of Figure 7-7(a) with a single 400-Ω resistance, as in Figure 7-7(b).

Step III. The two 200-Ω resistors and the 400-Ω resistor of Figure 7-7(b) are in series and may be replaced by the single 800-Ω equivalent resistance in Figure 7-7(c).

Step IV. Once again, two 800-Ω resistors in parallel are equivalent to the single 400-Ω resistance of Figure 7-7(d).

Step V. Finally, the simple series circuit of Figure 7-7(d) has a total resistance of

$$R_{in} = 100 + 400 + 100 = \mathbf{600 \; \Omega}$$

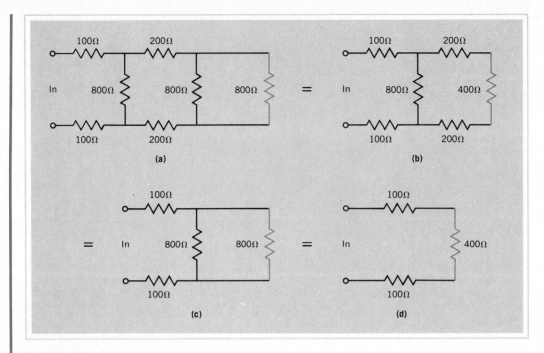

FIGURE 7-7
Equivalent Circuits for Example 7-2(b).

7-3 KIRCHHOFF'S-LAWS METHOD

Kirchhoff's laws provide us with an alternate method of solving series-parallel circuits, leaving them in their original form and not reducing them to a simple series or parallel circuit by substituting equivalent resistances.

KIRCHHOFF'S-LAW SOLUTION TO EXAMPLE 7-1
Referring to Figure 7-4, from Kirchhoff's current law,

$$I_1 = I_2 + I_3$$

But from Ohm's law, $I_1 = V_1/R_1$, etc.

$$\therefore \frac{V_1}{R_1} = \frac{V_2}{R_2} + \frac{V_3}{R_3} \quad \text{and} \quad \frac{V_1}{12} = \frac{V_2}{10} + \frac{V_3}{40}$$

Since R_2 and R_3 are in parallel, $V_2 = V_3$, and from Kirchhoff's voltage law,

$$V_2 = V_3 = E - V_1$$

Substitution in the preceding equation gives

$$\frac{V_1}{12} = \frac{100 - V_1}{10} + \frac{100 - V_1}{40}$$

Multiplying through by the least common denominator (120) to clear the fractions

$$10V_1 = 1200 - 12V_1 + 300 - 3V_1$$

Collecting the terms gives

$$25V_1 = 1500$$

from which

$$V_1 = 60 \text{ V}$$

$$\therefore V_2 = V_3 = E - V_1 = 100 - 60 = 40 \text{ V}$$

We may now determine the various currents by Ohm's law: $I_1 = V_1/R_1$, etc.

The information given in Example 7-1 was such that we had a choice of method in its solution. However, with the data given for Example 7-3, we cannot readily solve for the equivalent resistance, which would enable us to reduce the circuit to a simple series circuit. In this case, we can again use Kirchhoff's laws to help us to set up algebraic equations from which we can obtain the required data.

EXAMPLE 7-3

A resistor passing a 20-mA current is in parallel with a 5-kΩ resistor. This combination is in series with another 5-kΩ resistor, the whole network being connected to a 500-V source. What is the resistance of the resistor which is passing the 20-mA current?

SOLUTION

With reference to the circuit diagram of Figure 7-8, from Kirchhoff's voltage law,

$$V_1 + V_2 = E$$

and, since V = IR,

$$I_1 R_1 + I_2 R_2 = E$$

FIGURE 7-8
Circuit Diagram for Example 7-3.

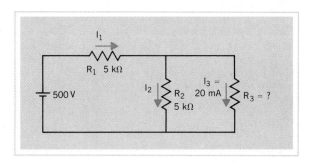

From Kirchhoff's current law,

$$I_2 = I_1 - 0.02$$

$$\therefore R_1 I_1 + R_2(I_1 - 0.02) = E$$

Substituting given values for R_1, R_2, and E,

$$5000I_1 + 5000(I_1 - 0.02) = 500$$

From which $I_1 = 60$ mA. Therefore,

$$I_2 = I_1 - 0.02 = 0.06 - 0.02 = 0.04 \text{ A} \quad \text{or} \quad 40 \text{ mA}$$

Since R_2 and R_3 are in parallel,

$$\therefore V_2 = V_3$$

But
$$V_2 = I_2 R_2 = 40 \text{ mA} \times 5 \text{ k}\Omega = 200 \text{ V}$$

and then
$$R_3 = \frac{V_3}{I_3} = \frac{200 \text{ V}}{20 \text{ mA}} = 10 \text{ k}\Omega$$

It is also possible to start the solution by substituting the appropriate V/R in the Kirchhoff's current-law equation, $I_1 = I_2 + 0.02$.

7-4 VOLTAGE-DIVIDER PRINCIPLE

In the series circuit of Figure 7-9, Kirchhoff's voltage law states that $E = V_1 + V_2 + V_3$. In other words, the total applied voltage is *divided* among the three resistors. Whereas we could measure only one voltage across the terminals of the source, we now have *six* possible combinations of terminals A, B, C, and D, across which we can connect a voltmeter, as shown in Figure 7-9. The series combination of R_1, R_2, and R_3, therefore, becomes a **voltage divider.**

We can solve for the six possible voltages in Figure 7-9 by using the long solution given in Example 6-3; that is, calculate R_T, find I from Ohm's law, and then calculate the voltage drop across each resistance from $V_1 = IR_1$, and so on. However, since our attention is directed to the *division* of voltage, we prefer the short solution, using the principle developed in Section 6-5. This is often referred to as the

Voltage-divider principle: In a series circuit, the ratio between any two voltage drops is the same as the ratio of the two resistances across which these voltage drops occur.

Given any resistor in a voltage divider,

$$\frac{V_x}{E} = \frac{R_x}{R_T}$$

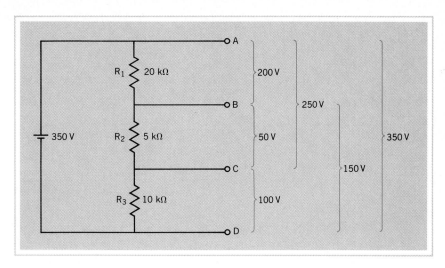

FIGURE 7-9
Voltage-Divider Principle.

from which

$$V_x = E\frac{R_x}{R_T} \tag{7-1}$$

Hence, Equation (7-1) becomes an algebraic statement of the voltage-divider principle.

EXAMPLE 7-4

What is the voltage between terminals B and D in the circuit of Figure 7-9?

SOLUTION

$$V_{BD} = E\frac{R_2 + R_3}{R_1 + R_2 + R_3}$$

$$= 350 \text{ V} \times \frac{(5 \text{ k}\Omega + 10 \text{ k}\Omega)}{(20 \text{ k}\Omega + 5 \text{ k}\Omega + 10 \text{ k}\Omega)}$$

$$= 350 \text{ V} \times \frac{15 \text{ k}\Omega}{35 \text{ k}\Omega} = 150 \text{ V}$$

We often require a continuously variable terminal voltage, rather than the fixed voltages provided by the simple voltage divider shown in Figure 7-9. This can be obtained by a voltage divider in which a sliding contact moves along either a wire-wound or a carbon-composition resistance element to form the **potentiometer** circuit shown in Figure 7-10.

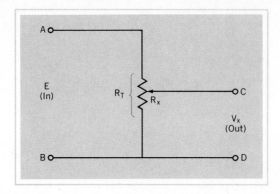

FIGURE 7-10
Voltage Divider or Potentiometer.

7-5 VOLTAGE DIVIDERS

The voltage-divider principle is widely used in electronic circuits where, for the sake of economy, the one voltage source must supply all the various voltages required by a piece of equipment. The series dropping resistor of Figure 7-11 provides the simplest method of obtaining the required voltage drop across a certain circuit element.

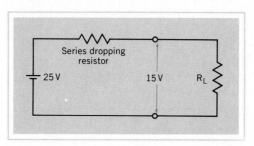

FIGURE 7-11
Series Dropping Resistor.

EXAMPLE 7-5

A portion of an electronic circuit requires an operating voltage of 15 V at a current drain of 20 mA. If the supply terminal voltage is 25 V, what value of series dropping resistor is required?

SOLUTION

From Ohm's law, we can represent this load circuit by a resistor, as in Figure 7-11.

$$R_L = \frac{V_L}{I_L} = \frac{15\ V}{20\ mA} = 750\ \Omega$$

From Kirchhoff's voltage law, the voltage drop across the series dropping resistor must be

$$V_D = E - V_L = 25 - 15 = 10\ V$$

Since this is a simple series circuit,

$$I_D = I_L = 20 \text{ mA}$$

$$\therefore \text{ dropping resistance} = \frac{V_D}{I_D} = \frac{10 \text{ V}}{20 \text{ mA}} = 500 \text{ } \Omega$$

To complete the design, we must know the minimum power rating that the chosen resistor must possess.

$$P = VI = 10 \text{ V} \times 20 \text{ mA} = 0.2 \text{ W}$$

We would probably select a $\frac{1}{2}$-W resistor in practice so that it would operate well below rated temperature, thus improving the reliability of the equipment.

The advantage of the simple series dropping resistor is that the current drain on the power supply is no greater than the current required by the circuit element in question. But this circuit has the disadvantage that any change in load resistance will cause appreciable change in the current through the series dropping resistor and, therefore, in the voltage drop across it. This, in turn, will allow appreciable change in the voltage supplied to the load.

In practical electronic circuits where R_L in Figure 7-11 is a transistor, it is quite possible for R_L to be infinitely high (when the transistor is biased to cutoff). Under such circumstances, the voltage drop across the load must rise to the full applied voltage. Even though the transistor is not drawing current, this high terminal voltage may be sufficiently high to cause voltage breakdown of the device. To prevent this breakdown from happening, in some voltage-divider designs we include a **bleeder** resistor[†] in parallel with the load, as in Figure 7-12. When the load resistance rises to infinity, the current through the bleeder maintains some current through the series dropping resistor. Consequently, there is still appreciable voltage drop across the series dropping resistor. Hence, the terminal voltage will not rise to full applied voltage.

In the design of power supplies for electronic equipment, a bleeder current of from 10 to 25% of the total current drawn from the source will provide sufficient protection against excessive terminal voltage under no-load conditions. With an increase in the bleeder current in relation to the normal load current, variations in load current will have even less effect on the load voltage. The improved **voltage regu-**

FIGURE 7-12
Simple Voltage Divider.

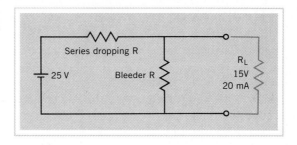

[†]So called because bleeder resistors may also serve to discharge filter capacitors in power supply units.

lation thus obtained is achieved at the expense of extra current drain from the source and extra heat produced in the voltage divider resistors. In designing voltage dividers for loads consisting of a single transistor, the voltage regulation is a more significant consideration than an extra few milliamperes of bleeder current.

EXAMPLE 7-6

Allowing a bleeder current of 50 mA, design a voltage divider to supply 15 V at 20 mA from a 25-V source. Calculate the open-circuit terminal voltage of the voltage divider.

SOLUTION

Since the bleeder resistor is in parallel with the load (Figure 7-12), the voltage drop across it is 15 V. Therefore,

$$\text{bleeder resistance} = \frac{V_L}{I_B} = \frac{15 \text{ V}}{50 \text{ mA}} = \mathbf{300 \ \Omega}$$

$$\text{power rating of bleeder} = V_L I_B = 15 \text{ V} \times 50 \text{ mA} = \mathbf{0.75 \ W}$$

Voltage drop across series dropping resistor,

$$V_D = E - V_L = 25 - 15 = 10 \text{ V}$$

and
$$I_D = I_B + I_L = 50 \text{ mA} + 20 \text{ mA} = 70 \text{ mA}$$

$$\therefore \text{ series dropping resistor} = \frac{V_D}{I_D} = \frac{10 \text{ V}}{70 \text{ mA}} = \mathbf{143 \ \Omega}$$

$$\text{power rating} = V_D I_D = 10 \text{ V} \times 70 \text{ mA} = \mathbf{0.7 \ W}$$

We would select at least 1-W resistors for this voltage divider.

Under open-circuit conditions, we can use the voltage-divider principle of Equation (7-1):

$$V_T = V_B = 25 \times \frac{300}{143 + 300} = \mathbf{16.9 \ V}$$

We recall that it is meaningless to talk about the voltage "at a point" in an electric circuit unless we add "with respect to some other point." Particularly in electronic circuits where we obtain several different voltages from one power supply, it would be very convenient to agree on one particular point in the circuit as a reference point for all voltage measurements. We usually connect this reference point electrically to the metal chassis or frame of the unit, which we represent by the graphic symbol shown in Figure 7-14. In many cases, we **ground** the chassis or frame electrically (to a water pipe or to a metal rod driven into the ground) so that there is zero potential difference between the chassis and the earth. We indicate that the reference point in the power supply voltage divider of Figure 7-13 is grounded by using the graphic symbol for **ground.** We may now say that the voltage at point A is $+250$ V with respect to ground.

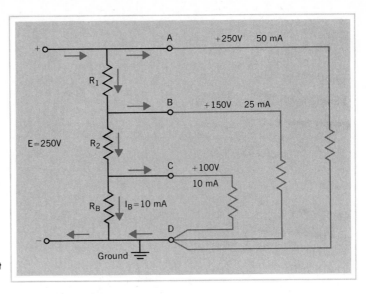

FIGURE 7-13
Voltage Divider with Multiple Output Voltages.

EXAMPLE 7-7
Design a voltage divider for the specifications shown in Figure 7-13.

SOLUTION
   Step I.   It will be easier to trace the various currents if we draw in resistors to represent the various loads given in the specifications. We should also draw in current-direction tracing arrows for each branch. Conventional current direction is shown in Figure 7-13. None of the 50-mA current drawn from terminal A flows through the voltage divider resistors. Hence, this current does not enter into our calculations.

   Step II.   Starting with the bleeder resistor, calculate the current through each resistor of the voltage divider and mark the circuit diagram of Figure 7-13 accordingly.
   The only current through R_B is the 10-mA bleeder current. R_2 passes bleeder current + the 10-mA current drawn by the 100-V load.

$$\therefore I_2 = 10 + 10 = 20 \text{ mA}$$

Finally, R_1 passes I_2 + the 25-mA current drawn by the 150-V load.

$$\therefore I_1 = 20 + 25 = 45 \text{ mA}$$

   Step III.   Calculate the resistance and power rating for each of the three resistors of the voltage divider.

$$R_B = \frac{V_B}{I_B} = \frac{100 \text{ V}}{10 \text{ mA}} = 10 \text{ k}\Omega$$

and

$$P_B = V_B I_B = 100 \text{ V} \times 10 \text{ mA} = 1 \text{ W}$$

The voltage drop across R_2 is the difference between the potentials at points B and C (with respect to ground)

$$\therefore V_2 = 150 - 100 = 50 \text{ V}$$

Hence,
$$R_2 = \frac{V_2}{I_2} = \frac{50 \text{ V}}{20 \text{ mA}} = 2.5 \text{ k}\Omega$$

and
$$P_2 = V_2 I_2 = 50 \text{ V} \times 20 \text{ mA} = 1 \text{ W}$$

(For a reliability margin, we would select 2-W resistors or even 5-W resistors for R_2 and R_B.)

Finally,
$$V_1 = 250 - 150 = 100 \text{ V}$$

Hence,
$$R_1 = \frac{V_1}{I_1} = \frac{100 \text{ V}}{45 \text{ mA}} = 2.22 \text{ k}\Omega$$

and
$$P_1 = V_1 I_1 = 100 \text{ V} \times 45 \text{ mA} = 4.5 \text{ W}$$

(Select a 10-W resistor.)

Transistor power supplies may require either the $+$ or the $-$ terminal to be grounded, depending on the type of transistor used. Or, more likely, a transistor circuit will require *both* $+$ and $-$ voltages with respect to the common chassis connection. We can obtain both positive and negative voltages (with respect to chassis) from the one power supply by connecting the appropriate tap on a voltage divider to chassis, as shown in Figure 7-14.

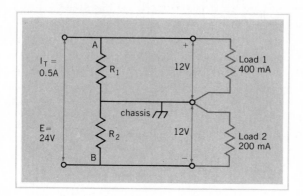

FIGURE 7-14
Voltage Divider for a Transistor Amplifier Power Supply.

EXAMPLE 7-8

The power supply for a transistor amplifier develops a full-load terminal voltage of 24 V dc. Design a voltage divider which will provide $+$ and -12-V outputs (with respect to chassis) when the current drain on the $+12$-V source is 400 mA and the drain on the -12-V source is 200 mA. The total current drain on the power supply is 0.5 A.

SOLUTION
Since R_1 is in parallel with Load 1, $V_1 = 12$ V. From Kirchhoff's current law applied to junction A,

$$I_1 = 500 \text{ mA} - 400 \text{ mA} = 100 \text{ mA}$$

$$R_1 = \frac{V_1}{I_1} = \frac{12 \text{ V}}{100 \text{ mA}} = \textbf{120 } \boldsymbol{\Omega}$$

and $$P_1 = V_1 I_1 = 12 \text{ V} \times 100 \text{ mA} = \textbf{1.2 W}$$

Similarly,

$$V_2 = 12 \text{ V} \quad \text{and} \quad I_2 = 500 \text{ mA} - 200 \text{ mA} = 300 \text{ mA}$$

$$\therefore R_2 = \frac{12 \text{ V}}{300 \text{ mA}} = \textbf{40 } \boldsymbol{\Omega} \quad \text{and} \quad P_2 = 12 \text{ V} \times 300 \text{ mA} = \textbf{3.6 W}$$

7-6 CURRENT-DIVIDER PRINCIPLE

In Example 7-6, the total current drawn from the voltage source *divides* between the load and the bleeder resistor. To assist us in the solution of problems in which we are concerned with the effects of current division between two parallel branches, we can apply the parallel-resistance principle developed in Section 6-12. This parallel circuit *dual* of the voltage-divider principle for series circuits is often called the

Current-divider principle: In a parallel circuit, the ratio between any two branch currents is the same as the ratio of the two conductances through which these currents flow.

To translate this principle into a useful algebraic equation, since V is common for two resistors in parallel,

$$\frac{I_1}{G_1} = \frac{I_2}{G_2} = \frac{I_T}{G_T} = \frac{I_x}{G_x} \tag{6-17}$$

from which $$I_x = I_T \frac{G_x}{G_T} \tag{7-2}$$

Since we are usually provided with data on the *resistance* of the parallel branches, rather than their conductances, we can substitute $G_x = 1/R_x$ and $G_T = 1/R_{eq}$ in Equation (7-2) giving

$$I_x = I_T \frac{R_{eq}}{R_x} \tag{7-3}$$

Note that the current ratio is the *inverse* of the *resistance* ratio. Hence, we may restate the current-divider principle:

In a parallel circuit, the ratio between any two branch currents is the inverse of their resistance ratio.

For two resistors in parallel,

$$R_{eq} = \frac{R_1 R_2}{R_1 + R_2} \tag{6-12}$$

$$\therefore I_1 = I_T \frac{R_1 R_2}{R_1(R_1 + R_2)}$$

and

$$I_1 = I_T \frac{R_2}{R_1 + R_2} \tag{7-4}$$

Similarly,

$$I_2 = I_T \frac{R_1}{R_1 + R_2} \tag{7-5}$$

EXAMPLE 7-9

Calculate the bleeder current in the voltage divider designed in Example 7-6 if the load resistance increases to 3 kΩ. See Figure 7-15.

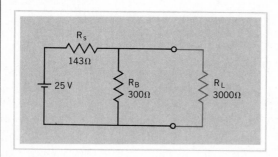

FIGURE 7-15
Circuit Diagram for Example 7-9.

SOLUTION

The equivalent resistance of the bleeder resistor and the load resistance in parallel is

$$R_{eq} = \frac{300\ \Omega \times 3\ k\Omega}{(300\ \Omega + 3\ k\Omega)} = 273\ \Omega$$

$$R_T = R_s + R_{eq} = 143 + 273 = 416\ \Omega$$

$$I_T = \frac{E}{R_T} = \frac{25\ V}{416\ \Omega} = 60.1\ mA$$

$$I_B = I_T \frac{R_L}{R_B + R_L} = 60.1\ mA \times \frac{3\ k\Omega}{3.3\ k\Omega} = 54.6\ mA$$

EXAMPLE 7-10

An ammeter has an internal resistance of 50 Ω and reads full scale when the current through it is 1.0 mA. What resistance of shunt must be placed in parallel with the meter so that it will read full scale with a circuit current of 1.0 A? See Figure 7-16.

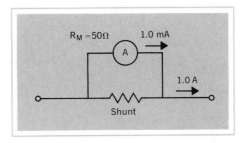

FIGURE 7-16
Circuit Diagram for Example 7-10.

SOLUTION

From Kirchhoff's current law,

$$I_{sh} = I_T - I_M = 1000 \text{ mA} - 1 \text{ mA} = 999 \text{ mA}$$

According to the current-divider principle,

$$\frac{R_{sh}}{R_M} = \frac{I_M}{I_{sh}}$$

$$\therefore R_{sh} = R_M \times \frac{I_M}{I_{sh}} = 50 \times \frac{1}{999} = 0.050\,05 \ \Omega$$

In Example 7-2, we calculated the input resistance of a typical **ladder** network, but we did not complete the circuit analysis by finding out how the output voltage and current compares with the input voltage and current. Now that we understand the current-divider principle, we can complete the task without resorting to a series of tedious Ohm's-law calculations.

EXAMPLE 7-11

With an input voltage of 204 mV applied to the ladder network of Figure 7-17 (and also Figure 7-5), what is the output voltage (a) with the output terminals open-circuit and (b) with a 600-Ω load resistance connected to the output terminals?

FIGURE 7-17
Circuit Diagram for Example 7-11.

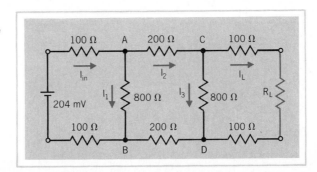

151

SOLUTION

(a) From the solution of Example 7-2(a),

$$R_{in} = 680 \ \Omega$$

Then
$$I_{in} = \frac{E}{R_{in}} = \frac{204 \ mV}{680 \ \Omega} = 300 \ \mu A$$

This current splits into two branches at node A. From the current-divider Equation (7-5) and Figure 7-6(b),

$$I_2 = 300 \ \mu A \frac{800 \ \Omega}{800 \ \Omega + 1200 \ \Omega} = 120 \ \mu A$$

Since the output terminals are open-circuit,

$$I_L = 0 \quad and \quad I_3 = I_2$$

Also, there will be zero voltage drop across the two right-hand 100-Ω resistors.

$$\therefore V_L = V_{CD} = I_2 R_{CD} = 120 \ \mu A \times 800 \ \Omega = \textbf{96 mV}$$

(b) From the solution of Example 7-2(b),

$$R_{in} = 600 \ \Omega$$

Then
$$I_{in} = \frac{E}{R_{in}} = \frac{204 \ mV}{600 \ \Omega} = 340 \ \mu A$$

This current splits into two branches at node A. From Figure 7-7(c), we note that each branch between nodes A and B has the same resistance.

$$\therefore I_2 = 0.5 \times 340 \ \mu A = 170 \ \mu A$$

At node C, I_2 splits into two branches. From Figure 7-7(a),

$$I_L = 0.5 \times 170 \ \mu A = 85 \ \mu A$$
$$\therefore V_L = I_L R_L = 85 \ \mu A \times 600 \ \Omega = \textbf{51 mV}$$

PROBLEMS

7-1. Calculate the equivalent resistance of each of the resistance networks in Figure 7-18.

7-2. What is the resistance measured from terminal R to ground in each of the resistance networks of Figure 7-19? (Note the calculator solution for Example 6-5A.)

7-3. What current is drawn from the voltage source in each of the circuits in Figure 7-20?

7-4. Determine the source current in each of the circuits in Figure 7-21.

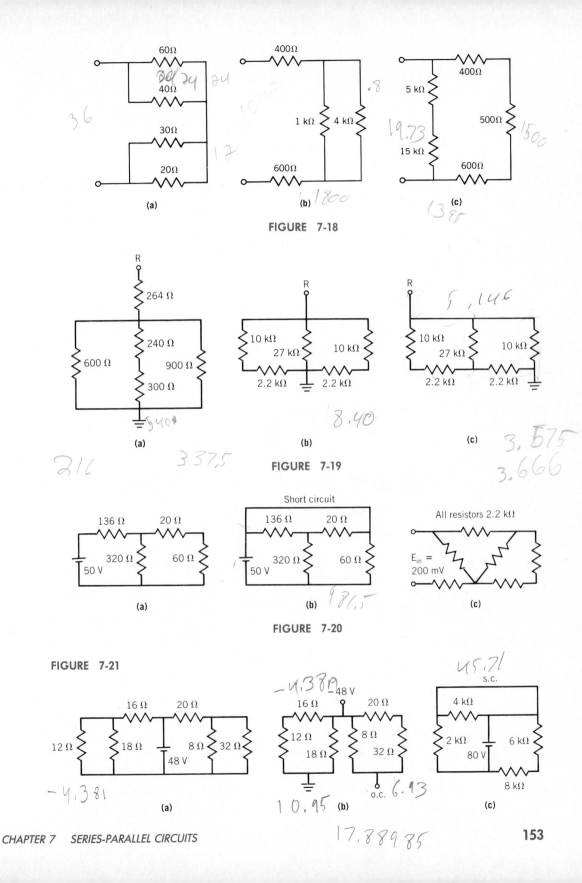

FIGURE 7-18

(a)   (b)   (c)

FIGURE 7-19

(a)   (b)   (c)

FIGURE 7-20

(a)   (b)   (c)

FIGURE 7-21

(a)   (b)   (c)

7-5. By connecting two leads to various terminals of the resistance network in Figure 7-22, six different values of equivalent resistance can be obtained. Calculate all six possibilities.

7-6. Calculate R_{AB}, R_{BC}, and R_{CA} in each of the networks in Figure 7-23.

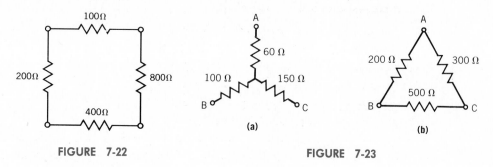

FIGURE 7-22

(a)

(b)

FIGURE 7-23

7-7. Determine the equivalent resistance of the circuit in Figure 7-24
   (a) With the switch open.
   (b) With the switch closed.

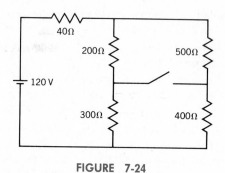

FIGURE 7-24

7-8. What is the voltage drop across the switch in the circuit of Figure 7-24 when the switch is open?

7-9. What is the input resistance of the ladder network in Figure 7-25
   (a) With a 75-Ω load resistor connected as shown?
   (b) With the load resistor disconnected?

FIGURE 7-25

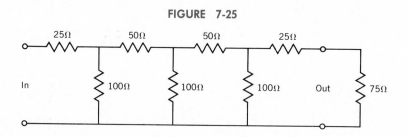

7-10. What voltage must be applied to the input terminals of the ladder network in Figure 7-25 to produce a 300-μW dissipation in the 75-Ω load resistor?

For Problems 7-11 to 7-16, draw fully labeled diagrams similar to Figure 7-1 and prepare tables similar to Table 7-1.

7-11. $E = 24$ V; $R_1 = 4$ Ω; $R_2 = 6$ Ω; $R_3 = 12$ Ω.

7-12. $P_T = 300$ W; $R_1 = 8$ Ω; $R_2 = 5$ Ω; $R_3 = 20$ Ω.

7-13. $E = 300$ V; $R_1 = 8$ Ω; $I_2 = 1$ A; $V_3 = 150$ V.

7-14. $E = 100$ V; $P_T = 75$ W; $R_2 = 100$ Ω; $I_3 = 0.5$ A.

7-15. $E = 180$ V; $R_1 = 5$ Ω; $R_2 = 15$ Ω; $I_3 = 4$ A.

7-16. $E = 250$ V; $R_1 + R_2 = 10$ kΩ; $V_2 = 100$ V; $I_3 = 5$ mA.

7-17. A 12-V generator has an internal resistance of 0.05 Ω. Two loads are connected in parallel to its terminals, one drawing a 12-A current and the other dissipating energy at the rate of 200 W. What is the terminal voltage of the generator with this load?

7-18. When the automatic gain control voltage of a radio receiver is measured with an electronic voltmeter whose input resistance is 10 MΩ, it reads 4.5 V. When an ordinary voltmeter whose input resistance is 10 kΩ is also connected across the AGC voltage source, both meters read 0.40 V. What is the internal resistance of the AGC voltage source?

7-19. What bleeder resistor is required to complete a voltage divider to give 300 V at 40 mA from a 500-V source if the series dropping resistor is 4 kΩ?

7-20. Design a voltage divider to supply 100 V at 80 mA from a 250-V source if the total drain on the source is to be 100 mA.

7-21. In the circuit of Figure 7-26; when $R_L = 400$ Ω, $I_L = 50$ mA; and when $R_L = 200$ Ω, $I_L = 75$ mA. Determine the resistances of R_1 and R_2.

7-22. Design a voltage divider to deliver 100 V at 1 mA from a 240-V source so that a 100% increase in load current causes only a 5% reduction in load voltage.

7-23. Figure 7-27 shows the focus control for a cathode-ray tube. In the one extreme position, the focus anode voltage is to be 600 V, at which the current drawn by the focus anode is 8 μA. At the other extreme, the focus anode voltage is to be 1200 V for a current of 12 μA. Calculate the resistance of R_1 and the focus control.

FIGURE 7-26

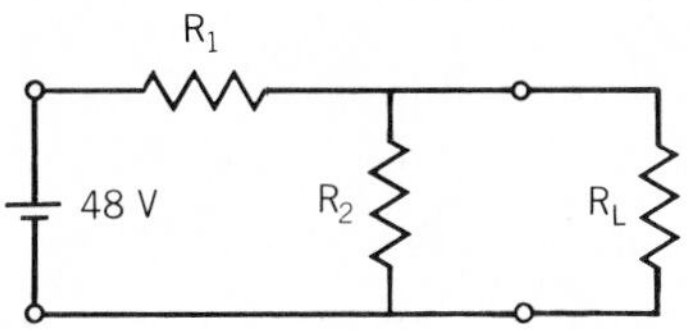

FIGURE 7-27

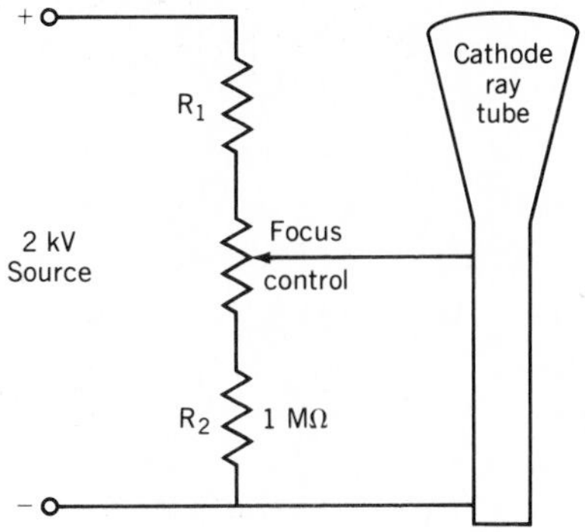

7-24. Design a voltage divider to supply 10 V at 10 mA and 25 V at 25 mA from a 40-V dc supply with a bleeder current (I_3 in Figure 7-28) of 10 mA.

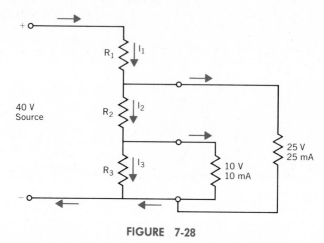

FIGURE 7-28

7-25. Assuming that load 2 in Figure 7-14 acts like a linear resistance, determine the voltage distribution (with respect to chassis) in Example 7-8 if load 1 becomes open-circuit.

7-26. A voltage divider consists of a 10-kΩ resistor with an adjustable tap. The tap is set so that it feeds 80 V at 5 mA to a load when the voltage divider is connected across a 200-V source. What is the resistance of the bleeder portion of the voltage divider?

7-27. A transistorized hi-fi amplifier system requires the following dc supplies: -25 V at 2.0 A, -12 V at 120 mA, and $+12$ V at 30 mA. All voltages are measured with respect to ground. Design a voltage divider which will permit this amplifier to be operated from a single dc supply which develops a terminal voltage of 40 V with a total current drain of 2.5 A.

7-28. The voltage-divider circuit in Figure 7-29 provides the transistor with the proper currents for operation as an amplifier. If $V_{BE} = 0.6$ V and $V_E = 1.1$ V, calculate V_C.

FIGURE 7-29

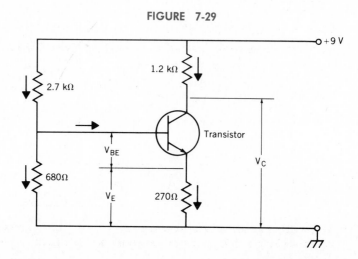

156

7-29. Calculate the current through each resistor in the circuit of Figure 7-30(a).

7-30. What value of resistor must be connected across the terminals of the constant-current source shown in Figure 7-30(b) to reduce the terminal voltage to 50 V?

7-31. In the circuit of Figure 7-31, what value of resistor R will draw a current equal to 25% of the source current?

7-32. In the circuit of Figure 7-31, what value of resistor R is required for the source current to be 200 mA?

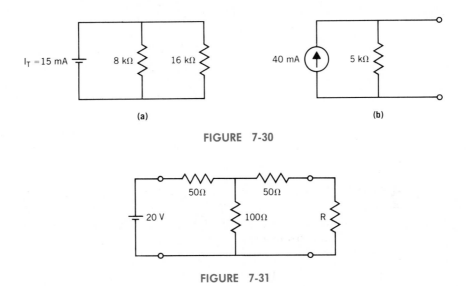

FIGURE 7-30

FIGURE 7-31

REVIEW QUESTIONS

7-1. With respect to series-parallel resistance networks, what is meant by an "equivalent resistance?"

7-2. What characteristics must an "equivalent resistance" possess?

7-3. What circuit rules allow us to replace R_2 and R_3 in Figure 7-32 with an equivalent resistance?

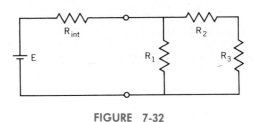

FIGURE 7-32

7-4. What advantage does solution of series-parallel resistance networks by writing Kirchhoff's-law equations have as compared with the "equivalent resistance" method?

7-5. Write one Kirchhoff's current-law and two Kirchhoff's voltage-law equations for the network of Figure 7-32.

7-6. What circuit condition must exist for the **voltage-divider principle** to be applied in network solutions?

7-7. The four-terminal potentiometer network of Figure 7-33 (sometimes called an "L" attenuator) is the volume-control circuit of an audio amplifier. Explain its operation.

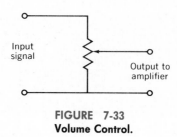

FIGURE 7-33
Volume Control.

7-8. What is the purpose of a **bleeder** resistor in a practical voltage-divider circuit?

7-9. Calculate the load voltage in Examples 7-5 and 7-6 if the load current decreases to 10 mA. Explain the difference in the two values.

7-10. Why does a decrease in the bleeder resistance of a given voltage divider improve the voltage regulation of the output of the voltage divider?

7-11. In Chapter 2, we noted that voltage is measured *between* two points in a circuit; yet in some electronic circuits the data might state that "The voltage at Point A is +12 V." Explain.

7-12. A dc power supply for an electronic amplifier has only two output terminals. Explain how it is possible to obtain simultaneously both positive and negative potentials with respect to the chassis of the amplifier.

7-13. What circuit conditions must exist for the **current-divider principle** to be applied in network solutions?

7-14. The current-divider principle is sometimes called the parallel-circuit *dual* of the voltage-divider principle. Explain the duality relationship.

7-15. Explain why the current-divider principle states that the ratio of the currents in two parallel branches is the *inverse* of their resistance ratio.

*7-16. Write a BASIC program which will ask the operator to enter values for E, R_1, R_2, and R_3 with respect to the series-parallel circuit of Figure 7-1. After these values have been entered, the program should then have the computer display the calculations in a table similar to Table 7-1.

*7-17. For a voltage divider similar to that shown in Figure 7-13, prepare a BASIC program which can be stored in permanent memory (cassette or diskette) which will ask the operator to input values for

1. Source voltage
2. Bleeder current
3. Output current from terminals A and D
4. Voltage and current for terminals B and D
5. Voltage and current for terminals C and D

The computer should then display in table format the resistance and power ratings of R_1, R_2, and R_B.

8

RESISTANCE NETWORKS

8-1 NETWORK EQUATIONS FROM KIRCHHOFF'S LAWS

At first glance, the **Wheatstone bridge** shown in Figure 8-1 may appear to be a simple series-parallel circuit. But no two resistors meet the requirements for either series or parallel components. Hence, we cannot use the *equivalent resistance* techniques of Chapter 7 to analyze a bridge circuit. There are two general methods we can use to solve such networks. The more recent method uses various *network theorems* to simplify the original network so that we can once more use equivalent-circuit techniques to analyze the complete circuit. This method is particularly suited to pocket calculators.

The second method is more traditional and is based on Kirchhoff's voltage and current laws, which we encountered in Sections 6-4 and 6-9. As fundamental *laws* of electric circuit behavior, Kirchhoff's laws must hold true, not only in the simple series and parallel circuits of the preceding chapters, but also in the most complex electric circuits we may encounter. Once we understand the significance of Kirchhoff's laws, we can write a series of equations (the algebraic statement of Kirchhoff's laws) which will enable us to analyze the behavior of the more sophisticated electric and electronic circuits we shall consider in later chapters and subsequent courses.

To apply the Kirchhoff's-law equations method, we need to learn a few basic rules which will permit us to write an orderly set of simultaneous equations. This structured approach makes this method particularly suited to the development of circuit-analysis software packages for computers. Hence, the traditional circuit analysis methods of this chapter are experiencing a resurgence in popularity.

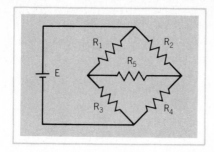

FIGURE 8-1
Wheatstone Bridge.

The disadvantage of these **mesh** and **nodal** analysis techniques in an introductory electric circuits course is that we can become more involved with plugging numbers into computer programs than with learning to understand the principles of electric circuit behavior. The equivalent-circuit and network-theorem methods of subsequent chapters accomplish the same results with less emphasis on mathematical formats.

An understanding of the principle of **source conversion** will allow us to simplify the final formats we use in writing Kirchhoff's-law equations for various resistance networks. Hence, we shall digress a bit into equivalent-circuit techniques for the next three sections of this chapter.

8-2 CONSTANT-VOLTAGE SOURCES

We have already discovered that most practical sources generate a terminal voltage which falls as we draw current from the source. We explained this effect in Section 6-6 by assuming that a practical voltage source consists of a constant-voltage source producing a constant E = open-circuit terminal V with an internal resistance in series with this constant-voltage source. Although we did not use the particular terminology at the time, we discovered an **equivalent circuit** for a practical source. This type of equivalent circuit is called a **constant-voltage source.** From a circuit-analysis point of view, the load in the circuit of Figure 8-2(a) "sees" the source as a constant-voltage

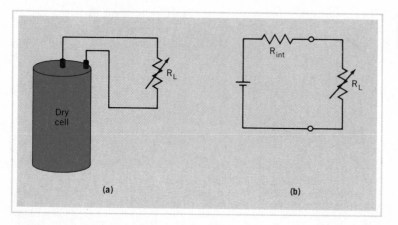

FIGURE 8-2
Equivalent Circuit for a Practical Voltage Source.

source which includes an internal series resistance, as shown in the equivalent circuit of Figure 8-2(b).

In replacing portions of a pencil-and-paper circuit diagram containing a practical voltage source with its equivalent constant-voltage source, we follow a mental procedure which duplicates a laboratory technique for finding out what is sealed into a "black box" that has only two terminals exposed. For example, let us consider a power supply which provides 16 volts dc to operate a transistor amplifier. As shown in Figure 8-3(a), this power supply is essentially a "black box" which derives its energy from a 120-V 60-Hz power line and delivers a dc output voltage from a pair of terminals. But as far as the load R_L is concerned, it can "see" only a source from which direct current may be drawn. Details of what happens inside the "black box" are of no significance to the electric circuit connected to its output terminals.

Given a voltmeter that draws negligible current, we can measure the **open-circuit terminal voltage** by disconnecting the load and connecting the voltmeter across the power supply output terminals. As we reduce R_L from an infinitely high value, current drawn from the power supply increases and the terminal voltage falls, as shown by the graph in Figure 8-3(c). In our laboratory experiment, we can reduce the load resistance to a point where the current drawn from the power supply is high enough to "blow" the protective fuse. In our pencil-and-paper circuit analysis, we are not hampered by such limitations, and we can proceed along the dashed portion of the graph until R_L = 0 ohms and we read the **short-circuit current.** With no knowledge

FIGURE 8-3
Equivalent Circuit for a DC Power Supply.

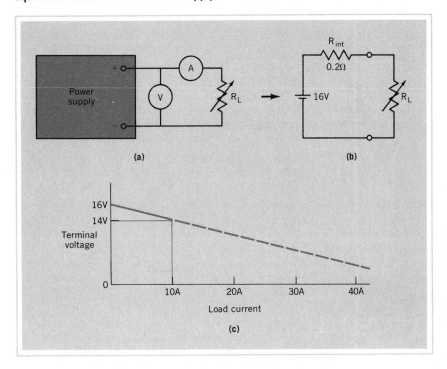

of the details of the internal operation of this particular power supply, all that we can state from the data shown in Figure 8-3(c) is that the power supply behaves as a constant-voltage source with an internal resistance of $R_{int} = \Delta V/\Delta I = \frac{2}{10} = 0.2$ ohm, as shown in Figure 8-3(b). Extending the graph of Figure 8-3(c) in a straight line until it cuts the x-axis, the short-circuit current is 80 amperes. Once we know the open-circuit voltage and the short-circuit current, the internal resistance becomes

$$R_{int} = \frac{\text{open-circuit terminal voltage}}{\text{short-circuit terminal current}} \qquad (8\text{-}1)$$

Hence, R_{int} in the constant-voltage source of Figure 8-3(b) is $V_{oc}/I_{sc} = 16$ V/80 A $= 0.2\ \Omega$.

8-3 CONSTANT-CURRENT SOURCES

As we have discovered on many occasions, there are always two ways of looking at the same property in electric circuits. For example, *resistance* is the ability of a circuit to *oppose* electric current. Looking at the same property from the opposite point of view, *conductance* is the ability of a circuit to *permit* electric current. From these definitions, it follows that conductance is the reciprocal of resistance, and $G = 1/R$. We select the point of view which is more convenient for the task at hand. In dealing with series circuits, we prefer to think and work in terms of resistance. But, in dealing with parallel circuits, it is often more convenient to switch our point of view and to think and work in terms of conductance.

Perhaps there is an opposite point of view of what the sealed-box power supply of Figures 8-3(a) and 8-4(a) might contain that will produce the same results in an external load. Our voltmeter/ammeter technique tells us that the sealed box has an open-circuit terminal voltage of 16 volts and a short-circuit terminal current of 80 amperes. Theoretically, the box could contain a special generator with a completely variable terminal voltage but which always produces a constant current equal to the short-circuit terminal current. For this constant current to flow when the terminals of the box are open-circuit, its internal resistance must now be shown in *parallel* with the terminals of the constant-current generator, as shown in Figure 8-4(b). To produce the required open-circuit terminal voltage, it follows that the value of R_{int} must be

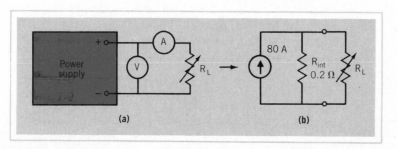

FIGURE 8-4
Constant-Current Source.

$$R_{int} = \frac{\text{open-circuit terminal voltage}}{\text{short-circuit terminal current}} \tag{8-1}$$

Note that this is the same equation we used to find the internal resistance of the constant-voltage source in Figure 8-3(b). The direction of the arrow in the symbol for a constant-current source shows the conventional current direction required to maintain the same polarity of voltage across R_L in Figure 8-4(b) as in Figure 8-4(a).

Suppose that we set R_L in the circuits shown in Figures 8-3(b) and 8-4(b) at 0.6 ohm. With the constant-voltage source of Figure 8-3(b), the voltage-divider principle tells us that

$$V_L = \frac{R_L}{R_{int} + R_L} \times V_{oc} = \frac{0.6}{0.2 + 0.6} \times 16\ V = 12\ V$$

and

$$I_L = \frac{V_L}{R_L} = \frac{12\ V}{0.6\ \Omega} = 20\ A$$

With the constant-current source of Figure 8-4(b), we switch to the current-divider principle.

$$I_L = \frac{R_{int}}{R_{int} + R_L} \times I_{sc} = \frac{0.2}{0.2 + 0.6} \times 80\ A = 20\ A$$

and

$$V_L = I_L R_L = 20\ A \times 0.6\ \Omega = 12\ V$$

For any value of load resistance we may choose, we find that the constant-voltage source of Figure 8-3(b) and the constant-current source of Figure 8-4(b) are exact equivalents, as "seen" by R_L. The constant-current equivalent source of Figure 8-4(b) is credited to an American engineer, Edward L. Norton, and is the **dual** of the constant-voltage source of Figure 8-3(b).

In most practical electric circuits, we are used to the concept of *opening* the circuit (disconnecting the load) to switch the circuit *off*. When we do so, we expect the current through the source, as well as the current in the load, to become zero. The power dissipation in the internal resistance of the source would also be zero under open-circuit conditions. This is the situation with a constant-voltage source. Hence, the equivalent circuit of Figure 8-3(b) represents the "real-world" situation in electric circuit diagrams.

It is difficult to imagine a current source existing in a practical electrical system. If we disconnect R_L in the circuit of Figure 8-4(b), the current through R_{int} must *increase* to the full constant-current value and the internal dissipation of the source becomes $V_{oc} \times I_{sc} = 1280\ W$. Theoretically, to switch the circuit of Figure 8-4(b) *off*, we *short-circuit* R_L so that the terminal voltage and the dissipation become zero. The constant-current equivalent source of Figure 8-4(b) is primarily a pencil-and-paper device to assist in network analysis.

Although we will not encounter current sources in electric power system networks, we do find devices which function as if they were current sources in low-power electronic circuits. For example, it is easier to analyze transistor circuits if we show a current source in their equivalent circuit. Also, it is possible to use electronic feedback circuits to construct a dc power supply which will maintain a constant load current with a terminal voltage which varies as the load resistance connected to it varies. Such constant-current power supplies are not as common as voltage-regulated power supplies which maintain a constant terminal *voltage* as the load resistances connected to them vary.

8-4 SOURCE CONVERSION

In developing either the constant-voltage source of Figure 8-3(b) or the constant-current source of Figure 8-4(b), we used the *same* "black-box" technique to determine R_{int}.

$$R_{int} = \frac{E_{oc}}{I_{sc}}$$

(8-1)

Hence, if we are given a constant-voltage source, as in Figure 8-5(a), R_x in the equivalent constant-current source of Figure 8-5(b) has the *same* value but appears in *parallel* with the ideal current source. Similarly, if we are given the constant-current source of Figure 8-5(c), R_x in the equivalent constant-voltage source of Figure 8-5(d) has the *same* value but appears in *series* with the ideal voltage source.

To be able to replace a constant-voltage source in a network with its constant-current equivalent, there must be some resistance R_x in *series* with an ideal voltage source. R_x may be an internal resistance of the source, or it may be one of the network resistors in series with the voltage source. Similarly, there must be some form of resistance in parallel with an ideal current source before we can replace it with its constant-voltage equivalent.

Given the constant-voltage source of Figure 8-5(a), from Equation (8-1), I_x for the equivalent constant-current source of Figure 8-5(b) is the short-circuit current

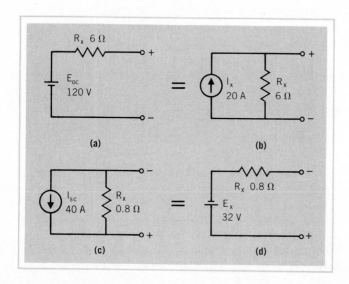

FIGURE 8-5
Source Conversion.

$$I_x = \frac{E_{oc}}{R_x} \tag{8-2}$$

And given the constant-current source of Figure 8-5(c), E_x for the equivalent constant-voltage source of Figure 8-5(d) is the open-circuit voltage

$$E_x = I_{sc}R_x \tag{8-3}$$

EXAMPLE 8-1

(a) A voltage source develops an open-circuit voltage of 120 V and has an internal resistance of 6 Ω. Determine the equivalent constant-current source.

(b) A current source develops a constant current of 40 A with an internal resistance of 0.8 Ω. Determine the equivalent constant-voltage source.

(c) Verify the conversion in part (a) by connecting a 12-Ω load to both sources.

SOLUTION

(a) R_x for the constant-current source is the same as for the voltage source. $R_x = 6\ \Omega.$

$$I_x = \frac{E_{oc}}{R_x} = \frac{120\ \text{V}}{6\ \Omega} = 20\ \text{A}$$

(b) R_x for the constant-voltage equivalent source is the same as for the current source. $R_x = 0.8\ \Omega.$

$$E_x = I_{sc}R_x = 40\ \text{A} \times 0.8\ \Omega = 32\ \text{V}$$

(c) When we connect a 12-Ω load to the voltage source of Figure 8-5(a),

$$I_L = \frac{E_x}{R_x + R_L} = \frac{120\ \text{V}}{(6 + 12)\Omega} = 6.67\ \text{A}$$

and $\qquad V_L = 6.67\ \text{A} \times 12\Omega = 80\ \text{V}$

When we connect a 12-Ω load to the current source of Figure 8-5(b),

$$V_L = I_x \times \frac{R_x \times R_L}{R_x + R_L}$$

$$= 20\ \text{A} \times \frac{6 \times 12}{6 + 12} = 80\ \text{V}$$

and $\qquad I_L = \frac{V_L}{R_L} = \frac{80\ \text{V}}{12\ \Omega} = 6.67\ \text{A}$

Once we are familiar with the concept, we can convert voltage sources to equivalent current sources, and vice versa, for network analysis purposes with no more effort than a bit of Ohm's-law mental arithmetic. Sometimes we may encounter resistance networks containing both voltage and current sources, particularly networks containing transistor equivalent circuits. Such networks are usually easier to solve if we use source conversion so that the network contains all voltage sources or all current sources.

8-5 KIRCHHOFF'S VOLTAGE-LAW EQUATIONS: LOOP PROCEDURE

As we discovered in Chapter 6, we can write two distinct types of equations for the same circuit: one set based on Kirchhoff's voltage law, and the other based on Kirchhoff's current law. Network equations based on Kirchhoff's voltage law are called **loop** or **mesh** equations. Network equations based on Kirchhoff's current law are called **nodal** equations. Each method has its advantages and disadvantages. To avoid confusion, we shall consider the two methods separately.

There are several ways of stating Kirchhoff's voltage law. In Section 6-4, we used the form.

In any complete electric circuit, the algebraic sum of the source voltages must equal the algebraic sum of the voltage drops.

This statement leads to equations in the form

$$E = V_1 + V_2 + V_3 \tag{8-4}$$

Some texts prefer to state the same law thus,

In any closed loop the algebraic sum of all voltage rises and voltage drops must equal zero.

This statement leads to an equation in which we must carefully record the sign of voltage drops and voltage rises.

$$E - V_1 - V_2 - V_3 = 0 \tag{8-5}$$

To enable us to set up our equations in a form that is suited to solution by determinants, we rearrange our original statement and say that

In any complete electric circuit (closed loop), the algebraic sum of the voltage drops must equal the algebraic sum of the source voltages.

This leads to equations in the form

$$V_1 + V_2 + V_3 = E \tag{8-6}$$

Perhaps the best way to develop our loop procedure is to refer to a specific example. We can start with an example which we can readily check by the equivalent-resistance techniques of Chapter 7 by removing the center resistor R_5 in the bridge circuit of Figure 8-1.

EXAMPLE 8-2

Determine the magnitude and polarity of the potential difference between the open-circuit terminals A and B in the bridge network of Figure 8-6.

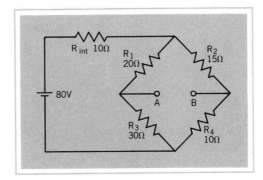

FIGURE 8-6
Bridge Network for Example 8-2.

SOLUTION

Step I. The first step is to draw a circuit diagram large enough for us to mark on tracing loops, polarities of voltages, and any other data we need in solving the network. We may change the *physical* location of components so long as we do not change their *electrical* connection. (Figure 8-7).

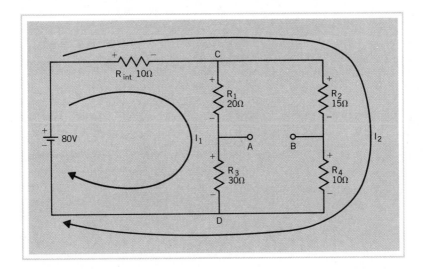

FIGURE 8-7
Circuit Diagram for Example 8-2.

Step II. We emphasize the polarity of the sources by marking the $+$ and $-$ terminals for two reasons: (1) if there is more than one source, reversing the polarity of one will completely alter all current and voltage relationships in the network; and (2) for our procedure, we draw current **tracing loops** on the circuit diagram, beginning at the $+$ terminal of the source and proceeding along a path through the network back to the − terminal of the source. These tracing loops will help us to write our Kirchhoff's voltage-law equations. We need sufficient tracing loops to include *all* components and to provide as many separate equations as there are unknowns. In this example, we require two tracing loops for the two unknown loop currents I_1 and I_2.

Step III. In Section 6-2, we considered the polarities of voltage drops and voltage rises in relation to current direction. Extending what is obvious in the simple circuit of Figure 8-8 to the tracing loops of Figure 8-7,

In any **passive** *component (such as a resistor) in an electric network, the tracing direction is from positive to negative through that component.*

Hence, once we have established the directions of tracing loops I_1 and I_2, we can mark the polarity of the various voltage drops on the circuit diagram of Figure 8-7.

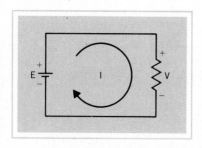

FIGURE 8-8
Polarity of Voltage Drop and Source Voltage in Relation to Direction of Current Tracing Loop.

Step IV. We are now ready to write our Kirchhoff's voltage-law equations. For the I_1 loop,

$$V_{R_{\text{int}}} + V_{R_1} + V_{R_3} = E$$

If we think now of I_1 as the current through the R_1, R_3 branch, from Ohm's law,

$$V_{R_1} = I_1 R_1 \quad \text{and} \quad V_{R_3} = I_1 R_3$$

Considering the junction point C in Figure 8-7, we see that the current I_1 *leaves* this point through R_1 and R_3, and the current I_2 *leaves* this point through R_2 and R_4. Hence, from Kirchhoff's current law, the current through R_{int} is $I_1 + I_2$. Thus,

$$V_{R_{\text{int}}} = (I_1 + I_2)R_{\text{int}}$$

The Kirchhoff's voltage-law equation for the I_1 tracing loop becomes

$$(I_1 + I_2)R_{\text{int}} + I_1 R_1 + I_1 R_3 = E$$

This is the standard form we shall use for writing our Kirchhoff's voltage-law equations, minimizing the number of unknowns by writing voltage drop as *IR* drops.

Step V. The Kirchhoff's voltage-law equation for the I_1 tracing loop becomes

$$10(I_1 + I_2) + 20I_1 + 30I_1 = 80 \text{ V}$$

Collecting terms, $\qquad\qquad 60I_1 + 10I_2 = 80 \qquad\qquad\qquad$ (1)

Similarly, for the I_2 tracing loop,

$$10(I_1 + I_2) + 15I_2 + 10I_2 = 80 \text{ V}$$

from which $\qquad\qquad\qquad 10I_1 + 35I_2 = 80 \qquad\qquad\qquad$ (2)

We then solve Equations (1) and (2) simultaneously:

Elimination method:

$$6 \times (2) \text{ gives} \quad 60I_1 + 210I_2 = 480$$
$$\textit{subtracting} \ (1) \quad 60I_1 + 10I_2 = 80$$
$$\overline{ 200I_2 = 400}$$

from which $\qquad\qquad\qquad I_2 = 2 \text{ A}$

Substituting this value in Equation (1), we obtain

$$I_1 = 1 \text{ A}$$

Determinant method:†

$$60I_1 + 10I_2 = 80 \qquad\qquad (1)$$
$$10I_1 + 35I_2 = 80 \qquad\qquad (2)$$

$$I_1 = \frac{\begin{vmatrix} 80 & 10 \\ 80 & 35 \end{vmatrix}}{\begin{vmatrix} 60 & 10 \\ 10 & 35 \end{vmatrix}} = \frac{2800 - 800}{2100 - 100} = \frac{2000}{2000} = 1 \text{ A}$$

$$I_2 = \frac{\begin{vmatrix} 60 & 80 \\ 10 & 80 \end{vmatrix}}{\begin{vmatrix} 60 & 10 \\ 10 & 35 \end{vmatrix}} = \frac{4800 - 800}{2000} = \frac{4000}{2000} = 2 \text{ A}$$

†The traditional algebraic procedure for solving simultaneous equations is the *Gaussian elimination* method in which equations with two or more unknowns are subtracted until a single equation with only one unknown is left. No fixed procedure is written down for this method since a certain amount of individual judgment is required in selecting the unknowns to be eliminated.

However, as more and more computation is turned over to computers, it is desirable to sacrifice flexibility for a standard procedure which we can use to solve *any* array of simultaneous equations. The procedure currently favored is the *determinant* method. For those whose background in mathematics does not include matrix algebra, a brief description of the determinant method is included in Appendix 1.

Step VI. We can now find the voltage drops across all components of the network. In this example, we need know only two of them.

$$V_{R_3} = I_1 R_3 = 1 \text{ A} \times 30 \text{ } \Omega = 30 \text{ V}$$

and

$$V_{R_4} = I_2 R_4 = 2 \text{ A} \times 10 \text{ } \Omega = 20 \text{ V}$$

To determine the potential difference between terminals A and B, we recall that a *voltage drop* in a resistance is just another way of expressing a *potential difference* between the two terminals of the resistance. Hence, terminal A in Figure 8-7 is 30 V more positive than junction D (the junction point of R_3 and R_4). Likewise, terminal B is 20 V more positive than junction D. Therefore,

Terminal A *is 10 V more positive than terminal* B.

At this point, let's consider how a personal computer might assist us in solving general resistance network examples. As specialized circuit-analysis software packages become more numerous, we can solve all sorts of examples without learning anything about loop equations or determinants or network theorems. These techniques are buried within the computer program. A typical circuit-analysis program asks us first of all to describe the network layout by identifying the type of circuit elements (resistor, voltage source, etc.) and the two specific nodes between which the element is connected. We then input numerical values for these elements and the computer proceeds to print out the complete solution of the network. These software packages are designed primarily to assist people in industry in optimizing a circuit design by rapidly printing out the results of many variations of circuit configuration and values. Since they do not assist in learning the principles of network analysis, we shall not pursue this type of computer program.

However, if we intend to solve quite a few examples involving simultaneous equations, it could be worthwhile setting up programs to solve second- and third-order determinants. We do this by following the convention for coefficients and the Cramer's rule formula developed in Appendix 1. The advantages of having such programs stored in permanent memory is that we can concentrate on setting up the Kirchhoff's-law equations from the network data without having to worry about arithmetic errors and the time required for individual calculations with a pocket calculator. A typical program that we could use with Example 8-1 might be

```
100 REM   SECOND ORDER DETERMINANTS
110 REM   SET UP TWO EQUATIONS IN THE FORM
120 PRINT  "AX + BY = K"
130 REM   A - SIGN IN THE EQUATION BELONGS TO THE COEFFICIENT
140 PRINT "ENTER COEFFICIENTS FOR EQUATION 1"
150 INPUT   A1,B1,K1
160 PRINT "ENTER COEFFICIENTS FOR EQUATION 2"
170 INPUT   A2,B2,K2
180 LET  D=A1*B2-A2*B1
190 LET  X=(K1*B2-K2*B1)/D
```

```
200 LET Y=(A1*K2-A2*K1)/D
210 PRINT "X = ";X
220 PRINT "Y = ";Y
230 END
```

For Example 8-2, this program produces the following result:

```
RUN
AX + BY = K
ENTER COEFFICIENTS FOR EQUATION 1
? 60, 10, 80
ENTER COEFFICIENTS FOR EQUATION 2
? 10, 35, 80
X = 1
Y = 2
```

If we add a meter between terminals A and B of the bridge circuit of Figure 8-6, the network becomes a Wheatstone bridge which is used for very accurate resistance measurements. To include the meter current in our Kirchhoff's-law equations, we now require one more tracing loop—I_3 in Figure 8-9. We have chosen a tracing loop for I_3 passing through R_1, R_M (the resistance of the meter), and R_4. We might have chosen a tracing loop for I_3 passing through R_2, R_M, and R_3. Since I_3 would then pass in the *opposite* direction through R_M, it follows that it would turn out to have the *opposite* sign in our solution. If a tracing loop current turns out to have a negative sign, we simply treat the current as a negative quantity when we add the loop currents *algebraically*.

Using the I_3 loop shown in Figure 8-9, the Kirchhoff's voltage-law equations for the three tracing loops are

$$\text{From loop 1,} \quad R_1(I_1 + I_3) + R_3I_1 = E$$

$$\text{From loop 2,} \quad R_2I_2 + R_4(I_2 + I_3) = E$$

$$\text{From loop 3,} \quad R_1(I_1 + I_3) + R_MI_3 + R_4(I_2 + I_3) = E$$

FIGURE 8-9
Solving Wheatstone Bridge by Loop Procedure.

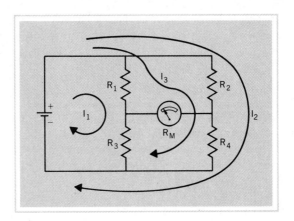

Substituting the given values of resistance and source voltage, we can solve these three simultaneous equations for the three loop currents from which

Total current is $\quad I_1 + I_2 + I_3$

Meter current is $\quad I_3$

Current through R_1 is $\quad I_1 + I_3$

Current through R_2 is $\quad I_2$

Current through R_3 is $\quad I_1$

Current through R_4 is $\quad I_2 + I_3$

EXAMPLE 8-3

If a galvanometer whose resistance is 50 Ω is connected between terminals A and B of the bridge circuit of Figure 8-7, what is the meter current?

SOLUTION

For the I_1 loop,

$$10(I_1 + I_2 + I_3) + 20(I_1 + I_3) + 30I_1 = 80 \text{ V}$$

for the I_2 loop,

$$10(I_1 + I_2 + I_3) + 15I_2 + 10(I_2 + I_3) = 80 \text{ V}$$

and for the I_3 loop,

$$10(I_1 + I_2 + I_3) + 20(I_1 + I_3) + 50I_3 + 10(I_2 + I_3) = 80 \text{ V}$$

Collecting terms,

$$60I_1 + 10I_2 + 30I_3 = 80 \tag{1}$$

$$10I_1 + 35I_2 + 20I_3 = 80 \tag{2}$$

$$30I_1 + 20I_2 + 90I_3 = 80 \tag{3}$$

Since we are asked only to determine the meter current, we need only solve for I_3. Thus, using the determinant method,

$$I_3 = \frac{\begin{vmatrix} 60 & 10 & 80 \\ 10 & 35 & 80 \\ 30 & 20 & 80 \end{vmatrix}}{\begin{vmatrix} 60 & 10 & 30 \\ 10 & 35 & 20 \\ 30 & 20 & 90 \end{vmatrix}}$$

$$= \frac{168\,000 + 24\,000 + 16\,000 - 84\,000 - 96\,000 - 8000}{189\,000 + 6000 + 6000 - 31\,500 - 24\,000 - 9000}$$

$$= \frac{20\,000}{136\,500} = 0.15 \text{ A}$$

8-6 NETWORKS WITH MORE THAN ONE VOLTAGE SOURCE

Even an electrical system as simple as a flashlight usually has more than one voltage source. To provide the lamp with a potential difference higher than that of a single battery, we connect two or more batteries in *series aiding*. Since both voltage sources in Figure 8-10(a) attempt to make current flow in the *same* direction through any load, the total effective source voltage becomes the *sum* of the individual source voltages. In pencil-and-paper electric circuits, we can replace the individual series-connected voltage sources with a single equivalent source. [The automobile storage battery in Figure 8-11(a) consists of six cells in series aiding.]

It is quite possible for us to load one of the flashlight (or radio or camera) batteries wrong-way-round. With such a *series opposing* connection [shown in Figure 8-10(b)] each source tries to make current flow in opposite directions in any load. As a result, the net terminal voltage for the equivalent source is the *difference* between the individual source voltages. We avoid series opposing connection of voltage sources in practical electric circuits for two reasons: (1) the equivalent source voltage is actually *reduced;* and (2), since the higher voltage source determines the current direction, current will flow the wrong way (charging) through the lower voltage source when we connect a load to the source in Figure 8-10(b).

Figure 8-11(a) shows a practical electrical system in which both a storage battery and a generator driven by an automobile engine provide current to a common load circuit. The terminals of these two voltage sources (each with its own internal resistance) are connected in *parallel*. If the auto engine stops, the generator emf must fall to zero and a heavy current—limited only by the two internal resistances in series—flows from the battery through the generator. In a practical automobile electrical system, we must arrange for the generator to be automatically disconnected from the rest of the system when the engine is stopped. Parallel connection of voltage sources requires special precautions.

The innocent-looking network of Figure 8-11(b) is the coupling circuit for a

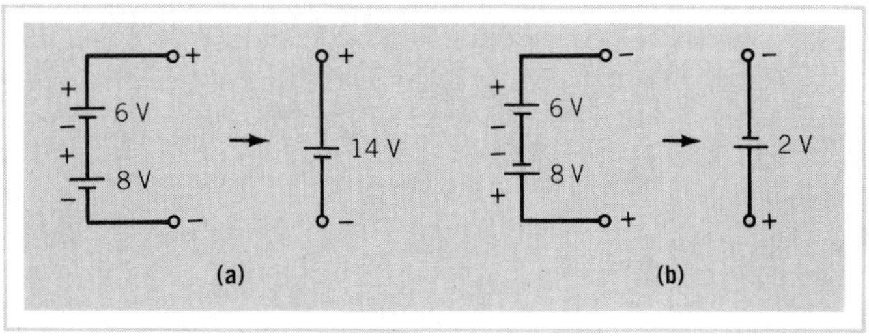

FIGURE 8-10
Voltage Sources: (a) Series Aiding; (b) Series Opposing.

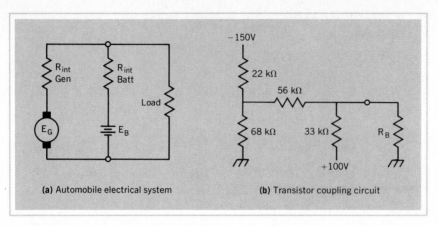

FIGURE 8-11
Examples of Networks with More than One Voltage Source.

transistor "flip-flop" circuit. Since dozens of stages of an electronic system are fed from a single power supply, we usually avoid cluttering circuit diagrams with long interconnecting leads by marking the dc supply voltages for individual stages, as shown in Figure 8-11(b). We can redraw this network in the form shown in Figure 8-15. It now becomes apparent that we have another example of a network containing two separate voltage sources. Since both networks in Figure 8-11 include more than one voltage source, we cannot solve them by simple Ohm's-law calculations.

8-7 LOOP EQUATIONS IN MULTISOURCE NETWORKS

The procedure we developed in Section 8-5 for writing Kirchhoff's voltage-law equations for closed tracing loops applies equally well to multisource networks. Again, we can illustrate the procedure and note possible points of difficulty by referring to specific examples. As our first example, we shall use the automobile electrical system of Figure 8-11(a).

EXAMPLE 8-4

An automobile generator with an internal resistance of 0.2 Ω develops an open-circuit voltage of 16.0 V. The storage battery has an internal resistance of 0.1 Ω and an open-circuit voltage of 12.8 V. Both sources are connected in parallel to a 1.0-Ω load. Determine the generator current, battery current, and load current.

SOLUTION 1

If we wish, we can redraw the circuit diagram of Figure 8-11(a) as shown in Figure 8-12 to make it easier to draw our tracing loops.

Writing a Kirchhoff's voltage-law equation for the generator loop I_G in Figure 8-12,

$$V_{RG} + V_L = E_G$$

But from Ohm's law, any voltage drop across a resistance is always equal to the product of the resistance and the current through it. Therefore,

$$V_{RG} = 0.2I_G$$

Since the diagram of Figure 8-12 shows both the I_G and the I_B loops going in the *same* direction through the load, from Kirchhoff's current law, the load current is

$$I_L = I_G + I_B \quad \text{and} \quad V_L = 1.0(I_G + I_B)$$

Hence, the Kirchhoff's voltage-law equation for the generator loop becomes

$$0.2I_G + 1.0(I_G + I_B) = E_G$$

or
$$1.2I_G + 1.0I_B = 16.0 \text{ V} \tag{1}$$

Similarly, the Kirchhoff's voltage-law equation for the battery loop becomes

$$0.1I_B + 1.0(I_G + I_B) = E_B$$

or
$$1.0I_G + 1.1I_B = 12.8 \text{ V} \tag{2}$$

Solving Equations (1) and (2) simultaneously,

Elimination method:

$$5 \times (1) \text{ gives} \quad 6I_G + 5.0I_B = 80$$
$$6 \times (2) \text{ gives} \quad 6I_G + 6.6I_B = 76.8$$

subtracting,
$$-1.6I_B = 3.2$$

from which
$$I_B = -2 \text{ A}$$

The negative answer for I_B simply tells us that the battery is not supplying current to the load, as we had supposed in setting up the direction for the tracing loop for I_B. However, the numerical answer is quite correct. In this example, the storage battery is *charging* at the rate of 2 A. To avoid having to change our tracing

FIGURE 8-12
Circuit Diagram for Example 8-4.

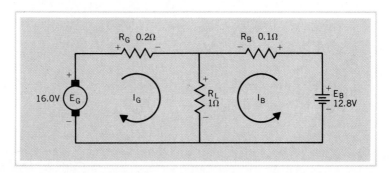

175

directions in Figure 8-12, we can carry on with the original tracing loops as long as we treat I_B as a negative quantity. Substituting this value of I_B in Equation (1),

$$1.2I_G + (1.0) \times (-2) = 16$$

or

$$1.2I_G - 2 = 16$$

from which

$$I_G = \textbf{15 A}$$

and

$$I_L = I_G + I_B = 15 + (-2) = \textbf{13 A}$$

Determinant method:

$$1.2I_G + 1.0I_B = 16.0$$

$$1.0I_G + 1.1I_B = 12.8$$

$$I_G = \frac{\begin{vmatrix} 16.0 & 1.0 \\ 12.8 & 1.1 \end{vmatrix}}{\begin{vmatrix} 1.2 & 1.0 \\ 1.0 & 1.1 \end{vmatrix}} = \frac{17.6 - 12.8}{1.32 - 1.0} = \frac{4.8}{0.32} = \textbf{15 A}$$

$$I_B = \frac{\begin{vmatrix} 1.2 & 16.0 \\ 1.0 & 12.8 \end{vmatrix}}{\begin{vmatrix} 1.2 & 1.0 \\ 1.0 & 1.1 \end{vmatrix}} = \frac{15.36 - 16.0}{0.32} = \frac{-0.64}{0.32} = \textbf{-2 A}$$

$$I_L = I_G + I_B = 15 - 2 = \textbf{13 A}$$

SOLUTION 2

The tracing loops we selected for Solution 1 are the normal choice from a circuit-operation point of view. However, other choices all lead to the same final solution. To illustrate the loop technique further, we can return to the circuit layout given in the original circuit diagram of Figure 8-11(a) and reproduced in Figure 8-13.

FIGURE 8-13
Circuit Diagram for Example 8-4, Second Solution.

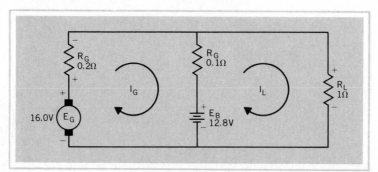

We select one tracing loop I_G from the $+$ terminal of the generator through the internal resistance of the generator and the battery and through the battery back to the $-$ terminal of the generator. The I_L tracing loop follows the same path as the I_B loop in Solution 1. But I_L is *not the same current* as I_B. In this case, the total load current is I_L, since this is the only tracing loop through R_L. The battery current in this solution is the discharge current I_L minus the charge current I_G.

Note that the voltage drop across R_B will be $R_B \times$ the *difference* of the two loop currents. In writing loop equations, the tracing loop we use is always the *positive* current direction; any current we encounter going *against* the tracing direction is negative. Hence, Kirchhoff's voltage-law equation for the I_L loop becomes

$$0.1(I_L - I_G) + 1I_L = 12.8 \text{ V}$$

from which
$$-0.1I_G + 1.1I_L = 12.8 \tag{1}$$

In writing the equation for the I_G loop, we note that I_G enters the $+$ terminal of the battery, just as it enters the $+$ terminal for voltage drops across resistors. Thus, E_B can appear as a positive term on the left-hand side of the equation. If we wish, we can think of E_B as a voltage source *bucking* the generator voltage and put E_B on the right-hand side of the equation as a *negative* quantity. The loop equation for the I_G loop becomes

$$0.2I_G + 0.1(I_G - I_L) + 12.8 = 16.0 \text{ V}$$

from which
$$0.3I_G - 0.1I_L = 3.2 \tag{2}$$

Once the loop equations are written, the rest of the solution is merely algebraic solution of simultaneous equations

$$-0.1I_G + 1.1I_L = 12.8 \tag{1}$$

$$0.3I_G - 0.1I_L = 3.2 \tag{2}$$

Generator current, I_G

$$I_G = \frac{\begin{vmatrix} 12.8 & 1.1 \\ 3.2 & -0.1 \end{vmatrix}}{\begin{vmatrix} -0.1 & 1.1 \\ 0.3 & -0.1 \end{vmatrix}} = \frac{-1.28 - 3.52}{+0.01 - 0.33} = \frac{-4.80}{-0.32} = \mathbf{15 \text{ A}}$$

Load current, I_L

$$I_L = \frac{\begin{vmatrix} -0.1 & 12.8 \\ 0.3 & 3.2 \end{vmatrix}}{D} = \frac{-0.32 - 3.84}{-0.32} = \frac{-4.16}{-0.32} = \mathbf{13 \text{ A}}$$

Battery current $= I_L - I_G = 13 - 15 = \mathbf{-2 \text{ A}}$

Again we find that the battery is charging, rather than discharging.

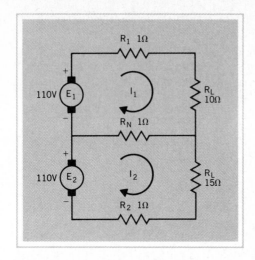

FIGURE 8-14
Three-Wire Distribution System.

For a second example, we can consider the Edison three-wire distribution system shown in Figure 8-14. Originally designed to reduce I^2R power loss in transmission lines, it provides us with a good example for our network analysis techniques.[†]

EXAMPLE 8-5
(a) Calculate the total power loss in the three 1-Ω conductors in the three-wire distribution system shown in Figure 8-14.
(b) Calculate the total power loss if the two loads were fed in parallel from a single 110-V source through a pair of 1-Ω conductors.

SOLUTION
(a) The voltage drop across the resistance of the neutral conductor is $1\ \Omega \times$ the *difference* of the two loop currents. As in the second solution for Example 8-4, tracing loop direction is the *positive* quantity and any current going *against* the tracing direction has a *negative* sign. Therefore, the Kirchhoff's voltage-law equation for the I_1 loop becomes

$$1I_1 + 10I_1 + 1(I_1 - I_2) = E_1$$

or
$$12I_1 - I_2 = 110 \tag{1}$$

[†]Because residential loads are many miles from generating stations, there is a costly loss of energy in the interconnecting transmission lines. We can reduce this loss if we can reduce the current required to feed a given load. Since $P = VI$, one method is to increase system voltage. For example, a 120-volt 1200-watt toaster draws a current of 10 amperes, but a 240-volt 1200-watt toaster requires a current of only five amperes. For safety reasons, it is not desirable to raise the operating voltage of household appliances too much. The three-wire distribution system reduces transmission line losses by arranging to have loop currents at least partially *cancel* in the *neutral* lead. If we arrange for equal values of R_L in Figure 8-14, the load is *balanced* and the neutral current is zero. The *unbalanced* load shown does increase power loss in the transmission lines, but not as much as operating both loads in parallel from a two-wire system with the same potential difference between lines.

For the I_2 loop,

$$1(I_2 - I_1) + 15I_2 + 1I_2 = E_2$$

or
$$-I_1 + 17I_2 = 110 \tag{2}$$

$$I_1 = \frac{\begin{vmatrix} 110 & -1 \\ 110 & 17 \end{vmatrix}}{\begin{vmatrix} 12 & -1 \\ -1 & 17 \end{vmatrix}} = \frac{1870 - (-110)}{204 - (+1)} = \frac{1980}{203} = 9.75 \text{ A}$$

$$I_2 = \frac{\begin{vmatrix} 12 & 110 \\ -1 & 110 \end{vmatrix}}{D} = \frac{1320 - (-110)}{203} = \frac{1430}{203} = 7.04 \text{ A}$$

Power in line 1, $\quad P_1 = I_1^2 R_1 = 9.75^2 \times 1 = 95$ W

Power in line 2, $\quad P_2 = I_2^2 R_2 = 7.04^2 \times 1 = 50$ W

Neutral current, $\quad I_N = I_1 - I_2 = 9.75 - 7.04 = 2.71$ A

Neutral power, $\quad P_N = I_N^2 R_N = 2.71^2 \times 1 = 7$ W

Total power loss in conductors,

$$P_T = P_1 + P_2 + P_N = 95 + 50 + 7 = \mathbf{152 \ W}$$

(b) If the two loads are connected in parallel, the equivalent load resistance is

$$R_{eq} = \frac{10 \times 15}{10 + 15} = 6 \ \Omega$$

Adding the resistance of two conductors, $R_T = 8 \ \Omega$. Hence, the circuit current becomes

$$I = \frac{E}{R_T} = \frac{110 \text{ V}}{8 \ \Omega} = 13.75 \text{ A}$$

Power in *each* line $= I^2 R = 13.75^2 \times 1 = 189$ W

Therefore, total power loss in conductors is

$$P_T = 2 \times 189 \text{ W} = \mathbf{378 \ W}$$

For our third example, we shall tackle the transistor coupling circuit shown in Figure 8-11(b). Even though the original circuit diagram does not show the actual voltage sources, we draw them in our working diagram of Figure 8-15 to enable us to draw complete tracing loops.

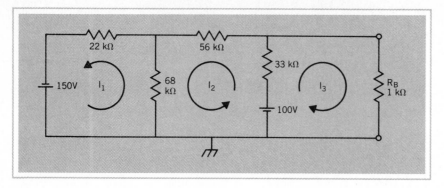

FIGURE 8-15
Circuit Diagram for Example 8-6.

EXAMPLE 8-6

If R_B in the circuit shown in Figures 8-11(b) and 8-15 is 1 kΩ, determine the magnitude and polarity of the voltage drop across R_B.

SOLUTION

We require three tracing loops to include all components in this circuit. Several combinations of loops are possible. Those selected in Figure 8-15 are the "natural" choice. Writing the three Kirchhoff's voltage-law equations from our loop procedure is quite straightforward. Since volts = milliamperes × kilohms, we can write the equation for the I_1 loop as

$$68(I_1 - I_2) + 22I_1 = 150 \text{ V} \tag{1}$$

(where I_1 and I_2 are in milliamperes). From the I_2 loop,

$$33(I_2 + I_3) + 56I_2 + 68(I_2 - I_1) = 100 \text{ V} \tag{2}$$

and from the I_3 loop,

$$33(I_3 + I_2) + 1I_3 = 100 \text{ V} \tag{3}$$

The tedious part of the solution is the numerical computation in solving third order simultaneous equations. However, we need only solve for I_3 to provide the requested information in this example. Collecting terms,

$$90I_1 - 68I_2 \qquad\quad = 150 \tag{1}$$
$$-68I_1 + 157I_2 + 33I_3 = 100 \tag{2}$$
$$33I_2 + 34I_3 = 100 \tag{3}$$

$$I_3 = \frac{\begin{vmatrix} 90 & -68 & 150 \\ -68 & 157 & 100 \\ 0 & 33 & 100 \end{vmatrix}}{\begin{vmatrix} 90 & -68 & 0 \\ -68 & 157 & 33 \\ 0 & 33 & 34 \end{vmatrix}}$$

$$= \frac{1\,413\,000 + 0 - 336\,600 - 0 - 297\,000 - 462\,400}{480\,420 + 0 + 0 - 0 - 98\,010 - 157\,216}$$

$$= \frac{317\,000}{225\,194} = 1.4 \text{ mA}$$

Hence, the voltage drop across R_B is

$$V_B = 1.4 \text{ mA} \times 1 \text{ k}\Omega = \mathbf{1.4 \text{ V}}$$

Since I_3 turned out to be a positive quantity, the top end of R_B is positive with respect to chassis.

8-8 MESH EQUATIONS

Now that we have a "feel" for writing equations which satisfy Kirchhoff's voltage law, we can turn our attention to a format procedure which is designed to simplify and speed up the task of writing the set of simultaneous equations we need to solve various resistance networks. In writing **mesh equations,**[†] the emphasis is on following a set of rules, rather than on the mental inspection of the network which we used with the loop procedure.

The price we pay for a short cut format is that mesh equations cannot handle some of the networks that we can solve with the loop procedure. We define a **mesh** as a closed loop within a network such that there are no other circuit elements within the loop. Therefore, the circuit diagram cannot contain any conductors crossing without there being an electrical junction at the point of crossing. In other words, the network diagram must be strictly two-dimensional or **planar.** Second, the mesh format requires all sources to be *voltage* sources. If there are any *current* sources, we must convert them to equivalent voltage sources before we use the mesh format. Third, we cannot arrange our current loops so that we only have to solve a single equation, as in Example 8-3.

After writing equations for voltage drops around the current loops in the loop

---

[†]Mesh equations were developed over a century ago by James Clerk Maxwell, the British physicist who worked out the electromagnetic theory of light and predicted the existence of radio waves before they were discovered by Heinrich Hertz in 1888.

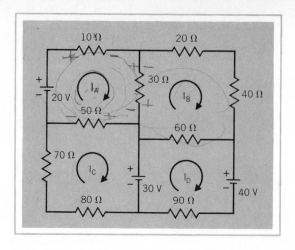

FIGURE 8-16
Mesh Network for Example 8-7.

procedure, we then rewrite the Kirchhoff's voltage-law equations to collect the current terms. In the mesh format, we go directly to this second step. To illustrate the mesh format, we can refer to the network shown in Figure 8-16. We have laid out our circuit diagram so that the network is clearly planar and the four meshes have no internal electrical components. The next step is to draw a current loop for each mesh. All mesh currents must be in the same direction—either clockwise or counterclockwise. The more popular choice is clockwise. The result is that, for any particular mesh current (e.g., I_A in Figure 8-16), all other currents through common circuit elements (I_B and I_C for mesh A) are in the *opposite* direction. This provides a standard format for every mesh equation. The mesh equation for mesh A has the form

$$R_T I_A - R_{AB} I_B - R_{AC} I_C = \Sigma E \qquad (8-7)$$

Since I_A passes through all resistors around mesh A, we insert the total resistance around mesh A for R_T. R_{AB} is the value of the resistor which is common to mesh A and mesh B—similarly for R_{AC}. ΣE is the algebraic sum of the voltage sources around the mesh. Where I_A passes from $-$ to $+$ *inside* a voltage source, this is its normal positive direction and that E is a positive quantity. Similarly, if the direction of I_A is from $+$ to $-$ inside a voltage source, that E would have a negative sign.

EXAMPLE 8-7

Find the magnitudes and polarities of the voltage drops across the 30-Ω, 50-Ω, and 60-Ω resistors in the network shown in Figure 8-16.

SOLUTION

Using the mesh format Equation (8-7) to write a Kirchhoff's voltage-law equation for each of the four meshes in Figure 8-16,

$$(10 + 30 + 50)I_A - 30I_B - 50I_C = 20 \text{ V}$$

$$(30 + 20 + 40 + 60)I_B - 30I_A - 60I_D = 0 \text{ V}$$

$$(80 + 70 + 50)I_C - 50I_A = -30 \text{ V}$$

$$(90 + 60)I_D - 60I_B = (30 - 40) \text{ V}$$

Collecting the terms,

$$90I_A - 30I_B - 50I_C = 20 \qquad (1)$$
$$-30I_A + 150I_B - 60I_D = 0 \qquad (2)$$
$$-50I_A + 200I_C = -30 \qquad (3)$$
$$-60I_B + 150I_D = -10 \qquad (4)$$

Unfortunately, the rules for solving fourth-order determinants are not as simple as those for second- and third-order determinants. We can get around this problem by eliminating one of the unknowns. From Equation (4),

$$I_D = \frac{60I_B - 10}{150}$$

Substituting in Equation (2),

$$-30I_A + 150I_B - (24I_B - 4) = 0$$

This leaves us with three simultaneous equations,

$$90I_A - 30I_B - 50I_C = 20 \qquad (1)$$
$$-30I_A + 126I_B = -4 \qquad (5)$$
$$-50I_A + 200I_C = -30 \qquad (3)$$

$$I_A = \frac{\begin{vmatrix} 20 & -30 & -50 \\ -4 & 126 & 0 \\ -30 & 0 & 200 \end{vmatrix}}{\begin{vmatrix} 90 & -30 & -50 \\ -30 & 126 & 0 \\ -50 & 0 & 200 \end{vmatrix}} = \frac{291\,000}{1\,773\,000} = 164.13 \text{ mA}$$

$$I_B = \frac{\begin{vmatrix} 90 & 20 & -50 \\ -30 & -4 & 0 \\ -50 & -30 & 200 \end{vmatrix}}{D} = \frac{13\,000}{1\,773\,000} = 7.33 \text{ mA}$$

$$I_C = \frac{\begin{vmatrix} 90 & -30 & 20 \\ -30 & 126 & -4 \\ -50 & 0 & -30 \end{vmatrix}}{D} = \frac{-193\,200}{1\,773\,000} = -108.97 \text{ mA}$$

$$I_D = \frac{60 \times 7.33 \times 10^{-3} - 10}{150} = -63.73 \text{ mA}$$

$$I_{30} = I_A - I_B = 164.13 \text{ mA} - 7.33 \text{ mA} = 156.8 \text{ mA}$$

$$\therefore V_{30} = 30 \ \Omega \times 156.8 \text{ mA} = \mathbf{4.7 \ V} \quad \text{positive at the top}$$

$$I_{50} = I_A - I_C = 164.13 \text{ mA} + 108.97 \text{ mA} = 273.1 \text{ mA}$$

$$\therefore V_{50} = 50 \ \Omega \times 273.1 \text{ mA} = \mathbf{13.7 \ V} \quad \text{positive on the right}$$

$$I_{60} = I_B - I_D = 7.33 \text{ mA} + 63.73 \text{ mA} = 71.06 \text{ mA}$$

$$\therefore V_{60} = 60 \ \Omega \times 71.06 \text{ mA} = \mathbf{4.3 \ V} \quad \text{positive on the right}$$

As an exercise in writing mesh equations, we can now repeat Examples 8-2 to 8-6 using the mesh format. Obviously, we should end up with the same numerical results.

EXAMPLE 8-2A

In the circuit diagram of Figure 8-7, there are two meshes. Mesh I_1 is the same as loop I_1. The second mesh is the closed loop containing R_3, R_1, R_2, and R_4 and no voltage source. The two mesh equations are

$$(10 + 20 + 30)I_1 - (20 + 30)I_2 = 80 \text{ V}$$

$$(30 + 20 + 15 + 10)I_2 - (20 + 30)I_1 = 0 \text{ V}$$

$$I_{R_3} = I_1 - I_2 \quad \text{and} \quad I_{R_4} = I_2$$

The remainder of the solution is the same as in Example 8-2.

EXAMPLE 8-3A

In the circuit diagram of Figure 8-9, there are three meshes. Mesh I_1 is the same as loop I_1. The second mesh is the closed loop containing R_1, R_2, and R_M. The third mesh contains R_3, R_M, and R_4. After collecting terms, the three mesh equations become

$$60I_1 - 20I_2 - 30I_3 = 80 \tag{1}$$

$$-20I_1 + 85I_2 - 50I_3 = 0 \tag{2}$$

$$-30I_1 - 50I_2 + 90I_3 = 0 \tag{3}$$

We need to solve for both I_2 and I_3, since $I_M = I_2 - I_3$.

EXAMPLE 8-4A

In the circuit diagram of Figure 8-12, we can write mesh-current equations by simply reversing the direction of I_B. The second solution for Example 8-4 using the circuit configuration of Figure 8-13 already satisfies the rules for the mesh format and will yield the same simultaneous equations when the terms are collected.

EXAMPLE 8-5A

The circuit layout of Figure 8-14 also satisfies the mesh format rules.

EXAMPLE 8-6A

Before we write the mesh equations for the circuit shown in Figure 8-15, all mesh currents must be in the same direction. Reversing I_3 so that all mesh currents are counterclockwise,

$$(22 + 68) \text{ k}\Omega \times I_1 - 68 \text{ k}\Omega \times I_2 = 150 \text{ V}$$

$$(33 + 56 + 68) \text{ k}\Omega \times I_2 - 68 \text{ k}\Omega \times I_1 - 33 \text{ k}\Omega \times I_3 = 100 \text{ V}$$

$$(1 + 33) \text{ k}\Omega \times I_3 - 33 \text{ k}\Omega \times I_2 = -100 \text{ V}$$

Collecting terms,

$$90 \text{ k}\Omega \times I_1 - 68 \text{ k}\Omega \times I_2 \qquad\qquad = \quad 150 \qquad (1)$$

$$-68 \text{ k}\Omega \times I_1 + 157 \text{ k}\Omega \times I_2 - 33 \text{ k}\Omega \times I_3 = \quad 100 \qquad (2)$$

$$-33 \text{ k}\Omega \times I_2 + 34 \text{ k}\Omega \times I_3 = -100 \qquad (3)$$

To solve for V_B, we need to solve only for I_3, which turns out to be -1.4 mA, since we reversed the direction of I_3.

8-9 KIRCHHOFF'S CURRENT-LAW EQUATIONS

The loop-current and mesh-equation methods of solving resistance networks are based on writing a sufficient number of Kirchhoff's voltage-law equations to include all circuit elements. The number of simultaneous equations required depends on the number of loops or meshes in the network. To solve the voltage-law equations, we substitute an *IR* drop for each voltage drop and solve for the various loop or mesh currents.

Kirchhoff's current law allows us to write a different set of equations for currents entering and leaving junction points or **nodes** in an electric circuit. Before we can write such current-law equations, we must be able to identify the nodes in a given network. In the network shown in Figure 8-17, we do not consider the junction of E

FIGURE 8-17
Identifying Nodes in a Network.

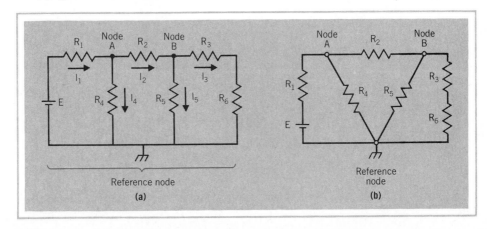

and R_1, and the junction point of R_3 and R_6 as nodes, since the current-law equation $I_1 = I_1$ is redundant. Also, as indicated by the alternate circuit layout of Figure 8-17(b), we consider all connections to the common grounded conductor in Figure 8-17(a) as a single node. Hence, there is a total of three nodes in the network of Figure 8-17.

Some texts prefer to state Kirchhoff's current law as a mathematical statement that the *algebraic* sum of currents at a node must equal zero. The difficulty in translating this statement into an equation is that we have to remember which currents are positive and which are negative. In setting up a nodal procedure, we prefer to state Kirchhoff's current law as

At any node in an electric circuit, the sum of the currents leaving the node equals the sum of the currents entering the node.

For the three nodes in the network of Figure 8-17, the above definition leads to the following equations.

$$\text{Node A:} \quad I_4 + I_2 = I_1 \tag{1}$$

$$\text{Node B:} \quad I_5 + I_3 = I_2 \tag{2}$$

$$\text{Reference node:} \quad I_1 = I_4 + I_5 + I_3 \tag{3}$$

Note that if we add Equations (1) and (2), we obtain Equation (3). Hence, the Kirchhoff's current-law equation for the third node is redundant in setting up the simultaneous equations. The number of Kirchhoff's current-law equations we need to solve any network is always one less than the total number of nodes. We label the redundant node as the **reference node.** We usually select as reference node one that is common to as many branch currents as possible. The *grounded* part of a network is a good choice for reference node, as shown in Figure 8-17.

The two equations for nodes A and B have five unknown currents. However, there are only *two* node voltages V_A and V_B with respect to the reference node in the network shown in Figure 8-17. To reduce the unknowns in the current-law equations to two, we substitute V/R ratios $(I = V/R)$ for each current in such a manner that we are solving for V_A and V_B. Once we know the node voltages, we can find any other unknown in the network by Ohm's law. Hence, in the network shown in Figure 8-17, $V_4 = V_A$ and $V_5 = V_B$.

For node A:

$$I_4 = \frac{V_A}{R_4}$$

$$I_2 = \frac{V_2}{R_2} = \frac{V_A - V_B}{R_2}$$

and
$$I_1 = \frac{V_1}{R_1} = \frac{E - V_A}{R_1}$$

Substituting these expressions in Equation (1) gives

$$\frac{V_A}{R_4} + \frac{V_A - V_B}{R_2} = \frac{E - V_A}{R_1}$$ (8-8)

For node B:

$$I_5 = \frac{V_B}{R_5}$$

$$I_3 = \frac{V_B}{R_3 + R_6}$$

and

$$I_2 = \frac{V_A - V_B}{R_2}$$

$$\therefore \frac{V_B}{R_5} + \frac{V_B}{R_3 + R_6} = \frac{V_A - V_B}{R_2}$$ (8-9)

Given numerical values for E and all resistors, the only unknowns are V_A and V_B.

EXAMPLE 8-4B
An automobile generator with an internal resistance of 0.2 Ω develops an open-circuit voltage of 16.0 V. The storage battery has an internal resistance of 0.1 Ω and an open-circuit voltage of 12.8 V. Both sources are connected in parallel to a 1-Ω load. Determine the load current.

SOLUTION
Although there are two nodes in Figure 8-18, except for a reversal of all signs, the same Kirchhoff's current-law equation applies to both nodes. Hence, we start the nodal solution by writing *one* Kirchhoff's current-law equation,

$$I_L = I_G + I_B$$

FIGURE 8-18
Circuit Diagram for Example 8-4B.

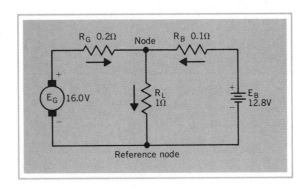

Note that we again assume that the battery is discharging. In Figure 8-18,

$$I_L = \frac{V_L}{R_L}, \quad I_G = \frac{E_G - V_L}{R_G} \quad \text{and} \quad I_B = \frac{E_B - V_L}{R_B}$$

Thus, we substitute V/R ratios for the currents in the Kirchhoff's current-law equation to obtain

$$\frac{V_L}{1} = \frac{16.0 - V_L}{0.2} + \frac{12.8 - V_L}{0.1}$$

Clearing the fractions by multiplying the equation by 0.1,

$$0.1\,V_L = 8.0 - 0.5\,V_L + 12.8 - V_L$$

from which

$$V_L = \frac{20.9}{1.6} = 13 \text{ V}$$

and

$$I_L = \frac{V_L}{R_L} = \frac{13 \text{ V}}{1 \ \Omega} = \textbf{13 A}$$

Since we only required one nodal equation in the solution of Example 8-4 B, we avoided the simultaneous equations of Example 8-4, which required two tracing loops. However, loop (and mesh) equations are free from algebraic fractions, since we substitute a *product*—the *IR* drop—for each voltage drop. In the nodal equation of Example 8-4 B, we substitute a V/R *ratio* for each current term, leaving us with an equation in fractional form. We can avoid the fraction produced by the expression $I = V/R$ if we switch our point of view from resistance to conductance, which is the reciprocal of resistance (Section 6-10). Now the equivalent expression becomes $I = VG$.

8-10 NODAL ANALYSIS

Nodal analysis is a circuit-analysis *format* which combines Kirchhoff's current-law equations with the source conversions we developed in Section 8-4. Converting all voltage sources to equivalent constant-current sources allows us to standardize the way we write the Kirchhoff's current-law equations. For the nodal analysis format, we consider source currents to flow *into* a node. If the arrows in a circuit diagram show that a source current actually flows out of a particular node, then that current is a *negative* quantity as far as that node is concerned. Similarly, we consider all resistor currents to flow *out of* a node. We are thus restating Kirchhoff's current law in the form

At any independent node, the algebraic sum of the resistor currents leaving the node equals the algebraic sum of the source currents entering the node.

This definition leads to Kirchhoff's current-law equations having the format

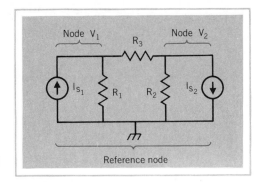

FIGURE 8-19
Circuit Diagram for Writing Nodal Equations.

$$I_{R_1} + I_{R_2} + \cdots = I_{S_1} + I_{S_2} + \cdots \tag{8-10}$$

The next step is to replace resistor currents on the left-hand side of Equation (8-10) with VG products. The voltage term depends on the two nodes between which the particular resistor is connected. When we write the nodal equation for node V_1 in the circuit of Figure 8-19, the current through R_1 becomes V_1G_1. But the voltage across R_3 is the algebraic difference between node voltages V_1 and V_2. To assure the correct sign for the current through R_3, we always subtract the adjacent node voltage (V_2) from the voltage "at" the node for which we are writing the equation (V_1). Hence the Kirchhoff's current-law equation for node V_1 becomes

$$V_1G_1 + (V_1 - V_2)G_3 = I_{S_1}$$

Collecting the voltage terms gives

$$(G_1 + G_3)V_1 - G_3 V_2 = I_{S_1} \tag{8-11}$$

Equation (8-11) is the standard format for writing a nodal equation for every independent node in the network. The conductance and current terms become the coefficients for determinant matrices.

We can write Equation (8-11) directly by noting that the positive voltage term on the left-hand side of the equation is the unknown voltage for the node in question multiplied by the sum of the conductances connected to that node. From this positive term, we must subtract a term for the node voltage at every adjacent node to which a resistor is connected from the node in question. This term consists of the adjacent node voltage multiplied by the conductance between the two nodes. Following these rules, we can write the nodal equation for node V_2 as[†]

$$(G_2 + G_3)V_2 - G_3V_1 = -I_{S_2}$$

Nodal analysis is particularly useful in analyzing networks where a common portion of the network is fed from several sources in parallel. Our automobile generator example is just such a network.

[†]Note the **duality** between the nodal format equation and Equation (8-7) for mesh format equations.

EXAMPLE 8-4C

An automobile generator with an internal resistance of 0.2Ω develops an open-circuit voltage of 16.0 V. The storage battery has an internal resistance of 0.1Ω and an open-circuit voltage of 12.8 V. Both sources are connected in parallel to a 1-Ω load. Determine the load current.

SOLUTION

Step I. Now that we are familiar with source conversion, we can skip the original circuit layout of Figure 8-18 and draw the equivalent circuit of Figure 8-20 with all sources converted to constant-current sources. The short-circuit current for the battery is

$$I_B = \frac{E_B}{R_B} = \frac{12.8 \text{ V}}{0.1 \text{ }\Omega} = 128 \text{ A}$$

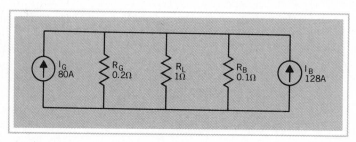

FIGURE 8-20
Circuit Diagram for Example 8-4C.

Similarly, the short-circuit current of the generator is

$$I_G = \frac{E_G}{R_G} = \frac{16.0 \text{ V}}{0.2 \text{ }\Omega} = 80 \text{ A}$$

Hence, we can fill in the required data on the circuit diagram of Figure 8-20 with nothing more than a bit of mental arithmetic. Note that the generator and battery currents in this *equivalent* circuit are not the same as the real currents in the original circuit.

Step II. To determine the conductance of each branch, we simply enter the value of the resistor into our calculator and press the reciprocal key. Hence,

$$G_G = \frac{1}{R_G} = 0.2 \text{ }\Omega\, \boxed{1/x} = 5 \text{ S} \qquad G_B = \frac{1}{R_B} = 0.1 \text{ }\Omega\, \boxed{1/x} = 10 \text{ S}$$

and

$$G_L = \frac{1}{R_L} = 1 \text{ S}$$

Step III. Although it might appear at first glance that there are six junction

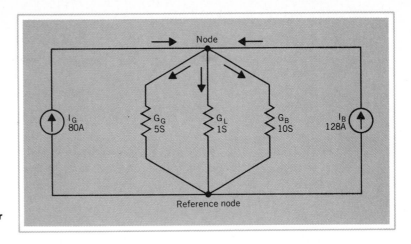

FIGURE 8-21
Equivalent Circuit for Example 8-4C.

points in Figure 8-20, there are only *two* significant nodes, because the *same* voltage appears across all parallel branches. For our nodal analysis, we can redraw the circuit in the form shown in Figure 8-21. If we label the lower node as the reference node, there is only one other independent node. Consequently, we require only one Kirchhoff's current law equation for the circuit of Figure 8-21. Using Equation (8-10), at the single independent node,

$$I_5 + I_1 + I_{10} = I_G + I_B$$

Or, we can go directly to the format of Equation (8-11) and substitute coefficients,

$$(G_5 + G_1 + G_{10})V = I_G + I_B$$

from which

$$V = \frac{80 + 128}{5 + 1 + 10} = \frac{208 \text{ A}}{16 \text{ S}} = 13 \text{ V}$$

and
$$I_L = VG_L = 13 \text{ V} \times 1 \text{ S} = \mathbf{13 \text{ A}}$$

Nodal analysis allowed us to solve Example 8-4C with not much more than mental arithmetic. This good fortune was due to a single independent node in the network. Hence, we did not need the simultaneous equations and determinants of the loop and mesh solutions. Unfortunately, many networks have more than one independent node. The transistor coupling circuit of Figures 8-11(b) and 8-15, and again in Figure 8-22, is such an example.

FIGURE 8-22
Circuit Diagram for Example 8-6B.

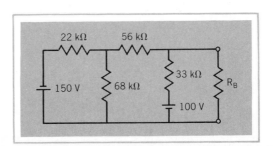

191

EXAMPLE 8-6B

If R_B in the circuit shown in Figure 8-22 is 1 kΩ, determine the magnitude and polarity of the voltage drop across R_B.

SOLUTION

Step I. In the circuit diagram of Figure 8-22, we can consider the 22-kΩ resistor to be the internal resistance of a 150-V constant-voltage source connected across the 68-kΩ resistor. Similarly, we can think of the 33-kΩ resistor as the internal resistance of a 100-V constant-voltage source connected across R_B. Therefore, when we convert the voltage sources to constant-current sources, we obtain the circuit of Figure 8-23 where

$$I_1 = \frac{150 \text{ V}}{22 \text{ k}\Omega} = 6.8 \text{ mA} \quad \text{and} \quad I_2 = \frac{100 \text{ V}}{33 \text{ k}\Omega} = 3.03 \text{ mA}$$

We must take care in marking the directions of I_1 and I_2 to be consistent with the polarities of the voltage sources they replace.

Step II. Converting each of the five resistances into its equivalent conductance, we obtain the values shown (in microsiemens) in Figure 8-23.

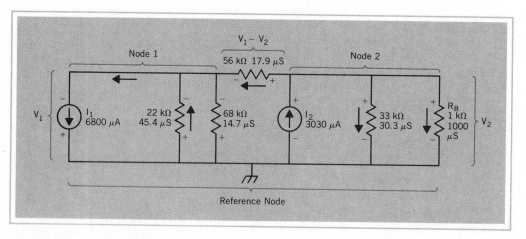

FIGURE 8-23
Equivalent Circuit for Example 8-6B.

Step III. The whole bottom conductor (chassis) in the circuit of Figure 8-23 becomes the reference node. This leaves two independent nodes—node 1 and node 2 in Figure 8-23. Each of these nodes has one source current "entering" the node and three resistor currents "leaving" the node.[†] To solve this network, we require

---

[†]At node 1, all currents turn out to be negative quantities since the source current actually *leaves* the node and resistor currents *enter* the node. If we solve for V_1, node 1 turns out to be negative with respect to chassis.

a nodal equation for each of the two independent nodes. Using the format of Equation (8-11), we note that the source current I_1 will be a negative quantity, since it *leaves*, rather than enters, node 1.

For node 1,

$$(45.4 + 14.7 + 17.9)V_1 - 17.9V_2 = -6800^\dagger$$

For node 2,

$$(17.9 + 30.3 + 1000)V_2 - 17.9V_1 = 3030^\dagger$$

Collecting terms in matrix form,

$$78V_1 - 17.9V_2 = -6800 \qquad (1)$$

$$-17.9V_1 + 1048.2V_2 = 3030 \qquad (2)$$

Step IV. To find the voltage drop across R_B, we need solve only for V_2.

$$V_2 = \frac{\begin{vmatrix} 78 & -6800 \\ -17.9 & 3030 \end{vmatrix}}{\begin{vmatrix} 78 & -17.9 \\ -17.9 & 1048 \end{vmatrix}} = \frac{237\,340 - 121\,720}{81\,744 - 320} = \frac{115\,620}{81\,424} = \mathbf{1.42\ V}$$

Hence, the voltage drop across R_B is 1.4 V and since it turned out to be a positive quantity, node 2 is positive with respect to chassis.

Although we had to solve simultaneous equations in Example 8-6B, they were only second-order equations, rather than the third-order simultaneous equations we encountered in the loop and mesh solutions.

In the Wheatstone bridge circuit of Figure 8-24(a), there are four nodes. Labeling one as a reference node, we would still need three nodal equations to solve the Wheatstone bridge by nodal analysis. Since this is no improvement on the three simultaneous equations required in the loop or mesh procedures, we would probably not select nodal analysis in this case. If we use the loop or mesh format to solve the circuit in Figure 8-24(b), we require *four* simultaneous equations. However, as shown in Figure 8-24(b), we would require only *one* nodal equation to solve the whole network by the nodal analysis procedure just outlined.

---

†Since microamperes = volts × microsiemens, we can avoid repetitive use of the exponent key of our calculator by expressing current in microamperes and conductance in microsiemens in this equation. We then enter the digits without the customary exponent for the prefix *micro*.

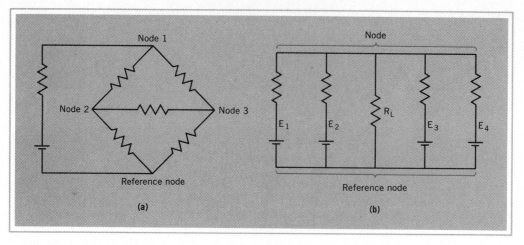

FIGURE 8-24
(a) Wheatstone Bridge; (b) Four Parallel Sources Supplying a Common Load.

8-11 THE SUPERPOSITION THEOREM

A characteristic of the Kirchhoff's-laws methods for solving resistance networks is that we have to solve a set of simultaneous linear equations. As long as we can solve networks with only two tracing loops or nodes, the numerical computation is not particularly onerous. But when we need three or more simultaneous equations, we start looking for alternative solutions. We can avoid simultaneous equations in multi-source networks by applying a fundamental principle of linear networks known as the **superposition theorem.** This theorem is useful only for networks containing more than one source and for determining the current through or voltage drop across *one* branch of these networks. However, in many cases, once we know the current through one branch of a network, we can solve the remainder of the network quite readily. We shall not worry about the proof of the superposition theorem, which may be stated thus:

> *The current that flows in any branch of a network of resistors resulting from the simultaneous application of a number of voltage sources distributed in any manner throughout the network is the algebraic sum of the component currents in that branch that would be caused by each source acting independently in turn while the others are replaced in the network by their respective internal resistances.*

In both the loop-current and nodal methods of solving resistance networks, the currents appearing in the simultaneous equations are not necessarily *actual* currents that we could measure at various points in the network. To obtain *actual* branch currents, we have to add the appropriate loop or mesh currents algebraically. Similarly, currents in a network with some of the sources inactive are not real currents.

Component currents calculated by the superposition theorem are only a means of calculating the actual current in a certain branch and do not represent currents that we can measure in a network. In using the superposition theorem, we must take care to use the component currents only for the purpose stated in the theorem.

EXAMPLE 8-4D
An automobile generator with an internal resistance of 0.2 Ω develops an open-circuit voltage of 16.0 V. The storage battery has an internal resistance of 0.1 Ω and an open-circuit voltage of 12.8 V. Both sources are connected in parallel to a 1.0-Ω load. Determine the load current.

SOLUTION
Because there are *two* sources, there will be *two* component currents through the load resistance. For the purpose of solving this example by the superposition theorem, we can redraw the circuit of Figure 8-11(a) and 8-12 as the two equivalent circuits of Figure 8-25.

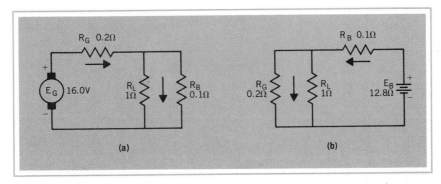

FIGURE 8-25
Equivalent Circuit for Example 8-4D.

Solving the first equivalent circuit of Figure 8-25(a) by the series-parallel resistance technique,

$$R_T = R_G + \frac{R_L \times R_B}{R_L + R_B} = 0.2 + \frac{1 \times 0.1}{1 + 0.1} = 0.29 \ \Omega$$

$$I_T = \frac{E_G}{R_T} = \frac{16.0 \ V}{0.29 \ \Omega} = 55 \ A$$

From the current divider principle, the first component of the load current is

$$I_{L_1} = I_T \times \frac{R_B}{R_L + R_B} = 55 \times \frac{0.1}{1.1} = 5 \ A$$

Similarly, for the circuit of Figure 8-25(b),

$$R_T = R_B + \frac{R_G \times R_L}{R_G + R_L} = 0.1 + \frac{0.2 \times 1}{0.2 + 1} = 0.267 \ \Omega$$

$$I_T = \frac{E_B}{R_T} = \frac{12.8 \ V}{0.267 \ \Omega} = 48 \ A$$

The second component of the load current is

$$I_{L_2} = I_T \times \frac{R_G}{R_G + R_L} = 48 \times \frac{0.2}{1.2} = 8 \ A$$

Since the direction for the component load currents is the *same* in Figures 8-25(a) and (b), the *actual* load current is the *sum* of the component currents.

$$\therefore I_L = I_{L_1} + I_{L_2} = 5 + 8 = \mathbf{13 \ A}$$

EXAMPLE 8-6C

If R_B in the circuit shown in Figure 8-12(b) is 1 kΩ, determine the magnitude and polarity of the voltage drop across R_B.

SOLUTION

Again there are two sources, hence the superposition theorem will yield two component currents. In the equivalent circuit of Figure 8-26(a), the 100-V source has been replaced by its zero-ohm internal resistance so that only the 150-V source contributes to the current through R_B. Again we can solve the circuit as a series-parallel network. However, in this case, we solve for R_T in two steps. First we replace the 56-kΩ and 33-kΩ resistors and R_B with their equivalent resistance, R_{eq} in Figure 8-26(b). For the equivalent circuits of Figure 8-26(a) and (b),

$$R_{eq} = 56 \ k\Omega + \frac{33 \ k\Omega \times 1 \ k\Omega}{33 \ k\Omega + 1 \ k\Omega} = 57 \ k\Omega$$

$$R_T = 22 \ k\Omega + \frac{68 \ k\Omega \times 57 \ k\Omega}{68 \ k\Omega + 57 \ k\Omega} = 53 \ k\Omega$$

$$I_T = \frac{E}{R_T} = \frac{150 \ V}{53 \ k\Omega} = 2.83 \ mA$$

We must now use the current-divider principle twice to find the component current through R_B. First, in the equivalent circuit of Figure 8-26(b), the current through R_{eq} is

$$I_{eq} = 2.83 \ mA \times \frac{68 \ k\Omega}{68 \ k\Omega + 57 \ k\Omega} = 1.54 \ mA$$

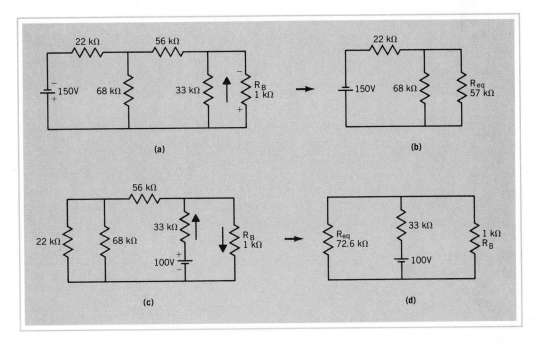

FIGURE 8-26
Equivalent Circuit for Example 8-6C.

In the circuit of Figure 8-26(a),

$$I_{B_1} = 1.54 \text{ mA} \times \frac{33 \text{ k}\Omega}{33 \text{ k}\Omega + 1 \text{ k}\Omega} = 1.49 \text{ mA}$$

Similarly, when the 100-V source operates alone, as in Figure 8-26(c), we start by replacing the 56-kΩ, 22-kΩ, and 68-kΩ resistors with an equivalent resistance. Solving for the second component current,

$$R_{eq} = 56 \text{ k}\Omega + \frac{22 \text{ k}\Omega \times 68 \text{ k}\Omega}{22 \text{ k}\Omega + 68 \text{ k}\Omega} = 72.6 \text{ k}\Omega$$

$$R_T = 33 \text{ k}\Omega + \frac{72.6 \text{ k}\Omega \times 1 \text{ k}\Omega}{72.6 \text{ k}\Omega + 1 \text{ k}\Omega} = 34 \text{ k}\Omega$$

$$I_T = \frac{E}{R_T} = \frac{100 \text{ V}}{34 \text{ k}\Omega} = 2.94 \text{ mA}$$

In Figure 8-26(d), we need only apply the current-divider principle once to find the second component of the current through R_B.

$$I_{B_2} = 2.94 \text{ mA} \times \frac{72.6 \text{ k}\Omega}{72.6 \text{ k}\Omega + 1 \text{ k}\Omega} = 2.90 \text{ mA}$$

We note from Figures 8-26(a) and (c) that the two component currents flow in *opposite* directions through R_B. Hence, the real current through R_B is

$$I_B = 2.90 - 1.49 = 1.41 \text{ mA}$$

and
$$V_B = 1.4 \text{ mA} \times 1 \text{ k}\Omega = \mathbf{1.4 \text{ V}}$$

Because the component current I_{B_2} is greater than I_{B_1}, the real current has the same direction through R_B as I_{B_2} in Figure 8-26(c). Thus, the top end of R_B is 1.4 V positive with respect to chassis.

Although we avoided simultaneous equations in Example 8-6C, we probably had to do just as much numerical computation to arrive at a solution as we did in Example 8-6, which we solved by loop equations.

PROBLEMS

NOTE For practice in circuit analysis techniques, solve problems with one method and check the solution with a different method.

8-1. As the current drawn from a dc power supply is increased from 600 mA to 800 mA, the terminal voltage decreases from 24 V to 23 V. Determine the equivalent constant-voltage source.

8-2. A voltage source has an open-circuit terminal voltage of 120 V and an internal resistance of 0.6 Ω. Determine the equivalent constant-current source.

8-3. Convert the voltage sources of Figure 8-27 into the equivalent constant-current sources.

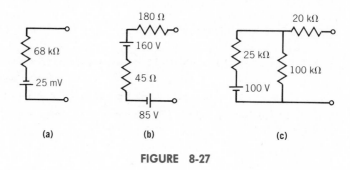

(a)   (b)   (c)

FIGURE 8-27

8-4. Replace the current sources shown in Figure 8-28 with equivalent constant-voltage sources.

FIGURE 8-28

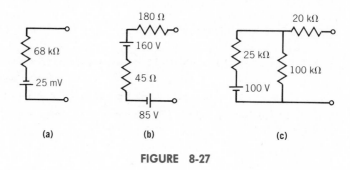

(a)   (b)   (c)

198

8-5. Determine the magnitude and polarity of the voltage drop across the 50-Ω resistor in the network shown in Figure 8-29(a).

8-6. Find the current drawn from each of the sources in the network shown in Figure 8-29(b).

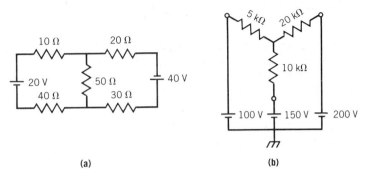

(a)   (b)

FIGURE 8-29

8-7. As the car slows down, the open-circuit terminal voltage of the generator in Example 8-4 drops to 14.0 V. What is the battery current?

8-8. (a) What must the open-circuit voltage of the generator be in Example 8-4 for the battery current to be zero?
     (b) What open-circuit voltage must the generator develop in Example 8-4 if the battery is to charge at a steady 10 A?

8-9. What will the neutral current be in the circuit of Example 8-5 if a second 15-Ω load is connected in parallel with the existing 15-Ω load?

8-10. Calculate the total power loss in the three 1-Ω conductors in the three-wire distribution system shown in Figure 8-14 if the polarity of source E_2 is reversed.

8-11. Two batteries are connected in parallel to feed a 12-Ω load. Battery A has an open-circuit voltage of 6.3 V and an internal resistance of 1.5 Ω. Battery B has an open-circuit voltage of 6.0 V and an internal resistance of 2.1 Ω. Find the current drain from each battery.

8-12. If the 12-Ω load in Problem 8-11 is replaced by a battery charger with an open-circuit voltage of 7.2 V and an internal resistance of 0.2 Ω, find the charging current of each battery.

8-13. What is the current through each voltage source in the circuit of Figure 8-30?

8-14. What is the magnitude and direction of the current through the 2.2-kΩ resistor in the circuit of Figure 8-30?

FIGURE 8-30

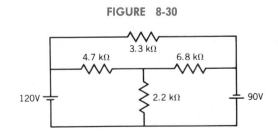

8-15. Determine the current through each of the five resistors in the circuit of Figure 8-31(a).

8-16. In the network shown in Figure 8-31(b), there is *no* electrical connection where the leads of the 20-Ω and 40-Ω resistors cross. Nevertheless, redrawing the network shows that it is a *planar* network. Determine the voltage drop across the 40-Ω resistor.

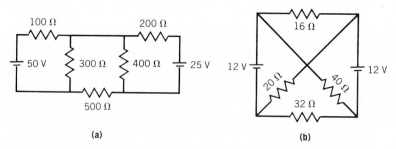

(a)   (b)

FIGURE 8-31

8-17. Determine the current through the 400-Ω resistor in the network of Figure 8-32(a).

8-18. Replace the 6-V source in the network of Figure 8-32(a) with an ammeter having a resistance of 0 ohms, and place the 6-V source in series with the 400-Ω resistor. What is the ammeter reading? Compare this with the current through the 400-Ω resistor in Problem 8-17. This is an example of the **reciprocity theorem**, which states that in any linear network with a single voltage source, when a voltage source in branch A is moved to branch B, then the current that was in branch B now appears in branch A.

8-19. Use loop or mesh equations to determine all eight branch currents in the network of Figure 8-32(b).

8-20. Use the superposition theorem to determine the voltage drop across the 40-kΩ resistor in Figure 8-32(b).

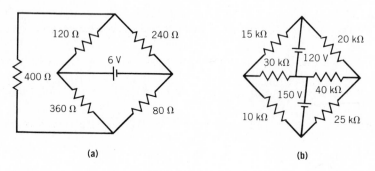

(a)   (b)

FIGURE 8-32

8-21. Find the current drawn from each of the sources in the network shown in Figure 8-29(b) when a 15-kΩ resistor is connected across the top of the network (from the + terminal of the 100-V source to the + terminal of the 200-V source).

8-22. The 56-kΩ resistor in the circuit of Figure 8-11(b) is replaced with a variable resistor which is adjusted until the voltage across R_B is exactly zero. What is the resistance of the variable resistor under these circumstances?

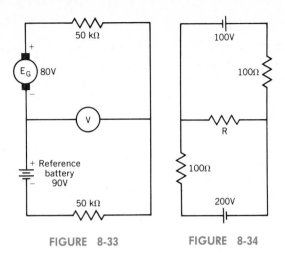

FIGURE 8-33   FIGURE 8-34

8-23. In the balance-detecting circuit of Figure 8-33, the voltmeter has a resistance of 100 kΩ. What is its reading?

8-24. The current through the unknown resistor in the circuit shown in Figure 8-34 is 0.5 A. What is its resistance?

8-25. What is the current through a 500-Ω load connected to the output terminals of the bridged-T attenuator of Figure 8-35 when the input voltage is 2 V?

8-26. If 360 V dc is applied to the input terminals of the lattice network of Figure 8-36, determine the current through a 48-Ω resistor connected to the output terminals.

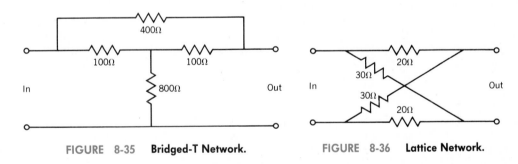

FIGURE 8-35   Bridged-T Network.   FIGURE 8-36   Lattice Network.

8-27. Determine the voltage drop across each of the three resistors in the network shown in Figure 8-37(a).

8-28. Determine the voltage drop across the 300-Ω resistor in the network shown in Figure 8-37(b).

FIGURE 8-37

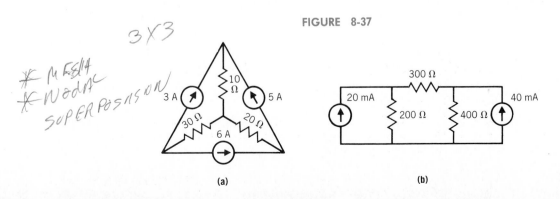

(a)   (b)   201

8-29. In the network shown in Figure 8-38
  (a) What is the voltage drop across the 2.2-kΩ resistor?
  (b) What is the voltage drop across the 4.7-kΩ resistor?

8-30. In the network shown in Figure 8-38, determine the voltage drop across the 10-kΩ resistor by the superposition theorem.

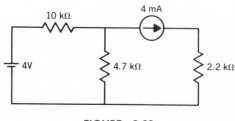

FIGURE 8-38

8-31. In the network shown in Figure 8-39(a), determine the current drawn from each of the voltage sources.

8-32. Determine the voltage drop across the 50-Ω resistor in the network shown in Figure 8-39(b).

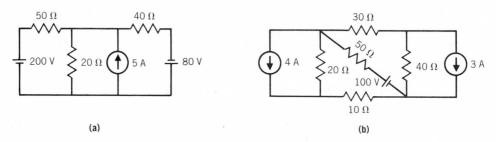

(a)   (b)

FIGURE 8-39

8-33. A pulse generator circuit acts as a switch in series with a 10-kΩ resistor, as shown in Figure 8-40. Calculate the output voltage of the interstage coupling network across a 470-kΩ resistor connected from the output terminal to ground
  (a) With the switch open.
  (b) With the switch closed.

FIGURE 8-40

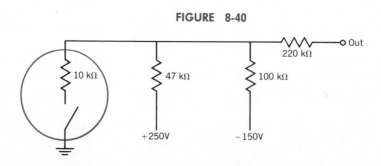

8-34. The circuit shown in Figure 8-40 is modified to the form shown in Figure 8-41. Calculate the voltage drop across a 470-kΩ resistor connected from the output terminal to ground
(a) With the switch open.
(b) With the switch closed.

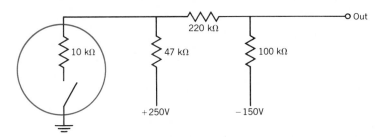

FIGURE 8-41

8-35. A trolleybus is halfway along a section of trolley wire which is fed at one end from a 600-V source and at the other end from a 596-V source. The internal resistances of the sources are negligible, but each trolley wire in the section has a resistance of 1.0 Ω. This trolleybus is the only one in the section at the moment and it draws a 50-A current from the trolley wires. What is the voltage drop across the trolleybus?

REVIEW QUESTIONS

8-1. Define a **constant-voltage source.**

8-2. Why is the terminal voltage of a constant-voltage source not necessarily constant as the load resistance is varied?

8-3. Describe a laboratory procedure for determining the internal resistance of an automobile storage battery.

8-4. What is the significance of the term **short-circuit current?**

8-5. How would you determine the short-circuit current of an automobile battery?

8-6. What is the significance of the term **open-circuit voltage?**

8-7. Define a **constant-current source.**

8-8. Why is the current drawn from a constant-current source not necessarily constant as the load resistance is varied?

8-9. What happens to the terminal voltage of a constant-current source when a load resistor is connected to it? Explain.

8-10. Compare equivalent constant-voltage and constant-current sources in terms of their open-circuit voltage and short-circuit current.

8-11. Describe the *two* methods for determining the internal resistance of constant-voltage and constant-current sources.

8-12. Why can we not replace the source shown in Figure 8-1 with an equivalent constant-current source?

8-13. Show that Equations (8-5) and (8-6) are numerically equal and, therefore, represent two different ways of stating Kirchhoff's voltage law.

8-14. Explain the procedure for determining the polarity of voltage drops around a current tracing loop, using Figure 8-7 as an example.

8-15. How does this procedure apply when determining the voltage drop across R in the circuit of Figure 8-34?

8-16. If, in solving Example 8-3, we had chosen the I_3 tracing loop (Figure 8-9) from the $+$ terminal of the source through R_2, R_M, and R_3 back to the $-$ terminal of the source, the solution for I_3 would have been -0.15 A. What is the significance of a negative result in solving loop equations?

8-17. Is it likely that I_1 or I_2 in Figure 8-9 would ever work out to be negative quantities? Explain.

8-18. Given the following data for the Wheatstone bridge of Figure 8-9, $R_1 = 40$ Ω, $R_2 = 10$ Ω, $R_3 = R_4 = 20$ Ω; what path would you select for the I_3 tracing loop in seeking to avoid a negative solution for I_3?

8-19. How do you write loop equations for the circuit of Figure 8-11(b) when no complete loops appear in the circuit diagram?

8-20. Loop equations are based on Kirchhoff's voltage law but the unknown quantities in the loop equations are *currents*. Explain the presence of current terms in voltage-law equations.

8-21. In the second solution to Example 8-4, how do we know that the battery current should be $I_L - I_G$ rather than $I_G - I_L$?

8-22. In distributing loads on a three-wire residential service, why is it desirable to have the two values of R_L in Figure 8-14 as nearly equal as possible?

8-23. Although the circuit diagram of Figure 8-31(b) shows conductors crossing with *no* electrical connection, the circuit is actually a *planar* circuit. Explain.

8-24. Why are we more likely to encounter negative current terms with the mesh format for writing Kirchhoff's voltage-law equations than with the loop procedure?

8-25. Many of the currents calculated with either the loop or mesh formats in a pencil-and-paper network cannot be measured in the real network. Explain.

8-26. How do we determine whether the source voltage in a mesh equation is a positive or negative quantity?

8-27. How do we determine the polarity of the voltage drop across a resistor in a network where the actual current is made up of two different pencil-and-paper mesh currents?

8-28. In writing loop equations, we insert branch currents in the basic Kirchhoff's voltage-law equation before we collect terms in the form of loop currents. With the mesh format, we skip the in-between step and write the term directly in terms of mesh currents. What features of the mesh format permit us to take this short cut?

8-29. Node-voltage equations are based on Kirchhoff's *current* law, but the unknown quantities in the equations are *voltages*. Explain the presence of voltage terms in current-law equations.

8-30. When we write node-voltage equations, we require one less equation than the total number of nodes. Explain.

8-31. What do we mean when we refer to "the voltage *at* node X"?

8-32. What factors would lead us to select nodal analysis in preference to loop or mesh analysis in solving a given network?

8-33. Under what circumstances are loop equations preferable to nodal analysis?

8-34. What advantage does the superposition theorem have over Kirchhoff's-law equations in network solutions?

8-35. What factors tend to limit the application of the superposition theorem in network analysis?

8-36. Are we likely to encounter negative solutions for component currents when applying the superposition theorem to a network problem? Explain.

8-37. Is it possible that we shall have to *subtract* component currents when finding the real branch current in a network by the superposition theorem? Explain.

8-38. To use mesh equations, we convert any current sources to equivalent constant-voltage sources. To use nodal analysis, we convert any voltage sources to equivalent constant-current sources. Do we have to make such conversions when using the superposition theorem?

8-39. A certain circuit-analysis computer program solves networks in terms of voltages between up to seven independent nodes and a reference node (node 0). The program asks the operator for the network configuration in terms of the names of the components connected between specific nodes. It assumes that the "TO" node is positive with respect to the "FROM" node and that component values are entered in basic units. The computer display informs us that its memory already contains the following network:

```
REF     NAME       FROM   TO    VALUE

 1      BATTERY      0      1      24
 2      BATTERY      0      4      20
 3      RESISTOR     2      1      50
 4      RESISTOR     0      2     100
 5      RESISTOR     3      2     200
 6      RESISTOR     0      3     100
 7      RESISTOR     3      4     300
 8      RESISTOR     4      1     500
```

Draw and label the circuit diagram for this network.

*8-40. Prepare a computer program that you can store for future use in solving third-order determinants.

9

EQUIVALENT-CIRCUIT THEOREMS

The advantages of the classical Kirchhoff's-law methods for solving resistance networks in Chapter 8 are (1) we do not have to tamper with the original configuration of the network, and (2) we can solve almost any network ranging from simple to very elaborate. The disadvantages are the large number of tedious computations and the many possibilities for writing an equation incorrectly or for making a numerical error. We learn these methods because they form the basis for computer circuit-analysis software packages. The universal application of Kirchhoff's laws allows the software to handle many different network configurations. And (after the original program debugging) computer programs can handle a very large number of computations rapidly and without error. Only the final solution is displayed by the computer.

In this chapter we shall investigate more contemporary methods used to simplify network calculations. Since these techniques involve substituting equivalent circuits for portions of the original network, they are not particularly adaptable to computer software packages. However, since they do involve thinking about how electric circuits behave, they are particularly suited to an introductory electric circuits course where our objective is to *understand* just how electric circuits must behave. With equivalent-circuit techniques, there are many ways of solving a particular network. Hence, an electronic calculator is better suited to most of the examples in this chapter than a personal computer.

9-1 EQUIVALENT CIRCUITS

One of the most useful circuit-analysis techniques is to simplify a circuit diagram by replacing several components with a simpler configuration which performs identically as far as the remainder of the circuit is concerned. We encountered this **equivalent-circuit** technique in Section 7-1, when we replaced series and parallel resistors with a single equivalent resistance, as shown in Figure 9-1. As far as the source is concerned, it can "see" no difference among the three circuits of Figure 9-1.

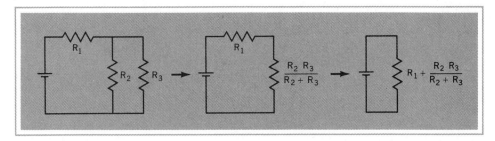

FIGURE 9-1
Equivalent Circuit for a Series-Parallel Circuit.

Figure 9-2(a) shows the complete circuit diagram for one stage of a transistor audio amplifier. Actually, this diagram contains two distinct circuits. A transistor requires certain dc potentials at its base, emitter, and collector to perform its function as an amplifying device. As far as direct current is concerned, the three capacitors are open circuits. Hence, the dc equivalent circuit becomes simply a voltage-divider system providing the necessary dc potentials, as shown in Figure 9-2(b).

Since the terminal voltage of the 16-volt dc power supply cannot vary at an audio frequency, the -16-volt dc terminal in the circuit of Figure 9-2(a) appears to be grounded as far as the audio signal is concerned. If the three capacitors have been selected properly, at a midrange audio frequency of 1 kilohertz they appear as short circuits by comparison with the resistances in the circuit. Consequently, for midrange audio frequencies, we can simplify the transistor amplifier circuit to the ac equivalent circuit shown in Figure 9-2(c).

The next step in the analysis of the performance of this circuit is to replace the transistor by an equivalent circuit composed of basic electrical properties. This is a chore which we must leave for a separate study of electronic circuits.[†]

9-2 THÉVENIN'S THEOREM

In Section 8-2, we adapted a laboratory "black-box" strategy to enable us to replace any practical voltage source in a network with a simple constant-voltage source (or constant-current source) which is an exact equivalent as far as the network is concerned. In this section, we shall expand this principle into one of the most useful of the circuit simplification theorems.

[†]We shall examine the equivalent circuit for a transistor briefly in Section 26-4.

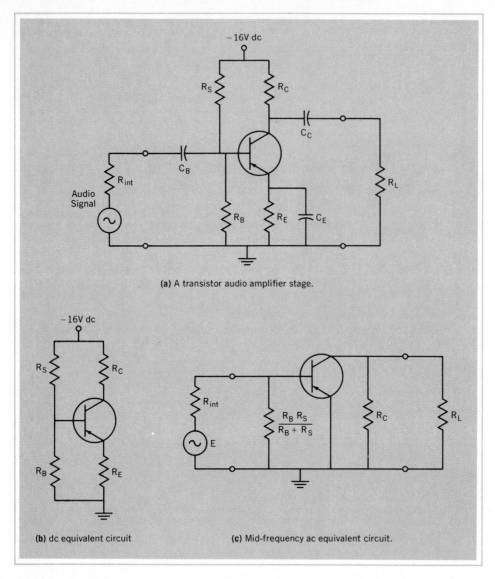

(a) A transistor audio amplifier stage.

(b) dc equivalent circuit

(c) Mid-frequency ac equivalent circuit.

FIGURE 9-2
Equivalent Circuits for a Transistor Audio Amplifier.

Suppose that the electric circuit of Figure 9-3(a) is sealed into a "black box" (represented by the shaded area) so that only the two terminals A and B are exposed. We can attempt to establish the nature of the contents of the box by first connecting a sensitive voltmeter across terminals A and B. The voltmeter will tell us that the box contains a voltage source, and, since the voltmeter draws negligible current, it shows the **open-circuit terminal voltage** of the source. As far as we can tell without opening the box, it contains an 80-volt source, as shown in the equivalent constant-voltage source of Figure 9-3(b).

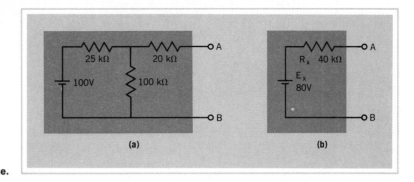

FIGURE 9-3
(a) Original Voltage Source;
(b) Thévenin-Equivalent Source.

Since we do not have to worry about blowing a fuse in our pencil-and-paper analysis, we can short-circuit terminals A and B through an ammeter having zero resistance. For the black box of Figure 9-3(a), we find that the short-circuit current is 2 milliamperes. Consequently, without any knowledge of the exact contents, the box appears to contain an 80-volt source which will permit a 2-milliampere short-circuit current. Hence, the source appears to have an internal resistance of

$$R_x = \frac{E_{oc}}{I_{sc}} = \frac{80 \text{ V}}{2 \text{ mA}} = 40 \text{ k}\Omega$$

as shown in Figure 9-3(b). We can check the equivalency by opening the box and solving the network it actually contains in Figure 9-3(a).

No matter how elaborate the network inside the sealed box may be, as long as it contains one or more voltage sources, all that we can determine about it without opening the box is that it is equivalent to a simple constant-voltage source with a single internal resistance in series with it. Any load connected to terminals A and B of the equivalent circuit of Figure 9-3(b) will draw exactly the same current with exactly the same voltage drop as it would if it were connected to terminals A and B of the original network of Figure 9-3(a).

Rather than leaving this demonstration as an interesting fundamental principle of electric circuits, Léon Charles Thévenin stated the principle in the form of a theorem that has become one of the most useful methods of simplifying electrical networks to assist in their pencil-and-paper analysis. We may state Thévenin's theorem as follows:

Any two-terminal network of fixed resistances and voltage sources may be replaced by a single voltage source having an equivalent voltage equal to the open-circuit voltage at the terminals of the original network and having an internal resistance equal to the resistance looking back into the network from the two terminals with all the voltage sources replaced by their internal resistances.

We can apply Thévenin's theorem to any of the resistance networks we have solved so far by treating one branch of the network as a *load* and the remainder of the network as a *two-terminal network containing one or more voltage sources*. Having decided which branch of the original network we are going to treat as a load, we *remove* it from the original network and place it in a Thévenin-equivalent circuit and apply the rules set forth in the theorem to determine the remainder of the equivalent circuit. Note that Thévenin uses an "ohmmeter" approach to determine the internal resistance "looking back into" the open-circuit terminals of the source network, rather than the short-circuit current method we used with our black-box technique. Either line of thought leads to the same answer for R_x. To illustrate the Thévenin method, we shall follow his instructions to determine the Thévenin-equivalent circuit for Figure 9-3(a).

EXAMPLE 9-1

What current will a 10-kΩ resistor draw when it is connected to a 100-V source through the "T" network shown in Figure 9-3(a)?

SOLUTION

Step I. Rather than connect the 10-kΩ resistor to the original circuit of Figure 9-3(a), we connect it to a Thévenin-equivalent source consisting of a constant-voltage source E_x with a series internal resistance R_x, as in Figure 9-4(b).

According to Thévenin's theorem, E_x of the equivalent circuit will be the open-circuit voltage between terminals A and B of Figure 9-3(a). Since an open circuit draws no current, there will be no voltage drop across the 20-kΩ resistor. Hence, the open-circuit terminal voltage will be the same as the voltage drop across the 100-kΩ resistor. Because terminal A is open-circuit, the 25-kΩ and the 100-kΩ resistors form a simple series circuit. Thus, the voltage-divider principle of Section 7-4 and a bit of mental arithmetic will provide us with the voltage drop across the 100-kΩ resistor, which is the open-circuit voltage between terminals A and B. This voltage, in turn, is E_x in the Thévenin-equivalent circuit.

$$E_x = \frac{100}{25 + 100} \times 100 \text{ V} = 80 \text{ V}$$

Step II. The next step in following the instructions stated in Thévenin's theorem is to replace the actual source in Figure 9-3(a) by its own internal resistance, which, in this case, is zero ohms. After we do this, the original circuit of Figure

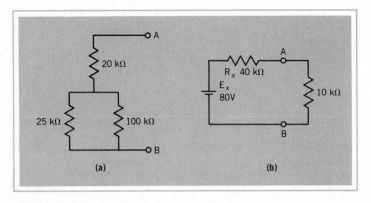

(a)

(b)

FIGURE 9-4
Equivalent Circuit for Step II of Example 9-1.

9-3(a) becomes the series-parallel resistance network of Figure 9-4(a). Using techniques shown in Figure 9-1, the internal resistance "looking back into" terminals A and B is

$$R_x = 20 \text{ k}\Omega + \frac{25 \text{ k}\Omega \times 100 \text{ k}\Omega}{25 \text{ k}\Omega + 100 \text{ k}\Omega}$$

$$= 20 \text{ k}\Omega + 20 \text{ k}\Omega = 40 \text{ k}\Omega$$

Step III. To complete the solution, we simply solve for the current through the 10-kΩ resistor in the Thévenin-equivalent circuit of Figure 9-4(b).

$$I = \frac{E_x}{R_T} = \frac{80 \text{ V}}{40 \text{ k}\Omega + 10 \text{k}\Omega} = 1.6 \text{ mA}$$

To illustrate further the procedure of Thévenin's theorem and, at the same time, to assess its usefulness, we can consider Example 7-3 which we solved originally by using Kirchhoff's-law equations.[†]

EXAMPLE 9-2

A resistor passing a 20-mA current is in parallel with a 5-kΩ resistor. This combination is in series with another 5-kΩ resistor, the whole network being connected to a 500-V source. What is the resistance of the resistor which is passing the 20-mA current?

SOLUTION

Step I. Having first drawn a circuit diagram for the circuit described in the example, we select the unknown resistance passing the 20-mA current as the load and *remove* it from the original circuit, as in Figure 9-5(a), and place it in the Thévenin-equivalent circuit of Figure 9-5(b).

Using the voltage-divider principle again, we can now state the open-circuit

FIGURE 9-5
Circuit Diagram for Step I of Example 9-2

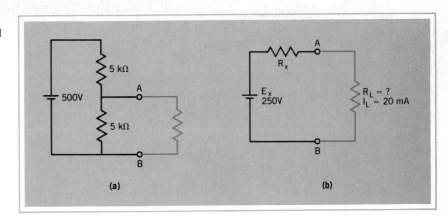

†See page 141.

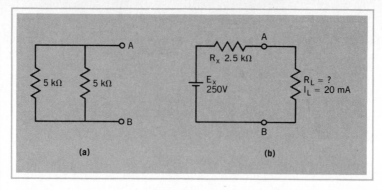

FIGURE 9-6
Equivalent Circuit for Step II of Example 9-2.

terminal voltage of the remaining circuit in Figure 9-5(a) by simple mental arithmetic. Hence, E_x for the Thévenin-equivalent circuit of Figure 9-5(b) is

$$E_x = \frac{5}{5 + 5} \times 500 \text{ V} = 250 \text{ V}$$

Step II. If we short out the original voltage source (and replace it with its internal resistance of zero ohms), the circuit of Figure 9-5(a) becomes the parallel-resistance network of Figure 9-6(a). This network is simple enough that we can again solve for the internal resistance "looking back into" terminals A and B by mental arithmetic. Thus,

$$R_x = \frac{5 \text{ k}\Omega \times 5 \text{ k}\Omega}{5 \text{ k}\Omega + 5 \text{ k}\Omega} = 2.5 \text{ k}\Omega$$

Step III. We can now solve the Thévenin-equivalent circuit of Figure 9-6(b) quite readily by Ohm's law.

$$R_T = \frac{E_x}{I} = \frac{250 \text{ V}}{20 \text{ mA}} = 12.5 \text{ k}\Omega$$

$$\therefore R_L = R_T - R_x = 12.5 \text{ k}\Omega - 2.5 \text{ k}\Omega = 10 \text{ k}\Omega$$

Thévenin's theorem allows us to solve a Wheatstone bridge network without resorting to the third-order simultaneous equations we required with the Kirchhoff's-laws solution of Chapter 8.

EXAMPLE 9-3
In the Wheatstone bridge circuit of Figure 8-1, R_1 and R_4 have a resistance of 300 Ω each. R_2 and R_3 each have a resistance of 150 Ω. R_5 is a galvanometer whose resistance is 50 Ω, and the source voltage is 100 V. What is the current in R_5?

SOLUTION
Step I. Since we are asked to find the current through only *one* branch of the

Wheatstone bridge, Thévenin's theorem allows us to *remove* R_5 (the meter) from the original circuit, as in Figure 9-7(a), to the Thévenin-equivalent circuit, as in Figure 9-7(b).

Step II. In the remaining circuit of Figure 9-7(a), from the voltage-divider principle,

$$V_3 = 100 \text{ V} \times \frac{150}{450} = 33.33 \text{ V}$$

and

$$V_4 = 100 \text{ V} \times \frac{300}{450} = 66.67 \text{ V}$$

From Kirchhoff's voltage law,

$$V_{AB} = V_3 - V_4 = 33.33 - 66.67 = -33.34 \text{ V}$$

(with terminal A *negative* with respect to terminal B). The open-circuit voltage V_{AB} becomes E_x of the Thévenin-equivalent circuit of Figure 9-7(b).

Step III. Assuming that the source has negligible internal resistance, we can short-circuit the generator symbol in Figure 9-7(a) to determine the internal resistance between terminals A and B. This becomes R_x for the Thévenin-equivalent circuit of Figure 9-7(b). With the generator short-circuited, R_1 is in parallel with R_3, and R_2 is in parallel with R_4. These two parallel circuits are then in series as far as terminals A and B are concerned. Therefore,

$$R_x = \frac{R_1 R_3}{R_1 + R_3} + \frac{R_2 R_4}{R_2 + R_4}$$

$$= \frac{300 \times 150}{300 + 150} + \frac{150 \times 300}{150 + 300} = 200 \ \Omega$$

FIGURE 9-7
Solving Wheatstone Bridge by Thévenin's Theorem.

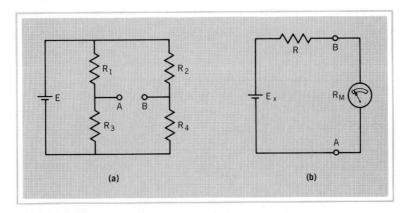

(a)　　　　　　　　　(b)

Step IV. In the Thévenin-equivalent circuit, the meter current then becomes

$$I_M = \frac{E_x}{R_x + R_M} = \frac{33.34 \text{ V}}{250 \text{ }\Omega} = 133 \text{ mA}$$

We can determine the current through any other one branch of the bridge circuit in exactly the same manner.

Thévenin's theorem applies to networks having more than one voltage source. Its usefulness depends on whether we can determine E_x without resorting to simultaneous equations. With only two sources, this is usually not a problem, as we can show with the network of Figure 9-8, which is similar to the automobile generator example of Chapter 8.

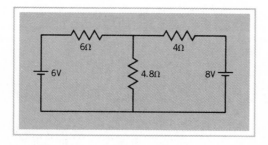

FIGURE 9-8
Network with Two Voltage Sources.

EXAMPLE 9-4
Determine the current in the 4.8-Ω resistor in the network shown in Figure 9-8.

SOLUTION
Step I. Remove the 4.8-Ω resistor from the original circuit of Figure 9-8 and place it in the Thévenin-equivalent circuit of Figure 9-9(b).

Step II. According to Kirchhoff's voltage law, the potential difference be-

FIGURE 9-9
Circuit Diagram for Example 9-4.

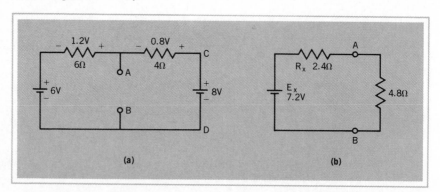

tween points C and D in Figure 9-9(a) must be the *same* via both paths. One path is the 8-V source. From inspection of the other path, it follows that there must be a 2-V drop across the 6-Ω and 4-Ω resistors in series. From the voltage-divider principle, these voltage drops become 1.2 V and 0.8 V, respectively, as shown in Figure 9-9(a).

When we trace the potential difference between points A and B by either path, the open-circuit voltage E_x becomes 7.2 V.

Step III. Replacing the sources by their internal resistances (zero ohms) results in the 6-Ω and 4-Ω resistors appearing in *parallel* between points A and B. Looking back into the source, we then find that

$$R_x = \frac{6 \times 4}{6 + 4} = 2.4 \ \Omega$$

Step IV. From the Thévenin-equivalent circuit of Figure 9-9(b),

$$I_L = \frac{E_x}{R_x + R_L} = \frac{7.2 \ \text{V}}{(2.4 + 4.8) \ \Omega} = 1 \ \text{A}$$

9-3 NORTON'S THEOREM

Figure 9-3 shows how we can reduce a network containing one or more voltage sources into a simple Thévenin-equivalent constant-voltage source. In Chapter 8, we found that we can readily convert a constant-voltage source into an exact equivalent constant-current source. Therefore, it follows that we should be able to reduce any network containing one or more voltage sources directly to a simple equivalent constant-current source, as shown in Figure 9-10. Norton's theorem combines Thévenin's theorem with constant-current source conversion into a single procedure. We may state Norton's theorem as follows:

Any two-terminal network of fixed resistances and voltage sources may be replaced by a single **constant-current** *source whose current is equal to the current drawn by a short circuit across the terminals of the original network and having in* **parallel** *with the constant-current source a resistance equal to the resistance looking back into the network from the two terminals.*

FIGURE 9-10
(a) Original Voltage Source;
(b) Equivalent Constant-Current Source.

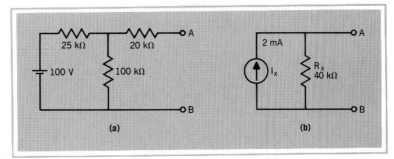

EXAMPLE 9-5

Determine the Norton-equivalent constant-current source for the network of Figure 9-10(a).

SOLUTION 1

Step I. Following the instructions of Norton's theorem, we place a short-circuit across terminals A and B in Figure 9-10(a). The total resistance "seen" by the 100-V source (in kilohms) is

$$R_T = 25 + \frac{100 \times 20}{100 + 20} = 41.67 \text{ k}\Omega$$

The current drawn from the 100-V source is

$$I_T = \frac{E}{R_T} = \frac{100 \text{ V}}{41.67 \text{ k}\Omega} = 2.4 \text{ mA}$$

From the current-divider principle,

$$I_x = I_{sc} = \frac{100}{100 + 20} \times 2.4 \text{ mA} = \textbf{2.0 mA}$$

Step II. Using the "ohmmeter" method of determining internal resistance, as stated in Norton's theorem, we replace the 100-V source with 0 Ω internal resistance. Then the resistance seen looking back into terminals A and B (in kilohms) is

$$R_x = 20 + \frac{25 \times 100}{25 + 100} = \textbf{40 k}\Omega$$

I_x and R_x are the required values for the Norton-equivalent constant-current source of Figure 9-10(b).

SOLUTION 2

Step I. We can avoid the current-divider calculations of Solution 1 by first finding the Thévenin-equivalent source of Figure 9-11(a). From Figure 9-10(a),

FIGURE 9-11
(a) Thévenin-Equivalent Constant-Voltage Source; (b) Norton-Equivalent Constant-Current Source.

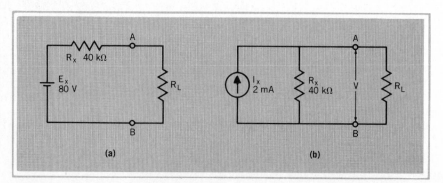

(a)   (b)

216

$$E_x = E_{oc} = \frac{100}{100 + 25} \times 100 \text{ V} = 80 \text{ V}$$

Step II. We find R_x by the same "ohmmeter" technique as in Solution 1. $R_x = 40 \text{ k}\Omega$.

Step III. Using source conversion to obtain the Norton-equivalent constant-current source of Figure 9-11(b), with reference to the Thévenin-equivalent source of Figure 9-11(a),

$$I_x = I_{sc} = \frac{E_x}{R_x} = \frac{80 \text{ V}}{40 \text{ k}\Omega} = 2 \text{ mA}$$

EXAMPLE 9-4A

Determine the current in the 4.8-Ω resistor in the circuit shown in Figure 9-8.

SOLUTION

Step I. Remove the 4.8-Ω resistor from the original circuit and place it in the Norton-equivalent circuit of Figure 9-12(b).

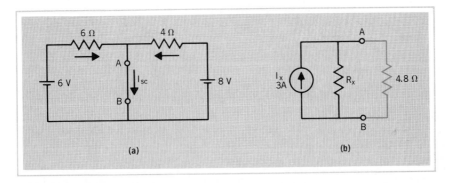

FIGURE 9-12
Circuit Diagram for Step II of Example 9-4A.

Step II. We replace the 4.8-Ω resistor in the original circuit by a short circuit, as shown in Figure 9-12(a). Drawing arrows for current directions shows that the short-circuit current is the *sum* (Kirchhoff's current law) of the currents resulting from placing the 6-Ω resistor directly across the 6-V source and placing the 4-Ω resistor directly across the 8-V source. Therefore, the short-circuit current, I_x in Figure 9-12(b) is

$$I_x = \frac{E_1}{R_1} + \frac{E_2}{R_2} = \frac{6}{6} + \frac{8}{4} = 3 \text{ A}$$

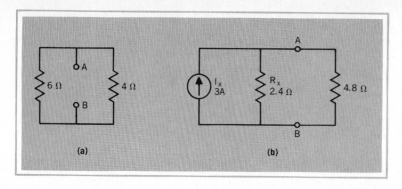

FIGURE 9-13
Circuit Diagram for Step III of
Example 9-4A.

Step III.   Replacing the sources by their internal resistance (zero ohms) gives the circuit of Figure 9-13(a). Looking back into the source, we find that R_x becomes the parallel equivalent of the 6-Ω and 4-Ω resistors, Hence,

$$R_x = \frac{6 \times 4}{6 + 4} = 2.4\ \Omega$$

Step IV.   Using the current-divider principle in the Norton-equivalent circuit of Figure 9-13(b),

$$I_L = \frac{2.4}{2.4 + 4.8} \times 3\ A = 1\ A$$

9-4 THÉVENIN'S THEOREM IN MULTISOURCE NETWORKS (MILLMAN'S THEOREM)

The general statement of Thévenin's theorem indicates that a two-terminal network containing *any number* of voltage sources can be replaced by a single voltage source with a single series resistance. If the original network contains only a single voltage source, finding the Thévenin-equivalent source is quite simple. Often we can handle *two* sources with little difficulty when, as shown in Figure 9-14, we can reduce the original network to a single loop by removing a resistor to create the two terminals.

FIGURE 9-14
Circuit Diagram for Example 9-6.

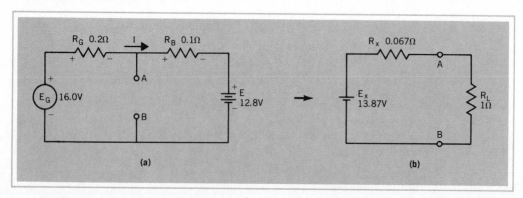

EXAMPLE 9-6

An automobile generator with an internal resistance of 0.2 Ω develops an open-circuit voltage of 16.0 V. The storage battery has an internal resistance of 0.1 Ω and an open-circuit voltage of 12.8 V. Both sources are connected in parallel to a 1-Ω load. Determine the load current.

SOLUTION

Step I. Removing the load resistance from the original circuit leaves the circuit shown in Figure 9-14(a), which we can solve by one simple Kirchhoff's voltage-law equation. Because the two sources are in opposition as far as current around the single closed loop of Figure 9-14(a) is concerned, the current direction is established by the source with the greater terminal voltage. Therefore,

$$E_G - E_B = IR_G + IR_B$$

$$16.0 - 12.8 = 0.2I + 0.1I$$

from which

$$I = 10.67 \text{ A}$$

and

$$V_{RG} = IR_G = 10.67 \times 0.2 = 2.13 \text{ V}$$

$$V_{RB} = IR_B = 10.67 \times 0.1 = 1.07 \text{ V}$$

If we inspect Figure 9-14(a) and note the polarities of the voltage drops, we find that the total potential difference between terminals A and B is either

$$E_x = E_G - V_{RG} = 16.0 - 2.13 = 13.87 \text{ V}$$

or

$$E_x = E_B + V_{RB} = 12.8 + 1.07 = 13.87 \text{ V}$$

Step II. If we short-circuit the generator and battery symbols in Figure 9-14(a) so that only their internal resistances remain in the circuit, we find that R_G and R_B are connected in parallel between terminals A and B. Consequently, following the procedure set out in Thévenin's theorem,

$$R_x = \frac{R_G \times R_B}{R_G + R_B} = \frac{0.2 \times 0.1}{0.2 + 0.1} = 0.067 \ \Omega$$

Step III. We can now solve the Thévenin-equivalent circuit of Figure 9-14(b) by Ohm's law.

$$R_T = R_x + R_L = 0.067 + 1 = 1.067 \ \Omega$$

$$I_L = \frac{E_x}{R_T} = \frac{13.87 \text{ V}}{1.067 \ \Omega} = \textbf{13 A}$$

If there are *three* or more voltage sources in parallel feeding a common load, as in Figure 9-15(a), we cannot find the Thévenin-equivalent voltage E_x directly without resorting to simultaneous equations. However, we can avoid such equations by converting *each* voltage source in the original network to its Norton-equivalent constant-

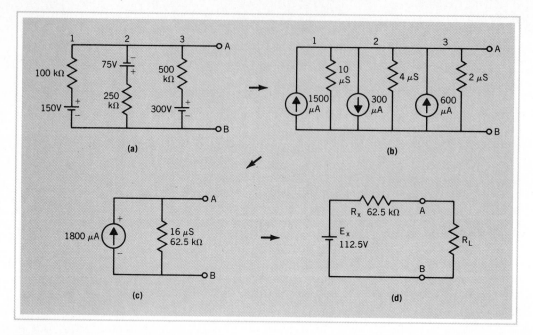

FIGURE 9-15
Equivalent Circuits for Example 9-7.

current source, as in Figure 9-15(b). It is then quite simple to reduce the three parallel constant-current sources to a single Norton-equivalent source, as in Figure 9-15(c). Finally, we convert back to the required Thévenin-equivalent source of Figure 9-15(d).

EXAMPLE 9-7

Find the Thévenin-equivalent source for the three sources in parallel shown in Figure 9-15(a).

SOLUTION

Step I. As for the Thévenin's-theorem procedure, the load is removed from the original circuit of Figure 9-15(a) and connected to the final Thévenin-equivalent circuit of Figure 9-15(d).

Step II. We now follow a procedure which is similar to the one we used in nodal analysis, the difference being that we have removed the load in this case. Replace each of the sources by its Norton-equivalent constant-current source. Again we substitute the internal *conductance* for internal resistance in the resulting parallel circuit. We also use microamperes and microsiemens as units (microamperes = volts × microsiemens).

$$I_1 = \frac{E_1}{R_1} = \frac{150 \text{ V}}{100 \text{ k}\Omega} = 1500 \ \mu A \quad \text{and} \quad G_1 = \frac{1}{R_1} = 10 \ \mu S$$

Similarly, $\quad I_2 = \dfrac{75 \text{ V}}{250 \text{ k}\Omega} = 300 \ \mu\text{A} \quad \text{and} \quad G_2 = 4 \ \mu\text{S}$

and $\quad I_3 = \dfrac{300 \text{ V}}{500 \text{ k}\Omega} = 600\mu\text{A} \quad \text{and} \quad G_3 = 2 \ \mu\text{S}$

In entering these data on Figure 9-15(b), we must make sure that the current direction for the constant-current source is consistent with the polarity of its equivalent constant-voltage source. Note the reversal of direction of I_2 with respect to I_1 and I_3.

Step III. Determining the Norton-equivalent source for Figure 9-15(b) is quite simple. It will have a single constant-current source equal to the *algebraic* sum of the three current sources of Figure 9-15(b). Hence, the Norton-equivalent constant-current source in Figure 9-15(c) is

$$I_x = I_1 + I_2 + I_3 = 1500 + (-300) + 600 = 1800 \ \mu\text{A}$$

The internal conductance in Figure 9-15(c) will be the sum of the three internal conductances in Figure 9-15(b).

$$G_x = 10 + 4 + 2 = 16 \ \mu\text{S}$$

Step IV. Finally, we convert the constant-current source of Figure 9-15(c) into the equivalent constant-voltage source of Figure 9-15(d).

$$R_x = \dfrac{1}{G_x} = \dfrac{1}{16 \ \mu\text{S}} = \textbf{62.5 k}\boldsymbol{\Omega}$$

and $\quad E_x = I_x R_x = 1800 \ \mu\text{A} \times 62.5 \text{ k}\Omega = \textbf{112.5 V}$

EXAMPLE 9-8
If R_B in the circuit shown in Figure 9-16 is 1 kΩ, determine the magnitude and polarity of the voltage drop across R_B.

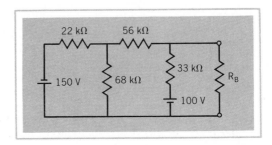

FIGURE 9-16
Circuit Diagram for Example 9-8

SOLUTION
Step I. Removing R_B from the original circuit leaves the circuit shown in Figure 9-17(a). We can replace everything to the left of the dashed line in Figure 9-17(a) by its Thévenin-equivalent source, as shown in Figure 9-17(b), where

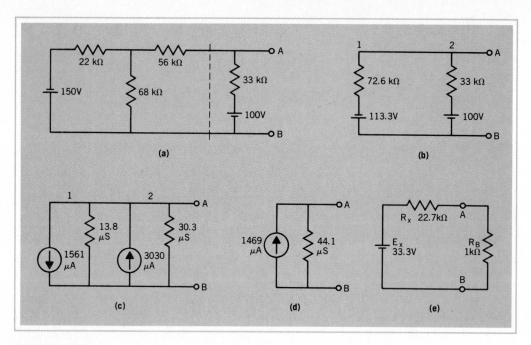

FIGURE 9-17
Equivalent Circuits for Example 9-8.

$$E_x = \frac{68 \text{ k}\Omega}{22 \text{ k}\Omega + 68 \text{ k}\Omega} \times 150 \text{ V} = 113.3 \text{ V}$$

and

$$R_x = \frac{22 \text{ k}\Omega \times 68 \text{ k}\Omega}{22 \text{ k}\Omega + 68 \text{ k}\Omega} + 56 \text{ k}\Omega = 72.6 \text{ k}\Omega$$

Step II. We now set out to find the single Thévenin-equivalent source to replace the two parallel sources of Figure 9-17(b), using either of the techniques just described. Converting each voltage source to its equivalent constant-current source,

$$I_1 = \frac{E_1}{R_1} = \frac{113 \text{ V}}{72.6 \text{ k}\Omega} = 1561 \ \mu\text{A} \quad \text{and} \quad G_1 = 13.8 \ \mu\text{S}$$

Similarly, $$I_2 = \frac{100 \text{ V}}{33 \text{ k}\Omega} = 3030 \ \mu\text{A} \quad \text{and} \quad G_2 = 30.3 \ \mu\text{S}$$

Step III. We can replace the circuit of Figure 9-17(c) with a single Norton-equivalent source. Noting the current directions,

$$I_x = I_1 + I_2 = -1561 + 3030 = 1469 \ \mu\text{A}$$

and $$G_x = G_1 + G_2 = 13.8 + 30.3 = 44.1 \ \mu\text{S}$$

222

Step IV. If we wish, we may solve for the answer directly from the Norton-equivalent circuit of Figure 9-17(d). Or we can carry on to the final Thévenin-equivalent circuit of Figure 9-17(e) where

$$R_x = \frac{1}{G_x} = \frac{1}{44.1\ \mu S} = 22.7\ k\Omega$$

and $$E_x = I_x R_x = 1469\mu A \times 22.7\ k\Omega = 33.3\ V$$

Step V. From the voltage-divider principle,

$$V_B = \frac{1\ k\Omega}{22.7\ k\Omega + 1\ k\Omega} \times 33.3\ V = 1.4\ V$$

The procedure we used in Examples 9-7 and 9-8 is sometimes referred to as **Millman's theorem.** Instead of going through Example 9-7 in four steps, we can summarize the procedure into two equations. From Figure 9-15,

$$R_x = \frac{1}{G_x} \qquad \text{and} \qquad G_x = G_1 + G_2 + G_3$$

Therefore, in the Thévenin-equivalent source for a group of parallel voltage sources,

$$R_x = \frac{1}{G_1 + G_2 + G_3 + \cdots + G_N} \tag{9-1}$$

Also, $$E_x = I_x R_x$$

But $$I_x = E_1 G_1 + E_2 G_2 + E_3 G_3$$

$$\therefore E_x = \frac{E_1 G_1 + E_2 G_2 + E_3 G_3 + \cdots + E_N G_N}{G_1 + G_2 + G_3 + \cdots + G_N} \tag{9-2}$$

Equations (9-1) and (9-2) represent the formal statement of Millman's theorem. Rather than work from memorized formulas, we should continue to use the separate steps outlined in Examples 9-7 and 9-8.

9-5 DEPENDENT SOURCES

The emf generated by the voltage sources we have encountered so far has been determined entirely by the internal energy-conversion process of the source. The open-circuit terminal voltage is, therefore, *independent of what happens elsewhere in a network containing the source*—hence the name *constant-voltage source* or **independent source.** Even though we know that the terminal voltage of a practical voltage source decreases as load current increases, we account for this effect in network diagrams by including an internal resistance in series with the ideal constant-voltage

TABLE 9-1

DEPENDENT SOURCES

Type of Source	Magnitude	Symbol
Voltage-controlled voltage source	$V = kV_1$	kV_1
Current-controlled voltage source	$V = kI_1$	kI_1
Current-controlled current source	$I = kI_1$	kI_1
Voltage-controlled current sources	$I = kV_1$	kV_1

source. From the above definition, the constant-current source of Section 8-3 is also an independent source. Most of the sources we encounter in practical electric circuits are independent sources.

By using electronic devices such as transistors, we can create an effect in an electric circuit which amounts to a voltage or current source whose magnitude is *dependent on* or *controlled by a voltage or current appearing in some other branch of the network*. We can represent such devices in our network diagrams by **dependent** or **controlled sources.** To distinguish dependent sources from independent sources in circuit diagrams, the most popular symbol at present is a diamond-shaped symbol with $+$ and $-$ polarity indicators for a voltage source and a current arrow for a current source. As shown in Table 9-1, there are four possible dependent sources.

Figure 9-18 shows a simple network containing a current-controlled voltage source in series with the more familiar 120-volt independent voltage source and its internal resistance. This dependent source is designed so that its terminal voltage is[†]

$$V = 5 \times I_1$$

By substituting various values for I_1 (as we change R_L), we soon discover that the dependent source adds a voltage which exactly offsets the IR drop across R_{int}. Consequently, V_L remains constant at 120 volts for any setting of R_L. We have created a voltage-regulated power supply.

We think of the CCVS in Figure 9-18 as a *source* because the relationship between the polarity of the voltage between its terminals and the current through it is that of a *source*. However, dependent sources differ from real energy sources. If we

[†]To be technically rigorous, we assign the unit *ohm* to the numerical constant $k = 5$ in this equation to satisfy the units for Ohm's law.

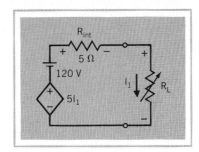

FIGURE 9-18
Network with a Dependent Source.

disconnect R_L in the circuit of Figure 9-18 so that $I_1 = 0$, the voltage across the CCVS disappears—unlike independent voltage sources. Hence, we use the symbol V rather than E for dependent sources. Although Kirchhoff's laws and network theorems hold true for networks containing dependent sources, we shall find some situations where we cannot use some of the methods to solve the network.

At first glance, it would seem that we could write two mesh equations based on Kirchhoff's current law to solve the network shown in Figure 9-19. But the rules for mesh equations require us to convert all current sources to equivalent voltage sources. We cannot convert the VCCS in Figure 9-19 to an equivalent VCVS because there is no resistor which we can use as R_x in parallel with the original current source. We can, however, "Thévenize" the portion of the network to the left of V_1 *without affecting* V_1 to create the single loop of Figure 9-20.

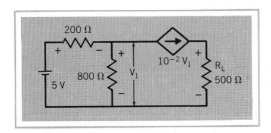

FIGURE 9-19
Circuit Diagram for Example 9-9.

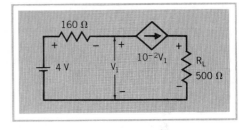

FIGURE 9-20
Equivalent Circuit for Solution 1 for Example 9-9.

EXAMPLE 9-9
Determine V_L in the network shown in Figure 9-19.

SOLUTION 1
In the equivalent circuit of Figure 9-20, from the VCCS,

$$I = 0.01 \ V_1$$

and from Ohm's law,

$$V_1 = 4 - 160 \ I$$

$$\therefore V_1 = 4 - 160 \times 0.01\, V_1 = \frac{4}{2.6} = 1.54 \text{ V}$$

$$I = 0.01\, V_1 = 15.4 \text{ mA}$$

and
$$V_L = IR_L = 15.4 \text{ mA} \times 500\ \Omega = 7.7 \text{ V}$$

We cannot use superposition theorem for this example since removing the 5-V source will drop V_1 to zero. Hence, I will also be zero. But we can convert the independent voltage source to its equivalent constant-current source and use nodal analysis.

SOLUTION 2

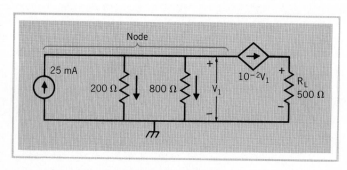

FIGURE 9-21
Equivalent Circuit for Solution 2 for Example 9-9.

In the equivalent circuit of Figure 9-21, the single nodal equation is

$$\left(\frac{1}{200} + \frac{1}{800} \right) V_1 = 25 \text{ mA} - 10^{-2}V_1$$

This reduces to
$$V_1 = \frac{25 \text{ mA}}{16.25 \text{ mS}} = 1.54 \text{ V}$$

Then
$$I_L = 10^{-2}V_1 = 0.01 \times 1.54 = 15.4 \text{ mA}$$

and
$$V_L = I_L R_L = 15.4 \text{ mA} \times 500\ \Omega = 7.7 \text{ V}$$

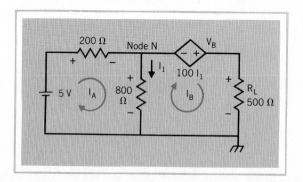

FIGURE 9-22
Circuit Diagram for Example 9-10.

In the circuit shown in Figure 9-22, we have replaced the VCCS of Figure 9-19 with a current-controlled voltage source. The superposition theorem still will not work. And we cannot replace the left-hand part of the network with its Thévenin-equivalent source because we will lose the I_1 which governs the voltage produced by the CCVS. However, we can use both mesh and nodal analysis techniques.

EXAMPLE 9-10
Determine V_L in the network shown in Figure 9-22.

SOLUTION 1
From Kirchhoff's voltage law,

$$(200 + 800)I_A - 800I_B = 5$$

$$(500 + 800)I_B - 800I_A = 100\ I_1 = 100(I_A - I_B)$$

Collecting the terms,

$$1000I_A - 800I_B = 5$$

$$900I_A - 1400I_B = 0$$

$$I_B = \frac{\begin{vmatrix} 1000 & 5 \\ 900 & 0 \end{vmatrix}}{\begin{vmatrix} 1000 & -800 \\ 900 & -1400 \end{vmatrix}} = \frac{-4500}{-680\,000} = 6.62 \text{ mA}$$

$$V_L = I_B R_L = 6.62 \text{ mA} \times 500\ \Omega = \textbf{3.31 V}$$

SOLUTION 2
Converting both voltage sources in the original network of Figure 9-22 into their Norton equivalents for the circuit of Figure 9-23;

Independent source:

$$I_A = \frac{5 \text{ V}}{200\ \Omega} = 25 \text{ mA}$$

FIGURE 9-23
Equivalent Circuit for Solution 2 for Example 9-10.

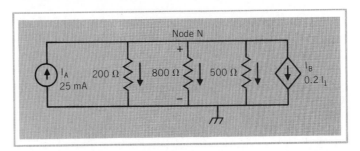

Dependent source:

$$I_B = \frac{V_B}{500}$$

But $\qquad V_B = 100\,I_1 \qquad \therefore I_B = 0.2\,I_1$

and $\qquad I_1 = \dfrac{V_N}{800} \qquad \therefore I_B = \dfrac{V_N}{4000}$

For the single independent node, the nodal equation is

$$\left(\frac{1}{200} + \frac{1}{800} + \frac{1}{500}\right)V_N = 25\ \text{mA} - \frac{V_N}{4000}$$

Bringing the term for the dependent source current to the left-hand side of the equation and using the calculator reciprocal key to clear the fractions,

$$V_N = 25\ \text{mA} \times 117.65\ \Omega = 2.94\ \text{V}$$

Note that V_N is *not* the same as V_L in the original network of Figure 9-22. Returning this value of V_N to the original network,

$$I_1 = \frac{V_N}{800} = \frac{2.94\ \text{V}}{800\ \Omega} = 3.676\ \text{mA}$$

and $\qquad V_B = 100 I_1 = 0.3676\ \text{V}$

Hence, $\qquad V_L = V_N + V_B = 2.94 + 0.37 = \mathbf{3.31\ V}$

In the solutions for Examples 9-9 and 9-10, we have not tried to include a dependent source in any Thévenin transformation. We can include dependent sources as long as we are careful how we determine R_x. If the controlling variable is within the portion of the network to be included in the Thévenin-equivalent source, we cannot use the shortcut method stated in Thévenin's theorem for finding R_x by "looking back" into the open-circuit terminals. We must use the short-circuit current of the "black-box" method of Section 8-2, where we found that

$$R_x = \frac{\text{open-circuit terminal voltage}}{\text{short-circuit terminal current}} \tag{8-1}$$

EXAMPLE 9-11

Removing the 500-Ω R_L from the network shown in Figure 9-22, determine the Thévenin-equivalent source which can replace the remainder of the network.

SOLUTION

For the open-circuit condition shown in Figure 9-24(a),

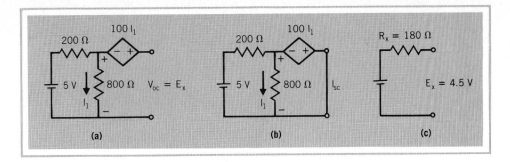

FIGURE 9-24
Dependent Source in a Thévenin Transformation.

$$I_1 = \frac{5\ V}{(200 + 800)\ \Omega} = 5\ mA$$

The dependent source voltage is

$$V = 100I_1 = 0.5\ V$$

$$\therefore V_{oc} = E_x = 5\ mA \times 800\ \Omega + 0.5\ V = \mathbf{4.5\ V}$$

To find the short-circuit current, we can write two mesh equations for the circuit of Figure 9-24(b).

$$(200 + 800)I_A - 800I_B = 5$$

and

$$800I_B - 800I_A = 100I_1 = 100(I_A - I_B)$$

Collecting the terms,

$$1000I_A - 800I_B = 5 \tag{1}$$

$$900I_A - 900I_B = 0 \tag{2}$$

$$I_{sc} = I_B = \frac{\begin{vmatrix} 1000 & 5 \\ 900 & 0 \end{vmatrix}}{\begin{vmatrix} 1000 & -800 \\ 900 & -900 \end{vmatrix}} = \frac{-4500}{-180\,000} = 25\ mA$$

$$R_x = \frac{E_{oc}}{I_{sc}} = \frac{4.5\ V}{25\ mA} = 180\ \Omega$$

To check E_x and R_x, from Figure 9-24(c),

$$V_L = \frac{500}{180 + 500} \times 4.5\ V = \mathbf{3.31\ V}$$

Note that R_x is *not* the same as the resistance "seen" looking back into the open-circuit terminals of Figure 9-24(a).

9-6 DELTA-WYE TRANSFORMATION

Another useful type of circuit transformation which replaces a given circuit with an equivalent circuit applies to the three-terminal resistance network of Figure 9-25. If we can state the conditions under which the "wye" circuit of Figure 9-25(a) is equivalent to the "delta" circuit of Figure 9-25(b), we can substitute the wye for the delta, and vice versa, in circuit simplification.

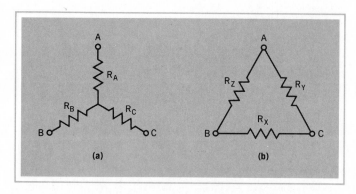

FIGURE 9-25
(a) Y-Network; (b) Δ-Network.

If we ignore terminal C for the moment, the resistance between terminals A and B must be the same for both networks. For the wye circuit of Figure 9-25(a), the circuit between A and B is a simple series circuit giving

$$R_{AB} = R_A + R_B$$

For the delta circuit of Figure 9-25(b), there are two branches in parallel between terminals A and B, giving

$$R_{AB} = \frac{R_Z(R_X + R_Y)}{R_Z + (R_X + R_Y)}$$

$$\therefore R_A + R_B = \frac{R_X R_Z + R_Y R_Z}{R_X + R_Y + R_Z} \tag{1}$$

Similarly,

$$R_B + R_C = \frac{R_X R_Y + R_X R_Z}{R_X + R_Y + R_Z} \tag{2}$$

and

$$R_C + R_A = \frac{R_Y R_Z + R_X R_Y}{R_X + R_Y + R_Z} \tag{3}$$

Subtracting (2) from (1) gives

$$R_A - R_C = \frac{R_Y R_Z - R_X R_Y}{R_X + R_Y + R_Z} \tag{4}$$

Adding (3) and (4) gives

$$2R_A = \frac{2R_Y R_Z}{R_X + R_Y + R_Z}$$

from which
$$R_A = \frac{R_Y R_Z}{R_X + R_Y + R_Z} \qquad (9\text{-}3)$$

Likewise,
$$R_B = \frac{R_X R_Z}{R_X + R_Y + R_Z} \qquad (9\text{-}4)$$

and
$$R_C = \frac{R_X R_Y}{R_X + R_Y + R_Z} \qquad (9\text{-}5)$$

These are the **delta to wye transformation equations.** In using them, we must be careful to associate the numerical values of the original delta circuit with their proper counterparts in the transformation equations. Note that there is a simple pattern to the three equations.

$$R_A = \frac{\text{product of the two } \Delta \text{ arms connected to terminal } A}{\text{sum of all three } \Delta \text{ arms}}$$

In using these equations with a calculator, we first calculate the sum $R_X + R_Y + R_Z$ $=$ and store this number in the calculator's memory. Then the calculation for R_A becomes simply $R_Y \times R_Z \div$ Memory recall .

We can also reduce Equations (1), (2), and (3) algebraically to solve for the equivalent delta circuit from a given wye circuit. The **wye to delta transformation equations** are usually stated as

$$R_X = \frac{R_A R_B + R_B R_C + R_C R_A}{R_A} \qquad (9\text{-}6)$$

$$R_Y = \frac{R_A R_B + R_B R_C + R_C R_A}{R_B} \qquad (9\text{-}7)$$

$$R_Z = \frac{R_A R_B + R_B R_C + R_C R_A}{R_C} \qquad (9\text{-}8)$$

Again the three equations have a pattern.

$$R_X = \frac{\text{sum of products of each } pair \text{ of } Y \text{ arms}}{\text{opposite } Y \text{ arm}}$$

In this case, we calculate the *numerator* and store this number in the calculator's memory to be divided by the appropriate denominator in using the wye to delta transformation equations.

Since Equations (9-6), (9-7), and (9-8) are more elaborate than Equations (9-3), (9-4), and (9-5), suppose we look for an alternative format. Since the delta circuit has *parallel* branches, we can *invert* Equation (9-6) and insert $1/G$ for each resistance.

$$G_X = \cfrac{\cfrac{1}{G_A}}{\cfrac{1}{G_A G_B} + \cfrac{1}{G_B G_C} + \cfrac{1}{G_C G_A}} = \cfrac{\cfrac{1}{G_A}}{\cfrac{G_A + G_B + G_C}{G_A G_B G_C}}$$

From which

$$G_X = \frac{G_B G_C}{G_A + G_B + G_C} \qquad (9\text{-}9)$$

We can establish similar equations for G_Y and G_Z. These equations are the **duals** of Equations (9-3), (9-4), and (9-5).

Inverting Equation (9-9) gives

$$R_X = R_B R_C (G_A + G_B + G_C) \qquad (9\text{-}10)$$

Similarly,

$$R_Y = R_A R_C (G_A + G_B + G_C) \qquad (9\text{-}11)$$

and

$$R_Z = R_A R_B (G_A + G_B + G_C) \qquad (9\text{-}12)$$

This *dual* form of the wye to delta transformation equations is particularly useful when dealing with complex impedances and admittances in ac circuits.

EXAMPLE 9-3A

In the Wheatstone bridge circuit of Figure 9-26(a), R_1 and R_4 have a resistance of 300 Ω each. R_2 and R_3 each have a resistance of 150 Ω. The source voltage is 100 V. What is the current through the galvanometer if its resistance is 50 Ω?

FIGURE 9-26
Solving Wheatstone Bridge by Delta-Wye Transformation.

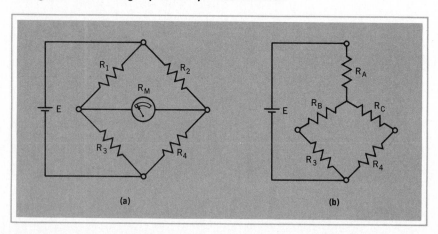

SOLUTION

Step I. Replace R_1, R_2, and R_M of the original bridge circuit of Figure 9-26(a) (which form a delta) with their equivalent wye circuit, as in Figure 9-26(b).

$$R_A = \frac{300 \times 150}{300 + 150 + 150} = 90 \ \Omega$$

$$R_B = \frac{300 \times 50}{500} = 30 \ \Omega$$

$$R_C = \frac{50 \times 150}{500} = 15 \ \Omega$$

Step II. The load on the source is now a simple series-parallel circuit with an equivalent resistance of

$$R_{eq} = R_A + \frac{(R_B + R_3) \times (R_C + R_4)}{(R_B + R_3 + R_C + R_4)}$$

$$= 90 + \frac{180 \times 315}{495} = 204.55 \ \Omega$$

$$\therefore I_T = \frac{E}{R_{eq}} = \frac{100 \ \text{V}}{204.55 \ \Omega} = 489 \ \text{mA}$$

Step III. From the current-divider principle,

$$I_3 = 489 \ \text{mA} \times \frac{315}{495} = 311 \ \text{mA}$$

and

$$I_4 = 489 \ \text{mA} \times \frac{180}{495} = 178 \ \text{mA}$$

Since R_3 and R_4 were not altered by the transformation, I_3 and I_4 are the *same* as for the original circuit. We can now return to the original circuit, and in Figure 9-26(a),

$$V_3 = I_3 R_3 = 311 \ \text{mA} \times 150 \ \Omega = 46.65 \ \text{V}$$

and

$$V_4 = I_4 R_4 = 178 \ \text{mA} \times 300 \ \Omega = 53.4 \ \text{V}$$

Step IV. From Kirchhoff's voltage law,

$$V_M = V_4 - V_3 = 53.4 - 46.65 = 6.75 \ \text{V}$$

(with the right-hand terminal positive with respect to the left-hand terminal)

and

$$I_M = \frac{6.75 \ \text{V}}{50 \ \Omega} = \textbf{135 mA}$$

9-7 SUBSTITUTION THEOREM

In Chapter 7 we discovered that we could *substitute* a single equivalent resistor for a branch in a network consisting of a combination of series and/or parallel resistors, as shown in Figures 9-27(a) and (b). By selecting (calculating) the appropriate value for this substitute resistor, the voltage drop across and current through the branch is identical with the original network. Hence, the rest of the network "sees" no difference with the substitution.

The substitution principle is such a fundamental property of electric circuits that we can expand it into the **substitution theorem.**

Any branch in a linear network can be replaced by any combination of circuit elements which will produce the same *voltage across and current through that branch.*

Figure 9-27(c), (d), and (e) shows several possible combinations which produce the *same* voltage and current as the circuit of Figure 9-27(a) and, therefore, will not affect the rest of the network. We should note that:

1. Since we have to know the voltage across and the current through the selected branch to make this type of substitution, this theorem is not useful for *solving* network examples. It is sometimes used by circuit designers to check or optimize a circuit design.
2. When we replace resistors by sources, current must flow the "wrong way"

FIGURE 9-27
Substitution Theorem.

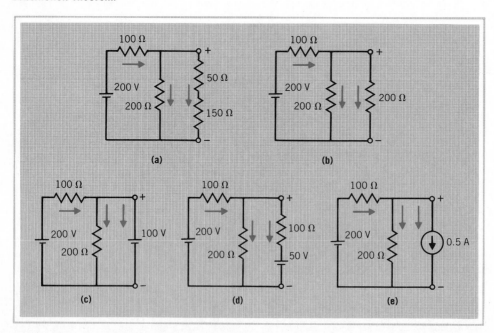

through voltage sources and the voltage drop must be the "wrong polarity" across current sources so that they act like the resistors they replace.

3. We cannot replace branches containing discharging sources by resistors alone (which cannot supply energy). In theory, we can call a voltage source a *negative resistance*.

As an exercise, check that the substitutions shown in Figure 9-27 do satisfy the substitution theorem.

9-8 RECIPROCITY THEOREM

We can also expand another principle based on a fundamental property of electric circuits into the **reciprocity theorem.**

In any linear network containing only a single voltage source, if that source located in branch A causes a current I in branch B, then moving that voltage source (but not its internal resistance) to branch B will cause a current I in branch A equal to the original current in branch B.

Although it applies to any network configuration, the reciprocity theorem is useful primarily in showing the bilateral property of the four-terminal networks we shall encounter in Chapter 26. There is also a version of the reciprocity theorem for a *current* source in branch A causing a *voltage drop* across branch B. Unlike the substitution theorem, the currents and voltages in the other branches of the network are *not* unchanged when we move the source from one branch to another. Check by determining the three currents within the four-terminal T-attenuator in both circuits shown in Figure 9-28.

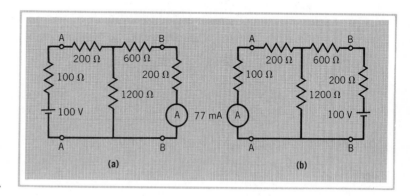

FIGURE 9-28
Reciprocity Theorem.

PROBLEMS

9-1. Determine the Thévenin-equivalent constant-voltage source for the two-terminal network of Figure 9-29.

9-2. Determine the Thévenin-equivalent source for the network of Figure 9-30.

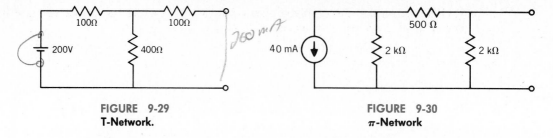

FIGURE 9-29
T-Network.

FIGURE 9-30
π-Network

9-3. Use Thévenin's theorem to determine the current through the 500-Ω resistor in the circuit of Figure 9-31.

9-4. Determine the current in the 100-Ω resistor in the circuit of Figure 9-31.

9-5. Determine the Norton-equivalent constant-current source for the two-terminal network of Figure 9-29.

9-6. Determine the Norton-equivalent source for the network of Figure 9-30.

9-7. What resistance must be connected to the source shown in Figure 9-29 to draw a 200-mA current through the unknown resistor? Use Thévenin's theorem.

9-8. What current will a 500-Ω load draw from the source shown in Figure 9-29? Use Norton's theorem.

9-9. Use Norton's theorem to determine the value of load resistance required to draw a 40-μA current from the source shown in Figure 9-30.

9-10. Use Thévenin's theorem to determine the current through a 5-kΩ load connected to the source shown in Figure 9-30.

9-11. What is the maximum power that can be developed in a load connected to the source shown in Figure 9-29?

9-12. What value of load resistance will dissipate energy at the rate of 100 mW when connected to the source shown in Figure 9-30? (Two answers)

9-13. Solve for the current drawn from the 20-V source in Figure 9-32 by replacing the portion of the circuit to the left of the dashed line with its Thévenin-equivalent source.

9-14. Determine the current through the center 50-Ω resistor in the network of Figure 9-32 by replacing each voltage source with its Norton-equivalent source. (Nodal analysis)

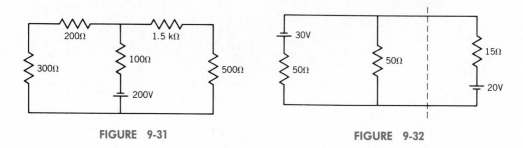

FIGURE 9-31

FIGURE 9-32

9-15. Find the voltage drop across the 25-Ω resistor in Figure 9-33 by
(a) Replacing the voltage source with an equivalent constant-current source.
(b) Replacing the current source with an equivalent constant-voltage source.

9-16. Find the current through the 100-Ω resistor in Figure 9-33 by leaving the sources in their original form and using the superposition theorem.

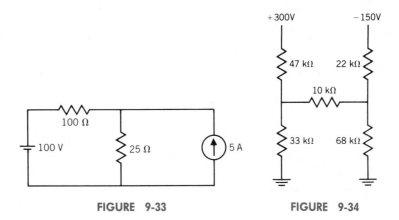

FIGURE 9-33

FIGURE 9-34

9-17. Determine the Thévenin-equivalent source as "seen" by the 10-kΩ resistor in the network shown in Figure 9-34.

9-18. Determine the Thévenin-equivalent source as "seen" by the 33-kΩ resistor in the network shown in Figure 9-34.

9-19. Determine the current through the 60-Ω resistor in the network of Figure 9-35.

9-20. Determine the voltage drop across the 5-kΩ resistor in the network of Figure 9-36.

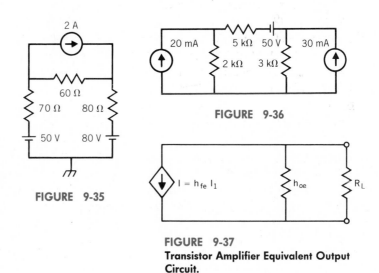

FIGURE 9-35

FIGURE 9-36

FIGURE 9-37
Transistor Amplifier Equivalent Output Circuit.

9-21. The output equivalent circuit of a typical transistor amplifier is shown in Figure 9-37. Calculate the output voltage if the input signal current $I_1 = 24$ μA, the forward current ratio $h_{fe} = 40$, the output admittance $h_{oe} = 5 \times 10^{-5}$ S, and the load resistance $R_L = 22$ kΩ.

9-22. Determine the Thévenin-equivalent source for the transistor output circuit of Problem 9-21.

9-23. Three voltage sources in a network ($E_1 = 40$ V, $R_1 = 2.2$ kΩ; $E_2 = 24$ V, $R_2 = 3.3$ kΩ; and $E_3 = -24$ V, $R_3 = 4.7$ kΩ) all feed a common component in parallel. Determine the Millman-equivalent voltage source.

9-24. A battery charger has an open-circuit voltage of 16.0 V and an internal resistance of 0.1 Ω. It is charging four storage batteries in parallel with open-circuit voltages and internal resistances as follows: $V_1 = 12.8$ V, $R_1 = 0.2$ Ω; $V_2 = 12.3$ V, $R_2 = 0.3$ Ω; $V_3 = 13.5$ V, $R_3 = 0.25$ Ω; and $V_4 = 12.6$ V, $R_4 = 0.15$ Ω. What is the total battery charger current?

9-25. Determine V_L in the circuit of Figure 9-38(a).

9-26. Determine V_L in the circuit of Figure 9-38(b).

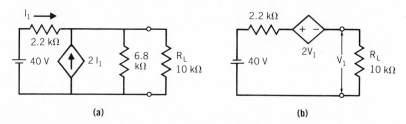

FIGURE 9-38

9-27. Find the Thévenin-equivalent voltage source for the network of Figure 9-39(a).

9-28. Find the Thévenin-equivalent voltage source for the network of Figure 9-39(b).

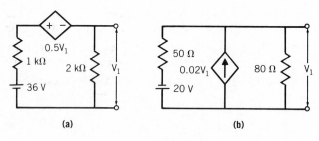

FIGURE 9-39

9-29. Determine V_L in the amplifier equivalent circuit of Figure 9-40.

9-30. A 2000-Ω feedback resistor is connected between the junction of the 200-Ω and 800-Ω resistors and the top of the VCCS in Figure 9-40. Determine V_L.

FIGURE 9-40

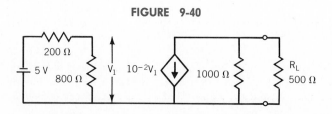

9-31. Given the π-network of Figure 9-41(a), use a delta-wye transformation to determine the equivalent T-network.

9-32. Given the T-network of Figure 9-41(b), use a wye-delta transformation to determine the equivalent π-network.

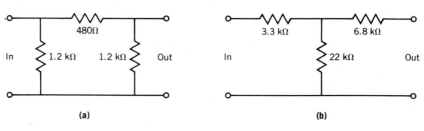

FIGURE 9-41
(a) π-Network; (b) T-Network.

9-33. Given the following data for the Wheatstone bridge of Figure 9-26(a), use a delta-wye transformation to determine the meter current. $E = 100$ V, $R_1 = 10\ \Omega$, $R_2 = 20\ \Omega$, $R_3 = 30\ \Omega$, $R_4 = 40\ \Omega$, and the meter resistance $R_M = 50\ \Omega$.

9-34. Given the following data for the Wheatstone bridge of Figure 9-26(a), use Thévenin's theorem to determine the meter current. $E = 200$ V, $R_1 = 10$ kΩ, $R_2 = 2.5$ kΩ, $R_3 = 5$ kΩ, $R_4 = 20$ kΩ, and $R_M = 25$ kΩ.

9-35. Use Thévenin's theorem to determine the current through a 500-Ω load connected to the output terminals of the bridged-T attenuator of Figure 9-42 when the input voltage is 2 V.

9-36. Use a delta-wye transformation to determine the current through a 300-Ω load connected to the output terminals of the bridged-T attenuator of Figure 9-42 when the input voltage is 4 V.

9-37. If the voltage applied to the four-terminal lattice network of Figure 9-43 is 360 V, what value of load resistor must be connected to the output terminals to draw a load current of 1 A? Use Thévenin's theorem.

9-38. Use a delta-wye transformation to determine what voltage must be applied to the input terminals of the lattice network of Figure 9-43 in order to draw a 2-A current from the source when a 20-Ω load is connected to the output terminals of the network.

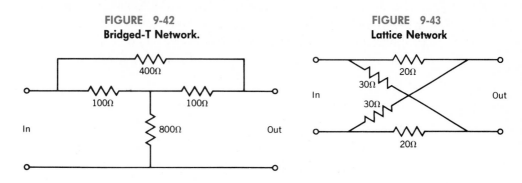

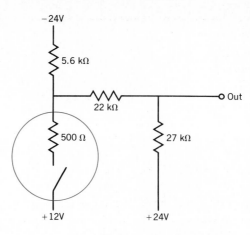

FIGURE 9-44

9-39. The switch in the pulse generator circuit of Figure 9-44 alternately closes for 1 ms and opens for 4 ms. Draw the open-circuit output voltage waveform (with respect to chassis) and determine its amplitude accurately.

9-40. A voltage source with zero internal resistance and an open-circuit voltage of 20 V is connected to the input terminals of the T-network of Figure 9-41(b). What is the reading on a milliammeter with zero resistance connected across the output terminals? What is the milliammeter reading when the voltage source is connected to the output terminals and the milliammeter is connected to the input terminals?

REVIEW QUESTIONS

9-1. What is meant by a **Thévenin-equivalent source?**

9-2. What is the difference between the "black-box" method and the Thévenin's-theorem method of determining the internal resistance of a Thévenin-equivalent source?

9-3. Show that the Thévenin-equivalent resistance, as found by the short-circuit method in the "black-box" experiment, is the same as that found by the procedure set forth in the statement of Thévenin's theorem.

9-4. Draw a Thévenin-equivalent circuit for the source shown in Figure 9-45 and express the equivalent voltage E_x and the equivalent internal resistance R_x in terms of the parameters of the original circuit.

9-5. If a load R_L is connected to the original circuit of Figure 9-45, derive a single equation for I_L.

9-6. In a Thévenin-equivalent circuit, $I_L = E_x/(R_x + R_L)$. Show that this is equal to the equation for I_L in Question 9-5, thereby verifying Thévenin's theorem.

9-7. What is meant by a **Norton-equivalent source?**

9-8. What happens to the terminal voltage of a Norton-equivalent source when a load resistor is connected to its terminals? Explain.

9-9. Attempt a solution for Problem 9-3 by Norton's theorem. Note that in this problem it is easier to find the Norton-equivalent source by first finding the Thévenin-equivalent source and then converting to the equivalent constant-current source.

9-10. Explain the polarity of the voltage drops across R_G and R_B in Figure 9-14(a).

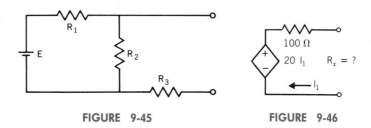

FIGURE 9-45   FIGURE 9-46

9-11. Show how you would use Millman's theorem to obtain the Thévenin-equivalent circuit of Figure 9-14(b).

9-12. The sources feeding the network shown in Figure 9-34 have negligible internal resistance. How do we go about replacing them with equivalent constant-current sources in applying Millman's theorem to the network?

9-13. The numerical constant for the current-controlled voltage source in Figure 9-18 is assigned the unit *ohm*. What unit is assigned to the numerical constant for
(a) A voltage-controlled voltage source?
(b) A voltage-controlled current source?

9-14. What restrictions are there on using the superposition theorem in a network containing dependent sources?

9-15. What restrictions are there on using Thévenin's and Norton's theorems in a network containing dependent sources?

9-16. What value of R_x is "seen" looking back into the terminals of the dependent voltage source of Figure 9-46? (*Hint:* Connect a fixed voltage source to the terminals and calculate the terminal voltage and current—the voltmeter-ammeter method of determining R_x.)

9-17. Derive the wye-delta transformation Equation (9-6).

9-18. Figure 9-26(b) shows a method for solving a Wheatstone bridge using a delta-wye transformation. Draw a diagram showing how we can solve the same bridge circuit using a wye-delta transformation.

9-19. Under what circumstances can we substitute a resistor for a voltage source in an electrical network?

9-20. The mathematical process of addition in electronic analog computers is performed by the circuit shown in Figure 9-47. By selecting several numerical combinations for E_1 and E_2, show that the open-circuit E_{out} is always directly proportional to $E_1 + E_2$. (All voltages are expressed with respect to ground.)

FIGURE 9-47

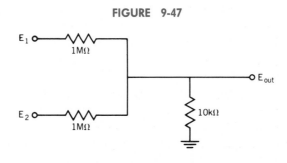

10

ELECTRICAL MEASUREMENT

We have investigated the interrelationships which must exist among applied voltages and resulting currents, voltage drops, and power in electric circuits ranging from simple series and parallel resistors to quite complex resistance networks. We discovered that all electric circuit behavior is governed by a few fundamental laws. To extend our *understanding* of the fundamentals of electric circuit behavior to practical examples, we must be able to *measure* the circuit parameters to which we have been assigning numerical values. In this chapter, we shall consider how the commonly used electrical measuring instruments obtain this numerical information. We shall not attempt to cover details of the construction, characteristics, and applications of this equipment. Such topics can be considered more effectively in a concurrent laboratory course on electrical measurement.

10-1 MECHANICAL TORQUE FROM ELECTRIC CURRENT

As we shall investigate in more detail in Chapter 13, when electric current flows in a conductor, it produces a magnetic field around the conductor. To assist us in visualizing the nature of a magnetic field, Michael Faraday suggested that we think of a magnetic field as being made up of **magnetic lines of force**. Figure 10-1(a) shows a cross section of an electric conductor in which the conventional current direction is away from us into the page (as indicated by the feathered end of an arrow). From this perspective, the electric current produces magnetic lines of force which appear as

242

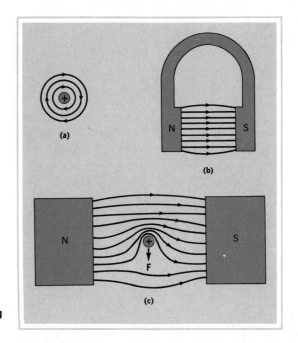

FIGURE 10-1
Force Acting on a Current-Carrying Conductor in a Magnetic Field.

concentric circles. Figure 10-1(b) shows a permanent magnet with its magnetic lines of force crossing the air gap between the N and S poles of the magnet.

If we now place the current-carrying conductor between the poles of the U-shaped permanent magnet of Figure 10-1(b), we cannot simply draw the magnetic field of the conductor superimposed on that of the permanent magnet. One of the basic rules of the concept of lines of force is that they cannot intersect. Therefore, the two component magnetic fields must combine to form a single composite magnetic field. Below the conductor in Figure 10-1(c), the direction of the lines of force created by the current in the conductor is *opposite* to the direction of the permanent magnet field. This results in a cancelling effect that reduces the flux density below the conductor. Above the conductor, the direction of both component fields is the *same*, thus increasing the flux density. We can think of the effect of the current in the conductor as causing the magnetic lines of force of the permanent magnet in the vicinity of the current-carrying conductor to *detour* above the conductor.

Since magnetic lines of force tend to become as short as possible and also tend to repel one another, the lines of force of the composite magnetic field of Figure 10-1(c) will attempt to straighten out and to space themselves uniformly in an effort to regain the flux pattern of Figure 10-1(b). To do so, the magnetic field must force the current-carrying conductor to move *downward*. The magnitude of this force is directly dependent on the strength of the two component magnetic fields. The strength of the permanent magnet field is represented by its flux density B. The strength of the magnetic field around the conductor is proportional to the current I in the conductor and also the length l of the conductor within the magnetic field of the permanent magnet. Since the international standard ampere is defined in terms of the force between parallel current-carrying electric conductors (Section 15-4),

$$F = BIl \tag{10-1}$$

where F is the force acting on the conductor in newtons, B is the flux density of the permanent magnet field in teslas, I is the current in the conductor in amperes, and l is the length in meters of the conductor perpendicular to the magnetic field.

Since the flux density produced by a permanent magnet remains constant with a given magnetic circuit, both B and l in Equation (10-1) are constant. Therefore, the force acting on the conductor is directly proportional to the current in it. If we can devise a means of indicating this mechanical force, we then have a device for measuring the current in the conductor. Figure 10-2 shows a spring scale, like those used for weighing fish, attached to the conductor. According to Hooke's law, the extension of the spring is directly proportional to the mechanical force applied to it. This force, in turn, is directly proportional to the electric current in the conductor.

A more practical form for a current-measuring instrument is based on two conductors (the two sides of a loop) supported in the field of a permanent magnet by a pivot X. If the current direction is away from us in conductor B, it must be toward us in the other side of the loop, conductor A. This current causes the composite magnetic field to be distorted as shown in Figure 10-3. The mechanical force acting downward on conductor B and upward on conductor A results in a turning force or **torque** that attempts to rotate the loop in a clockwise direction about its pivot. This torque is counterbalanced by the countertorque produced by a flat spiral spring fastened from the pivot to some stationary point. Since the countertorque of the hairspring is directly proportional to the angle through which the pivot is rotated (Hooke's law), the angle through which the loop rotates is directly proportional to the current

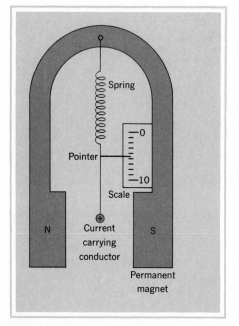

FIGURE 10-2
Simple Current-Measuring Device

FIGURE 10-3
Developing a Mechanical Torque from Electric Current.

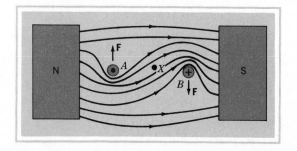

through it. Since torque equals force times the length of the lever arm, the torque produced by each side of the loop will be

$$T = Fr$$

where T is the torque in newton meters, F is the force acting on the conductors in newtons [Equation (10-1)], and r is the radius of the loop in meters.

The total torque contributed by *both* sides of the loop will be

$$T = 2Fr = Fd$$

where d is the diameter of the loop in meters.

If the loop consists of more than one turn of wire, each side of the loop can be thought of as a stranded conductor, with each strand carrying a current equal to the current in the coil. Therefore, the effective current in the loop will be

$$I_{eff} = NI$$

where N is the number of turns of wire in the pivoted coil.

We can modify Equation (10-1) to apply to a loop in a magnetic field.

$$\therefore T = BIlNd \qquad (10\text{-}2)$$

EXAMPLE 10-1

What torque is developed by a 1-mA current flowing in a 120-turn coil wound on a 1-cm × 1-cm square form? The flux density in the air gap is 0.1 T.

SOLUTION

$$T = BIlNd = 0.1 \text{ T} \times 1 \text{ mA} \times 1 \text{ cm} \times 120 \times 1 \text{ cm}$$

$$= 1.2 \times 10^{-6} \text{ N} \cdot \text{m}$$

Equation (10-2) is the basic **electric-motor** equation. But because we are using a countertorque spring to prevent the loop from rotating freely, we can adapt the electric-motor principle to the **moving-coil** meter movement shown in Figure 10-4. This movement is sometimes referred to as a D'Arsonval movement, after the man who first patented a current-measuring device based on the moving-coil principle. Because Edward Weston developed the moving-coil instrument from the delicate laboratory apparatus of D'Arsonval to the reasonably rugged panel type of meter in use today, the moving-coil movement is also known as the Weston movement.

Since B, l, N, and d are constant for a given meter movement, the torque is directly proportional to the current in the moving coil; and, since the countertorque of the spiral spring is directly proportional to its angular extension, the angular displacement of the pointer is directly proportional to the current in the moving coil. If we adjust the spiral spring so that the coil occupies the position shown in Figure 10-3 when the current is zero, the movement is called a **galvanometer.** A galvanometer can indicate the flow of direct current in either direction since a clockwise torque will

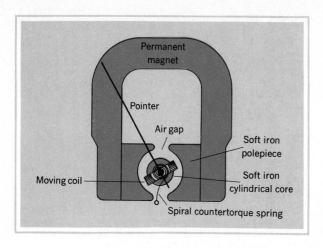

FIGURE 10-4
Moving-Coil Movement.

move the coil (and the pointer) in one direction from center, and a counterclockwise torque will move it in the other direction. We usually use a current meter or **ammeter** in a circuit in which the current direction does not change. By readjusting the stationary end of the spiral spring, we can set the moving coil and pointer in the position shown in Figure 10-4. Current through the coil in the proper direction will cause the pointer to be deflected across the scale, but reversed meter leads will cause the pointer to read off scale to the left, coming up against a mechanical stop.

10-2 THE AMMETER

By using magnetic circuits that produce a high flux density in the air gap and delicate countertorque springs, we can build commercial movements which will read full scale with a current of 20 microamperes or less through the moving coil. The same spiral springs that provide the countertorque are used to connect the moving coil to the meter terminals. By using less turns of a larger wire size on the moving coil and a stronger spiral spring, moving-coil movements can be built which require currents in the order of several milliamperes to obtain full-scale deflection. If the physical size of the meter is limited to about eight centimeters in diameter, the maximum wire size which can be conveniently used for the moving coil limits the meter to about 30 milliamperes for full-scale deflection. To use the moving-coil instrument as an ammeter for larger currents, we must apply our knowledge of resistors in parallel.

For the following example, we shall use a moving-coil movement which reads full scale when one milliampere flows through its moving coil. A one-milliampere movement is a reasonable compromise between the more sensitive 50-microampere movements used in electronic-circuit measurements and the more rugged 5-milliampere movements used for measurements in electric power circuits. The moving coil of the 1-milliampere movement in question has a resistance of 27 ohms. It is usually customary to include a **calibrating resistor** in series with the moving coil to bring the total resistance of the movement up to some convenient round number, as shown in Figure 10-5. It is common practice to select this calibrating resistor so that the full-scale current through the movement and its calibrating resistor produces a 50-mV drop between the meter terminals. For the movement in question,

$$R_T = \frac{V}{I} = \frac{50 \text{ mV}}{1 \text{ mA}} = 50 \; \Omega$$

and, therefore, the resistance of the calibrating resistor will be

$$R_C = 50 - 27 = 23 \; \Omega$$

The calibrating or **swamping** resistor is included in the case of the meter and is made from precision resistance wire having a small negative temperature coefficient to offset the positive temperature coefficient of the moving coil.

To use the one-milliampere movement to measure a current in the order of 0.6 ampere, as is the case in the circuit of Figure 10-6, we must arrange the meter circuit so that the movement and its **shunt** resistor form two parallel branches with 999 times as much current flowing through the shunt as through the meter movement. Since the shunt and the movement are in parallel, the *same* voltage drop appears across both and their current ratio is inversely proportional to their resistance ratio. Therefore,

$$V_{sh} = V_M$$

from which

$$I_{sh} \times R_{sh} = I_M \times R_M$$

and

$$R_{sh} = \frac{1}{999} \times R_M = \frac{50}{999} = 0.050\,05 \; \Omega$$

If the current flowing through the lamp in Figure 10-6 is sufficient to make the movement read full scale, the current in the moving-coil branch will be 1 milliampere.

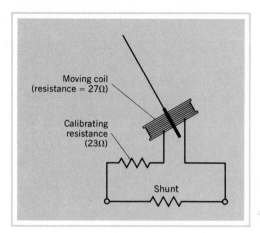

FIGURE 10-5
Moving-Coil Ammeter with Shunt Resistor.

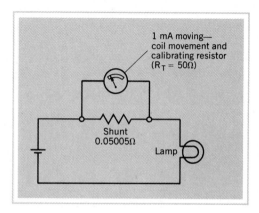

FIGURE 10-6
Measuring Electric Current

Therefore the current through the shunt resistor will be 999 milliamperes and the current through the lamp is $1 + 999$ mA $= 1$ ampere. As long as this movement and shunt are used together, we can interpret the meter scale as being calibrated for 1 ampere full scale.

If the lamp current happens to be 0.6 ampere, this total current will split so that one part (0.6 mA) will flow through the moving coil and 999 parts (0.599 4 A) will flow through the shunt. Therefore, we can again interpret the 0.6 scale marking of the basic 1-milliampere movement as indicating 0.6 ampere. For currents up to about 25 amperes, the shunt is physically small enough to be placed inside the meter case. For larger currents, external shunt resistors are required.

Connecting an ammeter into an electric circuit will have some slight effect on the circuit since we are adding a little extra resistance to the circuit. The total resistance of the meter and its shunt in Figure 10-6 is 0.05 ohm. The hot resistance of the lamp is in the order of 200 ohms. Therefore, the error in circuit current due to adding the ammeter to the circuit is only 0.05 part in 200, or 0.025%. In practice, the main problem with making current measurements is that we must place the ammeter in *series* with the circuit whose current is to be measured. Unless the meter is to be left in the circuit permanently, this involves opening the circuit in order to include the meter and then reconnecting the circuit when we remove the meter. Because of its low resistance (0.05 Ω for the 1-ampere meter), we must guard against connecting an ammeter *across* a source.

When an ammeter is to be permanently connected in a circuit, a single range is quite satisfactory. But for a test ammeter, it would be desirable to have several ranges so that an unknown current can be read on a range which provides a reading somewhere around two-thirds full scale where the meter calibration is most accurate. We can accomplish this by using several shunts, as shown in Figure 10-7(a). The switch must be of the *make-before-break* type so that the meter cannot be damaged by being in the circuit without a shunt as we change the range.

The Ayrton shunt of Figure 10-7(b) eliminates the possibility of the meter being in circuit without a shunt. This advantage is gained at the price of a slightly higher meter resistance. The Ayrton shunt also provides us with an excellent opportunity to apply our resistance network theory to a practical circuit.

FIGURE 10-7
(a) Multirange Shunt; (b) Ayrton Shunt.

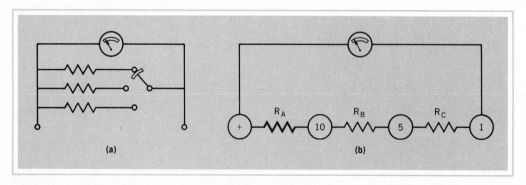

EXAMPLE 10-2

Using a 1-mA movement with a total resistance of 50 Ω in the circuit shown in Figure 10-7(b), determine R_A, R_B, and R_C so that when the common + terminal and terminal 1 are used, the meter is a 1-A meter; when the common + terminal and terminal 5 are used, the meter is a 5-A meter; and when the common + terminal and terminal 10 are used, the meter is a 10-A meter.

SOLUTION 1

On the 1-A range, $R_A + R_B + R_C$ forms the shunt, and it must take 999 parts of the total current when the meter movements take 1 part. Therefore, as in the example of the simple shunt,

$$R_A + R_B + R_C = \frac{1}{999} \times 50 = 0.050\,05 \ \Omega \tag{1}$$

On the 5-A range, $R_A + R_B$ forms the shunt and R_C is in series with the moving coil. For the 5-A range, 4999 parts of the total current must flow through the shunt when the meter movement passes 1 part. Therefore,

$$R_A + R_B = \frac{50 + R_C}{4999} \tag{2}$$

On the 10-A range, R_A is the shunt and R_B and R_C are in series with the meter. For the 10-A range, R_C takes 9999 parts of the total current, while 1 part flows through the meter and R_B and R_C. Therefore,

$$R_A = \frac{50 + R_B + R_C}{9999} \tag{3}$$

The solution now becomes an algebra problem involving solution of the simultaneous equations (1), (2), and (3) for the values of R_A, R_B, and R_C.

$$4999 \times (1) \text{ gives} \quad 4999R_A + 4999R_B + 4999R_C = 250.2$$
$$(2) \text{ gives} \quad \underline{4999R_A + 4999R_B - R_C = 50}$$
Subtracting, $\qquad\qquad\qquad\qquad\qquad\qquad\qquad 5000R_C = 200.2$

$$R_C = \frac{200.2}{5000} = 0.040\,04 \ \Omega$$

$$9999 \times (1) \text{ gives} \quad 9999R_A + 9999R_B + 9999R_C = 500.45$$
$$(3) \text{ gives} \quad \underline{9999R_A - R_B - R_C = 50}$$
Subtracting, $\qquad\qquad\qquad\qquad 10\,000R_B + 10\,000R_C = 450.45$

Substituting for R_C, $\qquad 10\,000R_B + 400.4 = 450.45,$

$$R_B = \frac{50.05}{10\,000} = 0.005\,005 \ \Omega$$

Substituting for R_B and R_C in Equation (1),

$$R_A + 0.005\,005 + 0.040\,04 = 0.050\,05$$

$$\therefore R_A = 0.050\,05 - 0.045\,045 = \mathbf{0.005\,005\ \Omega}$$

SOLUTION 2

We can avoid the simultaneous equations of Solution 1 by equating the voltage drops across the two parallel branches. On the 1-A range,

$$I_{sh}(R_A + R_B + R_C) = I_M R_M$$

from which $\quad R_A + R_B + R_C = \dfrac{0.001}{0.999} \times 50 = 0.050\,05\ \Omega$

as in Solution 1. On the 5-A range,

$$R_{sh} = R_A + R_B = (R_A + R_B + R_C) - R_C = 0.050\,05 - R_C$$

and the resistance of the meter branch is $50 + R_C$. Hence

$$I_{sh}(0.050\,05 - R_C) = I_M(50 + R_C)$$

Solving for R_C,

$$R_C = \frac{I_{sh}(R_A + R_B + R_C) - I_M R_M}{I_{sh} + I_M}$$

$I_M R_M$ remains the same for all ranges $= 0.001 \times 50 = 0.05$ V. Note that the denominator becomes the full-scale current for the range. Hence

$$R_C = \frac{4.999 \times 0.050\,05 - 0.05}{5.000} = \mathbf{0.040\,04\ \Omega}$$

Similarly,

$$R_B + R_C = \frac{9.999 \times 0.050\,05 - 0.05}{10.000} = 0.045\,045$$

from which $\quad R_B = 0.045\,045 - 0.040\,04 = \mathbf{0.005\,005\ \Omega}$

and $\quad R_A = 0.050\,05 - 0.045\,045 = \mathbf{0.005\,005\ \Omega}$

10-3 THE VOLTMETER

The same moving-coil movement forms the basis of a practical **voltmeter.** As we have already noted, if our 1-milliampere movement has a total resistance of 50 ohms, a voltage drop of $V = IR = 0.001 \times 50 = 0.05$ V will appear across it when it is

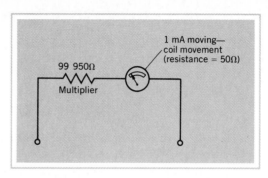

FIGURE 10-8
Simple Voltmeter.

reading full scale. If we now connect a 99 950-ohm resistor in series with the meter (as in Figure 10-8) to bring the total resistance up to 100 kilohms, the voltage we must apply to produce a 1-milliampere current through the meter (and full-scale deflection) is $V = IR = 0.001 \times 100\,000 = 100$ volts. If we apply a 50-volt potential difference to the voltmeter, the current through it will be $I = V/R = 50/100\,000$ ampere $= 0.5$ mA and the meter will read half scale. Therefore, we can provide this particular voltmeter with a linear scale marked from 0 to 100 volts. The series resistor which is required to convert the basic moving-coil movement into a voltmeter is called a **multiplier resistor.**

To measure open-circuit voltage, a perfect voltmeter should draw no current. Since current must flow in a moving-coil movement to obtain a reading, a 50-microampere basic movement will make a much more accurate voltmeter for use in high-resistance circuits than the 1-milliampere movement. The resistance of a standard 50-microampere movement and its calibrating resistor is 1000 ohms. Therefore, to obtain a 100-volt full-scale range, the total resistance must be $R_T = 100/0.000\,05$ ohms $= 2$ megohms; and the multiplier resistance must be $2\,000\,000 - 1000 = 1\,999\,000$ ohms. To obtain a 500-volt scale with a 50-microampere movement, the total resistance must be ten megohms, and to obtain a 1-volt full-scale range, $R_T = 1/0.000\,05$ ohms $= 20$ kilohms.

No matter what full-scale voltage we select for the 50-microampere basic movement, the total resistance required is 20 kilohms for *each* volt of the full-scale reading. Therefore, when we use a 50-microampere movement as a voltmeter, it has a **sensitivity** of **20,000 ohms per volt.** We determine voltmeter sensitivity by dividing the full-scale current into 1 volt. We find the total resistance of a meter movement and the multiplier resistor for a given range by multiplying the full-scale voltage by the voltmeter sensitivity. Therefore, when we use a 1-milliampere movement as a voltmeter, its sensitivity is $1/0.001 = 1000$ ohms per volt, and the total resistance required for a 500-V range will be 500×1000 ohms $= 500$ kilohms.

Multirange-voltmeter circuits are quite straightforward, as shown in Figure 10-9.

FIGURE 10-9
Multirange Voltmeter.

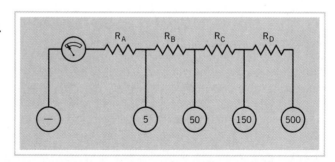

251

EXAMPLE 10-3

A 1-mA full-scale moving-coil movement with a resistance of 50 Ω is to be made into a voltmeter with scales of 5 V, 50 V, 150 V, and 500 V. Calculate R_A, R_B, R_C, and R_D in the circuit of Figure 10-9.

SOLUTION

Voltmeter sensitivity = $1/0.001$ = 1000 ohms per volt. Therefore, the total resistance for the 5-V range = 5 kΩ, and

$$R_A = 5000 - 50 = \mathbf{4950 \ \Omega}$$

The total resistance for the 50-V range = 50 kΩ, and

$$R_B = 50\,000 - 5000 = \mathbf{45 \ k\Omega}$$

The total resistance for the 150-V range = 150 kΩ, and

$$R_C = 150 - 50 = \mathbf{100 \ k\Omega}$$

The total resistance for the 500-V range = 500 kΩ, and

$$R_D = 500 - 150 = \mathbf{350 \ k\Omega}$$

The voltmeter is a very useful instrument for checking the operating conditions in an electric circuit. With suitable test leads, we can clip a voltmeter *across* any part of an electric circuit without having to disconnect or disturb the circuit being checked, as is the case when we test with an ammeter.

10-4 VOLTMETER LOADING EFFECT

Since an ammeter is connected in *series* with a circuit in which the current is to be measured, an ideal ammeter should have zero ohms internal resistance. If the internal resistance of an ammeter is comparable to the resistance of the circuit under test, adding the ammeter to the circuit will increase the total resistance of the circuit appreciably. Hence, the current indicated by the ammeter will be less than the circuit current when the ammeter is shorted out. However, as we noted in Section 10-2, in most applications the ammeter resistance is sufficiently small in comparison with the circuit resistance that the loading effect of the ammeter is negligible.

Since a voltmeter is connected in *parallel* with the portion of the circuit in which the voltage drop is to be measured, an ideal voltmeter should draw zero current and should, therefore, have an infinitely high internal resistance. Unfortunately, the moving-coil type of voltmeter must draw some current to obtain a pointer deflection. The smaller the current required for full-scale deflection, the more sensitive the voltmeter and the smaller the loading effect it will have on the circuit being tested. As noted in Section 10-3, voltmeter sensitivity is expressed in ohms per volt of full-scale voltage, rather than in terms of full-scale current drain. We can show the effect of voltmeter loading in practical circuits by the following numerical example.

EXAMPLE 10-4

(a) What is the actual voltage drop across the 200-kΩ resistor in Figure 10-10 with no meter in the circuit?

(b) What will a 20 000-ohm-per-volt meter with 150 V full scale read when connected across the 200-kΩ resistor?

(c) What will a 1000-ohm-per-volt-meter with 150 V full scale read when connected across the 200-kΩ resistor?

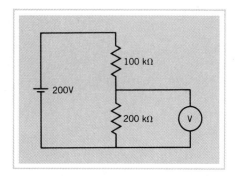

FIGURE 10-10
Effect of Voltmeter Sensitivity.

SOLUTION

(a) From the voltage-divider principle,

$$V = \frac{200 \text{ k}\Omega}{300 \text{ k}\Omega} \times 200 \text{ V} = \textbf{133.3 } \textit{V}$$

(b) A 20 000-ohm-per-volt meter with a 150-V range has a total resistance of $20 000 \times 150 \ \Omega = 3$ MΩ. This 3-MΩ resistance is in parallel with the 200-kΩ resistor, thus reducing the resistance of the parallel combination to

$$R = \frac{200 \times 3000}{200 + 3000} = 187.5 \text{ k}\Omega$$

and

$$V = \frac{187.5 \text{ k}\Omega}{287.5 \text{ k}\Omega} \times 200 \text{ V} = \textbf{130.4 V}$$

(c) A 1000-ohm-per-volt meter on a 150-V range has a total resistance of $1000 \times 150 \ \Omega = 150$ kΩ. Therefore, the 200-kΩ resistor and the voltmeter in parallel have an equivalent resistance of

$$R = \frac{200 \times 150}{200 + 150} = 85.7 \text{ k}\Omega$$

and

$$V = \frac{85.7 \text{ k}\Omega}{185.7 \text{ k}\Omega} \times 200 \text{ V} = \textbf{92.8 V}$$

When measuring voltage in electric power circuits where fairly large currents are flowing, an extra five mA does not alter the circuit conditions appreciably. Thus, the

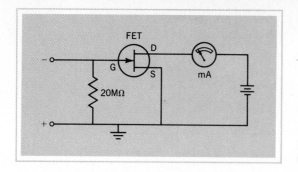

FIGURE 10-11
Electronic Voltmeter Principle.

more rugged and less sensitive voltmeters are quite satisfactory. However, in electronic circuits where the normal circuit current is very small, an insensitive voltmeter can alter circuit conditions considerably. Even the 20 000-ohm-per-volt meter used in part (b) of Example 10-4 introduced an error in excess of 2%. Consequently, the simple moving-coil type of voltmeter is not suitable for voltage measurement in high-resistance electronic circuits.

We can still use the simple moving-coil milliammeter to measure voltage by measuring the *drain* current of a field-effect transistor. Over the operating range, the drain current of the FET in the simplified circuit of Figure 10-11 is inversely proportional to the voltage applied between the *gate* and *source* terminals of the FET with the polarity shown. Hence, we can calibrate the milliammeter scale in terms of dc input voltage. As long as the gate of the FET is negative with respect to the source terminal, the gate/source *pn*-junction is *reverse*-biased and the gate current is negligible. Hence, the internal resistance of the **electronic voltmeter** is determined by the value of the gate return resistor in Figure 10-11, which can be several megohms.

10-5 THE POTENTIOMETER

Although the electronic voltmeter is quite satisfactory for most test purposes in high-resistance electronic circuits, its accuracy is not great enough for measuring the open-circuit voltage of sources with small terminal voltages and high internal resistances, such as the cell used to measure the pH of a chemical solution. We can still use a moving-coil movement for this purpose by making use of an accurately known reference voltage and a **potentiometer.**

When the galvanometer switch in Figure 10-12 is in the center open-circuit position, no current is drawn from the sliding tap B of the potentiometer. Hence, the current through R_{AB} is the *same* as the current through R_{BC}. Consequently,

$$\frac{V_{BC}}{R_{BC}} = \frac{V_{AC}}{R_{AC}}$$

But both V_{AC} and R_{AC} are fixed. Therefore, the open-circuit output potential difference V_{BC} of the potentiometer is directly proportional to the resistance R_{BC}. We can construct an experimental potentiometer by fastening a length of uniform-resistance wire to a one-meter ruler (see Figure 10-24). The open-circuit output voltage would then be

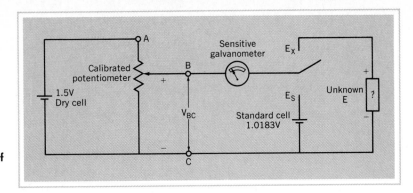

FIGURE 10-12
Potentiometer Method of Measuring an Unknown Source Voltage.

directly proportional to the *length* of wire representing R_{BC}. Commercial potentiometers are usually precision wire-wound resistors with the variable sliding contact arranged so that the ratio of R_{BC} to R_{AC} exactly equals the percentage of the full-scale rotation of the calibrated dial. Although we do not know the absolute magnitude of V_{BC} accurately (since the terminal voltage of the dry cell is not precise), we can *compare* two different values of V_{BC} quite accurately by the ratio of the two potentiometer pointer readings.

To determine the magnitude of the unknown source voltage in Figure 10-12, we carefully adjust the potentiometer until there is no trace of galvanometer deflection when we throw the galvanometer switch to the upper (E_x) position. If there is no meter deflection, there must be no current through the moving coil and, therefore, no potential difference across the meter movement. This, in turn, means that $V_{BC} = E_x$. And, because the meter current is zero, we are measuring the *open-circuit* voltage of the unknown source, regardless of its internal resistance.

Since we do not know the absolute magnitude of V_{BC}, we must compare it with a known reference voltage. The Weston standard cell is a precision laboratory standard intended for use only with potentiometers. It cannot be used to operate a lamp and still serve as a standard. With proper treatment, the Weston cell develops an open-circuit voltage of 1.0183 volts. Having recorded the potentiometer scale reading when $V_{BC} = E_x$, we adjust the potentiometer for zero meter deflection when we throw the switch to the lower (E_s) position. We can now calculate E_x, knowing that the ratio of the two open-circuit voltages equals the ratio of the two potentiometer readings.

EXAMPLE 10-5

The potentiometer circuit of Figure 10-12 gives zero meter deflection in the E_x position for a potentiometer scale reading of 478. Balance is obtained in the E_s position when the potentiometer reading is 662. What is the open-circuit voltage of the unknown source?

SOLUTION

$$\frac{E_x}{E_s} = \frac{478}{662}$$

$$\therefore E_x = 1.0183 \times \frac{478}{662} = \mathbf{0.735 \ V}$$

10-6 RESISTANCE MEASUREMENT

We can determine the resistance of the unknown resistor in Figure 10-13 by applying Ohm's law to the readings obtained from the voltmeter and ammeter. This system of resistance measurement has the disadvantage of having to connect the unknown resistance into a special circuit. However, we can use this method in the electrical laboratory as a means of determining very small resistances such as the armature resistance of a motor, particularly if we want the resistance to be measured with its normal current through it.

With the voltmeter connected as shown in Figure 10-13, the ammeter indicates the sum of the currents through R_x and the voltmeter. Unless the voltmeter current is an insignificant fraction of the current through R_x, the V/I ratio taken from the meter readings (which is the equivalent resistance of the two parallel branches) will be lower than the actual value of R_x. We can check for voltmeter loading by watching the ammeter reading as we disconnect the volt-meter. If there is any noticeable decrease in the current, we then connect the voltmeter on the source side of the ammeter. The ammeter now reads only the current through R_x. But the voltmeter now reads the sum of the voltage drops across the ammeter and R_x. However, if R_x is large enough to show a noticeable voltmeter loading effect, the voltage drop across the ammeter will be insignificant in comparison with V_x and we can neglect ammeter loading effect. If there is no noticeable change in current when we disconnect the voltmeter from its original position across R_x, we reconnect it in its original position, as shown in Figure 10-13.

For quick checks of circuit resistance, we need a simpler apparatus than that of Figure 10-13. We can once more use the basic moving-coil movement to construct an **ohmmeter.** The simple ohmmeter shown in Figure 10-14(a) consists of a 1-milliampere movement, a 4.5-volt battery, and sufficient resistance to permit a current of 1 milliampere when we short-circuit the ohmmeter terminals. A portion of the total resistance is adjustable so that we can calibrate the meter to read exactly full scale when we connect the two test leads together. In this example, the total resistance (including the meter movement) must be

$$R_T = \frac{E}{I} = \frac{4.5 \text{ V}}{1 \text{ mA}} = 4.5 \text{ k}\Omega$$

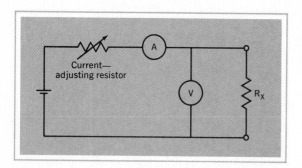

FIGURE 10-13
Measuring Resistance with a Voltmeter and an Ammeter.

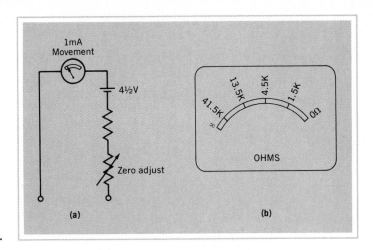

FIGURE 10-14
Simple Ohmmeter.

EXAMPLE 10-6

In preparing the scale for the ohmmeter of Figure 10-14, determine what resistance values must be marked at full scale, center scale, one-quarter of full scale, and one-tenth of full scale.

SOLUTION

As we have already noted, the total internal resistance of the ohmmeter is adjusted so that the meter reads exactly full scale when the test leads are short-circuited. This represents connecting the test leads to a zero-ohm resistor. Therefore, the full-scale point on the scale must be marked **zero ohms.**

For the meter to read half scale, we must connect the test leads to a resistor which has the same value as the total ohmmeter resistance so that the total resistance in the series loop is now doubled, thus reducing the current to 0.5 mA. Therefore, the center scale point must be marked **4.5 kΩ.**

$$I = \frac{E}{R_M + R_x} = \frac{4.5 \text{ V}}{4.5 \text{ k}\Omega + 4.5 \text{ k}\Omega} = 0.5 \text{ mA}$$

For one-quarter of full scale, the current must be 0.25 mA. Therefore, the total resistance in the loop must be

$$R_T = \frac{4.5 \text{ V}}{0.25 \text{ mA}} = 18 \text{ k}\Omega$$

and

$$R_x = 18 \text{ k}\Omega - 4.5 \text{ k}\Omega = \textbf{13.5 k}\Omega$$

Finally, the resistance value for the one-tenth of full-scale mark will be

$$R_x = \frac{4.5 \text{ V}}{0.1 \text{ mA}} - 4500 \ \Omega = \textbf{40.5 k}\Omega$$

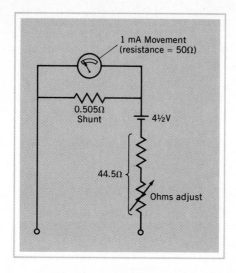

FIGURE 10-15
Low-Range Ohmmeter.

As shown in Figure 10-14(b), the scale of an ohmmeter is nonlinear as far as the spacing of the ohms values is concerned. But since a few ohms one way or the other make a greater *percentage* difference to a low resistance than to a high resistance, this nonlinear scale is not a disadvantage. The simple ohmmeter of Figure 10-14 is limited in its usefulness in that it is not accurate for very low or very high resistances, since these appear at the ends of the scale. For use in electronic circuit testing where we need a high-range ohmmeter, we would start with a 50-microampere movement rather than a 1-milliampere movement. If we use a 50-microampere movement in Figure 10-14(a), the total internal resistance of the ohmmeter will be

$$R_T = \frac{E}{I} = \frac{4.5 \text{ V}}{50 \ \mu A} = 90 \text{ k}\Omega$$

and, therefore, the center-scale reading of the meter will also be 90 kilohms.

From these examples, we note that the centerscale resistance mark of the ohmmeter is inversely proportional to the full-scale current of the meter. Therefore, we can convert the basic ohmmeter into a low-range ohmmeter by placing a shunt across the moving coil, as shown in Figure 10-15.

EXAMPLE 10-7

Design an ohmmeter using a 1-mA movement with an internal resistance of 50 Ω and a 4.5-V battery to read 45 Ω center scale rather than 4.5 kΩ.

SOLUTION

The total internal resistance of the ohmmeter must be 45 Ω for the meter to read full scale with only the 45 Ω of the meter in circuit and half scale when the total loop resistance is

$$R_x + R_M = 45 + 45 \ \Omega$$

Therefore, full-scale current must be

$$I = \frac{E}{R} = \frac{4.5 \text{ V}}{45 \text{ }\Omega} = 0.1 \text{ A}$$

Since the meter movement passes 1 mA at full scale, the shunt current must be 99 mA and the shunt resistance

$$R_{sh} = \frac{1}{99} \times 50 = \textbf{0.505 } \Omega$$

The total resistance of the meter movement and shunt in parallel is

$$R = \frac{50 \times 0.505}{50 + 0.505} = 0.500 \text{ }\Omega$$

Therefore, the total resistance of the series resistor and the "ohms adjust" rheostat will be

$$R_s = 45 - 0.5 = \textbf{44.5 } \Omega$$

As a test instrument, the ohmmeter has the same advantage as the voltmeter in that we can connect it *across* a portion of a circuit without disconnecting any wiring. However, since the ohmmeter is calibrated on the basis of its own voltage source (the 4.5-volt battery), we must make sure that no other voltage source is acting in the circuit while we measure its resistance.

When we wish to measure a very high resistance such as the leakage between the two conductors of a cable, as in Figure 10-16, we can apply our knowledge of electric circuit theory to use a voltmeter and a separate voltage source to form a high-resistance ohmmeter. This system has the added advantage that we can measure resistance under actual applied-voltage conditions and thus check for voltage break-down of an insulator.

FIGURE 10-16
Using a Voltmeter as a High-Resistance Ohmmeter.

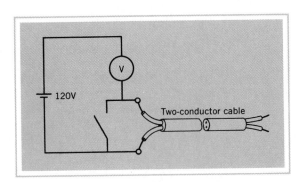

EXAMPLE 10-8

The voltmeter in Figure 10-16 is a 20 000-ohm-per-volt meter with a 150-V scale. When the switch is closed, it reads 120 V and when the switch is open it reads 10 V. What is the leakage resistance of the cable insulation?

SOLUTION

The resistance of the voltmeter is

$$R_M = 20\,000\ \Omega/V \times 150\ V = 3\ M\Omega$$

The voltage drop across the leakage resistance of the cable is $V = 120 - 10 = 110\ V$. And since the resistance of the voltmeter and the leakage resistance of the cable form a simple series circuit,

$$\frac{R_x}{R_M} = \frac{V_x}{V_M}$$

$$\therefore R_x = 3\ M\Omega \times \frac{110\ V}{10\ V} = 33\ M\Omega$$

For precision resistance measurements, we can use various forms of the Wheatstone bridge shown in Figure 10-17. If we adjust the circuit of Figure 10-17 in such a way that there is absolutely no deflection of the galvanometer when we close the switch, the voltage drops across R_x and R_A must be exactly the same, so that no potential difference appears across the galvanometer to cause current in its moving coil. With no current through the meter movement, $I_x = I_y$. Hence,

$$V_x = I_x R_x \quad \text{and} \quad I_x = \frac{E}{R_x + R_y}$$

Therefore,

$$V_x = \frac{E R_x}{R_x + R_y}$$

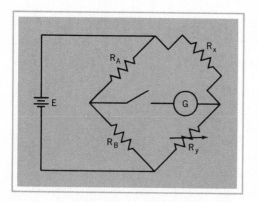

FIGURE 10-17
Measuring Resistance with a Wheatstone Bridge.

Similarly,
$$V_A = \frac{ER_A}{R_A + R_B}$$

Therefore, for perfect balance,

$$\frac{ER_x}{R_x + R_y} = \frac{ER_A}{R_A + R_B}$$

Cancelling out E and cross-multiplying gives

$$R_x R_A + R_x R_B = R_x R_A + R_A R_y$$

Canceling out $R_x R_A$,
$$R_x R_B = R_y R_A$$

or
$$R_x = \frac{R_A R_y}{R_B} \tag{10-3}$$

For the bridge to be balanced, the product of one pair of opposite arms must equal the product of the other pair of opposite arms of the bridge.

Since E does not appear in Equation (10-3), the magnitude of the source voltage used with a bridge circuit has no effect on the accuracy of the measurement. The source merely causes a deflection of the galvanometer pointer if the bridge is not properly balanced. With precision resistors for R_A, R_B, and R_y, one or more of which must be adjustable so that we can obtain a balance, we can then use Equation (10-3) to determine the value of R_x to a high degree of accuracy.

EXAMPLE 10-9
The Wheatstone bridge circuit of Figure 10-17 is balanced when $R_A = 1.0\ \Omega$, $R_B = 50\ \Omega$, and $R_y = 17\ \Omega$. What is the resistance of R_x?

SOLUTION

$$R_x = \frac{R_A R_y}{R_B} = \frac{1.0 \times 17}{50} = 0.34\ \Omega$$

10-7 MOVING-COIL MOVEMENT IN AC CIRCUITS

Since the magnetic field produced by the *permanent* magnet of the moving-coil movement must maintain a fixed direction, it follows that current flowing one way through the moving coil develops a clockwise torque, and current flowing in the other direction develops a counterclockwise torque. With the spiral spring adjusted so that the zero-current position of the pointer is center-scale (a galvanometer), the movement can indicate current in either direction. If we place the instrument in a circuit where the current builds up first in one direction, drops back to zero, and then builds up in the other direction, the torque will change from instant to instant, as shown in Figure 10-18(a). A negative value of torque on this graph simply indicates that the torque is

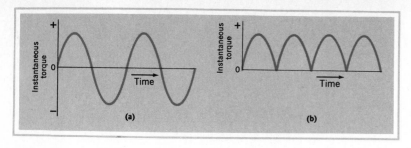

FIGURE 10-18
Instantaneous Torque Developed in a Moving-Coil Movement: (a) by an Alternating Current; (b) by a Rectified Alternating Current.

attempting to rotate the pointer in the opposite direction to that produced by a positive value of torque. If this variation in current is quite slow (once a second), the pointer will be able to follow the changing torque by swinging to the right and then to the left of zero. Since the pointer is never stationary, it is difficult to obtain a reading having any significance.

If the current direction alternates many times a second, the mechanical inertia of the moving-coil assembly is too great for the pointer to follow the instantaneous variations in torque (and in current through the moving coil). Therefore, the pointer must take up a position dictated by the *average* value of the changing torque. The average value of the sine wave in Figure 10-18(a) is zero. The pointer must then remain at the zero mark on the scale, both for the center-zero galvanometer and for those movements with spiral springs adjusted to place the zero mark at the left-hand end of the scale. Because of their permanent magnets, moving-coil movements are basically *direct-current* instruments.

The advantage of using batteries as the voltage sources in the resistance networks we have encountered in preceding chapters is that we do not have to consider the *time* element in our calculations. The *direct* currents in the various branches of dc networks have constant magnitudes and directions. In practical electrical systems, however, most voltage sources generate an *alternating* emf which reverses its polarity periodically at rates ranging from several times a second (power-line frequencies) to many millions of times a second (radio frequencies).[†] Hence, we require measuring devices which do not have the limitations of the basic moving-coil movement.

In Section 3-8, we discovered that semiconductor *pn*-junctions permit current to flow freely in one direction but have a high resistance to current in the opposite direction. We can use a **bridge rectifier** composed of four such *pn*-junctions to reverse the leads of a moving-coil meter movement automatically for each half-cycle of alternating current, as shown in Figure 10-19. The arrow of the graphic symbol for a semiconductor rectifier indicates the direction of minimum resistance for conventional current. The current through the meter movement, and consequently the instantaneous torque, now has the form shown in Figure 10-18(b). As a result, the pointer deflection will be proportional to the *average* value of the alternating current.[‡]

[†]See Section 17-2 for a detailed description of an alternating voltage.

[‡]As defined in Section 17-10. If the meter is to be used in sine-wave ac circuits, the fixed $1.1:1$ ratio between rms and average values of a sine wave (form factor) allows us to calibrate the meter scale to read rms values.

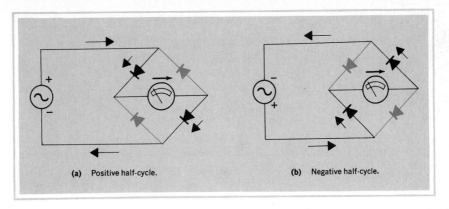

(a) Positive half-cycle. (b) Negative half-cycle.

FIGURE 10-19
Moving-Coil Movement with a Bridge Rectifier.

To use the rectifier-type meter as an ammeter, we require an ammeter shunt resistor across the input terminals of the bridge rectifier in Figure 10-19. Since the moving coil is not a pure resistance, the split in current between the shunt and the meter depends on the frequency of the alternating current. Consequently, a rectifier-type ammeter is accurate only for a specific frequency at which it is calibrated. Hence, moving-coil movements are seldom used in ac ammeters.

Since the multiplier resistance of a voltmeter is in *series* with the moving coil, slight variations in coil impedance are not a problem. Therefore, the rectifier and moving-coil movement ac voltmeter is widely used in **multimeters** in which a single meter movement is used for both dc and ac voltage measurements.

10-8 THE MOVING-IRON MOVEMENT

A more economical movement for use primarily in alternating-current circuits is the **moving-iron** movement, shown in Figure 10-20. It consists of a stationary solenoid with two soft-iron vanes within the coil. One vane is stationary, but the other is fastened to the pivot which carries the pointer and the counter-torque spring. Since the

FIGURE 10-20
Moving-Iron Movement.

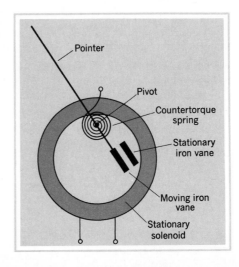

263

soft-iron vanes are located at the point where the flux density of the magnetic field created by the current through the solenoid is maximum, they become temporary magnets. Since the S pole of one vane is adjacent to the S pole of the other (the two N poles are also adjacent), the pivoted vane is *repelled* by the stationary vane. This repulsion is proportional to the current in the solenoid. Figure 10-20 shows only one of several possible shapes in which the soft-iron vanes may appear. The scale of this moving-iron instrument is not as linear as that of the moving-coil instrument; it is cramped at the low-current end of the scale. Scale linearity can be improved by the use of specially shaped soft-iron vanes.

We can calibrate the moving-iron movement in terms of *direct* current through the solenoid and sometimes use it as an inexpensive direct-current instrument where we do not require the greater accuracy of the moving-coil instrument. However, the moving-iron instrument was developed primarily for measuring *alternating* currents at power-line frequencies. When the direction of the current in the solenoid is reversed, the N and S poles of *both* vanes reverse, and the moving vane is still *repelled* by the stationary vane. Therefore, when alternating current flows in the solenoid, the instantaneous torque developed by the moving-iron vane is in the form of pulses which are always in the same direction, as shown in Figure 10-18(b). Because the mechanical inertia of the whole pointer and moving-vane assembly is too great to allow it to follow the instantaneous variations in torque, the pointer will indicate the *average* value of the torque. When calibrating the scale of a moving-iron instrument for alternating current, we must allow for hysteresis losses in the iron vanes. Therefore, the calibration of an ac moving-iron instrument is accurate only for power-line frequencies, not for direct current or high-frequency alternating currents.

Since the stationary solenoid of the moving-iron movement can be wound with a few turns of heavy wire or many turns of fine wire in order to produce the required magnetic field for full-scale deflection, the range of full-scale current readings available is much greater than for moving-coil instruments. The basic ammeter movement for ac power circuits has a five-ampere full-scale reading. Shunts similar to those used with moving-coil movements are used for larger currents. Although it is not possible to produce moving-iron movements that are as sensitive as moving-coil instruments, we can use one with a solenoid consisting of many turns of fine wire along with suitable series multiplier resistors to form a voltmeter for use in ac circuits.

10-9 THE ELECTRODYNAMOMETER MOVEMENT

As we noted in the preceding section, the fixed direction of the magnetic field of the permanent magnet is responsible for limiting the moving-coil movement to dc (or rectified ac) circuits. If we replace the two poles of the *permanent* magnet with an *electromagnet* consisting of two stationary coils, as shown in Figure 10-21, we have an **electrodynamometer** movement in which we can reverse the direction of the stationary magnetic field at will. Since it contains no iron, we can use the electrodynamometer as the basic movement for either dc or ac instruments. We can use it to measure *alternating* current by connecting the stationary and moving coils in series. Whenever the current in the moving coil reverses its direction, the magnetic field produced by the stationary coil also reverses. Therefore, the resulting torque tends to

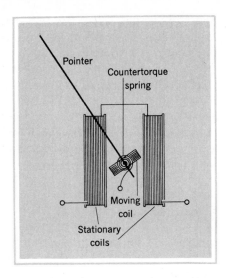

FIGURE 10-21
Electrodynamometer Movement.

move the pointer in a clockwise direction, regardless of the direction of the current in the coils. However, for voltmeter and ammeter applications, the electrodynamometer movement is too costly to compete with either the moving-coil movement for dc measurement or the moving-iron movement for ac measurement.

In developing Equation (10-2) for the torque of a moving coil, we noted that the force acting on a conductor is proportional to the *product* of the stationary magnetic field strength and the current in the moving coil. Therefore, the scale reading of an electrodynamometer movement is proportional to the product of the stationary-coil and moving-coil currents. If the wire on the stationary coils is heavy enough to be connected in series with a load, as in Figure 10-22, the stationary-coil current is the same as the load current. If the moving coil with a multiplier resistor in series with it is connected across the load, the moving-coil current is directly proportional to the load voltage. Therefore, the scale reading of the electrodynamometer movement is proportional to the product of the load current and voltage. Since $P = IV$, we have a direct-reading **wattmeter.** If we connect a wattmeter to a load in an ac circuit, both load current and load voltage reverse their direction the same number of times per second, thus producing a net positive torque. Therefore, we can use the electro-

FIGURE 10-22
Electrodynamometer Movement as a Wattmeter.

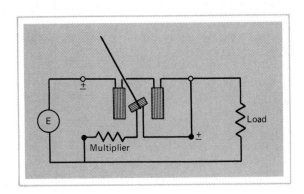

dynamometer movement as a wattmeter in either dc or ac circuits. Although we can calculate the power in a load by taking separate ammeter and voltmeter readings, a direct-reading wattmeter is a very useful instrument, particularly in ac circuits.

If we reverse the leads to *either* the current coil *or* the voltage coil when we connect a wattmeter into a circuit, a *negative* torque, which moves the pointer *down scale,* will be developed. To assist us in connecting a wattmeter properly, it is customary in practice to identify one end of each coil with $\pm$ marks, as shown in Figure 10-22. Standard connection requires that we connect the identified terminal of the current coil to the generator and the identified terminal of the voltage coil to the line containing the current coil, as shown in Figure 10-22.

10-10 DIGITAL METERS

We now call watches with hands *analog* watches to distinguish them from the newer *digital* watches. Similarly, the meter movements we have considered in this chapter are *analog* instruments. The term *analog* pertains to data in the form of some *continuously* variable quantity such as the angular position of watch hands or the angular displacement of a meter pointer across the scale of the meter. A *digital* device, on the other hand, displays data as discrete numerals with no in-between positions.

For many years, the most common piece of electrical test equipment has been the **multimeter** or VOM (volt-ohm-milliammeter) in which a multiposition switch combines the dc milliammeter ranges of Figure 10-17(b), the dc and ac voltmeter ranges of Figures 10-9 and 10-19, and the ohmmeter ranges of Figures 10-14 and 10-15. The multimeter's single moving-coil movement requires a multiplicity of scales which are not always easy to read accurately. As improved production techniques reduce the size and cost of integrated-circuit components, digital multimeters are rapidly gaining in popularity. They have the advantages of a high input resistance like the electronic voltmeter of Figure 10-11, greater accuracy than moving-coil meters, and nonambiguous readout of the numerical value of the measured quantity.

Digital meters usually operate by *counting* a number of pulses from an accurate oscillator circuit. These pulses are fed to the digital counter through a gate circuit which switches the flow of pulses on and then off after a precise time interval. This time interval is determined by an analog to digital converter circuit that develops a gating pulse which is directly proportional in width (time duration) to the magnitude of the electrical quantity being measured. The number of oscillator pulses getting by the gate, and therefore the numerals displayed on the digital readout, are calibrated to represent the voltage, current, or resistance being measured to well within 1% accuracy.

PROBLEMS

10-1. What force in newtons acts on a conductor which is perpendicular to a magnetic field whose flux density is 0.8 T if 15 cm of the conductor are in the magnetic field and the current in the conductor is 5 A?

10-2. The countertorque spring of a certain moving coil instrument develops a torque of 10^{-5} newton meter when the pointer is at full scale. The flux density in the air gap is 5×10^{-2} T, and there are 80 turns of wire wound on a form whose dimensions are 2 cm parallel to the pivot and 1.5 cm at right angles to the pivot. What current must flow through the coil for half-scale deflection?

10-3. The full-scale voltage drop across a 5-mA movement is 50 mV. Calculate the resistance of the shunt required if the movement is to be used as a 1-A full-scale ammeter.

10-4. If the shunt calculated in Problem 10-3 is used with a 1-mA movement having a total resistance of 50 Ω, what current would its full-scale reading represent?

10-5. Calculate an Ayrton shunt to provide 100-mA, 1-A, and 10-A ranges with a 50-μA movement whose total resistance is 1000 ohms.

10-6. Calculate an Ayrton shunt for 30-mA, 100-mA, 300-mA, and 1-A ranges with a 1-mA movement whose total resistance is 50 ohms.

10-7. Calculate the multiplier resistors required to form a voltmeter with 10-V, 50-V, 100-V, and 500-V ranges, using the meter movement of Problem 10-5.

10-8. The resistance of the movement for a 200 ohm-per-volt voltmeter is 10 ohms. Calculate the multiplier resistors required for 30-V, 150-V, and 1500-V ranges.

10-9. The terminal voltage of a 12-V source whose internal resistance is 100 kΩ is checked by a 20 000-ohm-per-volt voltmeter on its 50-V range. What is the voltmeter reading?

10-10. The bias voltage of a FET is applied to its gate terminal through a 470-KΩ resistor. If a 1000-ohm-per-volt meter on its 10-V range measures the voltage at the gate terminal (with respect to ground) as 0.8 V, what will it be when the voltmeter is removed if there is negligible gate current?

10-11. Resistor R_2 in the voltmeter shown in Figure 10-23 is open-circuit, thus its value cannot be determined. However, R_1 and R_3 can be measured as 97 kΩ and 300 kΩ, respectively. What value of resistance for R_2 is required to repair the voltmeter?

10-12. Using the data from Problem 10-11, calculate the meter resistance, full-scale current, and voltmeter sensitivity for the voltmeter shown in Figure 10-23.

10-13. In the potentiometer circuit of Figure 10-12, the Weston standard-cell voltage equals the potentiometer output voltage when R_{BC} is 71.0% of full-scale resistance. To obtain zero galvanometer deflection with the switch in the E_x position, we must set the potentiometer at 34.5% of full-scale resistance. What is the open-circuit voltage of the unknown source?

10-14. We want to use the potentiometer in Figure 10-12 to produce an open-circuit output voltage V_{BC} of 1.0 V. We first calibrate the potentiometer with a Weston standard cell, which requires a potentiometer setting of 654 to produce a balance. What potentiometer setting is required for the desired 1.0-V output?

FIGURE 10-23
Multirange Voltmeter.

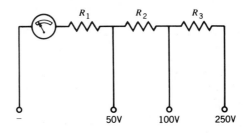

10-15. What is the center-scale reading of a basic ohmmeter constructed with a 50-μA movement and a $1\frac{1}{2}$-V battery?

10-16. What battery voltage would be required to use the same scale on a 1-mA movement?

10-17. What resistance must be connected to the terminals of the ohmmeter in Problem 10-15 for the pointer to read 20% of full-scale deflection?

10-18. Design an ohmmeter with 1000 Ω appearing at center scale, using a $1\frac{1}{2}$-V battery and a 50-μA movement having a resistance of 1000 ohms.

10-19. The insulation resistance of a solenoid is checked by connecting one terminal of a 1-kV source to the solenoid and the other terminal of the source through a 20 000-ohm-per-volt voltmeter having a 1500-V full-scale reading to the iron core of the solenoid. If the voltmeter reads 360 V, what is the insulation resistance?

10-20. What would the meter read in Problem 10-19 if the leakage resistance between the solenoid winding and the core dropped to 0.5 MΩ?

10-21. In the Wheatstone bridge circuit of Figure 10-17, balance is attained when $R_A = 5$ kΩ, $R_B = 1000$ Ω, and $R_y = 42$ kΩ. What is the resistance of R_x?

10-22. What value must R_y be set at to obtain a balance in the Wheatstone bridge of Figure 10-17 if $R_A = 1000$ Ω, $R_B = 1200$ Ω, and $R_x = 600$ Ω?

10-23. When the Wheatstone bridge in Figure 10-17 is balanced, V_A is 2.4 V and I_B is 0.24 A. If the total current drawn from the battery is 300 mA, what is the resistance of R_x?

10-24. Figure 10-24 is a variation of the Wheatstone bridge known as the **slide-wire bridge.** The slide wire is a length of uniform resistance wire 1 m long. The standard resistor $R_s = 24$ Ω. What is the value of R_x if the bridge is balanced when the slider is 37 cm from the end that is connected to the standard resistor?

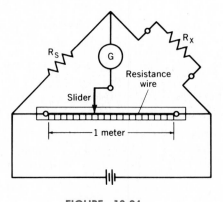

FIGURE 10-24
Slide-Wire Bridge.

10-25. Figure 10-25 is a variation of the Wheatstone bridge known as the **Varley loop.** It is used for locating grounds on long transmission lines. To accomplish this, the far ends of two identical conductors are connected together. Assume negligible resistance between the ground on the switch and the fault ground on the transmission line. With the switch in the *line* position, the bridge balances with $R_A = 100$ Ω, $R_B = 1000$ Ω, and $R_C = 200$ Ω. When the switch is thrown to the *ground* position, R_A and R_B are unchanged but R_C is changed to 150 Ω to obtain balance. How far is the ground from the end of the cable at which the apparatus is located?

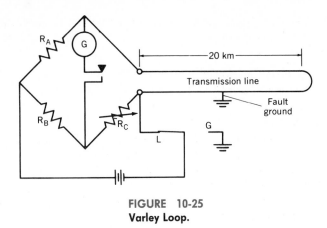

FIGURE 10-25
Varley Loop.

REVIEW QUESTIONS

10-1. Draw a sketch showing the composite field when an electric conductor with a conventional current direction coming out of the page is placed in an air gap in a magnetic circuit in which the lines of force pass from left to right across the air gap.

10-2. Explain why the composite magnetic field in Question 10-1 has the shape you have shown.

10-3. Explain why the composite magnetic field in Figure 10-3 has the pattern shown.

10-4. Why is the force acting on a coil of wire in a magnetic field directly proportional to the number of turns in the coil?

10-5. Why is the flat spiral spring essential in all the meter movements described in this chapter?

10-6. What is the relationship between Hooke's law and scale markings on a moving-coil movement?

10-7. What would be the effect on the operation of a moving-coil movement if the soft-iron cylinder (shown in Figure 10-4) were left out during the assembly of the instrument?

10-8. What is a **galvanometer?**

10-9. Why is a galvanometer used in the circuit of Figure 10-17 rather than a microammeter?

10-10. What is the purpose of the **calibrating resistor** used with moving-coil movements?

10-11. Suggest an advantage in arranging that moving-coil movements have a 50-mV drop across their terminals when full-scale current flows through them. (Refer to Problem 10-4.)

10-12. Show that

$$R_S = R_M \left(\frac{I_M}{I_T - I_M} \right)$$

where R_S is the resistance of an ammeter shunt, R_M is the resistance of the moving-coil movement and its calibrating resistor, I_M is the current in the moving coil required for full-scale deflection, and I_T is the desired full-scale ammeter reading.

10-13. Figure 10-26 shows the right and wrong ways of connecting an ammeter with an external shunt into a circuit. What would be the effect of connecting the circuit incorrectly as in Figure 10-26(a)?

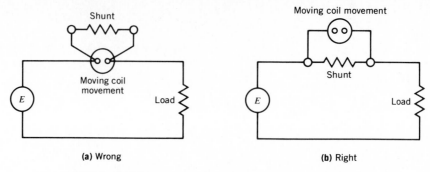

(a) Wrong

(b) Right

FIGURE 10-26
Connecting an Ammeter with an External Shunt into a Circuit.

10-14. Why is the insertion resistance of a multirange ammeter equipped with an Ayrton shunt greater than that of an ammeter which uses a separate shunt for each range?

10-15. The basic moving-coil movement is a *current*-indicating device. How is it possible to calibrate it as a voltmeter?

10-16. What is the significance of **voltmeter sensitivity?**

10-17. Why is a 20 000-ohm-per-volt meter more suitable than a 1000-ohm-per-volt meter for checking electronic circuitry?

10-18. Why is a 200-ohm-per-volt meter satisfactory for use with electrical machinery?

10-19. Why is voltmeter loading effect more of a problem in practical measurements than ammeter insertion resistance?

10-20. Why is the loading effect of an electronic voltmeter considerably less than that of a 20 000-ohm-per-volt meter?

10-21. With a precision potentiometer, we can readily measure an unknown voltage to a three-figure accuracy. Does the galvanometer scale have this degree of accuracy? Explain.

10-22. With the switch of the potentiometer circuit in Figure 10-12 in the E_x position, when we rotate the potentiometer dial so that R_{BC} increases, the meter deflection decreases but does not reach zero. Explain the condition which would give this effect.

10-23. What effect will a 2% decrease in the dry-cell voltage in the potentiometer circuit of Figure 10-12 have on the accuracy of the instrument? Explain.

10-24. Discuss the possible errors that might occur when using the voltmeter/ammeter method of determining the resistance of a 50-kΩ resistor. (Draw a circuit diagram.)

10-25. What precaution must be observed when checking the resistance of an electric circuit with an ohmmeter?

10-26. Lay out a scale (see Figure 10-14) for a simple ohmmeter, using a 50-μA movement and a $1\frac{1}{2}$-V battery.

10-27. Lay out a scale for the low-resistance ohmmeter shown in Figure 10.27.

10-28. What would be the effect of connecting an ammeter *across* a load?

10-29. What would be the effect of connecting a voltmeter in *series* with a load?

10-30. What circuit information could be obtained from the voltmeter reading in Question 10-29?

10-31. Which of the following movements would you select for the Wheatstone bridge circuit of Figure 10-17: a 50-μA movement or a 5-mA movement? Explain your selection.

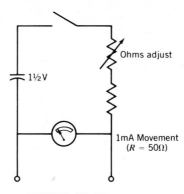

FIGURE 10-27
Low-Resistance Ohmmeter.

10-32. What effect does the accuracy of the galvanometer scale calibration have on the accuracy of the resistance measurements made with a Wheatstone bridge?

10-33. Why does a moving-coil movement read zero in an ac circuit?

10-34. Explain why a bridge rectifier "automatically reverses the leads of the meter movement for each half-cycle of alternating current."

10-35. A typical multirange volt-ohm-milliammeter (VOM) has the following ranges:
Volts dc: 2.5 V, 10 V, 50 V, 250 V, 1000 V
Volts ac: 2.5 V, 10 V, 50 V, 250 V, 1000 V
Ohms: $R \times 1, R \times 10, R \times 1000, R \times 10\,000$
Milliamperes dc: 100 μA, 1 mA, 10 mA, 50 mA, 250 mA, 1 A, 10 A
Why is there no provision for measuring alternating current?

10-36. Why is the pointer of a moving-iron movement able to indicate a steady reading in an ac circuit?

10-37. Why is it possible to use the same electrodynamometer-type wattmeter in both dc and ac circuits?

*10-38. Design a personal computer program which will print out the three resistor values in the Ayrton shunt of Figure 10-7(b) upon entering meter current for full-scale deflection, meter resistance, and the three desired current ranges.

CAPACITANCE AND INDUCTANCE

*THERE ARE THREE BASIC PROPERTIES that an electric circuit may possess. In all the electric circuits we have considered so far, we have encountered only one property—the tendency of the circuit to oppose electric current through it. This property is called **resistance**. No matter how complex an electric circuit may be, there are only two additional properties that it may possess. A second property tends to oppose any change in the electric current through it. This property is called **inductance**. The third property tends to oppose any change in the potential difference or voltage between two points in an electric circuit. This property is called **capacitance**.*

In Part 3, we shall investigate the nature of these two additional properties and discover their behavior in direct-current circuits. Since inductance depends on the magnetic fields associated with an electric current, we shall also review the principles of magnetism. Chapter 14 is an optional chapter of more interest to students proceeding to electric power systems options.

11

CAPACITANCE

11-1 STATIC ELECTRICITY

Since capacitance is associated with potential difference, we must first review some of the basic concepts concerning potential difference that we developed in Chapter 2. We noted that atoms and molecules in their normal state are electrically neutral. To create an electric potential difference, we must first *separate* some valence electrons from one group of neutral atoms—thus leaving them with a deficiency of electrons or a net positive electric charge—and then add these surplus electrons to a second group of atoms thus providing them with a net negative electric charge. To *maintain* a continuous electric current in a dc circuit, we employ electric "charge separators," such as batteries, which maintain a potential difference between their terminals by continuous expenditure of chemical energy.

But electric charge separation by friction between electric insulating materials was discovered many centuries before the earliest electric batteries were invented. Electric charges that are separated by friction between insulating materials are known as **static electricity**. The study of the behavior of these charges is called **electrostatics**.

Although static electricity is often no more than an annoyance encountered as a result of the electric charge separation that takes place when shoe leather is rubbed against a carpet (particularly when the air is dry), we shall find that the behavior of capacitance in electric circuits is based on the principles of electrostatics demonstrated by Charles Coulomb some 200 years ago.

274

11-2 THE NATURE OF AN ELECTRIC FIELD

In Section 2-1, we noted that Coulomb, using a delicate torsion balance, was able to measure the force that exists between two electrically charged pith balls. The result of these measurements led to

Coulomb's law of electrostatic force: The electric force between two point electric charges is directly proportional to the product of the two charges and inversely proportional to the square of the spacing between the charges.

We can now add units to Equation (2-1), giving

$$F = k\frac{Q_1 Q_2}{s^2} \tag{11-1}$$

For F to represent electric force in newtons, Q_1 and Q_2 the two charges in coulombs, and s the spacing between the charges in meters, the constant of proportionality k has a numerical value of 8.99×10^9.[†]

We have already observed that electric force requires no physical contact between the two electric charges. Hence, we describe electric force as a *field* force. We can extend this field concept by examining what happens when a tiny positive charge (such as a hydrogen ion which consists of one proton) is placed in the vicinity of a fixed positive charge. We now know that the repulsion of the *like* charges will cause the proton to move radially away from the fixed positive charge. We say that the tiny positive charge is acted upon by the **electrostatic** (or, simply **electric**) **field** of the fixed positive charge. We can show this electric field diagrammatically by drawing the paths along which the tiny + charge will move away from the fixed charge, as shown in Figure 11-1(a). It follows that the electric field surrounding a fixed negative charge

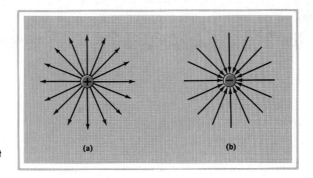

FIGURE 11-1
Electric Fields Surrounding Point Electric Charges.

[†]Strictly speaking, the constant k also has to have a unit such that other units in Equation (11-1) cancel out leaving only newtons. Using the dimensional analysis techniques of Section 1-7, the unit for k becomes $N \cdot m^2/C^2$.

will *attract* the tiny + charge, resulting in the diagrammatic representation of Figure 11-1(b). We can say that

An electric field is that region in which an electric charge is acted upon by an electric force.

In studying electric circuits, we are not particularly concerned with the electric fields surrounding isolated electric charges in space. Since we are concerned with electric potential difference between two points in an electric circuit, we are more interested in the electric field between two unlike electric charges. We can create such a field by connecting two parallel metallic plates to the two terminals of a voltage source, as shown in Figure 11-2. The left-hand plate becomes positively charged, since the + terminal of the battery removes some electrons from it; and the right-hand plate becomes negatively charged as the − terminal of the battery forces a surplus of electrons into it. If we now place a tiny positively charged particle between the plates, it will move toward the right-hand plate. If the particle has a negative charge, it will move toward the left-hand plate. Again, we can draw lines to show the paths followed by the tiny charged particle. These lines represent the direction of action of the electric force on the charged particle. Hence, we can think of an electric field as consisting of invisible electric lines of force, where

An electric line of force represents the path along which a weightless electrically-charged particle will move from one electrically-charged plate to the other.

Since a positively charged particle can travel only toward the right-hand plate and a negatively charged particle can travel only toward the left-hand plate, we can consider that electric lines of force possess *direction*. To agree with the conventional direction for electric current, we assume that

The direction of an electric line of force is the direction in which a positively-charged particle will travel between the two plates.

From the sketches in Figures 11-1, 11-2, and 11-3, we note that

Electric lines of force always begin and end on charged bodies and always meet a charged surface at right angles to that surface.

Although we are concerned mainly with the electric field between parallel plates, two other electric field patterns are of interest to us—the electric field between parallel conductors, as shown in Figure 11-3(a), and the electric field between concentric conductors, as shown in Figure 11-3(b).

As we have noted, a positively-charged particle in the electric field between the two parallel plates will be repelled by the + plate and attracted by the − plate. Because we developed the concept of an electric field to show the force that will be exerted on charged particles, we must be able to express the **intensity** or **strength** of the electric field numerically.

The letter symbol for electric field intensity is the Greek letter E (capital epsilon).

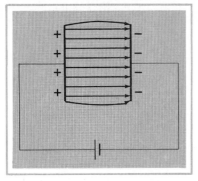

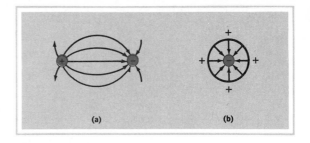

FIGURE 11-3
Electric Field between (a) Parallel Conductors, and (b) Concentric Conductors.

FIGURE 11-2
Electric Field between Parallel Plates.

To distinguish the symbol for electric field intensity from the symbol for applied voltage, we can use **boldface** type. Boldface type is used to call attention to *vector* quantities which possess *direction* as well as magnitude. No special unit is required for electric field strength or intensity since we can express it in terms of the charge on the particle and the force acting on the particle.

An electric field has an intensity of one unit at a certain point if a force of one newton is exerted on an electric charge of one coulomb located at that point.

We may express this in equation form as

$$E = \frac{F}{Q} \tag{11-2}$$

where E is the electric field intensity in **newtons per coulomb,** *F is the force acting on the charge in newtons, and Q is the magnitude of the charge in coulombs.* Since we show electric field intensity as a vector quantity on the left-hand side of Equation (11-2), we should also show the vector nature of *F* on the right-hand side.

In the special case of the electric field between parallel plates, the electric field is quite uniform between the plates. Therefore, a constant force is exerted on a charged particle as it moves from one plate to the other. (As the repulsion of one plate becomes weaker with an increase in distance, the attraction of the other plate increases as the distance decreases.) As the charge moves the complete distance from one plate to the other, the work done is

$$W = Fs \tag{2-3}$$

Since we established the magnitude of the volt in terms of the work done in moving a charge through a potential difference,

$$W = QV \tag{2-4}$$

$$\therefore Fs = QV$$

Substituting for F from Equation (11-2),

$$EQs = QV$$

$$(11\text{-}3)$$

from which

$$E = \frac{V}{s} \quad \text{(parallel plates)}$$

where E is the electric field intensity in **newtons per coulomb** *or* **volts per meter,** *V is the potential difference between the plates in volts, and s is the spacing between the plates in meters.*

This expression for electric field intensity leads to an interesting characteristic of electric fields. According to Kirchhoff's voltage law, when we apply 400 volts to the two parallel plates of Figure 11-4, there must be a 400-volt PD between them. Since the electric lines of force begin at the $+$ plate and end at the $-$ plate, this 400-volt potential difference appears across each line of force. The voltmeter in the circuit diagram draws zero current. We connect one voltmeter lead to the bottom plate, which is shown connected to ground; we may now refer all our voltage readings to ground without having to say "with respect to" for every point at which we locate the probe connected to the other voltmeter terminal.

When we touch the voltmeter probe to the top plate, the meter will read 400 volts. As we move the voltmeter probe one quarter of the way down from the top plate, the meter reads 300 volts; when the probe is halfway between the plates, the meter reading is 200 volts, and so on. Therefore, in the case of parallel plates, the total potential difference is distributed uniformly along the length of the electric line of force as if it were a resistance voltage divider. From this characteristic, we can speak of the **voltage gradient** in volts per meter. Because of the uniform electric field between parallel plates, in this case the voltage gradient is numerically equal to the electric field intensity. Therefore, we can express E in volts per meter.

Since every electric line of force has the total potential difference distributed along its length, we can mark a point on each line of force having the same potential difference with respect to ground. Joining these points produces what is known as an **equipotential surface** for the electric field. Equipotential-surface "maps" are very useful in applying the theory of electrostatics to the design of cathode-ray tubes. Equipotential surfaces for parallel plates and parallel conductors are shown by the colored lines in Figure 11-5.

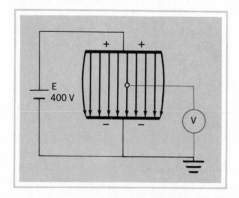

FIGURE 11-4
Showing the Voltage Gradient along an Electric Line of Force.

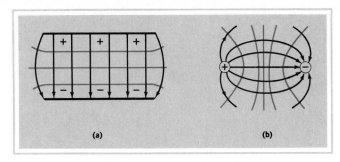

FIGURE 11-5
Equipotential Surfaces in the Electric Field between (a) Parallel Plates, and (b) Parallel Conductors.

(a)   (b)

If we think of an electric line of force joining every surplus electron in the negatively charged plate to every atom which is deficient an electron in the positively charged plate, the number of electric lines of force is directly proportional to the electric charge. The total number of electric lines of force in an electric field is called the **electric flux**.

The letter symbol for electric flux is the Greek letter ψ (psi).

Since the electric flux is directly proportional to the charge, we need no special unit for electric flux. We simply state that

$$\psi \equiv Q \tag{11-4}$$

where ψ is the electric flux in MKS units, and Q is the charge on each plate in coulombs.

The letter symbol for electric flux density is D.

Since the distribution of electric lines of force is quite uniform for the special case of two parallel plates, the electric flux density is also uniform and becomes simply

$$D = \frac{\psi}{A} \quad \text{(parallel plates)} \tag{11-5}$$

11-3 ELECTROSTATIC INDUCTION

Figure 11-6 shows an electric conductor in the electric field between two parallel plates. The conductor is shorter than the distance between the plates, so there is no electrical connection between it and the plates. Because there are countless free electrons drifting around in a conducting material, some of these electrons will be attracted by the + plate and repelled by the − plate. As a result, a surplus of electrons (negative charge) will appear at the end of the conductor closer to the + plate and a deficiency of electrons (positive charge) at the end nearer the − plate. Since there is no electrical connection between the conductor and the plates, these two equal and opposite charges that appear when the conductor is placed in the electric field are said to be **electrostatically induced** charges.

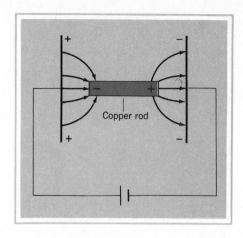

FIGURE 11-6
Electrostatically-Induced Charges.

Because a net movement of electrons in one direction constitutes an electric current, there is a slight surge of current in the conductor while the charges take up their positions. But since the conductor is insulated from the plates, there can be no continuous current. And since $V = IR$, when the **displacement** current ceases, there can be no voltage drop between the ends of the conductor. This means that there must be a redistribution of the electric field between the two parallel plates as shown in Figure 11-6.

11-4 DIELECTRICS

As a final step in this brief study of electrostatics, we shall place a piece of insulating material between the charged plates of Figure 11-7. Insulating materials exposed to electric fields are known as **dielectrics**. Although the atomic structure of an insulator is such that it is very difficult for electrons to move from one atom to another, the electrons orbiting around each nucleus will be attracted by the + plate and repelled

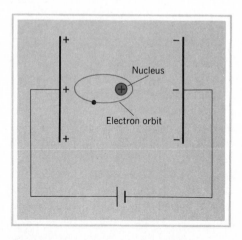

FIGURE 11-7
Effect of an Electric Field on an Atom of a Dielectric.

TABLE 11-1

Dielectric	Average Dielectric Strength (kilovolts per millimeter)
Air	3
Barium-strontium titanate (ceramic)	3
Porcelain (ceramic)	8
Transformer oil (organic liquid)	16
Bakelite (plastic)	16
Paper	20
Rubber	28
Teflon (plastic)	60
Glass	120
Mica	200

by the − plate. As a result, the orbits of the electrons in each atom of the dielectric will be displaced as shown for the single atom in Figure 11-7. The net effect is that, although electrons cannot move from atom to atom in an insulator, each individual atom behaves like the conductor in Figure 11-6. The center of the positive charge of each atom (the nucleus) no longer coincides with the center of the negative charge (the center of the electron orbit). The effect of the positive charge of the atom being closer to one side of the atom and the negative charge being closer to the other is called **polarization** of the atom.

The extent to which the orbits of the electrons in the atoms of the dielectric are distorted (and the extent to which the atoms become polarized) depends on the intensity of the electric field. As we increase the intensity of the electric field by increasing the potential difference between the two parallel plates, we reach a value at which so much force is exerted on the orbital electrons that they are torn free of their orbits, causing a breakdown of the dielectric. The dielectric then becomes a conductor and the two plates are short-circuited. Some dielectrics are able to withstand a much greater electric field intensity than others. The field intensity required to break down a dielectric is called its **dielectric strength**. The dielectric strength of some materials can vary considerably as a result of variations in manufacturing. The values of dielectric strength listed in Table 11-1 are typical average values for some of the common dielectric materials.

After exposing a dielectric to a high electric field intensity in Figure 11-7, we remove the source and momentarily connect the parallel plates together to allow their charges to become neutralized. The stress on the electron orbits of the dielectric is, therefore, removed and they can return to their normal positions. When we remove the connection between the two plates after the initial discharge, we find that with some dielectrics a small potential difference still appears between the two plates. This indicates that the electron orbits in the dielectric did not instantly return to their original positions. Therefore, even though the plates were connected together, the residual electron orbit displacement in the dielectric held an induced charge in the plates in the same manner that charges were induced into the conductor in Figure 11-6. This effect is called **dielectric absorption**.

11-5 CAPACITORS

So far, we have considered only what goes on in the insulated space between a pair of charged parallel plates. While this is important in a study of such devices as cathode-ray tubes, at the moment we are interested in the behavior of electric circuits. Therefore, it is time for us to consider what effect including a component consisting of a pair of parallel conducting plates with a dielectric between them has on the rest of the circuit. Such a component is called a **capacitor**.[†]

A capacitor may be constructed in the parallel-plate form by supporting two insulated metal plates with air between them, or by coating the two sides of a ceramic disk with metal, as in Figure 11-8(a). Some capacitors consist of interleaved parallel plates with either air or mica dielectric, as in Figure 11-8(b). Others consist of two long strips of aluminum foil rolled up with two strips of dielectric, as in Figure 11-8(c). Originally, waxed paper was used as the dielectric for such tubular capacitors. Modern tubular capacitors use thin films of a plastic such as polyethylene. Capacitors constructed by the methods shown in Figure 11-8 range in physical size from about the size of a match head in miniaturized electronic circuitry to industrial capacitors as large as a 10-liter gasoline can. Their capacitance is, however, limited to less than 1 microfarad.

An ingenious means of obtaining a reasonably large capacitance in a fairly compact space employs the electrolysis principle we investigated in Chapter 3. We can replace the plastic film dielectric in the construction of Figure 11-8(c) with a thin gauze soaked in an aluminum hydroxide solution. Like any electrolyte, the aluminum hydroxide solution becomes a conductor by dissociating into positive and negative ions. When first assembled, such an **electrolytic capacitor** is an electrical short circuit. To make it into a capacitor, it is necessary to **form** a dielectric by passing a direct current through the electrolyte between the two aluminum foils. The negative hydroxide ions are attracted to the positive foil where they chemically react with the aluminum to form a very thin layer of aluminum oxide which becomes the dielectric of the capacitor.

The forming current soon diminishes to a very small value and we now have a capacitor where one plate is the positive aluminum foil, the other is the electrolyte in contact with the negative foil, and the dielectric is the thin oxide layer. Electrolytic capacitors are used where capacitances in the range 1 to 100 microfarads are required.

However, there are some limitations in the circuit application of electrolytic

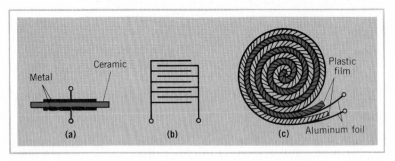

FIGURE 11-8
Construction of Capacitors.

[†]Sometimes referred to by the now obsolete term **condenser**.

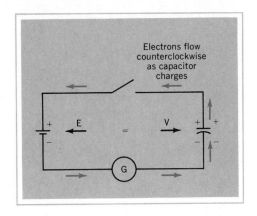

Electrons flow counterclockwise as capacitor charges

FIGURE 11-9
Charging a Capacitor.

capacitors. They must always be used in such a way that the polarity of the potential difference between their terminals is the same as the polarity of the forming voltage. Reversing the polarity of the operating voltage reverses the electrolytic action and tends to remove the oxide film and turn the capacitor into an electrolytic conductor in the process. Hence, the leads of electrolytic capacitors are clearly marked with (+) and (−) polarity indicators or color-coded wires. Electrolytic capacitors are not generally suited to alternating-current circuits.[†]

In order to obtain a high capacitance, the manufacturer keeps the oxide film of electrolytic capacitors as thin as possible. Hence the breakdown voltage of electrolytic capacitors is often quite low and is designated on the capacitor as the **working voltage** that cannot be exceeded without danger of dielectric breakdown. And, as we would expect, this type of dielectric is not a perfect insulator. There is always some leakage current through an electrolytic capacitor.

The development of semiconductor devices has introduced another interesting type of capacitor which is particularly useful in manufacturing microminiature circuit modules. As shown in Figure 3-16(a), a *reverse-biased pn*-junction acts as a nonconductor. Under these conditions, we have the basic requirements for a capacitor—two conducting bodies with a nonconducting depletion layer between them. As long as we maintain a reverse bias, a *pn*-junction can be used as a capacitor. We can even change the effective thickness of the dielectric by changing the magnitude of the reverse bias voltage. The resulting "voltage-variable capacitor," often called a **varactor**, is the heart of most automatic frequency control circuits in radio and television receivers. However, if the polarity of the potential difference between the terminals of a semiconductor capacitor ever reverses, it immediately becomes a short circuit.

In considering the manner in which a capacitor behaves as part of an electric circuit, we represent a capacitor by a standard graphic symbol which somewhat resembles the parallel plates of a physical capacitor. Table 1-1 shows two current styles for the capacitor symbol. The parallel-plate form is preferred internationally; but for the past 25 years, North American practice has moved toward the second style with one *curved* plate, as shown in Figure 11-9. We shall adopt this symbol for the remainder of this book.

[†]It is possible to use two electrolytic capacitors connected in series opposing as a starting capacitor for ac motors.

11-6 CAPACITANCE

When we first connected a pair of parallel plates to a voltage source in Figure 11-2, the charges could not appear on the plates *instantly*.[†] Electrons had to *flow* from the battery onto the right-hand plate to give it a negative charge, and a similar number of electrons had to flow into the battery from the left-hand plate to give it a positive charge. Since the current is the same in all parts of a series circuit, until we close the switch in the circuit of Figure 11-9, there can be no charging of either plate of the capacitor, even though the bottom plate is permanently connected to the − terminal of the battery. If neither plate possesses any charge, there will be no electric field between them. And since $E = V/s$, there will be no potential difference between plates.

When we close the switch in Figure 11-9, the **galvanometer** will register a sudden flow of electrons flowing counterclockwise around the circuit. Since electrons cannot flow right through the insulated area between the plates, a negative charge builds up on the bottom plate and a positive charge builds up on the top plate. As the charges build up, the intensity of the electric field increases and a potential difference builds up between the plates. According to Kirchhoff's voltage law, the potential difference between the two plates must stop building up when it is equal to the potential difference of the source. The galvanometer must, therefore, read zero after the inital charging surge is completed.

If, after the charging surge in Figure 11-9 is over, we suddenly double the applied voltage, there must be another similar surge as the capacitor plates charge sufficiently to raise the potential difference between them to equal the new voltage. We find that there is a fixed ratio between the potential difference between any given pair of insulated conducting plates and the charge required to establish this potential difference. Therefore, for any given pair of insulated conductors,

$$\frac{Q}{V} = \text{a constant} \qquad (11\text{-}6)$$

If we now open the switch in Figure 11-9, there is no longer any electric circuit connection between the two plates. And thus there is no way for the surplus electrons in the bottom plate to get to the atoms in the top plate (which are deficient some electrons) in order to neutralize the charges. As a result, the pair of parallel plates (a capacitor) *stores* electric charge when a potential difference is applied between the plates. Since Q/V is a constant, as long as charges remain on the plates, a potential difference exists between them. Therefore, we can say that

A capacitor is an electric circuit component constructed for the purpose of storing an electric charge with a potential difference between its terminals.

To determine just how much charge a capacitor can store when we connect it to a given voltage source, we must determine the factors that govern the numerical value of the

---

[†]Because of the attraction between unlike charges, the charges are concentrated at the surface of each plate facing the other plate. Therefore, we speak of the charge *on* the plate rather than the charge *in* the plate.

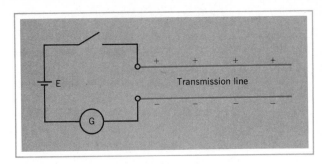

FIGURE 11-10
Capacitance between Parallel Conductors.

constant in Equation (11-6). This numerical constant, which represents the ability of a capacitor to store an electric charge, is called its **capacitance.**[†]

The letter symbol for capacitance is C.
The farad is the SI unit of capacitance.[‡]
The unit symbol for farad is **F.**
An electric circuit has a capacitance of one farad when a charge of one coulomb is required to raise the potential difference by one volt.

$$\therefore C = \frac{Q}{V} \qquad (11\text{-}7)$$

where C is the capacitance in farads, Q is the charge in coulombs, and V is the potential difference in volts.

To satisfy the specifications for the SI units, the basic unit of capacitance had to be defined in terms of *one* coulomb and *one* volt. But the farad represents a much greater capacitance than appears in practical circuits. In practice, the largest unit size used to express capacitance is the **microfarad** (1 μF = 10^{-6} farad). In many cases, even the microfarad is too large and we use the **picofarad** (1 pF = 10^{-12} farad). As North America becomes more accustomed to SI unit prefixes, we also encounter the **nanofarad** (1 nF = 10^{-9} farad). However, when substituting for C in equations, we must remember that these equations were set up in terms of the *basic* unit of capacitance, the *farad*.

Note that in defining the farad we spoke in general terms of the capacitance of a *circuit*. This was done purposely so that we would not be suggesting that capacitance appears in a circuit only when we connect a capacitor into the circuit. We find exactly the same conditions existing in the circuit of Figure 11-10 when we connect a voltage source to a pair of conductors which form an open-circuit transmission line. Capacitance, therefore, is an electric circuit property possessed by *any* pair of conductors that are insulated from each other. A capacitor is merely a device for *lumping* this property into a small physical space.

One of the characteristics of capacitance is its ability to store electric charge. But

---

[†]The ability of a capacitor to store a charge at a certain potential difference is sometimes referred to as its **capacity.** However, in present-day usage, the term capacitance (which rhymes with resistance and inductance) has replaced the older form—capacity.

[‡]Named in honor of Michael Faraday.

we would like to have a definition for capacitance that is in keeping with our definition for resistance. We have already noted that a capacitor cannot charge instantly. It takes time, even if only a fraction of a second, for the charge to build up. Consequently the potential difference across a capacitor cannot appear at the instant the switch is closed. It takes *time* for the potential difference across a capacitor to rise. And when we open the switch, the potential difference across the capacitor does not disappear. It remains until we supply a conducting path for electrons to flow from the negatively-charged plate to the positive plate. Therefore, since the time element is involved in any change in the potential difference between the plates of a capacitor, we may think of the capacitance of an electric circuit as opposing any change in the potential difference across it. (Such is not the case for resistance.) Therefore, we can say that

Capacitance is that property of an electric circuit that opposes any change in the voltage across that circuit.

Resistance is that property of an electric circuit that opposes electric current through that circuit.

11-7 FACTORS GOVERNING CAPACITANCE

In setting up equations for electric field intensity and electric flux, we considered only the case of the uniform electric field between two parallel plates. Therefore, we shall develop only those factors governing the capacitance of parallel plates.[†] We can consider all the capacitors shown in Figure 11-8 as parallel-plate capacitors. Equation (11-7), $C = Q/V$, is an *electrical* expression defining the magnitude of the farad. In order to determine the *physical* factors governing the capacitance of a circuit, we can solve for those factors governing the magnitude of the electric field, since the electric flux is directly proportional to the charges on the plates [Equation (11-4)] and the electric field intensity is directly proportional to the potential difference between the plates [Equation (11-3)].

If we double the area of the plates, we have room for twice as many electric lines of force with the same voltage across them. Hence, the charge that a capacitor can hold at a given potential difference is doubled, and since $C = Q/V$, the capacitance is doubled. Therefore,

The capacitance of parallel plates is directly proportional to their area.

To assist us in visualizing the effect of the spacing between the plates on their capacitance, we can refer back to Figure 11-6. When we placed an insulated conductor between the two plates, the positively charged plate attracted the electrons in the conductor toward the end closer to the positively charged plate. The negatively charged plate repelled some electrons from the end of the conductor closer to the negatively charged plate. The two parallel plates of a capacitor have a similar effect on each other. The positively charged plate helps to attract a greater number of surplus electrons from the − terminal of a voltage source onto the negatively charged plate

---

[†]Formulas for the capacitance of parallel conductors, concentric conductors, etc., are found in electrical and electronics handbooks.

of the capacitor. The negatively charged plate, in turn, helps to drive a greater number of electrons out of the positively charged plate into the + terminal of the source. As we bring the plates closer together, these forces of attraction and repulsion increase and the capacitor is able to store a greater charge when connected to a given voltage source. Therefore,

The capacitance of parallel plates is inversely proportional to their spacing.

Note the similarity between the dimensional factors governing the conductance of an electric conductor and those governing the capacitance of a capacitor.

$$G \propto \frac{A}{l} \qquad \text{(from)} \quad (4\text{-}4)$$

$$C \propto \frac{A}{s} \qquad (11\text{-}8)$$

where C is capacitance (farads), A is area of the parallel plates (square meters), and s is the spacing between the plates (meters).

We also noted that the conductance (and resistance) of a metallic conductor depends on the type of conductor material. Similarly, if we slide a sheet of glass or mica between the plates of the capacitor in Figure 11-9 after the charging current has ceased, the galvanometer would show an additional surge of charging current. Since the charge Q increased but the potential difference V did not change, replacing the air dielectric with either glass or mica as a dielectric has increased the capacitance of the capacitor. Therefore,

The capacitance of parallel plates is dependent on the type of dielectric between the plates.

To express the effect of different types of conductor materials, we defined conductivity of a material as the conductance between parallel faces of a one meter cube of the material. Similarly, we can express the "capacitivity" of different dielectric materials by measuring the capacitance between two parallel plates 1 square meter in area and spaced 1 meter apart. Since capacitance is *directly* proportional to the charge it can store at a given potential difference [Equation (11-7)], capacitance is directly proportional to the ability of the dielectric to *permit* the setting up of electric lines of force. Hence,

The capacitance between opposite faces of a unit length and cross section of a dielectric material is called the permittivity of the dielectric.
The letter symbol for permittivity is the Greek letter ϵ (epsilon).

Just as
$$R = \rho \frac{l}{A} \qquad (4\text{-}4)$$

and
$$G = \sigma \frac{A}{l}$$

it follows from the definition of permittivity that

$$C = \epsilon \frac{A}{s} \qquad \text{(parallel plates)} \qquad (11\text{-}9)$$

where C is the capacitance in farads, ϵ is the absolute permittivity in MKS *units, A is the area of each plate in square meters, and s is the spacing between the plates in meters.*

To determine the unit of permittivity, we transpose Equation (11-9):

$$\epsilon = C\frac{s}{A} = \text{farads} \times \frac{\text{meters}}{\text{meters}^2} = \textbf{farads/meter}$$

Substituting for C in the equation above from the defining Equation (11-7) for capacitance,

$$\epsilon = \frac{Q}{V} \times \frac{s}{A} = \frac{Q}{A} \times \frac{s}{V}$$

But from Equations (11-4) and (11-5), for parallel plates,

$$\frac{Q}{A} = D \qquad (11\text{-}10)$$

and from Equation (11-3), for parallel plates,

$$\frac{V}{s} = E \qquad (11\text{-}3)$$

By making these substitutions, we can also express permittivity in terms of the electric flux density and the intensity of the electric field set up in the dielectric between the plates of the capacitor.

$$\epsilon = \frac{D}{E} \qquad (11\text{-}11)$$

where ϵ is the absolute permittivity of a given dielectric in farads per meter, D is the electric flux density in coulombs per square meter, and E is the electric field intensity in volts per meter.

It is not absolutely essential to have a dielectric material between the plates of a capacitor in order to set up an electric field. If we place two parallel plates in a vacuum, an electric field will be set up between them when we apply a voltage to the plates. The ratio of electric flux density to electric field intensity will be

$$\frac{D}{E} = 8.85 \times 10^{-12} = \epsilon_v \qquad (11\text{-}12)$$

where ϵ_v is called the permittivity of free space in farads per meter.

When we do place a dielectric between the plates of a capacitor, the electron orbits of the atoms are forced off-center, as shown in Figure 11-7. As a result, the negative charge of each atom of the dielectric is a little closer to the positive plate and the positive charge of each atom of the dielectric is a little closer to the negative plate. This has the same effect on the capacitance as moving the parallel plates a little closer together. Therefore, the absolute permittivity of most solid and liquid dielectrics is considerably greater than that of free space. Because of their low density, the absolute permittivity of air and other gases is very close to that of free space. If we substitute Equation (11-12) in Equation (11-9), the capacitance of a parallel-plate *air dielectric* capacitor becomes

$$C = \frac{8.85\ A}{10^{12}\ s} \tag{11-13}$$

where C is the capacitance in farads, A is the area of each plate in square meters, and s is the spacing between the plates in meters.

11-8 DIELECTRIC CONSTANT

Rather than state the absolute permittivity of liquid and solid dielectrics, it has become the custom to prepare tables stating the ratio between the absolute permittivity of a given dielectric and the absolute permittivity of free space. This ratio is called the **relative permittivity** or the **dielectric constant** of that material.

$$k = \frac{\epsilon_r}{\epsilon_v} \tag{11-14}$$

where k is the dielectric constant, ϵ_r is the absolute permittivity of a given dielectric, and ϵ_v is the absolute permittivity of free space.

Therefore, the equation for the capacitance of *any* parallel-plate capacitor becomes.

$$C = \frac{8.85kA}{10^{12}s} \qquad \text{(parallel plates)} \tag{11-15}$$

where C is the capacitance in farads, k is the dielectric constant, A is the area of each plate in square meters, and s is the spacing between plates in meters.

The dielectric constant of a given type of material can vary considerably as a result of variation in manufacturing. Table 11-2 gives typical average values of dielectric constant for those dielectric materials listed in Table 11-1. On comparing Tables 11-1 and 11-2, we note that there is no correlation between dielectric strength and dielectric constant. In manufacturing a capacitor for use in high-voltage circuits, often we are not able to choose a dielectric with a high dielectric constant because its dielectric strength is too low. Distilled water has a dielectric constant of 80, but its dielectric strength is so low that it is not practical to build water dielectric capacitors.

TABLE 11-2

Dielectric	Average Dielectric Constant
Air	1.0006
Barium-strontium titanate (ceramic)	7500
Porcelain (ceramic)	6
Transformer oil	4
Bakelite (plastic)	7
Paper	2.5
Rubber	3
Teflon (plastic)	2
Glass	6
Mica	5

Some ceramics such as barium-strontium titanate have molecules which **polarize** very easily when exposed to an electric field. As a result, they have dielectric constants which are extremely high in comparison with the average dielectric materials. Development of these ceramics during the past thirty years has made it possible to manufacture capacitors which are much more compact for a given capacitance and voltage rating than the older mica and waxed-paper dielectric capacitors. (See Figure 11-8.) These ceramics are called **ferroelectric** dielectrics. Oddly enough, the term **ferroelectric** does not imply any iron content in such materials. It results from a comparison with a similar behavior of ferrous compounds in *magnetic* circuits. Just as iron permits a much greater magnetic flux density for a given magnetizing force than nonferrous materials, a ferroelectric dielectric permits a much greater electric flux density for a given electric field strength than most dielectrics [Equation 11-11]. The dielectric *constant* of ferroelectric materials varies slightly with changes in electric field intensity.

EXAMPLE 11-1

A capacitor consists of a disk of barium-strontium titanate 1 cm in diameter and 0.2 mm thick, silver-plated on each side. Using the average values of dielectric strength and dielectric constant given in Tables 11-1 and 11-2 calculate (a) its capacitance and (b) the maximum voltage that should be applied to it.

SOLUTION

(a)
$$C = \frac{8.85kA}{10^{12}s} = \frac{8.85 \times 7500 \times \pi \times (0.5 \text{ cm})^2}{10^{12} \times 0.2 \text{ mm}}$$

$$= 2.6 \times 10^{-8} \text{ F} = \mathbf{0.026 \ \mu F}$$

(b)
$$V_m = 3 \text{ kV/mm} \times 0.2 \text{ mm} = \mathbf{600 \ V}$$

11-9 CAPACITORS IN PARALLEL

As even the graphic symbols in Figure 11-11(a) suggest, connecting capacitors in parallel is like increasing the area of the plates of a single capacitor. Therefore, the

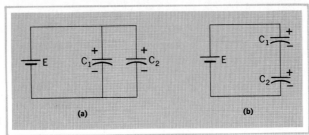

FIGURE 11-11
Capacitors: (a) in Parallel; (b) in Series.

total capacitance is greater than that of either one. Since both capacitors have to charge from the same battery, the total charge drawn from the battery is

$$Q_T = Q_1 + Q_2 \qquad\qquad (11\text{-}16)$$

Since Figure 11-11(a) is a simple parallel circuit,

$$E = V_1 = V_2$$

$$\therefore \frac{Q_T}{E} = \frac{Q_1}{V_1} + \frac{Q_2}{V_2}$$

and

$$C_T = C_1 + C_2$$

Therefore,

When we connect capacitors in parallel, the total capacitance is the sum of all the individual capacitances.

$$C_T = C_1 + C_2 + C_3 + \text{etc.} \qquad\qquad (11\text{-}17)$$

EXAMPLE 11-2
In the circuit of Figure 11-11(a), C_1 and C_2 have capacitances of 0.01 μF and 0.05 μF, respectively. What single capacitance can be used to replace C_1 and C_2?

SOLUTION
$$C_T = C_1 + C_2 = 0.01 \ \mu\text{F} + 0.05 \ \mu\text{F} = \mathbf{0.06 \ \mu F}$$

11-10 CAPACITORS IN SERIES

When we connect capacitors in series, the charging action will be similar to that shown in Figure 11-6. The bottom plate of C_1 in Figure 11-11(b) and the top plate of C_2 will be charged by electrostatic induction. And, as the graphic symbols suggest, the effect is much the same as increasing the spacing between the plates of a single capacitor. The total capacitance is less than that of either one, but the combination is capable of withstanding a higher total potential difference than either one by itself. Since Figure 11-11(b) is a simple series circuit,

$$E = V_1 + V_2$$

Since the current is the same in all parts of a simple series circuit,

$$Q_T = Q_1 = Q_2$$

$$\therefore \frac{E}{Q_T} = \frac{V_1}{Q_1} + \frac{V_2}{Q_2}$$

or

$$\frac{1}{C_{eq}} = \frac{1}{C_1} + \frac{1}{C_2}$$

Therefore,

The equivalent capacitance of series capacitors is

$$C_{eq} = \frac{1}{1/C_1 + 1/C_2 + 1/C_3 + \text{etc.}} \tag{11-18}$$

EXAMPLE 11-3

In the circuit of Figure 11-11(b), C_1 and C_2 have capacitances of 0.01 μF and 0.05 μF, respectively. What single capacitance can be used to replace C_1 and C_2?

SOLUTION

$$C_{eq} = \frac{1}{(1/0.01 \ \mu F) + (1/0.05 \ \mu F)} = 8.33 \times 10^{-3} \ \mu F$$

$$= 0.008\,33 \ \mu F$$

Equation (11-18) is easy to use with the reciprocal key of a calculator by first calculating the denominator and then reading the reciprocal. We can also reduce the calculator operations by omitting the exponent for *micro* in entering the capacitance. The display result will automatically be in microfarads. Hence, the calculator chain becomes

$$.01 \ 1/x \ + \ 0.5 \ 1/x \ = \ 1/x$$

With only two capacitors in *series,* we can use the same simplification that we used for two resistors in *parallel.* Equation (11-18) then becomes

$$C_{eq} = \frac{C_1 C_2}{C_1 + C_2} = \frac{0.01 \times 0.05}{(0.01 + 0.05)} = 8.33 \times 10^{-3} \ \mu F = 0.008\,33 \ \mu F$$

To avoid compound fractions such as Equation (11-18), we sometimes use a technique we developed when we considered parallel resistors in Section 6-9. Since connecting *more* resistors in parallel *reduces* the equivalent resistance, we consider the reciprocal property—the *conductance* of the various branches. Since connecting *more* capacitors in series *reduces* the equivalent capacitance, we can again switch our thinking to the reciprocal property. Because capacitance is proportional to the ability

292

of a dielectric to *permit* the setting up of electric lines of force in the dielectric [Equations (11-9) and (11-11)],

Elastance is the opposition to the setting up of electric lines of force in an electric insulator or dielectric.

The letter symbol for elastance is S.

From the definition,

$$S = \frac{1}{C} \tag{11-19}$$

where S is elastance in reciprocal farads,[†] and C is capacitance in farads. When we substitute in Equation (11-18), for capacitors in series,

$$S_T = S_1 + S_2 + S_3 + \text{etc.}$$

Since
$$Q_1 = C_1 V_1 \quad \text{and} \quad Q_2 = C_2 V_2,$$

and, since $Q_1 = Q_2$ in a series circuit,

$$C_1 V_1 = C_2 V_2$$

from which

$$\frac{V_1}{V_2} = \frac{C_2}{C_1} \tag{11-21}$$

When capacitors are connected in series, the ratio between any two potential differences is the *inverse* ratio of their capacitances.

EXAMPLE 11-4

A 0.01-μF capacitor and a 0.04-μF capacitor are connected first in parallel and then in series to a 500-V source. (In each case, the capacitors are discharged before the connections are made.)

(a) What is the total capacitance in each case?

(b) What is the total charge in each case?

(c) What is the charge on each capacitor and the potential difference across each capacitor in each case?

SOLUTION

(a) Parallel:
$$C_T = C_1 + C_2 = 0.01 + 0.04 = \mathbf{0.05 \ \mu F}$$

Series:
$$C_{eq} = \frac{C_1 C_2}{C_1 + C_2} = \frac{0.01 \times 0.04}{0.01 + 0.04} = \mathbf{0.008 \ \mu F}$$

[†] The International System of units does not provide a special name for the reciprocal of the farad. In some textbooks, the unit of elastance is called the **daraf** (*farad* spelled backwards).

(b) Parallel: $\quad Q_T = C_T V_T = 0.05 \times 500 = 25 \ \mu C$

Series: $\quad Q_T = C_{eq} V_T = 0.008 \times 500 = 4 \ \mu C$

(c) Parallel: $\quad V_1 = V_2 = E = 500 \ V$

$$Q_1 = C_1 V_1 = 0.01 \times 500 = 5 \ \mu C$$

$$Q_2 = C_2 V_2 = 0.04 \times 500 = 20 \ \mu C$$

Series: $\quad Q_1 = Q_2 = Q_T = 4 \ \mu C$

$$V_1 = \frac{Q_1}{C_1} = \frac{4}{0.01} = 400 \ V$$

$$V_2 = \frac{Q_2}{C_2} = \frac{4}{0.04} = 100 \ V$$

PROBLEMS

11-1. What force of attraction (newtons) will act on a single electron located 0.5 cm from a point charge of $+5 \ \mu C$?

11-2. Two electric charges, one of which is four times as large as the other, develop a mutual force of repulsion of 16 newtons when they are located 12 cm apart. Determine the magnitude of each charge.

11-3. Two parallel plates 1 cm apart are connected to a 500-V source. What force will be exerted on a free electron between the plates?

11-4. If 2×10^{10} electrons are removed from one parallel plate and added to the other, what is the total electric flux between the plates?

11-5. If moving 2×10^{10} electrons from one parallel plate to the other produces a potential difference between the plates of 180 V, what is the capacitance of the plates?

11-6. How many electrons must be removed from one plate of a 270-pF capacitor and added to the other to raise the voltage between the plates to 420 V?

11-7. A neutralizing capacitor in a radio transmitter consists of two aluminum disks, each 10 cm in diameter and $\frac{1}{2}$ cm apart with air between them. What is its capacitance?

11-8. If the electric flux density between the parallel plates of the capacitor in Problem 11-7 is $1.0 \ \mu C/m^2$, what is the voltage between the plates?

11-9. An interleaved parallel-plate capacitor consists of 10 sheets of mica 1 cm long $\times$ 0.5 cm wide and 0.025 mm thick. Eleven sheets of aluminum foil are pressed firmly between the mica sheets and connected as shown in Figure 11-8(b), What is the capacitance of this capacitor?

11-10. Two sheets of aluminum foil 2.5 cm wide and 1.0 m long and two sheets of 0.1 mm thick waxed paper 3.0 cm wide and 1.0 m long are rolled as shown in Figure 11-8(c) to form a tubular capacitor. The waxed paper has a dielectric strength of 6 kV/mm and a dielectric constant of 3. What is the capacitance and voltage rating of this capacitor?

11-11. A 10-μF and a 40-μF capacitor are connected in parallel to a 400-V source.
(a) What is the total capacitance?
(b) What is the magnitude of the charge stored by each capacitor?

11-12. The $10\text{-}\mu\text{F}$ capacitor in the circuit of Figure 11-12 is completely discharged. Draw a graph showing how the voltage across the $10\text{-}\mu\text{F}$ capacitor changes with time as the switch is operated for five complete cycles. Label the exact voltage after each cycle of switch operation. (Assume no leakage.)

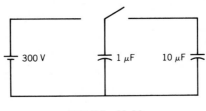

FIGURE 11-12

11-13. A $10\text{-}\mu\text{F}$, a $20\text{-}\mu\text{F}$, and a $40\text{-}\mu\text{F}$ capacitor are connected in series to a 400-V source.
   (a) What is the equivalent capacitance?
   (b) What is the magnitude of the charge stored by each capacitor?
   (c) What is the voltage across each capacitor?

11-14. After charging, the three capacitors in Problem 11-13 are disconnected from the source and from each other. They are then connected in parallel, with + plate connected to + plate. What is the voltage across each capacitor?

11-15. A sheet of glass 0.3 cm thick is placed between the plates of the neutralizing capacitor of Problem 11-7. What is its capacitance? (*Hint:* Capacitors in series.)

11-16. If the potential difference between the plates of the capacitor in Problem 11-15 is 1 kV, what is the voltage gradient in the air and in the glass?

11-17. Find the equivalent capacitance and the voltage across each of the capacitors in the circuit of Figure 11-13.

11-18. Find the equivalent capacitance and the voltage across each of the capacitors in the circuit of Figure 11-14.

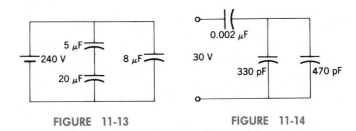

FIGURE 11-13   FIGURE 11-14

11-19. Given three capacitors with capacitances of 0.01 μF, 0.02 μF, and 0.05 μF, respectively, and a rated working voltage of 400 V for each capacitor, what is the highest voltage that can safely be applied to the series combination of the three capacitors?

11-20. What is the total elastance of a combination of the three capacitors of Problem 11-19 connected (a) in series and (b) in parallel?

REVIEW QUESTIONS

11-1. Why must energy be expended by an electric "charge separator," which removes electrons from neutral atoms?

11-2. How does charge separation in an electric battery differ from charge separation by rubbing a glass rod with a silk cloth?

11-3. Coulomb's law of electrostatic force states that electric force is proportional to the product of two electric charges; yet, Figure 11-1 shows an electric field for a *single* isolated electric charge. Explain the purpose of sketches such as those in Figure 11-1.

11-4. Why does a potential difference exist between two electric conductors possessing unlike charges?

11-5. Why are we more concerned with the electric field between unlike charges in electric circuits than we are with the repulsion between like charges which we can demonstrate in electrostatics?

11-6. What is the significance of the arrows on the sketch of electric lines of force between two parallel plates?

11-7. What characteristic of an electric field is referred to by the term **electric field intensity**?

11-8. What is meant by the **potential gradient** of an electric field?

11-9. How are **electric field intensity** and **potential gradient** related in the case of charged parallel plates?

11-10. What factors govern the **flux density** of the electric field between charged parallel plates?

11-11. Draw a sketch showing the equipotential surfaces in the electric field between two charged concentric conductors.

11-12. Draw a sketch showing the equipotential surfaces in the electric field of Figure 11-6.

11-13. Why must the + and − charges at the two ends of the insulated conductor in Figure 11-6 be equal?

11-14. Define a **displacement current.**

11-15. What is meant by a **polarized** atom?

11-16. What type of atoms can become polarized?

11-17. Why is the strength of a dielectric expressed in terms of the **voltage gradient** of an electric field rather than the total potential difference?

11-18. Define **dielectric absorption.**

11-19. Why must the ratio between the charge on a given pair of insulated conductors and the potential difference between them be a constant?

11-20. Define the term **capacitance** in terms of Q and V.

11-21. Define the term **capacitor** in terms of Q and V.

11-22. Why can we define capacitance as that property of an electric circuit that opposes any change in voltage across that circuit?

11-23. Why does the spacing between two parallel plates have an effect on their capacitance?

11-24. Why does the voltage applied to two parallel plates not have an effect on their capacitance?

11-25. Why is it possible to have capacitance between two parallel plates when the space between them is evacuated?

11-26. Why does the presence of an insulating material between the parallel plates increase their capacitance as compared with that of the same plates in a vacuum?

11-27. Since we are concerned with electric circuits, we defined the **farad** in terms of the charge on the conducting *plates* and the voltage between the *plates*. Define the **farad** in terms of flux density and intensity of the electric *field* in the dielectric between the plates.

11-28. Distinguish between the **absolute permittivity** and the **relative permittivity** of a given dielectric.

11-29. Why is it possible to define the **dielectric constant** of a given insulating material as the ratio between the capacitance of a pair of parallel plates with that material as a dielectric and the same parallel plates with air as a dielectric?

11-30. When two capacitors are connected in series, one plate of each has no electrical connection to any voltage source. How is it possible for these capacitors to become charged?

11-31. Why must capacitors in series all take the same charge?

11-32. A 0.1-μF 400-V capacitor and a 0.05-μF 200-V capacitor are connected in series. What is the maximum voltage that can be safely applied to this combination?

11-33. Some of the older electrolytic capacitors packaged in aluminum cans have been known to explode when connected across a dc voltage source such that the insulated terminal is negative with respect to the can. Explain what happens.

11-34. Explain why a motor-starting capacitor consisting of two electrolytic capacitors connected in series opposing can be used safely in an ac circuit.

*11-35. Design a personal computer program which will request the following input data: area of a capacitor plate specified in square centimeters, thickness of the dielectric specified in millimeters, and dielectric constant. After these data have been entered, the computer should display the capacitance of a simple parallel-plate capacitor specified in picofarads.

12

CAPACITANCE IN DC CIRCUITS

Now that we know something about capacitance as a basic property of electric circuits, we can examine the behavior of practical electric circuits when capacitance is present intentionally in the form of capacitors, or as a byproduct of other circuit configurations, such as a reverse-biased *pn*-junction. In either case, we represent circuit capacitance in our schematic diagrams by the standard graphic symbol shown in Figure 12-1.

12-1 CHARGING A CAPACITOR

Strictly speaking, when we speak of *charging* a capacitor, we should think in terms of the *charge* stored by it in *coulombs*. But $Q = CV$, and for a given capacitor, C is a constant based on such physical factors as the area of the plates, the distance between plates, and the type of dielectric. Therefore, the charge is directly proportional to the potential difference between the plates. Since we have voltmeters for reading potential difference directly, it is more convenient to speak of the charging of a capacitor in terms of the potential difference built up between its plates. This point of view is consistent with our definition of capacitance.

At the instant we close the switch in the circuit of Figure 12-1, there can be no potential difference across the capacitor, since it takes *time* for electrons to flow into the bottom plate and out of the top plate in order to build up a potential difference between the plates. But Kirchhoff's voltage law states that there must be a voltage drop in the circuit equal to the applied voltage when the switch is closed. Although no resistance is shown, we shall assume that the resistance of the wiring and the

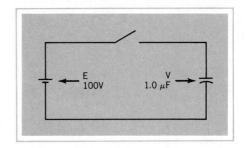

FIGURE 12-1
Charging a Capacitor without a Current-Limiting Resistor.

internal resistance of the battery add up to 0.5 ohm. Therefore, at the instant we close the switch, since there is no potential difference across the capacitor, there must be an IR drop in the circuit equal to the applied voltage. To produce this voltage drop, the current which must flow the instant we close the switch is

$$\text{initial } I = \frac{E}{R} = \frac{100}{0.5} = 200 \text{ A}$$

Since current is the rate of flow of electrons past a certain point in a circuit, a 200-ampere current will charge the plates of the capacitor very rapidly, and the potential difference across the capacitor will take only a few microseconds to rise to 100 volts. When the potential difference across the capacitor equals the applied voltage, current must cease. Therefore, if we charge a capacitor by connecting it directly across a voltage source, there is a very short duration pulse of very high current. This is rather rough on the source so, in practice, we prefer to charge capacitors through a resistance which limits the peak value of the initial current surge.

At this point, we must note an important convention regarding the letter symbols we use to represent electrical properties in equations. The resistance of an electric circuit is dependent on such physical factors as length, cross section, and type of material. Capacitance depends on such physical factors as area of the conducting surfaces, spacing between the surfaces, and the type of dielectric material between the conducting surfaces. The numerical values of these circuit parameters are essentially independent of *time*. Consequently, we represent resistance and capacitance by upper-case letter symbols. In the circuits we have considered in preceding chapters, the source voltage, the resulting current, the voltage drop, and the power have also been independent of any exact instant in time. Accordingly, we have represented all these **steady-state** values with uppercase letter symbols. But, in considering the charging of a capacitor, we note that the magnitude of the charging current and also the potential difference across the capacitor change from instant to instant and depend on how long the switch has been closed. We represent such **instantaneous** values of voltage and current by *lowercase* letter symbols. Note that we have always represented **time** itself by the lowercase t.

Lowercase letter symbols represent instantaneous values which are dependent on the exact instant in time for their numerical value.

Uppercase letter symbols represent steady-state values which are not dependent on the exact instant in time for their numerical value.

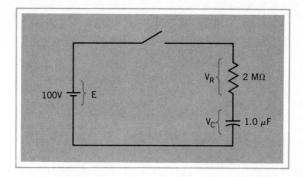

FIGURE 12-2
Charging a Capacitor through a Current-Limiting Resistor.

All the time the switch is closed in the circuit of Figure 12-2, the sum of the potential difference across the capacitor and the *IR* drop across the resistor must equal the applied voltage. Therefore,

$$E = v_C + iR \qquad (12\text{-}1)$$

Again, at the instant we close the switch, there will be no potential difference across the capacitor. Hence, the initial current must be such that the *IR* drop in the circuit is equal to the applied voltage. Therefore,

$$\text{initial } I = \frac{E}{R} \qquad (12\text{-}2)$$

The initial current in the circuit of Figure 12-2 is only 50 microamperes. This current immediately starts to charge the capacitor. Thus, a potential difference starts to build up across the capacitor. To satisfy Kirchhoff's voltage law, the *IR* drop across the resistor must decrease. Consequently, the instantaneous current (the dependent variable of our electric circuits) must become smaller. As a result, the potential difference across the capacitor must rise more slowly. Therefore, the *rate of rise* of the instantaneous voltage across the capacitor depends on the magnitude of the instantaneous current, which depends on how much *IR* drop must appear across the resistance of the circuit, which, in turn, depends on the instantaneous voltage across the capacitor, which depends on how long the switch has been closed. Before we can sort out these interdependences, we must consider for a moment just how we can express rate of rise or, in more general terms, *rate of change* of voltage numerically.

12-2 RATE OF CHANGE OF VOLTAGE

To do this, we can assume the existence of some special electric circuits in which the potential difference between two points in these circuits is represented by the graphs of Figure 12-3. When we close the switch in the circuits represented by the graphs of Figure 12-3(a), the potential difference increases as a linear function of time (rather than immediately taking on a steady-state value, as in the case of simple resistance circuits). Since these are straight-line graphs, we can read the rate of change of voltage directly from the graph by noting the number of volts increase in potential difference

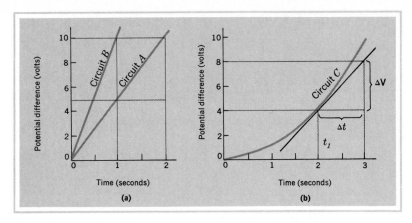

FIGURE 12-3
Graphical Determination of Rate of Change of Voltage.

during 1 second. For circuit A, the voltage rises from 0 to 5 volts in the first second and from 5 to 10 during the second second. In either case, the *rate of change* of voltage is 5 volts per second. Similarly, the rate of change of voltage in circuit B is 10 volts per second. Note that the graph for circuit B is twice as steep as the graph for circuit A. Hence, the rate of change of voltage is represented by the *slope* of the graph of voltage as a function of time.

Circuit C represents an electric circuit in which the magnitude of the potential difference between two specified points is a function of the *square* of the elapsed time. Since this graph is not a straight line, the rate of change of voltage is not the same for all time intervals. We must select some specific time t_1 at which we want to know the rate of change of voltage. By drawing a *tangent* to the graph at time t_1 in Figure 12-3(b), we can now read the rate of change of voltage at time t_1 by noting the change in voltage ΔV for any convenient change in time Δt (because the tangent is a straight line). Hence, the rate of change of voltage at time t_1 is $\Delta V/\Delta t$ volts per second. In practice, we adopt the calculus symbol for *rate of change*.[†]

The symbol for rate of change of voltage is $\mathrm{d}v/\mathrm{d}t$ *volts/second.*

[†]Calculus provides us with a more precise method of expressing *rate of change* of voltage by differentiating the equation for the instantaneous voltage as a function of time. The equation for the instantaneous voltage in circuit A in Figure 12-3(a) is

$$v = 5t$$

By differentiating, we obtain the rate of change of voltage

$$\frac{\mathrm{d}v}{\mathrm{d}t} = 5 \text{ V/s}$$

Similarly, in circuit B

$$v = 10t \quad \text{and} \quad \frac{\mathrm{d}v}{\mathrm{d}t} = 10 \text{ V/s}$$

Circuit C is just as easy to solve.

Since $$v = t^2, \quad \frac{\mathrm{d}v}{\mathrm{d}t} = 2t \text{ V/s}$$

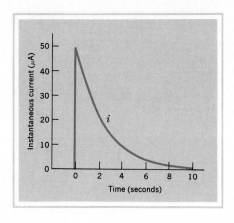

FIGURE 12-4
Instantaneous Charging Current.

When we close the switch in the circuit of Figure 12-2, current must flow instantly at its initial value and then become progressively smaller as the voltage across the capacitor rises, as shown by the graph of Figure 12-4. Since resistance is a physical constant, the instantaneous voltage drop across it must always be proportional to the instantaneous current through it (Ohm's law). Thus, the graph of the IR drop across the resistor will have the same shape as the instantaneous current graph of Figure 12-4.

Since the initial current in the CR circuit of Figure 12-2 is only 50 microamperes, it will take appreciable time for the capacitor to charge, that is, for the voltage across the capacitor to rise to the same value as the applied voltage. Because we know the magnitude of the initial current, we can determine the *initial* rate of change of voltage across the capacitor. Since

$$Q = CV \qquad (11\text{-}7)$$

and
$$Q = It \qquad (2\text{-}2)$$

$$CV = It$$

And since
$$\text{initial } I = \frac{E}{R} \qquad (12\text{-}2)$$

$$CV = \frac{E}{R} \times t$$

from which
$$\text{initial } \frac{dv}{dt} = \frac{E}{CR} \qquad (12\text{-}3)$$

Since Equation (12-3) contains all the independent variables in the circuit of Figure 12-2, no matter what numerical values we choose, the graph of the actual instantaneous voltage across the capacitor must start to rise with the same slope and at the same instant in time as the dashed line repesenting the initial rate of change of capacitor voltage in Figure 12-5. The slope of the actual potential difference must then become progressively more gradual until it finally ends up tangent to the dashed line representing the applied voltage.

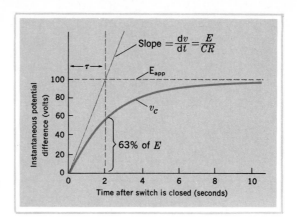

FIGURE 12-5
Rise of Potential Difference across a Capacitor Charging through a Resistance.

12-3 TIME CONSTANT

If we assume for the moment that the potential difference across the capacitor can rise at a steady rate equal to its initial rate until it equals the applied voltage, we can quite readily derive a convenient expression for the charging time of the capacitor. This time interval is called the **time constant** of the CR circuit. On the basis of the assumption above, we can define time constant thus:

The time constant of a CR circuit is the time it would take the potential difference across the capacitor to rise to the same value as the applied voltage if it were to continue to rise at its initial rate of change for the whole time interval.
The letter symbol for time constant is the Greek letter τ (tau).

From Figure 12-5, if the rate of change of voltage remains constant, the final potential difference across the capacitor becomes simply

$$V = \tau \times \text{initial } \frac{dv}{dt}$$

But $\qquad\qquad V = E \qquad$ and $\qquad$ initial $\dfrac{dv}{dt} = \dfrac{E}{CR}$ $\hfill$ (12-3)

from which $\qquad\qquad\qquad \tau = CR \text{ seconds}$ $\hfill$ (12-4)

where τ is the length of time in seconds defined as a time constant, C is the capacitance of the capacitor in farads, and R is the resistance through which the capacitor charges in ohms. Since the farad is not a practical size of unit, we can still make Equation (12-4) balance if we express capacitance in microfarads and resistance in megohms.

Note that the time constant of a CR circuit is directly proportional to the capacitance. If we double the capacitance, twice the charge must flow in order to raise the potential difference across the capacitor to the applied voltage. But if the resistance is not changed, the initial current remains unchanged, and, therefore, it will take twice as long for the capacitor to charge. If we double the resistance, the initial current is cut in half, and it will again take twice as long for the potential difference of a given capacitor to rise until it equals the applied voltage. Note also that the magnitude of the applied voltage has no effect on the time constant. If we double the applied voltage, the capacitor must store twice the charge, but the initial current will be twice as great. Therefore, it will take exactly the same time for the potential difference of the capacitor to reach the applied voltage.

Because of the fixed relationship between the solid exponential graph of Figure 12-5, representing the actual rise in potential difference across the capacitor, and the dashed lines from which we solved for the time constant, the actual charging time of a CR circuit bears a fixed relationship to the time constant. The instantaneous potential difference across a capacitor will always be 63% of the applied voltage one time constant after the switch is closed. For all practical purposes,

The potential difference across the capacitor will reach a steady-state value equal to the applied voltage after a time interval equal to five time constants has elapsed.

EXAMPLE 12-1

(a) What is the initial rate of change of potential difference across the capacitor in Figure 12-2?

(b) How long will it take the capacitor in Figure 12-2 to charge to a potential difference of 100 V?

SOLUTION

(a) $$\text{Initial } \frac{dv}{dt} = \frac{E}{CR} = \frac{100 \text{ V}}{1 \ \mu\text{F} \times 2 \text{ M}\Omega} = 50 \text{ V/s}$$

(b) $$\tau = CR = 1 \ \mu\text{F} \times 2 \text{ M}\Omega = 2 \text{ s}$$

Since $$100 \text{ V} = E = \text{full charge}$$

$$\therefore t = 5\tau = 5 \times 2 \text{ s} = 10 \text{ s}$$

12-4 GRAPHICAL SOLUTION FOR INSTANTANEOUS POTENTIAL DIFFERENCE

Because of the fixed relationship between the time constant and the time it actually takes the potential difference of a capacitor to rise to the same value as the applied voltage, we can use the exponential curve of Figure 12-5 to solve graphically for the instantaneous potential difference of any CR circuit at any instant in time. To do this, we carefully replot the exponential curve of Figure 12-5 with a horizontal (time) axis calibrated in terms of time constants rather than seconds and with a vertical (instantaneous voltage) axis marked in percentage of the applied voltage, as shown in Figure 12-6.

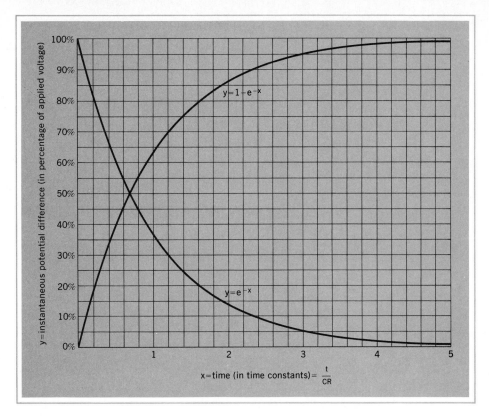

FIGURE 12-6
Universal Exponential Curves for the Graphical Solution of the Charge and Discharge of Capacitors in Direct-Current Circuits.

EXAMPLE 12-2
(a) What is the instantaneous potential difference across the capacitor in the circuit of Figure 12-2 three seconds after the switch is closed?
(b) How long will it take the potential difference across the capacitor to rise from zero to 55 V?

SOLUTION
(a)
$$\tau = CR = 1 \ \mu F \times 2 \ M\Omega = 2 \ s$$

$$\therefore 3 \ s \ \text{represents} \ \frac{3 \ s}{2 \ s} = 1.5 \ \tau$$

From Figures 12-6 and 12-7(a), when $t = 1.5\tau$,

$$v_C = 77\% \ \text{of E} = 0.77 \times 100 \ V = 77 \ V$$

(b)
$$55 \ V \ \text{represents} \ \frac{55 \ V}{100 \ V} \times 100\% = 55\% \ \text{of} \ E$$

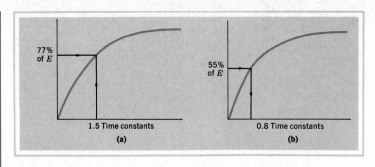

FIGURE 12-7
Using the Graph of Figure 12-6 to Solve Example 12-2.

From Figures 12-6 and 12-7(b), when $v_C = 55\%$ of E,

$$t = 0.8 \text{ time constant} = 0.8 \times 2 \text{ s} = \mathbf{1.6 \text{ s}}$$

12-5 DISCHARGING A CAPACITOR

If we assume that the switch in the circuit of Figure 12-8(a) has been in position 1 for a length of time equal to at least five time constants, the potential difference across the capacitor will be equal to the applied voltage, 200 volts. If we now place the switch in the open-circuit position 2, the potential difference across the capacitor will remain at 200 volts for some considerable time since there is no conducting path between the plates except a bit of leakage through the dielectric. At the instant we throw the switch to position 3, the potential difference across the capacitor will still be 200 volts since it takes *time* for the surplus electrons to flow from the bottom plate to the top plate of the capacitor. To satisfy Kirchhoff's voltage law, the initial current will have to be such that a 200 volt IR drop will be developed across the resistance of the conductor

FIGURE 12-8
Discharging a Capacitor.

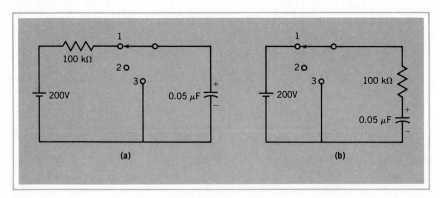

PART III   CAPACITANCE AND INDUCTANCE

connecting the two plates together. Therefore, the initial current is very high and the capacitor discharges very rapidly.

If we wish to reduce the magnitude of the initial discharge current, we can change the circuit to that shown in Figure 12-8(b) so that, in switch position 3, a resistance of 100 kilohms is connected across the charged capacitor. The initial current is now limited to

$$\text{initial } I = \frac{V_{\text{init}}}{R} = \frac{200 \text{ V}}{100 \text{ k}\Omega} = 2 \text{ mA} \tag{12-5}$$

Since the charge stored in the capacitor is being released at a slower rate $i = dq/dt$, it will take much longer for the capacitor to discharge fully.

The potential difference across the capacitor is at its maximum value just as the switch is thrown to position 3. Therefore, the initial current is the maximum value that the instantaneous discharge current can have and, as a result, the voltage across the capacitor starts to fall fairly rapidly. But, as the charge diminishes, the voltage across the capacitor must also decrease and, according to Kirchhoff's voltage law, the IR drop must have the same value as the potential difference of the capacitor. Therefore, the current must decrease and as a result, the capacitor will discharge more slowly. This interdependence among the potential difference, the IR drop, the current, and the rate of decrease of the potential difference results in the exponential discharge of Figure 12-9(a). The current in the 100-kilohm resistor flows in the opposite direction when the capacitor of Figure 12-8(b) is discharging with the switch in position 3 to that when the capacitor is charging in switch position 1. Consequently, the polarity of the IR drop across the resistance is reversed when the capacitor discharges. This reversal is indicated in the graph of the instantaneous current and IR drop in Figure 12-9(b) by using the negative vertical axis to plot instantaneous current or instantaneous voltage drop across the resistance.

With the same procedure we used in deriving Equation (12-3), the initial rate of change of the potential difference across the capacitor as it starts to discharge becomes

FIGURE 12-9
Instantaneous Voltage and Current as a Capacitor Discharges.

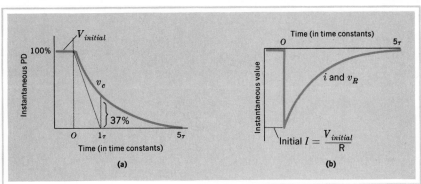

$$\frac{dv}{dt} = -\frac{V_{init}}{CR} \qquad (12\text{-}6)$$

The $-$ sign is a result of the discharge current flowing through the resistance in a direction opposite from that followed during the charging of the capacitor. Therefore, the rate of change of voltage is a *negative* quantity, which indicates that the voltage across the capacitor is decreasing rather than increasing. If we define the **discharge time constant as the time it would take for the capacitor to discharge completely *if* it were to continue to discharge at the initial rate of fall of potential difference for the whole time interval,** Equation (12-6) reduces to

$$\tau = CR \qquad (12\text{-}4)$$

Therefore, **we determine the discharge time constant in exactly the same manner as the charge time constant.** Again, the exponential discharge curve bears a fixed relationship to the time constant. In a time interval equal to one time constant, the potential difference of the capacitor will have dropped to 37% of its initial value. Or, if we wish, we can say that the capacitor loses 63% of its initial potential difference in a time interval of one time constant. For practical purposes, the time required for the capacitor to completely discharge will again be five time constants. In Figure 12-8(b),

$$\tau = CR = 0.05\ \mu F \times 100\ k\Omega = 5\ ms$$

and total discharge time = 5×5 ms = 25 ms.

EXAMPLE 12-3

After the switch has been in position 1 for a considerable period of time,

(a) What is the initial rate of change of potential difference across the capacitor in the circuit of Figure 12-8(b) when the switch is placed in position 3?

(b) What will the potential difference across the capacitor be 7 ms after the switch is placed in position 3?

(c) How many milliseconds will elapse after the switch is placed in position 3 before the instantaneous voltage drop across the 100-kΩ resistor reaches 110 V?

SOLUTION

(a) $\qquad\qquad$ Initial $\dfrac{dv}{dt} = -\dfrac{V_{init}}{CR}$

$$= -\frac{200\ V}{0.05\ \mu F \times 100\ k\Omega} = -40\ kV/s$$

(b) $\qquad\qquad \tau = CR = 0.05\ \mu F \times 100\ k\Omega = 5\ ms$

$$\therefore 7\ ms\ represents\ \frac{7\ ms}{5\ ms} = 1.4\ \tau$$

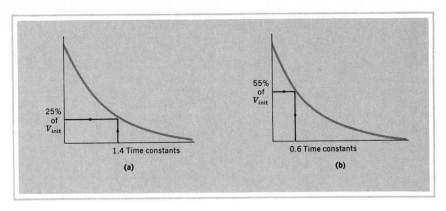

FIGURE 12-10
Using the Graph of Figure 12-6 to Solve Example 12-3.

From Figures 12-6 and 12-10(a), when $t = 1.4\tau$,

$$v_C = 25\% \text{ of } V_{init} = 0.25 \times 200 \text{ V} = \textbf{50 V}$$

(c) When the switch is in position 3, the resistor is connected directly across the capacitor. From Kirchhoff's voltage law,

$$v_C + v_R = 0 \quad \text{from which} \quad v_R = -v_C$$

$$110 \text{ V represents } \frac{110}{200} = 55\% \text{ of } V_{init}$$

From Figures 12-6 and 12-10(b), when $v_C = 55\%$ of V_{init},

$$t = 0.6\tau = 0.6 \times 5 \text{ ms} = \textbf{3 ms}$$

12-6 CR WAVESHAPING CIRCUITS

Suppose we replace the manually operated switches of Figures 12-2 and 12-8 with a motor-driven switch, as in Figure 12-11(a), which rotates at 500 revolutions a second. Consequently, the moving contact dwells for one millisecond on each stationary contact. The graph for the output voltage appears as shown in Figure 12-11(b), and the circuit of Figure 12-11(a) constitutes a simple **square-wave** generator. Since the polarity of the output voltage does not reverse, we describe it as a **pulsating dc** waveform. Although mechanical "choppers" are sometimes used to generate *intermittent* direct current from a steady dc source, the switching action is usually performed by transistors in the pulse circuits we encounter in radar, television, and digital computers.

In Figure 12-12, we have connected a CR network across the output of the square-wave generator. In Figure 12-12(a), the output voltage is v_C; and, in Figure

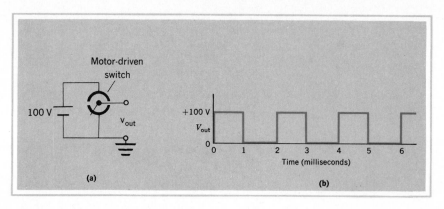

FIGURE 12-11
Simple Square-Wave Generator.

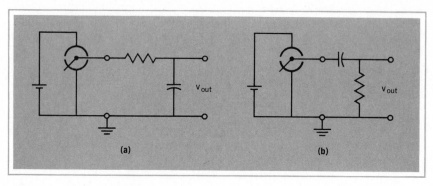

FIGURE 12-12
Simple *CR* Waveshaping Networks.

12-12(b), the output voltage is v_R. If the time constant of the *CR* network is short compared with the duration of each half-cycle of the square wave, the surge of charging and discharging current will be completed each half-cycle before the switch changes position. In this case, it is a fairly simple matter to sketch the output-voltage waveform.

Suppose that the *CR* networks of Figure 12-12 have time constants of 0.1 millisecond. Five time constants will then be 0.5 millisecond, and the charging and discharging current ceases halfway through each half-cycle of the square-wave input voltage. If we follow through the charge and discharge cycle of the capacitor, we find that the output voltages in the circuits of Figure 12-12 have the waveforms shown in the graphs of Figure 12-13.

However, if the time constant of the *CR* network is sufficiently long that charging or discharging of the capacitor is not completed during a half-cycle of the square-wave input, calculation of the waveform becomes a bit of a chore, as illustrated by Example 12-4.

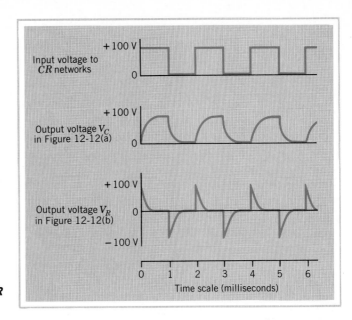

FIGURE 12-13
**Voltage Waveform Graphs for the *CR*
Networks of Figure 12-12.**

EXAMPLE 12-4
Starting from position 1, the switch in the circuit of Figure 12-14 dwells for 1 ms in each position. Calculate v_C 3 ms after the switch first moves to position 2.

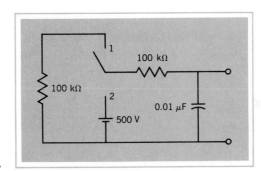

FIGURE 12-14
Circuit Diagram for Example 12-4.

SOLUTION
Step 1.   Starting from position 1, the capacitor is completely discharged. In position 2, the charging time constant is

$$\tau = CR = 0.01 \ \mu F \times 100 \ k\Omega = 1 \ ms$$

Since the switch remains in position 2 for 1 ms, the capacitor will charge for a period of time equal to one time constant. Hence, v_C will rise to 63% $\times$ 500 V = 315 V

Step II. As the switch returns to position 1, the discharging time constant becomes

$$\tau = CR = 0.01 \ \mu\text{F} \times (100 \ \text{k}\Omega + 100 \ \text{k}\Omega) = 2 \ \text{ms}$$

While the switch is in position 1, the capacitor will discharge for a period of time equal to one-half a time constant.

From Figure 12-6, v_C will drop to 60% of its initial value. Therefore, at the end of the second millisecond,

$$v_C = 0.6 \times 315 \ \text{V} = 189 \ \text{V}$$

As we start the third millisecond, the capacitor starts to charge again. But, this time, it starts from an initial potential difference of 189 V. We have a choice of two lines of thought in handling the rise in voltage across a partially-charged capacitor.

Step IIIA. 189 V represents $\frac{189}{500} = 38\%$ of E. On the charge curve of Figure 12-6, if the capacitor had started its charging from zero, a time interval of 0.48 time constant would have elapsed [Figure 12-15(a)] for the voltage across the capacitor to reach 189 V.

Since 1 ms = 1 charging time constant, one millisecond after the switch returns to the *charge* position, the potential difference across the capacitor will be the same as if it had charged from zero for a period of $0.48 + 1.0 = 1.48\tau$. From Figure 12-6, when $t = 1.48\tau$,

$$v_C = 77\% \text{ of } E = 0.77 \times 500 \ \text{V} = \textbf{385 V}$$

Step IIIB. The alternative method is to think in terms of the *additional* potential difference to which the capacitor can charge.

Since $E = 500$ V and the voltage across the capacitor is already 189 V, the

FIGURE 12-15
Graphical Solution of Step III of Example 12-4.

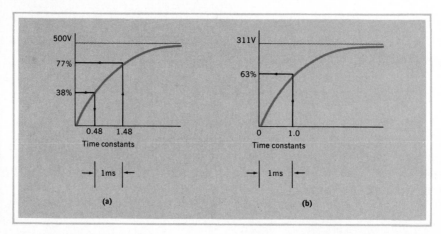

PART III   CAPACITANCE AND INDUCTANCE

maximum *additional* applied voltage is only 500 V − 189 V = 311 V.

In a charging time interval of one time constant, the voltage can rise by 63% of this additional E.

$$\therefore \text{ additional } v_C = 0.63 \times 311 \text{ V} = 196 \text{ V} \quad [\text{Figure 12-15(b)}]$$

Therefore, the final voltage across the capacitor is its *initial* voltage plus its *additional* increase in potential difference.

$$\therefore v_C = 189 \text{ V} + 196 \text{ V} = \mathbf{385 \text{ V}}$$

12-7 CALCULATOR SOLUTION FOR INSTANTANEOUS POTENTIAL DIFFERENCE

When we require greater accuracy than the universal exponential graph of Figure 12-6 provides, we may solve for instantaneous voltage across a capacitor by using the equations from which the curves of Figure 12-6 were plotted. Since a purely algebraic solution does not provide an opportunity to visualize what is going on in the circuit, it is well worthwhile accompanying any calculator solution with a freehand graphic solution to check any serious error in the more accurate algebraic solution. We start with the Kirchhoff's voltage-law equation for the basic circuit of Figure 12-2.

$$E = iR + v_C \qquad (12\text{-}1)$$

Because the instantaneous current is the instantaneous rate of charging,

$$i = \frac{dq}{dt}$$

And since the instantaneous charge is the product of the capacitance and the instantaneous PD across it, $q = Cv_C$. Therefore,

$$i = C \frac{dv_C}{dt} \qquad (12\text{-}7)$$

Substituting for i in Equation (12-1),

$$E = CR \frac{dv_C}{dt} + v_C \qquad (12\text{-}8)$$

Since Equation (12-8) is a differential equation, to solve for the instantaneous potential difference v_C, we must resort to some routine calculus. The solution of Equation (12-8) is developed in Appendix 2-2. Hence, **the instantaneous voltage across a capacitor *charging* from 0 volts is**

$$v_C = E(1 - e^{-x}) \qquad (12\text{-}9)$$

where e $= 2.718$ *(the base of natural logarithms) and* $x = t/CR$.

Since $\qquad \tau = CR$ seconds, $\qquad x = \dfrac{t}{\tau}$

We can again think of x as representing elapsed time measured in **time constants.**

Rather than repeat the solution of the differential Equation (12-1) to find the instantaneous charging current i_C, we can simply substitute the expression [Equation (12-9)] we obtained for the instantaneous PD across the capacitor directly into Equation (12-1). Therefore,

$$E = iR + E(1 - e^{-x})$$

from which $\qquad i_C = \dfrac{E}{R} e^{-x} \qquad\qquad\qquad$ (12-10)

When the switch in the circuit of Figure 12-8(b) is in position 3, there is no voltage source in the discharge loop. Hence, the applied voltage E is zero. Consequently, the IR drop across the resistance and the potential difference across the capacitor must be numerically equal at all instants in time in order to satisfy Kirchhoff's voltage law. Therefore,

$$0 = iR + v_C = CR \frac{dv_C}{dt} + v_C$$

Solving for v_C, **the instantaneous voltage across a *discharging* capacitor is**

$$v_C = V_0 e^{-x} \qquad\qquad\qquad (12\text{-}11)^\dagger$$

where V_0 *is the initial voltage across the capacitor,* e $= 2.718$ (the exponential function key of a calculator), *and* $x = t/CR$.

Equations (12-9) and (12-11) are the algebraic expressions from which the two curves of Figure 12-6 used in the graphical solution of CR examples were prepared.

EXAMPLE 12-2A

(a) What is the instantaneous voltage across the capacitor in the circuit of Figure 12-2 three seconds after the switch is closed?

(b) How long will it take the voltage across the capacitor to rise from 0 V to 55 V?

SOLUTION

(a) $\qquad\qquad x = \dfrac{t}{CR} = \dfrac{3 \text{ s}}{1 \ \mu\text{F} \times 2 \text{ M}\Omega} = 1.5$

$$v_C = E(1 - e^{-x}) = 100 \times (1 - e^{-1.5}) = 77.7 \text{ V}$$

---

†See Appendix 2-2 for solution.

The calculator chain for this computation becomes

$$100 \; \boxed{\times} \; \boxed{(} \; 1 \; \boxed{-} \; 1.5 \; \boxed{+/-} \; \boxed{e^x} \; \boxed{)} \; \boxed{=}$$

enter value of x converts to $-x$ solves for e^{-x}

When you press the $\boxed{e^x}$ key, the calculator displays the value of e^{-x} as 0.22313.

(b) $$55 \text{ V} = 100 \text{ V}(1 - e^{-x})$$

from which $e^{-x} = 0.45$.
    Solving for $-x$ with a calculator, we use the $\boxed{\textbf{ln}}$ key.

Hence, $$-x = 0.45 \; \boxed{\textbf{ln}} \; = -0.7985$$

from which $$x = 0.8 = \frac{t}{CR} \quad (\text{Press} \; \boxed{+/-} \; \text{key})$$

Hence, $$t = 0.8 \times 2 \text{ s} = \textbf{1.6 s}$$

EXAMPLE 12-4A
Starting from position 1, the switch in the circuit of Figure 12-14 dwells for 1 ms in each position. Calculate v_C 3 ms after the switch first moves to position 2.

SOLUTION
    Step I. When the switch first moves to position 2, the capacitor will charge from 0 V toward 500 V. At any instant,

$$v_C = E(1 - e^{-x}) = 500(1 - e^{-x})$$

After 1 ms,

$$x = \frac{t}{CR} = \frac{1 \text{ ms}}{0.01 \; \mu\text{F} \times 100 \text{ k}\Omega} = 1$$

(Hence, x represents a time interval of one time constant.)

$$\therefore v_C = 500(1 - e^{-1})$$
$$= 500 \times (1 - 0.3679) = 316 \text{ V}.$$

    Step II. As the capacitor discharges in position 1,

$$v_C = V_0 e^{-x} = 316 \times e^{-x}$$

After 1 ms,

$$x = \frac{t}{CR} = \frac{1 \text{ ms}}{0.01 \; \mu\text{F} \times 200 \text{ k}\Omega} = 0.5$$

$$\therefore v_C = 316 \times e^{-0.5}$$

$$= 316 \times 0.6065 = 192 \text{ V}$$

Again, we can solve the additional charge of a partially charged capacitor by either of the lines of thought we used in Steps IIIA and IIIB of the solution for Example 12-4. For the moment, we will use the longer A method.

Step III. When the switch returns to position 2, v_C starts at 192 V. The length of time it would take the capacitor to charge from 0 V to 192 V is

$$192 = 500(1 - e^{-x})$$

from which

$$e^{-x} = \frac{500 - 192}{500} = 0.616$$

Using the **ln** key of a calculator,

$$x = 0.616 \ \boxed{\text{ln}} \ \boxed{+/-} \ = 0.4845$$

Since

$$x = \frac{t}{CR}$$

$$\therefore t = 0.4845 \times 1 \text{ ms} = 0.4845 \text{ ms}$$

One millisecond later, the capacitor will reach a voltage equivalent to having charged for a total time interval of $0.4845 + 1.0 = 1.4845$ ms. Hence,

$$x = \frac{t}{\tau} = \frac{1.4845 \text{ ms}}{1 \text{ ms}} = 1.4845$$

and

$$v_C = 500(1 - e^{-1.4845})$$

$$= 500 \times (1 - 0.2266) = \mathbf{387 \text{ V}}$$

EXAMPLE 12-5

After being open for a considerable period of time, the switch in the circuit shown in Figure 12-16 is closed at $t = 0$ s and is opened again at $t = 1$ s. What is the voltage across the capacitor at $t = 2$ s?

SOLUTION

The discharge circuit when the switch is open is quite straightforward. However, to calculate the charge time constant when the switch is closed, we need Thévenin's theorem.

Step I. Considering the capacitor to be the load, we remove it from the original circuit of Figure 12-16(a) and place it in the Thévenin-equivalent circuit of Figure 12-16(b). Once again, mental arithmetic is all that is required to tell us that E_x and R_x for the equivalent circuit of Figure 12-16(b) are 400 V and 160 kΩ, respectively.

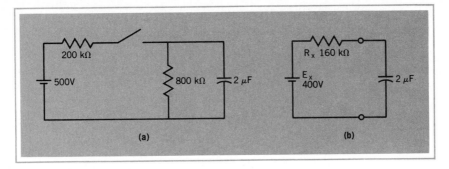

FIGURE 12-16
Thévenin-Equivalent Circuit of a Capacitor Charging Circuit.

Step II. Charging time constant is, therefore,

$$\tau = CR_x = 2\ \mu F \times 0.16\ M\Omega = 0.32\ s$$

$$\therefore 1\ s = \frac{1.0}{0.32} = 3.125\ \tau$$

Since the switch was originally open, the capacitor starts to charge with zero volts between its plates. From the algebraic solution of Equation (12-9), after charging for 1 s,

$$v_C = 400(1 - e^{-3.125}) = 400(1 - 0.044) = 382.4\ V$$

Step III. Discharging time constant is simply

$$\tau = CR_{dis} = 2\ \mu F \times 800\ k\Omega = 1.6\ s$$

$$\therefore 1\ s = \frac{1.0}{1.6} = 0.625\ \tau$$

From Equation (12-11), at $t = 2s$,

$$v_C = 382.4 \times e^{-0.625} = 382.4 \times 0.535 = \mathbf{205\ V}$$

BASIC computer language includes a function which allows us to include e^{-x} in a calculation by entering EXP($-X$). We can solve Examples 12-2A and 12-4A by using a personal computer in the calculator mode, or by writing a simple program which requests input of values for the independent variables E, T, C, and R, and then prints out the solution for

```
V=E-E*EXP(-T/(C*R))
```

A more useful application of the computer's abilities would employ its ability to repeat a series of calculations rapidly with small increments in one of the independent variables (*time* in this case). Hence, we can program a computer to prepare the table from which we can draw the universal exponential graph of Figure 12-6 accurately. Just for fun (because the resulting graph is not accurate enough for graphical solution of examples) we can also program the computer to draw the charging curve for a CR circuit on either a video display or a hard-copy printer.

12-8 TRANSIENT RESPONSE

Now that we are familiar with the way a capacitor charges and discharges in a dc circuit, and have algebraic procedures for calculating instantaneous voltages and currents, we can consider a third method of looking at the response of CR networks to opening and closing switches in direct-current circuits. Let's consider the charging of a partially charged capacitor, which we encountered in Step III of Example 12-4. In Step IIIB, we discovered that the final instantaneous capacitor voltage is equal to the initial voltage plus a percentage of the *additional* voltage based on elapsed time measured in time constants. Adapting this statement to Equation (12-9) gives

$$v_C = V_0 + (V_F - V_0)(1 - e^{-x}) \qquad (12\text{-}12)$$

where V_0 is the initial capacitor voltage and V_F is the steady-state voltage to which the

FIGURE 12-17
Transient Response of a dc CR Network.

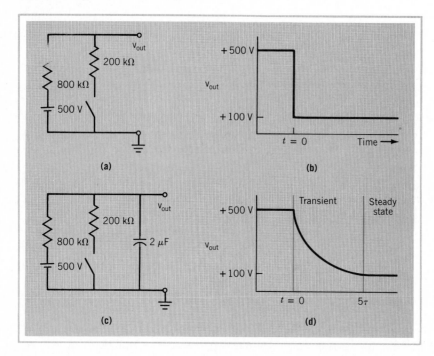

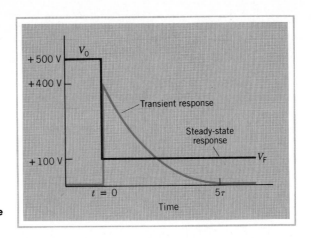

FIGURE 12-18
Separating v_{out} into Steady-State and Transient Components.

capacitor can charge (after an interval of at least five time constants). Carrying out the multiplication,

$$v_C = V_0 + V_F - V_0 - V_F e^{-x} + V_0 e^{-x}$$

Collecting the terms gives us a **universal equation for instantaneous capacitor voltage** in a dc CR network.[†]

$$v_C = V_F + (V_0 - V_F)e^{-x} \tag{12-13}$$

where v_C is the instantaneous voltage across the capacitor, V_0 is the initial voltage across the capacitor, and V_F is the final steady-state voltage after an interval of at least five time constants has elapsed.

In the purely resistive network of Figure 12-17(a), when we close the switch at $t = 0$, the output voltage of the two-terminal network instantly drops from 500 V to its new steady-state value of 100 V, as shown by the time graph of Figure 12-17(b). But in the CR network of Figure 12-17(c), the output voltage is the voltage across the capacitor which takes five time constants to discharge from its initial voltage V_0 of 500 V to its final steady-state voltage V_F of 100 V, as shown by the time graph of Figure 12-17(d).

Equation (12-13) represents the instantaneous output voltage as being made up of two components: the **steady-state response** (V_F) of Figures 12-17(b) and 12-18, and the **transient response** $(V_0 - V_F)e^{-x}$ shown in color in Figure 12-18. This transient response exists only for the short interval from $t = 0$ to $t = 5\tau$. Therefore

A transient is the time interval during which a network's instantaneous voltages and currents adjust from one steady state to another.

[†]We can also derive Equation (12-13) by considering a charged capacitor *discharging* toward a steady-state voltage other than zero.

EXAMPLE 12-6

If the switch in the two-terminal network of Figure 12-17(c) has been open for several minutes, what is the instantaneous output voltage 0.5 s after the switch is closed?

SOLUTION

From inspection of the network,

$$V_0 = 500 \text{ V} \quad \text{and} \quad V_F = 100 \text{ V}$$

The *discharge* time constant is governed by the Thévenin-equivalent resistance.

$$\therefore x = \frac{0.5 \text{ s}}{2 \text{ } \mu F \times 0.16 \text{ M}\Omega} = 1.5625$$

$$v_{out} = V_F + (V_0 - V_F)e^{-x}$$

$$= 100 + (500 - 100)e^{-1.5625}$$

$$= 100 + 400 \times 0.2096 = \textbf{184 V}$$

In the simple *CR* charging circuit of Figure 12-2, for the capacitor, $V_0 = 0$ and $V_F = 100$ V. Since $v_R = E - v_C$, for the resistor, $V_0 = 100$ V and $V_F = 0$. With this reversal of V_0 and V_F, *as long as we are careful in specifying V_0 and V_F for each steady-state situation, Equation (12-13) also applies to the instantaneous voltage drop across series-charging and shunt-discharging resistors.*

EXAMPLE 12-2B

What is the instantaneous voltage across
(a) The capacitor?
(b) The series charging resistor in the circuit of Figure 12-2 3 s after the switch is closed?

SOLUTION

$$x = \frac{t}{CR} = \frac{3 \text{ s}}{1 \text{ } \mu F \times 2 \text{ M}\Omega} = 1.5$$

(a) For the capacitor, $V_0 = 0$ and $V_F = 100$ V

$$v_C = V_F + (V_0 - V_F)e^{-x}$$

$$= 100 \text{ V} + (0 - 100 \text{ V})e^{-1.5}$$

$$= 100 \text{ V} - 100 \text{ V} \times 0.22313 = \textbf{77.7 V}$$

(b) For the resistor, $V_0 = 100$ V and $V_F = 0$

$$v_R = V_F + (V_0 - V_F)e^{-x}$$
$$= 0 + (100 \text{ V} - 0)e^{-1.5}$$
$$= 100 \text{ V} \times 0.22313 = \mathbf{22.3 \text{ V}}$$

Check: $E = v_C + v_R = 77.7 \text{ V} + 22.3 \text{ V} = 100 \text{ V}$

One advantage of the universal transient response point of view is that we can use the single Equation (12-13) in place of both Equations (12-9) and (12-11) and the procedure we used in Step III of Example 12-4A for determining the voltage across a partially charged capacitor.

EXAMPLE 12-4B

Starting from position 1, the switch in the circuit of Figure 12-14 dwells for 1 ms in each position. Calculate v_C 3 ms after the switch first moves to position 2.

SOLUTION

Step I. When the switch first moves to position 2,

$$V_0 = 0 \quad \text{and} \quad V_F = 500 \text{ V}$$

Just before the switch moves back to position 1,

$$x = \frac{1 \text{ ms}}{0.01 \ \mu\text{F} \times 0.1 \text{ M}\Omega} = 1.0$$
$$\therefore v_C = 500 + (0 - 500)e^{-1.0}$$
$$= 500 - 500 \times 0.3679 = 316 \text{ V}$$

Step II. When the switch first moves to position 1,

$$V_0 = 316 \text{ V} \quad \text{and} \quad V_F = 0$$

Just before the switch moves back to position 2,

$$x = \frac{1 \text{ ms}}{0.01 \ \mu\text{F} \times 0.2 \text{ M}\Omega} = 0.5$$
$$\therefore v_C = 0 + (316 - 0)e^{-0.5}$$
$$= 0 + 316 \times 0.6065 = 192 \text{ V}$$

Step III. When the switch moves back to position 2,

$$V_0 = 192 \text{ V} \quad \text{and} \quad V_F = 500 \text{ V}$$

At the end of the 3-ms interval,

$$x = \frac{1\text{ ms}}{1\text{ ms}} = 1.0$$

and

$$v_C = 500 + (192 - 500)e^{-1.0}$$
$$= 500 - 308 \times 0.3679 = \mathbf{387 \text{ V}}$$

12-9 ENERGY STORED BY A CAPACITOR

While a capacitor is charging, a displacement current flows in the circuit. As a result, energy is taken from the voltage source. Some of this energy is converted into heat as the charging current flows through the series current-limiting resistance in the circuit. This accounts for only half of the energy drawn from the source. The other half is stored by the capacitor. Energy is expended as the charging current forces a charge into a capacitor against the mounting opposition of the potential difference building up between the two plates. When we disconnect a charged capacitor from the source, its potential difference remains at a value equal to the original applied voltage for a considerable period of time, provided there is negligible leakage of the charge through the dielectric between the plates.

Since no current is required to maintain the potential difference between the plates of a charged capacitor, we can consider that capacitance *stores* electric energy in a **static** form. We think of this energy as being stored in the electric field which builds up between the plates of the capacitor and which produces a stress on the orbits of electrons bound to the molecules of the dielectric, as shown in Figure 11-7. We might compare the storing of electric energy by a capacitor to the storing of a charge of air under pressure in a steel cylinder. Energy is expended in forcing the air into the cylinder against the mounting back pressure. This stored energy can be recovered as desired by using the compressed air to operate a paint spray, pump up an automobile tire, etc. Similarly, we can recover the electric energy stored by a charged capacitor at will, as we shall discover in Section 12-10.

We can plot the graph of the *rate* at which a capacitor stores energy, since instantaneous power represents this rate, and $p = vi$. By calculating the product of the instantaneous potential difference across the capacitance and the instantaneous charging current for each division along the horizontal (time) axis of the graph of Figure 12-19, we can plot a graph of the rate at which the capacitance stores energy from instant to instant. As indicated by the shaded vertical line in Figure 12-19, this instantaneous power input to the capacitor reaches its peak value at the same instant in time that the graphs of instantaneous charging current and potential difference cross, that is, when the instantaneous voltage has risen to one-half of its final value and the instantaneous charging current has dropped to one-half of its initial value. From Figure 12-6, peak power input occurs 0.6 time constant after the capacitor commences to charge.

After a period of time equal to five time constants, the charging current has ceased, and the instantaneous power has, therefore, decreased to zero. The capacitor has now stored all the energy it is capable of storing at the particular potential difference applied to it. Since the total energy is the product of the average power and the time during which the capacitor is receiving energy, we can represent the energy

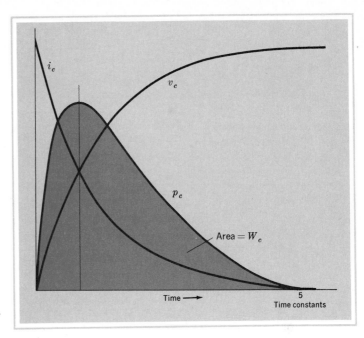

FIGURE 12-19
Graphical Representation of the Energy Stored by a Capacitor.

stored by the capacitance by the *area* contained by the instantaneous power graph in Figure 12-19.

Solving for the area under the instantaneous power graph of Figure 12-19 requires basic integral calculus, as outlined in Appendix 2-3. However, we can simplify the task of determining how much energy a capacitor can store by returning to the notion of a capacitor charging at a *constant* rate equal to the initial rate of change of voltage. To achieve this, we connect the capacitor to a *constant-current* source whose terminal voltage can vary but whose current output always remains constant at the initial value of $I = E/R$, as shown in Figure 12-20(a). Since the voltage across the capacitor now rises at a *constant* rate from zero to its final value of $V = E$, as shown in Figure 12-20(b), the *average* voltage during this time interval must be one-half of V. For this linear charging condition,

FIGURE 12-20
Determining the Energy Stored by a Capacitor.

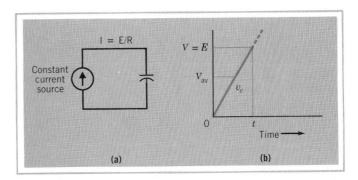

$$W = Pt \quad \text{becomes} \quad W = V_{av} \times I \times t$$

But
$$I \times t = Q \tag{2-2}$$

where Q is the charge stored by the capacitor after a time interval equal to t seconds. As shown in Figure 12-20(b), after t seconds, the voltage across the capacitor will have risen to V volts,

$$\therefore Q = CV \tag{11-7}$$

Since $V_{av} = \frac{1}{2}V$, $W = Pt$ becomes

$$W = \frac{1}{2}V \times C \times V = \frac{1}{2}CV^2 \tag{12-14}$$

where W is the energy stored by the capacitance in joules, C is the capacitance in farads, and V is the potential difference across the capacitance in volts.

EXAMPLE 12-7

How much energy is stored in the 500-pF high-voltage filter capacitor of a television receiver when it is charged to 15 kV?

SOLUTION

$$W = \frac{1}{2}CV^2 = \frac{1}{2} \times 500 \text{ pF} \times (15 \text{ kV})^2 = 5.63 \times 10^{-2} \text{ J} = \mathbf{56.3 \text{ mJ}}$$

12-10 CHARACTERISTICS OF CAPACITIVE DC CIRCUITS

The ability of a capacitor to store electric charge and, as a consequence, to oppose a rapid change in potential difference between its terminals, makes the capacitor a particularly useful circuit component in practical dc circuits. Its ability to oppose any instant change in the voltage across it makes the capacitor very suitable for filtering or smoothing any fluctuations in the output from a pulsating voltage source such as a rectifier, as shown in Figure 12-21(a). When the instantaneous source voltage drops below the voltage across the capacitor, the capacitor maintains the load voltage by discharging some of its stored energy into the load. The capacitor will then recover

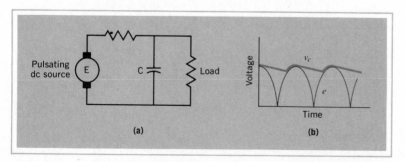

FIGURE 12-21
Filtering Voltage Fluctuations with a Capacitor.

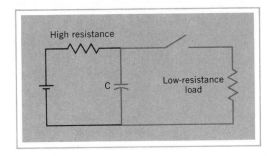

FIGURE 12-22
Obtaining Short-Duration, High-Current Pulses from a Reservoir Capacitor.

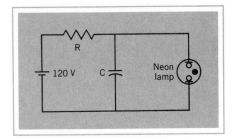

FIGURE 12-23
Using a *CR* Time Constant for Timing Purposes.

this charge the next time that the instantaneous source voltage rises to its peak value, as shown in Figure 12-21(b). This application of a capacitor is similar to the use of a reservoir tank with an air compressor used with a paint spray. Since the pressure of the air in the tank cannot change without the flow of the stored air, the tank maintains a fairly constant pressure at the paint spray in spite of the fluctuations in the pressure developed by the piston action of the compressor. (The compressed air tank analogy provides a fairly accurate means of visualizing the charge and discharge action of capacitance, with the friction of the air hose representing resistance.)

The energy-storage ability of a capacitor is very useful in cases where we wish to obtain a very high current for a short period of time without subjecting the source to a severe current surge. With a circuit patterned after the basic circuit of Figure 12-22, the capacitor charges slowly from the source and then discharges rapidly when we connect the low-resistance load across it. The peak current drawn from the source is limited by the series resistor. This system is used to obtain a one microsecond pulse for radar transmitter magnetrons with a peak power of over a million watts during the pulse. This system is also used in electric spot welding of aluminum where a short-duration surge of very high current is required to make the weld. The same system is used on a smaller scale in making the delicate welds which hold vacuum-tube elements in place. In this case, the circuit of Figure 12-22 is used, not because of its ability to limit the peak current drawn from the source, but because we are able to control very precisely the amount of energy in each discharge pulse by setting the potential difference to which the capacitor is allowed to charge.

The instantaneous potential difference across the capacitor of a given *CR* combination is an exponential function of elapsed time and will always rise along exactly the same curve. Therefore, we can use a series *CR* circuit as the basis for a very accurate timing device for time intervals of from a small fraction of a second to several seconds. In the simple circuit of Figure 12-23, the neon lamp "strikes" (glows) when the potential difference across it reaches 100 volts and extinguishes when the voltage drops below 20 volts. When a neon lamp is not struck, the gas is an insulator and we may consider the lamp to be an open circuit. When the lamp strikes, the gas becomes ionized and becomes an excellent electric conductor with the free electrons knocked out of the gas molecules flowing in one direction and the positively-charged gas ions moving in the other direction. The resistance of the neon lamp under these conditions

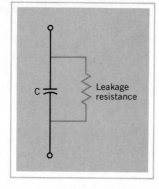

FIGURE 12-24
Equivalent Circuit of a "Leaky" Capacitor.

is only around 100 ohms. If the time constant of the series CR circuit in Figure 12-23 is several seconds, it will take several seconds for the potential difference across the capacitor to rise to the striking voltage of the neon lamp. The low resistance of the ionized neon will then quickly discharge the capacitor down to the extinguishing voltage of the lamp, whereupon the resistance of the neon lamp again becomes extremely high and the capacitor starts to recharge. The time interval between the short flashes from the neon lamp is, therefore, dependent on the time constant of C and R.

In Section 12-6, we noted one of the more significant applications of capacitance characteristics to waveshaping circuits. Before we complete this section, we should also note that capacitance can have undesirable characteristics in some instances. So far, we have assumed that there is negligible leakage of charge through the dielectric of capacitors. This is essentially true for practical purposes with such dielectrics as ceramic, mica, and plastic film. However, it is not true for electrolytic capacitors of the type described in Section 11-5. We can represent capacitor leakage in practical circuits by an appropriate high resistance in parallel with the capacitance of the capacitor in a circuit diagram, as shown in Figure 12-24.

FIGURE 12-25
Effect of Stray Capacitance.

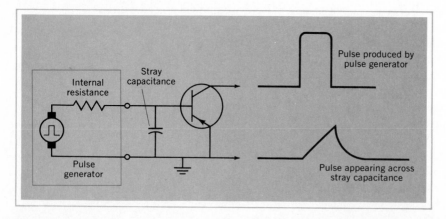

Capacitance in electric circuits is not limited to capacitors. As we noted in Figure 11-10, capacitance can exist between any pair of electric conductors which are at different potentials. In many circuit applications, this **stray capacitance** is small enough that its effect on the behavior of the circuit is negligible. But in modern digital computer circuits, where timing pulse voltages must rise and fall in a time interval of a few nanoseconds, the tendency of stray capacitance to oppose such rapid changes in voltage can have a serious effect on circuit performance. In the circuit of Figure 12-25, the capacitance across the *pn*-junctions of the transistor is directly in parallel with the input to the amplifier (even though it would not appear in the actual circuit diagram). Hence, the input signal to the transistor is the potential difference across this stray capacitance. Since this capacitance must charge and discharge through the internal resistance of the pulse generator, if the time constant is more than a small fraction of the pulse duration, there will be appreciable distortion of the desired pulse.

PROBLEMS

12-1. When the switch is first closed in the circuit shown in Figure 12-26(a)
   (a) What is the initial current?
   (b) What is the initial voltage across the capacitor?
   (c) What is the initial rate of rise of voltage across the capacitor?
   (d) What is the charging time constant?
   (e) How long will it take the capacitor to charge?

12-2. In the circuit shown in Figure 12-26(b), $v_C = 0$ at $t = 0$.
   (a) What is the initial charging current?
   (b) What is the initial rate of change of voltage across the capacitor?
   (c) What is the charging time constant?
   (d) What is the voltage across the capacitor at $t = \tau$?
   (e) What is the charge on the capacitor at $t = \tau$?

12-3. Using the universal exponential curves of Figure 12-6, for the circuit of Figure 12-26(a) determine
   (a) The voltage across the capacitor 5 s after the switch is closed.
   (b) The instantaneous charging current 2 s after the switch is closed.
   (c) The length of time for the voltage across the capacitor to reach 200 V.
   (d) The length of time for the voltage across the capacitor to rise from 50 V to 150 V.

FIGURE 12-26

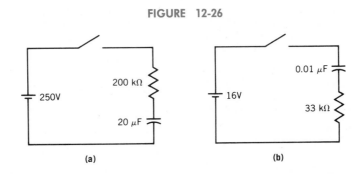

(a)   (b)

12-4. Using the universal exponential curves of Figure 12-6, for the circuit of Figure 12-26(b) determine
(a) How long it will take the capacitor to charge to 16 V.
(b) How long it will take the capacitor to charge to 10 V.
(c) The instantaneous voltage across the capacitor at $t = 0.5$ ms.
(d) The instantaneous charging current at $t = 0.25$ ms.
(e) How long it will take the voltage across the capacitor to rise from 4 V to 12 V.

12-5. A 40-μF capacitor has been charged to a voltage of 500 V.
(a) How long will it take to discharge the capacitor through a 470-kΩ resistor?
(b) What is the maximum value of the discharge current?

12-6. The capacitor of Problem 12-5 is discharged by shorting its terminals with an electric conductor having a resistance of 0.4 Ω.
(a) What is the peak discharge current?
(b) How long will it take to discharge the capacitor?

12-7. A 40-μF capacitor, which has been charged to a voltage of 500 V, is discharged through a 470-kΩ resistor. Using the universal exponential curves, determine
(a) How long it will take the voltage across the capacitor to drop to 50 V.
(b) The voltage across the capacitor 10 s after it starts to discharge.
(c) The instantaneous discharge current 20 s after it starts to discharge.
(d) How long it will take the voltage across the capacitor to drop from 400 V to 100 V.

12-8. If the resistor in the circuit of Figure 12-23 is 5 MΩ and the capacitor has a capacitance of 8 μF, calculate the time interval between flashes of the neon lamp. The lamp strikes at 100 V, and its resistance then becomes 100 Ω. It extinguishes when the voltage across it drops to 20 V, and its resistance then becomes infinitely high.

12-9. Repeat Problem 12-3 using algebraic expressions for instantaneous values of voltage and current.

12-10. Repeat Problem 12-4 using algebraic equations to solve for instantaneous voltage and current.

NOTE Use any method to solve the remaining problems.

12-11. After a considerable period of time in position 2, the switch in the circuit of Figure 12-27 is thrown to position 1 at $t = 0$.
(a) Calculate i_C and dv_C/dt at $t = 0$.
(b) How long must the switch remain in position 1 for v_C to rise to 100 V?
(c) The switch is thrown to position 1 at $t = 5$ min. Calculate i_C and dv_C/dt as the switch makes contact in position 2.
(d) If the switch returns to position 1 at $t = 5.5$ min, calculate v_C at $t = 10$ min.

12-12. After a considerable period of time in position 1, the switch in the circuit of Figure 12-28 is thrown to position 2 for 1 ms and is then returned to position 1.
(a) Calculate v_C 1 ms after the switch returns to position 1.
(b) Draw a graph of the output voltage at terminal A with respect to ground for this 2-ms interval.

12-13. Commencing from the open position, the switch in the circuit of Figure 12-29 closes for 1 ms and then opens. Calculate the voltage across the capacitor 1 ms after the switch reopens.

12-14. Repeat the procedure of Problem 12-13 for the circuit of Figure 12-30.

12-15. The switch in the network of Figure 12-31 has been in position 1 long enough for the initial charging transient to have been completed. What is the instantaneous output voltage (with respect to ground) (a) 2 s and (b) 4 s after the switch is thrown to position 2?

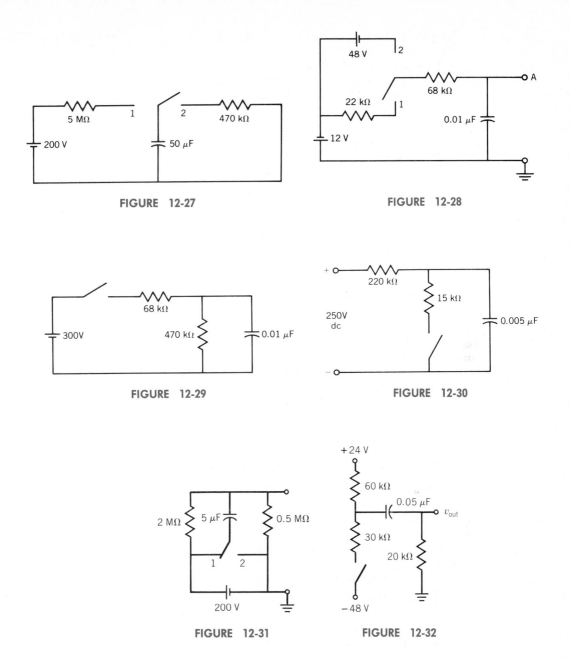

FIGURE 12-27

FIGURE 12-28

FIGURE 12-29

FIGURE 12-30

FIGURE 12-31

FIGURE 12-32

12-16. Assuming that the charging transient is completed before the switch is operated in each case, what is the instantaneous output voltage (with respect to ground) in the network of Figure 12-32
(a) 2 ms after the switch closes?
(b) 2 ms after the switch opens?

12-17. Calculate the energy stored by the capacitors in the circuits of Figure 12-26(a) and (b) when they are fully charged.

12-18. Calculate the energy stored by the capacitor in Problem 12-5
    (a) Before it starts to discharge.
    (b) After it has been discharging for one time constant.

12-19. A 50-μF capacitor charged to a potential difference of 32 V is discharged through a 20-kΩ resistor. How much energy is dissipated by the resistor as the capacitor is completely discharged? If the capacitor had been discharged through a 680-Ω resistor, how much energy would have been dissipated by the resistor?

12-20. In Problem 12-8, how much energy is transferred to the neon lamp with each flash?

12-21. The switch in the circuit of Figure 12-33 is alternately thrown to position 1 for 1 ms and then to position 2 for 9 ms.
    (a) Draw an accurate graph of the voltage between terminals A and C for an interval of 30 ms.
    (b) Superimpose (in another color) on the graph of part (a), a graph of the instantaneous voltage between terminals B and C during the same time interval. Label the magnitude of the instantaneous voltage at each instant of switch operation.
    (c) Draw a separate graph of the voltage drop between terminals A and B for the same time interval.

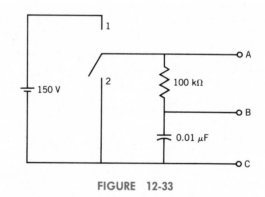

FIGURE 12-33

FIGURE 12-34

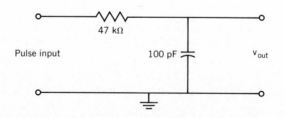

Input voltage waveform

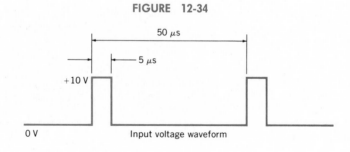

Pulse input

PART III   CAPACITANCE AND INDUCTANCE

12-22. Repeat Problem 12-21 with the switch remaining 1 ms in each position. Show a time interval of 10 ms.

12-23. A pulse generator having negligible internal resistance feeds a 20-kHz pulse waveform, shown in detail in Figure 12-34, into the CR waveshaping network shown.
(a) Draw an accurate graph of two complete cycles of v_{out}.
(b) Repeat with the positions of the capacitor and resistor reversed.

12-24. A 10-kΩ resistor is connected in series with the capacitor in the circuit of Figure 12-34. Draw an accurate graph of two complete cycles of v_{out} across the series combination of the capacitor and the 10-kΩ resistor.

REVIEW QUESTIONS

12-1. What factors determine the initial charging current of a capacitor?

12-2. What factors determine the rate of change of potential difference between the plates of a capacitor?

12-3. When a discharged capacitor is connected to a voltage source, the current changes instantly from zero to a value of E/R. Why then can the voltage across the capacitor not change instantly?

12-4. Account for the exponential shape of the graph of v_C in Figure 12-5.

12-5. Why must the rate of change of voltage across a capacitor diminish as the charge on its plates increases?

12-6. What is the advantage of using the **time constant** of a CR circuit as a unit of time in determining the instantaneous voltage?

12-7. Why does the initial charge on a capacitor have no effect on the length of time required for the capacitor to discharge through a given resistance?

12-8. Draw an accurately labeled graph of the instantaneous voltage across the capacitor in the circuit of Figure 12-35 plotted against time as the double-pole, double-throw switch is thrown to the opposite position.

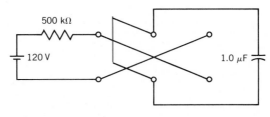

FIGURE 12-35

12-9. Figure 12-6 shows a graph of the algebraic equation $y = 1 - e^{-x}$. Why do we use this equation to solve for v_C rather than the simple exponential equation $y = e^{-x}$?

12-10. In the circuit of Figure 12-29, we use the Thévenin-equivalent resistance in the *charge* time constant, but not in the *discharge* time constant. In the circuit of Figure 12-30, we use the Thévenin-equivalent resistance in the *discharge* time constant, but not in the *charge* time constant. Explain the difference.

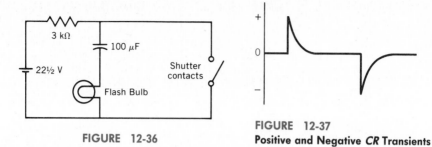

FIGURE 12-36

FIGURE 12-37
Positive and Negative *CR* Transients

12-11. Why are the charge and discharge time constants the same in the circuit of Figure 12-31?

12-12. Using the techniques we employed in producing Figure 12-19, plot a graph of the instantaneous power in a resistor when a charged capacitor is connected across it.

12-13. An air-dielectric capacitor is charged from a voltage source and then disconnected. The plates are then moved twice as far apart. Assuming no leakage of the charge, account for the difference in the energy stored by the capacitor before and after moving the plates.

12-14. Instead of changing the spacing between the plates in Question 12-13, a dielectric with a dielectric constant of 2.0 is placed between the plates after the capacitor is disconnected from the source. Account for any change in the energy stored by the capacitor.

12-15. The circuit of Figure 12-12(b) resembles the *CR* coupling used between stages of an electronic amplifier. However, the output waveform shown in Figure 12-13(c) is not a faithful replica of the input waveform of Figure 12-13(a). What conditions must exist for the *CR* coupling circuit to transmit the square wave with minumum distortion?

12-16. Figure 12-36 shows the circuit diagram of a BC-type (battery-capacitor) photoflash unit for standard flash bulbs. Explain its operation and advantages.

12-17. When steady-state voltages are suddenly switched in a dc *CR* network, the transient current always starts from zero and returns to zero, having either a positive or negative direction, as shown by the graph of Figure 12-37. Does this mean that the current in resistors of the network must be zero except during the transient? Explain.

12-18. Draw superimposed graphs (in different colors) to show how a graph of v_C is obtained by combining a sudden change in steady-state voltage with a transient having the form of either the positive or negative transient shown in Figure 12-37.

*12-19. Design a computer program to print out a labeled table of x and e^{-x} for values of x ranging from 0 to 5 in steps of 0.25.

*12-20. Design a computer program to display a labeled graph of the function $V = 100*(1 - EXP(-X))$ for values of X ranging from 0 to 5.

MAGNETISM

13-1 ELECTRICITY AND MAGNETISM

When the physical scientists of some two centuries ago investigated the *effects* of the natural phenomenon called "electricity," they had no real knowledge of the *nature* of electricity. Nevertheless, on the basis of observed effects, they were able to state the basic laws governing the behavior of electric circuits and to establish suitable units of measurement. Our present-day knowledge of the structure of atoms has allowed us to explain the effects of electricity observed by Coulomb (Figure 2-1) in terms of the electric force of attraction between the unlike electric charges which are an elemental property of the electrons and protons in all atoms. Knowledge of the *nature* of electricity has enabled us to manipulate the electrical properties of matter in order to create such devices as semiconductors, synthetic insulators, and dielectrics with amazingly high dielectric constants, which were not even dreamed of in the days of Ohm and Ampère.

In Chapter 2, we considered two natural phenomena—gravitation and electricity. We found that both phenomena were revealed through a force which can act without physical contact. Hence, we define both electric force and gravitational force as **field forces,** and we measure both forces in newtons. **Magnetism** is a third form of natural phenomenon, the *effects* of which were observed almost two thousand years ago by Chinese mariners, who used a crude form of magnetic compass for navigation. If we bring a magnet close to a compass needle (which itself is a tiny magnet carefully pivoted at its center of gravity), as in Figure 13-1, we find that the compass needle will be deflected so that one end of the needle always points to one end of the magnet. As we move the magnet, the compass needle follows the motion. Once again, the

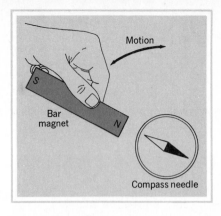

FIGURE 13-1
Detecting the Presence of a Magnetic Field.

phenomenon is revealed by a force which acts without physical contact. Consequently, magnetic force is a *field* force which also can be measured in newtons.

Although magnetism and electricity are often considered to be independent natural phenomena, in 1819, Hans Christian Oersted showed that magnetic fields and electric current are closely associated. Oersted discovered that a magnetic compass needle always takes up a position perpendicular to a current-carrying electric conductor placed near the compass needle. Although we do not as yet know as much about the nature of magnetism as we know about the nature of electricity, modern theories of magnetism suggest that all magnetic properties stem from the *motion* of electrons in an atom. (Electric properties stem from the *position* of electrons in an atom, as shown by the energy-level diagrams of Figures 3-3 and 3-4.) As we have already discovered, electric *current* also consists of electrons in motion, but, in this case, the motion is detached from the parent atom.

The third property of an electric circuit, **inductance,** is dependent on the relationship between magnetic field forces and electric current (just as capacitance depends on the relationship between electric field in a dielectric and potential difference). Before we can proceed with an investigation of inductance in electric circuits, we must be able to identify magnetic properties and express their magnitudes in suitable units.

The laws of magnetism were developed when the CGS system of units was the universal standard. The CGS magnetic units honor such pioneers as William Gilbert, Karl Friedrich Gauss, James Clerk Maxwell, and Hans Christian Oersted. There has been considerable reluctance to abandoning the CGS units in favor of the presently accepted MKS units, which are part of the International System of Units. There has also been a tendency in North America to use a hybrid set of magnetic units which combined CGS or MKS units with the English unit of linear measurement—the inch. We can avoid all this confusion by using the currently preferred SI units exclusively in this text.

13-2 THE NATURE OF A MAGNETIC FIELD

In studying the effect of electric force, we thought of the space surrounding an electric charge as constituting an electric **field.** We defined this electric field in terms of the region surrounding an electric charge in which a second elemental electric charge is acted upon by an electric force. Similarly,

334

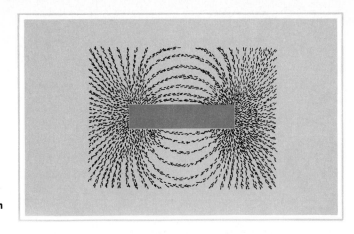

FIGURE 13-2
Pattern Formed by Sprinkling Iron
Filings over a Bar Magnet.

A magnetic field is that region in which a magnetic material is acted upon by a magnetic force.

In Chapter 11, we found it convenient to visualize an electric field as consisting of invisible electric lines of force, which are, in fact, the paths along which elemental positive electric charges move as a result of being acted upon by the electric force. Although we have not discovered a magnetic equivalent of an elemental positive electric charge, Michael Faraday suggested that we still think of a magnetic field as being made up of **magnetic lines of force.**

We can outline the magnetic field surrounding a bar magnet by placing it under a piece of paper sprinkled with fine particles of a magnetic material such as iron filings, as in Figure 13-2. We note that the filings tend to gather around two areas, one at each end of the magnet. These areas where the effect of the magnetic field is concentrated are called the **poles** of the magnet. By applying Faraday's magnetic lines of force concept to the magnetic field detected by the iron filings in Figure 13-2, we can sketch the magnetic field of a bar magnet in the form shown in Figure 13-3. If we could invent such a thing as an isolated magnetic pole (a concept used in some theories of magnetism),

FIGURE 13-3
Diagram of the Magnetic Field
around a Bar Magnet.

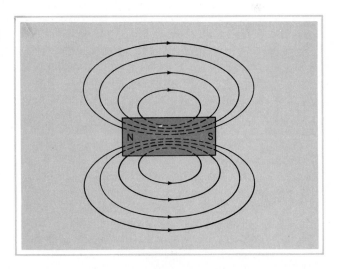

A magnetic line of force represents the path along which a theoretical isolated magnetic pole would move from one pole of a magnet to the other.

In Sections 13-11 and 13-12, we shall investigate the properties of magnetic materials which permit a **permanent magnet** to produce the magnetic field illustrated in Figures 13-2 and 13-3. For the moment, we are more concerned with the characteristics of the imaginary magnetic lines of force which make up a magnetic field.

13-3 CHARACTERISTICS OF MAGNETIC LINES OF FORCE

If we explore the magnetic fields of various shapes of magnets with a small compass needle, as in Figure 13-1, or with iron filings, as in Figure 13-2, we become aware of several important characteristics of magnetic lines of force. Since all the electromagnetic effects which we shall be studying can be traced directly to these characteristics, it is important that we learn them thoroughly at this point.

Magnetic lines of force possess direction.

Combining Figures 13-3 and 13-1, we can say that the compass needle in Figure 13-1 aligns itself with the magnetic lines of force. If we now bring the opposite pole of the bar magnet near the compass needle in Figure 13-1, we find that the compass needle will again align itself with the magnetic lines of force, but it will now have the *opposite* end of the needle pointing at the pole of the bar magnet. For the compass needle to be able to distinguish one pole of the bar magnet from the other, we must think of the magnetic lines of force as possessing a positive direction. Since magnetic lines of force are hypothetical, this direction will have to be arbitrarily assigned. However, since the earth itself has a magnetic field, we can identify the poles of a magnet more specifically. If we suspend the bar magnet by a thread tied around its center of gravity, it will act like a compass needle and align itself with the earth's magnetic field. One pole of the magnet will always point north and is, therefore, referred to as the **north pole** of the magnet. Similarly, the other pole is called the **south pole.** Having identified the poles of a magnet, we can now state that the positive direction of the magnetic lines of force in the region around a magnet shall be considered as being *from* the north pole of the magnet *toward* the south pole.

Magnetic lines of force always form complete loops.

They do not begin at the north pole of a magnet and end at the south pole but continue between south pole and north pole inside the magnet to form complete closed loops. We can show this by cutting the magnet of Figure 13-4(a) in two. As we pull the two sections apart, we can detect the presence of the magnetic lines of force in the gap between the two halves. New north and south poles are formed on each side of the gap, as shown in Figure 13-4(b). No matter how many times we cut the magnet, each section has its own north and south poles. We cannot continue the subdivision to a point where we find an isolated magnetic pole.

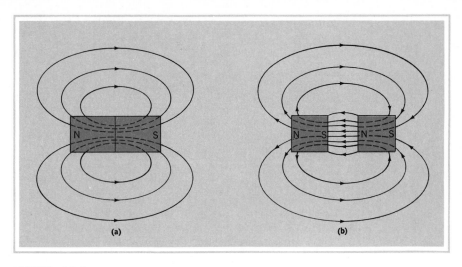

FIGURE 13-4
Showing that Magnetic Lines of Force Form Complete Loops by Continuing from South Pole to North Pole within the Magnet.

Magnetic lines of force represent a tension along their length which tends to make them as short as possible.

This effect would also be noticed as we separate the two halves of the magnet in Figure 13-4(b). In pulling the two halves apart, we would be lengthening the lines of force, and we would notice a definite attraction between the two pieces as the magnetic lines of force try to pull them together again in an attempt to shorten the lines of force. The analogy of a magnetic field composed of rubber bands sometimes helps in grasping this idea. Since the two pieces of the one magnet in Figure 13-4(a) became separate magnets in Figure 13-4(b), this tendency for magnetic lines of force to become as short as possible accounts for the familiar statement that **unlike poles of a magnet attract one another.**

Magnetic lines of force repel one another.

We can see this effect by examining the magnetic field around a bar magnet (Figure 13-4). The magnetic lines of force tend to diverge as we move away from the poles rather than to converge or even to remain parallel. As a result of this mutual repulsion,

Magnetic lines of force cannot intersect but must always form individual closed loops.

This mutual repulsion of magnetic lines of force accounts for the effect shown in Figure 13-5 when like magnetic poles approach one another.

Like poles of magnets repel one another.

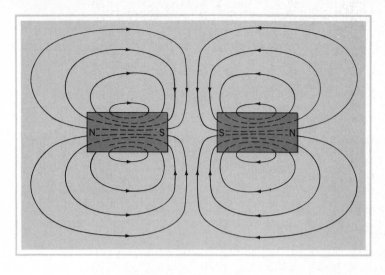

FIGURE 13-5
Repulsion of Like Poles of Magnets.

The combination of these basic characteristics is responsible for the shape of the magnetic field of a magnet. The magnetic lines of force are kept in equilibrium by each line tending to shrink its loop length and at the same time being prevented from doing so by the repulsion of the next smaller line-of-force loop. The magnetic field around a uniform magnet will be symmetrical unless there are other magnetic materials within the field. (Compare Figures 13-3 and 13-5.)

If we place a piece of soft iron in the field of a magnet, we find that some of the magnetic lines of force alter their normal paths to include the iron bar in their circuit, as shown in Figure 13-6. This is because iron is a better "conductor" of magnetic lines of force than air. The net result of the change in the pattern of the magnetic field is that the total number of lines of force is increased. Therefore,

Magnetic fields always tend to arrange themselves in such a manner that the maximum number of lines of force is set up.

While it is in the field of the main magnet, the piece of soft iron in Figure 13-6 exhibits the properties of a magnet. It takes on a south pole where the direction of the

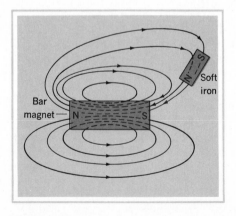

FIGURE 13-6
Magnetic Lines of Force Detour to Include Soft-Iron Bar in Their Circuit.

magnetic lines of force is toward the soft iron, and a north pole where their direction is away from the iron. The piece of soft iron has become a **temporary** magnet by **induction.** It will lose most of its magnetism when we remove it from the field of the main magnet.

We have noted the attraction of unlike poles as magnetic lines of force tend to become as short as possible. The magnetic lines of force which detour through the soft-iron bar in Figure 13-6 can become shorter if the iron object approaches the nearer pole of the main magnet. This accounts for the attraction of unmagnetized ferromagnetic objects such as nails and steel washers to the closer pole of a magnet.

13-4 MAGNETIC FIELD AROUND A CURRENT-CARRYING CONDUCTOR

So far we have discussed only the magnetic field around a permanent magnet. The disadvantage of this type of magnet is that we cannot turn the magnetic field on and off at will. However, Oersted discovered that a magnetic field is produced around an electric conductor whenever current flows through it. Since we can control an electric current quite readily, we, therefore, have a means of controlling the production of a magnetic field.

By exploring the magnetic field around a current-carrying conductor with iron filings or small compass needles, as in Figure 13-7, we find that the lines of force form concentric circles in a plane which is at right angles to the axis of the conductor. As the current increases from zero, a magnetic line of force loop of infinitesimal radius forms at the center of the conductor. As the current increases, the radius of the

FIGURE 13-7
(a) With No Current, Compass Needles All Indicate Earth's Magnetic Field; (b) When Current Flows, Compass Needles Indicate Circular Magnetic Lines of Force around the Conductor.

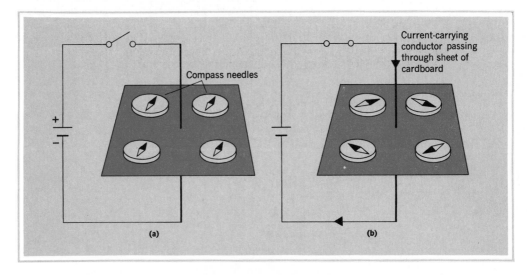

magnetic line of force increases and another loop forms at the center of the conductor. As the result of mutual repulsion of magnetic lines of force, this new loop forces the first line of force to expand still further. As long as the current is *increasing,* a whole series of closed-loop magnetic lines of force *expand* outward from the center of the conductor. When the current becomes steady, the magnetic field becomes stationary. Because of the tendency for magnetic lines of force to shorten like rubber bands, they will be concentrated at the center of the conductor and spaced farther and farther apart as we move outward from the center of the conductor. When the current is turned off, this tension causes each line of force to collapse back toward the center of the conductor.

Although the magnetic field around a straight current-carrying conductor has no north and south poles, the individual lines of force must still have direction. This direction can be checked by the compass needle in Figure 13-7(b). Reversing the direction of the current through the conductor causes all the compass needles to reverse their positions. We can determine the direction of the magnetic field around a straight current-carrying conductor by the right-hand rule illustrated in Figure 13-8.

Grasp the conductor (figuratively speaking) with the right hand so that the thumb points in the conventional current direction; the fingers circling the conductor then point in the direction of the magnetic lines of force around the conductor.

If we form the current-carrying conductor into a loop, as in Figure 13-9, we note that the magnetic lines of force all pass through the center of the loop in the *same* direction. For a given current, the total number of lines of force has not changed; therefore, we have strengthened the magnetic field by concentrating it into a smaller physical area.

FIGURE 13-8
Right-Hand Rule for Determining Direction of Magnetic Lines of Force around a Straight Current-Carrying Conductor.

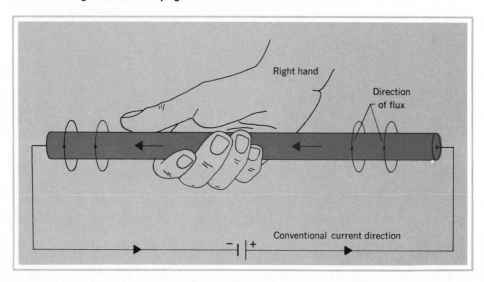

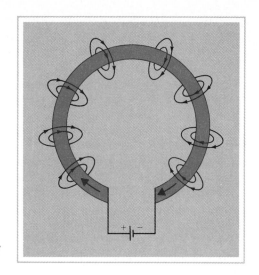

FIGURE 13-9
Concentrating the Magnetic Field by Forming the Conductor into a Loop.

We can concentrate the magnetic field of the conductor still further by winding the conductor around a cardboard tube, as in Figure 13-10, to form a **solenoid.** Because the current in the adjacent turns of the solenoid is traveling in the same direction around the circumference of the coil, the solenoid behaves as a single loop of a stranded conductor, each strand carrying a current equal to the actual solenoid current. This will result in a magnetic field around the solenoid having the pattern shown in Figure 13-10. Since this magnetic field pattern is similar to that of the bar magnet of Figure 13-3, we have produced an **electromagnet.**

FIGURE 13-10
Magnetic Field Pattern of a Current-Carrying Solenoid.

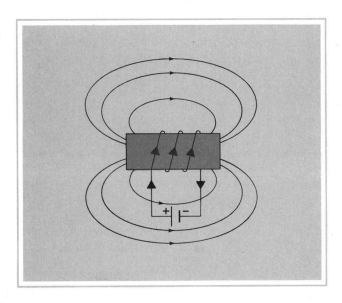

We can adapt the right-hand rule for the direction of the magnetic lines of force around a straight conductor to apply to a solenoid, as illustrated in Figure 13-11.

Grasp the solenoid with the right hand so that the fingers follow the conventional current direction around the circumference of the solenoid; the thumb then points in the direction of the magnetic lines of force through the center of the solenoid.

Because we can think of the solenoid as an electromagnet, we can think of the end of the solenoid where the direction of the magnetic lines of force is away from the coil (Figure 13-11) as being the north pole of the electromagnet. Therefore, the thumb in this hand rule also points to the north pole end of the electromagnet.

As shown in Figure 13-5 and 13-4(b), interaction between the magnetic fields of two permanent magnets produces a magnetic force between them. Similarly, a magnetic force exists between two parallel current-carrying electric conductors. In Figure 13-12(a), equal currents flow in opposite directions in the cross-sectional view of two parallel conductors. The tail feathers of an arrow on the left-hand conductor represents a conventional current direction *into* the page, and the point of the arrow on the right-hand conductor represents a conventional current direction *out* of the page (Check the direction of the magnetic lines of force with Figure 13-8.) Between the two conductors in Figure 13-12(a), the magnetic lines of force all have the same direction. Since they tend to repel one another, there is a resultant magnetic force of *repulsion* between the two conductors.

In Figure 13-12(b), where the conventional current direction in both conductors is into the page, the magnetic lines of force between the conductors have *opposite* directions. As a result, the magnetic lines of force which would have come midway between the two conductors can become shorter by combining to form single loops encompassing both conductors. The tendency of these lines of force to become still

FIGURE 13-11
Right-Hand Rule for Determining Direction of the Magnetic Lines of Force through a Current-Carrying Solenoid.

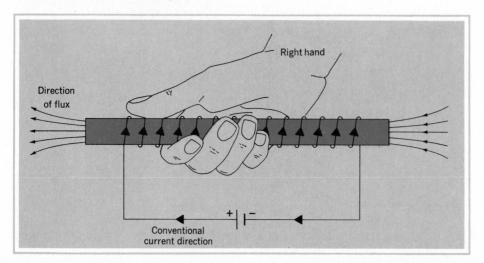

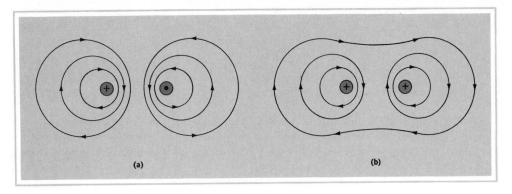

FIGURE 13-12
Magnetic Field around Parallel Current-Carrying Conductors.

shorter results in a magnetic force of *attraction* between the two conductors. The magnitude of these magnetic forces depends on the length and spacing of the two conductors and on the magnitude of the electric current in them.

In Section 1-2, we stated that the **ampere** is the fourth basic unit which extends the meter-kilogram-second system of units to include electrical measurement. If we now define the ampere in terms of the *magnetic* force in *newtons* between two *current-carrying* electric conductors, we can include magnetic measurements in the MKSA units, which form the basis of the International System of Units. Hence, the international ampere is the constant current that, if maintained in two straight parallel conductors that are of infinite length and negligible cross section and are separated from each other by a distance of 1 meter in a vacuum, will produce between these conductors a force equal to 2×10^{-7} newton per meter of length.

13-5 MAGNETIC FLUX

In keeping with the designation we used in dealing with electric fields, when we wish to refer collectively to the total *number* of lines of force in a magnetic field, we use the term **magnetic flux.** Therefore, **magnetic flux** and **magnetic lines** (plural) **of force** are synonymous.

The letter symbol for magnetic flux is the Greek letter Φ (phi).

For current to flow in an electric circuit, the circuit must form a closed loop. Since magnetic lines of force must always form closed loops, we can refer to the path we can trace around these loops as a **magnetic circuit.** Although there is no *flow* associated with a magnetic circuit comparable to the flow of electrons which constitutes the current in an electric circuit, we shall discover many similarities between current in an electric circuit and flux in a magnetic circuit. Since electricity and magnetism are closely related natural phenomena, we find that for every magnetic property there is a comparable electrical property. We note these comparisons as we proceed to develop suitable units of measurement for magnetic properties.

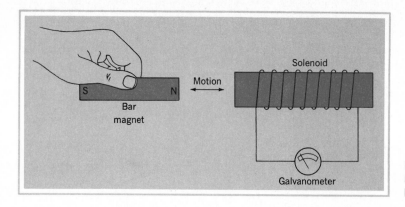

FIGURE 13-13
Faraday's Demonstration of Electromagnetic Induction.

Magnetic flux in a magnetic circuit is the counterpart of electric current in an electric circuit.

Because of the close relationships between magnetism and electricity, in establishing a unit of magnetic flux, we should base it on already established electrical units. In Figure 13-10, we noted that when we caused an electric current to flow in a solenoid, we established a magnetic field around it. Michael Faraday replaced the battery illustrated in Figure 13-10 with a sensitive current-indicating **galvanometer,** as shown in Figure 13-13. Faraday then thrust a permanent magnet into the solenoid. As he did so, he noticed a momentary deflection of the meter pointer. Therefore, not only does an electric current produce a magnetic field, but when we cause a magnetic field to *link* an electric circuit, we also cause an energy conversion from the *moving* **magnetic field to** *induce* an electromotive force by moving electric charge carriers from one end of the coil to the other. As we noted in Section 2-10, the induced emf, in turn, produces a potential difference between the ends of the coil which (under open-circuit conditions) is numerically equal to the induced emf. If the electric circuit is a closed loop (as it is through the galvanometer in Figure 13-13), the induced voltage, in turn, causes current to flow in the electric circuit. Faraday discovered that the induced voltage appeared only when he moved the magnet with respect to the electric circuit and that the faster he moved the magnet, the greater the deflection of the meter pointer. We shall pursue the consequences of this **electromagnetic induction** on the behavior of electric circuits in Chapter 15. But for the moment, Faraday's discovery gives us a means of specifying the basic unit of magnetic flux in terms of the emf induced into a *single-turn* loop (similar to Figure 13-9) as we cause the flux *linking* (passing through the center of) the loop to build up from zero.

The weber is the SI unit of magnetic flux.[†]
If the flux linking a single-turn coil builds up from zero at a rate which will induce an average emf of one volt in the coil, at the end of one second the flux linking the coil will have a magnitude of one weber.
The unit symbol for weber is **Wb.**

[†]The SI unit of magnetic flux was named in honor of Wilhelm Eduard Weber, who established many of the fundamental theories of magnetism.

13-6 MAGNETOMOTIVE FORCE

Just as an electric current cannot flow in an electric circuit until we connect it to a voltage source, magnetic flux (magnetic lines of force) cannot be established until a **magnetomotive force** is produced.

Magnetomotive force in a magnetic circuit is the counterpart of electromotive force in an electric circuit.

With the electromagnet of Figure 13-10, magnetic flux appears only when electric current flows in the solenoid. Therefore, magnetomotive force (mmf) must be a direct result of electric current. We can, therefore, establish a unit of mmf on the basis of the electric current in a *single-turn* coil of wire. Hence,

The ampere is the SI unit of magnetomotive force.

Since adding turns of wire to a solenoid produces the same effect as adding extra strands to a single-turn coil, each strand carrying the input current to the coil, the *effective* amperes of magnetomotive force is the product of the coil current and the number of turns in the coil.

The letter symbol for magnetomotive force is F_m.[†]

$$F_m = NI \qquad (13\text{-}1)$$

where F_m is the magnetomotive force in amperes (effective), N is the number of turns of wire in the coil, and I is the actual current through the coil in amperes.

So that we do not forget that the total magnetomotive force is the product of actual coil current and the number of turns of wire in the coil, we can say that

The ampere-turn is the practical unit of magnetomotive force.

Equation (13-1) becomes the defining equation for magnetomotive force. We shall use the practical unit—the ampere-turn—for clarity in this text.

13-7 RELUCTANCE

For a given electric circuit, Ohm discovered a constant proportionality between the applied voltage and the resulting current. He termed this constant the **resistance** of the electric circuit. Similarly, for a given magnetic circuit (of nonferrous materials), there is a constant proportionality between the magnetomotive force and the resulting magnetic flux. This constant is called the **reluctance** of the magnetic circuit.

Reluctance in a magnetic circuit is the counterpart of resistance in an electric circuit.

[†]Although current standards for letter symbols avoid the use of script letters, the script letter $\mathcal{F}$ is sometimes used to distinguish magnetomotive force from the magnetic force of attaction and repulsion.

From the so-called "Ohm's law for magnetic circuits,"

Reluctance is the opposition of a magnetic circuit to the establishing of magnetic flux. The letter symbol for reluctance is R_m.[†]

From the definition of reluctance,

$$R_m = \frac{F_m}{\Phi} \tag{13-2}$$

where R_m is reluctance, F_m is magnetomotive force in ampere-turns, and Φ is the magnetic flux in webers.

For present purposes, we can express reluctance in **ampere-turns per weber,** as indicated by Equation (13-2). The SI unit of reluctance is the *reciprocal henry.*

13-8 PERMEABILITY

In dealing with parallel electric circuits, we found it more convenient to think in terms of the conductance of the circuit elements rather than in terms of their resistance. Similarly, in dealing with magnetic circuits, we shall find it more convenient to think in terms of the ability of magnetic circuits to permit the setting up of magnetic lines of force rather than in terms of their opposition to magnetic flux.

Permeance in a magnetic circuit is the counterpart of conductance in an electric circuit.

Consequently,

Permeance is a measure of the ability of a magnetic circuit to permit the setting up of magnetic lines of force.
The letter symbol for permeance is P_m[‡]*.*

From the definition of permeance,

$$P_m = \frac{1}{R_m} \tag{13-3}$$

From Equation (13-3), we may express permeance in **webers per ampere-turn.** The SI unit of permeance is the *henry.*

In comparing the magnetic properties of various materials, we would then compare the permeance of sections of each material with unit length and cross-sectional area. Permeance per unit length and cross-sectional area of a material is called its **permeability.** As such, **permeability** is a figure indicating the ability of a

---

[†]Although current standards for letter symbols avoid the use of script letters, the script letter $\mathscr{R}$ is sometimes used to distinguish reluctance from resistance.

[‡]Although current standards for letter symbols avoid the use of script letters, the script letter $\mathscr{P}$ is sometimes used to distinguish permeance from power.

material to permit the setting up of magnetic lines of force, whereas **permeance** is a measure of the ability of a given *magnetic circuit* to permit the setting up of magnetic lines of force. Since permeability relates to the magnetic property of materials, it also has a counterpart relating to the property of materials in *electric* fields. In Section 11-6, we defined *permittivity* of a dielectric in terms of its ability to permit the setting up of *electric* lines of force.

The letter symbol for permeability is the Greek letter μ (mu).

Unfortunately, the letter symbol for the permeability of magnetic materials is the same as the prefix for *micro*. However, micro as a prefix is not required in dealing with magnetic circuits.

From the definition of permeability

$$\mu = P_m \frac{l}{A} = \frac{l}{R_m A} \tag{13-4}$$

where μ is the permeability, P_m is permeance in webers per ampere-turn, R_m is reluctance in ampere-turns per weber, l is the length of the magnetic circuit in meters, and A is the cross-sectional area in square meters.

For present purposes, we can express permeability in terms of **webers per ampere-turn for a one-meter cube.** In SI, permeability is expressed in *henrys per meter*.

13-9 FLUX DENSITY

Permeability is a measure of the ability of a *material* to permit the setting up of magnetic lines of force. We set up Equation (13-4) so that permeability could be expressed in terms of permeance of a cube of that material having unit length and cross-sectional area. Therefore, in thinking of permeability, we are not dealing with total flux. Rather, we are interested in the *flux per unit cross-sectional area.*

Flux per unit cross-sectional area is called flux density.
The letter symbol for flux density is B.
The tesla is the SI unit of flux density.[†]
The unit symbol for tesla is T.

$$B = \frac{\Phi}{A} \tag{13-5}$$

where B is flux density in teslas (or webers per square meter), Φ is the total flux in a magnetic circuit in webers, and A is the cross-sectional area of the magnetic circuit in square meters.

Equation (13-5) automatically defines a **tesla** as being equal to a **weber per square meter** ($T = Wb/m^2$).

[†]Named in honor of the inventor, Nikola Tesla, who immigrated from Hungary to the United States where he was responsible for much of the early design of rotating electric machines.

13-10 MAGNETIC FIELD INTENSITY

Since permeability is based on a *cube* of the magnetic material, we are not concerned with the magnetomotive force required for the full length of the magnetic circuit. Rather, we are interested in the magnetomotive force required to create a certain flux density in a unit length of the magnetic circuit.

Magnetomotive force per unit length is called magnetic field intensity.

It is sometimes called magnetic field strength, magnetizing force, or mmf gradient.

The letter symbol for magnetic field intensity is H.

From the definition of magnetizing force,

$$H = \frac{F_m}{l} \qquad (13\text{-}6)$$

where H is the magnetic field intensity in ampere-turns per meter,[†] F_m is the magnetomotive force in ampere-turns, and l is the length of the magnetic circuit in meters.

Now that we have defined permeability, flux density, and magnetic field intensity, we can derive a very useful relationship among these three magnetic quantities.

Since

$$R_m = \frac{F_m}{\Phi} \qquad \text{and} \qquad \therefore P_m = \frac{\Phi}{F_m}$$

substituting in Equation (13-4) gives

$$\mu = \frac{\Phi}{F_m} \times \frac{l}{A} \qquad \text{or} \qquad \mu = \frac{\Phi}{A} \times \frac{l}{F_m}$$

But

$$\frac{\Phi}{A} = B \qquad \text{and} \qquad \frac{F_m}{l} = H$$

$$\therefore \mu = \frac{B}{H} \qquad (13\text{-}7)$$

where μ is permeability in henrys per meter (or in webers per ampere-turn for a 1-meter cube), B is flux density in teslas (or in webers per square meter), and H is magnetic field intensity in amperes per meter (ampere-turns per meter).

To confirm the SI unit for permeability, we commence by doing a dimensional analysis of the units on the right-hand side of Equation (13-7).

[†]Strictly speaking, magnetomotive force is expressed simply in effective *amperes*. Hence, in SI, magnetic field intensity is expressed in *amperes per meter*.

$$\mu = \frac{Wb}{m^2} \times \frac{m}{A}$$

In Section 15-5, we shall discover that the *henry* is equivalent to webers per ampere. Making this substitution then gives permeability in henrys per meter.

Comparing Equation (13-7) with Equation (11-11),[†] we note that

Magnetic flux density in a magnetic circuit is the counterpart of electric flux density in a dielectric in an electric circuit.

Also,

Magnetic field intensity in a magnetic circuit is the counterpart of electric field intensity in an electric circuit.

and,

Permeability in a magnetic circuit is the counterpart of permittivity in an electric circuit.

13-11 DIAMAGNETIC AND PARAMAGNETIC MATERIALS

In Chapter 11, we noted that the behavior of materials in the electric field between parallel conducting plates in an electric circuit determined the permittivity of the material, which, in turn, governed the magnitude of the electric flux set up between the plates by a given potential difference. Similarly, the behavior of materials in magnetic fields determines their permeability; which, in turn, governs the magnetic flux that can be established by a given magnetomotive force. However, the behavior of materials in magnetic circuits is not as simple as the polarization of atoms of a dielectric shown in Figure 11-7. In Sections 13-11 to 13-18, we shall consider the effects of various materials in magnetic circuits, using currently accepted atomic theories of magnetism to explain these effects.

Like electric force and gravitational force, magnetic force is a *field* force which can act in free space. Hence, magnetic lines of force can exist in a vacuum. Just as we used the ratio of electric flux density to electric field intensity in free space as a reference value for permittivity [Equation (11-12)], we can use the ratio of magnetic flux density to magnetizing force in free space as a reference value for permeability.

The permeability of free space is

$$\mu_v = 4\pi \times 10^{-7} \text{ H/m} \tag{13-8}$$

Most substances have no more effect on a magnetic circuit than does free space. Such materials are **nonmagnetic** and have the same permeabilty as free space. A few

[†]Page 288.

materials have a permeability slightly less than that of free space. They appear to offer a slight opposition to magnetic lines of force as compared with free space. Such materials are called **diamagnetic** materials. Silver, copper, and hydrogen show slight diamagnetic tendencies. A few materials have a permeability just slightly greater than that of free space. These materials are, therefore, very slightly magnetic. Such materials are called **paramagnetic** materials. Platinum, aluminum, and oxygen are inclined to be paramagnetic. In both cases, these effects are so slight that, for practical purposes, we can assume that both diamagnetic and paramagnetic substances have the same permeability as free space.

We can explain the magnetic properties of materials in terms of electromagnetism and the Bohr model of an atom. We have already noted that electric current generates a magnetomotive force. We even measure mmf in *ampere*-turns. Since **electrons** are the charge carriers in metallic conductors, it follows that every electron traveling in an orbit around the nucleus of an atom behaves like a tiny electric current flowing in a single-turn loop. Consequently, every orbital electron generates a tiny magnetomotive force acting in a direction perpendicular to the plane of its orbit. We find it convenient to speak of the **magnetic moment** of these orbital electrons where magnetic moment is defined as the product of the current flowing in a single-turn loop and the area of the loop.

Although we might expect the Bohr model of the hydrogen atom with its single electron to have a magnetic moment, such is not the case. In addition to traveling in an orbit around the nucleus, the electron apparently *spins* on its axis, just as the earth rotates on its axis once a day as it revolves around the sun. If we visualize the electron as a spherical rotating charge, each tiny portion of that charge represents a microscopic electric current flowing in a circular path around the axis of spin. Hence, every electron develops a *spin* magnetic moment acting along the axis of spin, as well as an *orbital* magnetic moment. The magnitude of the spin magnetic moment is the same for all electrons in all atoms and is equal to the orbital magnetic moment of the electron of the Bohr hydrogen atom in its lowest energy level. This becomes the unit of magnetic moment for atomic purposes and is called the Bohr **magneton.**

When the orbital and spin magnetic moments of all the electrons of an atom are added together (vectorially), we find that most atoms have zero net magnetic moment. We can state that the net magnetic moment of all completely filled electron shells of an atom is always zero. Paramagnetic materials have atoms in which some shell is only partially filled in such a manner that these atoms possess a small net magnetic moment. These atoms tend to behave individually as weak permanent magnets. When we expose paramagnetic materials to an external magnetic field, their atoms tend to align their net magnetic moments with the magnetic field, thus slightly increasing the total magnetomotive force and the resulting flux. As a result, paramagnetic materials appear to have a permeability slightly greater than that of free space.

In paramagnetism, the external magnetic field acts on the net magnetic moment of the atom as a whole. In diamagnetism, we are concerned with the effect of an external magnetic field on the magnetic moments of the individual electrons. Even in an atom which has no net magnetic moment, interaction between the individual magnetic moments of the orbital electrons and the external magnetic field will induce some change in these individual magnetic moments. Therefore, the atom as a whole will now possess a slight overall magnetic moment. But, as we shall discover when we investigate magnetic induction in later chapters, the induced magnetic moment is always in such a direction as to *oppose* the magnetic field which is causing the distortion of the atom's magnetic equilibrium. Hence, the diamagnetic effect of a material tends to *reduce* the strength of the external magnetic field. All atoms have a tendency to behave diamagnetically, but, in some materials, the diamagnetic effect is masked by the paramagnetism displayed by the slight *net* magnetic moment of the atoms.

As we have noted, both diamagnetic and paramagnetic effects are so slight that, for practical purposes, we may consider such materials to be nonmagnetic.

13-12 FERROMAGNETIC MATERIALS

Some substances, including iron and its compounds and, to a lesser extent, nickel and cobalt, have permeabilities many hundreds of times greater than that of free space. These substances are called **ferromagnetic** since iron is the significant material in this group. These materials owe their magnetic properties to the special positions they occupy in the periodic table of elements. An iron atom has 26 orbital electrons, two in the first shell, eight in the second shell, 14 in the third shell, and two valence electrons in the fourth shell. Although the first two shells are completely filled, there is room for four more electrons in the third shell. It so happens that the four missing electrons must have *cumulative* spin magnetic moments in order to give a completely-filled third shell zero net magnetic moment. Hence, iron atoms do possess appreciable net magnetic moment due to their partially filled third shells.

In an unmagnetized state, the magnetic moments of the atoms in a given piece of iron have a random overall orientation so that the net mmf for the piece of iron as a whole is zero. The magnetic properties of a ferromagnetic material depend on the manner in which the magnetic moments of the iron molecules align themselves with an external magnetic field. This alignment process represents **magnetization** of the material. Magnetization of both paramagnetic and ferromagnetic materials depends on the alignment of magnetic moments with an external magnetic field. Whereas the weak magnetic moments of paramagnetic atoms allow them to respond *individually* to the external magnetic field, the net magnetic moments of ferromagnetic materials respond *collectively* in groups of 10^{15} (more or less) adjacent atoms forming magnetic **domains.** The magnetic moments of all atoms in a domain must act in the same direction and if any magnetic moment in the domain rotates, *all* magnetic moments in the domain must rotate together. This accounts for the permeabilities of ferromagnetic materials being very much greater than those of paramagnetic materials.

Iron crystals are shaped like cubes and each crystal usually contains many magnetic domains. These domains have their magnetic axes aligned with one of the edges of the cube rather than along one of the diagonals. In demagnetized iron crystals, one-third of the domains are aligned in each of the three planes which are parallel to the faces of the cube. Half of the domains in each plane have their magnetic axes in one direction, and the other half are in the opposite direction.

If we place a single crystal of iron within the magnetic field of a solenoid in such a manner that the magnetic lines of force are parallel to one edge of the crystal, it is fairly easy for all the domains to align themselves with this magnetizing field. This

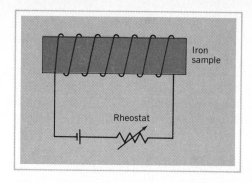

FIGURE 13-14
Magnetizing an Iron Sample.

greatly increases the number of lines of force which can be set up by the original mmf, and the permeability of the iron appears to be very high. If, however, the crystal is located so that a *diagonal* of the crystal is parallel to the magnetic lines of force, it is much more difficult for the magnetic axes of the domains to rotate into alignment with the flux since the diagonal is not one of the natural axes of the crystal. Before the magnetic axes of the domains can rotate, those domains with magnetic axes most nearly aligned with the external magnetic field increase in size at the expense of neighboring domains. This helps to increase the overall magnetic field intensity to a point where the domains can finally switch the orientation of their magnetic axes into alignment with the flux. The permeability of these crystals appears to be appreciably less, although it is still many hundreds of times greater than that of free space, since the alignment of the domains along any axis will greatly increase the total flux.

In a typical sample of iron, the orientation of the crystals is quite random. If we place this sample inside a solenoid, as in Figure 13-14, and gradually increase the current in the coil, those crystals whose *faces* are parallel to the magnetic field will become magnetized first. Then, as the magnetizing force increases, those crystals whose *diagonals* are parallel to the magnetic lines of force will become magnetized. If we increase the current through the solenoid to a point where all the domains are aligned with the magnetic field, any further increase in current can only increase the total flux by the same amount as if the iron were not present. The iron is now said to be completely **saturated** and its permeability has dropped to that of any nonmagnetic material. The current which is required to saturate a sample of iron within the solenoid depends on the type and grade of iron used.

The permeability of ferromagnetic materials also depends to some extent on temperature. As we raise the temperature of a sample of iron, we reach a temperature (about 770°C for iron) known as the **Curie temperature** at which all ferromagnetic properties disappear and the material behaves as a paramagnetic material. This phenomenon is thought to be brought about by thermal agitation becoming so great that the magnetic domains no longer exist. The magnetic moments of the atoms must now respond *individually* to an external magnetic field.

13-13 PERMANENT MAGNETS

When the current in the magnetizing solenoid of Figure 13-14 is switched off, there is a tendency for the various magnetic domains of the iron sample to return to their original orientations. Some types of iron (soft irons) have a very low magnetic

retentivity; that is, most of the domains do return to an orientation in which the **residual magnetism** or **remanence** is practically nil when the current is switched off. Such materials would be classified as **temporary magnets.** Residual magnetism in ferromagnetic materials is the magnetic counterpart of dielectric absorption noted in Section 11-4.

In some other types of iron (hard steels), a large percentage of the domains retain the orientation they were forced to take in the strong magnetic field produced by the energizing coil. These materials become **permanent magnets** of the type used in Figures 13-1 to 13-6. Permanent magnets maintain a magnetomotive force that cannot readily be adjusted like the mmf of an electromagnet.

Although the steel bar magnets shown in Figures 13-1 to 13-6 are satisfactory for laboratory demonstrations, the permanent magnets used in modern electric meter movements, loudspeakers, etc., are made of a very hard steel alloy containing aluminum, nickel, and cobalt (alnico). Because of its hardness, an alnico magnet is usually cast in the desired shape and is subjected to a very high magnetic flux density produced by an electromagnet while it is cooling.

In Section 11-6, we noted the fairly recent development of a "ferroelectric" ceramic material with an extremely high dielectric constant. At the same time, a group of ceramics containing iron oxides have been developed which display prominent ferromagnetic properties. Among these **ferrites** are some which make excellent permanent magnets. A typical "ceramic magnet" is formed by the usual ceramic technique of grinding barium carbonate and ferric oxide into particles the size of a magnetic domain, pressing the powder into the desired shape, and firing it in an oxygen atmosphere. Unlike iron magnetic materials but like ceramics, ferrites are poor electric conductors.

13-14 MAGNETIZATION CURVES

Although the magnetic field intensity in Figure 13-14 is an independent variable over which we have direct control by varying the current in the solenoid which produces the mmf, the permeability of a given sample of iron varies considerably as we increase the current in the coil. The exact value of permeability depends on what percentage of the magnetic domains have aligned their magnetic moments with the magnetic field. This percentage, in turn, is dependent on the flux density. Even though we can prepare a graph, as in Figure 13-15, showing how the permeability of a given sample of iron varies with flux density, we are faced with the interdependence which becomes evident when we rearrange Equation (13-7).

FIGURE 13-15
Manner in Which the Permeability of Cast Steel Varies with Flux Density.

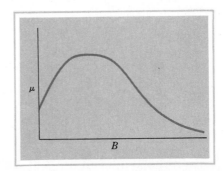

353

$$B = \mu H$$

For a given magnetic field intensity H, the flux density B depends on the permeability μ, which, in turn, depends on the flux density B. Such a situation results in a "cut-and-try" solution for flux density.

To avoid this situation, we can eliminate one of the dependent variables by disregarding permeability and concerning ourselves only with the manner in which the flux density of a certain type of iron varies with the magnetic field intensity. Since B and H as units of measurement were both based on a cube with *unit* dimensions, the **magnetization curve** (or simply ***BH* curve**) produced by plotting a graph of the manner in which B varies with H, is dependent only on the *type* of iron and not on its dimensions.

We can divide the magnetization curve of Figure 13-16 into three sections. As we gradually increase the current in the solenoid of Figure 13-14 from zero, we note that the flux density increases slowly at first. The region of the **lower knee** of the *BH* curve indicates the condition where domains whose magnetic axes are most nearly aligned with the applied magnetic field are *growing* in size at the expense of neighboring domains. The steep portion of the *BH* curve is the region in which the magnetic axes of the domains are *switching* to act along whichever axes of the crystals are closest to the direction of the applied magnetic field. Finally, when the magnetic field is great enough to increase the flux density beyond the **upper knee** of the *BH* curve, the magnetization process is entering a region where the magnetic field intensity is great enough to force the magnetic axes of the domains to align themselves with the external magnetic field rather than with the preferred crystal axes. Finally, for large values of magnetic field intensity, when all the domains are aligned with the magnetic field, the *BH* curve becomes a straight line with a gradual slope. This slope shows the same increase in B for a given increase in H as for any nonmagnetic material.

It is difficult to mark the exact point on the *BH* curve of Figure 13-16 where **saturation** occurs. Theoretical saturation is when all the domains are finally aligned

FIGURE 13-17
***BH* Curve for Iron Sample in Which All Crystals Have the Same Orientation.**

FIGURE 13-16
***BH* Curve for a Typical Cast-Steel Sample.**

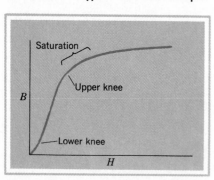

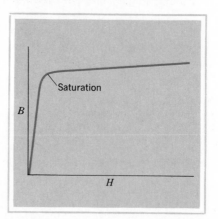

with the magnetic field. But this requires considerable magnetic field intensity for some types of iron. *Practical* saturation may be defined as the flux density beyond which it is *impractical* to magnetize a certain magnetic material. For some applications, this point is the upper knee in the *BH* curve where the steep, linear portion of the slope ends. In high-quality audio transformers, magnetic field intensity must be kept below this point, since any swing in *H* past this point does not permit a linear transfer of the audio signal from a current waveform in the coil to a flux waveform in the magnetic circuit. But, to reduce the cost of power transformers, we can reduce the amount of iron required if we design them so that their magnetic cores are magnetized to a point slightly beyond the upper knee of the *BH* curve.

The gradual saturation of most types of iron (as shown in Figure 13-16) is due to the random orientation of the iron crystals. If iron can be manufactured so that all crystals have the *same* orientation, it follows that the flux density will rise very steeply with an increase in magnetic field intensity, since all the magnetic domains will be acted upon in the same manner. At a certain value of magnetic field intensity, the iron will saturate (as shown in Figure 13-17) because *all* domains align themselves with the magnetic field at this particular value of magnetic field intensity. Such a type of iron is available for use in magnetic amplifiers, magnetic tapes for magnetic recording, magnetic memory cores for electronic computers, and the cores of high-quality audio transformers.

13-15 PERMEABILITY FROM THE *BH* CURVE

As we noted in Figure 13-15, the permeability of a ferromagnetic material is far from constant. It is quite dependent on flux density. However, we can determine permeability at a given flux density without being supplied with tables or graphs similar to Figure 13-15. We can apply the relationship stated in Equation (13-7) to the *BH* curve supplied for a given type of magnetic material. The manner in which we apply this $\mu = B/H$ relationship to a *BH* curve will depend on the conditions under which the magnetic circuit is to be operated.

If the current in the energizing solenoid is to be kept constant, we can locate the appropriate value for magnetic field intensity and the resulting value of flux density on the *BH* curve, as shown in Figure 13-18(a). The **normal permeability** for this flux density is simply

$$\mu = \frac{B}{H} \tag{13-7}$$

In some magnetic circuits, the magnetic field intensity will vary between two limits [*A* and *B* in Figure 13-18(b)]. This situation occurs when a fluctuating direct current flows in the coil, as in many audio transformers. In this case, we are interested in the permeability over a limited operating range. This value of permeability is called the **incremental permeability** and is determined as shown in Figure 13-18(b).

$$\mu_\Delta = \frac{\Delta B}{\Delta H} \tag{13-9}$$

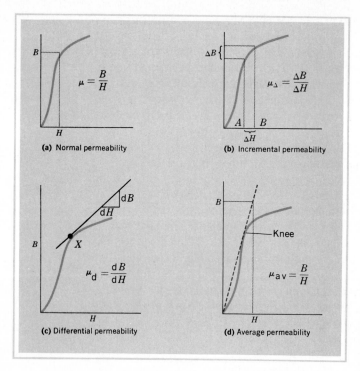

(a) Normal permeability $\quad\mu = \dfrac{B}{H}$

(b) Incremental permeability $\quad\mu_\Delta = \dfrac{\Delta B}{\Delta H}$

(c) Differential permeability $\quad\mu_d = \dfrac{dB}{dH}$

(d) Average permeability $\quad\mu_{av} = \dfrac{B}{H}$

FIGURE 13-18
Determining Permeability from a *BH* Curve.

where μ_Δ is the incremental permeability, ΔB is the difference between the maximum and minimum flux densities, ΔH is the difference between the maximum and minimum values of magnetic field intensity.

If alternating current flows in the solenoid, the magnetic field intensity will be constantly changing over wide limits. In this case, we are likely to be concerned with the permeability as the magnetic field intensity *swings through* a certain value, rather than as it maintains that value. This particular value of permeability is called the **differential permeability** for a particular point on the *BH* curve. We determine differential permeability from the *slope* of the magnetization curve at the point in question. To calculate the slope of the *BH* curve at point X in Figure 13-18(c), we draw a tangent to the curve at point X and then construct a right-angled triangle with the vertical side representing a certain increase in flux density, the horizontal side representing the equivalent increase in magnetic field intensity, and the tangent forming the hypotenuse. Then dB and dH are determined in the same manner as for incremental permeability, and the differential permeability at point X becomes

$$\mu_d = \frac{dB}{dH} \tag{13-10}$$

In some electronic circuits where the current is in the order of microamperes, the magnetic field intensity is so small that it is difficult to read the B/H ratio from the magnetization curve. However, the value of permeability for very small values of H and the associated values of B is quite important and is called **initial permeability**

(μ_0). Initial permeability could also be defined as the differential permeability when the magnetic field intensity is zero.

We can simplify the problem of solving magnetic circuits in which the magnetic field intensity and, consequently, the permeability is continually changing by determining an **average permeability** value for the type of iron being used. The average permeability is obtained by drawing a straight line through the origin of the BH graph and the upper knee of the curve as in Figure 13-18(d). Because this line is *straight,* we may pick *any* point on this line to read off the corresponding values of B and H from which μ_{av} is determined. As long as the flux density in a magnetic circuit is kept below saturation, this average permeability is reasonably accurate.

13-16 HYSTERESIS

Although the BH curves of Figures 13-16 to 13-18 represent the manner in which the flux density of an *unmagnetized* sample of iron rises as the value of H is increased, they do *not* represent the manner in which the flux density drops as the magnetic field intensity is decreased. Most magnetic materials have some **retentivity.** When the magnetizing force is returned to zero, the residual magnetism of the iron will produce an appreciable value of flux density. In order to get rid of this residual flux, it is necessary to pass some current through the solenoid in the *opposite* direction. The amount of *negative* magnetic field intensity or magnetizing force required to demagnetize a particular sample of iron is called the **coercive force.**

Figure 13-19 shows the complete cycle of magnetization of a piece of iron as we pass an *alternating* current through the solenoid. If we assume that the sample started from an unmagnetized state, as we increase the current from zero to maximum in one direction through the solenoid (positive value of H), the flux density increases along the customary magnetization curve OA. As we reduce the current to zero, the flux density decreases from saturation to a residual flux density at point B. To bring the flux

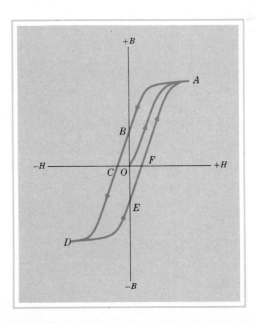

FIGURE 13-19
Typical Hysteresis Loop.

357

density in the iron to zero, we must reverse the current through the coil and increase it to a value represented by OC. As we now increase the current in this direction, the iron will again saturate at point D, with the magnetic domains all lined up in the opposite direction to that represented at point A. Again, as we reduce the current to zero, the flux density in the iron will decrease only to point E. Again we must reverse the current direction to bring the flux density to zero at point F. Increasing the current to its maximum value in the positive direction completes the cycle by saturating the iron and returning the flux density to point A in the graph of Figure 13-19.

As shown in Figure 13-19, when the magnetic field intensity is due to an alternating current, the flux density of ferromagnetic materials tends to *lag* the magnetic field intensity which creates it. This lagging of the magnetization of the iron behind the magnetic field intensity is called **hysteresis.** The graph which shows this lag, as in Figure 13-19, is called a **hysteresis loop.**

Since only ferromagnetic materials have residual magnetism, nonmagnetic materials show no hysteresis effect. For nonmagnetic materials, $B = \mu H$, where μ is a constant $(4\pi \times 10^{-7})$. Therefore, the flux density in a nonmagnetic material is always directly proportional to the magnetic field intensity, regardless of its variation.

Hysteresis is due to some of the magnetic domains in a ferromagnetic material not wanting to return to their original orientation and having to be forced to do so by a certain amount of reversed magnetic field intensity (coercive force). Whenever motion is accomplished against an opposing force, an energy transfer must take place. This energy is taken from the source of alternating voltage, which is responsible for the magnetic domains having to change their orientation, and is transferred to the molecules of the ferromagnetic material in the form of heat. The higher the frequency of the alternating current in the solenoid, the more rapidly the magnetic domains have to change their alignment and, therefore, the greater the **hysteresis loss.**

The greater the retentivity of a particular type of iron, the greater the coercive force that is required to demagnetize it. This means an increased opposition by the magnetic domains to reorientation, which results in a greater transfer of energy into heat. Therefore, hysteresis loss is also proportional to the retentivity of the iron. The iron selected for use in the magnetic circuits of transformers which are continually subject to an alternating mmf should have an absolute minimum retentivity if hysteresis loss is to be kept at a low value. Consideration of Figure 13-19 will show us that as the residual flux density of a sample of iron increases, the area within the hysteresis loop increases. Therefore, the area within the hysteresis loop for a given iron sample is a useful indication of its hysteresis loss.

Hysteresis can be used to advantage in some magnetic circuit applications. Certain ferrites composed of oxides of magnesium, manganese, and iron have an almost rectangular hysteresis loop, as shown in Figure 13-20. As the magnetic field intensity H is gradually increased in a positive direction, the flux density in the core remains unchanged until the magnetic field intensity reaches a critical level, at which the core suddenly demagnetizes and *saturates* in the positive direction. The same thing happens when the magnetic field intensity is gradually increased in the negative direction. Consequently, these ferrite cores will always be saturated in either the positive or negative direction. These two stable states of ferrite switching cores make tiny beads of ferrite threaded on a matrix of slender wires very useful for memory banks in digital computers.

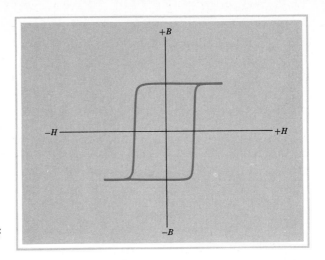

FIGURE 13-20
Rectangular Hysteresis Loop of a Ferrite Switching Core.

13-17 EDDY CURRENT

In section 13-5, we found that moving a magnetic field with respect to an electric conductor induced a voltage into the conductor. This situation exists inside a solenoid which is energized by an alternating current. If we place a metal core inside the solenoid, any motion of the magnetic field will induce a voltage in it. In electromagnetic induction, the direction of the magnetic lines of force, the motion of the magnetic lines of force with respect to the electric conductor, and the flow of electric current are all at right angles to one another.

As we increase the current in the solenoid, more magnetic lines of force must be set up. This crowding of more lines of force into the existing space causes the lines of force already present to move outward along a radius from the axis of the solenoid. The direction of the lines of force themselves is parallel to the axis of the solenoid. The motion of these lines of force out and in along the various radii when an alternating current flows in the solenoid will then induce a voltage into the metal core of the solenoid. As we have already noted, the current resulting from this induced voltage must be at right angles to both the radius and the axis of the solenoid.

The only possible manner in which this can be achieved is for current to flow in a circle around the circumference of the metallic core, as shown in Figure 13-21. This current is called an **eddy current.** Eddy current is produced in any core material

FIGURE 13-21
Eddy Current in a Solid Metal Core.

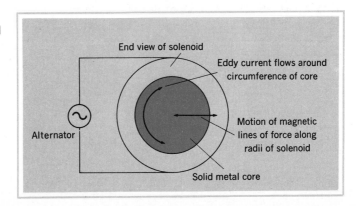

359

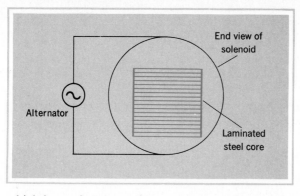

FIGURE 13-22
Cross Section of a Laminated Core.

which is an *electric* conductor. The core does not have to be a magnetic material. A brass core can become very hot due to eddy current produced as a result of a high-frequency alternating current in the solenoid. This heat comes from the customary I^2R power loss as the eddy current flows through the resistance of the core material. Since the resistance of iron is comparatively high, the eddy-current loss in a solid iron core would rule out its use with alternating magnetomotive force.

Eddy-current loss can be greatly reduced by replacing the solid core with one made up of thin **laminations** which are insulated from each other with a thin coat of varnish. The laminations are oriented as shown in Figure 13-22 so that the direction in which eddy current would have to flow is from lamination to lamination through the varnish. Because of the high resistance of the varnish, the eddy current in a laminated core is very small and the accompanying eddy-current power loss is, in many cases, negligible.

13-18 MAGNETIC SHIELDING

Although there is no "insulator" for magnetic lines of force, it is possible to shield a certain space such as a sensitive meter movement or a cathode-ray tube from stray magnetic fields by enclosing this space in a cylinder of high-permeability iron. As we noted in discussing the characteristics of magnetic lines of force, they will go out of their way to include the lower reluctance path through the iron as part of their circuit. This leaves the space within the cylinder comparatively free of magnetic lines of force, as shown in Figure 13-23.

FIGURE 13-23
Magnetic Shielding.

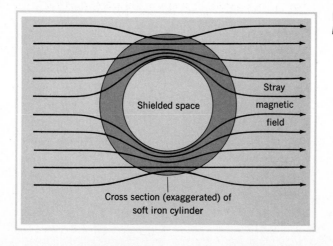

360

PROBLEMS

13-1. Draw the magnetic lines of force around the conductor in Figure 13-24(a) and indicate their direction.

13-2. Mark the conventional current direction on the conductor in Figure 13-24(b).

13-3. Draw the magnetic field around the coil in Figure 13-25(a) and indicate its direction.

13-4. Mark the conventional current direction on the coil of Figure 13-25(b).

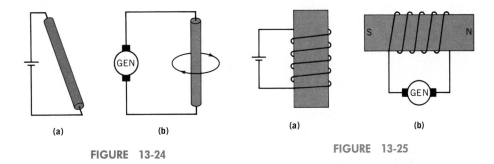

(a)   (b)   (a)   (b)

FIGURE 13-24   FIGURE 13-25

13-5. What is the reluctance of a magnetic circuit in which a total flux of 2×10^{-3} Wb is set up by a 5-A current flowing in a solenoid consisting of 200 turns of wire?

13-6. What current must be passed through a 500-turn solenoid to produce a total flux of 1.2 mWb in a magnetic circuit whose reluctance is 2×10^6 At/Wb?

13-7. The magnetic circuit in Problem 13-5 has a uniform cross-sectional area of 5 cm^2 and an average path length of 25 cm. What is the permeability of the core material?

13-8. The core of a solenoid consists of a brass cylinder 10 cm in length and 2 cm in diameter. What is its reluctance?

13-9. How many turns of wire are there in a solenoid if a 500-mA current through it produces a total flux of 8×10^{-4} Wb in a magnetic circuit whose reluctance is 5×10^5 At/Wb?

13-10. The magnet in Figure 13-4 has a rectangular cross section of 1×2 cm. If the spacing between the two sections in Figure 13-4(b) is 1 mm, what is the reluctance of the air gap?

13-11. A piece of iron 10 cm in length and having a rectangular cross section of 1×2 cm has a reluctance of 1.25×10^6 At/Wb. What is the permeability of the iron?

13-12. A magnetic field intensity of 2000 At/m produces a flux density of 1.0 T in a certain type of iron. What is its permeability at this flux density?

NOTE Refer to the typical magnetization curves of Figure 14-4 in solving the remaining problems.

13-13. What flux density is created in sheet steel by a magnetizing force of 2800 At/m?

13-14. What magnetic field intensity is required in order to maintain a constant flux density of 1.5 T in cast steel?

13-15. What is the normal permeability of cast steel when the flux density remains at 1.5 T?

13-16. What is the average permeability of sheet steel?

13-17. Find the incremental permeability of cast iron between flux densities of 0.30 T and 0.35 T.

13-18. Determine the differential permeability of cast steel for a magnetic field strength of 1600 At/m.

REVIEW QUESTIONS

13-1. Draw a diagram of the magnetic field around a horseshoe-shaped permanent magnet.

13-2. With reference to the characteristics of magnetic lines of force, show why a nail is attracted to either pole of a horseshoe-shaped permanent magnet.

13-3. Show with sketches how it is possible to determine the direction of the current in an electric conductor with the aid of a compass needle.

13-4. Why is the flux density of a magnetic field greater in the center of a current-carrying solenoid than around a straight wire carrying the same current?

13-5. How is the magnitude of the **weber** established?

13-6. Define the term **reluctance.**

13-7. What is the distinction between **magnetomotive force** and **magnetizing force?**

13-8. Define the term **permeability.**

13-9. Account for the upper knee on the *BH* curve of ferromagnetic materials.

13-10. Draw a *BH* curve for aluminum.

13-11. What is the distinction between **differential** and **incremental** permeability?

13-12. Compare permanent and temporary magnets in terms of **coercive force.**

13-13. Define the term **hysteresis.**

13-14. Laminating the iron core of a transformer greatly reduces the eddy-current losses. What effect does the laminating process have on hysteresis losses? Explain.

13-15. Some magnetic circuits consist of powdered iron pressed into form with a ceramic binder. Compare such a core with a solid core of the same type of iron in terms of
    (a) **permeability,**
    (b) **hysteresis,** and
    (c) **eddy current.**

14

MAGNETIC CIRCUITS

The design of *magnetic* circuits is usually of more interest to students who will be studying *heavy-current* or power systems options in electrical technology. For students who will be limiting their studies to *light-current* or electronics options, solution of the magnetic circuit examples of this chapter may be treated as optional.

14-1 PRACTICAL MAGNETIC CIRCUITS

Since magnetic fields are produced around current-carrying electric conductors, it follows that magnetic fields are associated with most electric apparatus. The operation of much of this apparatus (for example, motors, generators, transformers, loudspeakers, and relays) depends on the magnetic fields associated with it. It is just as important to be able to control the path and the strength of the magnetic fields in such devices as it is to control the flow of electric current in them. By making use of the very high permeability of ferromagnetic materials, we can construct **magnetic circuits** to control the path of the magnetic lines of force; and, by application of the relationship stated by the so-called "magnetic Ohm's law" of Equation (13-2), we can determine the magnetomotive force required to develop the desired field strength.

In some magnetic circuits, such as the loudspeaker and the meter movement in Figure 14-1, the required magnetomotive force is supplied by permanent magnets. But, in most electric machines, the mmf is produced by current-carrrying coils of wire (solenoids, field windings, armatures, operating coils, etc.). The shape and the location of the resulting magnetic field depends, in turn, on the reluctance of the magnetic circuit and the available magnetomotive force. We can determine the magnetomotive force we require to produce a certain flux in a given magnetic circuit in

363

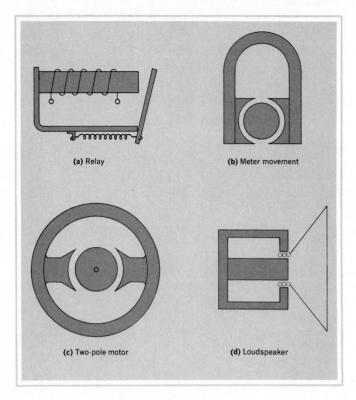

(a) Relay

(b) Meter movement

(c) Two-pole motor

(d) Loudspeaker

FIGURE 14-1
Typical Magnetic Circuit Cross Sections.

much the same manner that we determined the applied voltage (resulting from an electromotive force) needed to produce a certain electric current in an electric circuit possessing a given resistance.

14-2 LONG AIR-CORE COILS

In Figure 13-10, we noted that the magnetic field around a coiled electric conductor resembled that of a bar magnet. If a solenoid having air or any other nonmagnetic material as a core is at least ten times as long as its diameter, we can calculate fairly accurately the relationship between the flux at the center of the coil and the current through it.

As shown in Figure 14-2, all the magnetic lines of force pass through the center of the coil and then complete their loops through the space outside the coil. Since the permeability of the air outside the coil is the same as that of the nonmagnetic core, about half of the lines of force do not travel the whole length of the core but leave and enter the sides of the coil because of the tendency of magnetic lines of force to repel one another and to become as short as possible. Therefore, our calculations will apply only to the center of the coil.

Figure 14-2 also shows that all the lines of force that are crowded into the small cross section at the center of the coil complete their loops through a very large cross-sectional area outside the coil. Since the reluctance of a magnetic circuit is inversely proportional to its cross-sectional area, the reluctance of the portion of the

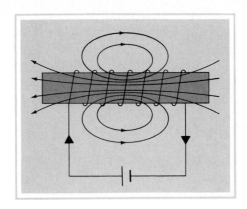

FIGURE 14-2
Magnetic Field around a Long Air-Core Coil.

magnetic path within the coil is much greater than that of the return portion outside the coil. Therefore, it is reasonable to assume that,

For long air-core coils, the total reluctance is approximately equal to the reluctance of the nonmagnetic core of the coil itself.

Transposing Equation (13-4) gives

$$R_m = \frac{l}{\mu A} \tag{14-1}$$

where R_m *is the reluctance of the nonmagnetic core in ampere-turns per weber (At/Wb),* l *is the length of the coil in meters,* μ *is the permeability of nonmagnetic materials* $(4\pi \times 10^{-7}$ H/m$)$, *and* A *is the inside cross-sectional area of the coil in square meters.*

Equation (14-1) is the magnetic circuit counterpart of

$$R = \rho \times \frac{l}{A} \quad (4\text{-}4) \qquad R = \frac{l}{\sigma A} \tag{6-16}$$

for electric circuits. Having calculated the reluctance of a magnetic circuit, we can readily determine the magnetomotive force needed to produce a required magnetic flux by the magnetic circuit counterpart of Ohm's law.

For electric circuits: $\qquad\qquad\qquad E = IR$ $\qquad\qquad\qquad$ (4-2)

For magnetic circuits: $\qquad\qquad\qquad F_m = \Phi R_m$ $\qquad\qquad\qquad$ (13-2)

EXAMPLE 14-1
An air-core coil 20 cm in length and 1 cm inside diameter has 1000 turns. What current must flow in the coil to develop a total flux of 10^{-6} Wb at the center of the coil?

SOLUTION
Step I. Solve for the reluctance of the core.

$$R_m = \frac{l}{\mu A}$$

$$\therefore R_m = \frac{20 \text{ cm}}{4\pi \times 10^{-7} \times \pi \times (0.5 \text{ cm})^2} = 2.026 \times 10^9 \text{ At/Wb}$$

There are two ways of setting up the calculator chain to solve this equation: (1) we can enter the data in the order given,

20 **EE** **+/−** 2 ÷ 4 ÷ **π** **EE** **+/−** 7 ÷ **π** ÷ 0.5 **EE** **+/−** 2 **x²** **=**

or (2) we can solve the denominator first, using the multiplication key, then using the reciprocal key to invert the denominator before we multiply by the numerator.

Step II. Solve for the magnetomotive force using Equation (13-2).

$$F_m = \Phi R_m = 10^{-6} \text{ Wb} \times 2.026 \times 10^9 \text{ At/Wb} = 2.026 \times 10^3 \text{ At}$$

Step III. Solve for the current in the coil using Equation (13-1).

$$I = \frac{F_m}{N} = \frac{2.026 \times 10^3 \text{ At}}{1000 \ t} = 2.03 \text{ A}$$

14-3 SIMPLE MAGNETIC CIRCUIT

If we place an iron core in the coil of wire in Figure 14-2, the high permeability of the iron will reduce the reluctance of the core well below that of the return path outside the core. Therefore, we can no longer assume that the total reluctance is very nearly that of the core itself. In this case, the reluctance of the return circuit outside the core is the major factor in determining the total reluctance of the magnetic circuit. This reluctance is considerably more difficult to calculate than that of the core with its finite dimensions. We usually avoid magnetic circuits in which the magnetic lines of force

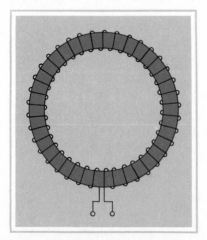

FIGURE 14-3
Construction of a Toroid Winding.

366

are allowed to find their own return path as best they can from one end of the core to the other.

By winding the coil completely around a ring-shaped core, forming a **toroid**-type winding, as shown in Figure 14-3, we confine the magnetic field to a path completely within the coil for both nonmagnetic and magnetic core materials.

14-4 LINEAR MAGNETIC CIRCUITS

In solving Example 14-1, we followed a procedure which is an exact counterpart of the procedure we used in solving simple electric circuits in Chapter 4. This procedure assumes that resistance in an electric circuit (and reluctance in a magnetic circuit) are dependent only on the physical properties of the circuit and are not in any way dependent on the applied voltage (or mmf) and the resulting current (or flux). In electric circuits, such resistors are called **linear** resistors. Similarly, in magnetic circuits, if the permeability μ in Equation (14-1) is independent of magnetic field intensity, the magnetic circuit is a **linear magnetic circuit.**

As shown in Figure 13-15, the permeability of ferromagnetic materials is very much dependent on the flux density. Hence, (except for a limited range at the hump of the graph in Figure 13-15) the reluctance of practical ferromagnetic circuits will change with variations in magnetomotive force and flux. Accordingly, we call these **nonlinear magnetic circuits.**

Linear magnetic circuits are, therefore, limited to nonmagnetic materials such as air, brass, aluminum, plastic, wood, etc. The solution of linear magnetic circuits follows much the same procedure as Example 14-1.

EXAMPLE 14-2

The brass core in Figure 14-3 has a circular cross section. The inside diameter of the core is 8 cm and the outside diameter is 12 cm. If there are 2000 turns in the winding, what total flux will be produced by a 1-A current in the coil?

SOLUTION

Step I. Some care is required in determining the correct values for the length and cross-sectional area of magnetic circuits. From the dimensions given, it follows that the diameter of the cross section of the core is 2 cm.

$$\therefore A = \pi \times (1 \text{ cm})^2 = \pi \times 10^{-4} \text{ m}^2$$

The length of the magnetic circuit will be the *average* path length. In this example, the inside circumference is $\pi \times 8$ cm and the outside circumference is $\pi \times 12$ cm. Hence, the average of these two extreme path lengths is $\pi \times 10$ cm.

Step II. $\quad R_m = \dfrac{l}{\mu A} = \dfrac{\pi \times 10 \text{ cm}}{4\pi \times 10^{-7} \times \pi \times 10^{-4} \text{ m}^2} = 7.958 \times 10^8 \text{ At/Wb}$

Step III. $\quad F_m = NI = 2000\, t \times 1 \text{ A} = 2000 \text{ At}$

Step IV. $\quad \Phi = \dfrac{F_m}{R_m} = \dfrac{2000 \text{ At}}{7.958 \times 10^8 \text{ At/Wb}} = \mathbf{2.5 \times 10^{-6} \text{ Wb}}$

14-5 NONLINEAR MAGNETIC CIRCUITS

In Section 4-9, we discovered that we could not provide data on nonlinear resistors by simply stating a resistance value in ohms, since their resistance is dependent on the current through them and/or the voltage drop across them. To work with nonlinear resistors, we are provided with data on their behavior in electric circuits in the form of a **volt-ampere** graph similar to those shown in Figures 4-7 to 4-9. Similarly, to solve nonlinear magnetic circuits, we must be provided with comparable data on the magnetic properties of the particular type of iron or ferrite we wish to use in our magnetic circuit. This information is usually published in the form of detailed *BH* graphs such as the curves of Figure 14-4.

FIGURE 14-4
Typical Magnetization Curves.

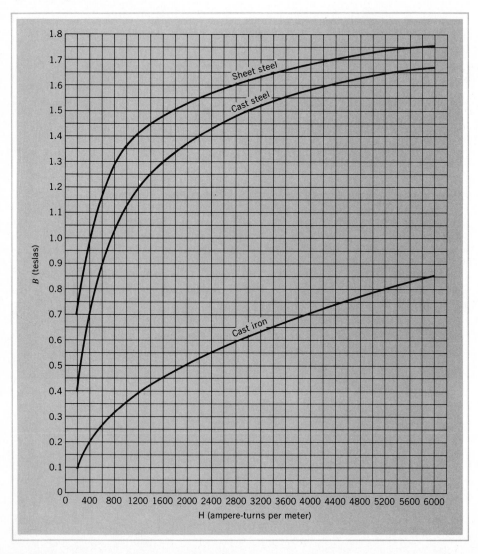

EXAMPLE 14-3

The cast-steel core in Figure 14-3 has a square cross section. The inside diameter of the ring is 12 cm, and the outside diameter is 15 cm. If there are 2000 turns in the coil, what current must flow through it to develop a total flux of 3×10^{-4} Wb?

SOLUTION 1

Step I. Determine the dimensions of the magnetic circuit:

Cross-sectional area: $\quad A = (1.5 \text{ cm})^2 = 2.25 \text{ cm}^2 = 2.25 \times 10^{-4} \text{ m}^2$

Average path length: $\quad l = \pi \times 13.5 \text{ cm} = 42.4 \text{ cm} = 0.424 \text{ m}$

Step II. Determine flux density:

$$B = \frac{\Phi}{A} = \frac{3 \times 10^{-4} \text{ Wb}}{2.25 \times 10^{-4} \text{ m}^2} = 1.333 \text{ T}$$

Step III. Determine permeability: From the magnetization curve for cast steel in Figure 14-4, a magnetic field intensity of 1750 At/m is required to produce a flux density of 1.33 T.

$$\therefore \mu = \frac{B}{H} = \frac{1.33 \text{ T}}{1750 \text{ At/m}} = 7.6 \times 10^{-4} \text{ H/m}$$

Step IV. $\quad R_m = \dfrac{l}{\mu A} = \dfrac{0.424 \text{ m}}{7.6 \times 10^{-4} \times 2.25 \times 10^{-4} \text{ m}^2} = 2.48 \times 10^6 \text{ At/Wb}$

Step V. $\quad F_m = \Phi R_m = 3 \times 10^{-4} \text{ Wb} \times 2.48 \times 10^6 \text{ At/Wb} = 744 \text{ At}$

Step VI. $\quad \therefore I = \dfrac{F_m}{N} = \dfrac{744 \text{ At}}{2000 \text{ t}} = 0.372 \text{ A} = 372 \text{ mA}$

Actually, the above procedure represents a rather roundabout way of solving magnetic circuits in which reference to *BH* curves is involved. Once we have determined the flux density, the magnetization curve tells us how much magnetic field intensity we require for each unit length of the magnetic circuit. We can, therefore, solve for the mmf directly without going through Steps III and IV. Hence, we completely bypass both permeability and reluctance in the second method of solving magnetic circuits.

SOLUTION 2

Step I. Dimensions of the magnetic circuit (as in Solution 1):

$$A = 2.25 \times 10^{-4} \text{ m}^2 \quad \text{and} \quad l = 0.424 \text{ m}$$

Step II. $$B = \frac{\Phi}{A} = \frac{3 \times 10^{-4} \text{ Wb}}{2.25 \times 10^{-4} \text{ m}^2} = 1.333 \text{ T}$$

Step III. From the BH curve for cast steel in Figure 14-4, $H = 1750$ At/m

Step IV. $$F_m = Hl = 1750 \text{ At/m} \times 0.424 \text{ m} = 742 \text{ At}$$

Step V. $$I = \frac{F_m}{N} = \frac{742 \text{ At}}{2000 \text{ t}} = 0.371 \text{ A} = \textbf{371 mA}$$

In a simple magnetic circuit where *all* the magnetomotive force is devoted to producing flux in the *uniform* cross section of the iron core, it is not difficult to solve for the flux when we are given the magnetomotive force.

EXAMPLE 14-4

A current of 0.5 A is passing through a coil consisting of 960 turns wound on a ring-type cast-steel core whose cross-sectional area is 12 cm^2 and whose average path length is 30 cm. What is the total flux in the core?

SOLUTION

$$F_m = NI = 960 \text{ t} \times 0.5 \text{ A} = 480 \text{ At}$$

$$H = \frac{F_m}{l} = \frac{480 \text{ At}}{30 \text{ cm}} = 1600 \text{ At/m}$$

From the BH curve for cast steel (Figure 14-4),

when $$H = 1600 \text{ At/m}, B = 1.3 \text{ T}$$

$$\therefore \Phi = BA = 1.3 \text{ T} \times 12 \text{ cm}^2 = \textbf{1.56} \times \textbf{10}^{-3} \textbf{ Wb}$$

14-6 LEAKAGE FLUX

In the preceding section, we wound the energizing coil along the full length of the magnetic circuit (toroid fashion) in an attempt to keep as much of the flux within the core as possible. This type of winding is expensive since it has to be wound by threading the wire through the center of the core. It is used only in cases where it is imperative that the number of lines of force traveling through the air outside the core be kept to a minimum. If a magnetic circuit is constructed as shown in Figure 14-5, the coil can be wound on a bobbin and then the magnetic core can be assembled. From our investigation of parallel resistors, we learned that *all* the current does not flow through the path of least resistance. Actually, the current splits up in proportion to the conductance of the parallel branches. When high-permeability iron is used for the full length of the magnetic circuit, the majority of the lines of force will follow the path established by the iron. But, since the air in the center of the core is not a magnetic insulator, and since magnetic lines of force have a tendency to try to become as short as possible, some of the flux will complete its circuit outside the core, as shown in Figure 14-5. These particular lines of force are called the **leakage flux.**

In designing a simple magnetic circuit similar to that of Figure 14-5, we can do

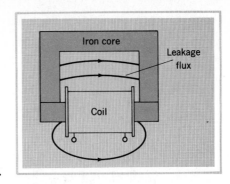

FIGURE 14-5
Leakage Flux in a Magnetic Circuit.

several things to keep the leakage flux to a minimum. If we keep the magnetic circuit compact, the length of a magnetic line of force taking the short cut through the air is not much shorter than the path length through the iron. The higher the permeance (the lower the reluctance) of the magnetic circuit, the greater will be the number of lines of force which stay within the magnetic circuit. High permeance is achieved by making sure that the cross-sectional area is great enough that the required flux density does not cause the core to saturate. As suggested by Figure 14-1, many practical magnetic circuits contain air gaps. The presence of the air gap greatly increases the total reluctance of the magnetic circuit. Accordingly, leakage flux will be more pronounced in such circuits.

In the magnetic circuit of Figure 14-5, the total flux through the top section of the circuit is less than that through the center of the coil, due to the leakage flux bypassing the top section. The effect of the leakage flux, then, is that we must increase the current in the coil slightly in order to obtain as much flux in the top section of the core as there would be if there were no leakage flux. The ratio between the total developed flux and the useful flux is known as the **leakage factor.** Since the leakage flux is only a very small percentage of the total flux, we may quite safely neglect it in making our magnetic circuit calculations in this chapter if a specific value of leakage factor is not given.

14-7 SERIES MAGNETIC CIRCUITS

Practical magnetic circuits are seldom as simple as the iron ring of Figure 14-3. The magnetic circuit of Figure 14-5 consists of two sections, a straight section and a U-shaped section in *series*. The two sections are in series since tracing around the complete loop of a line of force takes us first through one section and then through the other. As we have already noted, the equation for reluctance in magnetic circuits resembles Ohm's law of electric circuits. Therefore, just as the total resistance of a series electric circuit is the sum of the individual resistances, it follows that

The total reluctance of a series magnetic circuit is the sum of all the individual reluctances.

For electric circuits:

$$R_T = R_1 + R_2 + R_3 + \text{etc.} \qquad (6\text{-}1)$$

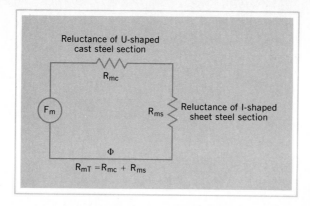

FIGURE 14-6
Schematic Representation of a Series Magnetic Circuit.

For magnetic circuits:

$$R_{mT} = R_{m1} + R_{m2} + R_{m3} + \text{etc.} \tag{14-2}$$

Graphic symbols are not provided for drawing "circuit diagrams" of magnetic circuits. However, since a series magnetic circuit can be compared with a series electric circuit, we might represent the series magnetic circuits of Figures 14-5 and 14-7 by the circuit diagram of Figure 14-6.

Using the relationship for total reluctance in a series magnetic circuit stated by Equation (14-2), we can solve the magnetic circuit of Example 14-5 by the first method that we used in solving Example 14-3. In many series magnetic circuits, the cross-sectional area of the various sections may not be the same, in which case we must find the flux density for each section. Also, the type of core material may vary from section to section.

EXAMPLE 14-5

What magnetomotive force must be developed by a coil wound on the magnetic circuit of Figure 14-7 to develop a total flux of 1.2×10^{-3} Wb? (The varnish on the laminations accounts for 10% of the lamination's thickness.)

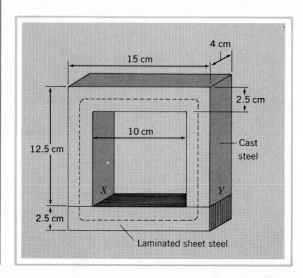

FIGURE 14-7
Magnetic Circuit for Example 14-5.

SOLUTION 1

Step I. Determine the reluctance of the U-shaped section.
Examination of the specifications shows that all three arms of the *cast*-steel section have the same width (2.5 cm).

$$\therefore A_c = 2.5 \text{ cm} \times 4 \text{ cm} = 10 \text{ cm}^2 = 10^{-3} \text{ m}^2$$

The *average* path is marked out by the dotted line in Figure 14-7. Its length can be obtained by averaging the outside length of the U-shaped section (12.5 + 15 + 12.5 = 40 cm) and the inside length (10 + 10 + 10 = 30 cm).

$$\therefore l_c = \frac{1}{2}(40 + 30) = 35 \text{ cm} = 0.35 \text{ m}$$

$$B_c = \frac{\Phi}{A_c} = \frac{1.2 \times 10^{-3} \text{ Wb}}{10^{-3} \text{ m}^2} = 1.2 \text{ T}$$

From the *BH* curves of Figure 14-4, for

$$B_c = 1.2 \text{ T in cast steel}$$

$$H_c = 1220 \text{ At/m}$$

$$\therefore \mu_c = \frac{B_c}{H_c} = \frac{1.2 \text{ T}}{1220 \text{ At/m}} = 9.84 \times 10^{-4} \text{ H/m}$$

$$R_{mc} = \frac{l}{\mu_c A_c} = \frac{35 \text{ cm}}{9.84 \times 10^{-4} \text{ H/m} \times 10^{-3} \text{ m}^2} = 3.557 \times 10^5 \text{ At/Wb}$$

Step II. Determine the reluctance of the I-shaped stack of laminations.

$$A_s = 2.5 \text{ cm} \times 4 \text{ cm} \times 90\% = 9 \text{ cm}^2 = 9 \times 10^{-4} \text{ m}^2$$

$$l_s = \tfrac{1}{2}(20 + 10) \text{ cm} = 15 \text{ cm} = 0.15 \text{ m}$$

$$B_s = \frac{\Phi}{A_s} = \frac{1.2 \times 10^{-3} \text{ Wb}}{9 \times 10^{-4} \text{ m}^2} = 1.333 \text{ T}$$

From the *BH* curves of Figure 14-4, for

$$B_s = 1.33 \text{ T in } sheet \text{ steel}$$

$$H_s = 900 \text{ At/m}$$

$$\therefore \mu_s = \frac{B_s}{H_s} = \frac{1.33 \text{ T}}{900 \text{ At/m}} = 1.48 \times 10^{-3} \text{ H/m}$$

$$R_{ms} = \frac{l_s}{\mu_s A_s} = \frac{15 \text{ cm}}{1.48 \times 10^{-3} \text{ H/m} \times 9 \times 10^{-4} \text{ m}^2} = 1.126 \times 10^5 \text{ At/Wb}$$

Step III. $\qquad R_{mT} = R_{mc} + R_{ms} = (3.557 \times 10^5) + (1.126 \times 10^5)$

$$= 4.683 \times 10^5 \text{ At/Wb}$$

and $\quad F_m = \Phi R_{mT} = 1.2 \times 10^{-3} \text{ Wb} \times 4.683 \times 10^5 \text{ At/Wb} = 562 \text{ At}$

Another characteristic of series electric circuits is that the total voltage drop is the sum of all the individual voltage drops. Similarly, in magnetic circuits,

The total magnetomotive force is the sum of the mmf's required for each section of the series magnetic circuit.

For electric circuits:

$$V_T = V_1 + V_2 + V_3 + \text{etc.} \tag{6-2}$$

For magnetic circuits:

$$F_{mT} = F_{m1} + F_{m2} + F_{m3} + \text{etc.} \tag{14-3}$$

Applying Equation (14-3) to Example 14-5, we may again completely bypass determination of permeability and reluctance in solving series magnetic circuits. Once we have determined the magnetic field intensity required for each section of the circuit, we can go right ahead and find the magnetomotive force required for that section.

SOLUTION 2
For the U-shaped section,

$$H_c = 1220 \text{ At/m} \qquad \text{(from Solution 1)}$$

$$\therefore F_{mc} = H_c l_c = 1220 \text{ At/m} \times 35 \text{ cm} = 427 \text{ At}$$

For the I-shaped section,

$$H_s = 900 \text{ At/m} \qquad \text{(from Solution 1)}$$

$$\therefore F_{ms} = H_s l_s = 900 \text{ At/m} \times 15 \text{ cm} = 135 \text{ At}$$

and $\qquad\qquad F_{mT} = F_{mc} + F_{ms} = 427 + 135 = 562 \text{ At}$

14-8 AIR GAPS

In all the magnetic circuits shown in Figure 14-1, the complete loop of any line of force is not wholly within the high permeability path provided by the iron. Every line of force has to cross an air gap to complete its loop. This gap is necessary in the case of motors, meters, and loudspeakers in order to provide space in which electric conductors can move. Even in stationary equipment like the magnetic circuit of Figure 14-7, an air gap may be inserted into the magnetic circuit at points X and Y between

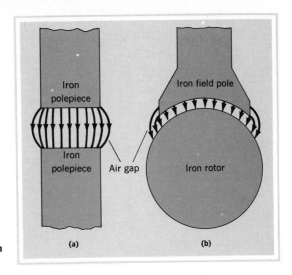

FIGURE 14-8
Fringing of Magnetic Lines of Force in an Air Gap.

(a)   (b)

the U-shaped and the I-shaped sections. This air gap increases the total reluctance of the circuit, to assist in preventing saturation of the iron portion of the magnetic circuit when a heavy current flows through the coil wound on the core. In such apparatus, the "air" gap is often a sheet of any nonmagnetic material such as fiberboard or brass, so that the length of the air gap can be accurately and rigidly set.

Since all the magnetic lines of force must pass through the air gaps in these circuits, the gaps are in *series* with the remainder of the circuit. Therefore, we can solve magnetic circuits with air gaps just like the series circuit of Example 14-5. Before we start to calculate the flux density in the air gap, we must note how the magnetic lines of force try to spread out as they pass through the air gap, as shown in Figure 14-8. This effect follows from the characteristic of magnetic lines of force in that they tend to repel one another. Since the air outside of the gap is just as capable of passing magnetic lines of force as the air within the gap, the lines of force take advantage of this opportunity to get farther apart as they cross the gap. This spreading apart of the magnetic lines of force as they cross an air gap is called **fringing.** Fringing results in the flux density in the air gap being slightly less than that in the adjacent iron sections of the magnetic circuit.

If the length of the air gap is small, the fringing is small, and we can make correction for it in magnetic circuit calculations by assuming that the cross-sectional dimensions of an air gap are greater than the cross-sectional dimensions of the adjacent iron by an amount equal to the length of the air gap.

EXAMPLE 14-6

What magnetomotive force must be developed by a coil wound on the magnetic circuit of Figure 14-7 to develop a total flux of 1.2×10^{-3} Wb if pieces of fiberboard 0.1 mm thick are inserted into the magnetic circuit at points X and Y? (Allow 10% of the lamination thickness for varnish.)

SOLUTION 1

Since the total flux is the same as in Example 14-5, the reluctance of the U-shaped and I-shaped steel sections will be the same as we have already calculated,

3.557×10^5 and 1.126×10^5 At/Wb, respectively. When we allow for fringing, the cross-sectional area of the air gap is

$$A_A = 2.51 \text{ cm} \times 4.01 \text{ cm} = 10.07 \text{ cm}^2 = 1.007 \times 10^{-3} \text{ m}^2$$

$$R_{mA} = \frac{l_A}{\mu_A A_A} = \frac{0.1 \text{ mm}}{4\pi \times 10^{-7} \text{ H/m} \times 1.007 \times 10^{-3} \text{ m}^2} = 7.9 \times 10^4 \text{ At/Wb}$$

Since each magnetic line of force has to pass through *both* air gaps, they are in *series*. Since both air gaps are identical, each has a reluctance of 7.9×10^4 At/Wb.

$$R_{mT} = (3.557 \times 10^5) + (1.126 \times 10^5) + 2(7.9 \times 10^4)$$

$$= 6.26 \times 10^5 \text{ At/Wb}$$

$$F_m = \Phi R_{mT} = 1.2 \times 10^{-3} \text{ Wb} \times 6.26 \times 10^5 \text{ At/Wb}$$

$$= 751 \text{ At}$$

SOLUTION 2

The mmf required for the iron will be the same as in Example 14-5, 562 At. For each air gap,

$$B_A = \frac{\Phi}{A_A} = \frac{1.2 \times 10^{-3} \text{ Wb}}{1.007 \times 10^{-3} \text{ m}^2} = 1.19 \text{ T}$$

$$H_A = \frac{B_A}{\mu_A} = \frac{1.19 \text{ T}}{4\pi \times 10^{-7} \text{ H/m}} = 9.47 \times 10^5 \text{ At/m}$$

$$F_{mA} = H_A l_A = 9.47 \times 10^5 \text{ At/m} \times 0.1 \text{ mm} = 94.7 \text{ At}$$

$$\therefore F_{mT} = 562 + (2 \times 94.7) = 751 \text{ At}$$

If we are given the magnetomotive force applied to a magnetic circuit which is made up of different sections in series, it is not quite so easy to solve for the total flux as it was in Example 14-4, because we do not know how much of the total mmf applies to each section. Although we examined the volt-ampere characteristics of nonlinear electric circuit elements such as the vacuum-tube and semiconductor diodes of Figure 4-9, we did not establish a procedure for solving series electric circuits containing nonlinear circuit elements. This topic is considered in courses on Electronic Circuits. Example 14-7 shows one method of tackling the problem in a series magnetic circuit. We start by making an educated guess at the answer and then solve the magnetic circuit by the procedures we used in Examples 14-5 and 14-6 to see if we arrive at the given value of magnetomotive force. If the first attempt is not close enough, we have to try again. This cut-and-try solution is included for general interest. We shall not pursue this type of solution further.

EXAMPLE 14-7

A current of 0.5 A is passing through a coil consisting of 960 turns wound on a cast-steel core whose cross-sectional dimensions are 5 cm $\times$ 2.5 cm and whose average path length is 30 cm. The continuity of this core is broken by a single air gap which is 0.25 mm in length. What is the total flux in the core?

SOLUTION
Area of air gap (allowing for fringing),

$$A_A = 5.025 \text{ cm} \times 2.525 \text{ cm} = 12.7 \text{ cm}^2$$

$$F_{mT} = NI = 960 \text{ t} \times 0.5 \text{ A} = 480 \text{ At}$$

Assume that, of this total mmf, 210 At is required for the air gap.

Then $\quad H_A = \dfrac{F_{mA}}{l_A} = \dfrac{210 \text{ At}}{0.25 \text{ mm}} = 8.4 \times 10^5 \text{ At/m}$

$$B_A = \mu_A H_A = 4\pi \times 10^{-7} \text{ H/m} \times 8.4 \times 10^5 \text{ At/m} = 1.056 \text{ T}$$

$$\therefore \Phi = B_A A_A = 1.056 \text{ T} \times 12.7 \text{ cm}^2 = 1.34 \times 10^{-3} \text{ Wb}$$

If this is so, the flux density in the iron must be

$$B_1 = \frac{\Phi}{A_1} = \frac{1.34 \times 10^{-3} \text{ Wb}}{12.5 \text{ cm}^2} = 1.07 \text{ T}$$

From the *BH* curve for cast steel (Figure 14-4), if

$$B_I = 1.07 \text{ T}$$

$$H_I = 885 \text{ At/m}$$

then $\qquad F_{mI} = H_I l_I = 885 \text{ At/m} \times 30 \text{ cm} = 266 \text{ At}$

$$\therefore F_{mT} = 210 + 266 = 476 \text{ At}$$

Since this is just slightly less than the given mmf, the total flux must be just slightly greater than $\mathbf{1.34 \times 10^{-3}}$ **Wb.**

14-9 PARALLEL MAGNETIC CIRCUITS

If we examine the magnetic circuits of Figure 14-1(c) and 14-1(d), we note that, although *all* the magnetic flux passes through some portions of the magnetic circuit, in other portions it splits, with half going one way and half going the other. In electric circuits, this is the condition that exists when two equal resistors are connected in *parallel*. Half the total current flows through each resistor and the *same* voltage drop appears across each branch. Designers have found that supplying more than one possible return path for magnetic lines of force appreciably reduces the leakage flux. Fortunately, in most designs, the parallel paths are limited to two symmetrical paths so that we can assume that the total flux splits equally.

EXAMPLE 14-8

If a 1000-turn coil is placed on the center leg of the shell-type core of Figure 14-9 to form a filter choke, what is the maximum current it can pass without the total flux in the core exceeding 2×10^{-3} Wb? The core is made of sheet-steel laminations stacked to a total thickness of 4 cm. (Allow 5% of the thickness for varnish.) The width of the outside legs of the core is 2.5 cm and that of the center leg is 5 cm. The average path length (see Figure 14-9) is 25 cm.

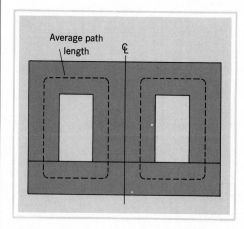

FIGURE 14-9
Magnetic Circuit for Example 14-8.

SOLUTION

We can divide the magnetic circuit down the centerline of Figure 14-9 and stack one section on top of the other without changing the total reluctance of the magnetic circuit. The cross-sectional area then becomes

$$A = 2 \times 2.5 \text{ cm} \times 4 \text{ cm} \times 95\% = 19 \text{ cm}^2$$

$$B = \frac{\Phi}{A} = \frac{2 \times 10^{-3} \text{ Wb}}{19 \text{ cm}^2} = 1.05 \text{ T}$$

From the *BH* curve for sheet steel (Figure 14-4), for

$$B = 1.05 \text{ T}$$

$$H = 460 \text{ At/m}$$

$$F_m = Hl = 460 \text{ At/m} \times 25 \text{ cm} = 115 \text{ At}$$

$$I = \frac{F_m}{N} = \frac{115 \text{ At}}{1000 \text{ t}} = 115 \text{ mA}$$

We could arrive at the same answer if we considered only *half* of the magnetic circuit accommodating only *half* of the total flux. The flux density in the iron would still be

$$B = \frac{\Phi}{A} = \frac{1 \times 10^{-3} \text{ Wb}}{9.5 \text{ cm}^2} = 1.05 \text{ T}$$

and again $F_m = 115$ At. This *same* mmf would cause the other 1×10^{-3} Wb to be set up in the other half of the magnetic circuit.

14-10 TRACTIVE FORCE OF AN ELECTROMAGNET

One of the characteristics of magnetic lines of force is that they represent a tension along their length which tends to make them as short as possible. If we pass current through the solenoid of the relay shown in Figure 14-10, a magnetic field will be set up in the iron. These magnetic lines of force must complete their loops via the air gap between the stationary pole piece and the moving arm of the relay. The tension along the length of the lines of force will, therefore, exert a force which attempts to close the air gap. In any magnetic circuit containing an air gap, a force will be exerted which tends to close the air gap. The force will depend on the number of lines of force crossing the air gap.

To determine the extent of this **tractive force,** let us assume that current is flowing in the solenoid of Figure 14-10 and that the air gap is closed. If we now apply a mechanical force F to the moving arm to pull the air gap open to a distance, l, we shall be doing mechanical work.

$$W = Fl$$

Since increasing the length of the air gap increases the total reluctance of the magnetic circuit, we must pass more current through the coil if we wish to maintain the same total flux. In doing so, we add extra energy to the magnetic circuit. If the total flux is kept constant, this extra energy is added to the flux in the air gap and will be the product of the flux in the air gap and the *average* mmf maintaining that flux in the air gap. As we force the air gap open, its length increases from zero to l, and, consequently, the magnetomotive force required to maintain flux in the air gap increases

FIGURE 14-10
Magnetic Circuit of a Simple Relay.

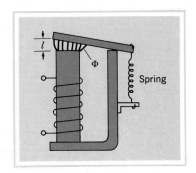

from zero to F_m. The average mmf will, therefore, have to be $\frac{1}{2}(0 + F_m) = \frac{1}{2}F_m$. The energy stored in the air gap will be

$$W = \tfrac{1}{2}F_m \Phi$$

But
$$F_m = Hl \quad \text{and} \quad \Phi = BA$$

$$\therefore W = \tfrac{1}{2}HlBA$$

But
$$H = \frac{B}{\mu}$$

$$\therefore W = l \times \frac{B^2 A}{2\mu}$$

This extra energy appearing in the air gap must equal the work done in pulling the moving arm away from the pole piece. Therefore,

$$Fl = l \times \frac{B^2 A}{2\mu}$$

and
$$F = \frac{B^2 A}{2\mu} \tag{14-4}$$

where F is the force acting to close the air gap in newtons, B is the flux density in teslas (webers per square meter), A is the cross-sectional area of the air gap in square meters, and μ is the permeability of air ($4\pi \times 10^{-7}$ henry per meter).

EXAMPLE 14-9

Figure 14-11 shows the cross section of an electromagnet used for lifting iron plates. The cross-sectional areas of both the center pole piece and the circular outside pole piece are 800 cm² each. The energizing solenoid produces a flux density of 1.2 T in each of the pole pieces. What is the maximum mass of boiler plate that can be supported by the magnet against the force of gravity?

SOLUTION

There are two air gaps in the complete magnetic circuit of Figure 14-11—one where the lines of force pass from the pole piece of the magnet into the boiler plate, and one where the lines of force pass from the plate back into the magnet. The total tractive force will be the sum of these two individual pulls. Therefore,

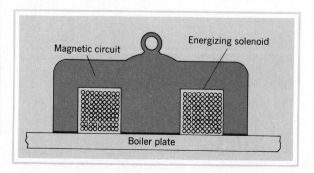

FIGURE 14-11
Cross Section of an Electromagnet.

$$F = 2 \times \frac{B^2A}{2\mu} = \frac{(1.2 \text{ T})^2 \times 800 \text{ cm}^2}{4\pi \times 10^{-7} \text{ H/m}} = 91\,670 \text{ N}$$

and $\quad m = \dfrac{F}{a} = \dfrac{91\,670 \text{ N}}{9.8 \text{ m/s}^2} = 9354 \text{ kg} = \textbf{9.35 Mg} \qquad$ (or tonnes)

For further examples of magnetic circuits, we may consider the electric meter movements of Section 10-1.

PROBLEMS

14-1. What flux will be developed at the center of an air-core coil passing a 600-mA current if its 750 turns occupy a length of 12 cm on a 1-cm diameter form?

14-2. A coil of wire 8 cm in length is wound on a $\frac{1}{2}$ cm diameter hardwood dowel. If there are 500 turns in the coil, what current must it pass to develop a flux of 5.0×10^{-7} Wb at the center of the coil?

14-3. How many turns of wire must be wound on a 2-cm-diameter Bakelite rod to produce 8×10^{-7} Wb at the center of the coil from a 500-mA current through the coil? The length of the coil is 20 cm.

14-4. The core for a low-loss inductor in a crossover filter is made from a piece of flat plastic stock 6 mm thick by 24 mm wide by 15 cm long formed into a closed loop; 1200 turns of wire are wound toroid-fashion on this form. What current must flow in the coil to develop a flux of 10^{-5} Wb in the core?

14-5. If the core in Problem 14-4 were a cast-iron ring with the same dimensions, what must the current be to produce a flux of 10^{-4} Wb?

14-6. The magnetic circuit of Figure 14-12 is made from a sheet-steel bar 15 cm long with a 12-mm × 20-mm rectangular cross section. It is bent into a ring, leaving an air gap 0.25 mm long. What current must flow through the 800-turn coil wound on this core in order to produce a total flux of 3×10^{-4} Wb in the air gap?

14-7. The horseshoe-shaped cast-steel electromagnet of Figure 14-13 is 3.0 cm in diameter and has an average path length of 30 cm. What current must flow through the windings of the electromagnet to develop a total flux of 1 mWb in the sheet-steel bar held by the magnet? the dimensions of the bar are 15 × 4 × 2 cm.

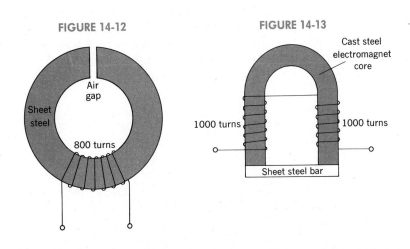

FIGURE 14-12

FIGURE 14-13

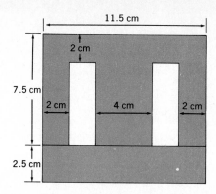

FIGURE 14-14

14-8. What is the total flux in the sheet-steel bar if the current in the coils of the electromagnet of Figure 14-13 is 0.5 A?

14-9. If a sheet of aluminum 1.5 mm thick is placed between the sheet-steel bar and the electromagnet of Figure 14-13, what current must flow in the coils to develop a total flux of 7.5×10^{-4} Wb in the sheet-steel bar if the leakage factor has now become 1.15?

14-10. If the sheet-steel laminations of Figure 14-14 are stacked to a total thickness of 2.5 cm, how many turns of wire must be wound on the center leg in order for a current of 100 mA to develop a total flux of 1.2×10^{-3} Wb in the center leg? Allow 5% of each lamination thickness for varnish. Assume no air gap and negligible leakage flux.

14-11. What force is exerted in the steel ring in Problem 14-6 as the magnetic field attempts to close the air gap?

14-12. What force is required to pull the sheet-steel bar away from the electromagnet in Problem 14-7?

REVIEW QUESTIONS

14-1. On what basis can we make the assumption that the total reluctance of a long air-core coil is approximately equal to the reluctance of the nonmagnetic core of the coil itself?

14-2. Describe a **torroid**-type winding on a magnetic circuit. What is its advantage?

14-3. What causes **leakage flux** to occur in magnetic circuits? State two means of reducing leakage flux in magnetic circuit design.

14-4. What characteristic of magnetic circuits allows us to calculate the total magnetomotive force required in a circuit made up of several different sections?

14-5. Many practical magnetic circuits contain an **air gap** in the form of a sheet of plastic or brass. Why do such materials constitute air gaps?

14-6. In an all-iron magnetic circuit, the relationship between the applied magnetomotive force and the resulting total magnetic flux is quite nonlinear, as indicated by the graphs of Figure 14-4. By purposely including an air gap in a magnetic circuit, there is a more linear relationship between applied mmf and total flux. Explain.

14-7. What causes **fringing** of the magnetic lines of force in an air gap?

14-8. If the two parallel paths for magnetic lines of force in the magnetic circuit of Figure 14-9 were not symmetrical, what complication would be introduced into our solution of an example such as Example 14-8?

14-9. Why does the tractive force increase as an iron bar is placed closer to an electromagnet with a constant current through its exciting coils?

14-10. Mechanical work is required to pull a soft-iron bar away from a U-shaped permanent magnet. Where does the energy go?

15

INDUCTANCE

In Chapter 11, we discovered that capacitance in an electric circuit depends on electric fields. Similarly, the third property of electric circuits, **inductance,** depends on the interaction of *magnetic fields* with electric circuits.

15-1 ELECTROMAGNETIC INDUCTION

We have already encountered the interaction of magnetic fields with electric circuits in defining the SI unit of magnetic flux—the weber. In Section 13-5, we noted Faraday's discovery of electromagnetic induction as he thrust a permanent magnet into a solenoid (Figure 13-13). Figure 15-1 shows another arrangement for demonstrating Faraday's discovery. As the conductor is moved downward through the magnetic lines of force, the galvanometer pointer swings one way from its center zero. As the conductor is moved upward, the galvanometer pointer swings in the opposite direction. But when there is no motion, there is no pointer deflection.

For current to flow through the galvanometer, there must be some voltage source in the loop consisting of the galvanometer and the conductor in the magnetic field. And, as experiment has shown, this voltage appears only when the conductor is *cutting across* the magnetic lines of force. Moving the conductor parallel to the lines of force will produce no deflection of the galvanometer pointer. In Figure 15-1, the conductor is moved while the magnetic field remains stationary; whereas, in Figure 13-13, the magnetic field is moved while the electric conductor remains stationary. In both cases, there is a cutting across magnetic lines of force by an electric conductor, and, in both cases, a voltage appears in the conductor. The generation of a voltage when an electric

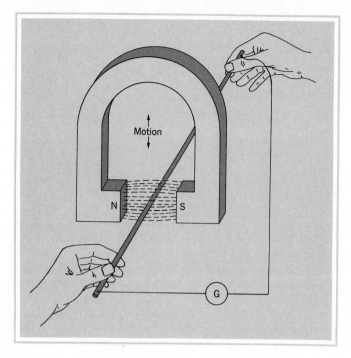

FIGURE 15-1
Demonstrating Electromagnetic Induction.

conductor cuts across magnetic lines of force is called **electromagnetic induction,** and the resulting voltage is called an **induced** voltage.

Another method of demonstrating electromagnetic induction is shown in Figure 15-2. In this example, *relative* motion between the magnetic lines of force and the electric conductor is achieved without mechanical motion of either the magnet or the electric conductor. As we reduce the resistance of the rheostat, the current in the left-hand **(primary)** winding increases. This causes an increase in the number of magnetic lines of force appearing in the iron core. Since the right-hand **(secondary)** winding is wound on the same core, the number of lines of force passing through the winding is increasing, and the deflection of the galvanometer pointer indicates that a voltage is being induced into the secondary winding. Since most of the magnetic lines of force are confined to the iron core, it is more usual for us to think of the flux and

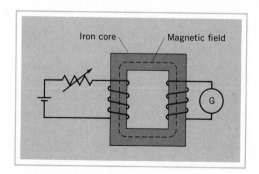

FIGURE 15-2
Demonstrating Mutual Induction.

384

linking the secondary winding rather than as cutting across the turns of the secondary winding.

When we increase the resistance of the rheostat, the current becomes smaller and the number of lines of force linking the secondary becomes less. At the same time, the pointer of the galvanometer swings in the opposite direction, indicating that the induced voltage has the opposite polarity. But as long as the current in the primary is steady, there is no deflection of the galvanometer pointer. The generation of a voltage in a secondary winding by a *changing* current in a primary winding is called **mutual induction.**

15-2 FARADAY'S LAW

One observation we can note in the demonstration of mutual induction in Figure 15-2 is that the faster the flux in the core builds up or collapses, the greater the deflection of the galvanometer pointer. When we consider the reason for this, we come to the conclusion which is referred to as

Faraday's law: The voltage induced in an electric circuit is proportional to the time rate of change of the flux of magnetic induction linked with the circuit.

We have already used Faraday's law in Section 13-5 to establish the SI unit of magnetic flux—the *weber*. To express *rate of change* of flux numerically, we adopt the convenient calculus derivative form which we discussed on page 301. Hence, from the definition of the weber,[†]

$$\text{voltage per turn (V)} = \frac{d\phi}{dt}\ (\text{Wb/s})$$

If the secondary winding in Figure 15-2 has more than one turn, the *same* changing flux links all turns and thus induces identical voltages in each turn. Since these "voltages per turn" are all in series, the total induced voltage is N times the voltage per turn. Hence, We may express Faraday's law in equation form as

$$e_T = N\frac{d\phi}{dt} \tag{15-1}$$

where e_T is the instantaneous value of the induced voltage between the terminals of the winding in volts, N is the number of turns in the winding, and $d\phi/dt$ is the rate of change of magnetic flux in webers per second.

Combining Equations (13-2) and (13-1),

$$\Phi = \frac{F_m}{R_m} = \frac{NI}{R_m}$$

[†]Note the use of lowercase ϕ to indicate *changing* flux.

The number of turns in the primary winding is fixed; and if we keep the flux density below the saturation point of the iron core, the reluctance of the magnetic circuit is reasonably constant. Therefore, the *rate of change of flux* is directly proportional to the *rate of change of current*. When we are dealing with electromagnetic induction due to a changing current rather than mechanical motion, we may restate Faraday's law:

The magnitude of the induced voltage is directly proportional to the rate of change of current.

15-3 LENZ'S LAW

We have already noted that the polarity of the induced voltage depends on the direction in which the conductor is moved across the magnetic lines of force as in Figure 15-1, or whether the current in the primary is increasing or decreasing as in Figure 15-2. We can apply a law stated by Heinrich Lenz to determine the exact polarity of the induced voltage. Lenz reasoned that the process of induction must conform to the well-known principle in physics that reaction is equal and opposite to action. Of several ways of stating Lenz's law, we shall use the following form:

Lenz's law: The polarity of the induced voltage must be such that any current resulting from it will develop a flux which tends to oppose any change in the original flux.

When we first close the switch in the mutual induction demonstration of Figure 15-3(a), the primary flux will increase from zero. Since a resistor is connected across the secondary winding to form a complete closed circuit, current will flow in the secondary winding when a voltage is induced in it by the rising primary flux. According to Lenz's law, the flux produced by this secondary current must try to oppose the increase in primary flux. Applying our right-hand rule to the primary winding, we find that the primary flux will have a clockwise direction around the core. To oppose an increase in this clockwise primary flux, the flux produced by the secondary current

FIGURE 15-3
Determining the Direction of an Induced Voltage by Lenz's Law.

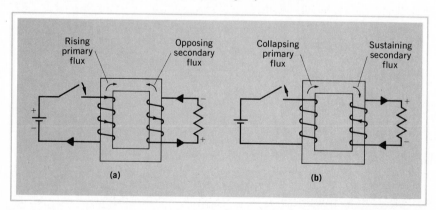

PART III   CAPACITANCE AND INDUCTANCE

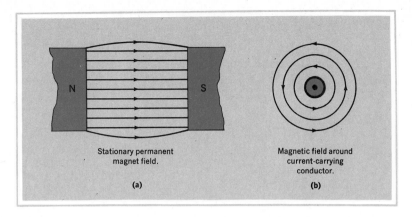

Stationary permanent
magnet field.

(a)

Magnetic field around
current-carrying
conductor.

(b)

FIGURE 15-4
**Component Magnetic Fields for the Electromagnetic Induction Demonstration
of Figure 15-1.**

must have a *counterclockwise* direction. Again applying the right-hand rule, this time
to the secondary winding, we can establish the direction in which the secondary
current must flow through the resistor, and thus determine the polarity of the induced
voltage in terms of the polarity of the voltage drop across the resistor.

When we open the switch, as in Figure 15-3(b), the primary current stops
flowing and the primary flux collapses. According to Faraday's law, this reduction in
flux will induce a voltage into the secondary, and, according to Lenz's law, the
secondary current must try to oppose the collapse of the primary flux. To help sustain
the primary flux, which has a clockwise direction in the core, the secondary current
must also produce a clockwise flux direction. Applying the right-hand rule, we obtain
the polarity shown in Figure 15-3(b). The polarity of the induced voltage when the
primary current is decreasing is, therefore, opposite to that induced when the primary
current is increasing.

We can also use Lenz's law to determine the direction of the induced voltage in
the demonstration of Figure 15-1. The galvanometer completes the circuit so that
current will flow in the conductor as we move it through the magnetic field. Current
in the conductor will, in turn, produce a magnetic field around the conductor. If we
consider the two magnetic fields separately, the field of the stationary magnet has the
form shown in Figure 15-4(a). If we assume for the moment that the conventional
current direction in the moving conductor is out of the page, the magnetic field around
the conductor has the form shown in Figure 15-4(b). Before we place the conductor
between the magnetic poles in the sketch and superimpose the two magnetic fields, we
must recall that magnetic lines of force can never intersect. Therefore, the two
component fields must combine to form a single composite magnetic field pattern, as
shown in Figure 15-5.

Above the conductor [Figure 15-4(b)], the direction of the magnetic lines of
force is *opposite* to that of the permanent magnet field [Figure 15-4(a)]. Consequently,
there is a cancelling effect which bends the resultant magnetic field away from the
conductor, as shown in Figure 15-5. Below the conductor, the direction of its magnetic
field is the *same* as that of the permanent magnet field. As a result, the flux density

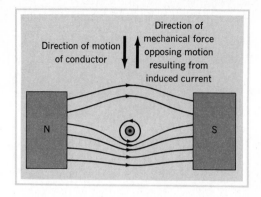

FIGURE 15-5
Using Lenz's Law to Determine the Direction of the Current Produced by an Induced Voltage.

is increased below the conductor in the composite flux pattern of Figure 15-5. The net effect of a conventional current out of the page in the conductor in Figures 15-1 and 15-5 is to warp the magnetic field produced by the permanent magnet so that the magnetic lines of force in the vicinity of the conductor *detour* below the conductor.

Another characteristic of magnetic lines of force is that they tend to become as short as possible and tend to repel one another. Therefore, the lines of force in the composite magnetic field of Figure 15-5 will attempt to straighten out and to space themselves uniformly, in order to regain the flux pattern of Figure 15-4(a). To accomplish this, the magnetic field will attempt to force the current-carrying conductor to move upward. According to the principle stated in Lenz's law, this force will *oppose* the actual motion of the conductor through the stationary magnetic field. Consequently, when the electric conductor in the demonstration shown in Figure 15-1 moves downward through the stationary magnetic field, the voltage induced into the conductor has a polarity such as to cause the current in the closed loop to have a direction out of the page.

15-4 SELF-INDUCTION

A voltage induced by a *changing* current is not restricted to mutual induction of a voltage in a secondary winding by current changing in a primary winding. When we first close the switch applying a voltage to the simple two-turn coil in Figure 15-6, the increasing flux produced by the rising current in turn A will induce a voltage into turn B. Similarly, the same rising current in turn B will induce a voltage into turn A. These voltages are in series and, according to Lenz's law, both will have a polarity such as

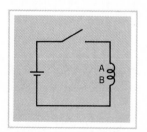

FIGURE 15-6
Self-Induction.

to oppose the increasing flux and the rising current. From Faraday's law equation (15-1), the total induced voltage again is $N \, d\phi/dt$. The generation of a voltage in an electric circuit by a *changing* current in the *same* circuit is called **self-induction.**

As we discovered in Section 2-8, generation of a potential difference requires a transfer of energy in order to perform the separation of electric charge—placing a surplus of electrons at one terminal and a deficiency at the other. This energy conversion constitutes an electromotive force in joules per coulomb, since it is responsible for generating the induced voltage. In older terminology, since the induced voltage has a polarity such as to *oppose* the force that creates it, the self-induced voltage is sometimes called a **counter emf,** or **cemf.** Since we can only measure the final induced voltage, we are inclined to forget the steps in its generation. When current is drawn from the voltage source in Figure 15-6, the voltage source is providing energy to build up a magnetic field around the coil. The rising magnetic field induces an emf *in* the turns of the coil. The emf, in turn, through electric charge-separation produces a potential difference *across* the coil. In modern terminology, we call this potential difference an **induced voltage.**

When we open the switch in Figure 15-6, the magnetic field collapses and again induces an emf in turns A and B. In this case, the polarity of the self-induced voltage is reversed as the coil tries to maintain current flow by having the polarity of its induced voltage such that it adds to the source voltage, thus raising the total voltage in an attempt to maintain current flow in the circuit.

15-5 SELF-INDUCTANCE

From the foregoing discussion, we find that the simple electric circuit of Figure 15-6 has the property of tending to oppose any *change* in current through it (or in the magnetic flux linking it). This property is called its **self-inductance.** Therefore,

Self-inductance is the property of an electric circuit that opposes any change in current in that circuit.

Because it will be quite apparent from the circuit diagram whether the induction action in question is *self*-induction or *mutual*-induction, it is customary to omit the *self* and simply speak of the **inductance** of an electric circuit. And since inductance is a property of electric circuits, we shall require a letter symbol to represent inductance and a unit of measurement.

The letter symbol for inductance is L.
The henry is the SI unit of inductance.[†]
The unit symbol for henry is H.

In the International System of Units, since we define the *ampere* in terms of the magnetic force between parallel current-carrying conductors, and the *weber* in terms

---

[†]Named in honor of the pioneer American physicist Joseph Henry.

of the voltage induced by a changing current, magnetic units are completely compatible with electrical units. We can define the unit of inductance either in terms of the change in flux resulting from a given change in current or in terms of the induced voltage resulting from a given change in current. The general definition of the henry is based on flux linkages. Hence,

An electric circuit has an inductance of one henry when a change in current of one ampere produces a change in total flux linkages of one weber.

As we noted in Section 13-6, the total flux linkages of a multiturn coil are N times as great as for a single turn carrying the same current. Hence, the defining equation for the henry becomes

$$L = N\frac{d\phi}{di} \tag{15-2}^{\dagger}$$

where L is the inductance of a circuit in henrys, N is the number of turns linked by the magnetic flux, and $d\phi/di$ is the change in flux for a given change in current in webers per ampere.

Rearranging Equation (15-1) gives

$$N = \frac{e_T}{d\phi/dt}$$

and rearranging Equation (15-2) gives

$$N = \frac{L}{d\phi/di}$$

from which

$$\frac{L}{d\phi/di} = \frac{e_T}{d\phi/dt}$$

and

$$L = e_T \times \frac{d\phi}{di} \times \frac{dt}{d\phi} = e_T \times \frac{dt}{di}$$

or

$$L = \frac{e_L}{(di/dt)} \tag{15-3}$$

where L is the inductance of a circuit in henrys, e_L is the voltage induced into the circuit in volts by current changing at the rate of di/dt amperes per second.

$\dagger$Since we have just defined the henry in terms of webers per ampere, we can now appreciate how we can express magnetic permeance (webers per effective ampere) in henrys, magnetic reluctance (effective amperes per weber) in *reciprocal* henrys, and magnetic permeability in henrys per meter on pages 346 and 347.

Thus we can also define the magnitude of the henry completely in electrical units without having to determine the magnetic flux.

An electric circuit has an inductance of one henry when current changing at the rate of one ampere per second induces a voltage of one volt into that circuit.

15-6 FACTORS GOVERNING INDUCTANCE

Even a straight wire possesses inductance. As the current through a conductor increases, magnetic lines of force first appear as tiny loops at the center of the conductor and then expand outward. As they expand, they cut across the copper of the conductor, thus inducing a voltage into the conductor. This induced voltage produced by the changing current is the evidence of the presence of inductance in the circuit. However, the inductance of a straight wire is so small that we can neglect it, except at very high radio frequencies.

If we wish to construct a circuit component in which the inductance effect is very pronounced, we would wind the conductor into a coil of many turns. From Equations (13-1) and (13-2),

$$\Phi = \frac{F_m}{R_m} = \frac{NI}{R_m}$$

Dividing both sides of the equation by t,

$$\frac{\Phi}{t} = \frac{N}{R_m} \times \frac{I}{t}$$

from which

$$\frac{d\phi}{dt} = \frac{N}{R_m} \times \frac{di}{dt}$$

Substituting in Equation (15-2) gives

$$e_T = \frac{N^2}{R_m} \times \frac{di}{dt}$$

And substituting this value for induced voltage into Equation (15-3),

$$L = \frac{N^2}{R_m} \qquad (15\text{-}4)$$

Hence, the inductance of a coil of wire is dependent on the number of turns and the reluctance of the magnetic circuit on which the coil is wound. A circuit component that is constructed for the express purpose of displaying the property of inductance is called an **inductor.** As Equation (15-4) indicates, we can greatly increase the inductance of an inductor by using an iron core to reduce the reluctance of the magnetic circuit. And since doubling the number of turns of wire in the coil of Figure 15-6 not only doubles the flux linking the coil for a given current (thus doubling the induced voltage per turn), but also doubles the number of turns that this flux links, the total induced voltage (and consequently the inductance) is proportional to the *square* of the number of turns.

Finally, substituting for the reluctance of the magnetic circuit of the inductor from Equation (14-1),

$$L = \frac{N^2 \mu A}{l} \tag{15-5}$$

where L is the inductance of the inductor in henrys, N is the number of turns in the coil, μ is the permeability of the magnetic circuit in henrys per meter, A is the cross-sectional area of the magnetic circuit in square meters, and l is the length of the magnetic circuit in meters (length of the coil for air-core coils).

In applying Equation (15-5) to practical inductors, we must realize that it is based on *all* the flux linking *all* the turns. In the case of iron-core coils and toroids, this equation is reasonably accurate. But, for long air-core coils where there is appreciable leakage flux, empirical formulas for inductance have been developed. We find these in electrical and radio handbooks.

EXAMPLE 15-1

2000 turns of wire are wound toroid-fashion on a cast-steel ring whose cross-sectional area is 2.25 cm^2, whose average path length is 42.4 cm, and whose permeability is 7.6×10^{-4} henrys per meter. (See Example 14-3.) What is the inductance of the toroid?

SOLUTION

$$L = \frac{N^2 \mu A}{l} = \frac{2000^2 \times 7.6 \times 10^{-4} \text{H/m} \times 2.25 \times 10^{-4} \text{ m}^2}{0.424 \text{ m}}$$

$$= 1.61 \text{ H}$$

(Note how dimensional units cancel.)

15-7 INDUCTORS IN SERIES

If we connect two inductors in series, as in Figure 15-7(a), the same current flows through both, and therefore, both will be subject to the same rate of change of current. For the present, we shall assume that the two inductors are physically located so that the magnetic field of one cannot induce a voltage into the other. Since inductance is directly proportional to the induced voltage, such mutual induction would alter the apparent self-inductance of the inductors. A changing current in the circuit of Figure

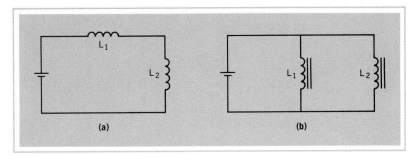

FIGURE 15-7
(a) Air-Core Inductors in Series; (b) Iron-Core Inductors in Parallel.

15-7(a) will induce a voltage of e_1 in L_1 and e_2 in L_2. The total induced voltage in the series circuit is

$$e_T = e_1 + e_2$$

From our second definition of the henry,

$$L_T = \frac{e_T}{di/dt} = \frac{e_1 + e_2}{di/dt} = \frac{e_1}{di/dt} + \frac{e_2}{di/dt} = L_1 + L_2$$

Therefore,

When there is no mutual coupling between inductors in series, the total inductance is the sum of the individual inductances.

$$L_T = L_1 + L_2 + L_3 + \text{etc.} \tag{15-6}$$

Note that the equation for inductors in series is similar to the equation for the total resistance of resistors in series.

15-8 INDUCTORS IN PARALLEL

If we now connect two inductors in parallel, as in Figure 15-7(b), each must have the same potential difference between its terminals. Since this voltage is produced by self-induction, the current in L_1 must change at the rate of di_1/dt, and current in L_2 must change at the rate of di_2/dt. Therefore, the total current must change at the rate of $(di_1 + di_2)/dt$.

$$\therefore L_{eq} = \frac{e_L}{\dfrac{di_1 + di_2}{dt}}$$

To simplify this equation, we invert both sides.

$$\frac{1}{L_{eq}} = \frac{\dfrac{di_1 + di_2}{dt}}{e_L} = \frac{\dfrac{di_1}{dt} + \dfrac{di_2}{dt}}{e_L} = \frac{\dfrac{di_1}{dt}}{e_L} + \frac{\dfrac{di_2}{dt}}{e_L}$$

$$\therefore \frac{1}{L_{eq}} = \frac{1}{L_1} + \frac{1}{L_2}$$

Therefore,

The equivalent inductance of inductors in parallel is

$$L_{eq} = \frac{1}{1/L_1 + 1/L_2 + 1/L_3 + etc.} \tag{15-7}$$

For only two inductors in parallel, this reduces to

$$L_{eq} = \frac{L_1 \times L_2}{L_1 + L_2} \tag{15-8}$$

Note again that we derived Equations (15-6), (15-7), and (15-8) on the assumption that no mutual induction takes place between the individual inductors. We shall leave these more complex calculations until Section 26-6.

PROBLEMS

15-1. Current changing at the rate of 200 mA/s in an inductor induces in it a voltage of 50 mV. What is its inductance?

15-2. What inductance must the collector load of a transistor have if collector current rising at a uniform rate from 2 mA to 7 mA in 50 μs induces in it a voltage of 20 mV?

15-3. What voltage will be induced into a 10-H inductor in which the current changes from 10 A to 7 A in 90 ms?

15-4. How long will it take the current in a 2-H inductor to rise from zero to 6 A if the voltage induced into the inductor is constant at 3 V?

15-5. If we can assume no leakage flux, what is the approximate inductance of an air-core coil of 20 turns with an inside diameter of 2 cm and a length of 2 cm?

15-6. How many turns must we wind on an air-core coil 3 cm long and 3 cm in diameter to obtain an inductance of 20 μH?

15-7. What is the inductance of the inductor in Example 14-7 of Chapter 14?

15-8. What is the inductance of the filter choke of Problem 14-10 of Chapter 14?

15-9. A 2-H inductor has 1200 turns. How many turns must we add to raise its inductance to 3 H?

15-10. A 500-mH inductor has 750 turns. Where must we place a tap to obtain an inductance of 250 mH between one end and the tap? What inductance is available if we use the other end and the tap?

15-11. Three inductors whose inductances are 30 mH, 40 mH, and 50 mH, respectively, are located so that there is no mutual induction between them. What is the effective inductance when the three coils are connected (a) in series and (b) in parallel?

15-12. What inductance must be placed in parallel with an inductance of 60 μH to reduce the net inductance to 15 μH?

REVIEW QUESTIONS

15-1. What is meant by **electromagnetic induction?**

15-2. What factors govern the magnitude of the emf induced into an electric conductor?

15-3. What factors govern the direction of the emf induced into an electric conductor?

15-4. Mark the conventional current direction in the sketches of Figure 15-8. The arrows show the direction of motion.

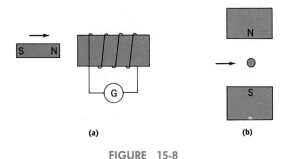

FIGURE 15-8

15-5. Mark the conventional current direction in both primary and secondary windings of Figure 15-9
(a) As the switch is first closed.
(b) As the switch is opened.

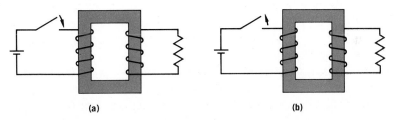

FIGURE 15-9

15-6. What is meant by **self-induction?**

15-7. How is the unit of inductance derived from previously established units?

15-8. Why is the self-inductance of an inductor proportional to the *square* of the number of turns?

15-9. Why does the inductance of an iron-core coil depend on the current through it?

15-10. What effect will the addition of an air gap to the magnetic circuit of a choke coil have on its inductance?

15-11. Why was it necessary to stipulate that there must be no magnetic coupling between the inductors in establishing the equation $L_T = L_1 + L_2 + L_3 +$ etc. for inductors in series?

15-12. Why is the total inductance of series inductors equal to the sum of the individual inductances?

16

INDUCTANCE IN DC CIRCUITS

16-1 CURRENT IN AN IDEAL INDUCTOR

To start our investigation of the behavior of inductance in dc circuits, we assume that the inductor of Figure 16-1 has no resistance. Such a situation is hypothetical since, in practice, there is no conductor material which has absolutely no resistance at room temperature. However, it will help us to determine what happens when we close the switch in Figure 16-1 if, for the moment, we do consider the resistance to be zero.

According to Kirchhoff's voltage law, all the time that the switch is closed in an electric circuit, a voltage drop exactly equal to the applied voltage must appear across the external circuit. If this circuit contains only *resistance,* this voltage drop is produced by current flowing through the resistance. Therefore, when we close the switch, the current in a resistance circuit must *instantly* assume a magnitude of $I = V/R = E/R$. But if there is no resistance in the circuit (as we have assumed in Figure 16-1), none of this voltage drop can be made up of IR drop.

In Section 15-4, we discovered that such a voltage is the result of an emf induced into the coil by a *changing* current in the coil. In Equation (15-3) we represented the induced voltage by the letter symbol e_L. Using e rather than v indicated a voltage resulting from an induced emf, and the lowercase e indicated that the magnitude of the voltage depended on the *rate of change* of current. However, we do not think of the inductor as a voltage source in Figure 16-1. Since the voltage between the terminals of an inductor exists only when there is a *changing* current in the coil, this induced voltage becomes the voltage drop required by Kirchhoff's voltage law. Thus, we can rewrite Equation (15-3) as

396

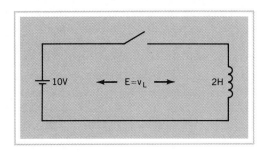

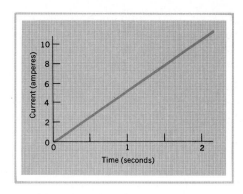

FIGURE 16-1
Simple Electric Circuit Containing Inductance.

$$v_L = L \times \frac{di}{dt} \qquad (16\text{-}1)$$

We can determine the current that must flow when we close the switch in Figure 16-1 by rearranging Equation (16-1) as

$$\frac{di}{dt} = \frac{v_L}{L}$$

Since the applied voltage is constant, the voltage across the inductor must also be constant, which allows us to represent it by the uppercase V_L. And since inductance depends on the number of turns and the reluctance of the magnetic circuit, the inductance of a given air-core coil is also constant. As a result, the V/L ratio is a constant. Hence, the rate of change of current must also be constant, which allows us to drop the d operator, giving

$$\frac{I}{t} = \frac{V}{L}$$

from which

$$I = \frac{V}{L} \times t$$

For the numerical values given in Figure 16-1,

$$I = \frac{10\ V}{2\ H} \times t = 5t\ A$$

Therefore, 0.5 second after we close the switch, the current will be 2.5 amperes; 1 second after we close the switch, it will be 5 amperes; at $t = 2$ s, the current must be 10 amperes; and so on, as shown by the graph of Figure 16-2.

Since the induced voltage in this case must remain constant, the rate of change of current must be a constant and, therefore, the graph of Figure 16-2 is a straight line. The *slope* of this graph of current plotted against time represents the *rate of change of current*. If we reduce the inductance of the circuit, the current must rise more rapidly in order to generate the required voltage. If the inductance is very small, the current must rise almost instantly to an infinitely high value. As the graph of Figure 16-2 indicates, the presence of appreciable inductance in the circuit acts to oppose this sudden change in current from zero to a very high value by forcing the current to rise at a slower, steady rate governed by the ratio V/L.

16-2 RISE OF CURRENT IN A PRACTICAL INDUCTOR

Because we do not have at our disposal any wire which has zero ohms resistance at room temperature, all practical inductors possess some resistance as well as inductance. Since there is only the one path for charge carriers to flow through, we must consider the practical inductor as consisting of inductance and resistance in *series,* as shown in Figure 16-3.

When we close the switch in the circuit of Figure 16-3, the voltage drop across the resistance of the coil plus the voltage drop across the inductance of the coil (resulting from the cemf generated in it by the rising current) must equal the applied voltage in order to satisfy Kirchhoff's voltage law. Therefore,

$$E = v_R + v_L$$

From Ohm's law,

$$v_R = iR$$

and from Equation (16-1),

$$v_L = L \frac{di}{dt}$$

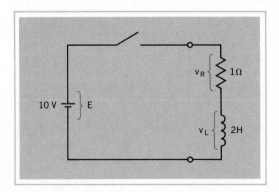

FIGURE 16-3
Simple Electric Circuit Containing a Practical Inductor.

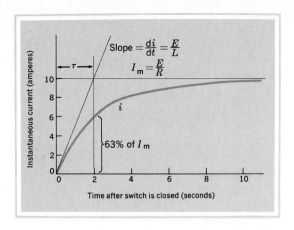

FIGURE 16-4
Rise of Current in a Direct-Current Circuit Containing Inductance and Resistance.

Note that we are using lowercase letters for current and voltage because, as we shall discover shortly, they change with time immediately after the switch is closed.

Substituting for v_R and v_L in the Kirchhoff's-law equation,

$$E = iR + L \frac{di}{dt} \qquad (16\text{-}2)$$

If we can arrange this equation to solve for the instantaneous current in the circuit, we can then determine the voltage drop across the resistance from Ohm's law. Subtracting v_R from E then gives us the inductive voltage drop. However, solving Equation (16-2) for i involves rather more than basic algebra.[†] Therefore, we shall set this equation aside for the moment and see if we can estimate the circuit behavior from our prior investigation of the rise in current in the circuit of Figure 16-1.

The inductance of the practical inductor and the applied voltage in Figure 16-3 are the same as for the theoretical inductor of Figure 16-1. Therefore, the current cannot rise at a more rapid rate than that shown in the graph of Figure 16-2, otherwise the induced voltage of the inductance alone would be greater than the applied voltage. Since there was no resistance in the circuit of Figure 16-1, there was no limit to the value of the instantaneous current. But, in the case of the practical inductor, it is not possible for the current to rise to a value such that the IR drop across the resistance exceeds the applied voltage, otherwise Kirchhoff's law again would not hold true. The greatest value that the instantaneous current can have is such that $v_R = E$. When this occurs, $v_L = 0$, and for the voltage across the inductance to be zero, $di/dt = 0$, and, therefore, the current must be a *steady* value which allows us to use the uppercase letter symbol;

$$\therefore I_m = \frac{E}{R} \qquad (16\text{-}3)$$

These two limits on the graph of the instantaneous current in a practical inductor are shown by the shaded lines in Figure 16-4.

Inductance opposes any *change* in current; consequently, as indicated by the graph of Figure 16-2, current cannot flow instantly when the switch is closed. It must build up from zero. At the instant we close the switch in Figure 16-3, the current

[†]See Appendix 2-4.

CHAPTER 16   INDUCTANCE IN DC CIRCUITS   **399**

through the resistance is zero and there will be no voltage drop across it. Therefore, at the instant the switch is closed, an inductive voltage equal to the applied voltage must appear across the inductance. Since $di/dt = v_L/L$,

$$\text{initial } \frac{di}{dt} = \frac{E}{L} \tag{16-4}$$

This is the same rate of change of current as for Figure 16-1 (since the resistance cannot make its presence known if there is no current through it). Hence, the graph of the instantaneous current in Figure 16-4 starts off with the same slope as in Figure 16-2.

However, since the current is rising in order to induce a voltage in the inductance, the voltage drop across the resistance must be rising. As a result, the induced voltage required of the inductance for $v_R + v_L = E$ must decrease; and consequently, the rate of change of current must become smaller. The slope of the instantaneous current curve in Figure 16-4, therefore, becomes more gradual. Although the current is now rising less rapidly, nevertheless it is still rising, and consequently, the inductive voltage must be decreasing. As v_L becomes smaller, the rate of change of current must become smaller. As a result, the instantaneous current curve in Figure 16-4 continues to become still more gradual. After an appreciable period of time, the current reaches its maximum value, which produces a voltage drop across the resistance equal to the applied voltage. Consequently, the current is no longer changing but has settled to the steady value given by Equation (16-3).

Therefore, when we close the switch in a dc circuit containing inductance and resistance in series, the current cannot instantly take on a value of E/R as would be the case if there were no inductance in the circuit. The effect of the inductance is to force the current to rise to this steady-state value along the colored curve in Figure 16-4. Once the current reaches this steady value, the effect of the inductance in the circuit disappears. Since $v_R = iR$, and R is a constant, the graph of the voltage drop across the resistance of the circuit must have the same shape as the instantaneous-current graph (Figure 16-5). We obtain the graph of the instantaneous voltage across the inductance by subtracting the instantaneous IR drop from the constant applied voltage.

$$v_L = E - v_R$$

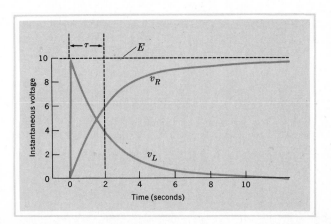

FIGURE 16-5
Instantaneous Voltage across the Resistance and Inductance.

16-3 TIME CONSTANT

When we close the switch in a dc circuit containing inductance it takes *time* for the current to rise to its steady-state value. Later on, we shall calculate this time element directly from the equation of the colored curve for instantaneous current in Figure 16-4. We can derive a simple and convenient method of indicating this time interval by assuming for the moment that the current can continue to rise at a *constant* rate equal to its initial rate of rise until it reaches the steady-state value, as shown by the shaded lines in Figure 16-4. This time interval is called the **time constant** of the *LR* circuit. On the basis of the above assumption, we can define time constant thus:

The time constant of an LR circuit is the time it would take the current to rise to its steady-state value if *it were to continue to rise at its initial rate of change for the whole time interval.*

The letter symbol for time constant is the Greek letter τ (tau).

If the rate of change of current remains constant (as shown by the straight shaded line in Figure 16-4), the steady-state current would be simply

$$I_m = \tau \times \text{initial } \frac{di}{dt}$$

But
$$I_m = \frac{E}{R} \tag{16-3}$$

and
$$\text{initial } \frac{di}{dt} = \frac{E}{L} \tag{16-4}$$

from which
$$\tau = \frac{L}{R} \tag{16-5}$$

where τ is the length of time in seconds defined as a time constant, L is the inductance of the circuit in henrys, and R is the resistance of the circuit in ohms.

Note that the time constant of an *LR* circuit (practical inductor) is directly proportional to the inductance. If we double the inductance, the initial rate of rise of current will only be half as great, and it will, therefore, take twice as long to reach the steady-state value. Also, the time constant is inversely proportional to the resistance. If we double the resistance, the steady-state current will be only half as much. So, for a given inductance, the current (rising at the same rate) will only take half as long to reach the steady-state value. Note also that the applied voltage has no effect on the time constant. If we double the applied voltage, we double the initial rate of rise of current E/L, but we also double the steady-state current E/R. Therefore, it will take exactly the same time to reach the steady-state value.

However, with resistance in the circuit, the instantaneous current does *not* continue to rise at its initial rate; it rises along an **exponential** curve, as shown by the colored graph in Figure 16-4. Nevertheless, this exponential curve must bear a fixed relationship to the two shaded straight lines in Figure 16-4. It starts at the same point and with the same slope as the initial rate of change of current line, and its slope gradually changes until it merges with the horizontal steady-state current line. Therefore, no matter what values of L and R are involved, in the time interval defined as a time constant, the instantaneous current actually rises to 63% of the steady-state value. If we wish, we may redefine time constant as the time it takes the instantaneous current in an *LR* circuit to reach 63% of its steady-state value. Although we might argue that the instantaneous current theoretically never does quite reach the steady-state value, we shall consider that, for practical purposes,

The instantaneous current will reach the steady-state value of E/R after a time interval equal to five *time constants has elapsed.*

EXAMPLE 16-1
(a) What is the steady-state current in the circuit of Figure 16-3?
(b) How long does it take the current to reach this value after the switch is closed?

SOLUTION

(a)
$$I_m = \frac{E}{R} = \frac{10 \text{ V}}{1 \text{ }\Omega} = 10 \text{ A}$$

(b)
$$\tau = \frac{L}{R} = \frac{2 \text{ H}}{1 \text{ }\Omega} = 2 \text{ s}$$

$$\therefore \text{ total rise time} = 5\,\tau = 5 \times 2 \text{ s} = 10 \text{ s}$$

FIGURE 16-6
Universal Exponential Curves for the Graphical Solution of the Rise and Fall of Current in Inductive Direct-Current Circuits.

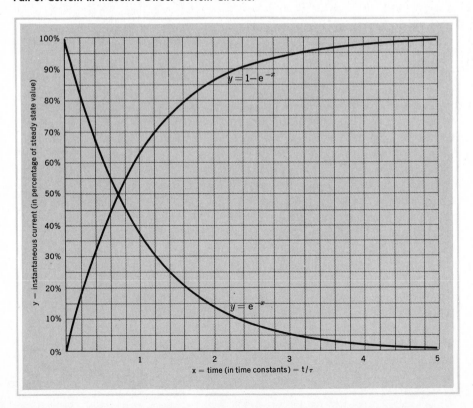

16-4 GRAPHICAL SOLUTION FOR INSTANTANEOUS CURRENT

Sometimes we wish to know what the instantaneous current is at a certain instant in time after the switch is closed or how long it takes the current to reach a certain percentage of its steady-state value. If we draw the instantaneous-current graph of Figure 16-4 accurately enough, we can solve such problems graphically from the exponential curve. We have already noted that, regardless of the numerical values of applied voltage, inductance, and resistance, the initial rate of rise of current will always be E/L; the final steady-state current will always be E/R; and, for all practical purposes, it will take a time equal to $5L/R$ to reach that steady-state value. This allows us to prepare a **universal exponential graph,** as in Figure 16-6, which will apply to *all* numerical values of E, L, and R. We calibrate the vertical (instantaneous current) axis of the graph in percentage of the final current and calibrate the horizontal (time) axis in time constants rather than in seconds.

EXAMPLE 16-2

(a) What is the instantaneous current in the circuit of Figure 16-3 1 s after the switch is closed?

(b) How long will it take the instantaneous current to rise from 0 A to 5 A?

SOLUTION

(a) We have already determined that the time constant of this circuit is

$$\tau = \frac{L}{R} = \frac{2\ \text{H}}{1\ \Omega} = 2\ \text{s}$$

Therefore, one second represents a time interval of 0.5 time constant on the universal graph of Figure 16-6. From this graph, 0.5 time constant after the switch is closed, as shown in Figure 16-7(a), the instantaneous current is

$$i = 39\% \times \frac{E}{R} = 0.39 \times \frac{10\ \text{V}}{1\ \Omega} = \textbf{3.9 A}$$

FIGURE 16-7
Using the Universal Exponential Graph to Solve Example 16-2.

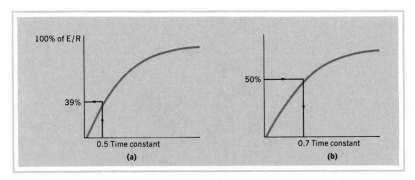

(b) Since $I_m = E/R = 10 \text{ V}/1 \, \Omega = 10 \text{ A}$,

$$5 \text{ A} = \frac{5}{10} \times I_m = 50\% \text{ of } I_m$$

The universal curve of Figure 16-6 shows that it takes 0.7 time constant [Figure 16-7(b)] for the instantaneous current to reach 50% of I_m. Therefore,

$$t = 0.7\tau = 0.7 \times 2 \text{ s} = 1.4 \text{ s}$$

EXAMPLE 16-3

In the circuit shown in Figure 16-8, switch S_2 is closed 1 s after switch S_1 is closed. What is the instantaneous current 1 s after switch S_2 is closed?

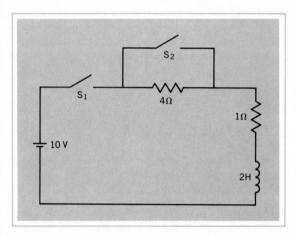

FIGURE 16-8
Circuit Diagram for Example 16-3.

SOLUTION

Step I. While S_1 is closed and S_2 is open, the total resistance in the circuit is

$$R_T = 4 + 1 = 5 \, \Omega$$

Time constant

$$\tau = \frac{L}{R} = \frac{2 \text{ H}}{5 \, \Omega} = 0.4 \text{ s}$$

$$\therefore 1 \text{ s} = \frac{1}{0.4} \times \tau = 2.5 \, \tau$$

From the curve of Fig. 16-6, when $t = 2.5\tau$,

$$i = 92\% \text{ of } \frac{E}{R} = 0.92 \times \frac{10 \text{ V}}{5 \, \Omega}$$

$$= 1.84 \text{ A}$$

There are two lines of thought that we may use to solve the second step which is the 1-s interval after S_2 is closed.

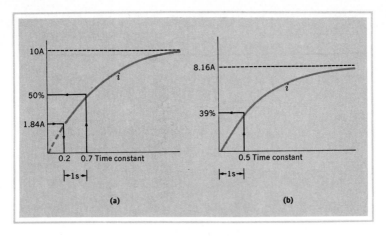

FIGURE 16-9
Graphical Solution of Step II of Example 16-3.

Step IIA. The first method is to consider the instantaneous current curve for the new circuit, which has only 1 Ω of resistance in circuit. The final steady-state current now becomes

$$I_m = \frac{E}{R} = \frac{10 \text{ V}}{1 \text{ }\Omega} = 10 \text{ A}$$

But at the instant S_2 is closed, i is already 1.84 A. Therefore, we must locate this point on the graph and mark off the additional 1 s from there [Figure 16-9(a)].

$$1.84 \text{ A} = \frac{1.84}{10} \times I_m = 18.4\% \text{ of } I_m$$

From the curve of Figure 16-6, when $i = 18.4\%$ of I_m,

$$t = 0.2\tau = 0.2\frac{L}{R} = 0.2 \times \frac{2 \text{ H}}{1 \text{ }\Omega} = 0.4 \text{ s}$$

Therefore, the instant in time we are concerned with will be 1 s later, or the 1.4-s mark on the curve of Figure 16-9(a). In terms of time constants,

$$1.4 \text{ s} = \frac{1.4}{2}\tau = 0.7\tau$$

Again, from Figure 16-6, when $t = 0.7\tau$,

$$i = 50\% \text{ of } I_m = 0.5 \times 10 \text{ A} = 5 \text{ A}$$

Step IIB. The alternative method requires only one reading from the universal exponential graph and is based on thinking in terms of the *additional* rise in current after S_2 is closed. Since we start Step II with 1.84 A already flowing in the inductance, and since $I_m = E/R = 10$ V/1 $\Omega = 10$ A, the maximum *additional* rise in current can be only $10 - 1.84 = 8.16$ A.

The time constant $= L/R = 2$ H/1 $\Omega = 2$ s; therefore, a time interval of 1 s after S_2 closes represents a time interval of 0.5 τ. From Figure 16-6, in half a time constant, the instantaneous current will rise 39% of the maximum *additional* current [Figure 16-9(b)].

$$\therefore additional \ i = 39\% \ of \ 8.16 \ A = 3.18 \ A$$

The instantaneous current 1 s after S_2 is closed will be 3.18 A greater than when S_2 is first closed. Therefore,

$$i = 1.84 + 3.18 = 5 \ A$$

16-5 CALCULATOR SOLUTION FOR INSTANTANEOUS CURRENT

The graphic solution for instantaneous current is sufficiently accurate for most purposes and has the advantage that it helps us to visualize the behavior of inductance in dc circuits. If we are required to determine the instantaneous current with greater accuracy, we must return to an algebraic solution based on Equation (16-2).

$$E = iR + L\left(\frac{di}{dt}\right) \tag{16-2}$$

Since Equation (16-2) is a differential equation, to solve for the instantaneous current i, we must once more resort to the calculus solution of Appendix 2-4, which yields

$$i = \frac{E}{R}(1 - e^{-x}) \tag{16-6}$$

where e $= 2.718$ *(the base of natural logarithms) and* $x = tR/L = t/\tau$.

Since $$\tau = \frac{L}{R}, \quad x = \frac{t}{\tau}$$

Therefore, we can think of x as representing elapsed time measured in **time constants.**

EXAMPLE 16-2A
(a) What is the instantaneous current in the circuit of Figure 16-3 1 s after the switch is closed?
(b) How long will it take the instantaneous current in this circuit to rise to 5 A?

SOLUTION

(a)

$$x = t \times \frac{R}{L} = 1s \times \frac{1\ \Omega}{2\ H} = 0.5$$

$$i = \frac{E}{R}(1 - e^{-x}) = \frac{10\ V}{1\ \Omega} \times (1 - e^{-0.5}) = 10 \times (1 - 0.6065) = \mathbf{3.93\ A}$$

The calculator chain for this computation becomes

$$10 \ \times \ (\ 1 \ - \ 0.5 \ +/- \quad e^x \ ) \ =$$

(b)

$$i = \frac{E}{R}(1 - e^{-x})$$

$$5 = \frac{10\ V}{1\ \Omega} \times (1 - e^{-x})$$

from which $e^{-x} = 0.5.$ $\qquad -x = \ln 0.5 = -0.693$

$$\therefore t = x \times \frac{L}{R} = 0.693 \times \frac{2\ H}{1\ \Omega} = \mathbf{1.39\ s}$$

EXAMPLE 16-3A

In the circuit shown in Figure 16-8, switch S_2 is closed 1 s after switch S_1 is closed. What is the instantaneous current 1 s after switch S_2 is closed?

SOLUTION

Step I. One second after S_1 is closed,

$$x = t \times \frac{R}{L} = 1\ s \times \frac{5\ \Omega}{2\ H} = 2.5$$

$$\therefore i = \frac{E}{R}(1 - e^{-x}) = \frac{10\ V}{5\ \Omega} \times (1 - e^{-2.5}) = 2 \times (1 - 0.0821) = 1.84\ A$$

Step IIA. Using the method illustrated by Figure 16-9(a),

$$i = \frac{E}{R}(1 - e^{-x})$$

$$1.84 = \frac{10\ V}{1\ \Omega} \times (1 - e^{-x})$$

$$\therefore e^{-x} = 0.816$$

from which $\qquad\qquad x = 0.203 = t \times \dfrac{R}{L}$

$$\therefore t = 0.203 \times \frac{2\ \mathrm{H}}{1\ \Omega} = 0.406\ \mathrm{s}$$

One second later, $\qquad t = 1.406\ \mathrm{s}$

and now $\qquad x = 1.406\ \mathrm{s} \times \dfrac{1\ \Omega}{2\ \mathrm{H}} = 0.703$

$$\therefore i = \frac{E}{R}(1 - e^{-x}) = \frac{10\ \mathrm{V}}{1\ \Omega} \times (1 - e^{-0.703}) = 10 \times (1 - 0.495) = \mathbf{5.05\ A}$$

Step IIB. Using the method illustrated in Figure 16-9(b),

Maximum *additional* current $= 10 - 1.84 = 8.16\ \mathrm{A}$

For a time interval of 1 s,

$$x = t \times \frac{R}{L} = 1\ \mathrm{s} \times \frac{1\ \Omega}{2\ \mathrm{H}} = 0.5$$

Actual additional current 1 s after S_2 is closed,

$$i = 8.16(1 - e^{-0.5}) = 8.16(1 - 0.6065) = 3.21\ \mathrm{A}$$

Therefore, the total instantaneous current 1 s after S_2 is closed is

$$i = 1.84 + 3.21 = \mathbf{5.05\ A}$$

The LR network of Figure 16-10(a) includes a 4-ohm **discharge resistor** in parallel with a practical inductor. We shall consider the function of this resistor in Section 16-7. In the meantime, to determine the time constant for the buildup of current in the inductor, we may consider the 4-ohm resistor to be part of the source network, as "seen" by the inductor. This allows us to use the Thévenin-equivalent circuit of Figure 16-10(b) to determine the rise of current in the inductor when the switch is first closed.

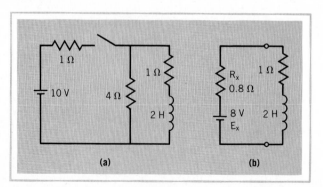

FIGURE 16-10
Thévenin-Equivalent Circuit for Determining Time Constant.

EXAMPLE 16-4

Calculate the instantaneous current in the inductor 1 s after the switch in the circuit of Figure 16-10(a) is first closed.

SOLUTION

Step I. Disconnecting the inductor from the rest of the circuit, the open-circuit voltage across the 4-Ω resistor is

$$E_x = 10 \text{ V} \times \frac{4}{4 + 1} = 8 \text{ V}$$

and

$$R_x = \frac{4 \times 1}{4 + 1} = 0.8 \ \Omega$$

Hence, the time constant is

$$\tau = \frac{L}{R_T} = \frac{2 \text{ H}}{(1 + 0.8) \Omega} = 1.111$$

Step II.

$$x = \frac{t}{\tau} = \frac{1 \text{ s}}{1.111 \text{ s}} = 0.9$$

and

$$i = \frac{E_x}{R_T}(1 - e^{-x}) = \frac{8 \text{ V}}{1.8 \ \Omega}(1 - e^{-0.9})$$

$$= 4.444 \text{ A}(1 - 0.4066) = \mathbf{2.64 \ A}$$

16-6 ENERGY STORED BY AN INDUCTOR

If we connect an ideal inductor to a voltage source having no internal resistance, the voltage generated by the inductance must remain constant and equal to the applied voltage; therefore, the current must rise at a constant rate, as in Figure 16-11(b). The

FIGURE 16-11
Determining the Energy Stored by an Inductor.

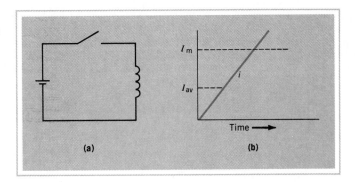

source is supplying electric energy to the ideal inductor at the rate of $p = Ei$. But, unlike resistance, inductance cannot convert this energy into heat or light. The energy that the source supplies is stored in the magnetic field, since the rising current forces the magnetic lines of force to expand against their natural characteristic of trying to become as short as possible. We can think of the storing of energy in an inductor as being similar to winding up the rubber band motor of a model airplane. As we apply torque to the propeller with our finger, we twist the rubber band and build up a countertorque in it. Since we are accomplishing motion against this countertorque, we are doing work and thus are storing energy in the rubber band. When we release the propeller, the energy stored in the rubber band is released, and it drives the propeller until the stored energy is all used up.

To determine the energy stored by an inductor, we must pick some final value of current [I_m in Figure 16-11(b)] and find out how much energy is stored as the current rises to this value. In resistance circuits, where the current and voltage do not change with a change in time, we found that $W = Pt = VIt$. Although the voltage remains constant in the circuit of Figure 16-11, the current steadily increases as time elapses. However, since the rate of change of current is constant, the *average* value of the instantaneous current as it rises from zero to I_m in Figure 16-11(b) will be $\frac{1}{2}I_m$. Therefore,

$$W = V \times \tfrac{1}{2}I_m \times t$$

But
$$V = L\frac{di}{dt} \tag{16-1}$$

and, since the rate of change of current is constant,

$$V = L \times \frac{I_m}{t}$$

Substituting gives
$$W = L \times \frac{I_m}{t} \times \tfrac{1}{2}I_m \times t$$

$$\therefore W = \tfrac{1}{2}LI_m^2 \tag{16-7}$$

In the case of the practical inductor, both the instantaneous voltage and current are changing and their rate of change is not constant. By first plotting accurate graphs of the instantaneous voltage across and current through the inductance on a piece of graph paper, we can then plot a graph of the instantaneous *rate* at which the inductance of the circuit stores energy from the relationship $p = vi$. By calculating instantaneous power for each division along the horizontal axis of the graph, we obtain the instantaneous power graph shown in Figure 16-12. When the current in a practical inductor reaches its steady-state value of $I_m = E/R$, the magnetic field ceases to expand, the voltage generated by the inductance has dropped to zero; hence, the instantaneous power $p = vi$ becomes zero. Therefore, the inductor increases the amount of energy stored in its magnetic field only while the current is building up to its steady-state value. As long as the current remains constant at this steady-state value, a constant amount of energy is stored by the magnetic field.

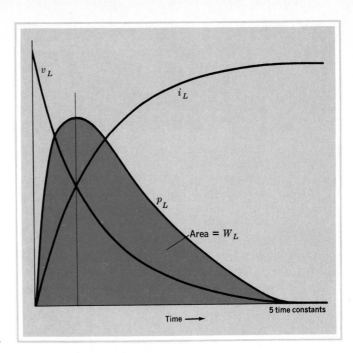

FIGURE 16-12
Graphical Representation of the Energy Stored by a Practical Inductor.

Although no additional energy is *stored* by the inductance of the practical inductor when I_m is reached, energy will be constantly dissipated by the resistance of the inductor at a steady rate of $P = I_m^2 R$. Energy is stored in an inductor in a *dynamic* form, and current must continue to flow for the magnetic field to be maintained. In this respect, energy storage by an inductor differs from energy storage by a capacitor, where the charging current ceases when the capacitor is charged. The electric field of a capacitor stores energy in a *static* form.

Note the similarity between Figure 16-12 and Figure 12-19. The significant difference is that the voltage curve for one graph becomes the current curve for the other. Since $p = vi$, the instantaneous power graph is the same in both sketches. Once again, the instantaneous rate of storage of energy p reaches its peak value when i_L rises to $\frac{1}{2}I_m$ and, at the same instant in time, v_L drops to $\frac{1}{2}E$.

Again, the *area* contained by the instantaneous power graph of Figure 16-12 represents the product of the *average power* and *time,* which, in turn, represents the *energy* stored by the inductance. By resorting to the basic integral calculus solution of Appendix 2-5 to determine this area, the energy stored in the magnetic field of the practical inductor of Figure 16-12 by the time the current reaches its steady-state value is

$$W = \tfrac{1}{2}LI^2 \tag{16-8}$$

where W is the stored energy in joules, L is the inductance in henrys, and I is the steady current in amperes. (I = E/R.)

EXAMPLE 16-5
What energy is stored by the magnetic field of an inductor whose inductance is 5 H and whose resistance is 2 Ω when it is connected to a 24-V source?

SOLUTION

$$I = \frac{E}{R} = \frac{24 \text{ V}}{2 \text{ }\Omega} = 12 \text{ A}$$

$$W = \tfrac{1}{2}LI^2 = \tfrac{1}{2} \times 5 \text{ H} \times (12 \text{ A})^2 = 360 \text{ J}$$

16-7 FALL OF CURRENT IN AN INDUCTIVE CIRCUIT

Interrupting the flow of current in an inductive circuit is like removing our finger after winding the rubber band motor of a model airplane; the stored energy is released. We have defined inductance as the circuit property that tends to oppose *any* change in current. If we examine the graphs of instantaneous current in this chapter, we note that it is not possible for the current to change instantly. It takes *time* for the current to change, even if only a few microseconds.

Therefore, at the instant we open the switch in the circuit of Figure 16-13, the current in the inductor must be the same as it was just before the switch was opened. To assist us in determining the effect of interrupting current flow in an inductive direct-current circuit, we shall give a specific value (500 kilohms) to the leakage resistance of the open switch. The steady-state current before the switch is opened is

$$I = \frac{E}{R} = \frac{12 \text{ V}}{2 \text{ }\Omega} = 6 \text{ A}$$

At the instant we open the switch, the current in the inductor must still be 6 amperes. Since this is a simple series circuit, this six-ampere current must flow through the leakage resistance of the switch, creating a voltage drop between the switch contacts of

$$V = IR = 6 \text{ A} \times 500 \text{ k}\Omega = 3\,000\,000 \text{ V}$$

This extremely high voltage drop is matched by the voltage generated in the turns of the inductor by the rapidly collapsing magnetic field. If we allow such a high voltage

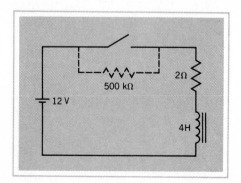

FIGURE   16-13
Interrupting Current in an Inductor.

to be induced in practice, not only will there be arcing of the switch contacts, but the insulation of the inductor will break down. Therefore, we must design dc circuits so that we never suddenly interrupt current in an inductor.

EXAMPLE 16-6

What is the voltage across the switch contacts in the circuit of Figure 16-14 at the instant that the switch is opened?

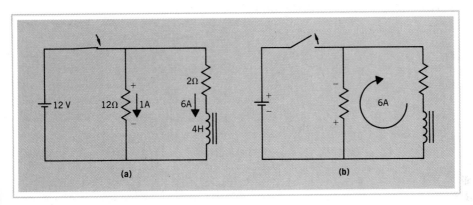

FIGURE 16-14
Circuit Diagram for Example 16-6.

SOLUTION

The steady-state current through the inductor is

$$I = \frac{E}{R} = \frac{12 \text{ V}}{2 \text{ }\Omega} = 6 \text{ A}$$

Although a 1-A current passes through the 12-Ω resistor with a conventional direction, as shown in Figure 16-14(a), this 1 A does *not* flow through the inductance of the circuit. When the switch is opened, the inductor will momentarily act as a generator as its magnetic field collapses. Across this generator, there is a total load of 12 Ω and 2 Ω in series. And since the current through the inductor cannot change instantly, a current of 6 A with a conventional direction, as shown in Figure 16-14(b), will flow through the 12-Ω resistor. This will create a voltage drop across the 12-Ω resistor of

$$v = iR = 6 \text{ A} \times 12 \text{ }\Omega = 72 \text{ V}$$

Between the switch contacts, we have the 72-V drop across the 12-Ω resistor and the 12-V potential difference of the battery. Their polarities are such that they *add*. Therefore, the total voltage across the switch contacts at the instant we open the switch is

$$v = 72 + 12 = \mathbf{84 \text{ V}}$$

According to Kirchhoff's voltage law, the initial induced voltage must equal the voltage drop across the total resistance at the instant we open the switch. Hence,

$$E_m = I_0 R_T \qquad (16\text{-}9)$$

where E_m is the peak voltage induced by the collapsing magnetic field, I_0 is the current through the inductor at the instant the switch is opened (which must be the same as the instant before the switch is opened), and R_T is the total resistance of the loop containing the inductor. In Example 16-6,

$$E_m = 6 \text{ A} \times 14 \ \Omega = 84 \text{ V}$$

According to Lenz's law, the direction of the induced voltage will always be such as to oppose any change in current. When the current rises, the direction of the induced voltage tends to cancel out some of the applied voltage. But when the source is removed, the magnetic field around the inductor collapses. The polarity of the voltage induced by a collapsing field is the opposite of that generated by a rising field. This reversed polarity tends to take the place of the original source to maintain current flow in the original direction through the inductor.

The 12-ohm resistor in the circuit of Figure 16-14 is called a **discharge resistor.** As the numerical results of Example 16-6 have shown, the presence of this discharge resistor in the circuit protects both the switch contacts and the insulation of the inductor from an excessively high voltage surge. Some such provision must always be made with inductive direct-current circuits which store appreciable energy, such as the field coils of dc motors and generators. Thyrite resistors are excellent for this purpose.

Since the induced voltage in Example 16-6 is much smaller than when no discharge resistor is used, the magnetic field must be collapsing at a much slower rate and, therefore, it takes longer to discharge the stored energy when a discharge resistor is included in the circuit. Since the magnetic field *is* collapsing and since the magnetic field strength depends on the instantaneous current in the windings, the current must start to decrease as soon as the switch is opened. In determining the rise time for the current in an inductive circuit, we used the term **time constant** based on a constant rate of change of current. Using the same technique, we may define **discharge time constant** as **the time it would take the current to drop to zero if it continued to drop at its initial rate for the whole time interval.** If the current continues to drop at a constant rate equal to its initial rate, from Figure 16-15,

$$\text{initial } \frac{di}{dt} = \frac{I_0}{\tau}$$

But $\qquad \text{initial } \dfrac{di}{dt} = \dfrac{E_m}{L} \qquad$ and $\qquad E_m = I_0 R_T \qquad (16\text{-}9)$

from which $\qquad\qquad\qquad\qquad \tau = \dfrac{L}{R_T} \qquad\qquad\qquad (16\text{-}5)$

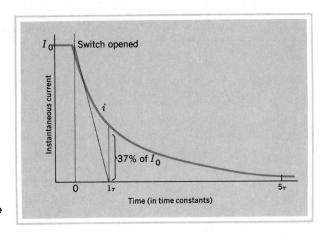

FIGURE 16-15
Fall of Current in an Inductive Direct-Current Circuit.

Therefore, **we determine the time constant of an *LR* circuit in exactly the same manner for both rise and fall of current.**

However, the current cannot fall at a constant rate. If it were to do so, the induced voltage would remain constant. But since the current is decreasing, the voltage drop across the resistance must also decrease. Because Kirchhoff's voltage law requires the induced voltage to equal the voltage drop across the resistance at every instant in time, the induced voltage must also decrease. As the induced voltage decreases the magnetic field must collapse more slowly. Thus, the rate of change of current must become progressively smaller with the passing of time. Since this sequence of events is similar to the rise of current when the switch was first closed, the instantaneous discharge graph (Figure 16-15) has the same basic exponential shape. Figure 16-6 includes an accurate graph for graphical solution of current decay in inductive dc circuits.

EXAMPLE 16-7

The switch in the circuit of Figure 16-14 is closed for 3 s and then opened. What is the instantaneous current in the inductor 0.25 s after the switch is opened?

SOLUTION
   Step I.

$$\text{Charge } \tau = \frac{L}{R} = \frac{4 \text{ H}}{2 \text{ } \Omega} = 2 \text{ s}$$

Expressing 3 s in time constants,

$$3 \text{ s} = \frac{t}{\tau} = \frac{3}{2} = 1.5 \text{ time constants}$$

From the graph for current *rise* in Figure 16-6, when $t = 1.5$ time constants, $i_L = 78\%$ of I_m.

Hence,
$$i_L = 0.78 \times \frac{E}{R} = 0.78 \times \frac{12\ V}{2\ \Omega} = 4.7\ A$$

Step II.

$$\text{Discharge } \tau = \frac{L}{R_T} = \frac{4\ H}{(12 \times 2)\Omega} = 0.286\ s$$

Expressing 0.25 s in time constants,

$$0.25\ s = \frac{t}{\tau} = \frac{0.25}{0.286} = 0.875 \text{ time constant}$$

From the graph for current *decay* in Figure 16-6, when $t = 0.88$ time constant, $i_L = 42\%$ of I_0.

Hence,
$$i_L = 0.42 \times 4.7\ A = 2\ A$$

A second method of preventing sparking of the switch or breaker contacts when we open the switch in the circuit of Figure 16-13 is shown in Figure 16-16. When we place a capacitor across the switch contacts, the potential difference across the capacitor is zero while the switch is closed. Since the voltage cannot rise instantly when we open the switch, there is sufficient time to move the switch contacts far enough apart that they do not spark.[†] Some of the energy released by the collapsing magnetic field of the inductor is transferred to the capacitor as it charges. The capacitor then partially discharges, building up a smaller magnetic field around the inductor in the opposite direction. Several cycles of diminishing energy transfer take place until the stored energy is all dissipated in the resistance of the load. We must leave the transient response of dc circuits containing both inductance and capacitance until we have investigated **resonance** in Chapter 24.

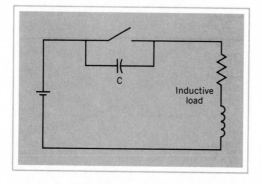

FIGURE 16-16
Supressing a Voltage Surge with a Capacitor.

[†]This principle is used to prevent sparking of the breaker points in an automobile ignition system.

16-8 CALCULATOR SOLUTION FOR INSTANTANEOUS DISCHARGE CURRENT

When we write the Kirchhoff's voltage-law equation for the circuit of Figure 16-14(b), the applied voltage in the circuit is zero. Therefore,

$$0 = iR_T + L\frac{di}{dt}$$

Solving this equation for the instantaneous current,[†]

$$i = I_0 e^{-x} \tag{16-10}$$

where I_0 is the initial current at the instant the switch is opened, $e = 2.718$, and $x = t(R_T/L) = t/\tau$.

EXAMPLE 16-7A

The switch in the circuit of Figure 16-14 is closed for 3 s and then opened. What is the instantaneous current in the inductor 0.25 s after the switch is opened?

SOLUTION

Step I. When the switch is closed,

$$x = t \times \frac{R}{L} = 3 \text{ s} \times \frac{2 \ \Omega}{4 \text{ H}} = 1.5 \text{ time constants}$$

$$\therefore i_L = \frac{E}{R}(1 - e^{-x}) = \frac{12 \text{ V}}{2 \ \Omega} \times (1 - e^{-1.5})$$

$$= 6 \text{ A} \times (1 - 0.223) = 4.66 \text{ A}$$

Step II. When the switch is opened,

$$x = t \times \frac{R_T}{L} = 0.25 \text{ s} \times \frac{14 \ \Omega}{4 \text{ H}} = 0.875 \text{ time constant}$$

$$\therefore i_L = I_0 \times e^{-x} = 4.66 \text{ A} \times e^{-0.875}$$

$$= 4.66 \times 0.417 = 1.94 \text{ A}$$

16-9 TRANSIENT RESPONSE

When we use either mechanical or electronic switches to change source voltage(s) suddenly in a dc circuit, a purely resistive network responds with instant changes in steady-state current. But LR networks respond by *delaying* the change to a new

[†]See Appendix 2-4.

steady-state current. We think of the time interval between the instant of switching ($t = 0$) and the final steady-state current ($t = 5\tau$) as the **transient response** of the LR network.

In the graphical solution for LR transients, we used separate $(1 - e^{-x})$ curve for rising current and e^{-x} curve for falling current, to help us to understand the behavior of inductance in dc circuits. This led to the two separate equations: Equation (16-6) for rising instantaneous current and Equation (16-10) for falling instantaneous current. However, we can use the transient concept to derive a single equation for any LR transient, as long as we are careful in specifying the initial steady-state current (I_0) just before the switch is operated, and the final steady-state current (I_F) which flows after the transient has ended.

To establish this general case, let's consider the situation in Step IIB of Example 16-3A (Figure 16-8), where closing S_2 attempts to change the current in the circuit from $I_0 = 1.84A$ at $t = 1$ s to $I_F = 10$ A. We noted that the instantaneous current during the transient is equal to the initial current plus a percentage of the *additional* current, based on the elapsed time measured in time constants. Expressing this statement in algebraic form,

$$i_L = I_0 + (I_F - I_0)(1 - e^{-x})$$ (16-11)

where I_0 is the initial current at the instant of switching and I_F is the final steady-state current (after an interval of at least five time constants has elapsed). Carrying out the multiplication,

$$i_L = I_0 + I_F - I_0 - I_F e^{-x} + I_0 e^{-x}$$

Collecting the terms gives us the **universal equation for instantaneous inductor current** in a dc LR network.

$$i_L = I_F + (I_0 - I_F)e^{-x}$$ (16-12)

where i_L is the instantaneous inductor current, I_0 is the initial inductor current, and I_F is the final steady-state inductor current after an interval of at least five time constants has elapsed.

Equation (16-11) expresses rising instantaneous current as *initial* current plus a percentage of the additional difference between initial and final current. In rearranging the terms, Equation (16-12) now expresses the instantaneous current during an LR transient in terms of the *final* steady-state current and an e^{-x} transient caused by the change in steady-state current.

EXAMPLE 16-8

In the LR circuit shown in Figure 16-17(a), the switch is closed at $t = 0$ and opened again at $t = 1.0$ s.
(a) What is the instantaneous current at $t = 1.0$ s?
(b) What is the voltage across the switch contacts at $t = 1.1$ s?
(c) Draw a graph of the voltage across the switch from $t = 0$ to $t = 5$ s.

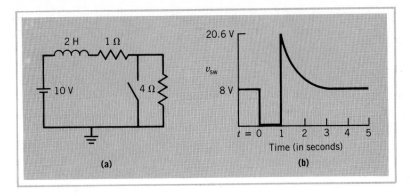

FIGURE 16-17
Transient Response in a DC *LR* Network.

SOLUTION
With the switch open, the steady-state current is

$$I_F = \frac{10 \text{ V}}{(4 + 1)\Omega} = 2 \text{ A}$$

and

$$\tau = \frac{L}{R_T} = \frac{2 \text{ H}}{5 \text{ }\Omega} = 0.4 \text{ s}$$

With the switch closed, the steady-state current is

$$I_F = \frac{10 \text{ V}}{1 \text{ }\Omega} = 10 \text{ A}$$

and

$$\tau = \frac{2 \text{ H}}{1 \text{ }\Omega} = 2.0 \text{ s}$$

(a) When the switch closes at $t = 0$, the initial current through the coil is 2 A. Hence, from $t = 0$ to $t = 1.0$ s,

$$i = I_F + (I_0 - I_F)e^{-t/\tau}$$
$$= 10 + (2 - 10)e^{-1.0/2.0}$$
$$= 10 - 8 \times 0.6065 = \mathbf{5.15 \text{ A}}$$

(b) At $t = 1.1$ s, the switch has been open for 0.1 s.

and

$$i = I_F + (I_0 - I_F)e^{-t/\tau}$$

becomes

$$i = 2 + (5.15 - 2)e^{-0.1/0.4}$$
$$= 2 + 3.15 \times 0.7788 = 4.453 \text{ A}$$

Since the switch is in parallel with the 4-Ω resistor,

$$v_{sw} = 4.453 \text{ A} \times 4 \ \Omega = \textbf{17.8 V}$$

(c) While the switch is open, the 2-A steady-state current through the 4-Ω resistor creates a steady voltage drop of 8 V. When the switch closes, the voltage across the resulting short circuit instantly becomes zero (in spite of a current transient through the switch).

When the switch opens at $t = 1.0$ s, the 5.15-A current creates a voltage drop of 5.15 A $\times$ 4 Ω = 20.6 V across the switch contacts, creating the transient shown in Figure 16-17(b). Since five time constants represent 2 s with the switch open, the transient will have disappeared by $t = 3$ s.

16-10 CHARACTERISTICS OF INDUCTIVE DC CIRCUITS

In summarizing the behavior of inductance in dc circuits, we should note that the most important characteristic is that the inductance develops an induced voltage which always tends to oppose any *change* in current. As a result of this characteristic, the current in inductive circuits cannot change instantly. A definite time interval is required for current to either rise or fall between any two instantaneous values. This time interval depends on the inductance and resistance of the circuit.

When we close the switch in an inductive circuit, it takes time for the current to reach its steady-state level. This delay is somewhat of an advantage in that it suppresses a sudden surge in the current drawn from the source. But when current in an inductive circuit is interrupted, an extremely high induced voltage transient is developed which can damage switch contacts and insulation unless we provide suitable circuit design so that current can decay gradually. Switches for dc circuits are designed so that their contacts move far apart as rapidly as possible to prevent arcing.

There are, however, some cases where we can use the very high induced voltage of a collapsing magnetic field to advantage. The ignition coil of an automobile engine is an example. Energy is stored gradually as current from the car battery builds up a magnetic field in the core of the coil. Then the current is suddenly interrupted by the breaker points, and a 10-kilovolt induced voltage, generated by the rapidly collapsing magnetic field, causes a spark between the points of the spark plug.

We can use the tendency of inductance to oppose any change in current in **filtering** or **smoothing** variations in load current that would otherwise occur when the load is fed from a dc source such as a rectifier system which provides a pulsating terminal voltage. As the source voltage in Figure 16-18 rises above its average value, the tendency for the current in the circuit to increase causes the inductance to store additional energy in its expanding magnetic field. Then, when the source voltage drops below average, the stored energy is released to help maintain a constant power input to the load.

In Chapter 12, we compared the energy storage ability of a capacitor with the *static* energy stored in a tank of compressed air. We may think of the energy storage ability of an inductor in terms of the *dynamic* energy stored by the *flywheel* of a piston engine.

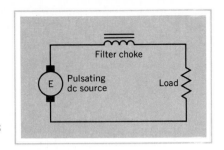

FIGURE 16-18
Filter Choke.

PROBLEMS

16-1. How long after the switch is closed will it take the current to reach its maximum value in a circuit whose resistance is 8 Ω and whose inductance is 2.4 H?

16-2. What is the resistance of a solenoid whose inductance is 4 H if it takes 0.2 s for the current to reach its maximum value?

16-3. What voltage must be applied to an 80-μH inductor if the initial rate of change of current is to be 10 000 A/s?

16-4. A 120-mH inductor has a resistance of 6 Ω. The initial rate of rise of current is 1000 A/s. What is the final current?

16-5. A choke with an inductance of 20 H and a resistance of 16 Ω is connected through a switch to a 24-V source.
   (a) What is the initial current when the switch is closed?
   (b) What is the initial rate of change of current when the switch is closed?
   (c) What is the final steady-state current?
   (d) How long does it take the current to reach its maximum value?

16-6. An inductor has an inductance of 20 H and a resistance of 500 Ω. In parallel with this coil is a 1000-Ω resistor. This combination is connected across a 120-V dc source whose internal resistance is negligible.
   (a) What initial current is drawn from the source when the switch is closed?
   (b) What is the initial rate of change of current in the coil?
   (c) What is the steady-state current in the coil?
   (d) How long does it take the current to reach its steady-state value?

16-7. (a) How long after the switch is closed will it take the current in the circuit of Problem 16-1 to reach 63% of its maximum value?
   (b) How long will it take the current to reach 50% of its maximum value?
   (c) If the current reaches a magnitude of 2.0 A in 0.5 s after the switch is closed, what is its maximum value?

16-8. (a) How long after the switch is closed will it take the current in the inductor of Problem 16-4 to reach a magnitude of 16 A?
   (b) What is the instantaneous current 10 ms after the switch is closed?

16-9. (a) How long after the switch is closed will it take the instantaneous current in the choke of Problem 16-5 to reach a magnitude of 1 A?
   (b) What is the instantaneous current 2 s after the switch is closed?

16-10. A 20-Ω resistor is connected in series with the choke of Problem 16-5, and this resistor is short-circuited 1 s after the switch is closed. What induced voltage is being developed by the inductance 1 s after the 20-Ω resistor is short-circuited?

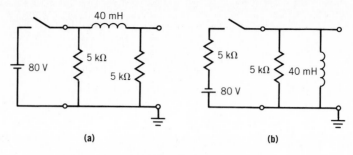

(a)   (b)

FIGURE   16-19

16-11. In the filter network of Figure 16-19(a), the 80-V source and the 40-mH inductor both have negligible internal resistance.
(a) What is the instantaneous output voltage 5 μs after the switch closes?
(b) If the switch remains closed for at least five time constants, how long after the switch opens will it be before the instantaneous output voltage becomes 50 V?

16-12. If the inductor in the circuit of Figure 16-19(b) has negligible resistance
(a) What is the instantaneous output voltage 5 μs after the switch closes?
(b) If the switch remains closed for at least five time constants, how long after the switch opens will it be before the instantaneous output voltage becomes $+50$ V (with respect to ground)?

16-13. (a) If the switch in the circuit of Figure 16-19(a) opens 10 μs after it closes, what is the instantaneous output voltage (with respect to ground) 10 μs after the switch opens?
(b) If the switch dwells alternately for 50 μs in each position, draw a labelled graph of two complete cycles of the output voltage waveform.

16-14. Repeat Problem 16-13 for the circuit of Figure 16-19(b).

16-15. (a) How much energy is stored by the choke in Problem 16-5 when the current reaches its steady-state value?
(b) At what rate is energy being drawn from the source when the current through the choke reaches its maximum value?

16-16. (a) How much energy is stored in the magnetic field of the inductor in Problem 16-6 when the current has reached its steady-state value?
(b) At what rate is electric energy being converted into heat when the current in the circuit has reached its maximum value?

16-17. The field coils of a dc motor have an inductance of 8 H, a resistance of 40 Ω, and are connected across a 120-V source.
(a) What maximum value of discharge resistor must be connected across the field coils if the voltage between the terminals of the coils must never exceed 200 V?
(b) How long will it take to discharge the energy stored in the field coils?
(c) How much energy is dissipated by the discharge resistor after the source is disconnected?

16-18. (a) What voltage will appear across the switch contacts when the switch in the circuit of Problem 16-6 is opened?
(b) How long does it take the coil to discharge its stored energy?
(c) If the switch is opened 1 s after it was first closed, how long does it take the current in the coil to drop 25 mA?

16-19. The switch in the circuit of Figure 16-20 is closed for 2 s and then opened. What is the instantaneous current in the inductor 1 s after the switch is opened?

16-20. Solve Problem 16-19 by replacing the source with an equivalent constant-current source.

422   PART III   CAPACITANCE AND INDUCTANCE

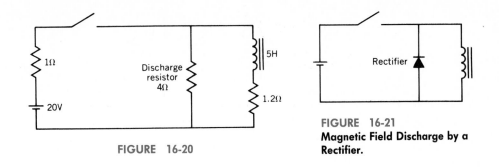

FIGURE 16-20

FIGURE 16-21
Magnetic Field Discharge by a Rectifier.

REVIEW QUESTIONS

16-1. Why is the graph of Figure 16-2 a straight line?

16-2. What is the significance of lowercase letter symbols?

16-3. Why are lowercase letter symbols used in representing the rise of current in inductive circuits?

16-4. Why must the rise of current follow the exponential curve of Figure 16-4?

16-5. What effect does the magnitude of the resistance of an inductor have on:
(a) The initial rate of change of current?
(b) The final steady-state current?
(c) The time it takes the current to reach its steady-state value?

16-6. What effect does the magnitude of the inductance of an inductor have on:
(a) The initial rate of change of current?
(b) The final steady-state current?
(c) The time it takes the current to reach its steady-state value?

16-7. Justify the following description of a time constant: "A time constant is the time it takes the current in an inductive circuit to rise to 63% of its final steady-state value."

16-8. Why does the value of the voltage applied to an inductive circuit have no effect on the time it takes the instantaneous current to reach a steady-state value?

16-9. Draw a detailed graph of the actual instantaneous current in Example 16-3.

16-10. Why is there no further increase in the energy stored in an inductor when the current has reached its steady-state value? Where is the energy supplied by the source going while a steady current flows in the circuit?

16-11. What is the effect of attempting to interrupt current in an inductive dc circuit instantly?

16-12. How does a discharge resistor prevent insulation breakdown in an inductor?

16-13. Why can we say that the current through an inductor the instant after the switch is opened is the same as the instant before the switch is opened?

16-14. Why is a **thyrite** type of varistor particularly suitable for use as a discharge resistor?

16-15. Some designers use a semiconductor rectifier which has a very high resistance to current in one direction and a very low resistance to current in the other direction in place of a discharge resistor, as shown in Figure 16-21. What are the advantages of such a circuit?

16-16. What effect does the magnitude of the resistance of the discharge loop of an inductive circuit have on:
(a) The initial current when the switch is opened?
(b) The initial rate of fall of current?
(c) The time it takes the current to fall to zero?

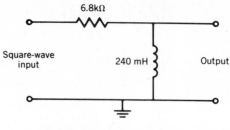

FIGURE 16-22
LR Waveshaping Circuit.

16-17. The ignition "tune-up" kit for an automobile engine contains a set of breaker points and a "condenser" to connect across the points. Why do these components need replacing from time to time to improve engine performance?

16-18. Derive Equation (16-12) by considering what happens in the circuit of Figure 16-8 when S_2 is opened but S_1 remains closed.

16-19. *LR* networks are sometimes used as waveshaping circuits in electronic pulse circuits. If a 1-kHz symmetrical square wave from a generator with negligible internal resistance is fed into the *LR* waveshaping network of Figure 16-22, draw an accurate graph of the output voltage waveform.

*16-20. Design a computer program identified as "INSTANTANEOUS CURRENT IN AN LR CIRCUIT" which will ask the operator to input values for initial current, final steady-state current, inductance, net resistance, and time interval. The program should then display the value of the instantaneous current in amperes.

*16-21. Design a computer program which will display an accurate, fully labeled graph for part (c) of Example 16-8.

PART **IV**

ALTERNATING CURRENT

IN ALTERNATING-CURRENT CIRCUITS, the magnitude and direction of the source voltage is continuously changing. To investigate the behavior of various resistance networks in the chapters of Part 2, we restricted voltage sources to such constant-voltage devices as batteries (and solar cells). The advantage of starting with direct-current circuits is that the time factor does not appear in our steady-state calculations—except when we want to know the total amount of energy dissipated in a certain time interval ($W = Pt$).

In certain chapters of Part 3, we encountered the effect of time on the instantaneous voltages and currents as capacitors and inductors charged and discharged their stored energy. However, after a short transient interval following the closing or opening of switches, even CR and LR circuits settle down to a steady-state condition where time is not a factor in a dc system.

In the remaining chapters of Parts 4 and 5, we encounter voltage sources in which the instantaneous generated emf continuously changes and produces an open-circuit terminal voltage whose magnitude and direction depend on the exact instant in time. Before we can work with the many forms of alternating-current networks in Part 5, we must investigate the rules for dealing with alternating currents in the chapters of Part 4.

17

ALTERNATING CURRENT

17-1 A SIMPLE ROTATING GENERATOR

The only practical voltage source we have considered so far is the battery. Because of the nature of the chemical action within it, the battery is able to maintain a constant terminal voltage for a considerable period of time. Although the battery is a convenient voltage source for flashlights and portable radios, it does not offer a suitable means of developing the large amounts of electric energy which we require for both home and industry.

In Chapter 15, we discussed a second method of generating a voltage, based on Faraday's discovery of electromagnetic induction. As an electric conductor is moved in such a manner that it cuts across magnetic lines of force, a voltage is induced in it. In order to maintain a potential difference by this method, we must keep the conductor in motion within a stationary magnetic field. We can achieve this by the arrangement shown in Figure 17-1. The electric conductor is in the form of a loop which can be rotated continuously within the magnetic field in the air gap between a north and a south pole. Electrical connection to the rotating loop is maintained through a pair of **sliprings** and **brushes**.

As the loop rotates, the two sides of the loop must always cut across the magnetic lines of force in opposite directions. As a result, the two induced voltages are connected in series so that the brushes receive the *sum* of the two voltages (just as two flashlight cells may be connected in series to double the applied voltage). If we connect a load resistor to the brushes of this simple generator, the induced voltage will cause current to flow in the closed circuit consisting of the loop and the load. This load current flowing in the rotating loop produces a magnetic field which, according to

426

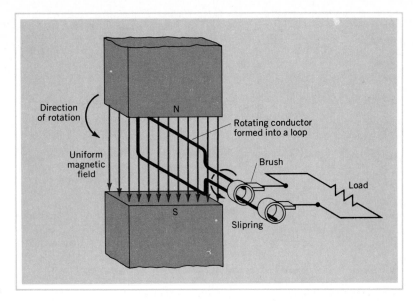

FIGURE 17-1
Simple AC Generator.

Lenz's law, reacts with the stationary magnetic field in such a manner as to *oppose* the rotation of the loop. Therefore, in order to maintain rotation, we must rotate the loop against this mechanical force opposing its rotation. Whenever motion is accomplished against an opposing force, mechanical energy is expended. Therefore, the electric energy that is converted into heat energy in the load resistance is generated directly at the expense of the mechanical energy maintaining the rotation of the loop.

17-2 THE NATURE OF THE INDUCED VOLTAGE

Let us consider the sequence of events as the conductor forming either side of the loop in Figure 17-1 rotates through a complete revolution (360°). First of all, it must cut across the magnetic lines of force in one direction and then move back across the same magnetic lines of force in the opposite direction. Since the induced voltage appearing at the brushes of this simple generator must reverse its polarity with each half-revolution, the basic rotating machine generator for converting mechanical energy into electric energy develops an **alternating** voltage.

Since most of our electric energy is generated by rotating machines, it is necessary for us to be able to cope with an *alternating* voltage in electric circuits just as readily as we have been able to determine the relationships among voltage, current, resistance, and power with the *direct* voltage from a battery, which we have been using up to this point. Figure 17-2(a) is a cross-sectional drawing of the simple generator of Figure 17-1. Only one side of the loop is shown in this sketch. The one conductor is shown in twelve positions during one rotation. In order to be able to relate the rotation of the conductor to numerical calculations, it is conventional to consider the normal or *positive* direction of rotation as being counterclockwise, starting from an angular position representing three o'clock.

At position 0, the motion of the conductor is parallel to the magnetic lines of

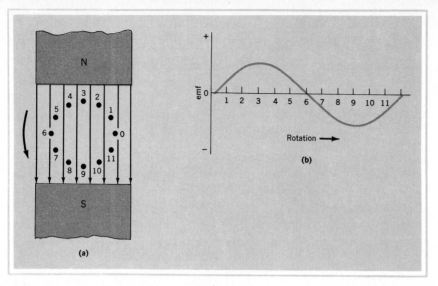

FIGURE 17-2
Nature of Induced Voltage.

force. Therefore, at that moment, no lines of force are being cut by the conductor. Hence, the induced voltage at the instant the conductor passes through point 0 is 0 volts. At position 1, the conductor has started to cut across the lines of force. Thus, appreciable voltage is induced into the conductor as it passes through point 1. At position 2, the conductor is cutting across the magnetic lines of force at an even steeper angle. Thus, the *rate* of cutting and, therefore, the voltage induced into the conductor as it passes through point 2 is even greater. At position 3, the conductor is cutting across the magnetic lines of force at right angles, which results in the greatest possible rate of cutting. Therefore, maximum voltage is induced into the conductor as it passes through point 3.

We can compare the manner in which the rate of cutting increases as the conductor rotates through the first 90° to four men walking along a sidewalk at the same velocity. As they reach a certain point, the first man continues to walk parallel to the street and, therefore, never reaches the other side. So his *rate of crossing the street* is zero. The second man makes a 30° turn and crosses the street at an angle. The third man makes a 60° turn and crosses the street at a steeper angle, while the fourth man makes a 90° turn and goes straight across the street. With all four men turning at the same instant and walking at the same speed, the fourth man has the shortest distance to go and, therefore, gets to the other side first. His rate of crossing the street is maximum even though his absolute velocity is the same as that of the other three. Number three man is the next to reach the other side, and number two man is the third to reach the opposite sidewalk. Their rates of crossing are proportionately less than that of number four man. Since the magnetic lines of force are parallel like the opposite sidewalks on a street, the same situation exists if we consider the actual direction in which the conductor is moving at each position.

From positions 3 to 6 in Figure 17-2, the *rate of cutting* across magnetic lines of force becomes progressively less, so the induced voltage at each position of the conductor gradually decreases back to zero volts as it passes through point 6, where

428

it is again momentarily moving parallel to the lines of force. From positions 6 to 12, the same sequence of events occurs, with the induced voltage rising to a maximum at position 9 and then dropping back to zero at position 12, except that the conductor is now cutting across the magnetic lines of force in the opposite direction. As a result, the polarity of the induced voltage is reversed during the second half of the revolution. The + sign on the graph of Figure 17-2(b) indicates the reference polarity of the terminal voltage. The − sign simply indicates that the polarity of the voltage is reversed with respect to the reference polarity. One complete rotation of the loop constitutes one **cycle**. The time taken to complete one cycle is called the **period** of the waveform.

17-3 THE SINE WAVE

In the preceding section, we observed the manner in which the alternating voltage induced into a conductor rotating in a uniform magnetic field varies continually as the angular position of the conductor changes. We can show the nature of this continual change best by plotting a graph showing the magnitude and polarity of the induced voltage at any position of the loop, as shown in Figure 17-2(b). To appreciate fully the manner in which electric circuits behave when the voltage applied to the system is of the alternating type, we must be able to determine exactly what the induced voltage is at any instant during the rotation of the conductor.

For such an analysis, we replace the cross-sectional sketch of Figure 17-2(a) with a much simpler diagram, as shown in Figure 17-3. The straight line OX represents the rotating conductor pivoted at point O (for origin). The end that is free to rotate is indicated by an arrowhead. We call this symbolic representation of the rotating conductor a **phasor**. As we have already noted, mathematical convention requires this phasor to rotate in a counterclockwise direction, starting from a position representing three o'clock, referred to as the *reference axis* in Figure 17-3. Instead of showing its position at twelve particular points, as in Figure 17-2(a), the position of the phasor at a certain instant in time is located by stating the angle ϕ through which it has rotated from the reference axis. Also, to keep the sketch as simple as possible, in Figure 17-3, we have not shown the magnetic lines of force, which are assumed to be uniformly spaced and in a vertical plane, as they were in Figure 17-2(a).

Faraday's law tells us that the voltage induced in a rotating conductor is directly proportional to the *rate* at which it cuts across the magnetic lines of force. If we assume a uniform flux density and a constant angular velocity, the rate of cutting will depend only on the angular position of the loop. At 0° (with respect to the reference axis), the loop moves parallel to the magnetic lines of force; hence, no voltage is

FIGURE 17-3
Phasor Representation of the Rotating Conductor.

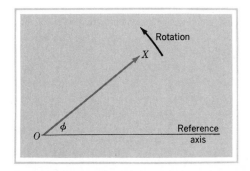

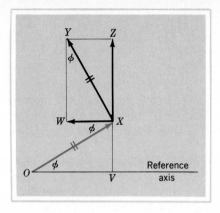

FIGURE 17-4
**Geometric Representation of the Rate
of Cutting of Magnetic Lines of Force.**

induced into the loop at that instant. As the loop swings around toward 90°, the rate of cutting increases, becoming maximum at 90°. To determine the exact value of the induced alternating voltage at any instant, we must be able to determine the exact rate of cutting of magnetic lines of force by the conductor at that instant with respect to the maximum rate of cutting of lines of force.

In Figure 17-4, OX represents the rotating conductor after it has traveled through an angular distance ϕ from the reference axis. Since OX is pivoted at O, the line XY drawn perpendicular to OX represents the direction of motion of the tip of the phasor (conductor) at that particular moment. If the angular velocity of the conductor is constant, the linear velocity of the tip of the phasor must also be constant. We can represent this by drawing $XY = OX = $ a constant for any value of ϕ. We can resolve the motion represented by XY into a vertical component of motion XZ, and a horizontal component XW. Since XZ is parallel to the magnetic lines of force, this component is ineffective in inducing any voltage. Since XW cuts across the magnetic lines of force at right angles, the induced voltage will be directly proportional to the length of XW. When ϕ becomes 90°, $XW = XY$. Therefore, the ratio of XW/XY in Figure 17-4 represents the ratio between the rate of cutting of lines of force at angle ϕ and the maximum rate of cutting.

From geometry:

Since $\qquad WX \parallel OV, \qquad \therefore \ \angle WXO = \angle XOV = \phi$

Since $\qquad \angle YXO = 90°, \qquad \therefore \ \angle WXY = 90° - \phi = \angle OXV$

Since $\qquad \angle XWY = 90°, \qquad \therefore \ \angle XYW = \phi$

Since $\qquad XY = OX, \qquad \therefore \ \triangle XOV \equiv \triangle XYW$

$$\therefore \quad \frac{XW}{XY} = \frac{XV}{OX}$$

As the loop rotates, the rate of cutting of magnetic lines of force at any instant is directly proportional to XV/OX. If OX remains constant, this ratio depends only on the magnitude of the angle ϕ. Since XV is the side of a right-angled triangle opposite the angle ϕ and OX is the hypotenuse, in trigonometry the ratio XV/OX is called the **sine** of the angle ϕ. We can determine accurately the ratio of XV/OX for any magnitude of the angle ϕ by using the **sin** key of an electronic calculator. Therefore,

The voltage induced into the rotating conductor at any point is directly proportional to the sine of the angle through which the loop has rotated from the reference axis.

This angle is called the **phase angle**.

In Figure 17-4, we determined by geometrical construction the relative magnitude of an induced voltage for an angle ϕ which is less than 90°. In its final form, the procedure amounted to drawing a perpendicular from the tip of the phasor (point X) to the horizontal reference axis (point V). As long as the length of the phasor OX remains constant, the magnitude of the induced voltage is directly proportional to the length of XV. Exactly the same geometrical construction applies to angular distances of more than 90° from the reference axis. In Figure 17-5(a), the phasor OX has rotated through an angle of 120°. When we draw a perpendicular from point X to the reference axis, we form a right-angled triangle in which the side XV is opposite an angle that is equal to $180° - \phi$ or 60°.

$$\therefore \sin 120° = \sin(180° - 120°) = \sin 60°$$

Similarly, in Figure 17-5(b), when $\phi = 210°$, the right-angled triangle has an angle opposite XV of $210° - 180° = 30°$. But, in this case, XV is *below* the reference axis. When we relate this diagram to the rotating loop of Figure 17-2, XV now represents a *negative* value of induced voltage, since the loop is moving in the opposite direction across the magnetic lines of force. Therefore, whenever XV is below the reference axis, it represents a negative quantity. So we must say, then, that sin 210° = −sin 30°.

The sine function is stored in the built-in memory of scientific calculators so that we can obtain the sine of *any* angle simply by pressing the sin key. BASIC language for personal computers also contains a SIN(X) function in its permanent memory. Most scientific calculators permit us to enter an angle in either degrees or radians. Since BASIC requires angles to be in radians, we must build a conversion step into any program providing for the input of angular data in degrees.

Since the voltage induced into a conductor rotating at a constant angular velocity in a uniform magnetic field is constantly varying, it is difficult for us to visualize its

FIGURE 17-5
Determining the Sine of Angles Greater than 90°.

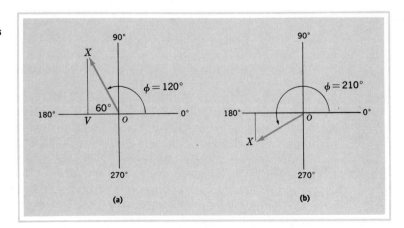

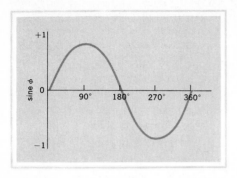

FIGURE 17-6
The Sine Curve.

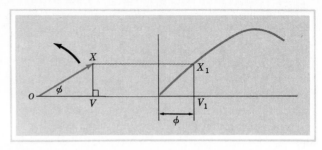

FIGURE 17-7
Preparing a Sine Curve by Geometric
Projection.

nature by a simple statement in words. Hence it is customary to plot a linear graph of the variation in the induced voltage over a complete cycle by expressing the angular distance ϕ as a linear distance along the horizontal axis of the graph and the induced voltage for each value of ϕ along the vertical axis of the graph, as in Figure 17-6. Since the induced voltage at any point during the rotation of the loop is directly proportional to the *sine* of the angle through which it has rotated from the reference axis, we can plot an accurate graph of the instantaneous variations in an alternating voltage by preparing a table of sine ϕ (at approximately 5° intervals) using a calculator **sin** key. We then plot these values on a sheet of graph paper and join the plots to form a smooth curve. A computer can calculate such a sine table automatically and then display the data in graph form on its output device. The resulting graph is called a **sine curve**. The alternating voltage that varies in accordance with this sine curve is called a **sine wave**.

Figure 17-7 shows another method of producing a sine curve. This method is a projection of the geometric construction we used in Figures 17-4 and 17-5. Since the length of phasor OX is constant, the altitude VX of the tip of the phasor is proportional to the sine of the angle ϕ, and, therefore, VX is directly proportional to the induced voltage. When we project VX to a position V_1X_1 on the linear graph at a distance along the horizontal axis representing the angular distance, the point X_1 will be a proper plot, as required to draw a sine curve. If we repeat this geometric projection at 5° intervals, we shall have sufficient plots to draw a complete sine curve.

17-4 THE PEAK VALUE OF A SINE WAVE

As we discovered in determining the nature of an alternating voltage, the greatest value that the induced voltage can attain occurs when $\phi = 90°$. At this point, the rate of cutting of magnetic lines of force is maximum. When $\phi = 270°$, the same maximum value is attained but the polarity of the induced voltage is reversed. The numerical value that the alternating voltage attains at these two angles is termed the **peak** value of the ac waveform.

The letter symbol for the peak value of an alternating source voltage is E_m (m for maximum).

432

This peak value is a numerical quantity in volts and does not have any polarity, since the same peak value in volts occurs at both 90° and 270°. The peak value is also independent of time, since, no matter where the rotating conductor may be at a certain instant, the peak value will appear as it passes the 90° and 270° points in its rotation. Therefore, the peak value is a constant for a given generator and does not have the sine-wave shape of the voltage induced from instant to instant.[†]

In the simple ac generator, the peak voltage will depend on just what the maximum rate of cutting of magnetic lines of force by the rotating loop happens to be. This rate, in turn, depends on (1) the total number of lines of force in the magnetic field, (2) the angular velocity of the loop, and (3) the number of turns of wire in the loop. From Faraday's law, when a conductor cuts across magnetic lines of force at the *rate* of one weber per second, an emf of one volt is induced in that conductor.

17-5 THE INSTANTANEOUS VALUE OF A SINE WAVE

The graphs of Figures 17-2, 17-6, and 17-7 show the manner in which the induced voltage changes from instant to instant as the loop rotates. Any value that is continually changing and thus is dependent on the exact instant in time for its numerical value is called an **instantaneous value**. Therefore, in representing the instantaneous value of the voltage induced in the rotating loop of the simple ac generator of Figure 17-1, we must use the lower case letter symbol e.

In the simple ac generator of Figure 17-1, during the first 180° of rotation of the loop, the voltage rose to a peak value in one direction and dropped back to zero; then, during the second 180° of rotation, it rose to a similar peak value but in the opposite direction and again dropped back to zero. Therefore, it took one complete revolution for the voltage to vary through the whole possible range of values. As the loop rotates past 360°, the same series of events recurs. We define this complete excursion of the instantaneous value of the induced voltage as one **cycle** of a **sine wave**. The time it takes for the instantaneous value to complete one cycle is called the **period** of the sine wave. The number of cycles completed in one second is called the **frequency** of the sine wave.

The letter symbol for frequency is f (lower case).

Until 1965, frequency was expressed in cycles per second. But people became lazy and stated frequency in "cycles." Consequently, the SI unit is now preferred in technical writing.

The hertz is the SI unit of frequency.[‡]
The unit symbol for hertz is Hz

Hertz and **cycles per second** are numerically synonymous.

[†]In mathematical terminology, we refer to E_m as a **scalar** quantity, that is, a quantity which has a numerical magnitude but which does *not* have any direction associated with it.

[‡]Named in honor of the German physicist Heinrich Hertz, who discovered "hertzian" or radio waves.

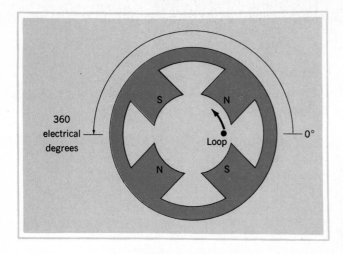

FIGURE 17-8
Relationship between Electrical and Mechanical Degrees in a Four-Pole Generator.

Although the instantaneous voltage is never constant, its exact magnitude at any instant is directly proportional to sin ϕ. Since the instantaneous voltage reaches its peak value at 90° and since sin 90° = 1, we can determine the instantaneous voltage for any particular instant in time as the loop rotates from

$$e = E_m \sin \phi \qquad (17\text{-}1)$$

In the simple **two-pole** generator of Figure 17-1, one complete mechanical revolution of the loop was required to generate one complete cycle of an electrical sine wave of voltage. But, in practical generators like the four-pole generator of Figure 17-8, one complete cycle of voltage will be generated as the loop moves through the magnetic flux under an N pole and then through the flux under an S pole. To use Equation (17-1), we must subdivide this cycle into 360 *electrical* degrees even though the loop has only rotated through 180 *mechanical* degrees. The relationship between *electrical* degrees and *mechanical* degrees will depend on the number of *pairs* of magnetic poles in the piece of rotating machinery. Sine waves of voltage can also be generated in **oscillator** circuits where there is no mechanical rotation. Again, to solve sin ϕ, a complete cycle must be subdivided into 360 *electrical* degrees.

EXAMPLE 17-1
At what speed must the shaft of a six-pole 60-Hz alternator be turned?

SOLUTION
With three pairs of poles, one mechanical revolution generates three electrical cycles. Therefore, the shaft speed is

$$\frac{60}{3} = 20 \text{ r/s} = 1200 \text{ r/min}$$

Our next problem is how to determine the angle ϕ in electrical degrees. We are usually given the frequency of the sine wave. Since frequency can be expressed in cycles *per second*, if we multiply frequency by the time elapsed since the instantaneous value passed through zero, we have the angular position expressed in cycles and fractions of complete cycles. Multiplying this product by 360 then gives us an angle representing the instantaneous position of the induced voltage in its cycle in electrical degrees. Therefore, we can express Equation (17-1) in a more useful form as

$$e = E_m \sin(360ft)°$$ (17-2)

where e is the instantaneous source voltage at a given instant in time, E_m is the peak value of the sine wave in question, sin is the sine of the angle expressed in electrical degrees, f is the frequency of the sine wave in hertz, and t is the elapsed time in seconds.

EXAMPLE 17-2

What is the instantaneous value of a 60-Hz sine wave whose peak voltage is 150 V 10 ms after the instantaneous voltage passes through 0 V?

SOLUTION

$$e = E_m \sin(360\ ft)°$$
$$= 150 \sin(360 \times 60 \times 0.01)°$$
$$= 150 \sin 216°$$

To solve this equation with a calculator, the chain becomes

$$150 \times 216\ \boxed{\textbf{sin}}\ = -88.17\ V$$

17-6 THE RADIAN

In dealing with the rotating conductor of Figure 17-1, we have been talking about angular *distance* and angular *velocity*. Although the Babylonian 360° system of angular measurement is very convenient for purposes of geometry and trigonometry, it is not suited to showing the relationship between the linear velocity of the conductor in Figure 17-1 as it moves around the circumference of a circle and its angular velocity. The linear velocity of the conductor governs the actual rate of cutting across magnetic lines of force. Therefore, for electrical purposes, it would be very useful to have a unit of angular distance which is related to the distance traveled by the free end of a rotating phasor. The **radian** is just such a unit.[†] By definition,

One radian is the angular distance through which the phasor travels when its free end (X in Figure 17-9) travels through a linear distance equal to the length (OX) of the phasor.

It is standard practice in electrical engineering to express the angular distance

[†]The **radian** is the SI unit for a plane angle.

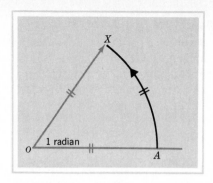

FIGURE 17-9
The Radian.

traveled by a rotating phasor in radians, and its angular velocity in radians per second. Since the rotating phasor in Figure 17-9 is the radius of a circle, and the circumference of a circle $= 2\pi r$, there are 2π radians in one complete cycle. Hence, in marking the x-axis of a time graph (such as Figure 17-6) in radians,

$$90° = \pi/2 \text{ rad} \qquad 180° = \pi \text{ rad}$$

$$270° = 3\pi/2 \text{ rad} \qquad 360° = 2\pi \text{ rad}$$

We must also convert Equation (17-2) from angular distance in degrees to angular distance in radians. Most engineering calculators will allow us to solve trigonometric functions in radians without having to convert back to degrees. If we do have to convert radians into degrees for some reason, the conversion factor becomes

$$2\ \pi \text{ rad} = 360°$$

$$\therefore \qquad 1 \text{ rad} \doteq 57.3° \qquad\qquad (17\text{-}3)$$

As in Equation (17-2), we usually express angular distance as the product of angular velocity and elapsed time. In radian measure, angular velocity is expressed in **radians per second**.

The letter symbol for angular velocity in radians per second is the Greek letter ω (omega).

Expressed in radian measure, Equation (17-2) becomes

$$e = E_m \sin \omega t \qquad\qquad (17\text{-}4)$$

Since there are 2π radians in a cycle,

$$\omega = 2\pi f \qquad\qquad (17\text{-}5)$$

Therefore, we write Equation (17-4) as

$$e = E_m \sin 2\pi f t \qquad\qquad (17\text{-}6)$$

In both Equation (17-4) and Equation (17-6), the angle is numerically expressed in radians.

EXAMPLE 17-3

Write the general equation for the instantaneous voltage of a 60-Hz generator whose peak voltage is 170 V.

SOLUTION

$$e = E_m \sin 2\pi ft$$

But

$$2\pi f = 2 \times 3.14 \times 60 = 377 \text{ rad/s}$$

$$\therefore e = 170 \sin 377t$$

Since 60 Hz is the standard power line frequency, we should memorize:

$$60 \text{ Hz} \doteq 377 \text{ rad/s} \qquad\qquad (17\text{-}7)$$

EXAMPLE 17-2A

What is the instantaneous value of a 60-Hz sine wave whose peak voltage is 150 V 10 ms after the instantaneous voltage passes through 0 V?

SOLUTION

$$e = E_m \sin 2\pi ft$$

$$= 150 \sin (2\pi \times 60 \times 0.01) \text{ rad}$$

$$= 150 \sin 1.2\pi \text{ rad}$$

We switch the calculator to the radian mode (usually by pressing a **d-r** key). The chain then becomes

$$150 \; \boxed{\times} \; (\; 1.2 \; \boxed{\times} \; \boxed{\pi} \;) \; \boxed{\sin} \; \boxed{=} \; -88.17 \text{ V}$$

17-7 INSTANTANEOUS CURRENT IN A RESISTOR

Now that we are acquainted with the sinusoidal nature of the instantaneous voltage generated by a basic ac voltage source, we can determine the nature of the current that will flow when we connect resistance to the generator terminals in the simple ac circuit of Figure 17-10. At a particular instant in time, the instantaneous voltage developed

FIGURE 17-10
Simple Alternating-Current Circuit.

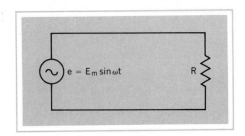

by the generator has a certain magnitude and a certain polarity. If we consider only this one particular instant, there is no difference between the simple ac circuit of Figure 17-10 and the equivalent dc circuit. The real difference between them is that, at the *next* instant in time, the magnitude of the instantaneous alternating voltage changes, but the applied voltage in the dc circuit continues on at a fixed magnitude. If we consider any particular *instant* in time, all the equations summarized at the end of Chapter 5 apply to the ac circuit of Figure 17-10.

Therefore, we can state that the instantaneous current through the resistance in Figure 17-10 will be

$$i = \frac{e}{R} \qquad (17\text{-}8)$$

where i is the instantaneous current through the resistance, e is the instantaneous voltage applied to the resistance, and R is the resistance of the circuit.

Since resistance is determined by such physical factors as type of material, length, and cross section, it is a constant for a given circuit at a given temperature [indicated by an upper case R in Equation (17-8)]. Because R is a constant in Equation (17-8), whatever the variation in instantaneous voltage, the instantaneous current must stay right in step in order to make Ohm's law hold true. If the instantaneous voltage is a sine wave, the instantaneous current must also be a sine wave, as shown in Figure 17-11. The instantaneous current must reach its peak value at the same instant that the instantaneous voltage becomes maximum. It will become zero at the same instant that the instantaneous voltage becomes zero, and it will reverse its direction at the same instant that the instantaneous voltage across the resistance reverses its polarity. Since the instantaneous current through a resistor must always be exactly in step with the voltage across the resistance, the current is said to be **in phase** with the voltage.

Because the general equation describing a sine-wave alternating applied voltage is $e = E_m \sin \omega t$, substituting in Equation (17-8) gives

$$i = \frac{E_m}{R} \sin \omega t \qquad (17\text{-}9)$$

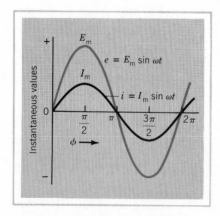

FIGURE 17-11
Instantaneous Current through a Resistor.

And since the peak value of the instantaneous current depends on the peak value of the applied voltage,

$$I_m = \frac{E_m}{R} \qquad\qquad (17\text{-}10)$$

$$\therefore i = I_m \sin \omega t \qquad\qquad (17\text{-}11)$$

where i is the instantaneous current through a resistor when a sine wave of voltage is applied, and I_m is the peak current as determined by the relationship $I_m = E_m/R$.

17-8 INSTANTANEOUS POWER IN A RESISTOR

If we still think in terms of instantaneous values, we may apply Equations (5-2), (5-3), and (5-4) from Chapter 5 to an ac circuit. Therefore,

$$p = vi = i^2 R = \frac{v^2}{R} \qquad\qquad (17\text{-}12)$$

where p is the instantaneous power in a resistor in watts, v is the instantaneous voltage drop across the resistance in volts, i is the instantaneous current through the resistance in amperes, and R is the resistance of the circuit in ohms.

As R is a constant for a given circuit, and since the instantaneous voltage and current in the basic circuit are *both* sine waves, the instantaneous power must be a sine-*squared* wave. To determine the nature of a sine-squared wave, we can plot a graph of instantaneous power in the same manner that we plotted the sine curve of Figure 17-6. If we calculate the *squares* of the sine ϕ plots on a linear graph, the instantaneous power in a resistor takes on the shape shown in Figure 17-12. Note that the instantaneous power in a resistor is always positive since squaring a negative quantity results in a positive quantity. We can interpret *positive* power as electric energy being converted into some other form of energy (heat in the case of a resistor).

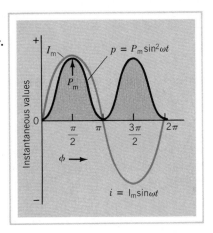

FIGURE 17-12
Instantaneous Power in a Resistor.

On this basis, *negative* power would represent some other form of energy being converted into electric energy. A resistor is not capable of doing this; therefore, the waveform of instantaneous power in a resistor has no negative component.

As Figure 17-12 indicates, the instantaneous power in the basic ac system is pulsating in nature, swinging from zero to maximum and back *twice* each cycle. If we consider an electric lamp as the resistance in Figure 17-10, as far as heating the filament is concerned, it does not matter which way the current flows through it. There will be one pulse of energy converted into heat and light on the one *half-cycle* of current and another similar pulse on the second half-cycle. If the lamp is operated from a 60-hertz source, it will increase in brilliance 120 times a second. Therefore, the instantaneous power pulsations are at *twice* the frequency of the voltage and current. This pulsating characteristic of instantaneous power is responsible for the flicker in 25-hertz lighting as used in some localities and for the vibration of small ac motors.

We can also show the pulsating nature of the power in an ac system by deriving an equation for instantaneous power.

By substituting $i = I_m \sin \omega t$ in Equation (17-12) we obtain

$$p = (I_m \sin \omega t)^2 \times R = I_m^2 R \times \sin^2 \omega t$$

And since peak values are maximum instantaneous values, from Equation (17-12)

$$P_m = I_m^2 R \tag{17-13}$$

$$\therefore p = P_m \sin^2 \omega t \tag{17-14}$$

where p is the instantaneous power in a resistance when a sine wave of current flows through it, and P_m is the peak power as determined by the relationship $P_m = I_m^2 R$.

From trigonometry, we can show that

$$\sin^2 \omega t = \tfrac{1}{2}(1 - \cos 2\omega t)^\dagger$$

$$\therefore p = \tfrac{1}{2}P_m - \tfrac{1}{2}P_m \cos 2\omega t \tag{17-15}$$

This relationship tells us three things about the instantaneous power waveform which we can check by referring to Figure 17-12.

1. In any right-angled triangle, the cosine of one acute angle is also the sine of the other acute angle. Therefore, the general shape of the cosine curve and the sine curve are the same. Consequently, the fluctuations in instantaneous power are sinusoidal in nature.

2. Since 2ω is twice the angular velocity of ω, the frequency of the sinusoidal variation is twice as great as the frequency of the instantaneous voltage and current.

---

†
$$\cos (A - B) = \cos A \cos B + \sin A \sin B$$

$$\cos (A + B) = \cos A \cos B - \sin A \sin B$$

If $A = B$, then $\cos 0° = \cos^2 A + \sin^2 A$ and $\cos 2A = \cos^2 A - \sin^2 A$. Since $\cos 0° = 1$, subtracting gives $1 - \cos 2A = 2 \sin^2 A$, from which $\sin^2 A = \tfrac{1}{2}(1 - \cos 2A)$.

3. Since the limits of the value of a cosine are $+1$ when $\phi = 0°$ and -1 when $\phi = 180°$, the value of the expression $\frac{1}{2}(1 - \cos 2\omega t)$ can vary between zero and $+1$. Therefore, the instantaneous power waveform for a resistance is always positive.

17-9 PERIODIC WAVES

So far, we have considered only one form of alternating current—the sine wave. Although the sine wave is by far the most important ac waveform and the only one we shall consider in detail in this book, we should note that the sine wave is one of a larger group of **periodic waves**. In electric circuits, a periodic wave refers to any time-varying function such as voltage, current, or power which continually repeats exactly the same sequence of values after a given time interval known as the **period** of the waveform. Figure 17-13 shows the graphs of the manner in which the instan-

FIGURE 17-13
Typical Periodic Waves.

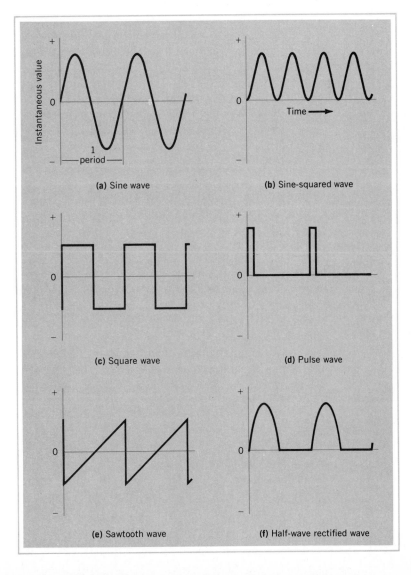

(a) Sine wave

(b) Sine-squared wave

(c) Square wave

(d) Pulse wave

(e) Sawtooth wave

(f) Half-wave rectified wave

taneous values of six typical periodic waves vary as a function of time. The graphs of the sine wave and sine-squared wave are repetitions of the two waves shown in Figure 17-12. In this case, however, we have marked the horizontal axis of the graph in elapsed time (milliseconds), rather than in terms of the angular position of the rotating phasor. As shown in Figure 17-13(a), the **period** of the sine wave is the length of time it takes the instantaneous voltage or current to complete one **cycle** of events. As we noted in Section 17-8, the instantaneous power waveform in a resistor has a sinusoidal component which is twice the frequency of the current and voltage waveforms. Hence, Figure 17-13(b) shows *four* complete cycles of a periodic wave. All other graphs in Figure 17-13 show *two* complete cycles.

The instantaneous values of a periodic wave change much too rapidly to be registered by the conventional electric meter movements of Chapter 10. However, the electron beam of the cathode-ray tube in an oscilloscope can quite readily respond to high frequency instantaneous voltages. If we apply the instantaneous voltage of a certain periodic wave [such as the sine wave of Figure 17-13(a)] to the vertical deflection system of an oscilloscope, the vertical motion of the electron beam will be exactly in step with these instantaneous values. This motion is still much too rapid for the eye to follow and all we would see is a vertical line proportional to the **peak to peak** value of the periodic wave. An important characteristic of the sawtooth periodic wave of Figure 17-13(e) is that the instantaneous voltage is a *linear* function of elapsed time. Hence, if we simultaneously apply to the horizontal deflection system of the oscilloscope a sawtooth wave having a period exactly twice that of the waveform fed to the vertical deflection system, the electron beam will trace out a duplicate of the graph of Figure 17-13(a).

17-10 THE AVERAGE VALUE OF A PERIODIC WAVE

By definition, an alternating current (or voltage) is one in which the *average* of the instantaneous values over a period of time is zero. Just by examining the waveforms of Figure 17-13, it is fairly obvious that the sine wave, square wave, and sawtooth wave are alternating currents (or voltages), since the instantaneous values are symmetrical about the horizontal axis of the graph. The sine-squared wave, pulse wave, and half-wave rectified wave do not qualify as alternating currents, since their instantaneous values are always of the same polarity.

We are quite familiar with the technique for *averaging* examination marks or determining batting averages in baseball. The same technique applies to determining the average value of any periodic wave. For complex periodic waves we can determine the average of the instantaneous values graphically by examining the *area* under a graph, as we did in Figures 12-19 and 16-12. But for the sine wave, we can obtain a more accurate average value by writing down the sine of 5°, 10°, 15°, and so on, for a complete cycle. To find the average value of any column of numbers, we add the column of figures and divide by the number of figures in the column. After going to all this trouble, the average value of a sine wave for a complete period is *zero*, since, for every positive value of instantaneous voltage or current during the one half-cycle, there is a similar negative value during the next half-cycle.

However, if we average *one-half cycle only*, we obtain a significant numerical constant. From the column of figures prepared as above from a sine table or calculator, we find that the average of sin ωt over one half-cycle is $22.9 \div 36 = 0.636$. If we average the half-cycle of a sine wave by integration (area under the curve method),[†] we arrive at a slightly more accurate result, expressing average values as $2/\pi$ or 0.637.

$$\text{Since } i = I_m \sin \omega t, \qquad \therefore I_{av} = I_m \times 0.637$$

Therefore, the *half*-cycle average value of a sine wave is

$$I_{av} = 0.637 I_m = \frac{2}{\pi} I_m \qquad \text{(17-16)}$$

or

$$E_{av} = 0.637 E_m = \frac{2}{\pi} E_m \qquad \text{(17-17)}$$

As we discovered in Section 17-8, the sine-squared wave of Figure 17-13(b) will have a *full*-cycle average value of one-half the peak value of the sine-squared wave. The pulse waveform will have an average value that depends on the ratio of the pulse duration to the time interval between pulses. The half-wave rectified sine wave of Figure 17-13(f) (see also Figure 3-8) will have a full-cycle average value which, in turn, is the average of $0.637 E_m$ for one half-cycle and zero for the next half-cycle, or $0.3183 E_m$.

17-11 THE RMS VALUE OF A SINE WAVE

Instantaneous values of alternating current and voltage are not satisfactory for most measurement purposes since the pointers of meters cannot follow the rapid changes in polarity. Nor are they suitable for calculation purposes because by the time we complete the computation, the instant to which the calculation applied has long since passed. Full-cycle average values of sine waves are zero, and half-cycle average values are only useful for such applications as rectification where we are concerned primarily with a half-cycle of a sine wave.

These problems do not arise in dc circuits since the voltage, current, and power in a resistive circuit remain constant for a considerable period of time. As far as the end result of converting electric energy into light is concerned, it makes no difference whether the lamp is lighted by a direct current or an alternating current flowing through it. Perhaps, then, we can establish *equivalent steady-state* values for alternating current and voltage which will allow us to use the same interrelationships among voltage, current, resistance, power, work, etc. that we use in dc circuits. We can determine this **equivalent dc** or **effective** value of an alternating current experimentally by finding out what direct current must flow through a given resistance to

[†]See Appendix 2-6 for calculus solution.

produce in a certain time interval the same amount of heat energy as when the resistor is connected to a source of alternating voltage.

To determine this equivalent dc value of an alternating current, we can conduct the same experiment algebraically with pencil and paper. The basis for comparison in the experiment is *the same amount of heat energy in a certain time interval*. We can translate this into *average work per unit time or* simply **average power**. To determine average power in a resistive ac circuit, we start with Equation (17-12),

$$p = i^2 R \qquad\qquad (17\text{-}12)$$

Using the averaging technique we discussed in Section 17-10, we compute the *square* of the instantaneous current at small time intervals over a full cycle and then calculate the average or *mean* value of all these i^2 values. Hence,

$$P_{av} = (\text{full-cycle average of } i^2) \times R$$

In a dc circuit, $P = I^2 R$. Since the effective values of an alternating current are to represent dc equivalent values, we can use the same letter symbols for both, that is, uppercase italic letters without subscripts. Therefore, in an ac circuit, $P = I^2 R$ where P is average power and I is the dc equivalent value of the alternating current. Solving for I by combining the equations above,

$$I = \sqrt{\frac{P}{R}} = \sqrt{\text{full-cycle average of } i^2}$$

Note that R cancels out. We can define the dc equivalent value of an alternating current in terms of the *process* we go through to determine its value. We find the square *root* of the *mean* value of the *squares* of the instantaneous current over a full cycle. **Root-mean-square** (usually abbreviated to **rms**), **effective** and **equivalent dc** are synonymous terms. At present time, **rms** is the preferred term, which we shall use for the remainder of the book.

For a sine wave of alternating current, we note that the instantaneous power in the graph of Figure 17-12 swings alternately and uniformly between zero and peak power. It appears quite reasonable to state, therefore, that *the average power in a resistor through which a sine-wave alternating current is flowing is one-half the peak power*.

We can also determine this relationship from the general Equation (17-14) for instantaneous power: $p = P_m \sin^2 \omega t$. As we have already shown, $\sin^2 \omega t = \frac{1}{2}(1 - \cos 2\omega t)$. If we average a sine or cosine function over a *complete* cycle, the average must be zero, since for every positive value during the first half-cycle, there is an equivalent negative value during the second half-cycle. Hence, when averaged over a complete cycle, $\cos 2\omega t$ becomes zero and Equation (17-14) becomes

$$P = 0.5\, P_m \qquad\qquad (17\text{-}18)$$

For a sine-wave alternating current in an ac circuit,

$$P = I^2R \quad \text{and} \quad P_m = I_m^2R$$

But
$$P = \tfrac{1}{2}P_m = \tfrac{1}{2}I_m^2R$$

$$\therefore I^2R = \tfrac{1}{2}I_m^2R, \text{ from which } I^2 = \tfrac{1}{2}I_m^2$$

and
$$I = \frac{I_m}{\sqrt{2}} = 0.707I_m \tag{17-19}$$

The rms (effective) value of a sine wave of current is $1/\sqrt{2}$ or 0.707 of the peak value.

The rms value of a sine-wave voltage should be such that the *average* power is the product of the *rms* voltage across and *rms* current through the resistance of the circuit, just as in the equivalent dc circuit. Therefore,

$$P = VI$$

But when a sine-wave current flows through a resistor,

$$P = 0.5P_m \tag{17-18}$$

Since peak instantaneous power occurs at the instant when both the instantaneous voltage across and the instantaneous current through a resistance circuit reach their peak values,

$$P_m = V_mI_m$$

As we have already determined,

$$I = \frac{I_m}{\sqrt{2}} \tag{17-19}$$

Substituting in Equation (17-18)

$$V \times \frac{I_m}{\sqrt{2}} = \tfrac{1}{2}V_mI_m$$

from which

$$V = \frac{V_m}{\sqrt{2}} = 0.707V_m \tag{17-20}$$

Note also that
$$V_m = \sqrt{2}\,V = 1.414V$$

Since rms values are used for the majority of ac circuit measurements and computations, we shall assume that **all currents and voltages are stated as rms values unless it is specifically stated otherwise.**

EXAMPLE 17-4

What is the peak voltage of the customary 120-V 60-Hz electric service?

SOLUTION

$$E_m = 1.414E = 1.414 \times 120 \text{ V} = \mathbf{170 \text{ V}}$$

A term we may encounter occasionally in working with rms and average values of an ac waveform is its **form factor**. Form factor is the ratio of the rms value to the half-cycle average value of an ac wave. For a sine wave,

$$\text{form factor} = \frac{I}{I_{av}} = \frac{0.707I_m}{0.637I_m} = 1.11 \tag{17-21}$$

PROBLEMS

17-1. If the peak voltage of a 60-Hz ac generator is 150 V, what is the instantaneous voltage after the loop has rotated through 135°?

17-2. If the peak voltage of a 1-kHz audio oscillator is 20 V, what is the instantaneous voltage when $\phi = 210°$?

17-3. A radio transmitter generates a sine-wave carrier with a frequency of 600 kHz. If the peak voltage fed to the antenna is 1 kV, what is the instantaneous voltage 2.0 μs after the instantaneous value passes through zero volts?

17-4. The instantaneous voltage from a 400-Hz aircraft generator is 95 V when $t = 1$ ms. What is the instantaneous voltage when $t = 2$ ms?

17-5. What is the instantaneous voltage after the loop of the generator in Problem 17-1 has rotated through 1 rad?

17-6. What is the instantaneous voltage from the audio oscillator in Problem 17-2 when $\phi = 5$ rad?

17-7. What is the instantaneous voltage after the loop of the generator in Problem 17-1 has rotated for 15 ms?

17-8. What is the instantaneous voltage from the audio oscillator of Problem 17-2 when $t = 600$ μs?

17-9. The instantaneous value of an 800-Hz sine-wave current is 5 A when $t = 0.5$ ms. What is the peak current?

17-10. If the instantaneous current drawn from a 50-Hz source is -263 mA when $t = 0.013$ s, what is the instantaneous current when $t = 0.017$ s?

17-11. Write the general voltage equation for a 400-Hz source whose peak voltage is 200 V.

17-12. A certain alternating voltage is described by the equation

$$e = 170 \sin 1570t$$

(a) What is the frequency of the sine wave?
(b) What is the instantaneous voltage when $t = 5$ ms?

17-13. What is the peak voltage in the European 230-V 50-Hz system?

17-14. What is the peak value of a 60-μV radio-frequency signal appearing at the input terminals of a radio receiver?

17-15. The equation of an alternating current is $i = 1.5 \sin 377t$. What rms voltage drop will appear across a 15-Ω resistor when this current flows through it?

17-16. Write the equation for the instantaneous voltage generated by a 117-V 400-Hz alternator.

17-17. If the symmetrical square wave of current shown in Figure 17-13(c) has a peak value of 40 mA, what is its rms value?

17-18. If the symmetrical sawtooth voltage waveform in Figure 17-13(e) has a peak voltage of 30 V, what is its rms value?

17-19. The average power of a 250-Ω soldering iron is 55 W.
(a) What is the rms current through the iron?
(b) What is the peak voltage across the iron?

17-20. What current will an ac ammeter show when connected in series with a 16-Ω load whose peak power is 288 W?

17-21. How long will it take a 10-Ω heater connected to a 120-V 60-Hz source to convert 1 kWh of electric energy into heat?

17-22. What is the "wattage" rating of a lamp whose hot resistance is 150 Ω when connected to a 117-V 25-Hz source?

17-23. The conductor in the simple generator of Figure 17-1 cuts across a total magnetic flux of 40 mWb during the first 180° of rotation. If the frequency of the ac output is 60 Hz, what is the half-cycle average induced voltage?

17-24. What is the rms value of the alternating voltage induced into the loop in Problem 17-23?

REVIEW QUESTIONS

17-1. What conditions govern the generation of a voltage by electromagnetic induction?

17-2. Why does the angular velocity of the rotating loop in Figure 17-1 tend to decrease when a load resistor is connected to the brushes?

17-3. Why is it not possible to maintain a constant polarity of induced voltage in the loop in Figure 17-1?

17-4. Why do we describe the instantaneous voltage waveform from the simple generator of Figure 17-1 as a **sine wave**?

17-5. With a diagram similar to Figure 17-5, show that $\sin 3\pi/4$ rad $= \sin \pi/4$ rad.

17-6. What effect has the angular velocity of the loop in Figure 17-1 on (a) the frequency of the voltage waveform? (b) the peak value of the induced voltage?

17-7. Why does doubling the number of turns in the loop double the peak voltage?

17-8. Why is the peak value of the induced voltage never a negative quantity?

17-9. Why is the answer to Example 17-2 a negative quantity?

17-10. What is the significance of a negative value of instantaneous current?

17-11. What is the significance of a negative value of instantaneous power?

17-12. What is meant by angular *distance*?

17-13. Why is the radian measure preferred in writing the general equation for an alternating current?

17-14. What is the significance of the term **sine wave** in describing the behavior of an electric circuit?

17-15. What effect does a variation in the peak value of an alternating current have on the waveform?

17-16. Why can we substitute instantaneous ac values in the various equations we derived for use in dc circuits containing a voltage source and a load resistor?

17-17. Why must the instantaneous current through a resistor have the same waveform as the instantaneous voltage across it?

17-18. What is meant by the statement that the current through a resistor is in phase with the voltage across it?

17-19. What factors govern the instantaneous value of the current through a resistor in an ac circuit?

17-20. Why is the instantaneous power input to a resistor never negative?

17-21. Why does the instantaneous power input to a resistor pulsate at *twice* the frequency of the applied voltage?

17-22. Why is the average power input to a resistor operating from a sine-wave source one-half the peak power input?

17-23. Why is it necessary to consider average *power* when deriving an rms value for alternating current and voltage sine waves?

17-24. The equation for instantaneous voltage in an ac system is sometimes written in the form $e = \sqrt{2}\, E \sin \omega t$. Suggest a reason for using this form.

17-25. We obtain the average value of the alternating power input to a resistive load by averaging the instantaneous power input over a complete cycle. Why can we not determine the average value of an alternating current in the same manner?

*17-26. Using a personal computer, determine the rms value of a sine wave by the process of finding the square root of the mean values of sine squared for one complete cycle.

*17-27. Design a computer program to display a graph of the instantaneous power when a sine-wave alternating current flows in a resistor.

18

REACTANCE

As we discovered in Chapter 17, when we apply a sine wave of voltage to a resistor, the instantaneous current in the circuit must be an ac sine wave exactly in phase with the applied voltage. We established this relationship on the basis that current is the dependent variable in the circuit and that, in an alternating current circuit too, it must automatically take on a magnitude such that Ohm's law will hold true. As long as we keep in mind the continuously-varying nature of a sine-wave alternating applied voltage, all the laws we have studied in dealing with dc circuits must also hold true for ac circuits. Any statement that would apply only for a special case and not for the general situation would not pass as a basic law of electric circuit theory.

Therefore, we were able to state Ohm's law in the form

$$i = \frac{v}{R} \tag{17-8}$$

where both v and i are instantaneous values. Although resistance is an electrical property of conductors, it is dependent on such physical factors as length, cross section, and type of material. Therefore, resistance cannot change in the short period of one cycle of a sine wave, and thus, as far as Equation (17-8) is concerned, resistance is a constant. Consequently, the instantaneous current must always be directly proportional to the instantaneous voltage.

Since the instantaneous current in a resistor reaches its peak value at the same instant as the instantaneous voltage across the resistance, we are also able to state Ohm's law in the form $I_m = V_m/R$, and, since the rms value of a sine wave is the dc equivalent value, which we will use for most of our ac circuit calculations, the most useful form of Ohm's law is simply

$$I = \frac{V}{R}$$

(18-1)

where I is the rms value of the sine-wave-current through a resistor, and V is the rms value of the sine-wave voltage across the resistor.

18-1 THE NATURE OF THE INSTANTANEOUS CURRENT IN AN IDEAL INDUCTOR

In setting out to determine the nature of the instantaneous current that must flow in the circuit of Figure 18-1, we assume that the resistance is so small that we can neglect it. In Chapter 16, we decided that we would think of the inductor as a load connected across a voltage source. Therefore, in order to satisfy Kirchhoff's voltage law, at every instant in time, the inductive voltage across the coil in Figure 18-1 must exactly equal the applied voltage. Hence,

$$v_L = e = E_m \sin \omega t$$

If there is no resistance in the circuit, any voltage drop that appears across the terminals of an inductor must be due to the voltage induced in the coil by a *changing* current through it. From Equation (16-1), we can express the instantaneous voltage that must appear across the inductor in Figure 18-1 as

$$v_L = L\frac{di}{dt}$$

Therefore,

$$L\frac{di}{dt} = E_m \sin \omega t$$

from which

$$\frac{di}{dt} = \frac{E_m}{L} \sin \omega t$$

(18-2)

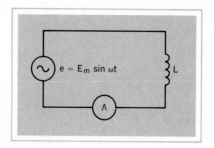

FIGURE 18-1

Inductance in an Alternating-Current Circuit.

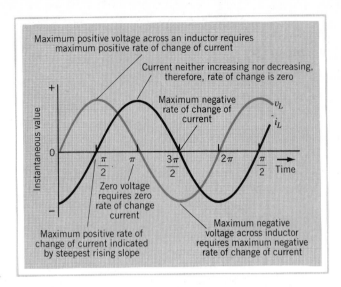

FIGURE 18-2
Instantaneous Current in an Inductor.

By integrating Equation (18-2), we can establish a mathematical expression for the instantaneous current through an ideal inductor in an ac circuit in one fell swoop.[†] However, to help us *understand* the behavior of inductance in ac circuits, we can determine the nature of the instantaneous current through the inductor in Figure 18-1 by applying principles we have already learned. Since Kirchhoff's voltage law requires the voltage across the inductor in Figure 18-1 to be exactly equal to the applied voltage at every instant in time, we can represent the instantaneous voltage v_L across the inductor by the colored sine curve of Figure 18-2. According to Faraday's law of electromagnetic induction, the magnitude of this inductive voltage at any instant is directly proportional to the *rate of change* of current through the coil at that instant. At the instant when we require maximum *positive* voltage across the inductor (at the $\pi/2$-rad point in Figure 18-2), the current must, therefore, be *changing* at the greatest *rate* in a *positive* direction. As we noted in discussing the rise of current through an inductor in Chapter 16, *rate of change of current* in a graph of current plotted against time is indicated by the *slope* of the current graph. Hence, maximum rate of change is indicated in the instantaneous current graph of Figure 18-2 by the steepest slope.

One-quarter cycle later in time (at the π-rad point in Figure 18-2), when the instantaneous voltage across the inductor must momentarily drop to zero, the current must stop *changing*. It does not matter how great the current may be, as long as it is neither rising nor falling, there will be no *change* in the number of magnetic lines of force linking the turns of the coil in order to induce a voltage in them. Therefore, at the π-rad point on the graph of Figure 18-2, the *slope* of the instantaneous current curve must become horizontal. Between $\pi/2$-rad and π-rad, the voltage across the inductor has to decrease along a sine curve, and thus the *rate of change* of current has to decrease in the same manner.

One-quarter cycle still later in time (at the $3\pi/2$-rad point in Figure 18-2), the instantaneous inductive voltage must become maximum negative, and, therefore,

[†]See Appendix 2-7.

the instantaneous current must again change at a maximum rate but, this time, in the opposite direction. If we draw an instantaneous current graph that fulfills the conditions of having a maximum positive rate of change when the instantaneous voltage across the inductor is maximum positive, neither rising nor falling when the inductive voltage is zero, having a maximum negative rate of change when the voltage across the inductor is maximum negative, and so on, we find that the instantaneous current through the inductor in Figure 18-1 must have the *sine-wave* shape shown by the black curve in Figure 18-2. But the instantaneous current sine wave reaches its positive peak $\pi/2$ radians later in time than the instantaneous voltage drop across the inductor, and it passes through zero, going in a negative direction, $\pi/2$ radians later in time than the inductive voltage drop. Therefore,

For a sine-wave voltage drop to appear across an ideal inductor, the current through it must be a sine wave which lags the inductive voltage drop by $\pi/2$ radians.

We can now write an equation for the instantaneous current in the circuit of Figure 18-1.

$$i_L = I_{\mathrm{m}} \sin\left(\omega t - \frac{\pi}{2}\right) \text{ rad} \qquad (18\text{-}3)$$

18-2 INDUCTIVE REACTANCE

When we connected a dc voltage source to a circuit containing inductance but no resistance in Chapter 16, the current had to keep rising at a constant rate in order to maintain a constant induced voltage in the inductor. But when we connect a sine-wave voltage source to the circuit of Figure 18-1, the instantaneous current must take on a sine-wave form, lagging the applied voltage sine wave by $\pi/2$ radians in order to produce the required sine-wave voltage drop across the inductance. If the current in the inductor is a sine wave, it must have some definite rms value; thus we would discover a constant reading on the ac ammeter in Figure 18-1. Therefore, for a given alternating voltage source and a given inductance, the ratio V/I is a constant.

In discussing Ohm's law of constant proportionality, we decided that a constant V/I ratio represented the opposition of the circuit to electric current, and we called this opposition the **resistance** of the circuit. We can, therefore, say that the constant V_L/I_L ratio in Figure 18-1 represents the opposition of the inductance to *alternating* current. However, we cannot call this opposition **resistance,** since the alternating current in a resistor is in phase with the voltage drop across it, whereas the current in an ideal inductor lags the voltage across it by $\pi/2$ radians. Furthermore, an inductor does not convert electric energy into heat as does a resistor. Therefore, we must use a different term which suggests *opposition* to alternating current. Thus,

Inductive reactance is the opposition of inductance to alternating current. The letter symbol for inductive reactance is X_L.

$$X_L = \frac{V_L}{I_L} \qquad (18\text{-}4)$$

Since inductive reactance is a V/I ratio just as is resistance, we can use the **ohm** as the unit of inductive reactance. Hence,

An ac circuit has an inductive reactance of one ohm when an alternating current having an rms value of one ampere flowing through the inductance creates an inductive voltage drop with an rms value of one volt across the inductance of the circuit.

18-3 FACTORS GOVERNING INDUCTIVE REACTANCE

If we assume that the shape of the applied voltage in the circuit of Figure 18-1 is to remain sinusoidal, there are only three factors which we can vary independently: the peak voltage or **amplitude** of the sine wave of applied voltage (hence, also its rms value), the frequency of the applied voltage, and the inductance. Let us consider in turn, the effect of each independent variable on the current in the circuit.

If we double the peak value of the applied voltage leaving the frequency constant, the maximum rate of change of current must be doubled in order to develop the required inductive voltage drop. Since the frequency is unchanged, the current will reach its maximum value in the same length of time; yet, at each instant in time, its slope is twice as steep as its original values. Therefore, the peak value of the current must also be twice as great as originally. Consequently, doubling the amplitude of the applied voltage also doubles the current. As we would expect, since $X_L = V_L/I_L$, there is no change in inductive reactance. Hence,

The magnitude of the applied alternating voltage has no effect on the inductive reactance of an ac circuit.

If we double the frequency leaving the peak value of the applied voltage the same, the instantaneous current in the inductance must have the *same* maximum rate of change. But, because the frequency is doubled, the instantaneous current must cease rising in half the original time. Therefore, since the current rises with the same maximum slope for half as long, the amplitude of the current sine wave is now only half as great as originally, even though the amplitude of the voltage waveform is unchanged. Consequently, since $X_L = V_L/I_L$, the inductive reactance has been doubled. And

Inductive reactance is directly proportional to frequency.

If we double the inductance, leaving the source unchanged, according to Equation (18-2), the maximum rate of change of current will have to be cut in half to develop the same voltage. Therefore, the amplitude of the currrent is again only half as great for the same amplitude of voltage. And

Inductive reactance is directly proportional to the inductance.

We can avoid the calculus solution for Equation (18-2) by applying a technique we used in Section 16-6 to determine the energy stored by an inductance.[†] We can substitute the *average* rate of change of current for (di/dt) and the *average* value for sin ωt. Since the current must rise from zero to I_m in one quarter-cycle, and since the **period** of one cycle is $1/f$ seconds,

$$\text{average } \frac{di}{dt} = \frac{I_m}{\frac{1}{4} \times (1/f)} = 4fI_m$$

Therefore, Equation (18-2) becomes

$$4fI_m = \frac{E_m \times 2/\pi}{L}$$

from which

$$\frac{E_m}{I_m} = 2\pi fL$$

and

$$X_L = 2\pi fL \tag{18-5}$$

where X_L is inductive reactance in ohms, f is frequency in hertz, and L is inductance in henrys.

Since

$$\omega = 2\pi f \tag{17-5}$$

$$X_L = \omega L \tag{18-6}$$

where ω is angular velocity in radians per second.

18-4 THE NATURE OF THE INSTANTANEOUS CURRENT IN A CAPACITOR

We can follow a similar procedure in order to determine the behavior of capacitance in ac circuits. If we connect a capacitor across a sine-wave voltage source, as in Figure 18-3, Kirchhoff's voltage law requires the voltage across the capacitor at every instant in time to be exactly the same as the applied voltage at that instant. But the only way that a voltage across a capacitor can change is for the capacitor to charge or discharge. Consequently, the capacitor in Figure 18-3 must charge and discharge in such a manner that the voltage across it is a sine wave equal to the applied voltage at every instant in time.

Since

$$q = Cv \tag{11-7}$$

$$\frac{dq}{dt} = C\frac{dv}{dt}$$

[†]See Appendix 2-7 for calculus solution.

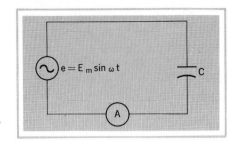

FIGURE 18-3
Capacitance in an Alternating-Current Circuit.

But by definition, $i = dq/dt$. Therefore,

$$i = C \frac{dv}{dt}$$ (18-7)

Since capacitance depends on such physical factors as the area of the plates and the dielectric constant, it cannot change in the short period of one cycle. Hence, as far as Equation (18-7) is concerned, capacitance is a constant. Accordingly, in determining the nature of the instantaneous current in Figure 18-3, we can state that it is always directly proportional to the *rate* at which the potential difference or voltage across the capacitor is changing.

Upon examining the colored sine curve of Figure 18-4 representing the instantaneous voltage across the capacitor, we note that the maximum voltage across the capacitor and the maximum *rate of change* of voltage occur $\pi/2$ radians apart in time. When the voltage across the capacitor is maximum, momentarily it is neither rising nor falling. Therefore, at this particular instant, the instantaneous current must be zero. Maximum rate of change of voltage occurs when the colored sine curve of Figure

FIGURE 18-4
Instantaneous Current in a Capacitor.

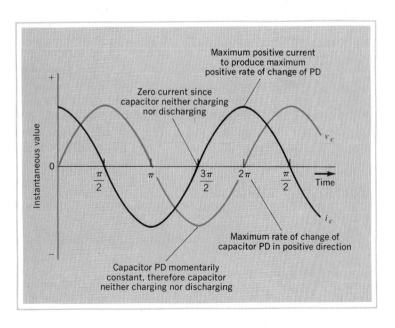

18-4 is steepest. The steepest slope of the instantaneous voltage sine wave occurs at the instant that the capacitor finishes discharging its stored charge and starts to recharge with the opposite polarity.[†] Therefore, the instantaneous current must be maximum in the positive direction when the voltage across the capacitor is changing from a negative polarity to a positive polarity. It will be maximum negative when the voltage changes from a positive polarity to a negative polarity, as shown in Figure 18-4.

When we plot an instantaneous current graph to fulfill the conditions of being zero when the voltage across the capacitance of the circuit is maximum and having a maximum positive value when the voltage is changing from a negative to a positive polarity, and so on, we find that the instantaneous current in the circuit of Figure 18-3 must follow a sine curve, as shown by the black graph of Figure 18-4. But the instantaneous current must be maximum positive at the instant that the voltage across the capacitance is just starting to build up in the positive direction. Then, $\pi/2$ radians later in time when the voltage across the capacitance has reached its positive peak, the instantaneous current has passed its positive peak and has dropped back to zero. Therefore,

For a sine-wave potential difference to be developed across a capacitor, the current through it must be a sine wave which leads the instantaneous voltage by $\pi/2$ radians.

Writing this statement in equation form,

$$i_c = I_m \sin\left(\omega t + \frac{\pi}{2}\right) \text{ rad} \tag{18-8}$$

18-5 CAPACITIVE REACTANCE

When we connect a capacitor across a sine-wave voltage source, the instantaneous current must take on a sine wave form leading the voltage sine wave by $\pi/2$ radians in order to produce a sine wave of potential difference across the capacitor. If the current in the capacitor is a sine wave, it must have some definite rms value, and we would discover a steady reading on the ac ammeter of Figure 18-3. A steady reading on the ammeter does not mean that current is flowing *through* the insulating dielectric of the capacitor. In an ac circuit, instantaneous current flows in one direction for one half-cycle and then in the opposite direction for the next half-cycle. (The average of one complete cycle of a sine wave of current is zero.) Therefore, the capacitor can

---

[†]If we consider the manner in which a capacitor is discharged by electrons flowing into the plate which has a deficiency of electrons and flowing out of the plate which has a surplus of electrons, we realize that if this flow continues in the same direction after the original charges have been neutralized, the plate which was originally positively charged due to a deficiency of electrons will start to accumulate a surplus of electrons, thus becoming negatively charged. Similarly, the plate which was originally negatively charged will lose more electrons than are required to neutralize the original charge, thus starting to build up a positive charge. Therefore, there will be no change in the direction of the current as the polarity of the voltage across the capacitor reverses.

alternately charge, discharge, recharge with the opposite polarity, etc., with the required **displacement current** for the charging and discharging process forming a sine-wave alternating current in the circuit.[†]

For a given alternating voltage source and a given capacitance, both the sine-wave voltage across and the sine-wave current "through" the capacitor have definite rms values. Therefore, the ratio V/I is a constant. Just as in the case of resistance and inductive reactance, this constant V/I ratio represents the opposition of the capacitor to alternating current. Therefore,

Capacitive reactance is the opposition of capacitance to alternating current. The letter symbol for capacitive reactance is X_C.

$$X_C = \frac{V_C}{I_C} \qquad (18\text{-}9)$$

An ac circuit has a capacitive reactance of one ohm when alternating current with an rms value of one ampere creates an alternating voltage with an rms value of one volt across the capacitance of the circuit.

18-6 FACTORS GOVERNING CAPACITIVE REACTANCE

Now that we have established that the displacement current charging and discharging a capacitor in an ac circuit is a sine wave, we can discover the factors which govern capacitive reactance by again following the procedure of independently varying the amplitude and frequency of the applied voltage and the value of the capacitance in the circuit of Figure 18-3.

If we double the amplitude of the applied voltage, leaving the frequency constant, the capacitor will have to store twice as much charge in the same time interval. Therefore, the peak charging current must be twice as great. Since $X_C = V_C/I_C$, if both V_C and I_C are doubled, the capacitive reactance remains unchanged. And

The magnitude of the applied alternating voltage has no effect on the capacitive reactance of an ac circuit.

If we double the frequency, leaving the peak value of the applied voltage the same, the capacitor must store the same charge but in half the time. Therefore, the peak charging current must be twice as great. Consequently, although V_C is unchanged, I_C is doubled, and, since $X_C = V_C/I_C$, the magnitude of the capacitive reactance has been cut in half. Thus,

*Capacitive reactance is **inversely** proportional to frequency.*

[†]Hence, the characteristic of capacitors is that they can "*pass*" alternating current but *block* direct current.

Finally, if we double the capacitance, leaving the source unchanged, according to Equation (11-7), at the peak of the voltage sine wave, the capacitor must have stored twice as much charge. Since the frequency is unchanged, the capacitor must acquire this greater charge in the same length of time. Therefore, the peak charging current must be twice as great. Again, V_C is unchanged and I_C is doubled. Hence,

Capacitive reactance is inversely proportional to the capacitance.

Again we can solve for reactance without resorting to calculus by considering the *average* rate of voltage rise.[†] Rewriting Equation (18-7) in the form

$$\frac{dv}{dt} = \frac{I_m \sin \omega t}{C} \tag{18-10}$$

we can substitute the *average* rate of voltage rise for dv/dt and the *average* value for $\sin \omega t$. Since the voltage across the capacitor must rise from zero to E_M in one-quarter of a cycle,

$$\text{average } \frac{dv}{dt} = 4fE_m$$

Therefore, Equation (18-10) becomes

$$4fE_m = \frac{I_m \times 2/\pi}{C}$$

from which

$$\frac{E_m}{I_m} = \frac{1}{2\pi fC}$$

and

$$X_c = \frac{1}{2\pi fC} \tag{18-11}$$

where X_C is capacitive reactance in ohms, f is frequency in hertz, and C is capacitance in farads.

Since

$$\omega = 2\pi f \tag{17-5}$$

when we substitute in Equation (18-11),

$$X_c = \frac{1}{\omega C} \tag{18-12}$$

[†]See Appendix 2-8 for calculus solution.

18-7 RESISTANCE, INDUCTIVE REACTANCE, AND CAPACITIVE REACTANCE

Resistance, inductive reactance, and capacitive reactance represent the opposition of the three basic properties of electric circuits to alternating current. Each is defined in terms of a V/I ratio and each is expressed in ohms. Although the instantaneous currents through all three components vary sinusoidally when a sine wave of voltage is applied across them, for a given circuit and a given frequency, the magnitude of this opposition (in ohms) is a *nonvarying* quantity which is independent of any particular instant in time. Resistance, inductive reactance, and capacitive reactance are *scalar* quantities represented by upper-case letter symbols.

However, there are significant differences in the behavior of resistance, inductance, and capacitance in ac circuits, which we must constantly keep in mind: (1) whereas resistance converts electric energy into heat, inductance and capacitance alternately *store* energy during the charging quarter-cycle, only to return this energy to the circuit during the next (discharging) quarter-cycle; (2) resistance is independent of frequency, but inductive reactance is *directly* proportional to frequency and capacitive reactance is *inversely* proportional to frequency; and (3) whereas the current through a resistor is in phase with the voltage across it, the current through an ideal inductor *lags* the voltage across it by 90° or $\pi/2$ radians and the alternating current "through" a capacitor *leads* the voltage across it by 90° or $\pi/2$ radians. These characteristics are summarized in Table 18-1.

TABLE 18-1

R, L, AND C IN AC CIRCUITS		
Resistance	$R = \dfrac{V_R}{I_R} = R$ ohms	I_R is in phase with V_R
Inductive Reactance	$X_L = \dfrac{V_L}{I_L} = \omega L$ ohms	I_L lags V_L by 90°
Capacitive Reactance	$X_C = \dfrac{V_C}{I_C} = \dfrac{1}{\omega C}$ ohms	I_C leads V_C by 90°

It is important, therefore, that we identify clearly the *type* of opposition in making statements about ac circuits. For example, the statement that *the reactance of circuit A is* 200 *ohms* is a correct statement. But it provides only part of the necessary information. As shown in Table 18-1, there is a considerable difference between an ac circuit containing *inductive* reactance and one containing *capacitive* reactance. Careful attention to letter symbols and subscripts (R, X_L, and X_C) will provide the necessary identification. To maintain the identification along with a numerical value expressed in ohms, we sometimes include the appropriate *type* of ohms designation, as shown in the following examples.

EXAMPLE 18-1

What is the resistance of a 660-W toaster operated from a 110-V 60-Hz source?

SOLUTION

$$R = \frac{E^2}{P} = \frac{(110 \text{ V})^2}{660 \text{ W}} = 18.3 \ \Omega \ (\text{resistive})$$

EXAMPLE 18-2

What inductive reactance develops a 40-V induced voltage when the current through it has an rms value of 80 mA?

SOLUTION

$$X_L = \frac{V_L}{I_L} = \frac{40 \text{ V}}{80 \text{ mA}} = 500 \ \Omega \ (\text{inductive})$$

EXAMPLE 18-3

What is the reactance of a 68-pF capacitor when it is used in a 4.5-MHz circuit?

SOLUTION

$$X_C = \frac{1}{2\pi fC} = \frac{1}{2 \times \pi \times 4.5 \text{ MHz} \times 68 \text{ pF}} = 520 \ \Omega \ (\text{capacitive})$$

Rather than considering dc and ac circuit-analysis techniques separately, some textbooks like to consider steady-state direct current as a special zero-frequency case of alternating current. At zero frequency, $\omega = 0$. Hence, $X_L = \omega L = 0$ and $X_C = 1/\omega C = \infty$; which only goes to show what we already know. The steady-state dc response of inductance is a short circuit, and the steady-state dc response of capacitance is an open circuit.

PROBLEMS

18-1. What is the inductive reactance of an ideal inductor which allows a 4-A current to flow when it is connected across a 117-V ac source?

18-2. What value of inductive reactance is required to limit current through it to 24 mA when the voltage across it is 35 V rms?

18-3. What value of capacitive reactance will allow a 3-mA alternating current to flow when it is connected to a 50-V 400-Hz source?

18-4. What is the reactance of a capacitor which allows a 7-μA current to flow when the voltage across it is 22 mV rms?

18-5. What is the 60-Hz reactance of a 4-H choke?

18-6. What is the reactance of a 250-mH radio-frequency choke at a frequency of 4.5 MHz?

18-7. What is the reactance of a 0.05-μF coupling capacitor at 400 Hz?

18-8. What is the reactance of a 27-pF capacitor at a frequency of 88 MHz?

18-9. At what frequency will a 0.5-H inductor have a reactance of 2000 Ω?

18-10. At what frequency will a 48-μH inductor have a reactance of 250 Ω?

18-11. What value of inductance will have a reactance of 1400 Ω at 475 kHz?

18-12. What value of inductance will draw a 160-mA rms current when connected to a 110-V 25-Hz source?

18-13. At what frequency will an 8-μF capacitor have a reactance of 160 Ω?

18-14. At what frequency will a 0.002-μF capacitor have a reactance of 80 Ω?

18-15. What value of capacitance is required if its 400-Hz reactance is to be 50 Ω?

18-16. The input resistance to a transistor audio amplifier is 47 kΩ. What value of coupling capacitor should be used if its reactance is to equal the input resistance at 20 Hz?

18-17. The solenoid of a magnetic contactor for starting an industrial electric motor draws 1.8 A from the 208-V 60-Hz source. Assuming negligible resistance, what is the inductance of the solenoid?

18-18. A capacitor is connected across a 6.6-kV 60-Hz transmission line for power factor correction. What capacitance is required for a 40-A reactive current?

18-19. What value of inductance is required in order that its reactance at 580 kHz be equal to the reactance of a 350-pF capacitor at that frequency?

18-20. At what frequency will the reactance of a 640-μH coil equal that of a 400-pF capacitor?

REVIEW QUESTIONS

18-1. Why must the instantaneous current sine wave be *exactly* $\pi/2$ radians out of phase with the applied voltage waveform across an ideal inductor?

18-2. Referring to the definition of inductance in Chapter 15, why must the current *lag* the voltage by $\pi/2$ radians, rather than lead it?

18-3. What is the advantage of expressing the phase difference between the voltage across and the current through an ideal inductor as $\pi/2$ radians, rather than 90°?

18-4. Why is it possible for us to label the *time* axis of the graph of Figure 18-2 in terms of *radians*?

18-5. What condition is imposed in expressing the elapse of time in radians?

18-6. Why does the meter in Figure 18-1 show a constant reading?

18-7. The term **resistance** is used in both dc and ac circuits. Why does the term **inductive reactance** not apply to dc circuits?

18-8. Why is it possible to express inductive reactance in *ohms*?

18-9. What is the significance of the symbol ω in the equation which states the factors governing inductive reactance?

18-10. With reference to the definition of capacitance in Chapter 11, why does the current *lead* the voltage across the capacitor by $\pi/2$ radians, rather than lag it?

18-11. When we connect the terminals of a capacitor to an ohmmeter, we obtain an infinitely high resistance reading. Nevertheless, the ammeter in Figure 18-3 can indicate a current of many amperes in the circuit. Explain.

18-12. The reactance of an inductor increases with an increase in frequency but the reactance of a capacitor decreases with an increase in frequency. Explain.

18-13. A single electrical component is sealed into a "black box" with a pair of terminals marked "1000 Ω." Suggest a laboratory procedure for determining whether the sealed box contains a resistor, a capacitor, or an inductor.

18-14. Explain how the polarity of the voltage across a capacitor can change without any change in the direction of the current "through" the capacitor.

19

PHASORS

Now that we know how resistance, inductance, and capacitance behave individually in an alternating current circuit, we can proceed to ac circuits containing both resistance and reactance. We require a type of algebra that can handle numbers in *two* dimensions, rather than the simple *linear* algebra we have been using for dc circuits. Such algebra is called **complex algebra,** not because it is difficult, but rather because it deals with **real** and **imaginary** numbers that exist in the two-dimensional *complex plane*. There are several forms of complex numbers. In more advanced ac theory courses, we use the exponential form of complex numbers based on Euler's theorem. However, for this introductory course, we can limit our complex algebra to the polar and rectangular coordinates of complex numbers. Students who are quite familiar with conversion from polar to rectangular coordinates, and vice versa, can proceed to Chapter 20.

19-1 ADDITION OF SINE WAVES

Figure 19-1 shows a simple ac circuit containing resistance and inductance in series. The sine-wave voltage source will cause a sine wave of current to flow in the circuit. Since this is a simple series circuit, the current must be the same in all parts of the circuit. The current in the inductance must have the same magnitude as the current in the resistance, must reach its peak magnitude at the same instant, and become zero at the same instant. In fact, the sine-wave currents in the inductance and the resistance are one and the same thing, and the letter symbol for current in a series circuit requires no subscript.

462

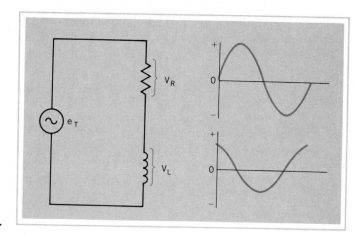

FIGURE 19-1
Simple AC Circuit.

The sine wave of current through the resistance will cause a sine-wave voltage drop across it. This sine wave of voltage is exactly *in phase* with the current sine wave. The instantaneous voltage across the resistance becomes maximum when the instantaneous current is maximum, etc.

In Chapter 18, we found that the instantaneous current through inductance has to *lag* the instantaneous voltage applied to it by $\pi/2$ radians. Conversely, then, the sine-wave voltage across the inductance portion of the circuit of Figure 19-1 must *lead* the current waveform through it by $\pi/2$ radians. Consequently, the sine-wave voltage across the inductance and the sine-wave voltage across the resistance are $\pi/2$ radians *out of phase*, with the instantaneous voltage across the inductance reaching its positive peak $\pi/2$ radians earlier in time than the instantaneous voltage across the resistance.

Kirchhoff's voltage law states that the sum of the voltage drops in a series circuit must equal the applied voltage. In dc circuits where the terminal voltage of the source is independent of *time*, we were able to find the sum of the voltage drops by simple arithmetic (linear algebra). Even when dealing with the charging of a capacitor in Chapter 12 and the rise of current in an inductor in Chapter 16, we were able to find the sum of the *instantaneous* voltages drops by simple arithmetic addition.

Our chore now is to find the *sum* of two out-of-phase sine-wave voltages appearing across the resistance and inductance of the circuit shown in Figure 19-1. As we discovered in Chapter 17, we can represent a sine-wave voltage either by its instantaneous value $v = V_m \sin \omega t$, or by its rms value $V = 0.707 V_m$.

The instantaneous value varies continuously and depends on an exact instant in *time*. Hence, we say that instantaneous values belong to the **time domain.** In the time domain, we use simple linear algebra and express angular distance (traveled by a rotating phasor in a given *time* interval) in radians.

We have already discovered that the rms value of a sine wave does not vary with the exact instant in time. However, we shall find that V_R and V_L also are 90° out of phase. To describe an alternating voltage by its rms value, we must include both magnitude and phase angle. $V \angle \phi$ is a complex number. We say that it belongs to the **frequency domain.** In the frequency domain, we use complex algebra and return to stating phase angle in *degrees*.

So that we may appreciate the advantages of being able to add rms values of sine waves, we shall first consider addition of their instantaneous values in the time domain.

19-2 ADDITION OF INSTANTANEOUS VALUES ON A LINEAR GRAPH

Since Kirchhoff's voltage law states that, at *every instant in time*, the algebraic sum of the various voltage drops must equal the algebraic sum of the applied voltages, we can then state that

$$e_T = v_R + v_L \tag{19-1}$$

at each instant in time. Therefore, we can add the *instantaneous* values of sine waves by simple algebraic addition. Although such an addition for only one instant in time does not give us the overall picture of the addition of complete sine waves, we can repeat the calculation at small intervals in angular distance over a complete cycle to determine the nature of the resultant sum. To assist us in this time-consuming chore, we can draw the sine curves for the two sets of instantaneous values we wish to add on a common *time* axis, as in Figure 19-2. Since the instantaneous voltage across the inductance leads the instantaneous voltage across the resistance by $\pi/2$ radians in the circuit of Figure 19-1 it must reach its positive peak $\pi/2$ radians earlier in time than the voltage across the resistance. We indicate the passage of time by distance to the right along the horizontal axis of the graph in Figure 19-2. And since angular velocity is constant for a given frequency, we can mark the horizontal axis of the graph in radians, rather than in milliseconds.

At $t = 0$, the instantaneous voltage across the resistance is zero; therefore, the instantaneous value of the total voltage is the same as the voltage across the inductance at that instant. As time elapses, the instantaneous voltage across the inductance is decreasing and the instantaneous voltage across the resistance is increasing. At ap-

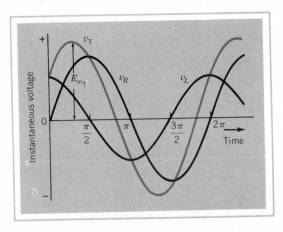

FIGURE 19-2
Addition of Instantaneous Values of Two Out-of-Phase Sine Waves on a Linear Graph.

proximately one radian, their sum is a maximum, and the resultant waveform reaches its peak value at this point. At $\pi/2$ radians, the instantaneous voltage across the inductance is zero, and the instantaneous value of the total voltage becomes the same as the instantaneous voltage across the resistance. At approximately $(1 + \pi/2)$ radians, the instantaneous values of the voltages across the resistance and the inductance are equal but opposite in polarity. Therefore, at this point, the instantaneous value of the sum must become zero.

As we draw the colored curve of Figure 19-2 through the plots we obtained by finding the simple algebraic sum of the instantaneous values at regular small intervals, we note that

The resultant sum of sine waves of the same frequency is also a sine wave of the same frequency.

We also note that, since the two sine waves we are adding do not reach their peak values at the *same* instant in time, the peak value of the resultant occurs at a time when the instantaneous values of the two component voltages are *less* than their peak values. Therefore,

$$E_{mT} \neq V_{mR} + V_{mL}$$

Since the rms value of any sine wave is 0.707 of the peak value,

$$|E_T| \neq |V_R| + |V_L|$$

The vertical bars enclosing the letter symbols indicate that the letter symbol represents the *magnitude* or *absolute value* of the quantity represented by the symbol.[†] The magnitude of a quantity is a simple positive numeral with no directional or angular data. Hence, the addition process represented by the above equations is simple algebraic addition. Therefore, although simple algebraic addition *does* apply to the instantaneous values of a sine wave, it does *not* apply to the peak and rms values if the sine waves being added are out of phase, as in Figure 19-2.

In the special case of sine waves that are exactly in phase, as in Figure 19-3, the resultant and all component sine waves do reach their peak values at the same instant in time. For *in-phase* sine waves,

$$E_{mT} = V_{m1} + V_{m2} \quad \text{and} \quad |E_T| = |V_1| + |V_2|$$

Therefore, we may add by simple arithmetic the rms values of sine waves that are exactly in phase.

With respect to the circuit of Figure 19-1, we can expand Equation (19-1) to

$$e_T = V_{mR} \sin \omega t + V_{mL} \sin(\omega t + \pi/2) \tag{19-2}$$

[†]The mathematical term for the absolute value of a quantity is the **modulus** of the quantity.

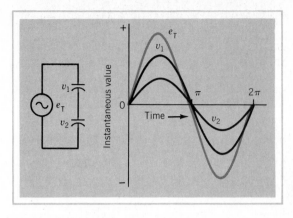

FIGURE 19-3
Addition of Instantaneous Values of Two In-Phase Sine Waves on a Linear Graph.

19-3 REPRESENTING A SINE WAVE BY A PHASOR DIAGRAM

Although we can determine the rms value of the resultant sine wave and its phase angle with respect to the component sine waves by means of the graphical addition of instantaneous values, as in Figures 19-2 and 19-3, the main feature of these graphs, which is apparent at a glance, is the sine-wave shape of the resultant. Once we accept that the resultant of the addition of sine waves of the same frequency is also a sine wave of the same frequency, the tedious procedure required to produce Figures 19-2 and 19-3 is not warranted. The original information from which we constructed the graphs of Figures 19-2 and 19-3 was simply their rms values and the phase angle between them. Rms value and phase angle will also serve to identify the resultant.

In Figure 17-3, we used a phasor as a means of drawing a simple sketch to indicate the position of a loop of wire as it cut across magnetic lines of force. The length of this phasor represented the maximum rate of cutting or E_m (which was a constant), and its position indicated the angular distance the loop had traveled in its cycle at that particular instant in time. If we draw phasors for the two sine-wave voltages in the circuit of Figure 19-1, the rotating phasor for the instantaneous voltage across the inductance will always stay exactly $\pi/2$ radians ahead of the rotating phasor for the instantaneous voltage across the resistance. Since the conventional direction for indicating this motion on a phasor diagram is counterclockwise, the rotating phasor for v_L will then be $\pi/2$ radians further counterclockwise than the phasor for v_R at every instant of time.

Because the phasors representing the voltages across the inductance and the resistance are always 90° apart, there is no motion of the one with respect to the other. Therefore,

We can represent sine waves of the same frequency by "stationary" phasors with the angle between the phasors representing the phase angle between the sine waves.

We can adopt a concept of stationary phasors by defining a **phasor quantity** as any quantity that is expressed in complex form, i.e., by an absolute magnitude and an angle. And since the rms value of a sine wave is always 0.707 of its peak value,

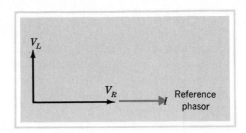

FIGURE 19-4
Phasor Diagram for the Circuit of Figure 19-1.

we might as well make the length of the phasors (magnitude) proportional to the rms value of the sine wave they represent. These ac phasors belong to the frequency domain where we usually express angles in degrees.

In the simple series circuit of Figure 19-1, the current is common to all components. Therefore, in constructing the phasor diagram for a series circuit (Figure 19-4), we draw a phasor representing the rms value of the current along the reference axis of the diagram (in the three o'clock direction). Since the voltage across the resistance is in phase with the current through it, we mark V_R in the same direction as the current phasor. Since the voltage across the inductance *leads* the current through it by 90°, we draw the phasor for V_L 90° counterclockwise from the reference axis.

We prepared the phasor diagram of Figure 19-4 much more rapidly than the instantaneous voltage graphs of Figure 19-2. Nevertheless, we can see the pertinent data regarding rms values and phase angles at a glance in Figure 19-4, whereas we have to derive these data from the information shown in Figure 19-2. If we can now locate the phasor for the resultant voltage on the phasor diagram of Figure 19-4 without excessive calculations, we shall have a very powerful aid to the solution of ac circuits.

19-4 LETTER SYMBOLS FOR PHASOR QUANTITIES

In studying the behavior of electric and magnetic circuits, we encounter three distinct types of numerical quantities.

Scalars are quantities which possess magnitude but have no direction or angular information. Ohms, henrys, and farads are scalars. In dc circuits, where voltage and current are not time-varying quantities, they can be treated as scalars. In ac circuits, *instantaneous* values of voltage and current are scalars. Scalars are represented in equations by lightface italic letter symbols (R, X_L, C, v, i, etc.). We add scalar quantities by simple algebraic addition.

Vectors are quantities which possess both magnitude and direction. Gravitational force, wind velocity, electric field intensity, magnetic flux density are all vector quantities. Although vector quantities may be represented by lightface italic letter symbols, if we wish to call attention to their *vector* characteristic, we use **boldface** italic letter symbols ($\boldsymbol{F}$, $\boldsymbol{v}$, $\boldsymbol{E}$, $\boldsymbol{B}$, etc.). The boldface type helps to remind us that vector quantities cannot be added by simple algebraic addition. Vector quantities are *complex* numbers representing magnitude and direction. Hence, vectors must be added by the rules for complex algebra.

Phasors are complex numbers also. The term **phasor quantity** refers specifically to steady-state values of quantities in alternating-current circuits which are complex numbers. The term **phasor** is used to avoid confusion with space *vectors* which are not time-varying functions. In technical journals, we are expected to know that rms ac quantities are phasor quantities—hence, the approved letter symbol for a phasor quantity is the familiar lightface italic letter symbol.[†] We should, therefore, represent an alternating voltage as a phasor quantity by the letter symbol V *where* $V = |V| \underline{/\phi}$.

To assist in distinguishing calculations requiring phasor algebra, many textbooks represent phasor quantities by boldface roman letter symbols. With this distinction, we can omit the vertical-bar solidus symbols of Section 19-2 and represent the magnitude of a phasor quantity by the standard lightface italic letter symbol. Hence, in the following chapters, we shall represent a phasor voltage by $\mathbf{V}$ *where* $\mathbf{V} = V \underline{/\phi}$.

Since we must use phasor algebra to add the two rms voltages shown in Figure 19-4, the equation that we could not write in Section 19-2 becomes

$$\mathbf{E}_T = \mathbf{V}_R + \mathbf{V}_L \tag{19-3}$$

19-5 PHASOR ADDITION BY GEOMETRICAL CONSTRUCTION

The resultant of two phasor quantities in an electrical phasor diagram (as in the vector diagram of forces in mechanics) is the diagonal from the origin to the opposite corner of a parallelogram with the two phasors as sides. If we draw the two phasors as accurately as possible on *polar* graph paper, we can draw the resultant with ruler and compass, as shown in Figure 19-5. Adjusting the compass to a radius equal to the length of $\mathbf{E}_1$, we trace an arc using the tip of the $\mathbf{E}_2$ phasor as the center. Then, adjusting the compass to a radius equal to the length of $\mathbf{E}_2$, we trace an arc using the tip of $\mathbf{E}_1$ as center. We then draw the resultant by joining the intersection of the two arcs to the origin with a straight line.

Applying this technique to the series circuit of Figure 19-1, we can quickly show the magnitude and phase relationships among $\mathbf{E}$, $\mathbf{I}$, $\mathbf{V}_R$, and $\mathbf{V}_L$. To illustrate the advantages of phasor diagrams, both the sine-wave graph of Figure 19-2 and the phasor diagram of Figure 19-4 are shown in Figures 19-6(a) and 19-6(b) respectively.

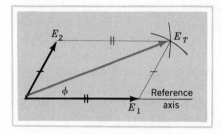

FIGURE 19-5
Geometrical Construction of the Resultant Phasor.

[†]ANSI/IEEE Standard 280-1985, *IEEE Standard Letter Symbols for Quantities Used in Electrical Science and Electrical Engineering*.

PART IV   ALTERNATING CURRENT

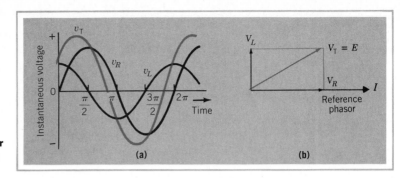

FIGURE 19-6
Sine-Wave Graph and Phasor Diagram for the *Same* Series Circuit of Figure 19-1.

Although we can determine (with some difficulty) the rms values and phase angles for I, V_R, V_L, and V_T from the graphical addition of sine waves in Figure 19-6(a), this information is clearly displayed by the phasor addition on the phasor diagram of Figure 19-6(b).

If we perform the geometrical construction carefully, we can measure the phase angle between the applied voltage E and the current I, and the rms values of the voltages to two-figure accuracy directly from the phasor diagram. Although geometrical solutions are not considered accurate enough for ac circuit problems in later chapters, the phasor diagram is a valuable guide in performing the more accurate complex algebra solutions. If we draw the phasors in their proper proportions and mark off reasonably accurate phase angles, we can construct very quickly a free hand phasor diagram that will give us an approximate answer to the problem. The phasor diagram also assists us in checking sine, cosine, and tangent functions of angles larger than 90°. It also shows us whether we have the correct sign for the numerical quantities used in the algebraic solution. Therefore, we *always* construct a phasor diagram in solving ac circuits to show us what we have to do and to provide a quick check for the more accurate complex algebra procedures.

EXAMPLE 19-1

Three sources of sine-wave voltages of the same frequency have the rms values and phase angles listed. $E_1 = 80$ V $\underline{/0°}$, $E_2 = 100$ V $\underline{/150°}$, and $E_3 = 40$ V $\underline{/45°}$. By geometrical construction on a phasor diagram solve for the total applied voltage when these sources are connected in series.

SOLUTION

Step I. Construct a phasor for each given voltage.

Step II. Construct a resultant phasor for any pair of phasors (E_1, and E_3 in Figure 19-7).

Step III. Construct a resultant phasor for the resultant of Step II and the remaining phasor.

Step IV. Measure the length and phase angle of the final resultant (approximately 79 V $\underline{/80°}$, in Figure 19-7).

CHAPTER 19   PHASORS   469

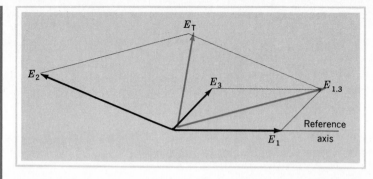

FIGURE 19-7

Solving Example 19-1 by Geometrical Construction on a Phasor Diagram.

19-6 ADDITION OF PHASORS WHICH ARE AT RIGHT ANGLES

We have already noted that, for the specific case of phasor quantities having a zero-degree angle between them, we can accomplish phasor addition by simple arithmetical addition (Figure 19-3). Similarly, when the angle between a pair of phasor quantities is exactly 180°, we can determine the phasor sum by simple arithmetical subtraction. In the specific case of two phasor quantities with an angle of exactly 90° between them, we can determine the phasor sum very accurately by making use of the relationships among the various sides and angles of a right-angled triangle as defined in trigonometry. Geometry also supplies a very useful relationship which applies to this specific case of two quantities with a 90° angle between them. Pythagoras' theorem states that the square of the hypotenuse of a right-angled triangle is equal to the sum of the squares of the other two sides.

In the simple series circuit of Figure 19-1, since the voltage drop across the resistance must be exactly in phase with the current and the voltage across the inductance must lead the current by exactly 90°, the two voltage phasors we wish to add are exactly 90° out of phase. If we solve for the resultant sum by geometrical construction, as in Figure 19-8(a) the parallelogram becomes a rectangle with side AC exactly equal to the phasor OB which represents the voltage across the inductance.

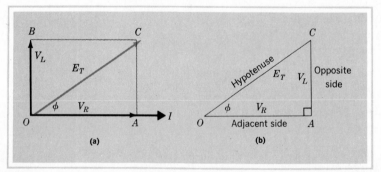

FIGURE 19-8

Addition of Phasors at Right Angles.

Since the triangle of Figure 19-8(b) contains exactly the same data as the conventional phasor diagram of Figure 19-8(a), which has all phasors starting from the same point of origin, we are allowed to use triangular diagrams in solving problems in phasor addition if we find that we can show data more clearly in this form.

Since E_T is a phasor quantity, it must possess angle information as well as magnitude information. Solving first for the angle between the total voltage and the current, from trigonometry,

$$\tan \phi = \frac{\text{opposite side}}{\text{adjacent side}} = \frac{AC}{OA} = \frac{V_L}{V_R}$$

Therefore, if we know the magnitude of the voltage across the inductance and the voltage drop across the resistance, we can determine the phase angle ϕ from trigonometric tangent tables or calculators with $\tan^{-1}$ keys.

$$\therefore \phi = \arctan \frac{V_L}{V_R} \tag{19-4}$$

where **arctan** means "the angle whose tangent is".[†]

We now have a choice of two methods of solving for the magnitude of the total voltage phasor OC. From trigonometry,

$$\sin \phi = \frac{\text{opposite side}}{\text{hypotenuse}} = \frac{AC}{OC}$$

from which

$$E_T = \frac{V_L}{\sin \phi} \tag{19-5}$$

or

$$\cos \phi = \frac{\text{adjacent side}}{\text{hypotenuse}} = \frac{OA}{OC}$$

from which

$$E_T = \frac{V_R}{\cos \phi} \tag{19-6}$$

The second method for determining the magnitude of the resultant is based on Pythagoras' theorem. In Figure 19-8(b),

$$OC^2 = OA^2 + AC^2$$

$$\therefore E_T = \sqrt{V_R^2 + V_L^2} \tag{19-7}$$

Note that it is conventional to state the resistance component before the reactive component in Equation (19-7). Since we require both magnitude and angle information to identify the exact nature of the total voltage, we may combine Equations (19-7) and (19-4) in the form

[†]**Arctan** usually appears on calculator keys as $\tan^{-1}$.

$$\mathbf{E}_T = \sqrt{V_R^2 + V_L^2} \ \bigg/ \arctan \frac{V_L}{V_R} \qquad (19\text{-}8)$$

EXAMPLE 19-2

If the voltage drop across the resistance in the circuit of Figure 19-1 is 15 V rms and the voltage across the inductance is 10 V rms, what are the magnitude of the total voltage and its phase angle with respect to the current?

TRIGONOMETRY SOLUTION

$$\phi = \arctan \frac{V_L}{V_R} = \arctan \frac{10}{15} = 33.7°$$

$$E_T = \frac{V_R}{\cos \phi} = \frac{15}{0.832} = 18 \text{ V}$$

or

$$E_T = \frac{V_L}{\sin \phi} = \frac{10}{0.555} = 18 \text{ V}$$

or

$$E_T = \sqrt{V_R^2 + V_L^2} = \sqrt{225 + 100} = 18 \text{ V}$$

$$\therefore \mathbf{E}_T = 18 \text{ V} \ \underline{/+33.7°}$$

The positive angle shows that the voltage *leads* the current.

CALCULATOR SOLUTION

We can solve for $\mathbf{E}_T$ with a calculator using trigonometry function keys according to the above steps. It is worth the time involved just to become familiar with the use of these keys.

However, calculators with a →P key can give us the solution in a single step when the two phasor quantities are at right angles. In Figure 19-8, V_R and V_L are the rectangular coordinates of $\mathbf{E}_T$. Entering 15 as the value for x, we then press the $x \leftrightarrow y$ key and enter 10 as the value for y. When we press the →P key, the display shows the magnitude of the resultant voltage 18.027 756. Pressing the $x \leftrightarrow y$ key once more then displays the phase angle 33.690 067°. We will understand this coordinate conversion procedure better after we have studied the next section.

19-7 EXPRESSING A PHASOR QUANTITY IN RECTANGULAR COORDINATES

When we describe a phasor by stating its magnitude and phase angle with respect to the reference axis, as we did in Example 19-2, we are stating its **polar coordinates;** that is, the coordinates that identify the phasor are its length along a radius from the origin and its angular distance from the reference axis. Polar coordinates are desirable for expressing the final answer to electric circuit calculations since the numerical quantities involved in this method of representation are those which we obtain in the laboratory with measuring instruments such as voltmeters, ammeters, and wattmeters.

By applying the basic trigonometric relationships to the phasor diagram of

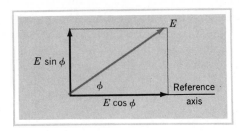

FIGURE 19-9
Determining the Rectangular Coordinates of a Phasor.

Figure 19-8, we were able to determine the polar coordinates of the total voltage from two components, one of which lies along the reference axis and one of which is perpendicular to the reference axis. Therefore, if we are given a phasor in polar form ($E \angle \phi$ in Figure 19-9), we can use the same basic trigonometry functions to express it in terms of one component along the reference axis and one component at right angles to the reference axis. We find the reference axis component (called the **real** component in mathematics) by multiplying the magnitude of the phasor by the *cosine* of its phase angle. We find the quadrature (perpendicular) component (called the **imaginary** component in mathematics) by multiplying the magnitude of the phasor by the *sine* of its phase angle.

We can, therefore, identify a phasor by two coordinates, one of which is the reference axis (or real) component and one of which is the quadrature (or imaginary) component. These two components are called the **rectangular coordinates** of the phasor. A rectangular coordinate must fall along a boundary between the four quadrants of the complex plane. In polar coordinate representation, we identify the boundaries between the quadrants in terms of Babylonian degrees, as shown in Figure 19-10(a). In rectangular coordinate representation, we identify the four possible directions for the coordinates by the complex plane *operators* shown in Figure 19-10(b).[†]

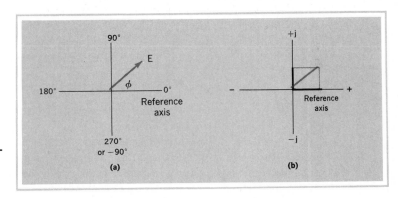

FIGURE 19-10
Boundaries between the Quadrants in (a) Polar Coordinates, and (b) Rectangular Coordinates.

[†]In the complex plane, multiplying a vector quantity along the reference axis by the j operator *rotates* it 90° in a counterclockwise direction. To rotate it another 90°, we simply multiply once more by the j operator. Hence, multiplying a vector quantity along the reference axis by j^2 changes its direction by 180°. In polar coordinates, reversing the direction of a phasor is the same as multiplying by −1. Hence $j^2 = -1$. It follows that $j = \sqrt{-1}$, which has no solution—thus the name *imaginary* for quantities preceded by the +j or −j operator. Multiplying the original vector quantity once more by j rotates it a further 90° to the 270° location in polar coordinates. Since $j^3 = j^2 \times j$ and $j^2 = -1$, then the complex plane equivalent of 270° is −j.

Hence, the rectangular coordinates of a phasor consist of a real number identified simply by + or − signs, and an imaginary number identified by either +j or −j operators. In writing the rectangular coordinates of a phasor, it is conventional to write the reference axis (real) component before the j (imaginary) component.

With reference to Figure 19-9, we can state the procedure for converting a phasor from polar coordinates to rectangular coordinates in equation form as

$$E \underline{/\phi} = +E \cos \phi + jE \sin \phi \qquad (19\text{-}9)$$

EXAMPLE 19-3

Express 18 V $\underline{/33.7°}$ in rectangular coordinates.

SOLUTION

Step I. Always draw a freehand phasor diagram as a quick means of identifying the correct operators to use with the rectangular coordinates (Figure 19-11).

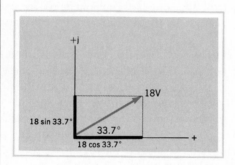

FIGURE 19-11
Phasor Diagram for Example 19-3.

Step IIA. Using sin and cos keys of a calculator:

$$\mathbf{E} = +E \cos \phi + jE \sin \phi = +18 \cos 33.7° + j18 \sin 33.7°$$

We calculate each coordinate separately.

$$18 \times 33.7 \; \cos \; = 15 \quad \text{and} \quad 18 \times 33.7 \; \sin \; = 10$$
$$\therefore \mathbf{E} = 15 + j10 \text{ V}$$

Step IIB. Using the →R coordinate conversion key: For coordinate conversion, the calculator can handle two unknowns. In polar coordinates, x represents the magnitude and y the phase angle. In rectangular coordinates, x represents the real component and y the imaginary or j component. The calculator automatically enters or displays the x component first. Hence, we enter

$$18 \; x{\leftrightarrow}y \; 33.7$$

The $x{\leftrightarrow}y$ key moves the x component into storage so that the y component can be entered. We have now entered the polar coordinates of $\mathbf{E}$. Pressing the →R key automatically converts x and y to rectangular coordinates and displays the x or real component. Pressing the $x{\leftrightarrow}y$ key then displays the j component.

EXAMPLE 19-4
State 4 A $\underline{/240°}$ in rectangular coordinates.

SOLUTION

Step I. Again a reasonably accurate freehand phasor diagram (Figure 19-12) shows the proper operators for the solution as a check on the signs displayed by the calculator.

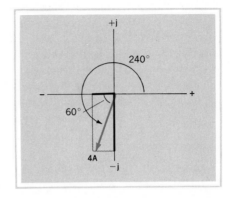

FIGURE 19-12
Phasor Diagram for Example 19-4.

Step II. Whether we use the $\boxed{\sin}$ and $\boxed{\cos}$ keys or the $\boxed{\rightarrow R}$ key, the calculator can automatically handle *any* angle and provide the correct + or − signs for the rectangular coordinates. Hence,

$$4 \quad \boxed{x \leftrightarrow y} \quad 240 \quad \boxed{\rightarrow R} \quad \text{displays} \quad -2.0$$

and pressing $\boxed{x \leftrightarrow y}$ again displays -3.464

$$\therefore \mathbf{I} = -2 - j3.46 \text{ A}$$

EXAMPLE 19-5
State 48 $\underline{/-45°}$ in rectangular coordinates.

SOLUTION

A *negative* angle is simply one measured *clockwise* from the reference axis. From Figure 19-13,

FIGURE 19-13
Phasor Diagram for Example 19-5.

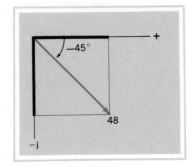

$$48 \underline{/-45^\circ} = +48 \cos 45^\circ - j48 \sin 45^\circ = +33.9 - j33.9$$

Again the calculator will handle the negative angle automatically if we enter

$$48 \quad \boxed{x \leftrightarrow y} \quad 45 \quad \boxed{+/-} \quad \boxed{\rightarrow R}$$

The procedure for converting from rectangular coordinates into polar coordinates is similar to that indicated by Equation (19-8). The magnitude of the polar quantity is simply the root of the sum of the squares of the reference axis rectangular coordinate and the j or quadrature rectangular coordinate. We find the proper phase angle for the resultant by drawing a phasor diagram of the rectangular coordinates. The angle contained within the triangle constructed on the horizontal axis is numerically equal to

$$\arctan \frac{\text{quadrature coordinate}}{\text{reference axis coordinate}}$$

From the position of the resultant in the phasor diagram, we can then determine the actual phase angle with respect to the reference axis.

EXAMPLE 19-6
State $-60 + j30$ volts in polar form.

TRIGONOMETRY SOLUTION
If we use the trigonometry function keys of the calculator, we must pay special attention to the phasor diagram of Figure 19-14. Since the $\tan^{-1}$ key of the calculator cannot distinguish between $30/-60$ and $-30/60$, it will display an angle of -26.565°. From Figure 19-14, we realize that the angle we want for the polar coordinates is

$$\phi = 180^\circ - 26.6^\circ = +153.4^\circ$$

$$V \text{ is } 60 \quad \boxed{x^2} \quad + \quad 30 \quad \boxed{x^2} \quad = \quad \boxed{\sqrt{x}}$$

which displays 67.1 V or

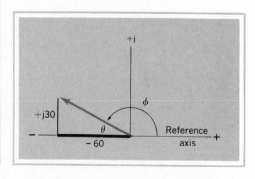

FIGURE 19-14
Phasor Diagram for Example 19-6.

$$V \text{ is } 60 \div 153.4 \boxed{\cos} = -67.1 \text{ V}$$

In polar coordinates, the magnitude of the phasor is always a positive quantity. Hence, we ignore the $-$ sign that cos 153.4° gives in the second solution.

$$\therefore \mathbf{V} = 67 \text{ V } \underline{/+153.4°}$$

CALCULATOR SOLUTION
Since the calculator knows that x is the real coordinate and y is the imaginary coordinate, when we use the $\boxed{\rightarrow P}$ key, the calculator will automatically locate the correct angle for polar coordinates.

$$60 \boxed{+/-} \boxed{x \leftrightarrow y} \ 30 \boxed{\rightarrow P}$$

displays 67.082 V. Pressing the $\boxed{x \leftrightarrow y}$ key then displays 153.43°.

19-8 PHASOR ADDITION BY RECTANGULAR COORDINATES

Although there are trigonometric procedures for adding phasors having angles between them or other than 0° or 90°, now that we can express phasor quantities in rectangular coordinates, we seldom need such procedures. When we are required to add several phasors with various phase angles, we first convert each one to its rectangular coordinates. We can now add all components in a horizontal (+ and −) direction by simple algebraic addition. Similarly, we can add all components falling in the vertical (+j and −j) direction by simple algebraic addition. This leaves us with one + or − coordinate and one +j or −j coordinate. These then are the rectangular coordinates of the sum of the individual phasors. All that remains is to convert these final rectangular coordinates into polar form. This procedure, as set forth in Example 19-7, is the conventional method of phasor addition for alternating-current circuit problems.

EXAMPLE 19-7
What is the sum of $\mathbf{E}_1 = 80 \text{ V } \underline{/+60°}$ and $\mathbf{E}_2 = 80 \text{ V } \underline{/-135°}$?

TRIGONOMETRY SOLUTION
Step I. Although we do not rely on geometrical construction for an accurate solution, we *always* construct a freehand phasor diagram (Figure 19-15), so we can check on any error in entering data into the calculator.

Step II. Converting each voltage into its rectangular coordinates,

$$\mathbf{E}_1 = 80 \cos 60° \quad + \text{ j}80 \sin 60° \quad = +40 \quad + \text{ j}69.28 \text{ V}$$

$$\mathbf{E}_2 = 80 \cos -135° + \text{ j}80 \sin -135° = -56.57 - \text{ j}56.57 \text{ V}$$

Step III. Adding gives $\qquad\qquad \mathbf{E}_T = -16.57 + \text{ j}12.71 \text{ V}$

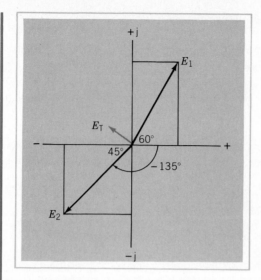

FIGURE 19-15
Phasor Diagram for Example 19-7.

Step IV. Converting the resultant into polar coordinates,

$$\theta = \arctan \frac{12.71}{-16.57} = -37.49°$$

From Figure 19-15,

$$\phi = 180° - 37.49° = +142.51°$$
$$E_T = \sqrt{16.57^2 + 12.71^2} = 20.88 \text{ V}$$
$$\therefore \mathbf{E_T} = \mathbf{20.9 \text{ V} \underline{/+142.5°}}$$

CALCULATOR SOLUTION
Most pocket calculators do not have enough storage to keep the solution to each step in the calculator. Hence, we jot down the coordinates after each conversion.

$$80 \text{ V}\underline{/+60°} = +40 \qquad + \text{j}69.2820 \text{ V}$$
$$80 \text{ V}\underline{/-135°} = -56.5685 - \text{j}56.5685 \text{ V}$$

Adding gives
$$\mathbf{E_T} = -16.5685 + \text{j}12.7135 \text{ V}$$

Converting to polar coordinates,

$$\mathbf{E_T} = \mathbf{20.884 \text{ V} \underline{/+142.5°}}$$

COMPUTER SOLUTION
Unless we have ready access to a previously prepared program for converting polar to rectangular coordinates and vice versa, a personal computer cannot compete with

478

an engineering calculator for this type of example. BASIC computer language is quite capable of performing the mathematics required to find E_T. However, BASIC does not provide for automatic coordinate conversion like the $\rightarrow$P and $\rightarrow$R keys of a calculator. We can use the computer in the calculator mode to perform the steps of the Trigonometry Solution, or we can write a simple program to add two phasor quantities. In the following sample program note that Line 160 recognizes that BASIC solves trigonometric functions in radians and does not store a value for π. Line 230 takes care of the dilemma in the Trigonometry Solution as to which quadrant arctan ATN(X) belongs.

```
100 REM     SUM OF TWO PHASORS
110 PRINT "ENTER MAGNITUDE AND ANGLE IN DEGREES OF PHASOR 1"
120 INPUT   M1, A1
130 PRINT "ENTER MAGNITUDE AND ANGLE IN DEGREES OF PHASOR 2"
140 INPUT   M2, A2
150 REM     CONVERT ANGLES TO RADIANS
160 PI = 3.1416: R1=A1*PI/180: R2=A2*PI/180
170 REM     ADD REAL X AND IMAGINARY Y COORDINATES
180 LET   XR=M1*COS(R1) + M2*COS(R2)
190 LET   YR=M1*SIN(R1) + M2*SIN(R2)
200 REM     CONVERT RESULTANT TO POLAR COORDINATES
210 LET   MR=SQR(XR^2 + YR^2)
220 LET   AR=180/PI*ATN(YR/XR)
230 IF   XR < 0 THEN AR=AR + 180
240 PRINT "MAGNITUDE OF RESULTANT IS ";MR
250 PRINT "ANGLE OF RESULTANT IS ";AR;" DEGREES"
260 END
```

19-9 SUBTRACTION OF PHASOR QUANTITIES

When expressed in rectangular coordinates, subtraction of two phasor quantities requires only simple algebraic subtraction of the reference axis and the quadrature components. We then convert resultant rectangular coordinates into polar form.

EXAMPLE 19-8
Subtract $4 - j5$ from $3 + j6$.

SOLUTION

$$3 + j6$$
$$\underline{4 - j5}$$

Subtracting: $\qquad -1 + j11$

Draw a simple phasor diagram of $-1 + j11$ to help us determine the phase angle of the polar form (Figure 19-16).

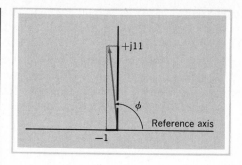

FIGURE 19-16
Phasor Diagram for Example 19-8.

$$-1 + j11 = \sqrt{1^2 + 11^2} \underline{/180 - \arctan 11/1}$$

$$= 11.05 \underline{/+95.2°}$$

We can also subtract by reversing the $+$ and $-$ signs in the subtrahend and then performing an algebraic *addition*.

$$
\begin{array}{r}
3 + j6 \\
-4 + j5 \\
\hline
\end{array}
$$

Adding gives $\qquad -1 + j11$

Again, the calculator chain 1 $+/-$ $x{\leftrightarrow}y$ 11 $\rightarrow$P $x{\leftrightarrow}y$ automatically provides the correct angle.

The second procedure in Example 19-8 gives us a clue as to how to go about subtracting phasors on a phasor diagram. If we locate $+4 - j5$ and $-4 + j5$ on a phasor diagram, as in Figure 19-17, we note that they represent phasors which are exactly equal in length but 180° apart in angular location. Therefore, we can use geometrical construction to subtract two phasor quantities by simply reversing the direction of the subtrahend phasor and then carrying out the customary geometrical construction for the addition of phasors on a phasor diagram.

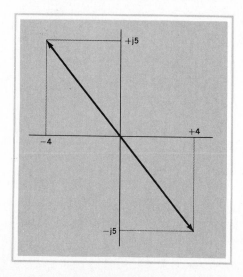

FIGURE 19-17
Reversing the Signs of Rectangular Coordinates.

480

EXAMPLE 19-9

Subtract $E_1 = 40\ \underline{/30°}$ from $E_2 = 50\underline{/135°}$ by geometrical construction on a phasor diagram.

SOLUTION

See Figure 19-18.

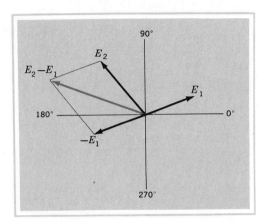

FIGURE 19-18
Subtraction of Phasors by Geometrical Construction.

19-10 MULTIPLICATION AND DIVISION OF PHASOR QUANTITIES

Multiplication and division of phasor quantities in *polar* form is the simplest of all the complex algebra processes.

To multiply phasor quantities expressed in polar form, multiply their magnitudes and algebraically add their phase angles.

$$\therefore |E_1|\underline{/\phi_1} \times |E_2|\underline{/\phi_2} = |E_1||E_2|\underline{/\phi_1 + \phi_2} \qquad (19\text{-}10)$$

To divide phasor quantities expressed in polar form, divide their magnitudes and algebraically subtract their phase angles.

$$\therefore |E_1|\underline{/\phi_1} \div |E_2|\underline{/\phi_2} = \frac{|E_1|}{|E_2|}\underline{/\phi_1 - \phi_2} \qquad (19\text{-}11)$$

EXAMPLE 19-10

Solve $80\underline{/40°} \times 11\underline{/-15°}$ and $80\underline{/40°} \div 11\underline{/-15°}$.

SOLUTION

$$80\underline{/40°} \times 11\underline{/-15°} = 80 \times 11\underline{/40 + (-15)} = \mathbf{880\underline{/25°}}$$

$$80\underline{/40°} \div 11\underline{/-15°} = \tfrac{80}{11}\underline{/40 - (-15)} = \mathbf{7.27\underline{/55°}}$$

Although the procedure for multiplying and dividing phasor quantities expressed in *rectangular* coordinates is quite straightforward, it does require skill in manipulating algebraic terms. Therefore, it is usually safer to carry out phasor multiplication and division by the simple polar method even though this often requires conversion from rectangular to polar coordinates. If our calculator has $\rightarrow P$ and $\rightarrow R$ keys, such conversion is the quickest way to carry out phasor multiplication and division. However, to illustrate the procedure for multiplication and division in rectangular form, we can refer to Example 19-11.

EXAMPLE 19-11

Solve $(8 + j6) \times (2 - j4)$ and $(8 + j6) \div (2 - j4)$.

SOLUTION

$$8 + j6$$
$$\underline{2 - j4}$$

Multiplying by 2, $\quad 16 + j12$

Multiplying by $-j4$, $\quad \underline{-j32 - j^2 24}$

Adding, $\quad 16 - j20 - j^2 24$

The next step is to take care of the term j^2. Since the operator j represents a rotation of $90°$ from the reference axis, j^2 represents a rotation of $180°$ from the reference axis. Therefore, the operator j^2 is the same as the operator $-$. If we wish, we can say $j^2 = -1$.

$$\therefore 16 - j20 - j^2 24 \quad \text{becomes} \quad 16 - j20 + 24$$

and the product becomes $40 - j20$.

The procedure for division is more involved. First, we set up the phasor quantities involved in fraction form. If we multiply both numerator and denominator of a fraction by the same quantity, we do not change the value of the fraction. In choosing this extra factor, we wish to get rid of any j term in the denominator. To eliminate the j term, the extra factor must have the same reference axis coordinate as the denominator and the same magnitude of quadrature coordinate as the denominator *but with the opposite sign.*[†]

$$\therefore (8 + j6) \div (2 - j4) = \frac{8 + j6}{2 - j4} = \frac{(8 + j6) \times (2 + j4)}{(2 - j4) \times (2 + j4)}$$

Completing the multiplication gives

$$\therefore (8 + j6) \div (2 - j4) = \frac{-8 + j44}{20}$$

[†]Such a term is called the **conjugate** of the denominator.

Now that there is no j term in the denominator, we can carry out the division quite readily.

$$\therefore (8 + j6) \div (2 - j4) = -0.4 + j2.2$$

PROBLEMS

Solve Problems 19-1, 19-2, and 19-3 by drawing reasonably accurate freehand sine curves on a sheet of graph paper and determining the resultant by the addition of instantaneous values.

19-1. A 200-V sine wave lags a 100-V sine wave by 45°. Find the rms value of the resultant.

19-2. A 2-A sine-wave current flows through a capacitor and an inductor in series. The reactance of the capacitor is 200 Ω and the reactance of the inductor is 150 Ω. Find the rms value of the total voltage.

19-3. The currents in the two branches of a parallel circuit are given as $i_1 = 80 \sin 377t$ and $i_2 = 50 \sin (377t - \pi/3)$. Solve graphically for the total current.

19-4. Draw phasor diagrams for Problems 19-1 and 19-2 and solve by geometrical construction.

19-5. Three sources of alternating voltage each generate 120 V rms, but $\mathbf{E}_2$ leads $\mathbf{E}_1$ by 120°, and $\mathbf{E}_3$ lags $\mathbf{E}_1$ by 120°. Draw a phasor diagram and determine the resultant voltage when all three sources are connected in series.

19-6. Draw a phasor diagram and determine the resultant voltage if the leads to source $\mathbf{E}_1$ in Problem 19-5 are reversed.

19-7. Find the resultant of the following phasors by geometrical construction: $80\underline{/+45°}$, $60\underline{/-135°}$, $40\underline{/-\pi/3}$ rad, $20\underline{/2}$ rad.

19-8. An induction motor draws a 10-A current from a 120-V 60-Hz source. This current lags the applied voltage by 30°. A capacitor in parallel with the motor draws a 5-A current from the source. Draw a schematic diagram, a phasor diagram, and solve for the total current drawn from the source by geometrical construction.

19-9. Convert the following phasor quantities into rectangular coordinates.
   (a) $25\underline{/36.9°}$
   (b) $36\underline{/140°}$
   (c) $178\underline{/300°}$
   (d) $60\underline{/-120°}$
   (e) $14.2\underline{/2}$ rad
   (f) $6.8\underline{/\pi/5}$ rad

19-10. Convert the following to polar coordinates.
   (a) $3 + j4$
   (b) $-14 - j10$
   (c) $12.7 + j0$
   (d) $-0.8 + j1.2$
   (e) $0 - j18$
   (f) $0.67 - j0.43$

Express all the following answers in polar coordinates.

19-11. Add $13 - j7$ and $11 + j18$.

19-12. Add $1.4 + j0.8$, $-0.9 + j2.1$, and $0 - j3.9$.

19-13. Add $170\underline{/200°}$ and $88\underline{/-75°}$.

19-14. Add $1.8\underline{/125°}$, $2.7\underline{/-157°}$, and $1.3\underline{/-66°}$.

19-15. Add $40 + j72$ and $60\underline{/100°}$.

19-16. Add $120 + j0$, $120\underline{/-120°}$, and $-60 + j104$.

19-17. Subtract $2.8\underline{/60°}$ from $3.1\underline{/45°}$.
   (a) By geometrical construction.
   (b) By complex algebra.

19-18. Add $120 + j0$, $-120\underline{/-120°}$, and $-60 + j104$.

19-19. Solve $4.1\underline{/-64°} \times 13\underline{/13°}$ and $4.1\underline{/-64°} \div 13\underline{/13°}$.

19-20. Solve $(3 + j4) \times (5 - j6)$ and $(3 + j4) \div (5 - j6)$.

19-21. Solve $\mathbf{Z}_{eq} = \dfrac{\mathbf{Z}_1\mathbf{Z}_2}{\mathbf{Z}_1 + \mathbf{Z}_2}$ if $\mathbf{Z}_1 = 40\ \Omega\underline{/30°}$ and $\mathbf{Z}_2 = 12\ \Omega\underline{/-45°}$.

19-22. Solve $\dfrac{(47 + j13)(-21 - j32)}{51\underline{/-111°} + 19\underline{/+70°}}$

REVIEW QUESTIONS

19-1. Rms values in an ac circuit are defined as dc equivalent values. Why then can we not add rms values by simple algebra as we can in a dc circuit?

19-2. Why is it possible to add instantaneous values in an ac circuit by simple algebra?

19-3. When writing an equation for the instantaneous value of the sum of two sine waves, how is the phase difference between them indicated?

19-4. What advantage does the phasor diagram have over the linear graph of instantaneous values plotted against time?

19-5. What information does a linear graph show that is not shown by a phasor diagram?

19-6. Why is a knowledge of complex algebra required in solving ac circuit problems?

19-7. What condition is required to be able to add rms values in an ac circuit either by phasor construction or by complex algebra?

19-8. Is it possible to add the instantaneous values of two sine waves if their frequencies are different? Explain.

19-9. What is the significance of an equation such as $|I_T| = |I_1| + |I_2|$?

19-10. Show with a phasor diagram that $\sin 210° = -\sin 30°$.

19-11. Show with a phasor diagram that $\cos 135° = -\cos 45°$.

19-12. What is meant by **algebraic addition?**

19-13. What is meant by **Pythagorean addition?**

19-14. What is meant by **polar coordinates?**

19-15. What is meant by **rectangular coordinates?**

19-16. Why is it possible to convert phasor quantities from one set of coordinates to the other?

19-17. What is the significance of the operator j in dealing with phasor quantities?

19-18. What is the advantage of expressing phasor quantities in rectangular coordinates?

484

19-19. Explain the procedure for subtracting phasor quantities by geometrical construction on a phasor diagram.

19-20. Is it possible to determine the product of two phasor quantities by geometrical construction on a phasor diagram? Explain.

*19-21. Given two phasors $I_1 = 40 \text{ mA} \angle +120°$ and $I_2 = 60 \text{ mA} \angle -30°$, either run the computer program for Example 19-7 or go through the steps indicated in the program to determine the resultant sum.

*19-22. Write a computer program to express in rectangular coordinates the quotient of two phasor quantities which are stated in rectangular coordinates.

20

IMPEDANCE

20-1 RESISTANCE AND INDUCTANCE IN SERIES

In Chapter 19, we noted that all rules and laws which apply to dc circuits must also apply to ac circuits, provided that we keep in mind the sinusoidally-varying nature of the voltage source in an ac system. The most important of the rules of series circuits is the one by which we define a series circuit.

The same current must flow in all components in a series circuit.

Therefore, the starting point in determining the behavior of all series ac circuits is to represent this common current by the reference phasor of a phasor diagram, as in Figure 20-1.

A second characteristic of series circuits is that the sum of the various voltage drops must equal the applied voltage. If we use instantaneous values, this becomes a simple algebraic sum. However, using instantaneous values is a tedious means of solving an ac circuit. To use the more practical *rms* values of voltage and current, we can restate this second characteristic of series circuits thus:

In a series ac circuit, the phasor *sum of the various voltage drops must equal the applied voltage.*

Since the voltage drop across the resistance must be in phase with the current, its phasor falls along the + rectangular coordinate axis in Figure 20-1. Similarly,

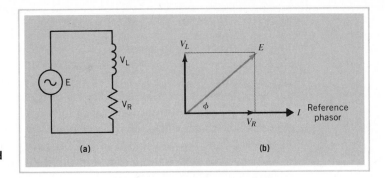

FIGURE 20-1
Phasor Diagram of Voltage and Current Relationships in a Series Circuit Containing Resistance and Inductance.

since the voltage across the inductance must lead the current by 90°, its phasor falls along the +j rectangular coordinate axis. Therefore, we may express the relationship among the voltages in a simple ac series circuit quite readily in rectangular coordinates as

$$\mathbf{E} = V_R + jV_L \tag{20-1}$$

Drawing in the resultant phasor on the phasor diagram of Figure 20-1, we can also express this relationship in polar form.

$$\mathbf{E} = \sqrt{V_R^2 + V_L^2} \;\bigg/\; \arctan \frac{V_L}{V_R} \tag{20-2}$$

Because current is the only parameter common to all components in a series circuit, in constructing the phasor diagram of Figure 20-1, we started with current as the reference phasor and solved by geometrical construction for the location of the phasor representing the total voltage or applied voltage. In the actual circuit, when the inductance and resistance are connected in series to a given source, the current (the dependent variable) must automatically take on a sine-wave shape and a magnitude and phase such that it will develop a voltage drop across the resistance and voltage across the inductance with a phasor sum equal to the applied voltage. Nevertheless, we always think of current as the reference phasor for series circuits, and the angle in Equation (20-2) is the **phase angle** of the applied voltage *with respect to* the common current.

The letter symbol for phase angle is the Greek letter φ (phi) or the Greek letter θ (theta).[†]

[†]The international standard symbol for phase angle is the Greek letter ϕ. North American textbooks have, in the past, shown a preference for the Greek letter θ. At one point in time, one of several North American standards used θ to represent the phase angle of the voltage *with respect to* the current, and ϕ to represent the phase angle of the current *with respect to* the voltage. Current North American standards permit the use of either ϕ or θ to represent the phase angle *between* current and voltage. In line with the present North American trend to adopt international standards as rapidly as is feasible, we shall use the Greek letter ϕ to represent phase angle in this text.

20-2 IMPEDANCE

Since the current in the series circuit of Figure 20-1 must automatically take on a sine-wave form of certain magnitude and phase with respect to the applied voltage, for a given ac circuit the ratio between the applied voltage and the current is a constant. Just as V_R/I represents the opposition of resistance to alternating current and V_L/I represents the opposition of inductance to alternating current, the ratio V_T/I or E/I represents the *total* opposition of the circuit to alternating current. In resistance the voltage and current are in phase, and in reactance the voltage and current are exactly 90° out of phase. But as we can see from the phasor diagram of Figure 20-1, the angle ϕ between the applied voltage and the current is between 0° and +90°. Therefore, we cannot call this total opposition either resistance or reactance.

The total opposition to current in an ac circuit is called impedance.
The letter symbol for impedance is Z.

$$\mathbf{Z} = \frac{\mathbf{E}}{\mathbf{I}} \tag{20-3}$$

where $\mathbf{Z}$ is the impedance of an ac circuit in ohms, $\mathbf{E}$ is the applied voltage in volts, and $\mathbf{I}$ is the current drawn from the source in amperes.

Since current is the dependent variable in the circuit, we must have some means of determining the impedance before we can solve for the current. Once we know the impedance, we can use Equation (20-3) to determine the current in the circuit. Because the current is common to all components in a series circuit, we can divide every term in Equation (20-1) by I. Hence,

$$\frac{\mathbf{E}}{I} = \frac{V_R}{I} + \mathrm{j}\frac{V_L}{I}$$

from which

$$\mathbf{Z} = R + \mathrm{j}X_L \tag{20-4}$$

Also, when we divide each term in Equation (20-2) by I,

$$\mathbf{Z} = \sqrt{R^2 + X_L^2} \Big/ \arctan \frac{X_L}{R} \tag{20-5}$$

Therefore,

The impedance of a series circuit is the phasor sum of the resistance and reactance.

Like resistance and reactance, impedance is a V/I ratio expressed in ohms. Like resistance and reactance, impedance is not a sinusoidally varying quantity. But unlike R, X_L, and X_C which do indicate the exact phase angle between V and I, the letter symbol Z gives no such indication. The statement $Z = 5$ kilohms tells us only the magnitude of the total opposition to electric current. There is no indication as to whether the impedance is largely inductive (with the voltage leading the current) or

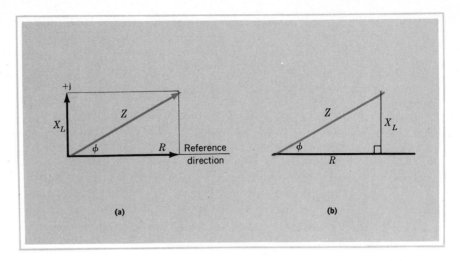

FIGURE 20-2
Impedance Diagram for a Series Circuit Containing Resistance and In-ductance.

capacitive (with the voltage lagging the current). To provide this information we must recognize that impedance is a complex number which fits the definition of a phasor quantity. Hence, Equation (20-3) can be expanded into polar coordinates as

$$\mathbf{Z} = \frac{|E|\underline{/\phi}}{|I|\underline{/0°}} = |Z|\underline{/\phi} \qquad (20\text{-}6)$$

We can represent impedance in the customary common-origin type of phasor diagram of Figure 20-2(a). However, to preserve a distinction between sinusoidal phasor quantities such as voltage and current on the one hand and nonsinusoidal phasor quantities measured in ohms on the other, we reserve the common-origin type of phasor diagram for voltage and current. In Section 19-6, we found that we could perform the same graphical addition of phasors with a triangular type of diagram. Hence, we use the format of Figure 20-2(b) and call it an **impedance diagram.** As indicated by Equation (20-4), resistance is always drawn in the reference direction and inductive reactance in the +j direction. It follows that capacitive reactance is drawn in the −j direction.

EXAMPLE 20-1
A solenoid having an inductance of 0.5 H and a resistance of 100 Ω is connected to a 120-V 60-Hz source. What is the magnitude and phase of the current with respect to the applied voltage?

SOLUTION
    Step I. In an ac circuit, we are more concerned with the *reactance* of the solenoid than its inductance. Therefore, the first step is to express its reactance at 60 Hz.

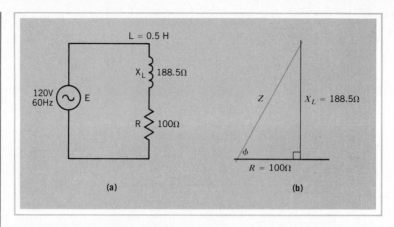

FIGURE 20-3
Impedance Diagram for Example 20-1.

$$X_L = \omega L = 377 \times 0.5 = 188.5 \ \Omega \ (\text{inductive})$$

Step II. Always draw a circuit diagram, labeling it with all pertinent data, and draw either a voltage phasor diagram or an impedance diagram to show the relationships that exist in the given circuit (Figure 20-3).

Step III. From the impedance diagram, the rectangular coordinates of the impedance are

$$\mathbf{Z} = R + jX_L = 100 + j\,188.5 \ \Omega$$

Step IVA. Solve for the impedance in polar form.

$$\phi = \arctan \frac{X_L}{R} = \arctan \frac{188.5}{100} = 62°$$

$$Z = \sqrt{R^2 + X_L^2} = \sqrt{100^2 + 188.5^2} = 213 \ \Omega$$

$$\therefore \mathbf{Z} = 213 \ \Omega \underline{/+62°}$$

Step IVB. With a calculator, we can bypass the trigonometry of rectangular to polar coordinate conversion and let the calculator do all the work.

$$100 \ \boxed{x \leftrightarrow y} \ 188.5 \ \rightarrow \text{P} \ 213.383 \ \boxed{x \leftrightarrow y} \ 62.054$$

Step VA. Solve for the magnitude of the current.

$$I = \frac{E}{Z} = \frac{120 \ \text{V}}{213 \ \Omega} = 0.563 \ \text{A}$$

And from Figure 20-3, this current *lags* the applied voltage by 62°.

Step VB. Treating the terms of Equation (20-6) as phasor quantities with the angle of the applied voltage as $0°$, we can solve for both magnitude and angle of the current with respect to the applied voltage by phasor division. Therefore,

$$I = \frac{E}{Z} = \frac{120 \text{ V}\underline{/0°}}{213 \text{ }\Omega\underline{/62°}} = 0.563 \text{ A }\underline{/-62°}$$

20-3 PRACTICAL INDUCTORS

In practical circuits, we are not able to obtain inductance alone since the wire used in the winding possesses some resistance. Therefore, the practical inductor is not a reactance with a $+90°$ angle. With only two leads from the choke in Figure 20-4(a), we cannot connect a voltmeter to read only the inductive voltage drop across its inductance or the *IR* drop across its resistance. We can read only the total voltage drop between the two leads. This voltage leads the current through the choke by something *less* than $90°$. If we are given the *impedance* of the choke in polar form with its proper angle, we can determine what the resistance and inductance components of the practical inductor are by conversion to rectangular coordinates.

EXAMPLE 20-2

The choke in the circuit of Figure 20-4(a) has an impedance of $80 \text{ }\Omega\underline{/80°}$. What is the total impedance when it is connected in series with a $75\text{-}\Omega$ resistor?

SOLUTION

Since we do not directly tackle the addition of two phasors with an $80°$ angle between them as in the phasor diagram of Figure 20-4(b), we must first express the impedance of the choke in rectangular coordinates.

FIGURE 20-4
Practical Inductor and Resistor in Series.

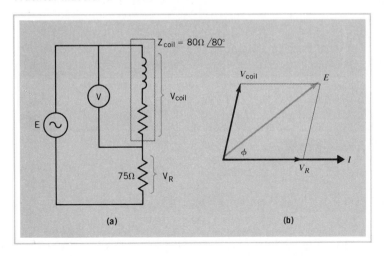

$$\therefore \mathbf{Z}_{coil} = 80 \cos 80° + j80 \sin 80° = 14 + j79 \; \Omega$$

The total impedance of the circuit in Figure 20-4(a) now becomes

$$\mathbf{Z}_T = 75 + 14 + j79 \; \Omega$$

Since the two resistance components are both along the + axis of an impedance diagram, we may add them by simple arithmetic.

$$\therefore \mathbf{Z}_T = 89 + j79 \; \Omega$$

Changing back to polar form,

$$\phi = \arctan \frac{79}{89} = 41.6°$$

$$Z_T = \sqrt{89^2 + 79^2} = 119 \; \Omega$$

$$\therefore \mathbf{Z}_T = 119 \; \Omega \; \underline{/+41.6°}$$

Since the $\rightarrow$ P and $\rightarrow$ R keys of a calculator conceal the actual process of converting from rectangular to polar coordinates, and vice versa, the solutions to examples will continue to show the "longhand" steps. Students have the option of using calculator short cuts and checking these results with the solutions shown. Now that we are familiar with the Pythagorean relationship among the magnitudes of resistance, reactance, and impedance, we might consider the shorter trigonometry procedure for determining Z with calculators not possessing $\rightarrow$ P and $\rightarrow$ R keys.

$$Z = \frac{R}{\cos \phi} \quad \text{or} \quad \frac{X}{\sin \phi} \tag{20-7}$$

While the angle ϕ is still displayed by the calculator from having solved $\arctan X/R$, press $\boxed{\cos}$ $\boxed{1/x}$ $\boxed{\times}$ and enter R. Pressing $\boxed{=}$ then displays Z.

The following example doesn't lend itself to shortcuts. An alternative solution can be based on the cosine rule of trigonometry rather than on a Pythagorean equation.

EXAMPLE 20-3

A coil and a resistor each have an 80-V drop across them when connected in series to a 120-V 60-Hz source. If they draw a 0.5-A current from the source, what is the resistance of the coil?

SOLUTION

Step I.

$$Z_T = \frac{E_T}{I} = \frac{120 \; V}{0.5 \; A} = 240 \; \Omega$$

492

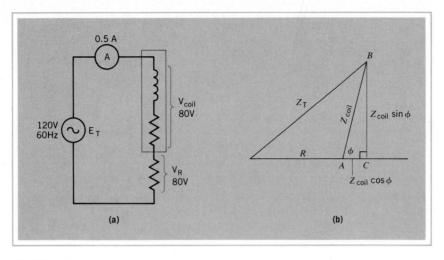

FIGURE 20-5
Impedance Diagram for Example 20-3.

$$Z_{coil} = \frac{V_{coil}}{I} = \frac{80 \text{ V}}{0.5 \text{ A}} = 160 \text{ } \Omega$$

$$R = \frac{V_R}{I} = \frac{80 \text{ V}}{0.5 \text{ A}} = 160 \text{ } \Omega$$

Step II. Since the coil possesses both resistance and inductance, the angle of its impedance ϕ must be something less than $+90°$, as shown in Figure 20-5.

$$\therefore \mathbf{Z}_{coil} = 160 \text{ } \Omega \underline{/\phi}$$

We can express this in rectangular coordinates as

$$\mathbf{Z}_{coil} = 160 \cos \phi + j160 \sin \phi$$

Step III.

$$\therefore \mathbf{Z}_T = 160 + 160 \cos \phi + j160 \sin \phi = 240 \text{ } \Omega$$

In polar form, this becomes

$$240 = \sqrt{(160 + 160 \cos \phi)^2 + (160 \sin \phi)^2}$$

Squaring both sides of the equation,

$$57\,600 = 25\,600 + 51\,200 \cos \phi + 25\,600 \cos^2 \phi + 25\,600 \sin^2 \phi$$

$$\therefore \cos \phi = \frac{32\,000 - 25\,600(\cos^2 \phi + \sin^2 \phi)}{51\,200} = \frac{5 - 4(\cos^2 \phi + \sin^2 \phi)}{8}$$

Step IV. But from triangle ABC in Figure 20-5(b),

$$\cos^2 \phi = \frac{AC^2}{AB^2} \quad \text{and} \quad \sin^2 \phi = \frac{BC^2}{AB^2}$$

$$\therefore \cos^2 \phi + \sin^2 \phi = \frac{AC^2}{AB^2} + \frac{BC^2}{AB^2} = \frac{AC^2 + BC^2}{AB^2}$$

But from Pythagoras' theorem, $AC^2 + BC^2 = AB^2$

$$\therefore \cos^2 \phi + \sin^2 \phi = 1$$

and
$$\cos \phi = \frac{5 - 4}{8} = 0.125$$

Step V. The resistance of the coil then becomes

$$R_{\text{coil}} = Z_{\text{coil}} \cos \phi = 160 \times 0.125 = 20 \ \Omega$$

20-4 RESISTANCE AND CAPACITANCE IN SERIES

In the circuit of Figure 20-6(a), the current again is common to all components. Since the potential difference across the capacitor *lags* the current through it by 90°, the phasor diagram shows the total-voltage phasor located in quadrant IV. From this phasor diagram, we can state that

$$\mathbf{E} = V_R - jV_C \tag{20-8}$$

or
$$\mathbf{E} = \sqrt{V_R^2 + V_C^2} \bigg/ -\arctan \frac{V_C}{V_R} \tag{20-9}$$

and
$$\mathbf{Z} = R - jX_C \tag{20-10}$$

or
$$\mathbf{Z} = \sqrt{R^2 + X_C^2} \bigg/ -\arctan \frac{X_C}{R} \tag{20-11}$$

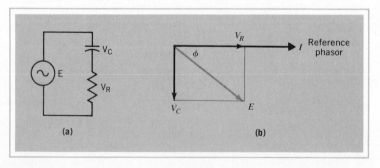

FIGURE 20-6

Phasor Diagram of Voltage and Current Relationships in a Series Circuit Containing Resistance and Capacitance.

Since *inductive* reactance is always drawn in a $+j$ direction in impedance diagrams and *capacitive* reactance in a $-j$ direction, in polar form, an inductive impedance always has a $+$ angle between 0 and $+90°$ and a capacitive impedance always has a $-$ angle between 0 and $-90°$.

EXAMPLE 20-4

What capacitance when connected in series with a 500-Ω resistor will limit the current drawn from a 48-mV 465-kHz source to 20 μA?

SOLUTION

$$Z = \frac{E}{I} = \frac{48 \text{ mV}}{20 \text{ } \mu A} = 2.4 \text{ k}\Omega$$

From Pythagoras' theorem with respect to Figure 20-7,

$$X_C = \sqrt{Z^2 - R^2} = \sqrt{2400^2 - 500^2} = 2350 \text{ } \Omega$$

since

$$X_C = \frac{1}{2\pi f C}$$

$$C = \frac{1}{2\pi f X_C} = \frac{1}{2 \times \pi \times 465 \text{ kHz} \times 2350 \text{ } \Omega} F = \quad 146 \text{ pF}$$

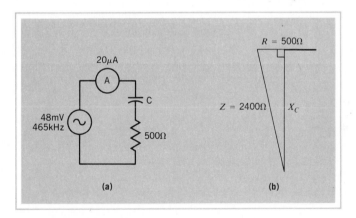

FIGURE 20-7
Impedance Diagram for Example 20-4.

20-5 RESISTANCE, INDUCTANCE, AND CAPACITANCE IN SERIES

We can express the sum of the various voltages in the series circuit of Figure 20-8 in rectangular form as

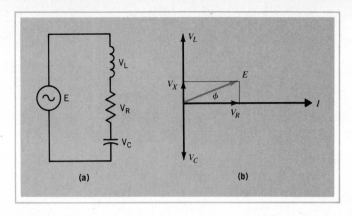

FIGURE 20-8
Phasor Diagram of Voltage and Current Relationships in a Series Circuit Containing Resistance, Inductance, and Capacitance.

$$\mathbf{E} = V_R + jV_L - jV_C \qquad (20\text{-}12)$$

The voltage across the inductance *leads* the common current by 90°, and the voltage across the capacitance *lags* the current by 90°. Therefore, these two voltages are exactly 180° out of phase. We can then add these two phasors by simple arithmetical subtraction, and Equation (20-12) becomes

$$\mathbf{E} = V_R + j(V_L - V_C) \qquad (20\text{-}13)$$

Unless the two reactive voltages are exactly equal, this leaves the circuit with an **equivalent reactive voltage** (V_X in Figure 20-8), which will be either inductive or capacitive depending on which of the original reactive voltages is the larger. Note that as far as a fixed-frequency generator is concerned in the circuit of Figure 20-8, the circuit appears to contain only inductance since V_L is greater than V_C. The overall circuit cannot be both inductive and capacitive at the same time. The current must *either* lead *or* lag the applied voltage; it cannot do both.

Therefore, converting Equation (20-13) into polar coordinates gives

$$\mathbf{E} = \sqrt{V_R^2 + V_X^2} \Big/ \arctan \frac{V_X}{V_R} \qquad (20\text{-}14)$$

where $V_X = V_L - V_C$ *(the net reactive voltage)*.

If we divide every term in Equations (20-13) and (20-14) by the common current, the impedance of series circuits containing R, L, and C becomes

$$\mathbf{Z} = R + jX_L - jX_C = R + j(X_L - X_C) \qquad (20\text{-}15)$$

or
$$\mathbf{Z} = \sqrt{R^2 + X_{eq}^2} \Big/ \arctan \frac{X_{eq}}{R} \qquad (20\text{-}16)$$

where $X_{eq} = X_L - X_C$ *(the net equivalent reactance)*.

EXAMPLE 20-5
What is the total impedance at 20 kHz of a series circuit consisting of a 1.5-mH inductance, a 100-Ω resistance, and an 0.08-μF capacitance?

SOLUTION

Step I. Again, in ac circuits we are interested primarily in the *reactance* of the inductor and the capacitor. Hence,

$$X_L = 2\pi f L = 2 \times \pi \times 20 \text{ kHz} \times 1.5 \text{ mH}$$
$$= 188.5 \ \Omega \text{ (inductive)}$$

and

$$X_C = \frac{1}{2\pi f C} = \frac{1}{2 \times \pi \times 20 \text{ kHz} \times 0.08 \ \mu\text{F}}$$
$$= 99.47 \ \Omega \text{ (capacitive)}$$

Step II. Draw a schematic diagram and either a phasor diagram or an impedance diagram, as in Figure 20-9(b).

Step III.

$$\mathbf{Z} = R + jX_L - jX_C = 100 + j188.5 - j99.5 = 100 + j89 \ \Omega$$

Step IV.

$$\phi = \arctan \frac{X_{eq}}{R} = \arctan \frac{89}{100} = +41.7°$$

$$Z = \frac{R}{\cos \phi} = \frac{100 \ \Omega}{\cos 41.7°} = 134 \ \Omega$$

$$\therefore \mathbf{Z} = 134 \ \Omega \ \underline{/+41.7°}$$

FIGURE 20-9
Impedance Diagram for Example 20-5.

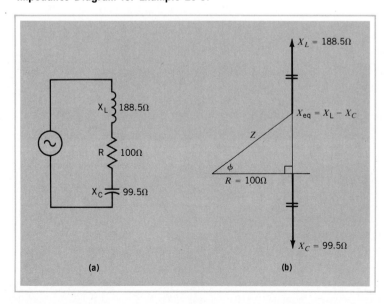

20-6 RESISTANCE, INDUCTANCE, AND CAPACITANCE IN PARALLEL

On examining Figure 20-10(a), we recall that the characteristic by which we define a parallel circuit is that

The same voltage appears across all parallel branches.

Therefore, in preparing a phasor diagram for any parallel circuit, we always use this common voltage as the reference phasor. If we are given the numerical value of this voltage, we can solve for the individual branch currents by Ohm's law. Since

$$I_R = \frac{E\,\underline{/0°}}{R\,\underline{/0°}}$$

then I_R is in phase with the reference phasor.

Since
$$I_L = \frac{E\,\underline{/0°}}{X_L\,\underline{/+90°}}$$

then I_L lags the reference voltage by 90°.

And since
$$I_C = \frac{E\,\underline{/0°}}{X_C\,\underline{/-90°}}$$

I_C leads the reference voltage by 90°, as shown in Figure 20-10. The total current in the parallel circuit then becomes the *phasor* sum of the branch currents. Therefore, from Figure 20-10,

$$\mathbf{I_T} = I_R - jI_L + jI_C = I_R + j(I_C - I_L) \tag{20-17}$$

FIGURE 20-10
Phasor Diagram of Voltage and Current Relationships in a Parallel Circuit Containing Resistance, Inductance, and Capacitance.

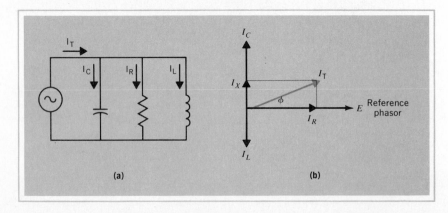

Converting Eqauation (20-17) to polar coordinates gives

$$\mathbf{I_T} = \sqrt{I_R^2 + I_X^2} \bigg/ \arctan \frac{I_X}{I_R} \qquad (20\text{-}18)$$

where $I_X = I_C - I_L$ (*the net reactive current*). We can now determine the equivalent impedance of a parallel circuit by using Equation (20-6).

$$\mathbf{Z}_{eq} = \frac{\mathbf{E}}{\mathbf{I_T}} \qquad (20\text{-}6)$$

This method of solving for the equivalent impedance of a parallel circuit is called the *total-current method*. Even if the exact value of the applied voltage is not known, we may assume any convenient value in order to solve for the equivalent impedance.

EXAMPLE 20-6
What is the equivalent impedance at 20 kHz of a circuit consisting of a 1.5-mH inductance, a 100-Ω resistance, and an 0.08-μF capacitance all connected in parallel?

SOLUTION
   Step I.

$$X_L = 2\pi f L = 2 \times \pi \times 20 \text{ kHz} \times 1.5 \text{ mH}$$
$$= 188.5 \ \Omega \text{ (inductive)}$$

$$X_C = \frac{1}{2\pi f C} = \frac{1}{2 \times \pi \times 20 \text{ kHz} \times 0.08 \ \mu\text{F}}$$
$$= 99.5 \ \Omega \text{ (capacitive)}$$

   Step II. Assume a convenient value of applied voltage. For example, let $\mathbf{E} = 200 \text{ V} \underline{/0°}$.

Then $\qquad I_R = \dfrac{E \underline{/0°}}{R} = \dfrac{200 \underline{/0°}}{100 \underline{/0°}} = 2 \text{ A} \underline{/0°}$

$$I_L = \frac{E \underline{/0°}}{+jX_L} = \frac{200 \underline{/0°}}{188.5 \underline{/+90°}} = 1.06 \text{ A} \underline{/-90°}$$

and $\qquad I_c = \dfrac{E \underline{/0°}}{-jX_C} = \dfrac{200 \underline{/0°}}{99.5 \underline{/-90°}} = 2.01 \text{ A} \underline{/+90°}$

   To facilitate the phasor division, we expressed the direction associated with R, X_L, and X_C in polar form rather than rectangular coordinate operators. Note also that dividing a quantity having the same direction as the reference phasor by a $+j$ quantity creates a $-j$ quantity, and vice versa.

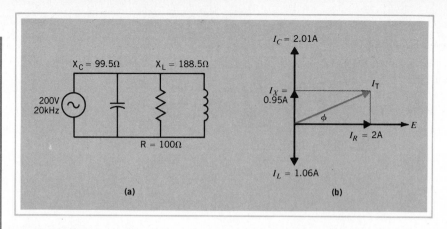

FIGURE 20-11
Phasor Diagram for Example 20-6.

Step III.   Prepare a circuit diagram and phasor diagram, as in Figure 20-11.

Step IV.

$$\mathbf{I_T} = I_R - jI_L + jI_C = 2 - j1.06 + j2.01 = 2 + j0.95 \text{ A}$$

Step V.

$$\phi = \arctan \frac{I_X}{I_R} = \arctan \frac{0.95}{2} = +25.4°$$

$$I_T = \frac{I_R}{\cos \phi} = \frac{2.0 \text{ A}}{\cos + 25.4°} = 2.214 \text{ A}$$

$$\therefore \mathbf{I_T} = 2.21 \text{ A}\underline{/+25.4°}$$

Step VI.

$$\mathbf{Z}_{eq} = \frac{\mathbf{E}}{\mathbf{I_T}} = \frac{200\underline{/0°}}{2.21\underline{/+25.4°}} = 90.5 \text{ }\Omega\underline{/-25.4°}$$

Note that the equivalent impedance is capacitive, since the capacitive branch draws a greater *current* from the source than does the inductive branch. When the *same* three components were connected in series in Example 20-5, the total impedance was inductive, since there was a greater *voltage* across the inductance than across the capacitance.

20-7 CONDUCTANCE, SUSCEPTANCE, AND ADMITTANCE

Although we listed the total-current method in solving for the equivalent resistance of parallel dc circuits, we preferred to use a method that did not require the in-between step of solving for total current. To do this, we found it necessary to change our point of view in dealing with parallel circuits. Rather than thinking in terms of the ability of a circuit to *oppose* current, we found that in dealing with parallel circuits we could obtain a better picture of the circuit behavior if we thought in terms of the ability of the various branches to *pass* current. This reciprocal property of resistance is called **conductance**.

Conductance is a measure of the ability of resistance to pass *electric current.*
The letter symbol for conductance is G.

The unit of conductance is the **siemens** as defined by the equation $G = 1/R$.
If we wish, we can write Equation (20-17) as

$$\frac{E}{\mathbf{Z}} = \frac{E}{R} - j\frac{E}{X_L} + j\frac{E}{X_C}$$

Since the *same* voltage appears across all components in a simple parallel circuit, this becomes

$$\frac{1}{\mathbf{Z}} = \frac{1}{R} - j\frac{1}{X_L} + j\frac{1}{X_C} \tag{20-19}$$

We already have a term and a letter symbol for the reciprocal of resistance, namely, **conductance**. We could certainly clean up the appearance of Equation (20-19) if we also had such reciprocals for impedance and reactance. We defined impedance as the total *opposition* to current in an ac circuit. Hence, its reciprocal,

Admittance is the overall ability of an electric circuit to pass *alternating current.*
The letter symbol for admittance is Y.

where $\mathbf{Y} = 1/\mathbf{Z}$ *siemens*. Similarly, the reciprocal of reactance,

Susceptance is the ability of inductance or capacitance to pass *alternating current.*
The letter symbol for susceptance is B.

where $B = 1/X$ *siemens.*
Therefore, Equation (20-19) becomes

$$\mathbf{Y} = G - jB_L + jB_C = G + j(B_C - B_L) \tag{20-20}$$

Note that when we divide a phasor quantity with a $+90°$ angle into $1\,\underline{/0°}$, the quotient has a $-90°$ angle, and vice versa. Therefore, the reciprocal of $+jX_L$ is $-jB_L$, and the reciprocal of $-jX_C$ is $+jB_C$. Using the now familiar relationship, we can express the polar coordinates of admittance as

$$\mathbf{Y} = \sqrt{G^2 + B_{eq}^2}\,\Big/\!\arctan\frac{B_{eq}}{G} \tag{20-21}$$

where $B_{eq} = B_C - B_L$ *(the net equivalent susceptance)*. Since $B_L = 1/X_L$,

$$B_L = \frac{1}{2\pi f L} = \frac{1}{\omega L} \qquad (20\text{-}22)$$

and

$$B_C = 2\pi f C = \omega C \qquad (20\text{-}23)$$

EXAMPLE 20-6A

What is the equivalent impedance at 20 kHz of a circuit consisting of a 1.5-mH inductance, a 100-Ω resistance, and an 0.08-μF capacitance all connected in parallel?

SOLUTION

Step I.

$$G = \frac{1}{R} = \frac{1}{100\ \Omega} = 0.01\text{ siemens}$$

$$B_C = 2\pi f C = 2 \times \pi \times 20\text{ kHz} \times 0.08\ \mu\text{F}$$
$$= 0.01\text{ S (capacitive)}$$

$$B_L = \frac{1}{2\pi f L} = \frac{1}{2 \times \pi \times 20\text{ kHz} \times 1.5\text{ mH}}$$
$$= 0.0053\text{ S (inductive)}$$

Step II. Prepare a current phasor diagram or preferably an admittance diagram for the circuit of Figure 20-11(a) and (Figure 20-12). Note that it is impossible to draw an *impedance* diagram using the actual values given in the circuit of Figure 20-11(a), since the equivalent impedance of a parallel circuit must be *smaller* than the resistance, and it is impossible for the hypotenuse of a right-angled triangle to be smaller than either of the two sides.

Step III.

$$\mathbf{Y} = G + \mathrm{j}(B_C - B_L) = 0.01 + \mathrm{j}(0.01 - 0.0053)$$
$$= 0.01 + \mathrm{j}0.0047\text{ S}$$

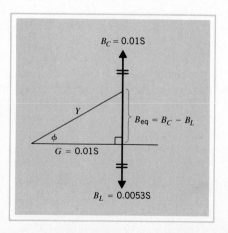

FIGURE 20-12
Admittance Diagram for Example 20-6A.

502

Step IV.

$$\phi = \arctan \frac{B_{eq}}{G} = \arctan \frac{0.0047}{0.01} = +25.2°$$

$$Y = \frac{G}{\cos \phi} = \frac{0.01 \text{ s}}{\cos 25.2°} = 0.011\ 05 \text{ S}$$

$$\therefore \mathbf{Y} = 11.05 \text{ mS} \underline{/+25.2°}$$

Step V.

$$\mathbf{Z}_{eq} = \frac{1}{\mathbf{Y}} = \frac{1 \underline{/0°}}{11.05 \text{ mS} \underline{/+25.2°}} = 90.5 \ \Omega \underline{/-25.2°}$$

20-8 IMPEDANCE AND ADMITTANCE

In the simple parallel circuit of Figure 20-10, we made sure that each branch contained only one type of opposition to alternating current. Hence, we were able to substitute G for $1/R$ and B for $1/X$. Suppose, however, that one of the branches had been a practical inductor possessing both resistance and inductive reactance, as in Figure 20-13. The current through this branch would now be inversely proportional to its **impedance**. Hence, if we want to use a reciprocal unit, it would have to be $\mathbf{Y}$ for $1/\mathbf{Z}$.

Using the resistance and reactance values given for the practical inductor in Figure 20-13, we *cannot* say that the conductance of this branch is

$$G = \frac{1}{R} = \frac{1}{30} = 0.333 \text{ S}$$

and the susceptance of the branch is *not*

$$B = \frac{1}{X} = \frac{1}{40} = 0.025 \text{ S}$$

Doing so would leave us with a conductance and susceptance in *series*, just as in the original circuit of Figure 20-13. But in Equation (20-20), conductance and susceptance are the rectangular coordinates of admittance, and we developed Equation (20-20) from *parallel* branches. We can illustrate the procedure for finding the conductance and susceptance of a practical inductor from the following example.

FIGURE 20-13
Finding the Conductance and Susceptance of a Practical Inductor.

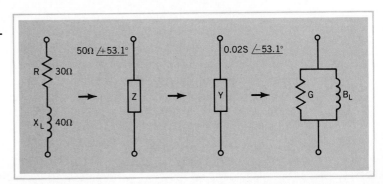

EXAMPLE 20-7

What is the conductance and susceptance of a practical inductor whose resistance is 30 Ω and whose inductive reactance is 40 Ω?

SOLUTION

Step I. Find the impedance of the inductor.

$$\mathbf{Z} = \sqrt{R^2 + X^2} \Big/ \arctan \frac{X}{R}$$

$$= \sqrt{30^2 + 40^2} \; \arctan \frac{40}{30} = 50 \; \Omega \underline{/+53.1°}$$

Step II. From the definition of admittance,

$$\mathbf{Y} = \frac{1}{\mathbf{Z}} = \frac{1}{50 \; \Omega \underline{/+53.1°}} = 0.02 \; S \underline{/-53.1°}$$

Step III. Since $\mathbf{Y} = G - jB$, we split $\mathbf{Y}$ into its rectangular coordinates.

$$G = Y \cos \phi = 0.02 \cos 53.1° = 0.02 \times 0.6$$

$$\therefore G = \mathbf{0.012 \; siemens}$$

And
$$B_L = Y \sin \phi = 0.02 \sin 53.1° = 0.02 \times 0.8$$

$$\therefore B_L = \mathbf{0.016 \; siemens}$$

Compare these answers with the *wrong* answers above. With a calculator, we can make the double conversion with a single chain as follows:

30 $\boxed{x \leftrightarrow y}$ 40 $\rightarrow$ P displays 50 = Z

Press $\boxed{1/x}$ to display 0.02 = Y

Press $\boxed{x \leftrightarrow y}$ to display 53.13° = ϕ

Press $\boxed{+/-}$ to change the sign of ϕ

Press $\boxed{\rightarrow R}$ and then $\boxed{x \leftrightarrow y}$ to display the recangular coordinates of $\mathbf{Y}$

$$0.012 - j0.016 \; S$$

Note that the rectangular coordinates of an **impedance** are the resistance and reactance components which always represent the *series* equivalent circuit, while the rectangular coordinates of an **admittance** are the conductance and susceptance components which always represent the *parallel* equivalent circuit. We first encountered these "similar but opposite" characteristics of series and parallel circuits when we compared series and parallel resistances in dc circuits in Section 6-8. Extending this comparison to ac circuits, we can summarize the reciprocal characteristics of series and parallel circuits in Table 20-1.

TABLE 20-1

SERIES AND PARALLEL AC CIRCUIT CHARACTERISTICS

Characteristic	Series components	Parallel components
Phasor diagram	Reference phasor	Reference phasor
Sum of phasors	$\mathbf{V_T} = \mathbf{V_1} + \mathbf{V_2} + \mathbf{V_3} +$ etc.	$\mathbf{I_T} = \mathbf{I_1} + \mathbf{I_2} + \mathbf{I_3} +$ etc.
Defining equation	$\mathbf{Z} = \dfrac{\mathbf{V_T}}{\mathbf{I}}$	$\mathbf{Y} = \dfrac{\mathbf{I_T}}{\mathbf{V}}$
Impedance diagram		Not possible
Admittance diagram	Not possible	
Rectangular form	$\mathbf{Z} = R + jX_{eq}$	$\mathbf{Y} = G + jB_{eq}$
Resistive component	R ohms	G siemens
Reactive component	$+ jX_{eq} = +j(X_L - X_C)$ where $X_L = \omega L$ $X_C = \dfrac{1}{\omega C}$	$+jB_{eq} = +j(B_C - B_L)$ where $B_C = \omega C$ $B_L = \dfrac{1}{\omega L}$
Polar form	$\mathbf{Z} = |Z| \; \underline{/\phi}$	$\mathbf{Y} = |Y| \; \underline{/\phi}$
Magnitude	$|Z| = \sqrt{R^2 + X_{eq}^2}$	$|Y| = \sqrt{G^2 + B_{eq}^2}$
Phase angle	$\phi = \arctan \dfrac{X_{eq}}{R}$	$\phi = \arctan \dfrac{B_{eq}}{G}$

PROBLEMS

Draw a circuit diagram and either a phasor diagram or an impedance diagram (series circuits) or admittance diagram (parallel circuits) with the solution for each problem.

20-1. Calculate the magnitude of the missing voltage in the ac circuits shown in Figure 20-14.

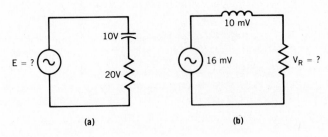

(a) (b)

FIGURE 20-14

20-2. The voltages across a capacitor and a resistor connected in series in an ac circuit are measured as 60 V and 80 V rms, respectively. What is the magnitude and phase angle of the total voltage with respect to the common current?

20-3. An inductor connected to a 120-V 60-Hz source has an IR drop of 72 V across the resistance of its winding. What is the magnitude and the angle of the voltage across its inductance with respect to the applied voltage?

20-4. Calculate the magnitude and angle of the open-circuit output voltage in the coupling circuits shown in Figure 20-15 if the applied voltage in both cases is 150 μV $\underline{/0°}$. [There are two possible answers for the circuit in Figure 20-15(b).]

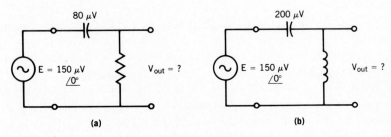

(a) (b)

FIGURE 20-15

20-5. When the operating coil of a contactor is connected to a 120-V 60-Hz source, it draws a 0.9-A current which lags the applied voltage by 72°. What is its impedance?
   (a) In polar coordinates?
   (b) In rectangular coordinates?

20-6. What are the magnitude and phase angle of the current that an $80 - j95\ \Omega$ impedance will draw from a 117-V ac source?

20-7. When a 400-Hz sine-wave voltage of 4 $V\underline{/0°}$ is applied to the voice coil of a loudspeaker, the resulting current is 0.5 $A\underline{/-20°}$. Express the impedance of the loudspeaker in both polar and rectangular coordinates.

20-8. The input circuit of a radio receiver draws a current of 3 $\mu A\underline{/+15°}$ from a 1-MHz signal generator when the terminal voltage is 225 $\mu V\underline{/0°}$. Express the input impedance of the receiver in rectangular coordinates.

20-9. Calculate the total impedance (polar coordinates) at 60 Hz of a 1.8-H inductance in series with a 500-Ω resistance.

20-10. Calculate the total impedance (polar coordinates) at 1 kHz of a 0.5-μF capacitor and a 500-Ω resistor in series.

20-11. What is the 1.5-MHz impedance (polar coordinates) of a 200-pF capacitor in series with a 1.2-kΩ resistor?

20-12. What two components in series will have a 2-kHz impedance of 5 $k\Omega\underline{/+60°}$?

20-13. What is the resistance of a 4-H choke whose impedance at 25 Hz has a magnitude of 750 Ω? What is the angle of the impedance?

20-14. At what frequency will a 0.02-μF capacitor and a 120-kΩ resistor in series have an impedance of 200 kΩ?

20-15. Determine the series equivalent circuit of a device that has an impedance of 2.4 $k\Omega\underline{/+30°}$ at a frequency of 200 Hz.

20-16. Determine the series equivalent circuit of a 4.5-MHz impedance of 150 $\Omega\underline{/-75°}$.

20-17. What is the total 60-Hz impedance (polar form) of a 1-H choke whose resistance is 100 Ω in series with a 10-μF capacitor?

20-18. Calculate the impedance (polar form) of an ac circuit consisting of a 680-Ω resistor, a 200-mH inductor, and a 0.1-μF capacitor in series at frequencies of (a) 400 Hz, (b) 1 kHz, and (c) 3 kHz.

20-19. At a frequency of 400 Hz, what value of capacitor must be connected in series with an impedance of 75 $\Omega\underline{/+60°}$ to form a total impedance of 75 $\Omega\underline{/-60°}$?

20-20. What is the total impedance when $Z_1 = 240$ $\Omega\underline{/+15°}$, $Z_2 = 480$ $\Omega\underline{/-45°}$, and $Z_3 = 1$ $k\Omega\underline{/+60°}$ are all connected in series?

20-21. When connected to a 12-V battery, a certain choke draws a current of 0.25-A. When connected to a 120-V 60-Hz source, it draws a current of 1.0 A. Determine the resistance and inductance of the choke.

20-22. When connected to a 117-V 25-Hz source, a certain choke draws a 5.0-A current. When connected to a 120-V 60-Hz source, the same choke draws a 3-A current. Determine the resistance and inductance of the choke.

20-23. A relay coil requires a 100-mA current through it in order to close the contacts of the relay. If it is operated from a dc source, the required voltage is 24 V. If it is operated from a 60-Hz source, the required voltage is 160 V. What value of a capacitance in series with the relay coil will allow its operation from a 120-V 60-Hz supply? (Two answers)

20-24. At a frequency of 400 Hz, what value of capacitance must be connected in series with an impedance of 75 $\Omega\underline{/+60°}$ to form
(a) An inductive impedance of 50 Ω?
(b) A capacitive impedance of 50 Ω?

20-25. What value of capacitor must be connected in series with a 560-Ω resistor to limit its dissipation to 5 W when connected to a 120-V 60-Hz source?

20-26. A projector using a special 120-V 300-W lamp is operated on the European 230-V 50-Hz system by connecting a choke in series with the lamp. If the resistance of the choke is 20 Ω, what must its inductance be in order to provide the correct voltage drop across the lamp?

20-27. A 15-W fluorescent lamp operates with a ballast inductance in series with it. The voltage across the lamp is measured as 56 V rms, and the total voltage across the ballast inductor is measured as 100 V rms when the total applied voltage is 120 V 60 Hz. What are the resistance and inductance of the ballast? (The lamp itself represents resistance only.)

20-28. A coil and capacitor are connected in series across a 400-Hz source. The voltage across the coil is measured as 48 V, across the capacitor as 52 V, and across the whole circuit as 6 V. The current in the circuit is 100 mA. Determine
(a) The inductance of the coil.
(b) The resistance of the coil.
(c) The capacitance of the capacitor.
(d) The total impedance in polar form.

20-29. Calculate the magnitude and the phase angle with respect to the applied voltage of the unknown current in the ac circuits shown in Figure 20-16.

20-30. If the magnitudes of the applied voltages in the circuits shown in Figure 20-16 are 75 V rms, calculate the admittance and impedance of each circuit in polar form.

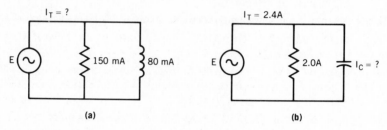

FIGURE 20-16

20-31. A 60-W incandescent lamp and a 44.2-μF capacitor are connected in parallel to a 120-V 60-Hz ac source.
(a) What is the total current drawn from the source?
(b) What is the total admittance in polar form?
(c) What is the equivalent impedance in polar form?

20-32. The bias circuit of an audio amplifier consists of a 2200-Ω resistor and a 0.2μF capacitor in parallel. At a frequency of 1 kHz
(a) What is the admittance of the bias circuit in polar form?
(b) What is the equivalent impedance of the bias circuit in rectangular coordinates?

20-33. Calculate the total admittance and equivalent impedance (polar coordinates) at 60 Hz of a 20-mS conductance in parallel with a 40-mS capacitive susceptance.

20-34. Calculate the total admittance and equivalent impedance (polar coordinates) at 45 MHz of a 300-μS conductance in parallel with a 30-pF capacitor.

20-35. Calculate the 60-Hz admittance of a 0.5-H inductance in parallel with a 12-mS conductance.

20-36. What is the equivalent impedance of a device whose admittance is $0.05 - j0.015$ S? Is this device inductive or capacitive?

20-37. What values of conductance and susceptance must be connected in parallel to form an admittance of 500 mS $\underline{/-30°}$?

20-38. What values of resistance and capacitance in parallel have a 465-kHz admittance of $40\ \mu S \underline{/+60°}$?

20-39. Determine the parallel equivalent circuit of a device that has an impedance of 2.4 k$\Omega \underline{/+30°}$ at a frequency of 200 Hz. (Compare with the answer to Problem 20-15.)

20-40. Determine the parallel equivalent circuit at 4.5 MHz of an impedance of 150 $\Omega \underline{/-75°}$. (Compare with the answer to Problem 20-16.)

20-41. Determine the series components which will have an equivalent admittance at 60 Hz of $20 - j80$ mS.

20-42. Determine the series components which will have an equivalent admittance at 25 MHz of $120 + j90$ S.

20-43. What value of capacitance must be connected in parallel with a 75-Ω resistor to form a total admittance at 500 Hz of 20 mS?

20-44. What value of resistance must be connected in parallel with an inductance of 1.5 H to produce a 60-Hz impedance of 400 Ω?

20-45. Calculate the admittance (polar form) of an ac circuit consisting of a 680-Ω resistor, a 200-mH inductor, and a 0.1-μF capacitor in parallel at frequencies of (a) 400 Hz, (b) 1 kHz, and (c) 3 kHz.

20-46. Calculate the impedance (polar form) of the circuit in Problem 20-45 (Compare with the answers to Problem 20-18.)

20-47. What is the total admittance (in polar coordinates) when an ac circuit consists of two parallel branches whose admittances are $36 + j24$ mS and $42 - j20$ mS?

20-48. What is the equivalent impedance of an ac circuit consisting of two parallel branches whose admittances are 64 $\mu S \underline{/+15°}$ and 48 $\mu S \underline{/-45°}$?

20-49. A given impedance draws a 5.0-A current from a 120-V 60-Hz source. The current through the impedance lags the voltage across it by 60°. What is the total current drawn from the source when a 100-μF capacitor is connected in parallel with the given impedance?

20-50. What is the equivalent impedance (polar form) when $\mathbf{Z}_1 = 240\ \Omega \underline{/+15°}$, $\mathbf{Z}_2 = 480\ \Omega \underline{/-45°}$, and $\mathbf{Z}_3 = 1$ k$\Omega \underline{/+60°}$ are all connected in parallel? (Compare with the answer to Problem 20-20.)

REVIEW QUESTIONS

20-1. Why is the additional term **impedance** required in dealing with the opposition of a circuit to alternating current?

20-2. Why must we add the resistance and reactance of a series circuit by the Pythagorean method in order to find the total impedance?

20-3. Why is it necessary to state an angle as well as a magnitude when an impedance is given in polar form?

20-4. Why does a + angle with an impedance always represent a circuit possessing some inductive reactance?

20-5. How is the angle of an impedance related to the voltage across and the current through the impedance?

20-6. Why is it not necessary to state an angle when an impedance is stated in rectangular coordinates?

20-7. What is the significance of a total impedance having a 0° angle?

20-8. Why must the angle of the impedance of a practical inductor be something less than $+90°$?

20-9. Expressing the impedance of a practical inductor in rectangular coordinates gives us the actual resistance and reactance components which form the practical inductor. Explain.

20-10. How do we go about determining the total impedance of a practical inductor and a resistor in series when the angle between them is less than $90°$?

20-11. Under what circumstances is it possible for the total impedance to decrease when a capacitor is added in series with a given impedance?

20-12. Under what circumstances is it possible for the rms current drawn from a certain source to have the same magnitude when an inductance is connected in series with a given impedance?

20-13. Explain the statement that an ac circuit containing both inductance and capacitance appears to the source only as either an inductive circuit or a capacitive circuit.

20-14. What is meant by **equivalent reactance**?

20-15. Why is quadrant I the capacitive quadrant of a parallel circuit phasor diagram, whereas quadrant IV is the capacitive quadrant of a series circuit phasor diagram?

20-16. Why can we not prepare an impedance diagram for a parallel circuit?

20-17. What is the advantage of thinking in terms of admittance, conductance, and susceptance in dealing with parallel ac circuits?

20-18. Why does B_L have a $-j$ operator for phasor addition, whereas X_L has a $+j$ operator?

20-19. What is meant by the **equivalent impedance** of a parallel ac circuit?

20-20. Explain why $\mathbf{Z} = 20 + j40 \; \Omega$ is *not* the equivalent impedance of $\mathbf{Y} = 50 - j25$ mS.

* 20-21. Write a computer program which will permit entering the rectangular coordinates of an impedance and then print out the rectangular coordinates of the equivalent admittance, or vice versa. Verify the program by using it to solve Example 20-7.

21

POWER IN ALTERNATING-CURRENT CIRCUITS

Most of the examples in this chapter deal with power in ac circuits at the standard North American power-line frequency of 60 hertz. This is not surprising when we recall that we defined **power** as the *rate of doing work* or the *rate of converting energy from one form to another*. Much of the terminology we encounter in this chapter has come from power-systems applications and has resisted a trend toward increased use of the mathematical terminology of complex algebra. The **decibel** of Section 21-10, on the other hand, is an indirect result of the invention of the telephone.

21-1 POWER IN A RESISTOR

In Chapter 17, we investigated the nature of the instantaneous power in a resistor in order in establish rms values for voltage and current in ac circuits. We can set a pattern for dealing with power in ac circuits in general by reviewing the procedure we used to determine the power (or rate of conversion of energy) in a resistor.

If we consider only one particular instant in time, all the power equations we used in direct-current circuits (Chapter 5) must apply to alternating-current circuits. Therefore,

$$p = vi \tag{21-1}$$

And when we plot the sine-wave voltage drop across and current through a resistor, as shown by the shaded curves of Figure 21-1, we can place a series of plots on a linear graph at small intervals, representing the product of v and i at those particular instants.

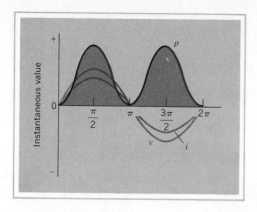

FIGURE 21-1
Instantaneous Power in a Resistor.

Joining these plots gives us the black curve of Figure 21-1 representing power input to a resistor.

Since the voltage drop across a resistor must be exactly in phase with the current through it, the instantaneous voltage and current reach their peak values simultaneously. Whenever i is a positive quantity, v is also a positive quantity; and whenever i is a negative quantity, v is also a negative quantity. Since the product of two negative quantities is a positive quantity, the instantaneous-power graph for a resistor is always positive. This indicates that a resistor must always convert electric energy into heat, never vice versa.

We also noted that the instantaneous power in a resistor is pulsating in nature, fluctuating sinusoidally between zero and a peak value of $P_m = V_m I_m$ twice during each complete cycle of voltage and current. Because of this smooth sinusoidal variation between zero and P_m, we concluded that, as far as doing work is concerned, we can consider the power in a resistive ac circuit as having an average, or dc equivalent value equal to $\frac{1}{2}P_m$.

We verified this conclusion by substituting the sine-wave form for the instantaneous voltage and current in Equation (21-1), giving

$$p = V_m I_m \sin^2 \omega t = P_m \sin^2 \omega t \tag{17-14}$$

From trigonometric half-angle relationships, this became

$$p = \tfrac{1}{2}P_m(1 - \cos 2\omega t) \tag{17-15}$$

Since the *average* of a cosine wave over a complete cycle is zero, from Equation (17-15) the average power $P = \frac{1}{2}P_m$.

In Chapter 17, we used this average value of power in a resistor to establish the rms value of a sine wave of voltage and current as 0.707 of its peak value. Therefore, it follows that the average power input to the resistance of an ac circuit is simply the product of the rms value of the voltage drop across and the rms value of the current through the resistance, and

$$P = V_R I_R \tag{21-2}$$

512

Because a resistor can convert energy only from an electric form into heat or light but never vice versa, the power in watts determined by Equation (21-2) used to be called the **true power** input to the resistance of an ac circuit. More recent terminology refers to the average power input to the resistance of an ac circuit as **real power** or **active power.** We shall appreciate the choice of such designations when we consider power in an ac circuit containing both resistance and reactance in Section 21-4. Note that the letter symbol for average power in an ac circuit is the same as that for power in a dc circuit.

Since $R = V_R/I_R$, just as in dc circuits,

$$P = V_R I_R = I_R^2 R = \frac{V_R^2}{R} \qquad (21\text{-}3)$$

where V_R is the rms voltage across only the resistance portion of the circuit and I_R is the rms current through only the resistance portion of the circuit.

EXAMPLE 21-1

A wattmeter shows an average power of 144 W in an ac circuit in which the rms current is 2 A. What is the resistance of this circuit?

SOLUTION

$$R = \frac{P}{I^2} = \frac{144 \text{ W}}{(2 \text{ A})^2} = 36 \text{ } \Omega$$

21-2 POWER IN AN IDEAL INDUCTOR

We can determine the nature of the instantaneous power in an ideal inductor by the same technique of plotting the product of the instantaneous voltage and current on a linear graph. But in preparing the shaded sine curves of Figure 21-2, we must note that the instantaneous current in an ideal inductor *lags* the voltage across it by $\pi/2$ radians. (Refer to Figure 18-2.)

FIGURE 21-2
Instantaneous Power in an Ideal Inductor.

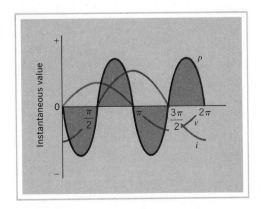

At the zero-radian point on the graph of Figure 21-2, the instantaneous voltage across the inductance of the circuit is zero. Therefore, between 0 and $\pi/2$ radians, the instantaneous voltage is increasing in a positive direction. To generate this voltage, the *rate of change* of current is also increasing in a positive direction as the instantaneous current decreases toward zero. During this interval the instantaneous current is actually flowing *against* the applied voltage as the magnetic field around the inductor collapses, thus returning the energy stored in its magnetic field to the source. As we would expect, the product of a positive v and a negative i gives a *negative* instantaneous power as energy is transferred from the inductance to the source. At $\pi/2$ radians, the current momentarily becomes zero; thus the instantaneous power is also zero.

Between $\pi/2$ and π radians, the instantaneous voltage across the inductor is still positive but decreasing in magnitude. The instantaneous current goes positive as it maintains its positive rate of change in order to induce a positive voltage in the inductance. The product $v \times i$ is now a positive quantity as the instantaneous current increases and energy is stored in the magnetic field of the inductor. At π radians when the instantaneous voltage becomes zero, the instantaneous power also momentarily becomes zero.

The instantaneous power again becomes negative between π and $3\pi/2$ radians when the instantaneous current is decreasing and the collapsing magnetic field again returns energy to the system. Between $3\pi/2$ and 2π radians both v and i have negative polarities creating a positive $v \times i$ product as the source supplies energy to rebuild the magnetic field of the inductor.

The resulting black curve of Figure 21-2 for instantaneous power in the inductance of an ac circuit is sinusoidal in shape at twice the frequency of the instantaneous voltage and current, just as in the resistance of an ac circuit. But in the inductance, the instantaneous power fluctuates equally between a positive maximum and a negative maximum as the ideal inductor alternately stores and then returns equal amounts of energy in the circuit. Therefore, **the *average* power in the inductance of an ac circuit is zero.**

Again we can substitute for v and i in Equation (21-1) using the expression for instantaneous current in an ideal inductor given in Equation (18-3). Hence,

$$p = v \times i = V_m \sin \omega t \times I_m \sin \left(\omega t - \frac{\pi}{2} \right)$$

From trigonometry, $\sin \theta - 90°$ is the same as $-\cos \theta$. Hence,

$$p = -V_m I_m \sin \omega t \cos \omega t$$

From the half-angle relationship, this becomes[†]

(21-4)

[†]
$$\sin (A + B) = \sin A \cos B + \cos A \sin B$$
$$\sin (A - B) = \sin A \cos B - \cos A \sin B$$

If $A = B$, since $\sin 0° = 0$, $\sin 2A = 2 \sin A \cos A$.

514

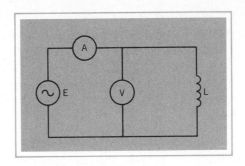

FIGURE 21-3
Reactive Power in an Ideal Inductor.

$$p = -\tfrac{1}{2}V_m I_m \sin 2\omega t$$

Hence, the instantaneous power in an ideal inductor is a sine wave at twice the frequency of the instantaneous voltage and current and starting off in the negative direction from $0°$. But when we average $\sin 2\omega t$ over a complete cycle, there is no nonvarying term in Equation (21-4) as there was in Equation (17-15). Hence, for an ideal inductor, the *average* power is zero.

For the resistance of an ac circuit, we can determine average power simply by multiplying the rms values of current and voltage [Equation (21-2)]. In the circuit of Figure 21-3, the meters show us the rms values of the voltage across and the current through an ideal inductor. But since the average power in the inductance is zero, $V_L I_L$ does *not* represent average power. However, this product is directly proportional to the amount of energy stored and returned by the inductor every time the current changes direction. Therefore,

The product of the rms voltage and current in an ideal inductor is called the reactive power of the inductor.
The letter symbol for reactive power is Q.[†]

$$\therefore Q = V_L I_L \tag{21-5}$$

We can now see why the preferred term for the average power in the resistance of an ac circuit is **active power.** Since the average power in the inductance of an ac circuit is zero, to avoid the possibility of mistaking reactive power for active power, we reserve the **watt** exclusively for active power in an ac circuit. Reactive power is simply the product of the voltage across and the current through the reactance. Therefore,

The voltampere (reactive), which is usually abbreviated to **var,** *is the unit of reactive power.*

Since
$$X_L = \frac{V_L}{I_L} \tag{18-4}$$

$$\therefore Q = V_L I_L = I_L^2 X_L = \frac{V_L^2}{X_L} \qquad \text{vars} \quad \text{(21-6)}$$

[†]Prior to 1968, reactive power was represented by the more descriptive letter symbol P_q.

EXAMPLE 21-2

What is the reactive power of an ideal inductor with an inductance of 0.5 H drawing 0.5 A from a 60-Hz source?

SOLUTION

$$X_L = \omega L = 377 \times 0.5 \text{ H} = 188.5 \ \Omega \text{ (inductive)}$$

$$Q = I_L^2 X_L = (0.5 \text{ A})^2 \times 188.5 \ \Omega = \textbf{47.1 vars}$$

21-3 POWER IN A CAPACITOR

Since the instantaneous current in a capacitor leads the instantaneous voltage across it by $\pi/2$ radians, we again find that the instantaneous power graph of Figure 21-4, although a sine wave at twice the frequency of the instantaneous voltage and current, goes alternately positive and negative for equal time intervals. As we would expect, on examining Figure 21-4 we find that the instantaneous power is positive while the potential difference across the capacitor is rising and the capacitor is taking energy from the source to store up a charge on its plates. While the voltage across the capacitor is decreasing, the capacitor is discharging the stored energy back into the system. During this $\pi/2$-radian interval, the instantaneous power is, therefore, a negative quantity.

Because an ideal capacitor must discharge as much energy between $\pi/2$ and π radians as it stored between 0 and $\pi/2$ radians, **the average power in the capacitance of an ac circuit is zero.** In this case, substitution in Equation (21-1) gives

$$p = \tfrac{1}{2} V_m I_m \sin 2\omega t \qquad (21\text{-}7)$$

Again we find that the product of the rms voltage across and current in the capacitance of an ac circuit does *not* represent average power.

The product of the rms voltage and current in a capacitor is called the reactive power of the capacitor.

$$\therefore Q = V_C I_C = I_C^2 X_C = \frac{V_C^2}{X_C} \qquad \text{vars} \quad (21\text{-}8)$$

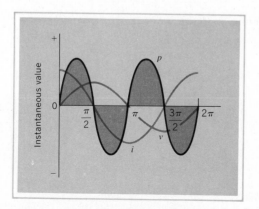

FIGURE 21-4
Instantaneous Power in a Capacitor.

EXAMPLE 21-3

What value of capacitance will have a reactive power of 50 vars when connected across a 120-V 60-Hz source?

SOLUTION

Since $Q = V_C^2/X_C$,

$$X_C = \frac{V_C^2}{Q} = \frac{(120 \text{ V})^2}{50 \text{ vars}} = 288 \ \Omega \text{ (capacitive)}$$

Since $X_C = 1/\omega C$,

$$C = \frac{1}{\omega X_C} = \frac{1}{377 \times 288 \ \Omega} = 9.2 \times 10^{-6} \text{ F} = \mathbf{9.2 \ \mu F}$$

21-4 POWER IN A CIRCUIT CONTAINING RESISTANCE AND REACTANCE

If an electric circuit consists of equal resistance and inductive reactance in series, as in Figure 21-5, the instantaneous current lags the applied instantaneous voltage by $\pi/4$ radians (45°). Therefore, when we plot the instantaneous power on the graph of Figure 21-6, it is a negative quantity between 0 and $\pi/4$ radians as the inductance is returning more energy to the system than the resistance is converting to heat. For the longer time interval between $\pi/4$ and π radians, the instantaneous power is positive. At the peak of this instantaneous power wave, both the resistance and the inductance are taking energy from the source.

The instantaneous-power graph for a circuit containing both resistance and reactance is again a sine wave at twice the frequency of the voltage and current. But, for the circuit of Figure 21-5, the instantaneous-power graph is neither all positive, as in the case of resistance, nor equally positive and negative, as in the case of reactance. Since the instantaneous-power graph is more positive than negative, there is an average power component which represents the active power input to the resistance

FIGURE 21-5
Apparent Power in an AC Circuit Containing Resistance and Inductive Reactance in Series.

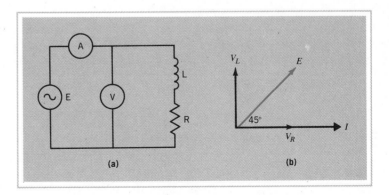

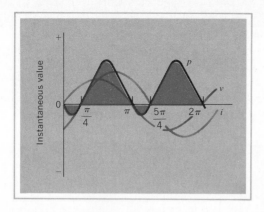

FIGURE 21-6
Instantaneous Power in an AC Circuit Containing Resistance and Inductive Reactance.

of the circuit. Part of the positive instantaneous power is offset by the negative instantaneous power to account for the reactive power of the reactance of the circuit. In the general case,

$$p = V_m I_m \sin \omega t \sin (\omega t + \phi) \tag{21-9}$$

where ϕ is the phase angle of the current with respect to the applied voltage. The phase angle of the rms current with respect to the rms voltage in the phasor diagram of Figure 21-5(b) is $-45°$. In the time graph of instantaneous values in Figure 21-6, the same phase angle is shown as $-\pi/4$ radians.

If we use the $\sin (A + B)$ relationship to substitute for $\sin (\omega t + \phi)$ in Equation (21-9),

$$p = V_m I_m \sin \omega t (\sin \omega t \cos \phi + \cos \omega t \sin \phi)$$

$$= \cos \phi (V_m I_m \sin^2 \omega t) + \sin \phi (V_m I_m \sin \omega t \cos \omega t)$$

We should recognize the expressions in parentheses as the instantaneous power in the resistance and reactance of the circuit respectively. Hence,

$$p = \cos \phi p_A + \sin \phi p_R \tag{21-10}$$

where p_A is the instantaneous power in the resistance of the circuit and p_R is the instantaneous power in the reactance of the circuit.

In the example above, the product of total rms voltage and total rms current represents neither active power in **watts** nor reactive power in **vars.** If we have only the voltmeter and ammeter readings to go by in the circuit of Figure 21-5, we have no means of determining the active and reactive power components of the circuit. Therefore,

The product of the total rms voltage and the total rms current in an ac circuit is called the **apparent power** *of the circuit.*

The letter symbol for apparent power is S. [†]

[†]Prior to 1968, apparent power was represented by the more descriptive letter symbol P_s.

$$\therefore S = V_T I_T \qquad \text{(21-11)}$$

At this point, we can only talk about the magnitude of the apparent power. In the next section, we find that, like impedance and admittance, we can treat S as a complex number with an angle as well as a magnitude. Because apparent power is neither active power in **watts** nor reactive power in **vars,** we must always express apparent power simply as the product of volts and amperes. Therefore,

The voltampere (VA) is the unit of apparent power.[†]

Since
$$\frac{V_T}{I_T} = Z \qquad \text{(20-3)}$$

$$\therefore S = V_T I_T = I_T^2 Z = \frac{V_T^2}{Z} \qquad \text{(21-12)}$$

EXAMPLE 21-4

The input impedance of a given transmitting antenna system is 500 $\Omega/\underline{+20°}$. A radio-frequency ammeter shows an input current of 8 A. What apparent power is the transmitter supplying to the antenna system?

SOLUTION

$$S = I^2 Z = (8 \text{ A})^2 \times 500 \ \Omega = 32\,000 \text{ VA} = \mathbf{32 \ kVA}$$

21-5 THE POWER TRIANGLE

Although total voltage times total current does not give us any indication of the active power in the circuit of Figure 21-5, we can find the active power from the product of V_R and I and the reactive power from the product of V_L and I. Since V_L leads V_R by 90° (Figure 21-5), we can state that

$$V_T = \sqrt{V_R^2 + V_L^2} \qquad \text{(20-2)}$$

Multiplying each side of equation (20-2) by the common current,

$$V_T I = \sqrt{(V_R I)^2 + (V_L I)^2}$$

from which
$$S = \sqrt{P^2 + Q^2} \qquad \text{(21-13)}$$

Therefore, we can use the sides of a right-angled triangle to represent the relationship among the three types of power in an ac circuit.

[†]**Load** in an electric circuit is defined as either a device that receives power or the power or apparent power to such a device. Hence, we sometimes find voltamperes referred to as the **load** on a source. **Voltampere** and **var** do not follow the customary pattern for naming SI units after pioneers in science. However, lacking other names for the SI units of apparent and reactive power, the International Electrotechnical Commission accepts the names **voltampere** and **var.**

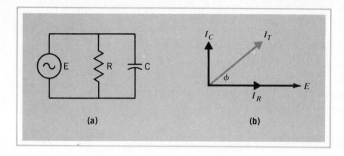

FIGURE 21-7
Apparent Power in an AC Circuit Containing Resistance and Capacitive Reactance in Parallel.

Similarly, in the parallel circuit of Figure 21-7, since the current in the capacitor branch must lead the current in the resistor branch by exactly 90°,

$$I_T = \sqrt{I_R^2 + I_C^2} \qquad (20\text{-}17)$$

Multiplying each side of Equation (20-17) by the common voltage,

$$V\,I_T = \sqrt{(VI_R)^2 + (VI_C)^2}$$

and again
$$S = \sqrt{P^2 + Q^2} \qquad (21\text{-}13)$$

Note that Equation (21-13) applies to both series and parallel circuits. Since power is the product of current (the common phasor of series circuits) and voltage (the common phasor of parallel circuits), we find that the *same* equations for power apply to series, parallel, and also combination series-parallel ac circuits. This knowledge will provide us with a simple method of solving series-parallel circuits without having to reduce them to series-equivalent or parallel-equivalent circuits, as we shall discover in Section 21-7.

Although instantaneous power does vary sinusoidally, it does so at *twice* the frequency of the voltage and current. Consequently, we cannot draw power phasors on the same phasor diagram with voltage and current. Therefore, we prefer not to represent the relationship among active power, reactive power, and apparent power by the conventional phasor diagram of the type shown in Figures 21-5(b) and 21-7(b). It is customary to represent this Pythagorean relationship of Equation (21-13) by means of right-angled triangles, as shown in Figure 21-8.

As we would expect, we draw that side of the triangle which represents active power horizontally. We draw the side representing reactive power in a vertical direction, with the right angle at the *right*-hand end of the side representing active power. Since Equation (21-13) applies to either a series or a parallel circuit, we can develop our **power triangle** from either an impedance diagram (Figure 20-2) or an admittance diagram (Figure 20-12). If we start with an impedance diagram, in which inductive reactance is drawn in the $+j$ direction, the power triangle appears as shown in Figure 21-8(a). But if we start from an admittance diagram for parallel circuits, in which inductive susceptance is drawn in the $-j$ direction, the power triangle has the form shown in Figure 21-8(b).

To avoid confusion between inductive and capacitive reactive power in our power triangles, we must select *one* of these formats for *all* ac circuits, whether series, parallel, or series-parallel. The International Electrotechnical Commission, which

520

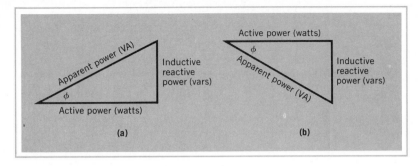

FIGURE 21-8
**Power Triangle: (a) Based on Common Current as Reference Axis;
(b) Based on Voltage as Reference Axis.**

establishes international standards for electrical units and symbols, prefers inductive reactive power to be shown in the $-j$ direction, as in Figure 21-8(b). Hence, it is conventional to think of the *voltage* of the source as the reference phasor in dealing with power in alternating-current circuits.

We can now add to the magnitude of the apparent power the angle ϕ shown in Figure 21-8(b). With the convention adopted for drawing power diagrams, it turns out that **S** has the same angle as the admittance of the circuit and the angle of the conjugate impedance (opposite sign) of the circuit. When we treat **S** as a complex number, we call it the **phasor power** of the circuit. Thus,

$$\mathbf{S} = P + jQ \tag{21-14}$$

We can now see where the terms **real power** for active power and **quadrature** (imaginary) **power** for reactive power originated. To satisfy the convention for the sign of ϕ, capacitive-reactive power has a $+j$ operator and inductive-reactive power has a $-j$ operator.

Since the voltage across the capacitance in a series circuit is 180° out of phase with the voltage across the inductance, the net reactive voltage is the *difference* between the two reactive voltages. Similarly, in a simple parallel circuit, the net reactive current is the *difference* between the capacitive and inductive branch currents. Therefore, since $Q = V_X I_X$, the net reactive power in any alternating-current circuit (either series or parallel) is the *difference* between the capacitive reactive power and the inductive reactive power.

If we examine Figure 21-2 closely, we note that the instantaneous power in an inductance is positive when the current is rising and building up a magnetic field around the inductor. But Figure 21-4 shows that, when the current is rising, the capacitance is discharging its stored energy back into the system. Therefore, in an ac circuit containing both inductance and capacitance, the capacitance always returns energy to the circuit when the inductance takes energy from the circuit, and vice versa. Consequently, a certain amount of energy can be traded back and forth between the inductance and the capacitance of the circuit and, as far as the source is concerned, the net reactive power required by the circuit is the *difference* between the capacitive reactive power and the inductive reactive power.

EXAMPLE 21-5

A circuit consists of two branches connected in parallel to a 120-V 60-Hz source. Branch I consists of a 75-Ω resistance and a 100-Ω inductive reactance in series. Branch II consists of a 200-Ω capacitive reactance. Determine the apparent power of the circuit.

SOLUTION 1

  Step I. In the circuit diagram of Figure 21-9,

$$\mathbf{Z}_1 = R + jX_L = 75 + j100 = 125 \ \Omega \underline{/+53.1°}$$

$$\therefore I_1 = \frac{E}{Z_1} = \frac{120 \text{ V}}{125 \ \Omega} = 0.96 \text{ A}$$

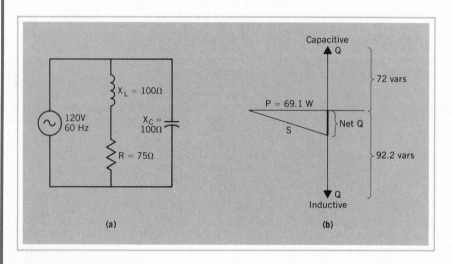

FIGURE 21-9
Diagram for Example 21-5.

  Step II. Active power of the circuit is

$$P = I_1^2 R = 0.96^2 \times 75 = 69.1 \text{ W}$$

Reactive power of the inductance is

$$Q = I_1^2 X_L = 0.96^2 \times 100 = 92.2 \text{ vars}$$

Reactive power of the capacitance is

$$Q = \frac{E^2}{X_C} = \frac{120^2}{200} = 72 \text{ vars}$$

Net reactive power of the circuit is

$$Q = 92.2 - 72 = 20.2 \text{ vars (inductive)}$$

Step III.

$$S = \sqrt{P^2 + Q^2} = \sqrt{69.1^2 + 20.2^2} = 72 \text{ VA}$$

SOLUTION 2

A somewhat similar procedure involves solving Steps II and III in terms of current rather than power.

Step I.

$$\mathbf{Z}_1 = R + jX_L = 75 + j100 = 125 \ \Omega \underline{/+53.1°}$$

$$\therefore \mathbf{I}_1 = \frac{E}{\mathbf{Z}_1} = \frac{120 \underline{/0°}}{125 \underline{/+53.1°}} = 0.96 \text{ A} \underline{/-53.1°}$$

$$\therefore \mathbf{I}_1 = 0.58 - j0.77 \text{ A}$$

Step II.

$$I_C = \frac{E}{-jX_C} = \frac{120 \underline{/0°}}{200 \underline{/-90°}} = 0.6 \text{ A} \underline{/+90°}$$

$$= 0 + j0.6 \text{ A}$$

$$\therefore \mathbf{I}_T = I_1 + I_C = (0.58 - j0.77) + (0 + j0.6)$$

$$= 0.58 - j0.17 \text{ A}$$

and

$$I_T = \sqrt{0.58^2 + 0.17^2} = 0.6 \text{ A}$$

Step III.

$$\therefore S = E_T \ I_T = 120 \times 0.6 = 72 \text{ VA}$$

21-6 POWER FACTOR

Since the hypotenuse of a right-angled triangle must be greater than either of the other two sides, the apparent power that a generator must supply to a reactive load is always greater than the active power that the load can convert into some other form of energy. Since most industrial loads possess appreciable inductive reactance, this relationship between the active power that the load can use and the apparent power that the source must supply is of considerable importance.

The ratio between the active power and the apparent power of a load in an ac circuit is called the power factor of the load.

Examining the power triangle of Figure 21-8(b), we note that the ratio of active power to apparent power is the cosine of the **power factor angle** ϕ between active power and app nt power. Tracing the construction of the power triangle back through admitta diagram to the phasor diagram for current and voltage, we find that the **power factor angle** is the same as the **phase angle** between voltage across and current through the load.

From the definition of power factor,

$$\text{Power factor} = \frac{P}{S} = \cos \phi \qquad (21\text{-}15)$$

In keeping with international practice, North American standards have dropped a letter symbol for power factor.[†] In equations,

The symbol for power factor is cos ϕ.

In keeping with our decision to show inductive reactive power in a $-j$ direction in power triangles, we distinguish between inductive and capacitive loads by stating that inductive loads always have a **lagging power factor** and that capacitive loads always have a **leading power factor.** Since the active power cannot be greater than the apparent power, the power factor cannot have a numerical value greater than unity. Since it is a simple ratio, power factor is always a positive dimensionless number. We may express power factor either as a decimal fraction or as a percentage.

EXAMPLE 21-6

What are the active power and power factor of a load whose impedance is $60\ \Omega\ \underline{/+60^\circ}$ when connected to a 120-V 60-Hz source?

SOLUTION 1

Step I.   Expressing the impedance in rectangular coordinates produces the equivalent of a series circuit consisting of resistance and reactance in series. Therefore, as shown in Figure 21-10,

FIGURE 21-10
Schematic Diagram for Example 21-6.

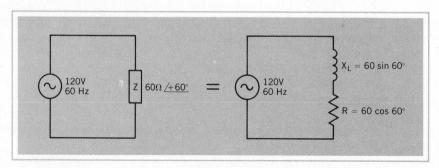

[†]Prior to 1968, power factor was represented by the letter symbol, F_p.

$$\mathbf{Z} = 60 \cos 60° + j60 \sin 60°$$

from which $R = 30\ \Omega$.

Step II.

$$I = \frac{E}{Z} = \frac{120\ \text{V}}{60\ \Omega} = 2.0\ \text{A}$$

$$\therefore P = I^2R = 4 \times 30 = \mathbf{120\ W}$$

Step III.

$$S = I^2Z = 4 \times 60 = 240\ \text{VA}$$

$$\therefore \cos \phi = \frac{P}{S} = \frac{120}{240} = \mathbf{0.5\ lagging}$$

Although Solution 1 follows the definition of power factor, it is a roundabout way of tackling the problem.

SOLUTION 2

Step I.   Power Factor $= \cos \phi = \cos 60° = \mathbf{0.5\ lagging}$

Step II.

$$S = \frac{E^2}{Z} = \frac{120^2}{60} = 240\ \text{VA}$$

Step III.   Since $\cos \phi = P/S$ (Figure 21-11),

$$P = S \cos \phi = 240 \times 0.5 = \mathbf{120\ W}$$

FIGURE   21-11
Power Triangle for Example 21-6.

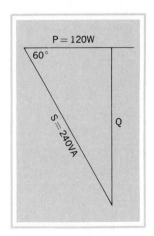

From the second solution, we note that we can express active power in terms of power factor and apparent power. This leads to the following very useful expression for active power in ac circuits.

$$P = E I \cos \phi \qquad (21\text{-}16)$$

Once we know the active power and apparent power of an ac load, we can determine reactive power by

$$Q = \sqrt{S^2 - P^2} \qquad (21\text{-}13)$$

Since $P = S \cos \phi$, this equation becomes

$$Q = S\sqrt{1 - \cos^2 \phi} \qquad (21\text{-}17)$$

Although our main concern in dealing with ac loads is the ratio between active power and apparent power, there are some instances when we wish to solve directly for the reactive power of the load. In these cases, it is useful to be able to express the ratio between reactive power and apparent power.

Reactive factor is the ratio between the reactive power and the apparent power of an ac load.

Once again, as shown by the power triangle of Figure 21-8(b), from the definition of reactive factor,

$$\text{Reactive factor} = \frac{Q}{S} = \sin \phi \qquad (21\text{-}18)$$

The symbol for reactive factor is sin ϕ.[†]

21-7 POWER FACTOR CORRECTION

We can appreciate the consequences of a low power factor in an industrial load by considering the state of affairs illustrated by Figure 21-12. Since the motor in Figure 21-12(a) has a 70% lagging power factor, the apparent power, as far as the source is concerned, is

$$S = \frac{P}{\cos \phi} = \frac{840}{0.7} = 1200 \text{ VA}$$

and, therefore, the current drain on the source is

$$I = \frac{S}{V} = \frac{1200 \text{ VA}}{120 \text{ V}} = 10 \text{ A}$$

[†]Prior to 1968, reactive factor was represented by the letter symbol F_q.

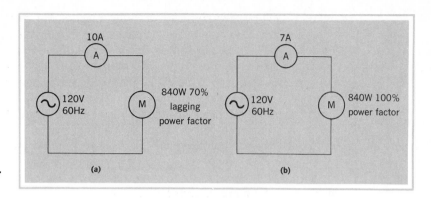

FIGURE 21-12
Effect of Load Power Factor.

If we now replace the original motor by one capable of developing an active power of 840 watts at 100% power factor, as in Figure 21-12(b), the power apparently drawn from the source will now be 840 VA. The current drawn by the second motor is only 7 amperes. Therefore, the 70% power-factor motor requires a current of 10 amperes to do the same work that the unity-power-factor motor can do by drawing only 7 amperes from the source.

The wire size used in the windings of the generator and the conductors connecting the load to the generator depends on the *current* they have to pass, and the copper losses in the system depend on the square of the *current*. Therefore, it is more economical for an electrical utility company to feed an 840-W 100% power-factor load than to feed an 840-W 70% lagging-power-factor load. Since a great many industrial loads possess considerable inductive reactance, these electrical utility companies are very much concerned with maintaining a high overall power factor in their systems.

As we have noted, an 840-W 70% lagging-power-factor motor draws three amperes more from the source than an equivalent 840-W 100% power-factor motor. As a power triangle shows, this additional current is involved in supplying the reactive power required by the inductive reactance of the motor. But, as shown in Figure 21-2, this portion of the current alternately transfers energy from the system into the magnetic field of the inductance and then returns the stored energy back to the system. Although the average power used by the inductance during this process is zero, this charge and discharge current has to flow through the generator windings and connecting conductors, thus forcing us to use a larger wire size for a given load power.

We can illustrate this charge and discharge or *reactive* current component by expressing the current of each of the motors in Figure 21-12 in reactangular coordinates. In the case of the 70% lagging-power-factor motor, the current lags the applied voltage by arccos $0.7 = 45.6°$.

$$\therefore \mathbf{I} = 10 \cos 45.6° - j10 \sin 45.6° = 7 - j7.14 \text{ A}$$

In the case of the 100% power-factor motor, the current is in phase with the applied voltage.

$$\therefore \mathbf{I} = 7 - j0 \text{ A}$$

When we express the load current in this manner, we note that both have the same *reference axis* component. It is the component that contributes to the active power of the load. The 70% lagging-power-factor load requires the additional *quadrature* component to take care of the inductance of the load.

We recall that in an ac circuit containing both inductance and capacitance, the capacitance returns energy to the system while the inductance takes energy from the system; and vice versa. If we then place a capacitor in the circuit containing the 70% lagging-power-factor motor, it should be possible for the capacitor and the inductance of the load to trade their reactive power back and forth without the reactive component of the current having to travel all the way from the source to the load and back. In practice, this would allow us to use lower power factor loads yet obtain the advantages of a high system power factor.

For the moment, we connect a 308-μF capacitor in *series* with the 70% lagging-power-factor motor, as in Figure 21-13. The impedance of the motor is $Z_M = V/I = 120/10 = 12\ \Omega$.

$$\phi = \text{arccos } 0.7 = 45.6°$$

$$\therefore \mathbf{Z}_M = 12\ \Omega\underline{/+45.6°}$$

$$= 12 \cos 45.6° + j12 \sin 45.6° = 8.4 + j8.6\ \Omega$$

$$X_C = \frac{1}{\omega C} = \frac{1}{377 \times 308\ \mu F} = 8.6\ \Omega \text{(capacitive)}$$

Because the motor and capacitor are in series,

$$\mathbf{Z}_T = 8.4 + j8.6 - j8.6 = 8.4 + j0\ \Omega$$

$$\therefore \mathbf{Z}_T = 8.4\ \Omega\underline{/0°}$$

Since the phase angle of the total impedance is now 0°, the current drawn from the generator is in phase with the source voltage. But

$$I = \frac{V}{Z_T} = \frac{120\ V}{8.4\ \Omega} = 14.3\ A$$

Therefore, connecting a capacitor in *series* with an inductive load to reduce the net reactive power does not *reduce* the current drain on the source. It *increases* it. Moreover, since the impedance of the motor itself is still 12 ohms, the voltage across the motor in the circuit of Figure 21-13 becomes

$$V_M = IZ_M = 14.3\ A \times 12\ \Omega = 172\ V$$

Although connecting a capacitor in *series* with the motor has raised the system power factor to 100%, it has served only to aggravate the original problem by increasing the current drawn from the source by applying excessive voltage to the motor. Therefore, it is not practical to correct for a low power factor by connecting a capacitor in *series* with an inductive load.

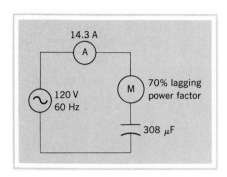

FIGURE 21-13
Effect of a Series Capacitor on a
Lagging Power-Factor Load.

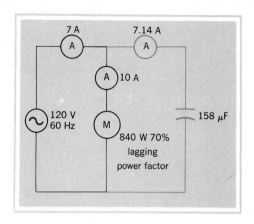

FIGURE 21-14
Power-Factor Correction.

Power factor correction or improvement consists of adding a capacitive reactive power to an ac circuit in such a manner that the apparent power drawn from the source is reduced without altering the current through or the voltage across the load itself.

One of the characteristics of a parallel circuit is that, although a change in one branch will affect the *total* current, it will *not* affect the current in the other branches. Therefore, in the circuit of Figure 21-14, we have connected a capacitor in parallel with the 840-W 70% lagging-power-factor motor. Since the motor is still connected directly across the 120-V source, its current is still 10 amperes, lagging the applied voltage by 45.6°. Because of its lagging power factor, the motor has a reactive power of $Q = VI \sin \phi = 120 \times 10 \times 0.714 = 860$ vars. If the capacitor in parallel with the motor also has a reactive power of 860 vars, they will be able to trade this reactive power back and forth between them, and, as far as the generator is concerned, the net reactive power in the circuit is zero. Since the voltage across and the current through the motor are unchanged, the active power is still 840 watts. Therefore, the overall apparent power that the generator must supply is 840 voltamperes, and the current drawn from the generator becomes seven amperes. Note that adding the capacitor to the circuit has decreased the current to seven amperes, while the motor current still remains at ten amperes. As far as the generator is concerned, the 70% lagging-power-factor motor and capacitor in parallel are the equivalent of a unity-power-factor motor.

Although we can solve for the required capacitance by noting that the capacitor current must equal the reactive component of the load current or that the susceptance of the capacitor must equal the susceptance component of the load admittance, the simplest procedure is to work in terms of reactive power, since power equations are independent of series or parallel connection of the circuit components.

To obtain an overall unity power factor with a lagging power factor in a load, a capacitor having a reactive power equal to the reactive power of the load is connected in parallel with the load.

EXAMPLE 21-7

What value of capacitance must be connected in parallel with a motor drawing 10 A at 70% lagging power factor from a 120-V 60-Hz source in order for the generator current to be minimum?

SOLUTION

Step I. The reactive power of the load is

$$Q = VI\sqrt{1 - \cos^2 \phi} = 120 \times 10\sqrt{1 - 0.7^2} = 857 \text{ vars}$$

Therefore, we must select a capacitor with a reactive power of 857 vars when connected across a 120-V 60-Hz source.

Step II. Since $Q = V^2/X_C$,

$$\therefore X_C = \frac{V^2}{Q} = \frac{(120 \text{ V})^2}{857 \text{ vars}} = 16.8 \ \Omega \text{ (capacitive)}$$

$$C = \frac{1}{\omega X_C} = \frac{1}{377 \times 16.8 \ \Omega} = 1.58 \times 10^{-4} \text{ F} = \textbf{158 } \boldsymbol{\mu}\textbf{F}$$

Note: As a check that *adding* an extra branch does *reduce* the total current drawn from the source without affecting the load current, draw an accurate phasor diagram for the circuit of Figure 21-14.

EXAMPLE 21-8

A fluorescent lamp and its ballast inductance draw a 1.0-A current at a 50% lagging power factor from a 120-V 60-Hz source. What is the overall power factor when a 26.5-μF capacitor is connected across the fixture?

SOLUTION

Step I. Prepare a power triangle for the lamp and its ballast inductance.

$$S = VI = 120 \text{ V} \times 1.0 \text{ A} = 120 \text{ VA}$$

$$P = S \cos \phi = 120 \text{ VA} \times 0.5 = 60 \text{ W}$$

$$Q = \sqrt{S^2 - P^2} = \sqrt{120^2 - 60^2} = 104 \text{ vars (inductive)}$$

Step II.

$$X_C = \frac{1}{\omega C} = \frac{1}{377 \times 26.5 \ \mu\text{F}} = 100 \ \Omega \text{ (capacitive)}$$

and

$$Q_C = \frac{V^2}{X_C} = \frac{120^2}{100} = 144 \text{ vars (capacitive)}$$

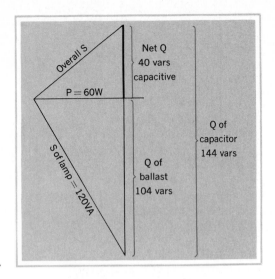

FIGURE 21-15
Power Triangle for Example 21-8.

Step III.   From the power triangle of Figure 21-15,

$$\text{Final } Q = 144 - 104 = 40 \text{ vars (capacitive)}$$
$$\text{Final } S = \sqrt{P^2 + Q^2} = \sqrt{60^2 + 40^2} = 72 \text{ VA}$$

Step IV.

$$\text{Overall } \cos\phi = \frac{P}{S} = \frac{60}{72} = \mathbf{83.2\% \text{ leading}}$$

EXAMPLE 21-9
What value of capacitance must be connected in parallel with a load drawing 1 kW at 70.7% lagging power factor from a 208-V 60-Hz source in order to raise the overall power factor to 91% lagging?

SOLUTION
   Step I.

$$\text{Original } S = \frac{P}{\cos\phi} = \frac{1000 \text{ W}}{0.707} = 1.41 \text{ kVA}$$
$$\text{Load } Q = \sqrt{S^2 - P^2} = \sqrt{1.41^2 - 1^2} = 1 \text{ kvar (inductive)}$$

Step II.

$$\text{Final } S = \frac{P}{\cos\phi} = \frac{1000 \text{ W}}{0.91} = 1.1 \text{ kVA}$$
$$\text{Overall } Q = \sqrt{S^2 - P^2} = \sqrt{1.1^2 - 1^2} = 456 \text{ vars (inductive)}$$

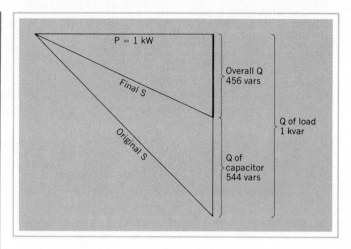

FIGURE 21-16
Power Triangle for Example 21-9.

Step III. Therefore, from the power triangle of Figure 21-16, the reactive power of the capacitor must be

$$Q_C = 1000 - 456 = 544 \text{ vars (capacitive)}$$

$$\therefore X_C = \frac{V^2}{Q_C} = \frac{208^2}{544} = 80 \ \Omega \text{ (capacitive)}$$

and

$$C = \frac{1}{\omega X_C} = \frac{1}{377 \times 80 \ \Omega} = 3.3 \times 10^{-5} \text{ F} = \mathbf{33 \ \mu F}$$

21-8 MEASURING POWER IN *AC* CIRCUITS

In a dc system, we seldom require wattmeters to indicate power input to a load, since we can obtain this information simply by multiplying the voltmeter reading by the ammeter reading. However, in an ac system, the product of the voltmeter and ammeter readings gives us only the *apparent*-power input to a load. Unless we have further data as to the power factor of the load or the phase angle between load voltage and load current, we cannot determine the *active*-power input to the load in this manner.

In Chapter 10, we discovered that an electrodynamometer movement produces a torque which is proportional to the product of the currents in the moving coil and the stationary coils. If we connect the electrodynamometer as shown in Figure 21-17, the *load current* flows through the stationary coils. The current through the moving coil and its series multiplier resistor is proportional to the *voltage across the load*. Hence the pointer displacement of the resulting wattmeter is proportional to the *product* of load current and load voltage, i.e., load *power*.

In direct-current circuits, this wattmeter reading is a real-power reading, since reactive power does not exist. To determine what the wattmeter will read in an

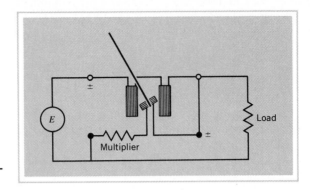

FIGURE 21-17
Measuring AC Power with a Watt-meter.

alternating-current circuit, we must examine the nature of the instantaneous torque produced by the sine-wave currents in the stationary and moving coils. We can do this by referring first to the voltage and current relationships for a resistance load, as shown in Figure 21-1.

The current through the current coil is represented by the shaded current curve, and the current through the voltage coil is represented by the shaded voltage curve. The instantaneous power curve in Figure 21-1 represents the product $v \times i$ and also represents the instantaneous torque developed by the electrodynamometer movement. Since the current in both coils reverses at the same instant, the instantaneous torque (like the instantaneous power) is always positive. However, the inertia of the pointer and voltage coil assembly will not allow the pointer to follow the variations in instantaneous torque due to the alternating current in the coils. Therefore, the pointer will take up a position which represents the average between zero and maximum instantaneous torque. The position of the pointer then represents the *average* power into the resistor.

If we now connect the electrodynamometer to an inductor, the currents in the two coils will be 90° out of phase, as indicated by the shaded curves in Figure 21-2. Therefore, the product of the two instantaneous fluxes will produce a clockwise torque for one quarter-cycle and a counterclockwise torque for the next quarter-cycle. As a result, the net meter deflection is zero, just as the average of the instantaneous power (the active power) in inductance is zero. If we connect the electrodynamometer to the circuit of Figure 21-5(a), the instantaneous torque can be represented by the instantaneous power curve in Figure 21-6. Now the clockwise torque is greater than the counterclockwise torque and the meter will give a reading. We can consider that the instantaneous torques due to the inductance average to zero, leaving a net positive torque representing the active power into the resistance of the circuit.

Therefore, when we connect an electrodynamometer movement with its current coils in series with a load and its voltage terminals across the load in an ac circuit, the inertia of the movement causes it to indicate the *average* instantaneous power or the *active* power input to the load. Consequently, we call this instrument a **wattmeter.** We should also note that the wattmeter reading can be represented by

$$P = EI \cos \phi \qquad (21\text{-}16)$$

where E is the voltage applied to the voltage terminals, I is the current flowing through the current terminals, and ϕ is the phase angle between the voltage and current applied to the wattmeter.

In the wattmeter movement, the required current through the moving voltage coil is determined by a *multiplier resistor* in series with the coil. If we replace this resistor with a *capacitor* having the required capacitive reactance at the power line frequency, the current through the voltage coil will have the same magnitude but now will be 90° out of phase with the voltage applied to the voltage terminals of the meter (since the capacitor current leads the applied voltage by 90°). If we connect this meter to a resistive circuit, the net torque is zero, since the voltage-coil current is now 90° out of phase with the current through the current coil. Similarly, the meter will now show a reading when connected to an inductive or capacitive load. This modification to the electrodynamometer has produced a meter which indicates *reactive power*. Hence, we call this meter a **varmeter.**

21-9 EFFECTIVE RESISTANCE

We can determine the resistance of an electric circuit by measuring the voltage across and the current through it. However, for the results to be accurate, we must be sure that the voltage across the resistor is due only to *IR* drop. We recall that, as the current through an electric conductor increases, closed loop magnetic lines of force expand outward from the center of the conductor. As a result, in addition to the *IR* drop, a small counter emf is induced in the copper that is being linked by the expanding magnetic field. In a dc circuit, where the current is not changing, these induced voltages do not appear. Therefore, when we make *direct* current measurements of the voltage across and current through a practical resistor, we are reasonably sure that the value of resistance determined from $R = V/I$ is the actual **ohmic resistance** of the resistor.

If we wish, we can also determine the resistance of a dc circuit by using a wattmeter and an ammeter. This is more in keeping with the techniques we use in establishing the rms or dc equivalent value of an alternating current. Since $R = V/I$, multiplying numerator and denominator by I gives

$$R = \frac{V}{I} = \frac{VI}{I^2} = \frac{P}{I^2}$$

Because the wattmeter reads the *product* of the voltage across and current through a dc circuit, we are essentially measuring the same parameters by both techniques.

However, in an ac circuit, the ratio between the voltage across and the current through a circuit component is, by definition, its **impedance,** rather than just its resistance. Since it is not possible for us to connect an ac voltmeter in such a manner that it indicates only the *IR* drop across the resistance of the component, we cannot use the voltmeter-ammeter method to determine the **effective resistance** of an ac circuit. But, as we noted in Chapter 20, the *same* current flows through both the resistance and the reactance of practical impedances. We can then accurately measure I_R in an ac circuit, even though we cannot accurately measure V_R. As we noted earlier in this chapter, the active power in an ac circuit represents the power input to the resistance of the circuit. Since a wattmeter measures only the active power in an ac circuit, we can determine the effective resistance of an ac circuit by the wattmeter-ammeter method, even though the voltmeter-ammeter technique is not valid. Hence, in both ac and dc circuits, we are correct in expressing resistance by the relationship

$$R = \frac{P}{I^2}$$

(21-19)

After going to this trouble to obtain a true resistance value in a practical ac circuit (rather than the total impedance of the circuit), we would discover that the **effective or ac resistance** of the circuit is appreciably higher than the plain **ohmic resistance** of the same circuit with only direct current through it. In addition to thinking of resistance as the property of an electric circuit that opposes electric current, we must now think of **resistance as the property of an electric circuit that dissipates electric energy.** There are several mechanisms for dissipating energy in an ac circuit that are not present in a dc circuit.

One such factor is apparent if we again consider the expanding and collapsing magnetic lines of force around an electric conductor in an ac circuit. There will be more changes in flux linkage at the center of the conductor than at its surface. Therefore, there is more opposition to alternating current in the center of the conductor than there is at the surface. As the frequency of the alternating current is increased, the rate of change of flux increases, which, in turn, increases the induced voltage at the center of the conductor. This effect is so pronounced that at radio frequencies almost all the current flows along the surface of the conductor. This is called **skin effect.** Since there is practically no current in the center of the conductor at radio frequencies, its effective cross-sectional area is greatly reduced, and, therefore, the effective resistance of the conductor is much greater than its dc ohmic resistance.

A second factor governing the effective resistance of an ac circuit is the dissipation of energy through radiation of electromagnetic waves from the magnetic and electric fields associated with the ac circuit. As we have discovered, these fields build up and collapse twice each cycle. Under ideal conditions, as much reactive power is returned to the electric circuit during the collapsing of the field as is taken from the circuit during the building up of the field. But as the frequency is increased, some of the energy escapes as a radiated electromagnetic wave; hence, less energy is returned to the circuit during the collapse of the field than was needed to build it up. These **radiation losses** increase the value of P in Equation (21-19) and, therefore, increase the effective resistance of the ac circuit.

Suppose that we now connect an ammeter and a wattmeter so that we can determine the effective resistance of a coil of wire in an ac circuit. Having made the necessary calculations, we now insert an iron core into the coil. Due to the increase in inductive reactance, both meter readings decrease. But when we calculate the effective resistance, we find that it is greater than before. We can blame this increase on the hysteresis of the iron core and also on eddy current in the core. Both hysteresis and eddy current cause energy supplied by the electric source to be converted into heat. This requires an increase in the active power input to the coil for a certain coil current. Since this increase in active power is registered by the wattmeter, the effect of hysteresis and eddy current shows up in our calculation of the effective resistance of the iron-core coil. Again, the higher the frequency, the greater the discrepancy between ohmic resistance and effective resistance.

Similarly, a capacitor which, when measured in a dc circuit has an infinitely high resistance, can cause an appreciable wattmeter reading in an ac circuit. This produces the effect of a resistance in parallel with the capacitor. The wattmeter, in this case, is indicating the transfer of energy into heat as the dielectric of the capacitor is stressed, first in one direction and then in the other, many times a second. Since this loss of energy by conversion to heat is similar to the hysteresis loss in a magnetic circuit, it is called **dielectric hysteresis.** It is interesting to note that, because magnetic and dielectric hysteresis losses and eddy current losses all tend to raise the temperature of the ac circuit, there will also be an additional increase in the ohmic resistance if the circuit has an appreciable positive temperature coefficient of resistance.

21-10 THE DECIBEL

In working with communications systems, we are often more interested in the power *gain* of amplifiers and the power *loss* of transmission lines and attenuators than we are in the actual power in the system. Power gain and power loss infer a *ratio* between output power and input power. And since determination of power gain and power loss was important in the early days of the telephone, it was desirable to express these ratios as *logarithms* in order to duplicate the nonlinear response of the human ear to sound waves. Consequently, a unit called the **bel**[†] was defined as the common logarithm (base of ten) of the ratio of output power to input power. The bel proved to be too large for practical purposes; hence, we usually express the power ratio in tenths of a bel, or **decibels.**

Power gain and power loss are expressed in decibels where the number of decibels is determined by

$$dB = 10 \log_{10} \frac{P_{out}}{P_{in}} \qquad (21\text{-}20)$$

EXAMPLE 21-10

What is the power gain of an amplifier in which an input power of 50 μW results in an output power of 20 W?

SOLUTION

$$\text{Power gain} = 10 \log_{10} \frac{20 \text{ W}}{50 \text{ } \mu\text{W}} = +56 \text{ dB}$$

On a calculator,

$$10 \times (\; 20 \div 50 \; \boxed{EE} \; +/- \; 6 \;) \; \boxed{log} \; =$$

[†]Named in honor of Alexander Graham Bell.

536

EXAMPLE 21-11

What is the power loss of the attenuator shown in Figure 21-18 when it is connected between a signal source and a 600-Ω load?

FIGURE 21-18
Attenuator Circuit.

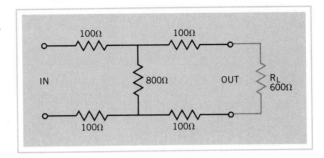

SOLUTION

Step I. Assume an rms load current of 1 mA. Then

$$P_{out} = I^2R_L = (1 \text{ mA})^2 \times 600 \ \Omega = 600 \ \mu\text{W}$$

Step II. Input resistance [see Example 7-2(b)],

$$R_{in} = 600 \ \Omega$$

From the current-divider principle if $I_L = 1$ mA, then

$$I_{in} = 2 \text{ mA}$$

Hence,
$$P_{in} = I^2R_{in} = (2 \text{ mA})^2 \times 600 \ \Omega = 2.4 \text{ mW}$$

Step III.

$$\text{Power loss} = 10 \ \log_{10}\frac{600 \ \mu\text{W}}{2.4 \text{ mW}} = -6 \text{ dB}$$

Since the logarithm of a quantity less than 1 is a negative number, a negative number of decibels automatically indicates a power *loss*.

Since decibels represent *power* ratios, if (and only if) the input and output resistance are the same, we can substitute V^2/R for P in Equation (21-20) with R canceling out. Hence,

$$\text{dB} = 10 \ \log_{10}\frac{P_{out}}{P_{in}} = 10 \ \log_{10}\frac{V^2_{out}}{V^2_{in}}$$

In logarithms, the log of a square is twice the log of the root quantity. Therefore,

$$\text{dB} = 20 \ \log_{10}\frac{V_{out}}{V_{in}} \quad (R_{out} = R_{in}) \quad\quad\quad (21\text{-}21)$$

The logarithmic nature of the decibel makes it possible for us to *add* decibels whereas we would have to *multiply* a chain of power gains and losses using only the basic power ratios. Every time we *double* power gain (or power loss), the ratio expressed in decibels changes by 3. For example, a 2 : 1 power ratio = 3 dB; a 4 : 1 power ratio = 2^2 = 6 dB; and an 8 : 1 power ratio = 2^3 = 9 dB.

PROBLEMS

21-1. A toaster draws a 6-A current from a 110-V 60-Hz source.
   (a) What is the active power of the toaster?
   (b) What is the peak value of the instantaneous power input to the toaster?

21-2. The voltage drop across the heater of a certain cathode-ray tube is 6.3 V rms when the current through it is 0.3 A. What is the average rate of conversion of electric energy into heat?

21-3. What is the average power dissipation of a 75-Ω resistor in which the sine-wave current reaches a peak value of 2.0 A?

21-4. What power rating must a 300-Ω dummy antenna resistor possess if the rms voltage drop across it is 96 V?

21-5. An inductance of 3.0 mH passes a 1.0-kHz sine-wave current of 20 mA.
   (a) What is the average power input to the inductor?
   (b) What is the reactive power?
   (c) What is the peak rate at which it can store energy?

21-6. A $26\frac{1}{2}$-μF capacitor is connected across a 120-V 60-Hz source.
   (a) What is the reactive power input to the capacitor?
   (b) What is the apparent power input?
   (c) What is the peak rate at which it can discharge its stored energy back into the circuit?

21-7. A solenoid having an inductance of 0.5 H and a resistance of 24 Ω is connected to a 120-V 60-Hz source.
   (a) What is the apparent power input to the solenoid?
   (b) What is the active power input to the solenoid?
   (c) What is the power factor of the solenoid?

21-8. The operating coil of a relay draws 100 mA from a 24-V 400-Hz source at a 15% lagging power factor.
   (a) What is the resistance of the coil?
   (b) What is the inductance of the coil?

21-9. What is the average power input to the transmitting antenna in Example 21-4?

21-10. The reactive power input to an impedance of 480 $\Omega\,/\!-60°$ is 300 mvars. At what rate is energy being dissipated by the impedance?

21-11. What value of resistance and inductance in series will draw the same current from a 120-V 60-Hz source as a 500-W 60% lagging-power-factor load?

21-12. What value of resistance and inductance in parallel will draw the same current from a 120-V 60-Hz source as a 500-W 60% lagging-power-factor load?

21-13. (a) What current must a 110-V alternator supply to operate a 2-kW 75% lagging-power-factor load?
   (b) What active power could the alternator supply for the same magnitude of current in its windings to a unity-power-factor load?

21-14. An industrial capacitor used in power factor correction is rated at 7.5 kvars when connected to a 208-V 60-Hz circuit. What is its capacitance?

21-15. An apparent power input of 25 kVA must be provided to the "work coil" of a radio-frequency induction heater in order to generate heat in the steel work piece at the rate of 5 kW. What is the reactive factor of the load?

21-16. Part of the "crossover" network used to feed the "woofer" and "tweeter" in a loudspeaker system consists of an 8-Ω resistor and a capacitor in parallel. The power input to these two components at 1 kHz is 500 mW at a 20% leading power factor. What is the capacitance of the capacitor?

21-17. An impedance coil having an 0.2 lagging power factor is connected in series with a 300-W lamp in order to supply the lamp with 120 V from a 208-V 60-Hz source. What is the voltage across the terminals of the impedance coil?

21-18. An induction motor which draws 2.0 A from a 120-V 60-Hz source at 0.8 lagging power factor and a 100-W lamp are connected in parallel. What is their overall power factor?

21-19. An induction motor draws 6.0 A at 0.8 lagging power factor from a 208-V 60-Hz source.
    (a) What value of capacitance must be placed in parallel with the motor to raise the overall power factor to unity?
    (b) What are the magnitudes of the final motor current, capacitor current, and line current?

21-20. What value of capacitance is required to produce an overall power factor of 0.96 lagging with the motor of Problem 21-19?

21-21. What is the overall power factor when a 50-μF capacitor is connected in parallel with the motor of Problem 21-19?

21-22. A synchronous motor capable of operating with a leading power factor draws 15 kW from a distribution transformer while driving an air compressor. The remainder of the load on the transformer is 80 kW at 0.85 lagging power factor.
    (a) How many kilovars of capacitive reactive power must the synchronous motor produce to raise the overall power factor to 0.96 lagging?
    (b) What is the reactive factor of the synchronous motor operating in this manner?

21-23. The power factor of a load on a 120-V 60-Hz source is raised from 0.707 lagging to 0.866 lagging by connecting a 53-μF capacitor across the load. What is the active power of the load?

21-24. The power factor of a load on a 120-V 60-Hz source is raised from 0.866 lagging to 0.966 leading by connecting a $110\frac{1}{2}$-μF capacitor in parallel with the load. What is the rms load current?

21-25. A 100-pF capacitor is connected in series with a coil having an inductance of 100 μH and a resistance of 20 Ω. At what frequency will this network have unity power factor?

21-26. The ratio between the reactive power and the active power of the coil in Problem 21-25 is called its Q. What is the Q of the coil at a frequency of 1 MHz?

21-27. What is the effective resistance of the transmitting antenna system in Example 21-4?

21-28. An 0.0005-μF radio transmitter capacitor has a 10% leading power factor at 500 kHz. What is the effective resistance of the capacitor (in parallel with its capacitance) at this frequency? At what rate is heat generated in this capacitor when the rms radio-frequency current "through" it is 3.0 A?

21-29. The ac resistance of an air-core solenoid is measured by the ammeter-wattmeter method. When the solenoid is connected to a 120-V 60-Hz source, the ammeter indicates 2.0 A and the wattmeter reads 48 W. When an iron core is inserted into the coil, the readings are 180 mA and 500 mW, respectively.
    (a) What is the effective resistance and inductance of the solenoid without the iron core?
    (b) What is the effective resistance and inductance of the coil with the iron core inserted?

21-30. We can also calculate effective resistance by employing a precision noninductive resistor instead of a wattmeter. An iron-core inductor draws a current of 1.0 A from a 120-V 60-Hz source. When a noninductive 100-Ω precision resistor is connected in series with the coil, the current becomes 0.6 A. What is the effective resistance and inductance of the iron-core inductor?

21-31. An input power of 12 mW is required to produce a power output of 30 W from an audio power amplifier. What is its power gain in decibels?

21-32. A broadcast studio feeds a 6-mW signal into a telephone line between the studio and the transmitter. If the signal power arriving at the transmitter is 950 μW, what is the power loss of the telephone line in dB?

21-33. What power must a TV signal generator feed into a 12-dB attenuator if the power output is to be 50 mW?

21-34. What signal voltage must be applied to the 10-kΩ input terminals of an amplifier having a power gain of 40 dB to produce a power output of 15 W?

21-35. By how many dB must the gain of an audio amplifier be increased if the output power is to be increased from 5 W to 25 W?

21-36. By how many times must the input voltage to a fixed-gain amplifier be increased in order to increase the output power by 15 dB?

REVIEW QUESTIONS

21-1. Why can we state that $p = vi$ in any ac circuit?

21-2. Why is the instantaneous power in resistance always a positive quantity?

21-3. Why can we state that the average power in the resistance in a sine-wave ac system is one-half the peak power?

21-4. Why is the product $V_R \times I_R$ sometimes called the **real power?**

21-5. Given an ac voltmeter and an ac ammeter, how would you determine the peak value of the instantaneous power as an ideal inductor builds up its magnetic field in an ac circuit?

21-6. Why is the average power of an ideal inductor zero?

21-7. What is the significance of the term **reactive power?**

21-8. What is the significance of the term **apparent power?**

21-9. What is the apparent power of an ideal resistor?

21-10. What is the apparent power of an ideal capacitor?

21-11. Why should we not express apparent power in watts?

21-12. Why can we say that the net reactive power of an ac circuit is the *difference* between the inductive reactive power and the capacitive reactive power?

21-13. Why is the total active power of a network always the sum of the individual active powers regardless of series or parallel connection of the components?

21-14. Why is the apparent power of an ac circuit the root of the sum of the squares of the active and reactive powers rather than the simple sum?

21-15. Why is the angle between the active power and apparent power in a power triangle the same as the angle between the total current and total voltage of a load?

21-16. What is the significance of the term **power factor** in dealing with industrial loads?

21-17. What is meant by a **leading power factor?**

21-18. Why are ac generators and transformers rated in terms of kilovoltamperes rather than kilowatts?

21-19. Of what practical use is the term **reactive factor?**

21-20. Derive an expression for reactive factor in terms of power factor.

21-21. Although placing capacitance in series with an inductive load raises the overall power factor, why can we not consider this to be a satisfactory means of power-factor correction?

21-22. What is the purpose of power-factor correction?

21-23. Why does adding capacitance in parallel with an inductive load achieve this purpose?

21-24. In determining the rms value of a sine-wave by laboratory experiment, we determine the equivalent direct current that will produce the same lamp brilliance. Why do we work with current rather than voltage?

21-25. Why does the effective resistance of an electric conductor increase as the frequency of the alternating current through it increases?

21-26. Some radio coils are wound from "Litz" wire, which consists of many strands of fine, separately insulated wire. Suggest why Litz wire is more suitable for this purpose than a solid conductor of the same gauge.

21-27. How do you account for the change in effective resistance of the solenoid in Problem 21-29? What resistance would you expect an ohmmeter to indicate for the iron-core solenoid?

21-28. What are the purpose and advantage of expressing power ratios in a communications system in logarithmic terms?

21-29. When the power *output* of amplifier is stated in decibels, it is also necessary to state a reference power (usually 1 mW). Explain.

21-30. What is the overall power gain of a system composed of a magnetic tape playback head feeding a preamplifier having a gain of 20 dB, which in turn feeds a tone-control network with an insertion loss of 10 dB, which in turn feeds a power amplifier with a gain of 40 dB, which in turn feeds a loudspeaker?

PART V

IMPEDANCE NETWORKS

ALTERNATING CURRENTS IN IMPEDANCE NETWORKS must obey all the basic laws of electric circuit behavior we encountered in the direct-current circuits in Part 2. To qualify as fundamental principles of electric circuit behavior, all dc circuit-analysis formats and theorems must also hold true for alternating-current circuits. The only difference is that steady-state ac quantities possess angles as well as magnitudes. Hence, the solution of numerical examples usually requires using the rules of complex algebra. This added dimension simply requires a little extra caution in performing the various algebraic routines on a calculator or personal computer.

In Chapters 22 and 23, we shall review the basic laws and circuit-analysis techniques we first encountered in Part 2 and adapt them to alternating-current examples. The remaining chapters deal with electric circuit characteristics which are unique to ac systems.

22

SERIES AND PARALLEL IMPEDANCES

22-1 RESISTANCE AND IMPEDANCE

Resistance and **impedance** both represent opposition to electric current. Both are expressed in **ohms.** However, whereas resistance is a specific property of both direct- and alternating-current circuits, we define impedance as the opposition of a portion of a circuit to *alternating* current. We commenced our consideration of circuit-analysis techniques with resistance networks because all numerical quantities in direct-current circuits are *scalar* quantities, which are limited to a numerical magnitude and, in some instances, a + or − polarity. Consequently, all numerical computation in Chapters 8 and 9 is based on *linear* algebra operations. As we discovered in Chapter 20, impedance is a phasor quantity in alternating-current circuits. Therefore, numerical computation for impedance networks is based on *phasor* algebra operations.

As long as we perform all numerical computation by phasor *algebra, we may apply to impedance networks all the relationships we developed in Part 2 for resistance networks.*

In this chapter, we shall consider the basic series and parallel impedance relationships as a means of becoming acquainted with the characteristics and limitations of phasor algebra before we tackle the more elaborate and multisource networks which require application of the network theorems we encountered in Chapters 8 and 9.

22-2 IMPEDANCES IN SERIES— KIRCHHOFF'S VOLTAGE LAW

In section 19-2, we noted that, if we consider only a particular *instant* in time (if time could stand still), that the circuit relationships in Figure 22-1(a) for that instant are exactly the same as in a direct-current circuit with the numerical values that prevail at that instant. Consequently, from Kirchhoff's voltage law, we are able to state that

$$e = v_T = v_R + v_L \tag{19-1}$$

where we perform the numerical operations by *linear* algebra.

To extend Kirchhoff's voltage law to the much more useful rms values for alternating-current circuit parameters, we must note that rms values are *phasor* quantities. We can write the Kirchhoff's voltage-law relationship for the circuit of Figure 22-1(a) in the form

$$\mathbf{E_T} = \mathbf{V}_R + \mathbf{V}_L \tag{19-3}$$

provided that we note that each term is a phasor and that we must perform the numerical computation by phasor algebra.

We can now proceed one step further and apply Kirchhoff's voltage law to the series circuit of Figure 22-1(b). Although we can write the Kirchhoff's voltage-law equation in terms of instantaneous values, we are concerned mainly with rms values; hence, we write the equation in the form

$$\mathbf{E} = \mathbf{V_T} = \mathbf{V}_1 + \mathbf{V}_2 + \mathbf{V}_3 \tag{22-1}$$

where (unlike the equivalent equation for a dc circuit) all terms are phasor quantities.

FIGURE 22-1
Kirchhoff's Voltage Law Applied to an Alternating-Current Circuit.

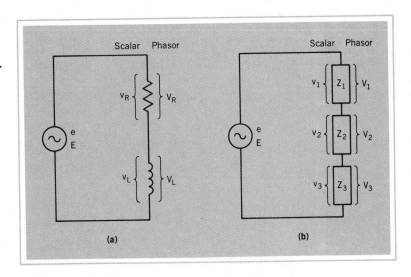

We can divide every term in Equation (22-1) by the common current in the series circuit of Figure 22-1(b), (using phasor division, of course), giving

$$\frac{\mathbf{V_T}}{\mathbf{I}} = \frac{\mathbf{V_1}}{\mathbf{I}} + \frac{\mathbf{V_2}}{\mathbf{I}} + \frac{\mathbf{V_3}}{\mathbf{I}}$$

From the definition of impedance, this becomes

$$\mathbf{Z_T} = \mathbf{Z_1} + \mathbf{Z_2} + \mathbf{Z_3} \qquad (22\text{-}2)$$

and we may state that

The total impedance of several impedances in series is their **phasor** *sum.*

EXAMPLE 22-1

What is the total impedance when $\mathbf{Z_1} = 60\ \Omega\underline{/+60°}$ and $\mathbf{Z_2} = 80\ \Omega\underline{/-45°}$ are connected in series?

SOLUTION

Step I. Draw a circuit diagram on which we can show all pertinent data for ready reference, and a fairly accurate freehand impedance diagram to indicate the approximate solution.

Step II. Express the impedances in rectangular form so that we can add their coordinates.

$$\mathbf{Z_1} = 60\cos 60° + \text{j}60\sin 60° = 30 \quad + \text{j}52$$
$$\mathbf{Z_2} = 80\cos 45° - \text{j}80\sin 45° = \underline{56.6 - \text{j}56.6}$$

Adding gives
$$\mathbf{Z_T} = 86.6 - \quad \text{j}4.6\ \Omega$$

FIGURE 22-2
Impedances in Series.

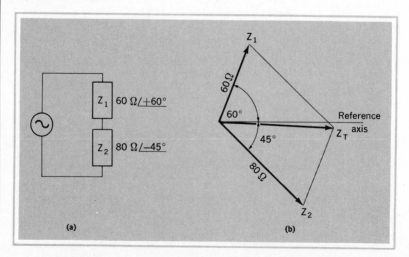

(a) (b)

Step III. Depending on the nature of any further computations, convert the total impedance into polar form. [For example, if the next step is to determine the current in the circuit of Figure 22-2(a) for a given E.]

$$\phi = \arctan \frac{-4.6}{86.6} = -3°$$

$$Z = \frac{86.6}{\cos -3°} = 86.7 \ \Omega$$

$$\therefore \mathbf{Z} = 86.7 \ \Omega \ \underline{/-3°}$$

EXAMPLE 22-2

Given $\mathbf{Z}_1 = 10\,000 + j15\,000 \ \Omega$, $\mathbf{Z}_2 = 4700 + j0 \ \Omega$, and $\mathbf{Z}_3 = 25\,000 - j10\,000 \ \Omega$. Express the total impedance of these three impedances in series in rectangular coordinates.

SOLUTION

All we need to do in this case is to perform the addition of rectangular coordinates.

$$\mathbf{Z}_1 = 10\,000 + j15\,000$$
$$\mathbf{Z}_2 = 4700 + j0$$
$$\mathbf{Z}_3 = 25\,000 - j10\,000$$

and

$$\mathbf{Z}_\mathrm{T} = 39\,700 + j5000 \ \Omega$$

We might be inclined to suggest from the simple solution of Example 22-2 that we should always express impedance in rectangular coordinates. However, we must not allow ourselves to be misled by first appearances. As we noted in Section 20-8 (see Table 20-1), the rectangular coordinates of an impedance represent a *series* resistance and reactance. Hence, the addition process of Example 22-2 applies only to *series* impedances.

As an application of Kirchhoff's voltage law to series ac circuits, we can now extend the **voltage-divider principle** of Section 7-4 to series impedances. Therefore, Equation (7-1) becomes

$$\mathbf{V}_x = \mathbf{E} \frac{\mathbf{Z}_x}{\mathbf{Z}_\mathrm{T}} \tag{22-3}$$

where every term is treated as a phasor quantity.

EXAMPLE 22-3

Given an applied voltage in the circuit of Figure 22-2(a) of 120 V $\underline{/0°}$, what is the voltage drop across $\mathbf{Z}_1$?

SOLUTION 1

By Ohm's law:

Step I. Having determined $\mathbf{Z}_T$ in Example 22-1,

$$\mathbf{I} = \frac{\mathbf{E}}{\mathbf{Z}_T} = \frac{120 \text{ V} \underline{/0°}}{86.7 \text{ } \Omega \underline{/-3°}} = 1.38 \text{ A} \underline{/+3°}$$

Step II.

$$\mathbf{V}_1 = \mathbf{I}\mathbf{Z}_1 = 1.38 \text{ A} \underline{/+3°} \times 60 \text{ } \Omega \underline{/+60°}$$

$$= 83 \text{ V} \underline{/+63°}$$

SOLUTION 2

By the voltage-divider principle: We can solve for $\mathbf{V}_1$ in a single step without first determining $\mathbf{I}$.

$$\mathbf{V}_1 = \mathbf{E}\frac{\mathbf{Z}_1}{\mathbf{Z}_T} = 120 \text{ V} \underline{/0°} \times \frac{60 \text{ } \Omega \underline{/+60°}}{86.7 \text{ } \Omega \underline{/-3°}}$$

$$= 83 \text{ V} \underline{/+63°}$$

22-3 IMPEDANCES IN PARALLEL— KIRCHHOFF'S CURRENT LAW

In working with resistors in parallel in Chapter 6, we noted that the *more* resistors we connect in parallel, the *less* is the equivalent resistance. We, therefore, found it to our advantage to switch our point of view in dealing with parallel circuits and to think in terms of the ability of the various branches to *conduct* electric current. Consequently, the solution of parallel circuits is a bit more involved than the solution of series circuits, because we must first make this conversion from *opposition* into *ability to conduct*.

In the solution of parallel ac circuits, it is even more important that we do our computation in numerical quantities that represent *ability to conduct*. (See Table 20-1.) We can use the same procedures in parallel ac circuits that we used in parallel dc circuits. There are two sets of parameters which are proportional to the ability of a circuit to conduct: (1) the total of all branch *currents*, and (2) the total of all branch *admittances*. In ac circuits, these must of course be *phasor* totals.

For parallel branches in an ac circuit, Kirchhoff's current law gives

$$\mathbf{I}_T = \mathbf{I}_1 + \mathbf{I}_2 + \mathbf{I}_3 + \text{etc.} \tag{22-4}$$

where all terms are phasor quantities. Dividing every term in Equation (22-4) by the *common* voltage across *parallel* branches, gives

$$\frac{\mathbf{I}_T}{\mathbf{V}} = \frac{\mathbf{I}_1}{\mathbf{V}} + \frac{\mathbf{I}_2}{\mathbf{V}} + \frac{\mathbf{I}_3}{\mathbf{V}}$$

From the definition of admittance, this equation becomes

548

$$\mathbf{Y_T} = \mathbf{Y_1} + \mathbf{Y_2} + \mathbf{Y_3} + \text{etc.} \qquad (22\text{-}5)$$

and we may state that

The total admittance of several parallel branches is the phasor *sum of the branch admittances.*

We can determine the equivalent *impedance* of a parallel ac circuit either from Equation (22-4) by the total-current method,

$$\mathbf{Z_{eq}} = \frac{\mathbf{E}}{\mathbf{I_T}} \qquad (22\text{-}6)$$

or from Equation (22-5) by the total-admittance method,

$$\mathbf{Z_{eq}} = \frac{1}{\mathbf{Y_T}} \qquad (22\text{-}7)$$

EXAMPLE 22-4

What is the equivalent impedance when $\mathbf{Z_1} = 60\ \Omega\underline{/+60°}$ and $\mathbf{Z_2} = 80\ \Omega\underline{/-45°}$ are connected in parallel?

SOLUTION 1

Total-current method:

Step I. Draw a schematic diagram and a current phasor diagram, as in Figure 22-3.

Step II. Assume a convenient value of applied voltage, e.g., $\mathbf{E} = 240\ V\underline{/0°}$.

FIGURE 22-3
Impedances in Parallel.

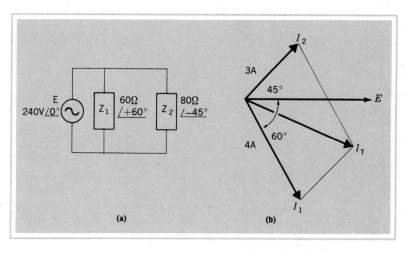

(a)　　　　　(b)

Step III. Determine the branch currents.

$$\mathbf{I}_1 = \frac{\mathbf{E}}{\mathbf{Z}_1} = \frac{240\,\underline{/0^\circ}}{60\,\underline{/+60^\circ}} = 4 \text{ A}\,\underline{/-60^\circ}$$

$$\mathbf{I}_2 = \frac{\mathbf{E}}{\mathbf{Z}_2} = \frac{240\,\underline{/0^\circ}}{80\,\underline{/-45^\circ}} = 3 \text{ A}\,\underline{/+45^\circ}$$

Step IV. Express the branch currents in rectangular coordinates so they can be added.

$$\mathbf{I}_1 = 4 \cos 60^\circ - j4 \sin 60^\circ = 2 \qquad - j3.464$$
$$\mathbf{I}_2 = 3 \cos 45^\circ + j3 \sin 45^\circ = 2.121 + j2.121$$

Adding gives
$$\mathbf{I}_T = 4.121 - j1.343 \text{ A}$$

Step V. Convert the total current to polar form.

$$\phi = \arctan \frac{-1.343}{4.121} = -18^\circ$$

$$I_T = \frac{4.121}{\cos -18^\circ} = 4.334 \text{ A}$$

$$\therefore \mathbf{I}_T = 4.334 \text{ A}\,\underline{/-18^\circ}$$

Step VI.

$$\mathbf{Z}_{eq} = \frac{\mathbf{E}}{\mathbf{I}_T} = \frac{240 \text{ V}\,\underline{/0^\circ}}{4.334 \text{ A}\,\underline{/-18^\circ}} = 55.4 \text{ } \Omega\,\underline{/+18^\circ}$$

SOLUTION 2
Total-admittance method:

Step I. as shown in Figure 22-4,

$$\mathbf{Y}_1 = \frac{1}{\mathbf{Z}_1} = \frac{1}{60 \text{ } \Omega\,\underline{/+60^\circ}} = 0.0167 \text{ S}\,\underline{/-60^\circ}$$

$$\mathbf{Y}_2 = \frac{1}{\mathbf{Z}_2} = \frac{1}{80 \text{ } \Omega\,\underline{/-45^\circ}} = 0.0125 \text{ S}\,\underline{/+45^\circ}$$

Step II.

$$\mathbf{Y}_1 = 0.0167 \cos 60^\circ - j0.0167 \sin 60^\circ = 0.008\,33 - j0.014\,43 \text{ S}$$
$$\mathbf{Y}_2 = 0.0125 \cos 45^\circ + j0.0125 \sin 45^\circ = 0.008\,84 + j0.008\,84 \text{ S}$$

Adding gives
$$\mathbf{Y}_T = 0.017\,17 - j0.005\,59 \text{ S}$$

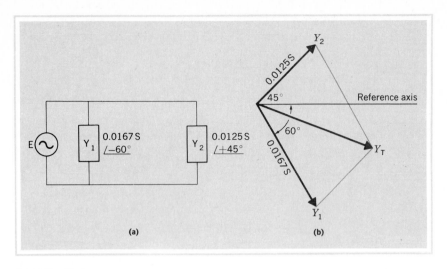

FIGURE 22-4
Admittances in Parallel.

Step III.

$$\phi = \arctan \frac{-0.005\ 59}{0.017\ 17} = -18°$$

$$Y_T = \frac{0.017\ 17}{\cos -18°} = 0.018\ 05 \text{ S}$$

$$\therefore \ \mathbf{Y}_T = 0.018 \text{ S } \underline{/-18°}$$

Step IV.

$$\mathbf{Z}_{eq} = \frac{1}{\mathbf{Y}_T} = \frac{1}{0.018 \text{ S } \underline{/-18°}} = 55.4\ \Omega \underline{/+18°}$$

Although Solution 1 may seem easier, since we can avoid switching our thinking to quantities expressed in siemens, Solution 2 is much preferred as we encounter more elaborate ac networks. We shall now drop the total-current procedure as a means to an end and work as much as possible in terms of admittance in parallel circuits. For the special case of *two* branches in parallel,

$$\mathbf{Y}_T = \mathbf{Y}_1 + \mathbf{Y}_2 = \frac{1}{\mathbf{Z}_1} + \frac{1}{\mathbf{Z}_2} = \frac{\mathbf{Z}_2 + \mathbf{Z}_1}{\mathbf{Z}_1 \mathbf{Z}_2}$$

Inverting this equation,

$$\mathbf{Z}_{eq} = \frac{\mathbf{Z}_1 \mathbf{Z}_2}{\mathbf{Z}_1 + \mathbf{Z}_2} \tag{22-8}$$

where all terms are treated as phasor quantities. Applying Equation (22-8) to Example 22-4, we obtain a shorter numerical computation *without adjusting our thinking to the admittance concept of parallel branches.*

SOLUTION 3

Step I. Solving the numerator,

$$\mathbf{Z}_1\mathbf{Z}_2 = 60\underline{/+60°} \times 80\underline{/-45°} = 4800\underline{/+15°}$$

Step II. Solving the denominator,

$$\mathbf{Z}_1 = 60 \ \Omega\underline{/+60°} = 30 \quad + j51.96$$
$$\mathbf{Z}_2 = 80 \ \Omega\underline{/-45°} = 56.57 - j56.57$$

Adding gives $\qquad \mathbf{Z}_1 + \mathbf{Z}_2 = 86.57 - j4.61 \ \Omega$

from which $\qquad \mathbf{Z}_1 + \mathbf{Z}_2 = 86.69 \ \Omega\underline{/-3°}$

Step III.

$$\frac{\mathbf{Z}_1\mathbf{Z}_2}{\mathbf{Z}_1 + \mathbf{Z}_2} = \frac{4800\underline{/+15°}}{86.69\underline{/-3°}}$$

$$\therefore \ \mathbf{Z}_{eq} = 55.4 \ \Omega\underline{/+18°}$$

EXAMPLE 22-5

Given $\mathbf{Z}_1 = 10\ 000 + j15\ 000 \ \Omega$, $\mathbf{Z}_2 = 4700 + j0 \ \Omega$, and $\mathbf{Z}_3 = 25\ 000 - j10\ 000 \ \Omega$. What equivalent impedance (rectangular form) does the source "see" when it feeds $\mathbf{Z}_1$, $\mathbf{Z}_2$, and $\mathbf{Z}_3$ all connected in parallel?

SOLUTION

Step I. Since we cannot work with the rectangular coordinates of impedance in dealing with parallel branches, we first convert to polar form.

$$\mathbf{Z}_1 = 10\ 000 + j15\ 000 = 18.03 \ k\Omega\underline{/+56.3°}$$
$$\mathbf{Z}_2 = \quad 4700 + j0 \quad = \quad 4.7 \ k\Omega\underline{/0°}$$
$$\mathbf{Z}_3 = 25\ 000 - j10\ 000 = 26.93 \ k\Omega\underline{/-21.8°}$$

Step II. Taking the reciprocal of each impedance,

$$\mathbf{Y}_1 = 55.47 \ \mu S \underline{/-56.3°}$$
$$\mathbf{Y}_2 = 212.8 \ \mu S \underline{/0°}$$
$$\mathbf{Y}_3 = 37.14 \ \mu S \underline{/+21.8°}$$

Step III. Converting to rectangular coordinates,[†]

$$\begin{aligned}
\mathbf{Y}_1 &= 30.77 - j46.15 \ \mu S \\
\mathbf{Y}_2 &= 212.8 + j0 \ \mu S \\
\mathbf{Y}_3 &= 34.48 + j13.79 \ \mu S \\
\hline
\mathbf{Y}_T &= 278.1 - j32.36 \ \mu S
\end{aligned}$$

Step IV. Converting $\mathbf{Y}_T$ to polar form,

$$\mathbf{Y}_T = 280 \ \mu S \underline{/-6.64°}$$

Step V. Taking the reciprocal of $\mathbf{Y}_T$,

$$\mathbf{Z}_{eq} = \frac{1}{\mathbf{Y}_T} = \frac{1}{280 \ \mu S \underline{/-6.64°}} = 3572 \ \Omega \underline{/+6.64°}$$

Step VI. Converting to rectangular coordinates,

$$\mathbf{Z}_{eq} = 3548 + j413 \ \Omega$$

Since the *same* voltage appears across parallel branches, we can establish the **current-divider principle** for ac circuits from the relationship

$$\mathbf{V} = \mathbf{I}_1\mathbf{Z}_1 = \mathbf{I}_2\mathbf{Z}_2$$

from which

$$\frac{\mathbf{I}_1}{\mathbf{I}_2} = \frac{\mathbf{Z}_2}{\mathbf{Z}_1} \qquad\qquad (22\text{-}9)$$

Hence, as we would expect, the split in current between two parallel branches in an ac circuit is *inversely* proportional to the impedance of the two branches. However, we find it more convenient in dealing with parallel branches in ac circuits to express the current-divider principle in terms of branch admittances. Starting with the common voltage for all branches in parallel,

$$\mathbf{V} = \frac{\mathbf{I}_x}{\mathbf{Y}_x} = \frac{\mathbf{I}_T}{\mathbf{Y}_T}$$

from which

$$\mathbf{I}_x = \mathbf{I}_T \frac{\mathbf{Y}_x}{\mathbf{Y}_T} \qquad\qquad (22\text{-}10)$$

[†]As we noted in Section 20-8, the rectangular coordinates of the admittance are *not* the reciprocals of the rectangular coordinates of the impedance.

where every term is treated as a phasor quantity.

Note that when we switch our thinking to admittance, the current-divider Equation (22-10) becomes the exact dual of the voltage-divider Equation (22-3). For the special case of only *two* parallel branches, we can substitute the reciprocal of Equation (22-8) for $\mathbf{Y}_T$ and express the current-divider principle in the form

$$\mathbf{I}_1 = \mathbf{I}_T \frac{\mathbf{Z}_2}{\mathbf{Z}_1 + \mathbf{Z}_2} \tag{22-11}$$

Equation (22-11) again allows us to minimize numerical computation in dealing with two parallel branches without adjusting our thinking to the admittance concept of parallel branches.

EXAMPLE 22-6

What current will a 10-k$\Omega\,\underline{/-45°}$ load draw from the Norton-equivalent source shown in Figure 22-5?

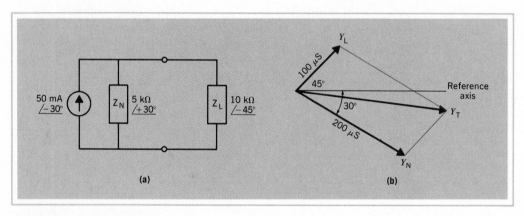

FIGURE 22-5
Diagrams for Example 22-6.

SOLUTION 1
By Ohm's law:

Step I.

$$\mathbf{Z}_{eq} = \frac{\mathbf{Z}_1 \times \mathbf{Z}_2}{\mathbf{Z}_1 + \mathbf{Z}_2} = \frac{5\ k\Omega\,\underline{/+30°} \times 10\ k\Omega\,\underline{/-45°}}{(4330 + j2500) + (7071 - j7071)}$$

$$= \frac{50 \times 10^6\,\underline{/-15°}}{(11\ 401 - j4571)} = \frac{50 \times 10^6\,\underline{/-15°}}{12\ 283\,\underline{/-21.85°}}$$

$$= 4.07\ k\Omega\,\underline{/+6.85°}$$

Step II.

$$\mathbf{V} = \mathbf{I}_T \mathbf{Z}_{eq} = 50 \text{ mA} \underline{/-30°} \times 4.07 \text{ k}\Omega \underline{/+6.85°}$$
$$= 203.5 \text{ V} \underline{/-23.15°}$$

Step III.

$$\mathbf{I}_L = \frac{\mathbf{V}}{\mathbf{Z}_L} = \frac{203.5 \text{ V} \underline{/-23.15°}}{10 \text{ k}\Omega \underline{/-45°}} = \mathbf{20.35 \text{ mA} \underline{/+21.85°}}$$

SOLUTION 2

By the current-divider principle:

Step I.

$$\mathbf{Y}_L = \frac{1}{\mathbf{Z}_L} = \frac{1}{10 \text{ k}\Omega \underline{/-45°}} = 100 \ \mu\text{S} \underline{/+45°} = 70.71 + j70.71 \ \mu\text{S}$$

$$\mathbf{Y}_N = \frac{1}{\mathbf{Z}_N} = \frac{1}{5 \text{ k}\Omega \underline{/+30°}} = 200 \ \mu\text{S} \underline{/-30°} = \underline{173.2 - j100 \ \mu\text{S}}$$

$$\mathbf{Y}_T = 243.91 - j29.29 \ \mu\text{S}$$

From which $\quad \mathbf{Y}_T = 245.7 \ \mu\text{S} \underline{/-6.85°}$

Step II. From Equation (22-10)

$$\mathbf{I}_L = \mathbf{I}_T \frac{\mathbf{Y}_L}{\mathbf{Y}_T} = 50 \text{ mA} \underline{/-30°} \times \frac{100 \ \mu\text{S} \underline{/+45°}}{245.7 \ \mu\text{S} \underline{/-6.85°}}$$
$$= \mathbf{20.35 \text{ mA} \underline{/+21.85°}}$$

22-4 SERIES-PARALLEL IMPEDANCES

As long as we remember that we are working with phasor quantities, the procedure for solving series-parallel impedance combinations is essentially the same as the procedure we developed for dc circuits in Section 7-1.

EXAMPLE 22-7

What total load is seen by the source in the circuit diagram of Figure 22-6(a)?

SOLUTION

Step I. Determine the single $\mathbf{Z}_{eq}$ which can replace the parallel combination of $\mathbf{Z}_2$ and $\mathbf{Z}_3$.

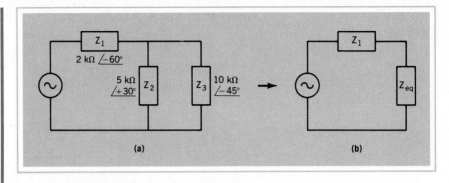

FIGURE 22-6
Circuit Diagrams for Example 22-7.

$$\mathbf{Z}_{eq} = \frac{\mathbf{Z}_2 \times \mathbf{Z}_3}{\mathbf{Z}_2 + \mathbf{Z}_3} = \frac{5\ k\Omega\ \underline{/+30°} \times 10\ k\Omega\ \underline{/-45°}}{(4330 + j2500) + (7071 - j7071)}$$

$$= \frac{50 \times 10^6\ \underline{/-15°}}{11\ 401 - j4571} = \frac{50 \times 10^6\ \underline{/-15°}}{12\ 283\ \underline{/-21.85°}}$$

$$= 4.07\ k\Omega\ \underline{/+6.85°}$$

Step II. Convert $\mathbf{Z}_1$ and $\mathbf{Z}_{eq}$ to rectangular coordinates. For the simple series circuit of Figure 22-6(a),

$$\mathbf{Z}_1 = 1000 - j1732\ \Omega$$

$$\mathbf{Z}_{eq} = \underline{4041 + \ \ j485\ \Omega}$$

$$\mathbf{Z}_T = 5041 - j1247\ \Omega = \mathbf{5193\ \Omega\ \underline{/-13.9°}}$$

EXAMPLE 22-8

If the output terminals of the π-network in Figure 22-7 are left open-circuit, what is the input impedance at 10 kHz?

SOLUTION
Step I. First we must determine the single impedance that can replace the

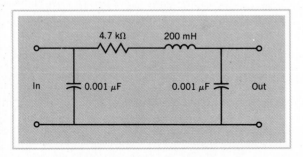

FIGURE 22-7
Pi-Network for Example 22-8.

resistor, inductor, and right-hand capacitor, all of which form a simple series circuit when no load is connected to the output terminals.

$$X_L = 2 \times \pi \times 10 \text{ kHz} \times 200 \text{ mH} = 12\ 566\ \Omega$$

$$X_C = \frac{1}{2 \times \pi \times 10 \text{ kHz} \times 0.001\ \mu\text{F}} = 15\ 915\ \Omega$$

$$\mathbf{Z}_s = R + j(X_L - X_C) = 4700 + j(12\ 566 - 15\ 915)\ \Omega$$

$$= 4700 - j3349\ \Omega = 5771\ \Omega\underline{/-35.47°}$$

Step II. We can now use any of the methods for determining the equivalent impedance of two impedances in parallel.

$$B_C = \omega C = 2 \times \pi \times 10 \text{ kHz} \times 0.001\ \mu\text{F} = 62.83\ \mu\text{S (capacitive)}$$

$$\mathbf{Y}_s = \frac{1}{\mathbf{Z}_s} = \frac{1}{5771\ \Omega\underline{/-35.47°}} = 173.3\ \mu\text{S}\underline{/+35.47°}$$

Step III.

$$\mathbf{Y}_T = B_C + \mathbf{Y}_s = +j62.83\ \mu\text{S} + (141.2 + j100.4)\ \mu\text{S}$$

$$= 141.2 + j163.23\ \mu\text{S} = 215.8\ \mu\text{S}\underline{/+49.1°}$$

Step IV.

$$\therefore\ \mathbf{Z}_{in} = \frac{1}{\mathbf{Y}_T} = \frac{1}{215.8\ \mu\text{S}\underline{/+49.1°}} = \mathbf{4634\ \Omega\underline{/-49.1°}}$$

22-5 EQUIVALENT CIRCUITS

We shall encounter many impedance networks which can be solved more readily if we can replace an elaborate network by a simple equivalent network. This essentially was the basis for the circuit-simplification pattern of solving series-parallel impedance networks in Section 22-4. The equivalent-circuit technique becomes particularly important in dealing with equivalent circuits for electronic amplifiers.

The complex nature of an impedance provides a procedure (in addition to the techniques we considered in Chapter 8) for substituting an equivalent parallel circuit for a series circuit, and vice versa. This procedure had no particular significance in dc circuits. Given an impedance sealed in a "black box" as in Figure 22-8(a), the source cannot distinguish between the possible circuits of Figures 22-8(b) and 22-8(c). Hence, the circuit of Figure 22-8(b) is equivalent of the circuit of Figure 22-8(c), and vice versa. As we noted in Section 20-8 (Table 20-1), Figure 22-8(b) represents the rectangular coordinates of the impedance of the black box in ohms; but to find the parallel equivalent circuit of figure 22-8(c), we must first find the rectangular coordinates of the equivalent admittance in siemens.

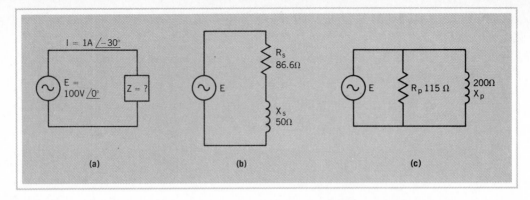

FIGURE 22-8
Series- and Parallel-Equivalent Circuits for an Inductive Load in an AC Circuit.

When we solve impedance network problems by plugging numbers into formulas, one of the most common sources of error is selecting the wrong equivalent circuit for the rectangular coordinates we intend to use. Again, we should note and understand thoroughly that

Whenever we express an impedance (ohms) in rectangular coordinates, we are automatically specifying an equivalent circuit consisting of a resistance (ohms) and reactance (ohms) in **series,** *as shown in Figure 22-9.*

Applying basic trigonometry functions to the impedance diagram of Figure 22-9,

$$R_s = Z \cos \phi \qquad (22\text{-}12)$$

and
$$X_s = Z \sin \phi \qquad (22\text{-}13)$$

Likewise,

Whenever we express an admittance (siemens) in rectangular coordinates, we are automatically specifying an equivalent circuit consisting of a conductance (siemens) and susceptance (siemens) in **parallel,** *as shown in Figure 22-10.*

FIGURE 22-9
Series-Equivalent Circuit of an Impedance.

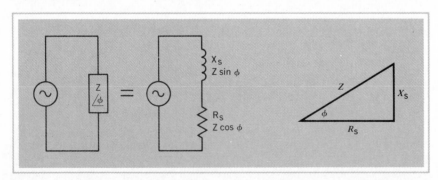

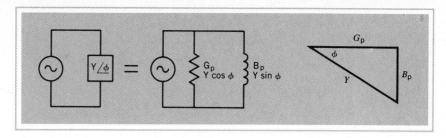

FIGURE 22-10
Parallel-Equivalent Circuit of an Admittance.

In schematic diagrams for practical circuits, we find that it is customary to state the values of resistors and the reactances of chokes in **ohms,** even though they may actually be connected in *parallel*. Therefore, in Figure 22-10, we would like to express the R_p and the X_p of the two parallel branches in ohms. From the admittance diagram of Figure 22-10,

$$G_p = Y \cos \phi \qquad (22\text{-}14)$$

But
$$G = \frac{1}{R} \quad \text{and} \quad Y = \frac{1}{Z}$$

$$\therefore \ R_p = \frac{Z}{\cos \phi} \qquad (22\text{-}15)$$

Similarly,
$$B_p = Y \sin \phi \qquad (22\text{-}16)$$

and
$$X_p = \frac{Z}{\sin \phi} \qquad (22\text{-}17)$$

The next step is to be able to replace the given series circuit of Figure 22-11(a) with an exact parallel-equivalent circuit with R_p and X_p in ohms, without having to perform all the in-between calculations every time. For a parallel circuit to be the exact equivalent of a given series circuit as far as the source is concerned, their impedances must be exactly the same both in magnitude and in angle.

FIGURE 22-11
Parallel Equivalent of a Series Branch.

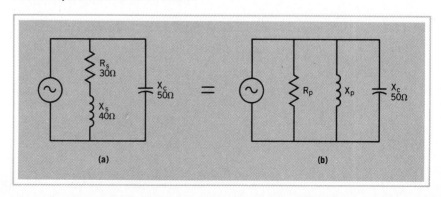

From Equation (22-12), $$\frac{R_s}{Z} = \cos \phi$$

and from Equation (22-15), $$\frac{Z}{R_p} = \cos \phi$$

$$\therefore \quad \frac{R_s}{Z} = \frac{Z}{R_p} \tag{22-18}$$

from which $$R_p = \frac{Z^2}{R_s} = \frac{(R_s^2 + X_s^2)}{R_s} \tag{22-19}$$

Similarly $$X_p = \frac{Z^2}{X_s} = \frac{(R_s^2 + X_s^2)}{X_s} \tag{22-20}$$

EXAMPLE 22-9

An ac circuit consists of two parallel branches, one consisting of a 30-Ω resistance and 40-Ω inductive reactance in series. The other branch consists of a 50-Ω capacitive reactance. Find the equivalent parallel circuit.

SOLUTION

$$R_p = \frac{R_s^2 + X_s^2}{R_s} = \frac{900 + 1600}{30} = 83.3 \ \Omega$$

$$X_p = \frac{R_s^2 + X_s^2}{X_s} = \frac{2500}{40} = 62.5 \ \Omega \ \text{(inductive)}$$

The equivalent susceptance of the parallel inductance and capacitance branches of the equivalent circuit of Figure 22-11(b) is

$$B_{eq} = B_C - B_L = \frac{1}{X_C} - \frac{1}{X_L} = \frac{1}{50} - \frac{1}{62.5} = 0.02 - 0.016$$

$$= 0.004 \ \text{S (capacitive)}$$

$$\therefore \ X_{eq} = \frac{1}{B_{eq}} = \frac{1}{0.004 \ \text{S}} = 250 \ \Omega \ \text{(capacitive)}$$

Therefore, the equivalent parallel circuit consists of an 83.3-Ω resistance in parallel with a 250-Ω capacitive reactance.

We shall find this type of conversion very useful in dealing with practical parallel-resonant circuits in Chapter 24.

Returning to Equation (22-18), we can derive the following equations for the series-equivalent components of a given parallel circuit.

$$R_s = \frac{R_p X_p^2}{R_p^2 + X_p^2} \tag{22-21}$$

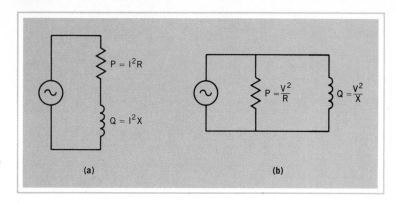

FIGURE 22-12
Series- and Parallel-Equivalent Circuits from Power Relationships.

and

$$X_s = \frac{R_p^2 X_p}{R_p^2 + X_p^2} \tag{22-22}$$

In Section 21-5, we discovered that the *same* power relationship [Equation (21-13)] applies to series, parallel, and also series-parallel networks. We used this knowledge to advantage in establishing a procedure for solving power-factor correction problems.

We can also add this useful relationship to our collection of equivalent-circuit techniques. Given the active power and reactive power for any ac network, as shown in Figure 22-12(a), P/I^2 and Q/I^2 automatically give us the *series*-equivalent circuit in ohms where

$$R_s = \frac{P}{I^2} \tag{22-23}$$

and

$$X_s = \frac{Q}{I^2} \tag{22-24}$$

Likewise, as shown in Figure 22-12(b), V^2/P and V^2/Q automatically give us the *parallel*-equivalent circuit in ohms where

$$R_p = \frac{V^2}{P} \tag{22-25}$$

and

$$X_p = \frac{V^2}{Q} \tag{22-26}$$

EXAMPLE 22-9A

An ac circuit consists of two parallel branches. One branch consists of a 30-Ω resistance and a 40-Ω inductive reactance in series. The other branch consists of a 50-Ω capacitive reactance [Figure 22-11(a)]. Find the equivalent parallel circuit.

SOLUTION
Assume a convenient source voltage. Let $E = 100$ V. The impedance of the series branch is

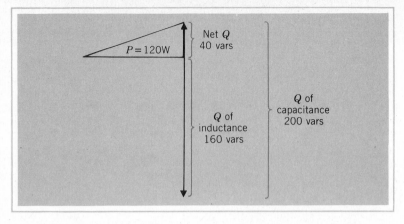

FIGURE 22-13
Power Triangle for Example 22-9A.

$$Z_s = \sqrt{30^2 + 40^2} = 50 \ \Omega$$

Hence, the current in each branch has a magnitude of

$$I = \frac{100 \text{ V}}{50 \ \Omega} = 2 \text{ A}$$

The active power is $\qquad P = I_s^2 R = 2^2 \times 30 = 120 \text{ W}$

The inductive reactive power is $\qquad Q_L = I_s^2 X_L = 2^2 \times 40 = 160 \text{ vars}$

The capacitive reactive power is $\qquad Q_C = \dfrac{V^2}{X_C} = \dfrac{100^2}{50} = 200 \text{ vars}$

As shown in the power triangle of Figure 22-13, the net reactive power is $200 - 160 = 40$ vars capacitive from which,

$$R_\mathrm{p} = \frac{V^2}{P} = \frac{100^2}{120} = 83.3 \ \Omega$$

and $\qquad X_\mathrm{p} = \dfrac{V^2}{Q} = \dfrac{100^2}{40} = 250 \ \Omega \text{ (capacitive)}$

Note that, in both of the equivalent-circuit techniques we have developed in this section, we have escaped *phasor* algebra because the relationships are based on *magnitudes* only.

22-6 SOURCE CONVERSION

Another useful equivalent-circuit technique is the conversion of a constant-voltage source to an equivalent constant-current source, and vice versa, which we developed in Section 8-4. Although *instantaneous* voltages and currents vary continuously in ac

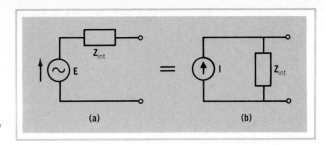

FIGURE 22-14
Alternating-Current Source Conversion.

circuits, rms values are represented by steady-state phasors. Hence, we can still speak of constant-voltage and constant-current sources in alternating-current networks. For any "black-box" ac source, Equation (8-1) becomes

$$\mathbf{Z}_{int} = \frac{\text{open-circuit rms voltage}}{\text{short-circuit rms current}} \qquad (22\text{-}27)$$

If we are given a constant-voltage ac source, as in Figure 22-14(a), the equivalent constant-current source of Figure 22-14(b) has the *same* internal impedance and a short-circuit current of

$$\mathbf{I}_{sc} = \frac{\mathbf{E}}{\mathbf{Z}_{int}} \qquad (22\text{-}28)$$

Similarly, if we are given the constant-current ac source of Figure 22-14(b), the equivalent constant-voltage source of Figure 22-14(a) has the *same* internal impedance and an open-circuit voltage of

$$\mathbf{E}_{oc} = \frac{\mathbf{I}}{\mathbf{Z}_{int}} \qquad (22\text{-}29)$$

In both cases, the phasor division provides us with an angle, as well as a magnitude, for the equivalent source.

Source conversion provides a third solution for Example 22-6.

EXAMPLE 22-6A
What current will a 10-k$\Omega\underline{/-45°}$ load draw from the Norton-equivalent source shown in Figure 22-5?

SOLUTION
We can replace the original constant-current source shown again in Figure 22-15(a) with the equivalent constant-voltage source of Figure 22-15(b).

$$\mathbf{E}_x = \mathbf{IZ}_x = 50 \text{ mA}\underline{/-30°} \times 5 \text{ k}\Omega\underline{/+30°} = 250 \text{ V}\underline{/0°}$$

$$\mathbf{Z}_T = \mathbf{Z}_x + \mathbf{Z}_L = 5 \text{ k}\Omega\underline{/+30°} + 10 \text{ k}\Omega\underline{/-45°}$$

$$= (4330 + j2500) + (7071 - j7071)$$

$$= (11\,401 - j4571) = 12\,283 \ \Omega\underline{/-21.85°}$$

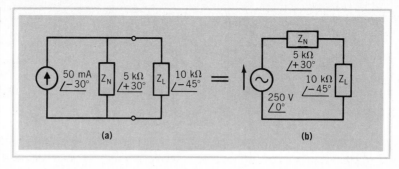

(a)   **(b)**

FIGURE 22-15
Source Conversion for Example 22-6A.

$$\mathbf{I}_L = \frac{\mathbf{E}_x}{\mathbf{Z}_T} = \frac{250 \text{ V} \underline{/0°}}{12\,283 \text{ } \Omega \underline{/-21.85°}} = 20.35 \text{ mA} \underline{/+21.85°}$$

22-7 MAXIMUM POWER TRANSFER

In Section 6-7, we considered the selection of a load resistance which would permit maximum power input to the load from a source having a fixed open-circuit voltage and a fixed internal resistance. We discovered that maximum power input occurs when $R_L = R_{int}$. We can now extend our discussion of maximum power transfer to alternating-current circuits in which the internal **impedance** of the source is partially reactive.

From our discussion of active power in Chapter 21, the power input to the load in the circuit in Figure 22-16 is

$$P_L = I^2 R_L$$

But

$$I = \frac{E}{Z_T} = \frac{E}{\mathbf{Z}_{int} + \mathbf{Z}_L} = \frac{E}{\sqrt{(R_{int} + R_L)^2 + (X_{int} + X_L)^2}}$$

Hence,

$$P_L = \frac{E^2 R_L}{(R_{int} + R_L)^2 + (X_{int} + X_L)^2} \tag{22-30}$$

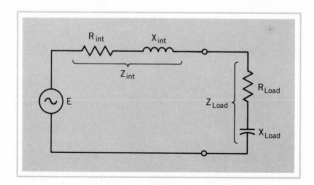

FIGURE 22-16
Maximum Power Transfer in an AC Circuit.

R_{int} and X_{int} remain fixed, depending on the characteristics of the source. For the moment, we leave R_L fixed and vary X_{Load}. From Equation (22-30) P_L will be maximum when the magnitude of the denominator of the expression is minimum. If we select a value of load reactance that is equal in magnitude but *opposite* in sign to the internal reactance, $(X_{int} + X_L)^2$ in Equation (22-30) becomes zero. Consequently, P_L will be maximum for the given value of R_L. Thus, as shown in Figure 22-16, if the source has an *inductive* internal impedance, we select a *capacitive* load reactance such that the net reactance of the circuit is zero.

Having cancelled out the reactance of the source, we now vary R_L just as we did in Section 6-7. Maximum power input to the load again requires that we select $R_L = R_{int}$. Hence, in an ac circuit,

Maximum power transfer occurs when the load impedance is the conjugate *of the internal impedance of the source.*

$\mathbf{Z_L}$ is the *conjugate* of $\mathbf{Z_{int}}$ when $R_L = R_{int}$ and the reactance of the load is equal in magnitude but *opposite* in sign to the reactance of the source.

EXAMPLE 22-10

A 50-MHz RF signal generator has an internal impedance of $60 + j45\ \Omega$. What value of load resistance and capacitance in parallel with R_L are required to obtain maximum power output from the signal generator?

SOLUTION
For maximum power transfer,

$$\mathbf{Z_L} = 60 - j45\ \Omega$$

$60\ \Omega$ and $45\ \Omega$ are the resistance and reactance, respectively, of a *series* circuit. From Equation (22-19),

$$R_p = \frac{R_s^2 + X_s^2}{R_s} = \frac{60^2 + 45^2}{60} = 93.75\ \Omega$$

and from Equation (22-20),

$$X_p = \frac{R_s^2 + X_s^2}{X_s} = \frac{5625}{45} = 125\ \Omega$$

$$\therefore C = \frac{1}{\omega X_C} = \frac{1}{2 \times \pi \times 50\ \text{MHz} \times 125\ \Omega} = 25.5\ \text{pF}$$

22-8 CIRCLE DIAGRAMS

When we wish to examine the effect of varying one of the parameters of an impedance network, for example, the reactance of the load in the circuit of Figure 22-16, we can avoid having to draw a whole series of phasor diagrams by resorting to a bit of basic geometry.

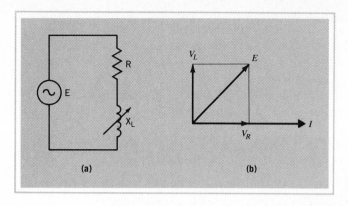

FIGURE 22-17
Simple Series Circuit with a Variable Reactance.

Since the current in a series circuit is common to all components, we have been using current as the reference phasor for series-circuit phasor diagrams, as in Figure 22-17(b). We then draw the total voltage as the resultant of the voltages across the resistance and the reactance of the circuit.

But, in practice, the applied voltage is a constant, and as either the resistance or reactance of the circuit is changed, the current (the dependent variable of most electric circuits) must change in such a manner that

$$V_R + jV_L = \mathbf{E}$$

We can redraw the phasor diagram with the applied voltage as the reference phasor, as shown in Figure 22-18. Since the applied voltage is constant, and since the voltage across the inductance must always lead the voltage across the resistance by exactly 90°, this phasor diagram will always be in the form of a right-angled triangle with a constant hypotenuse. We can show that, under these circumstances, the junction of the V_R and V_L phasors must be on the circumference of a circle with the $\mathbf{E}$ phasor as a diameter.[†]

In the circuit of Figure 22-17(a), the resistance remains constant but the inductive reactance can change. Since $I = V_R/R$, and since R is constant, the current phasor in Figure 22-18 will have the same angle and same proportionate magnitude

[†]$\triangle ABC$ in Figure 22-19,

$$\angle CAB + \angle ABC + \angle BCA = 180°$$
$$\therefore \angle CAB + \angle ABD + \angle DBC + \angle BCA = 180°$$

But since $AD = DB = DC$,

$$\angle CAB = \angle ABD$$

Similarly,
$$\angle BCA = \angle DBC$$
$$\therefore 2(\angle ABD + \angle DBC) = 180°$$

from which
$$\angle ABC = 90°$$

Conversely, whenever
$$\angle ABC = 90°$$
$$AD = DB = DC = \text{radii of a circle}$$

566

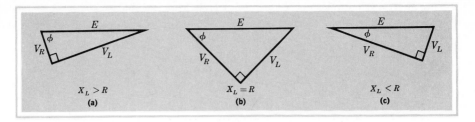

FIGURE 22-18
Effect of Varying X_L in a Series Circuit Connected to a Fixed Source Voltage.

as the V_R phasor. Therefore, as shown in Figure 22-20, in a series circuit containing a fixed resistance and a variable reactance, the tip of the current phasor must also lie on the circumference of a circle having the applied voltage as the reference phasor and a diameter of $I_{max} = E/R$, because the current will be greatest when the reactance drops to zero. Such is the situation when the source has a constant magnitude but a variable frequency. We use the circular locus of the current phasor as the frequency changes in the analysis of resonant circuits and transmission lines.

Figure 22-21 shows a simple series circuit in which the inductive reactance remains constant and the resistance varies. This condition will be encountered in analyzing the operating characteristics of induction motors as the mechanical load on the motor is changed. The current will be maximum when the resistance is zero. Hence, $I_{max} = E/X_L$, lagging the applied voltage by 90° as shown. As the resistance increases, the current decreases and, at the same time, the phase angle decreases. Again, since the applied voltage is constant, and since V_R and V_L must always be at right angles, the tip of the current phasor must be on the circumference of a circle, with E/X_L as the diameter.

From Figure 22-21(b),

$$\frac{I}{I_{max}} = \cos(90° - \phi) = \sin \phi \qquad (22\text{-}31)$$

FIGURE 22-20
Circle Diagram for Current in a Series Circuit with a Fixed Resistance and a Variable Reactance.

FIGURE 22-19

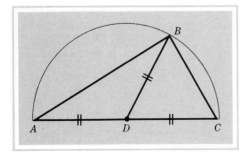

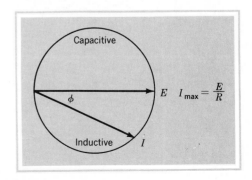

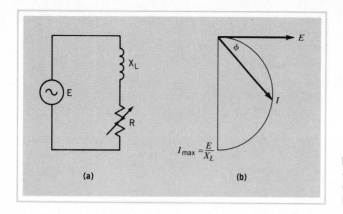

FIGURE 22-21
Circle Diagram for Current in a Series Circuit with a Fixed Inductive Reactance and a Variable Resistance.

EXAMPLE 22-11

A series circuit with a fixed reactance has an 0.5 lagging power factor when it draws a 10-A current from a 120-V 60-Hz source. What is the maximum true power for this circuit?

SOLUTION

$$\phi = \arccos 0.5 = 60°$$

$$I_{max} = \frac{I}{\sin 60°} = \frac{10}{0.866} = 11.55 \text{ A}$$

$$P = EI \cos \phi$$

where E is the applied voltage and I cos ϕ is the in-phase rectangular coordinate of the current. As the circle diagram of Figure 22-21(b) shows, $I \cos \phi$ will be maximum when $\phi = 45°$. Therefore, when $\phi = 45°$,

$$I = I_{max} \sin \phi = 11.55 \times 0.7071 = 8.165 \text{ A}$$

$$\therefore P_{max} = EI \cos \phi = 120 \times 8.165 \times 0.7071 = \mathbf{693 \text{ W}}$$

PROBLEMS

Draw schematic and phasor diagrams wherever applicable.

22-1. What is the total impedance when 400 $\Omega\underline{/+15°}$ and 800 $\Omega\underline{/+60°}$ are connected in series?

22-2. What is the total impedance when $\mathbf{Z}_1 = 5 \text{ k}\Omega\underline{/+30°}$, $\mathbf{Z}_2 = 4.7 \text{ k}\Omega\underline{/-45°}$, and $\mathbf{Z}_3 = 2.2 \text{ k}\Omega\underline{/+75°}$ are connected in series?

22-3. What is the total admittance when the two components of Problem 22-1 are connected in parallel?

22-4. What is the equivalent impedance when the three impedances of Problem 22-2 are connected in parallel?

568

22-5. A coil having a resistance of 100 Ω and an inductance of 0.5 H, a 60-μF capacitor, and a 30-Ω resistor are all connected in series to a 60-Hz source whose open-circuit voltage is 208 V $\underline{/0°}$.
(a) What are the magnitude and phase of the current through the coil?
(b) What are the magnitude and phase of the voltage across the coil?

22-6. What is the overall power factor when the three components of Problem 22-5 are connected in parallel?

22-7. What values of capacitance and resistance in series will form a 1-kHz impedance of 600 $\Omega \underline{/-30°}$?

22-8. What values of capacitance and resistance in parallel will form a 1-kHz impedance of 600 $\Omega \underline{/-30°}$?

22-9. What values of inductance and resistance in parallel will be equivalent at 60 Hz to a 0.5-H inductance and a 120-Ω resistance in series?

22-10. What values of capacitance and resistance in series will be equivalent at 400 Hz to a 0.02-μF capacitor and 6800-Ω resistor in parallel?

22-11. What value of impedance must be connected in series with a 60-$\Omega \underline{/+45°}$ impedance for the current drawn from a 120-V source to be 1 A lagging the source voltage by 30°?

22-12. What value of impedance must be connected in parallel with a 10-k$\Omega \underline{/-60°}$ impedance for the current from a 1.5-V 100-kHz source to be 150 μA $\underline{/0°}$?

22-13. Determine the magnitude and phase of the two capacitor currents in the circuit of Figure 22-22(a).

22-14. Find the magnitude and phase of the voltage across the capacitor in the circuit of Figure 22-22(b).

22-15. Find the magnitude and phase of the current in the output resistor in the circuit of Figure 22-23(a).

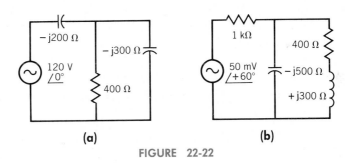

(a)   (b)

FIGURE 22-22

FIGURE 22-23

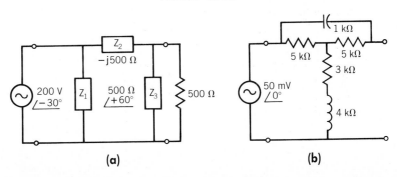

(a)   (b)

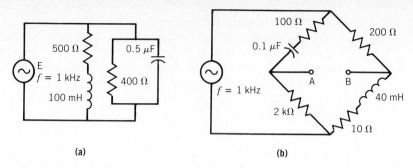

(a)

(b)

FIGURE 22-24

22-16. Find the magnitude and phase of the current drawn from the voltage source in the circuit of Figure 22-23(b)
(a) With the output terminals open-circuit,
(b) With the output terminals short-circuit.

22-17. Find the magnitude and phase of the open-circuit output voltage in the circuit of Figure 22-23(b).

22-18. Find the magnitude and phase of the voltage source required in the circuit of Figure 22-24(a) for the total drain on the source to be 50 mA $\underline{/+45°}$.

22-19. Find the magnitude and phase of the open-circuit voltage between terminals A and B in the Hay bridge circuit shown in Figure 22-24(b), if the source voltage is 50 V $\underline{/+30°}$.

22-20. If the source voltage in Figure 22-24(b) is 15 V $\underline{/0°}$, what is the total source current with terminals A and B (a) open circuit and (b) short circuit.

22-21. Reduce the circuit of Figure 22-25 to the simplest possible series circuit containing a resistance and a reactance for (a) $f = 60$ Hz and (b) $f = 100$ Hz.

22-22. Reduce the circuit of Figure 22-25 to the simplest possible parallel circuit for $f = 50$ Hz.

22-23. Reduce the circuit of Figure 22-26 to the simplest possible series circuit.

22-24. Reduce the circuit of Figure 22-26 to the simplest possible parallel circuit.

22-25. Determine the input impedance (rectangular coordinates) for the network in Figure 22-27.

22-26. Determine the input admittance (rectangular coordinates) for the network in Figure 22-27.

22-27. If the open-circuit voltage of the RF signal generator in Example 22-10 is 240 mV, what is the maximum power output?

FIGURE 22-25   FIGURE 22-26

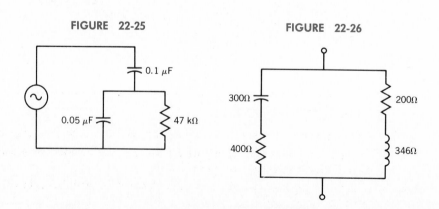

PART V   IMPEDANCE NETWORKS

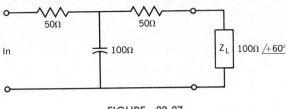

FIGURE 22-27

22-28. What is the power output in Problem 22-27 if we remove the 25-pF capacitor?

22-29. A television signal generator has an internal impedance of 50 $\Omega\underline{/-30°}$ at 83.25 MHz. Determine the series components of a dummy antenna load designed for maximum power transfer.

22-30. A transmitter operating on an assigned frequency of 1850 kHz delivers maximum power to an antenna when the antenna is tuned to represent a load consisting of a 200-pF capacitance in parallel with a 160-Ω resistance. Express the internal impedance "seen" at the output terminals of the transmitter in rectangular coordinates.

22-31. A circuit consisting of a 0.75-H inductance, a 40-Ω resistance, and a 26.5-μF capacitance in series is connected to a source which remains constant at 120 V, but whose frequency can vary from 25 Hz to 60 Hz. Draw an accurate diagram showing the manner in which the magnitude and phase of the current vary as the frequency is varied through its range.

22-32. Draw an accurate circle diagram showing how the open-circuit output voltage E_0 varies as the rheostat is varied from zero to 500 kΩ in the phase shifter shown in Figure 22-28.

22-33. In an ac circuit, $\mathbf{Z}_1$ consists of a 50-Ω capacitive reactance and $\mathbf{Z}_2 = 50\ \Omega\underline{/+30°}$. Determine the equivalent impedance of $\mathbf{Z}_1$ and $\mathbf{Z}_2$ in series and of $\mathbf{Z}_1$ and $\mathbf{Z}_2$ in parallel. Determine the relationship that must exist between any two impedances so that their parallel equivalent impedance is the same as their series equivalent impedance.

FIGURE 22-28
Phase Shifter.

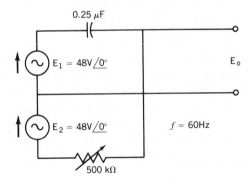

REVIEW QUESTIONS

22-1. Write a statement for Kirchhoff's voltage law as it applies to alternating-current circuits.

22-2. Given the magnitudes of two impedances in ohms, why can we not calculate the total impedance of these two impedances in series?

22-3. Under what conditions could the total impedance of two impedances in series be less than that of either one by itself?

22-4. Under what circumstances is it (a) more convenient and (b) less convenient, to state impedances in rectangular coordinates?

22-5. Write a statement for the voltage-divider principle as it applies to alternating-current circuits.

22-6. What is the advantage in thinking in terms of total **admittance** when dealing with parallel ac circuits?

22-7. Show how Equation (22-8) is derived.

22-8. Derive Equation (22-11) from Equation (22-10).

22-9. What conditions must be fulfilled for a series circuit to be the exact equivalent of a given parallel circuit?

22-10. What is the significance of equivalent circuits in analyzing the performance of electric and electronic circuits?

22-11. Given a resistance and reactance in series with $R_s = 4X_s$, in the equivalent parallel circuit, $X_p = 4 R_p$. Explain this inversion.

22-12. A coil and capacitor connected in series to a given source represent an inductive impedance. When the same two components are connected in parallel to the same source, the total impedance becomes capacitive. Explain.

22-13. If we vary the frequency of the source in the circuit of Figure 22-8(a), we can in fact determine whether the "black box" contains circuit (b) or circuit (c). Explain how this is achieved.

22-14. Why is it possible to derive either a series- or a parallel-equivalent circuit from the *same* power triangle?

22-15. Show that Equation (22-23) is the same as $R_s = Z \cos \phi$.

22-16. Show that Equation (22-26) is the same as

$$X_p = \frac{Z}{\sin \phi}$$

22-17. What is meant by the **conjugate** of an impedance?

22-18. What is the significance of conjugate impedance in ac networks?

22-19. Why does the tip of the current phasor move around the circumference of a circle when the resistance alone is varied in an ac circuit containing resistance and reactance in series.

22-20. Draw a circle diagram for the output voltage of the phase shifter of Figure 22-29 as the variable resistance is increased from zero to 500 kΩ.

*22-21. Circle diagrams are alternatives to tedious repetitive calculations in ac networks containing a variable impedance. Computers can do such repetitive calculations rapidly and accurately. Write a computer program to print out a table of magnitudes and phase angles which will verify the circle diagram you produced for the phase shifter in Problem 22-32.

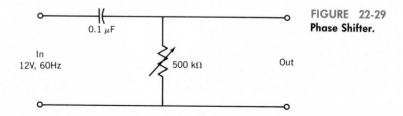

FIGURE 22-29
Phase Shifter.

23

IMPEDANCE NETWORKS

In Chapter 8, we encountered resistance networks which we could not solve by simple series-parallel circuit relationships. We developed two general lines of attack on such networks: (1) we can write a set of simultaneous **loop** or **nodal** equations for the networks based on Kirchhoff's voltage and current laws, respectively; or (2) we can use network theorems to simplify the original networks to a level where we can tackle them by simple series-parallel relationships.

As long as we perform all numerical computations by *phasor* algebra, we follow exactly the same procedures in dealing with impedance networks in ac circuits. Unfortunately, this switch from linear algebra to phasor algebra greatly increases the amount of numerical computation. To minimize the possibility of errors in performing these calculations on an electronic calculator, we need to develop an understanding of the various network analysis techniques so that we can select the one that is most straightforward and involves the minimum of computation for a particular network.

The advantage of a personal computer is that errors are limited to incorrectly entering requested data—once a suitable program has been verified. The disadvantage is that too much time is required to write (and debug) a special program for each example involving complex algebra. From this chapter onward, for a personal computer to be a useful aid in solving alternating-current networks, we require access to a suitable circuit-analysis software package. Due to the variations in such programs, we must leave specific instruction in their application to the manuals provided with the software.

23-1 LOOP EQUATIONS

We can restate Kirchhoff's voltage law from Section 8-5, worded to suit the phasor quantities we encounter in ac circuits.

In any complete ac circuit (closed loop), the phasor sum of the voltage drops must equal the phasor sum of the source voltages.

This leads to equations in the form

$$\mathbf{V}_1 + \mathbf{V}_2 + \mathbf{V}_3 = \mathbf{E} \tag{23-1}$$

where every term is a phasor quantity. As in direct-current circuits, we have a choice of two methods of writing Kirchhoff's voltage-law equations: **loop** equations and **mesh** equations. The loop procedure more closely follows our statement of Kirchhoff's voltage law and is probably easier to *understand*. Mesh equations provide a format for setting up essentially the same Kirchhoff's voltage-law equations by following a few simple rules.

In writing *loop* equations, we substitute an $\mathbf{IZ}$ voltage drop across each impedance for each $\mathbf{V}$ term in Equation (23-1). Since the various values of $\mathbf{Z}$ are usually known, we end up with a set of simultaneous equations in which loop currents are the unknown quantities.

If we could omit $\mathbf{Z}_4$ in the network shown in Figure 23-1, we would have a simple T-network which we could solve by the series-parallel circuit techniques of

FIGURE 23-1
Bridged-T Network.

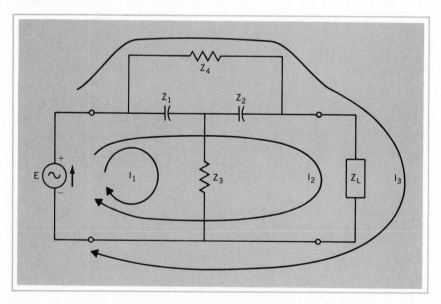

Chapter 22. The addition of Z_4 turns the network into a bridge circuit in which it is quite likely that there will be a different current through each of the five impedances. We can avoid having to write *five* simultaneous equations by constructing a set of tracing loops from the + terminal of the source to its − terminal. Hence, instead of writing I_L for the current through Z_L in Figure 23-1, we immediately write $Z_L(I_2 + I_3)$ as our substitution for V_L. We write the *sum* of I_2 and I_3 since both tracing loops pass in the same direction through Z_L.

Because the current in a dc circuit always flows in the same direction, establishing the direction of the tracing loops presents no problem. We simply assumed a direction from the + terminal of the source, through the network, and back to the − terminal of the source. However, in an ac circuit, the instantaneous current reverses its polarity twice each cycle. Nevertheless, we must adopt some method of marking the direction of an ac source, since reversing the leads of an ac source will change the angle of the source voltage by 180°, just as reversing the leads of a dc source will invert the polarity of the source voltage. In Figure 23-1, the lower lead is common to both the source and the load. In nodal analysis, we call this the **reference node.** It is customary to express as a positive number the voltage appearing at the upper terminal of the source *with respect to* the reference node (which is sometimes grounded).

To make this choice quite clear, particularly in multisource networks, there are two methods of marking the direction for ac sources. We can use + and − signs, as shown in Figure 23-1. In an ac circuit, these signs tell us that the angle given for E is measured from the + terminal *with respect to* the − terminal. Hence, we should mark the positive direction for tracing loops through the external network from the + terminal to the − terminal of the source. Obviously, these + and − signs do not indicate instantaneous polarity.

The second method is to draw an arrow alongside the source, also shown in Figure 23-1. This arrow tells us that the angle given for E is measured from the head of the arrow *with respect to* the other terminal. We can also take the direction for tracing loops through the external network by projecting the arrow around the loop. To avoid possible confusion from using + and − signs in an ac circuit diagram, we shall use the arrow method in the following chapters.

There are several possible combinations of tracing loops which we can draw for the network of Figure 23-1. Using those shown in Figure 23-1, we can write the three loop equations from Kirchhoff's voltage law.

For Loop I_1,

$$Z_1(I_1 + I_2) + Z_3 I_1 = E$$

For Loop I_2,

$$Z_1(I_1 + I_2) + Z_2 I_2 + Z_L(I_2 + I_3) = E$$

For Loop I_3,

$$Z_4 I_3 + Z_L(I_2 + I_3) = E$$

Collecting terms,

$$(\mathbf{Z}_1 + \mathbf{Z}_3)\mathbf{I}_1 + \mathbf{Z}_1\mathbf{I}_2 = \mathbf{E}$$
$$\mathbf{Z}_1\mathbf{I}_1 + (\mathbf{Z}_1 + \mathbf{Z}_2 + \mathbf{Z}_L)\mathbf{I}_2 + \mathbf{Z}_L\mathbf{I}_3 = \mathbf{E}$$
$$\mathbf{Z}_L\mathbf{I}_2 + (\mathbf{Z}_4 + \mathbf{Z}_L)\mathbf{I}_3 = \mathbf{E}$$

EXAMPLE 23-1

Given $\mathbf{E} = 20 \text{ V} \underline{/0°}$, $\mathbf{Z}_1 = \mathbf{Z}_2 = -j5 \text{ k}\Omega$, $\mathbf{Z}_3 = \mathbf{Z}_4 = (20 + j0) \text{ k}\Omega$, and $\mathbf{Z}_L = (5 + j5) \text{ k}\Omega$ for the bridged-T network of Figure 23-1, find the current through $\mathbf{Z}_L$.

SOLUTION

Although we could proceed to substitute in the equations we wrote for the tracing loops of Figure 23-1, we would have to solve for both $\mathbf{I}_2$ and $\mathbf{I}_3$ in order to determine $\mathbf{I}_L$. If we select a different set of tracing loops, as shown in Figure 23-2, we need solve only for $\mathbf{I}_3$.

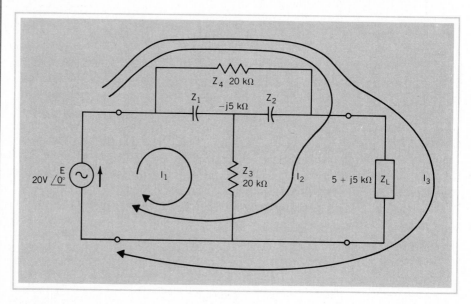

FIGURE 23-2
Circuit Diagram for Example 23-1.

Step I. Writing the three loop equations,

$$\mathbf{Z}_1\mathbf{I}_1 + \mathbf{Z}_3(\mathbf{I}_1 + \mathbf{I}_2) = \mathbf{E}$$
$$\mathbf{Z}_4(\mathbf{I}_2 + \mathbf{I}_3) + \mathbf{Z}_2\mathbf{I}_2 + \mathbf{Z}_3(\mathbf{I}_2 + \mathbf{I}_1) = \mathbf{E}$$
$$\mathbf{Z}_4(\mathbf{I}_3 + \mathbf{I}_2) + \mathbf{Z}_L\mathbf{I}_3 = \mathbf{E}$$

Collecting terms,

$$(\mathbf{Z}_1 + \mathbf{Z}_3)\mathbf{I}_1 + \mathbf{Z}_3\mathbf{I}_2 = \mathbf{E}$$

$$\mathbf{Z}_3\mathbf{I}_1 + (\mathbf{Z}_2 + \mathbf{Z}_3 + \mathbf{Z}_4)\mathbf{I}_2 + \mathbf{Z}_4\mathbf{I}_3 = \mathbf{E}$$

$$\mathbf{Z}_4\mathbf{I}_2 + (\mathbf{Z}_4 + \mathbf{Z}_L)\mathbf{I}_3 = \mathbf{E}$$

Step II. If we substitute for $\mathbf{Z}$ in *kilohms* and for $\mathbf{E}$ in *volts,* the current will be in *milliamperes*. And since the loop equations show that we shall be adding impedances, we might as well make our substitutions in rectangular coordinates. Therefore,

$$(20 - j5)I_1 + (20 + j0)I_2 = 20 + j0$$

$$(20 + j0)I_1 + (40 - j5)I_2 + (20 + j0)I_3 = 20 + j0$$

$$(20 + j0)I_2 + (25 + j5)I_3 = 20 + j0$$

Step III. Since we now have to *multiply* the diagonals of the determinant matrices, we use the $\rightarrow$ P key of our calculator to write the impedances in the matrices in polar form.

$$I_3 = \frac{\begin{vmatrix} 20.61\underline{/-14.04°} & 20\underline{/0°} & 20\underline{/0°} \\ 20\underline{/0°} & 40.31\underline{/-7.13°} & 20\underline{/0°} \\ 0 & 20\underline{/0°} & 20\underline{/0°} \end{vmatrix}}{\begin{vmatrix} 20.61\underline{/-14.04°} & 20\underline{/0°} & 0 \\ 20\underline{/0°} & 40.31\underline{/-7.13°} & 20\underline{/0°} \\ 0 & 20\underline{/0°} & 25.5\underline{/+11.31°} \end{vmatrix}}$$

$$\mathbf{I}_3 = \frac{16\,616\underline{/-21.17°} + 0 + 8000\underline{/0°} - 0 - 8244\underline{/-14.04°} - 8000\underline{/0°}}{21\,185\underline{/-9.86°} + 0 + 0 - 0 - 8244\underline{/-14.04°} - 10\,196\underline{/+11.31°}}$$

$$= \frac{(15\,495 - j6000) - (7998 - j2000)}{(20\,872 - j3628) - (7998 - j2000) - (9998 + j2000)}$$

$$= \frac{7497 - j4000}{2876 - j3628} = \frac{8497\underline{/-28.1°}}{4630\underline{/-51.6°}} = 1.835 \text{ mA}\underline{/+23.5°}$$

$$\therefore \mathbf{I}_L = 1.84 \text{ mA}\underline{/+23.5°}$$

In the single-source network of Example 23-1, we started all tracing loops from the "positive" terminal of the source. This keeps the number of negative terms in the loop equations to a minimum. In Example 23-1, we needed three loops to include all circuit elements in such a way that we could combine loop currents to determine six different currents in the six different branches of the network. If there is more than one voltage source in a network, at least one tracing loop starts and ends with each source. Again, the source direction determines the loop current direction and there must be sufficient loop equations to determine actual currents in all branches. In Example 23-2, there are two sources and only three different branches. Hence, we can determine all actual currents from two loop current equations.

EXAMPLE 23-2

Two 60-Hz alternators are connected in parallel to feed a load whose impedance is $20 \ \Omega \underline{/-15°}$. Alternator 1 develops an open-circuit voltage of 120 V and has an internal impedance of $10 \ \Omega \underline{/+30°}$. Alternator 2 has an open-circuit voltage of 117 V which lags the voltage of alternator 1 by a constant 15°. Alternator 2 has an internal impedance of $8 \ \Omega \underline{/+45°}$. Find the load current.

SOLUTION

Step I. Writing the two loop equations for the circuit of Figure 23-3,

For the $\mathbf{I}_1$ loop,

$$\mathbf{Z}_1 \mathbf{I}_1 + \mathbf{Z}_L (\mathbf{I}_1 + \mathbf{I}_2) = \mathbf{E}_1$$

And for the $\mathbf{I}_2$ loop,

$$\mathbf{Z}_2 \mathbf{I}_2 + \mathbf{Z}_L (\mathbf{I}_2 + \mathbf{I}_1) = \mathbf{E}_2$$

Collecting terms,

$$(\mathbf{Z}_1 + \mathbf{Z}_L)\mathbf{I}_1 + \mathbf{Z}_L \mathbf{I}_2 = \mathbf{E}_1$$
$$\mathbf{Z}_L \mathbf{I}_1 + (\mathbf{Z}_2 + \mathbf{Z}_L)\mathbf{I}_2 = \mathbf{E}_2$$

FIGURE 23-3
Circuit Diagram for Example 23-2.

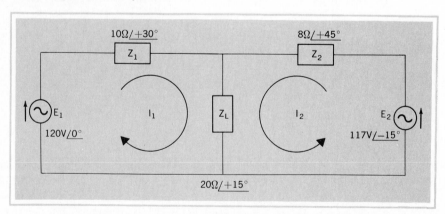

$$\mathbf{I}_1 = \frac{\begin{vmatrix} \mathbf{E}_1 & \mathbf{Z}_L \\ \mathbf{E}_2 & (\mathbf{Z}_2 + \mathbf{Z}_L) \end{vmatrix}}{\begin{vmatrix} (\mathbf{Z}_1 + \mathbf{Z}_L) & \mathbf{Z}_L \\ \mathbf{Z}_L & (\mathbf{Z}_2 + \mathbf{Z}_L) \end{vmatrix}} \quad \text{and} \quad \mathbf{I}_2 = \frac{\begin{vmatrix} (\mathbf{Z}_1 + \mathbf{Z}_L) & \mathbf{E}_1 \\ \mathbf{Z}_L & \mathbf{E}_2 \end{vmatrix}}{\begin{vmatrix} (\mathbf{Z}_1 + \mathbf{Z}_L) & \mathbf{Z}_L \\ \mathbf{Z}_L & (\mathbf{Z}_2 + \mathbf{Z}_L) \end{vmatrix}}$$

Step II.

$$\mathbf{Z}_1 + \mathbf{Z}_L = 10\underline{/+30°} + 20\underline{/+15°} = (8.66 + j5) + (19.32 + j5.176)$$
$$= 27.98 + j10.176 = 29.77 \ \Omega\underline{/+19.99°}$$

$$\mathbf{Z}_2 + \mathbf{Z}_L = 8\underline{/+45°} + 20\underline{/+15°} = (5.657 + j5.657) + (19.32 + j5.176)$$
$$= 24.98 + j10.833 = 27.23 \ \Omega\underline{/+23.45°}$$

$$\mathbf{I}_1 = \frac{\begin{vmatrix} 120\underline{/0°} & 20\underline{/+15°} \\ 117\underline{/-15°} & 27.23\underline{/+23.45°} \end{vmatrix}}{\begin{vmatrix} 29.77\underline{/+19.99°} & 20\underline{/+15°} \\ 20\underline{/+15°} & 27.23\underline{/+23.45°} \end{vmatrix}} = \frac{3268\underline{/+23.45°} - 2340\underline{/0°}}{810.6\underline{/+43.44°} - 400\underline{/+30°}}$$

$$= \frac{(2998 + j1301) - (2340 + j0)}{(588.6 + j557.4) - (346.4 + j200)} = \frac{658 + j1301}{242.2 + j357.4}$$

$$= \frac{1458\underline{/+63.17°}}{431.7\underline{/+55.88°}} = 3.377 \ \text{A}\underline{/+7.29°}$$

$$\mathbf{I}_2 = \frac{\begin{vmatrix} 29.77\underline{/+19.99°} & 120\underline{/0°} \\ 20\underline{/+15°} & 117\underline{/-15°} \end{vmatrix}}{D} = \frac{3483\underline{/+4.99°} - 2400\underline{/+15°}}{D}$$

$$= \frac{(3470 + j303) - (2318 + j621)}{D} = \frac{1152 - j318}{D}$$

$$= \frac{1195\underline{/-15.43°}}{431.7\underline{/+55.88°}} = 2.768 \ \text{A}\underline{/-71.31°}$$

Step III. From Figure 23.3,

$$\mathbf{I}_L = \mathbf{I}_1 + \mathbf{I}_2 = 3.377 \ \text{A}\underline{/+7.29°} + 2.768 \ \text{A}\underline{/-71.31°}$$
$$= (3.35 + j0.429) + (0.887 - j2.622)$$
$$= 4.237 - j2.193 = \mathbf{4.77 \ \text{A}\underline{/-27.4°}}$$

23-2 MESH EQUATIONS

The rules for writing the optional mesh equations are the same for both dc and ac circuits. However, in ac networks, both the dependent and independent variables are phasor quantities.

1. We must convert any current sources in the network to equivalent constant-voltage sources before laying out the circuit diagram.
2. We cannot use mesh equations unless we can lay out the circuit diagram so that there are no electrical components *within* any closed loop "mesh" (a *planar* circuit diagram).
3. We draw and identify a current loop for every mesh so that mesh currents are either all clockwise or all counterclockwise. (We shall choose clockwise.) As a result, when we write the mesh equations, the current for the identified mesh is a *positive* quantity and all currents in adjacent meshes with mutual components are *negative* quantities.
4. If the direction of the mesh current is the *same* as the direction indicated for a source in the mesh, the source voltage is a *positive* quantity. If the directions are opposite, the source voltage is a *negative* quantity. (There may be no source in some meshes.)
5. Since the object of these rules is to set up the simultaneous equations directly in matrix format (as in the collecting-of-terms step in the *loop* procedure), the coefficient for the identified positive mesh current is the *sum* of the impedances around the mesh. The coefficients for the negative adjacent mesh currents are the impedances of the mutual components. Thus the mesh equation for the I_A mesh in Figure 23-4 has the form

$$(\mathbf{Z}_1 + \mathbf{Z}_3)\mathbf{I}_A - \mathbf{Z}_3\mathbf{I}_B - \mathbf{Z}_1\mathbf{I}_C = +\mathbf{E} \tag{23-2}$$

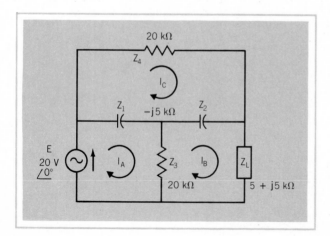

FIGURE 23-4
Circuit Diagram for Example 23-1A.

EXAMPLE 23-1A

Given $\mathbf{E} = 20 \text{ V}\underline{/0°}$, $\mathbf{Z}_1 = \mathbf{Z}_2 = -j5 \text{ k}\Omega$, $\mathbf{Z}_3 = \mathbf{Z}_4 = (20 + j0) \text{ k}\Omega$, and $\mathbf{Z}_L = (5 + j5) \text{ k}\Omega$ for the bridged-T network of Figures 23-1 and 23-4, find the current through $\mathbf{Z}_L$.

SOLUTION

Step I. Substituting for $\mathbf{Z}$ in kilohms and solving for $\mathbf{I}$ in milliamperes, and following the mesh format rules, the three mesh equations for $\mathbf{I}_A$, $\mathbf{I}_B$ and $\mathbf{I}_C$, respectively, are

$$(20 - j5)\mathbf{I}_A - (20 + j0)\mathbf{I}_B - (0 - j5)\mathbf{I}_C = +20\underline{/0°}$$

$$-(20 + j0)\mathbf{I}_A + (25 + j0)\mathbf{I}_B - (0 - j5)\mathbf{I}_C = 0$$

$$-(0 - j5)\mathbf{I}_A - (0 - j5)\mathbf{I}_B + (20 - j10)\mathbf{I}_C = 0$$

Step II. Having written the required mesh equations in a single step, solution of the three simultaneous equations now follows the same procedure as the loop method. To find $\mathbf{I}_L$, we need solve only for $\mathbf{I}_B$. Using the $\rightarrow$ P key of the calculator, the matrices become

$$\mathbf{I}_B = \frac{\begin{vmatrix} +20.62\underline{/-14.04°} & +20\underline{/0°} & -5\underline{/-90°} \\ -20\underline{/0°} & 0 & -5\underline{/-90°} \\ -5\underline{/-90°} & 0 & +22.36\underline{/-26.57°} \end{vmatrix}}{\begin{vmatrix} +20.62\underline{/-14.04°} & -20\underline{/0°} & -5\underline{/-90°} \\ -20\underline{/0°} & +25\underline{/0°} & -5\underline{/-90°} \\ -5\underline{/-90°} & -5\underline{/-90°} & 22.36\underline{/-26.57°} \end{vmatrix}}$$

$$= \frac{0 + 500\underline{/-180°} + 0 - 0 - 0 + 8944\underline{/-26.57°}}{11\,524\underline{/-40.61°} - 500\underline{/-180°} - 500\underline{/-180°} - 625\underline{/-180°} - 515.4\underline{/-194.04°} - 8944\underline{/-26.57°}}$$

$$= \frac{(-500 + j0) + (8000 - j4000)}{(8750 - j7500) - (-1625 + j0) - (-500 + j125) - (8000 - j4000)}$$

$$= \frac{7500 - j4000}{2875 - j3625} = \frac{8500\underline{/-28.1°}}{4627\underline{/-51.6°}}$$

$$\mathbf{I}_L = \mathbf{I}_B = 1.84 \text{ mA}\,\underline{/+23.5°}$$

EXAMPLE 23-2A

Two 60-Hz alternators are connected in parallel to feed a load whose impedance is $20 \Omega\underline{/-15°}$. Alternator 1 develops an open-circuit voltage of 120 V and has an internal impedance of $10 \Omega\underline{/+30°}$. Alternator 2 has an open-circuit voltage of 117 V which lags the voltage of alternator 1 by a constant 15°. Alternator 2 has an internal impedance of $8 \Omega\underline{/+45°}$. Find the load current.

SOLUTION

We can use the circuit diagram of Figure 23-3 for mesh equations simply by reversing the direction of the $\mathbf{I}_2$ loop.

Step I. The mesh equations for $\mathbf{I}_1$ and $\mathbf{I}_2$, respectively, are

$$(\mathbf{Z}_1 + \mathbf{Z}_\mathrm{L})\mathbf{I}_1 - \mathbf{Z}_\mathrm{L}\,\mathbf{I}_2 = +\mathbf{E}_1$$

$$-\mathbf{Z}_\mathrm{L}\,\mathbf{I}_1 + (\mathbf{Z}_2 + \mathbf{Z}_\mathrm{L})\mathbf{I}_2 = -\mathbf{E}_2$$

$$\mathbf{I}_1 = \frac{\begin{vmatrix} \mathbf{E}_1 & -\mathbf{Z}_\mathrm{L} \\ -\mathbf{E}_2 & \mathbf{Z}_2 + \mathbf{Z}_\mathrm{L} \end{vmatrix}}{\begin{vmatrix} \mathbf{Z}_1 + \mathbf{Z}_\mathrm{L} & -\mathbf{Z}_\mathrm{L} \\ -\mathbf{Z}_\mathrm{L} & \mathbf{Z}_2 + \mathbf{Z}_\mathrm{L} \end{vmatrix}} \qquad \mathbf{I}_2 = \frac{\begin{vmatrix} \mathbf{Z}_1 + \mathbf{Z}_\mathrm{L} & +\mathbf{E}_1 \\ -\mathbf{Z}_\mathrm{L} & -\mathbf{E}_2 \end{vmatrix}}{\begin{vmatrix} \mathbf{Z}_1 + \mathbf{Z}_\mathrm{L} & -\mathbf{Z}_\mathrm{L} \\ -\mathbf{Z}_\mathrm{L} & \mathbf{Z}_2 + \mathbf{Z}_\mathrm{L} \end{vmatrix}}$$

Step II. Comparing these mesh equations with the loop equations of Example 23-2 after collecting terms, we find the same coefficients but with some $-$ signs due to reversing the direction of $\mathbf{I}_2$. Solving for $\mathbf{I}_1$ and $\mathbf{I}_2$,

$$\mathbf{I}_1 = 3.377 \text{ A}\underline{/+7.29°} \qquad \mathbf{I}_2 = 2.768 \text{ A}\underline{/+108.69°}$$

Step III. From Figure 23-3, with both mesh currents clockwise,

$$\mathbf{I}_\mathrm{L} = \mathbf{I}_1 - \mathbf{I}_2 = 3.377 \text{ A}\underline{/+7.29°} - 2.768 \text{ A}\underline{/+108.69°}$$

$$= (3.35 + \mathrm{j}0.429) - (-0.887 + \mathrm{j}2.622)$$

$$= 4.237 - \mathrm{j}2.193 = 4.77 \text{ A}\,\underline{/-27.4°}$$

23-3 SUPERPOSITION THEOREM

In networks containing more than one source, we can avoid the simultaneous equations of the loop procedure by applying the **superposition theorem.** We can restate the theorem to embrace the phasor quantities encountered in ac circuits.

The current that flows in any branch of a network of impedances resulting from the simultaneous application of a number of voltage sources distributed in any manner throughout the network is the phasor sum of the component currents in that branch that would be caused by each source acting independently in turn while the others are replaced in the network by their respective internal impedances.

Since Example 23-1 has only one source, the superposition theorem cannot help us with the bridged-T network problem. However, we can use the superposition theorem to avoid the simultaneous equations of Example 23-2.

EXAMPLE 23-2B

Two 60-Hz alternators are connected in parallel to feed a load whose impedance is $20\ \Omega\underline{/+15°}$. Alternator 1 develops an open-circuit voltage of 120 V and has an

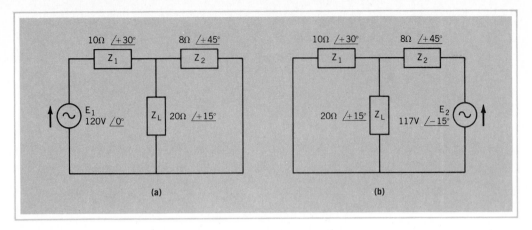

FIGURE 23-5
Circuit Diagram for Example 23-2B.

internal impedance of $10 \; \Omega\underline{/+30°}$. Alternator 2 has an open-circuit voltage of 117 V which lags the voltage of alternator 1 by a constant 15°. Alternator 2 has an internal impedance of $8 \; \Omega\underline{/+45°}$. Find the load current.

SOLUTION
   Step I.   We consider $\mathbf{E}_1$ to be acting alone in the network. All that remains of $\mathbf{E}_2$ in the circuit of Figure 23-5(a) is its internal impedance, $\mathbf{Z}_2$.

$$\mathbf{Z}_T = \mathbf{Z}_1 + \frac{\mathbf{Z}_2 \mathbf{Z}_L}{\mathbf{Z}_2 + \mathbf{Z}_L}$$

$$= (8.66 + j5) + \frac{(5.657 + j5.657) \times (19.319 + j5.176)^\dagger}{(5.657 + j5.657) + (19.319 + j5.176)}$$

$$= (8.66 + j5) + \frac{(80.01 + j138.57)}{(24.98 + j10.83)}$$

$$= (8.66 + j5) + (4.72 + j3.5) = (13.38 + j8.5) \; \Omega$$

$$= 15.85 \; \Omega\underline{/+32.43°}$$

$$\therefore \mathbf{I}_1 = \frac{\mathbf{E}_1}{\mathbf{Z}_T} = \frac{120\underline{/0°}}{15.85\underline{/+32.43°}} = 7.571 \; A\underline{/-32.43°}$$

Using the current-divider principle to find the first component of the load current, from Figure 23-5(a),

$$\mathbf{I}_L = \mathbf{I}_1 \frac{\mathbf{Z}_2}{\mathbf{Z}_2 + \mathbf{Z}_L} = 7.571\underline{/-32.43°} \times \frac{8\underline{/+45°}}{27.23\underline{/+23.44°}}$$

$$= 2.224 \; A\underline{/-10.87°}$$

†As an alternative to repeated use of the $\rightarrow$ P and $\rightarrow$ R keys, we can, if we wish, carry out the phasor multiplication and division in rectangular coordinates, using the procedure shown in Example 19-11.

Step II. We now consider $\mathbf{E}_2$ to be acting alone in the network giving the circuit of Figure 23-5(b),

$$\mathbf{Z}_T = \mathbf{Z}_2 + \frac{\mathbf{Z}_1 \mathbf{Z}_L}{\mathbf{Z}_1 + \mathbf{Z}_L}$$

$$= (5.657 + j5.657) + \frac{(8.66 + j5) \times (19.319 + j5.176)}{(8.66 + j5) + (19.319 + j5.176)}$$

$$= (5.657 + j5.657) + \frac{(141.42 + j141.42)}{(27.98 + j10.18)}$$

$$= (5.657 + j5.657) + (6.088 + j2.839) = (11.745 + j8.496) \ \Omega$$

$$= 14.50 \ \Omega \underline{/+35.88°}$$

$$\therefore \mathbf{I}_2 = \frac{\mathbf{E}_2}{\mathbf{Z}_T} = \frac{117\underline{/-15°}}{14.5\underline{/+35.88°}} = 8.069 \ A\underline{/-50.88°}$$

The second component of the load current is

$$\mathbf{I}_L = \mathbf{I}_2 \frac{\mathbf{Z}_1}{\mathbf{Z}_1 + \mathbf{Z}_L} = 8.069\underline{/-50.88°} \times \frac{10\underline{/+30°}}{29.77\underline{/+19.99°}}$$

$$= 2.710 \ A\underline{/-40.87°}$$

Step III. The actual load current is the phasor sum of the two component currents. Hence,

$$\mathbf{I}_L = 2.224\underline{/-10.87°} + 2.710\underline{/-40.87°}$$

$$= (2.184 - j0.419) + (2.049 - j1.773)$$

$$= (4.233 - j2.192) \ A = \mathbf{4.77} \ \mathbf{A}\underline{/-27.4°}$$

The superposition theorem was developed as a circuit-analysis tool long before the concept of equivalent constant-current sources became popular. Nevertheless, we can still use the superposition principle in networks containing current sources. To illustrate the procedure without a lot of complex algebra, we can consider the simple network of Figure 23-6(a).

EXAMPLE 23-3

Assuming that the ac milliammeter in the network of Figure 23-6(a) has zero impedance, calculate its reading.

SOLUTION

Step I. For $\mathbf{E}_A$ to act alone in the network, $\mathbf{I}_B$ becomes zero by leaving an *open* circuit in place of its symbol, as shown in Figure 23-6(b). The first component of the meter current is

$$I_1 = \frac{E_A}{Z_1 + Z_2} = \frac{20 \text{ V}\underline{/0°}}{(2 + j4) \text{ k}\Omega} = 4.472 \text{ mA}\underline{/-63.43°}$$

Step II. For I_B to act alone, E_A becomes zero by placing a short circuit in place of its symbol, as shown in Figure 23-6(c). Using the current-divider principle, the second component of the meter current is

$$I_2 = \frac{Z_1}{Z_1 + Z_2} \times I_B = \frac{2 \text{ k}\Omega\underline{/0°}}{4.472 \text{ k}\Omega\underline{/+63.43°}} \times 40 \text{ mA}\underline{/+30°}$$

$$= 17.89 \text{ mA}\underline{/-33.43°}$$

Step III. From the directions of the component currents in Figures 23-6(b) and (c),

$$I_M = I_1 + I_2 = (2 - j4) + (14.93 - j9.86) = 16.93 - j13.86 \text{ mA}$$

$$\therefore I_M = 21.88 \text{ mA}$$

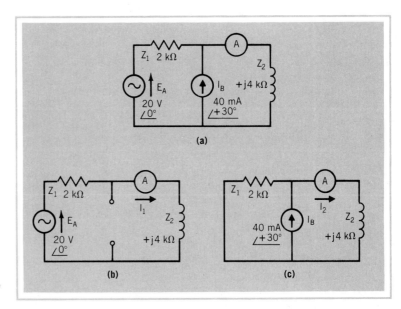

FIGURE 23-6
Circuit Diagram for Example 23-3.

23-4 THÉVENIN'S THEOREM

Because of the considerable amount of numerical computation involved when we solve networks in their original form, techniques which will permit us to replace the original circuit with a simplified equivalent circuit are particularly significant in the analysis of alternating-current circuit behavior. Since Thévenin's theorem allows us to replace a fairly complex impedance network containing one or more voltage sources with a single source and its internal impedance, this theorem has become one of the

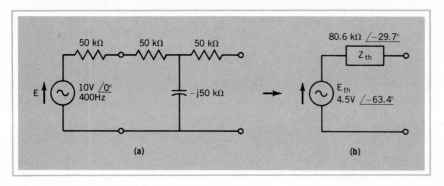

FIGURE 23-7
AC Circuit Simplification by Thévenin's Theorem.

most useful of our circuit-analysis techniques. When we apply it to ac circuits, we may state Thévenin's theorem thus,

Any two-terminal network of fixed impedances and voltage sources may be replaced by a single voltage source having an equivalent voltage equal to the open-circuit voltage at the terminals of the original network and having an internal impedance equal to the impedance looking back into the network from the two terminals with all voltage sources replaced by their internal impedances.

Thévenin's theorem is particularly useful in analyzing the performance of electronic amplifiers. We can represent the *active* components of the circuit (transistors) by Thévenin-equivalent sources. Then, as shown in Figure 23-7, we can incorporate the coupling network into the signal source by a Thévenin transformation.

EXAMPLE 23-4
Given the signal source and four-terminal network of Figure 23-7(a), determine the Thévenin-equivalent source.

SOLUTION
Step I. With no load connected to the output terminals, the impedance across the original source is

$$\mathbf{Z} = 50 \text{ k}\Omega + 50 \text{ k}\Omega - j50 \text{ k}\Omega = 111.8 \text{ k}\Omega\underline{/-26.56°}$$

$$\mathbf{I} = \frac{\mathbf{E}}{\mathbf{Z}} = \frac{10 \text{ V}\underline{/0°}}{111.8 \text{ k}\Omega\underline{/-26.56°}} = 89.44 \ \mu A\underline{/+26.56°}$$

Therefore, the open-circuit voltage is

$$\mathbf{E}_{th} = \mathbf{V}_C = \mathbf{I}X_C = 89.44 \ \mu A\underline{/+26.56°} \times 50 \text{ k}\Omega\underline{/-90°}$$
$$= 4.47 \text{ V}\underline{/-63.44°}$$

Step II. Shorting out the alternator symbol leaving only its 50-kΩ internal resistance in the circuit, we now look back into the open-circuit output terminals to find Z_{th}. Since all impedances are in kilohms, we can simplify the following calculation by working in kilohms.

$$Z_{th} = (50 + j0) + \frac{(100 + j0) \times (0 - j50)}{(100 + j0) + (0 - j50)}$$

$$= (50 + j0) + \frac{-j5000}{(100 - j50)} = (50 + j0) + (20 - j40)$$

$$= (70 - j40)\ k\Omega = \mathbf{80.62\ k\Omega \underline{/-29.74°}}$$

By being able to remove the load from the source network, Thévenin's theorem provides an alternate solution for Example 23-1.

EXAMPLE 23-1B

Given the source and bridged-T network of Figure 23-8(a), what current will flow through a $(5 + j5)$-kΩ load connected to this source?

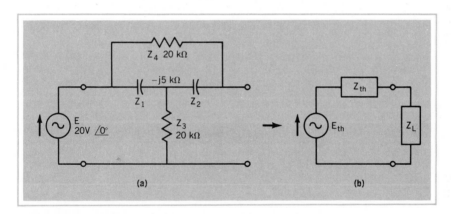

FIGURE 23-8
Circuit Diagram for Example 23-1B.

SOLUTION

Step 1. With the load removed from the original network, we can solve for the total impedance now seen by the source using series-parallel impedance techniques. To assist us in writing the necessary equations, we can rearrange the circuit of Figure 23-8(a) into the form shown in Figure 23-9(a) and again work directly in kilohms.

$$Z_T = Z_3 + \frac{Z_1(Z_4 + Z_2)}{Z_1 + (Z_4 + Z_2)} = (20 + j0) + \frac{-j5(20 - j5)}{-j5 + (20 - j5)}$$

$$= (20 + j0) + \frac{-25 - j100}{20 - j10} = (20 + j0) + (1 - j4.5)$$

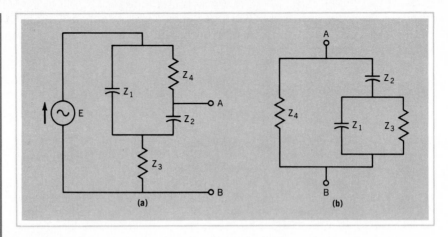

FIGURE 23-9
Determining Thévenin-Equivalent Voltage and Impedance.

$$\mathbf{Z}_T = (21 - j4.5) \text{ k}\Omega = 21.48 \text{ k}\Omega\underline{/-12.09°}$$

$$\mathbf{I}_T = \frac{\mathbf{E}}{\mathbf{Z}_T} = \frac{20 \text{ V}\underline{/0°}}{21.48 \text{ k}\Omega\underline{/-12.09°}} = 931.2 \text{ }\mu\text{A}\underline{/+12.09°}$$

From the current-divider principle, the current through $\mathbf{Z}_4$ and $\mathbf{Z}_2$ in Figure 23-9(a) is

$$\mathbf{I}_2 = \mathbf{I}_T \frac{\mathbf{Z}_1}{\mathbf{Z}_1 + (\mathbf{Z}_4 + \mathbf{Z}_2)} = \mathbf{I}_T \frac{-j5 \text{ k}\Omega}{(20 - j10)\text{l}\Omega}$$

$$= 931.2 \text{ }\mu\text{A}\underline{/+12.09°} \times \frac{5 \text{ k}\Omega\underline{/-90°}}{22.36 \text{ k}\Omega\underline{/-26.57°}}$$

$$= 208.2 \text{ }\mu\text{A}\underline{/-51.34°}$$

Step II. Calculating the voltage drops across $\mathbf{Z}_2$ and $\mathbf{Z}_3$,

$$\mathbf{V}_2 = \mathbf{I}_2 \mathbf{Z}_2 = 208.2 \text{ }\mu\text{A}\underline{/-51.34°} \times 5 \text{ k}\Omega\underline{/-90°}$$

$$= 1.041 \text{ V}\underline{/-141.34°}$$

$$\mathbf{V}_3 = \mathbf{I}_T \mathbf{Z}_3 = 931.2 \text{ }\mu\text{A}\underline{/+12.09°} \times 20 \text{ k}\Omega\underline{/0°}$$

$$= 18.624 \text{ V}\underline{/+12.09°}$$

$$\therefore \mathbf{E}_{th} = \mathbf{V}_2 + \mathbf{V}_3 = 1.041 \text{ V}\underline{/-141.34°} + 18.624 \text{ V}\underline{/+12.09°}$$

$$= (-0.813 - j0.650) + (18.211 + j3.901)$$

$$= (17.398 + j3.251) \text{ V} = 17.7 \text{ V}\underline{/+10.58°}$$

Step III.　From Figure 23-9(b), we can first find the equivalent of $\mathbf{Z}_3$ and $\mathbf{Z}_1$ in parallel. Working in kilohms,

$$\mathbf{Z}_{eq} = \frac{\mathbf{Z}_3 \times \mathbf{Z}_1}{\mathbf{Z}_3 + \mathbf{Z}_1} = \frac{20 \times (-j5)}{20 - j5} = \frac{-j100}{20 - j5}$$

$$= (1.176 - j4.706) \text{ k}\Omega$$

$$\mathbf{Z}_{th} = \frac{\mathbf{Z}_4 \times (\mathbf{Z}_2 + \mathbf{Z}_{eq})}{\mathbf{Z}_4 + (\mathbf{Z}_2 + \mathbf{Z}_{eq})} = \frac{20 \times (-j5 + 1.176 - j4.706)}{20 + (-j5 + 1.176 - j4.706)}$$

$$= \frac{23.53 - j194.12}{21.176 - j9.706} = (4.39 - j7.15) \text{ k}\Omega$$

Step IV.　Returning to the Thévenin-equivalent circuit of Figure 23-8(b),

$$\mathbf{I} = \frac{\mathbf{E}_{th}}{\mathbf{Z}_{th} + \mathbf{Z}_L} = \frac{17.7 \text{ V}\underline{/+10.58°}}{(4.39 - j7.15) \text{ k}\Omega + (5 + j5) \text{ k}\Omega}$$

$$= \frac{17.7 \text{ V}\underline{/+10.58°}}{(9.39 - j2.15) \text{ k}\Omega} = \frac{17.7 \text{ V}\underline{/+10.58°}}{9.64 \text{ k}\Omega\underline{/-12.92°}}$$

$$= \mathbf{1.84 \text{ mA}\underline{/+23.5°}}$$

Thévenin's theorem also allows us to escape the simultaneous equations in the multisource network of Example 23-2. When we remove $\mathbf{Z}_L$ in the circuit of Figures 23-3 and 23-10(a), we are left with a simple series circuit from which we can determine the Thévenin-equivalent source feeding $\mathbf{Z}_L$ in Figure 23-10(b).

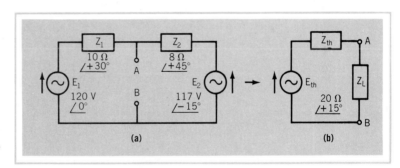

FIGURE 23-10
Circuit Diagrams for Example 23-2C.

EXAMPLE 23-2C
Two 60-Hz alternators are connected in parallel to feed a load whose impedance is $20 \text{ }\Omega\underline{/-15°}$. Alternator 1 develops an open-circuit voltage of 120 V and has an internal impedance of $10 \text{ }\Omega\underline{/+30°}$. Alternator 2 has an open-circuit voltage of 117 V which lags the voltage of alternator 1 by a constant 15°. Alternator 2 has an internal impedance of $8 \text{ }\Omega\underline{/+45°}$. Find the load current.

SOLUTION

Step I. Writing a single Kirchhoff's voltage law equation for the remaining loop in Figure 23-10(a),

$$(\mathbf{Z}_1 + \mathbf{Z}_2)\mathbf{I} = \mathbf{E}_1 - \mathbf{E}_2$$

$$\mathbf{I} = \frac{120 \text{ V}\underline{/0°} - 117 \text{ V}\underline{/-15°}}{10 \text{ }\Omega\underline{/+30°} + 8 \text{ }\Omega\underline{/+45°}}$$

$$= \frac{(120 + \text{j}0) - (113 - \text{j}30.28)}{(8.66 + \text{j}5) + (5.657 + \text{j}5.657)}$$

$$= \frac{7 + \text{j}30.28}{14.317 + \text{j}10.657} = \frac{31.078\underline{/+76.98°}}{17.85\underline{/+36.66°}} = 1.741 \text{ A}\underline{/+40.32°}$$

The voltage drop across $\mathbf{Z}_1$ is

$$\mathbf{V}_1 = \mathbf{I}\mathbf{Z}_1 = 1.741 \text{ A}\underline{/+40.32°} \times 10 \text{ }\Omega\underline{/+30°}$$

$$= 17.41 \text{ V}\underline{/+70.32°}$$

$$\mathbf{V}_{AB} = \mathbf{E}_1 - \mathbf{V}_1 = 120 \text{ V}\underline{/0°} - 17.41 \text{ V}\underline{/+70.32°}$$

$$= (120 + \text{j}0) - (5.863 + \text{j}16.393)$$

$$\therefore \mathbf{V}_{th} = \mathbf{V}_{AB} = 114.137 - \text{j}16.393 = 115.3 \text{ V}\underline{/-8.173°}$$

Step II. The Thévenin-equivalent resistance as "seen" looking back into terminals A and B in Figure 23-10(a) consists of $\mathbf{Z}_1$ and $\mathbf{Z}_2$ in parallel. Hence,

$$\mathbf{Z}_{th} = \frac{\mathbf{Z}_1 \times \mathbf{Z}_2}{\mathbf{Z}_1 + \mathbf{Z}_2} = \frac{80 \text{ }\Omega^2\underline{/+75°}}{17.85 \text{ }\Omega\underline{/+36.66°}}$$

$$= 4.482 \text{ }\Omega\underline{/+38.34°} = 3.515 + \text{j}2.780 \text{ }\Omega$$

Step III. In the Thévenin-equivalent circuit of Figure 23-10(b),

$$\mathbf{I}_L = \frac{\mathbf{E}_{th}}{\mathbf{Z}_{th} + \mathbf{Z}_L} = \frac{115.3 \text{ V}\underline{/-8.173°}}{(3.515 + \text{j}2.780) + (19.319 + \text{j}5.176) \text{ }\Omega}$$

$$= \frac{115.3 \text{ V}\underline{/-8.173°}}{22.834 + \text{j}7.956 \text{ }\Omega} = \frac{115.3 \text{ V}\underline{/-8.173°}}{24.18 \text{ }\Omega\underline{/+19.21°}}$$

$$= 4.77 \text{ A}\underline{/-27.4°}$$

23.5 NORTON'S THEOREM

In Section 8-2, we used a "black-box" technique to find the open-circuit terminal voltage and internal resistance of a two-terminal source network. Starting with the same source network, Norton showed that the black box could contain an equivalent constant-current source. The same black-box "measurements" of E_{oc} and I_{sc} yield either a constant-voltage source of E_{oc} and $R_x = E_{oc}/I_{sc}$, or a constant-current source of I_{sc} and $R_x = E_{oc}/I_{sc}$.

After we find the Thévenin-equivalent constant-voltage source for a two-terminal network of voltage sources and fixed impedances, we can carry circuit simplification one more step by converting to the Norton-equivalent constant-current source. We combine these two steps into a single procedure called Norton's theorem. For alternating-current circuits, Norton's theorem becomes

Any two-terminal network of fixed impedances and voltage sources may be replaced by a single **constant-current** *source whose current is equal to the current drawn by a short circuit across the terminals of the original network and having in* **parallel** *with the constant-current source an impedance equal to the impedance looking back into the network from the two terminals.*

Converting a constant-voltage source to an equivalent constant-current source is an application of Norton's theorem. Norton's theorem is particularly useful when a selected branch in a network is fed from several voltage sources in parallel. Hence, Norton's theorem provides an even shorter solution for Example 23-2C than Thévenin's theorem.

EXAMPLE 23-2D

Two 60-Hz alternators are connected in parallel to feed a load whose impedance is $20\ \Omega/{+15°}$. Alternator 1 develops an open-circuit voltage of 120 V and has an internal impedance of $10\ \Omega/{+30°}$. Alternator 2 has an open-circuit voltage of 117 V which lags the voltage of alternator 1 by a constant 15°. Alternator 2 has an internal impedance of $8\ \Omega/{+45°}$. Find the load current.

SOLUTION

Step I. Determine the short-circuit current of each alternator.

$$\mathbf{I}_1 = \frac{\mathbf{E}_1}{\mathbf{Z}_1} = \frac{120\ \text{V}/0°}{10\ \Omega/{+30°}} = 12\ \text{A}/{-30°}$$

$$\mathbf{I}_2 = \frac{\mathbf{E}_2}{\mathbf{Z}_2} = \frac{117\ \text{V}/{-15°}}{8\ \Omega/{+45°}} = 14.63\ \text{A}/{-60°}$$

In drawing Figure 23-11, we made sure that the tracing directions for the constant-current sources matched the tracing directions for the alternators in Figure 23-3.

FIGURE 23-11
Circuit Diagram for Examples 23-2D and 23-2E.

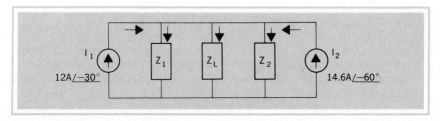

Hence, the *sum* of the two Norton currents divides among the three impedance branches in the equivalent circuit of Figure 23-11.

$$\therefore \mathbf{I}_T = \mathbf{I}_1 + \mathbf{I}_2 = 12 \text{ A}\underline{/-30°} + 14.6 \text{ A}\underline{/-60°}$$

$$= (10.39 + j6) + (7.31 - j12.67) = (17.7 - j18.67) \text{ A}$$

$$= 25.73 \text{ A}\underline{/-46.53°}$$

Step II.

$$\mathbf{Y}_T = \mathbf{Y}_1 + \mathbf{Y}_2 + \mathbf{Y}_3 = 0.1 \text{ S}\underline{/-30°} + 0.05 \text{ S}\underline{/-15°} + 0.125 \text{ S}\underline{/-45°}$$

$$= (0.0866 - j0.05) + (0.0483 - j0.0129) + (0.0884 - j0.0884)$$

$$= (0.2233 - j0.1513) \text{ S} = 0.2697 \text{ S}\underline{/-34.12°}$$

From the current-divider principle [Equation (22-10)],

$$\mathbf{I}_L = \mathbf{I}_T \frac{\mathbf{Y}_L}{\mathbf{Y}_T} = 25.73 \text{ A}\underline{/-46.53°} \times \frac{0.05 \text{ S}\underline{/-15°}}{0.2697 \text{ S}\underline{/-34.12°}}$$

$$= \mathbf{4.77 \text{ A}}\underline{\mathbf{/-27.41°}}$$

We used Example 23-3 to show that the superposition theorem applies to both voltage and current sources. However, when we encounter mixed sources in a network, as in Figures 23-6(a) and 23-12(a), it is usually simpler to convert to either one or the other. In the network of Figure 23-12(a), if we convert the constant-current source into a constant-voltage source, we have to include $\mathbf{Z}_2$ and the milliammeter in the transformation. Hence, the meter current in the new circuit would not be the same as in the original. But we can replace the constant-voltage source on the left of Figure 23-12(a) with its Norton-equivalent constant-current source, as shown in Figure 23-12(b).

FIGURE 23-12
Circuit Diagrams for Example 23-3A.

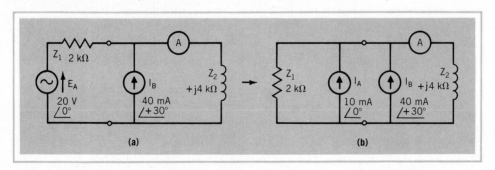

EXAMPLE 23-3A

Assuming that the ac milliammeter in the network of Figure 23-12(a) has zero impedance, calculate its reading.

SOLUTION

Step I. The Norton-equivalent source on the left of Figure 23-12(b) has the *same* internal impedance $\mathbf{Z}_1$ and

$$\mathbf{I}_A = \frac{\mathbf{E}_A}{\mathbf{Z}_1} = \frac{20 \text{ V}\underline{/0°}}{2 \text{ k}\Omega\underline{/0°}} = 10 \text{ mA}\underline{/0°}$$

Step II. Writing a single Kirchhoff's current-law equation for the network of Figure 23-12(b),

$$(\mathbf{Y}_1 + \mathbf{Y}_2)\mathbf{V} = \mathbf{I}_A + \mathbf{I}_B$$

$$\mathbf{V} = \frac{10 \text{ mA}\underline{/0°} + 40 \text{ mA}\underline{/+30°}}{500 \text{ }\mu\text{S}\underline{/0°} + 250 \text{ }\mu\text{S}\underline{/-90°}}$$

$$= \frac{(10 + j0) + (34.64 + j20)}{(500 + j0) + (0 - j250)} = \frac{44.64 + j20}{500 - j250}$$

$$\therefore \mathbf{V} = \frac{48.92 \text{ mA}}{559 \text{ }\mu\text{S}} = 87.5 \text{ V}$$

Step III.

$$I_M = \frac{V}{Z_2} = \frac{87.5 \text{ V}}{4 \text{ k}\Omega} = 21.88 \text{ mA}$$

23-6 NODAL ANALYSIS

Adapting Kirchhoff's current law to alternating-current circuits, we can state that

The phasor sum of the currents leaving any junction point or node in an ac circuit must equal the phasor sum of the currents entering that node.

The nodal equivalent of the loop procedure requires us to mark the directions of all branch currents leaving and entering independent nodes, as shown in Figure 23-11. From inspection, we can then write a Kirchhoff's current-law equation.

$$\mathbf{I}_{Z_1} + \mathbf{I}_{Z_L} + \mathbf{I}_{Z_2} = \mathbf{I}_1 + \mathbf{I}_2$$

We then replace impedance currents with $\mathbf{I} = \mathbf{V} \times \mathbf{Y}$, the product of the voltage across each impedance and its admittance. Writing the Kirchhoff's current-law equa-

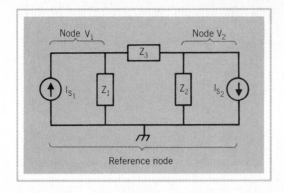

FIGURE 23-13
Circuit Diagram for Writing Nodal Equations.

tion is quite straightforward in the circuit of Figure 23-11, since all currents leaving the upper node are impedance currents and all currents entering the node are source currents. This is not the case in the circuit of Figure 23-13, and it is a bit more difficult to write suitable nodal equations by inspection. Consequently, we usually use a nodal analysis *format* which is the *dual* of mesh equations.

For the nodal analysis format, we can restate Kirchhoff's current law in the form,

> *The phasor sum of the* impedance *currents leaving any node in an ac circuit must equal the phasor sum of the* source *currents entering that node.*

Regardless of the actual directions of source currents, Kirchhoff's current-law nodal equations always have the format,

$$\mathbf{I}_{Z_1} + \mathbf{I}_{Z_2} + \mathbf{I}_{Z_3} + \cdots = \mathbf{I}_{S_1} + \mathbf{I}_{S_2} + \cdots \qquad (23\text{-}3)$$

If an *actual* source current *leaves* the node, we enter it as a *negative* quantity in Equation (23-3). The signs for impedance currents are taken care of automatically by the rules for substituting voltages in the **VY** products which replace each impedance current.

1. For nodal equations to form the *dual* of mesh equations, we start by converting all voltage sources to equivalent constant-current sources to provide the source-current terms for the right-hand side of Equation (23-3).

2. We select a point in the network which is common to as many branches as possible and label it "reference node" in the circuit diagram. We label all other independent nodes with a **V** symbol and a suitable subscript. (The dependent variables in nodal analysis are the voltages between independent nodes and the reference node.) We can redraw our diagram to show junction point nodes, as in Figure 8-21, or mark the nodes with colored brackets, as in Figure 23-13.

3. In writing a Kirchhoff's current-law equation for node $\mathbf{V}_1$ in Figure 23-13, the voltage across an impedance connected between that node and the reference node ($\mathbf{Z}_1$ in Figure 23-13), is simply the node voltage $\mathbf{V}_1$ (with respect to the reference node). For an impedance connected between the node in question and an adjacent independent node ($\mathbf{Z}_3$ in Figure 23-13), the voltage across such an

impedance always has the form $(V_x - V_A)$ where V_x is the node voltage in question and V_A is the adjacent node voltage (with respect to reference node). Following this format automatically takes care of the sign for the impedance current *leaving* the node in question. We can now write Kirchhoff's current-law equations for the two independent nodes in the circuit of Figure 23-13.

For node V_1,

$$V_1 Y_1 + (V_1 - V_2)Y_3 = I_{S_1} \tag{1}$$

For node V_2,

$$V_2 Y_2 + (V_2 - V_1)Y_3 = -I_{S_2} \tag{2}$$

Collecting the terms into matrix form,

$$(Y_1 + Y_3)V_1 - \qquad Y_3 V_2 = I_{S_1} \tag{23-4}$$
$$-Y_3 V_1 + (Y_2 + Y_3)V_2 = -I_{S_2} \tag{23-5}$$

4. By noting the structure of Equations (23-4) and (23-5), we can skip Step 3 and write nodal equations directly in matrix form. The left-hand side of nodal equations consists of one positive term and any number of negative terms. The positive term is the node voltage in question and its coefficient is the phasor sum of *all* admittances connected to that node. Each negative term is an adjacent node voltage with the admittance between that adjacent node and the node in question as its coefficient. The right-hand side of the nodal equation is the phasor sum of all source currents *entering* the node in question.

EXAMPLE 23-2E

Two 60-Hz alternators are connected in parallel to feed a load whose impedance is $20\ \Omega/{+15°}$. Alternator 1 develops an open-circuit voltage of 120 V and has an internal impedance of $10\ \Omega/{+30°}$. Alternator 2 has an open-circuit voltage of 117 V which lags the voltage of alternator 1 by a constant 15°. Alternator 2 has an internal impedance of $8\ \Omega/{+45°}$. Find the load current.

SOLUTION

Step I. Replacing each source by its equivalent constant-current source and maintaining the same direction for the currents produces the equivalent circuit of Figure 23-11 where

$$I_1 = \frac{E_1}{Z_1} = \frac{120\ V/\underline{0°}}{10\ \Omega/{+30°}} = 12\ A/\underline{-30°}$$

and
$$I_2 = \frac{E_2}{Z_2} = \frac{117\ V/\underline{-15°}}{8\ \Omega/{+45°}} = 14.63\ A/\underline{-60°}$$

Step II. If we take the lower conductor in the circuit diagram of Figure 23-11 as the reference node, the upper conductor represents a single independent node. Hence we need only one nodal equation.

$$(\mathbf{Y}_1 + \mathbf{Y}_L + \mathbf{Y}_2)\mathbf{V} = \mathbf{I}_1 + \mathbf{I}_2$$

from which

$$\mathbf{V} = \frac{12 \text{ A}\underline{/-30°} + 14.63 \text{ A}\underline{/-60°}}{0.1 \text{ S}\underline{/-30°} + 0.05 \text{ S}\underline{/-15°} + 0.125 \text{ S}\underline{/-45°}}$$

$$= \frac{(10.39 - j6) + (7.31 - j12.67)}{(0.0866 - j0.05) + (0.0483 - j0.0129) + (0.0884 - j0.0884)}$$

$$= \frac{17.7 - j18.67}{0.2233 - j0.1513} = \frac{25.73 \text{ A}\underline{/-46.53°}}{0.2697 \text{ S}\underline{/-34.12°}}$$

$$= 95.4 \text{ V}\underline{/-12.41°}$$

Step III.

$$\mathbf{I}_L = \frac{\mathbf{V}}{\mathbf{Z}_L} = \frac{95.4 \text{ V}\underline{/-12.41°}}{20 \text{ Ω}\underline{/+15°}} = \mathbf{4.77 \text{ A}\underline{/-27.41°}}$$

There are much simpler methods than nodal analysis for solving the network of Figure 23-13. If we convert both current sources to equivalent constant-voltage sources, we have a simple single-loop circuit which we can easily solve by Ohm's law. However, as a further example, we shall complete the solution by nodal analysis.

EXAMPLE 23-5

Given the following parameters for the network of Figure 23-13, determine the magnitude of the current in Z_3 by nodal analysis.

$$\mathbf{I}_{S_1} = 50 \text{ mA}\underline{/0°} \qquad \mathbf{I}_{S_2} = 100 \text{ mA}\underline{/+90°}$$

$$\mathbf{Z}_1 = +j50 \text{ Ω} \qquad \mathbf{Z}_2 = -j25 \text{ Ω} \qquad \mathbf{Z}_3 = 50\text{-Ω resistor}$$

SOLUTION

Step I. Converting impedances to admittances,

$$\mathbf{Y}_1 = -j20 \text{ mS} \qquad \mathbf{Y}_2 = +j40 \text{ mS} \qquad \mathbf{Y}_3 = 20\text{-mS conductance}$$

(These are the values we mark on the circuit diagram of Figure 23-13.)

$$\mathbf{Y}_1 + \mathbf{Y}_3 = (20 - j20)\text{mS} = 28.3 \text{ mS}\underline{/-45°}$$

$$\mathbf{Y}_2 + \mathbf{Y}_3 = (20 + j40)\text{mS} = 44.7 \text{ mS}\underline{/+63.4°}$$

Step II. The nodal equation for node $\mathbf{V}_1$ becomes

$$28.3\underline{/-45^\circ}\ \mathbf{V}_1 - 20\underline{/0^\circ}\ \mathbf{V}_2 = 50\underline{/0^\circ}{}^\dagger$$

The nodal equation for node $\mathbf{V}_2$ becomes

$$44.7\underline{/+63.4^\circ}\ \mathbf{V}_2 - 20\underline{/0^\circ}\ \mathbf{V}_1 = -100\underline{/+90^\circ}{}^\dagger$$

Step III.

$$\mathbf{V}_1 = \dfrac{\begin{vmatrix} 50\underline{/0^\circ} & -20\underline{/0^\circ} \\ -100\underline{/+90^\circ} & 44.7\underline{/+63.4^\circ} \end{vmatrix}}{\begin{vmatrix} 28.3\underline{/-45^\circ} & -20\underline{/0^\circ} \\ -20\underline{/0^\circ} & 44.7\underline{/+63.4^\circ} \end{vmatrix}} = \dfrac{2236\underline{/+63.4^\circ} - 2000\underline{/+90^\circ}}{1265\underline{/+18.4^\circ} - 400\underline{/0^\circ}}$$

$$= \dfrac{1000\underline{/0^\circ}}{894\underline{/+26.6^\circ}} = 1.12\ \text{V}\underline{/-26.6^\circ}$$

$$\mathbf{V}_2 = \dfrac{\begin{vmatrix} 28.3\underline{/-45^\circ} & 50\underline{/0^\circ} \\ -20\underline{/0^\circ} & -100\underline{/+90^\circ} \end{vmatrix}}{D} = \dfrac{-2828\underline{/+45^\circ} + 1000\underline{/0^\circ}}{D}$$

$$= \dfrac{2236\underline{/-116.6^\circ}}{894\underline{/+26.6^\circ}} = 2.5\ \text{V}\underline{/-143.1^\circ}$$

Step IV.

$$\mathbf{V}_1 - \mathbf{V}_2 = (-2 - j1.5) - (1 - j0.5) = (-3 - j1)\ \text{V}$$
$$= 3.16\ \text{V}\ \underline{/-161.6^\circ}$$

Step V.

$$I_{Z_3} = \frac{\mathbf{V}_1 - \mathbf{V}_2}{Z_3} = \frac{3.16\ \text{V}}{50\ \Omega} = 63\ \text{mA}$$

If we attempt to solve the bridged-T network of Example 23-1 by nodal analysis, we would discover that there are *three* independent nodes. Therefore, we would require three nodal equations requiring third-order determinants. Second, we would find that we cannot convert the voltage source to an equivalent constant-current source, since the internal impedance of the given source is zero.

---

†Since milliamperes/millisiemens $=$ volts, the nodal equations balance without repeated use of the prefix *milli* when entering admittance and current coefficients into the calculator.

23-7 DELTA-WYE TRANSFORMATION

By substituting the impedance letter symbol for the resistance symbol, we can apply to ac circuits the equations we developed in Section 9-6 for converting a given delta network into an equivalent wye network, and vice versa. Since the derivation of these equations is the same as for resistance networks, we shall simply state the equations at this point. In using these transformation equations, we must remember that we must treat each $\mathbf{Z}$ symbol as a *phasor* quantity.

$$\mathbf{Z}_A = \frac{\mathbf{Z}_Y \mathbf{Z}_Z}{\mathbf{Z}_X + \mathbf{Z}_Y + \mathbf{Z}_Z} \tag{23-6}$$

$$\mathbf{Z}_B = \frac{\mathbf{Z}_X \mathbf{Z}_Z}{\mathbf{Z}_X + \mathbf{Z}_Y + \mathbf{Z}_Z} \tag{23-7}$$

$$\mathbf{Z}_C = \frac{\mathbf{Z}_X \mathbf{Z}_Y}{\mathbf{Z}_X + \mathbf{Z}_Y + \mathbf{Z}_Z} \tag{23-8}$$

$$\mathbf{Z}_X = \frac{\mathbf{Z}_A \mathbf{Z}_B + \mathbf{Z}_B \mathbf{Z}_C + \mathbf{Z}_C \mathbf{Z}_A}{\mathbf{Z}_A} \tag{23-9}$$

$$\mathbf{Z}_Y = \frac{\mathbf{Z}_A \mathbf{Z}_B + \mathbf{Z}_B \mathbf{Z}_C + \mathbf{Z}_C \mathbf{Z}_A}{\mathbf{Z}_B} \tag{23-10}$$

$$\mathbf{Z}_Z = \frac{\mathbf{Z}_A \mathbf{Z}_B + \mathbf{Z}_B \mathbf{Z}_C + \mathbf{Z}_C \mathbf{Z}_A}{\mathbf{Z}_C} \tag{23-11}$$

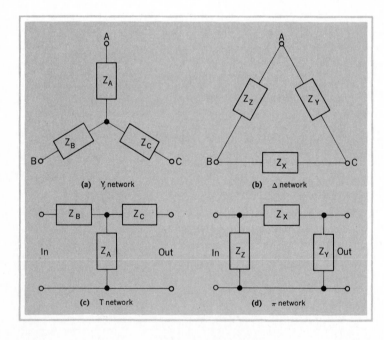

FIGURE 23-14
Delta-Wye Transformation.

(a) Y network
(b) Δ network
(c) T network
(d) π network

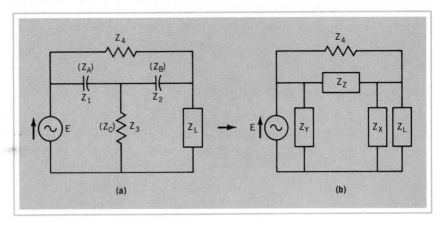

FIGURE 23-15
Circuit Diagram for Example 23-1C.

In Section 9-6, we developed an alternate form for the wye to delta transformation equations based on the *dual* format of the delta to wye equations. This alternate form is particularly useful in ac circuits, where we must use *phasor* algebra, since they involve fewer computations.

$$\mathbf{Z}_X = \mathbf{Z}_B \mathbf{Z}_C (\mathbf{Y}_A + \mathbf{Y}_B + \mathbf{Y}_C) \qquad (23\text{-}12)$$

$$\mathbf{Z}_Y = \mathbf{Z}_A \mathbf{Z}_C (\mathbf{Y}_A + \mathbf{Y}_B + \mathbf{Y}_C) \qquad (23\text{-}13)$$

$$\mathbf{Z}_Z = \mathbf{Z}_A \mathbf{Z}_B (\mathbf{Y}_A + \mathbf{Y}_B + \mathbf{Y}_C) \qquad (23\text{-}14)$$

These transformation equations were developed originally for use in three-phase ac power-distribution networks having the wye and delta forms of Figure 23-14(a) and (b). However, many of the four-terminal coupling networks encountered in electronic circuitry have a *common* input and output terminal (which is often grounded). In such coupling circuits, we can often identify the wye network in the form of the T-network of Figure 23-14(c). Similarly, the π-network of Figure 23-14(d) is essentially the same as the delta network of Figure 23-14(b). Therefore, we may also use the Equations (23-6) to (23-14) for T to π transformations, and vice versa. This transformation provides an additional means of solving the bridged-T network of Example 23-1. If we replace the T-network consisting of $\mathbf{Z}_1$, $\mathbf{Z}_2$, and $\mathbf{Z}_3$ in Figure 23-15(a) with the equivalent π-network, we can solve the resulting circuit of Figure 23-15(b) as a series-parallel impedance network.

EXAMPLE 23-1C
Given $\mathbf{E} = 20$ V $\underline{/0°}$, $\mathbf{Z}_1 = \mathbf{Z}_2 = -j5$ kΩ, $\mathbf{Z}_3 = \mathbf{Z}_4 = (20 + j0)$ kΩ, and $\mathbf{Z}_L = (5 + j5)$ kΩ for the bridged-T network of Figure 23-15(a), find the current through $\mathbf{Z}_L$.

SOLUTION
    Step I. Labeling $\mathbf{Z}_1$, $\mathbf{Z}_2$, and $\mathbf{Z}_3$ as $\mathbf{Z}_A$, $\mathbf{Z}_B$, and $\mathbf{Z}_C$, respectively, in Figure

23-15(a), we can then label the π-network of Figure 23-15(b) with the appropriate subscripts. These must correspond according to the exact relationships shown in Figure 23-14; that is, Z_X does *not* connect to the same terminal as Z_A, and so forth around the network.

Note that we need not solve for Z_Y, since it is directly across the source and does not affect the current through the load.

Step II. Working out the common numerator for Equations (23-9) and (23-11) and again working directly in kilohms,

$$Z_A Z_B + Z_B Z_C + Z_C Z_A = -j5 \times (-j5) + (-j5) \times 20 + 20 \times (-j5)$$
$$= -25 + (-j100) + (-j100) = -25 - j200$$
$$= 201.56 \,\underline{/-97.12°}$$

$$\therefore Z_X = \frac{Z_A Z_B + Z_B Z_C + Z_C Z_A}{Z_A} = \frac{201.56 \,\underline{/-97.12°}}{5 \,\underline{/-90°}} = 40.31 \text{ k}\Omega \,\underline{/-7.12°}$$

$$\text{and } Z_Z = \frac{Z_A Z_B + Z_B Z_C + Z_C Z_A}{Z_C} = \frac{201.56 \,\underline{/-97.12°}}{20 \,\underline{/0°}} = 10.08 \text{ k}\Omega \,\underline{/-97.12°}$$

Note that Z_Z appears to have a *negative* resistance component. Therefore, in this case, we cannot construct an *actual* equivalent π-network, using only passive circuit elements. Nevertheless, we can still use the pencil-and-paper transformation as a means of solving a network problem.

Step III. Excluding Z_Y, the load on the source in Figure 23-15(b) is

$$Z_T = \frac{Z_4 \times Z_Z}{Z_4 + Z_Z} + \frac{Z_X \times Z_L}{Z_X + Z_L}$$

$$= \frac{20 \,\underline{/0°} \times 10.08 \,\underline{/-97.12°}}{(20 + j0) + (-1.25 - j10.0)} + \frac{40.31 \,\underline{/-7.12°} \times 7.07 \,\underline{/+45°}}{(40 - j5) + (5 + j5)}$$

$$= \frac{201.56 \,\underline{/-97.12°}}{18.75 - j10.0} + \frac{284.99 \,\underline{/+37.88°}}{45 + j0}$$

$$= \frac{201.56 \,\underline{/-97.12°}}{21.25 \,\underline{/-28.07°}} + \frac{284.99 \,\underline{/+37.88°}}{45 \,\underline{/0°}}$$

$$= 9.49 \,\underline{/-69.05°} + 6.33 \,\underline{/+37.88°}$$

$$= (3.39 - j8.86) + (5.0 + j3.89) = 8.39 - j4.97$$

$$= 9.75 \text{ k}\Omega \,\underline{/-30.64°}$$

Step IV. The current drawn from the source by these four impedances is

$$\mathbf{I_T} = \frac{\mathbf{E}}{\mathbf{Z_T}} = \frac{20 \text{ V } \underline{/0°}}{9.75 \text{ k}\Omega \ \underline{/-30.64°}} = 2.05 \text{ mA } \underline{/+30.64°}$$

Step V. Using the current-divider principle,

$$\mathbf{I_L} = \mathbf{I_T}\frac{\mathbf{Z}_X}{\mathbf{Z}_X + \mathbf{Z}_L} = 2.05 \text{ mA } \underline{/+30.64°} \times \frac{40.31 \text{ k}\Omega \ \underline{/-7.12°}}{45 \text{ k}\Omega \ \underline{/0°}}$$

$$= \mathbf{1.84 \text{ mA }} \underline{/+23.5°}$$

23-8 ALTERNATING-CURRENT BRIDGES

If we replace the resistance symbols of the basic Wheatstone bridge with impedance symbols, we obtain the general ac bridge circuit of Figure 23-16. A brief consideration of ac bridges at this point will serve the dual purpose of acquainting us with the method by which we can obtain precision measurement of ac circuit parameters, and of providing us with a source of practical ac network problems.

For the null indicator in Figure 23-16 to read zero, the voltage across its terminals must be zero. For this to be so, the voltage across $\mathbf{Z}_1$ and $\mathbf{Z}_3$ must be exactly the same both in magnitude and phase. Therefore,

$$\mathbf{V}_1 = \mathbf{V}_3 \quad \text{and} \quad \mathbf{I}_1\mathbf{Z}_1 = \mathbf{I}_3\mathbf{Z}_3$$

Because no current flows through the null indicator when the bridge is balanced,

$$\mathbf{I}_1 = \frac{\mathbf{E}}{\mathbf{Z}_1 + \mathbf{Z}_2} \quad \text{and} \quad \mathbf{I}_3 = \frac{\mathbf{E}}{\mathbf{Z}_3 + \mathbf{Z}_4}$$

Hence,

$$\frac{\mathbf{E}\mathbf{Z}_1}{\mathbf{Z}_1 + \mathbf{Z}_2} = \frac{\mathbf{E}\mathbf{Z}_3}{\mathbf{Z}_3 + \mathbf{Z}_4}$$

from which

$$\mathbf{Z}_1\mathbf{Z}_4 = \mathbf{Z}_2\mathbf{Z}_3 \tag{23-15}$$

FIGURE 23-16
General Form of an AC Bridge.

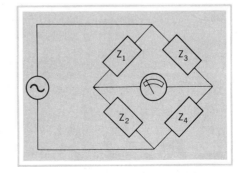

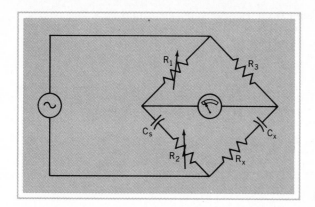

FIGURE 23-17
Capacitance-Comparison Bridge.

This is known as the **general bridge equation** and must hold true for any special form of an ac bridge when it is properly balanced to give zero voltage across the null indicator.

The bridge arrangement of Figure 23-17 is called a **capacitance comparison bridge** since the unknown capacitance C_x and its loss resistance R_x are to be balanced against the standard capacitor C_s. In this particular case, the terms of the general bridge equation become

$$\mathbf{Z}_1 = R_1 \qquad \mathbf{Z}_2 = R_2 - j\frac{1}{\omega C_s}$$

$$\mathbf{Z}_3 = R_3 \qquad \mathbf{Z}_4 = R_x - j\frac{1}{\omega C_x}$$

Substituting in Equation (23.15) gives

$$R_1 R_x - j\frac{R_1}{\omega C_x} = R_2 R_3 - j\frac{R_3}{\omega C_s}$$

As an impedance diagram will show, altering the magnitude of a reactance can have no effect on the magnitude of the total resistance of the circuit because these two quantities are at right angles. Therefore, the j term on the left of the above equation must equal the j term on the right of the equation. Similarly, the two horizontal or *real* components must be equal. Hence, we can separate equations expressed in rectangular coordinates into two equations.

$$R_1 R_x = R_2 R_3$$

from which
$$R_x = \frac{R_2 R_3}{R_1} \tag{23-16}$$

and
$$\frac{R_1}{\omega C_x} = \frac{R_3}{\omega C_s}$$

from which
$$C_x = C_s \frac{R_1}{R_3} \tag{23-17}$$

Note that in deriving Equations (23-16) and (23-17), the angular velocity term disappeared. Therefore, the accuracy of the capacitance comparison bridge is not dependent on the accuracy of the frequency of the source. None of the ac bridges is dependent on the magnitude of the applied voltage for an accurate reading when balanced, since this term disappeared in deriving Equation (23-15).

Although we can construct a similar inductance comparison bridge to determine accurately an unknown inductance, standard inductors are expensive and bulky. The **Maxwell bridge** circuit of Figure 23-18 allows us to balance the bridge by using a standard capacitance for comparison purposes. In this case, Branch 1 contains parallel components. Thus, it is more convenient to identify the four arms as

$$\mathbf{Y}_1 = \frac{1}{R_1} + j\omega C \qquad \mathbf{Z}_2 = R_2$$

$$\mathbf{Z}_3 = R_3 \qquad\qquad \mathbf{Z}_4 = R_x + j\omega L_x$$

Substituting $1/\mathbf{Y}_1$ for $\mathbf{Z}_1$ in Equation (23-15) gives

$$\mathbf{Z}_4 = \mathbf{Y}_1 \mathbf{Z}_2 \mathbf{Z}_3$$

Substituting the given values for the Maxwell bridge gives

$$R_x + j\omega L_x = \frac{R_2 R_3}{R_1} + j\omega C_s R_2 R_3$$

Again separating the real and j terms into separate equations,

$$R_x = \frac{R_2 R_3}{R_1} \tag{23-18}$$

$$L_x = C_s R_2 R_3 \tag{23-19}$$

The Maxwell bridge is best suited to coils in which R_x is an appreciable fraction of ωL_x. As Equation (23-18) indicates, if R_x is very small, R_1 must be very large in

FIGURE 23-18
Maxwell Bridge.

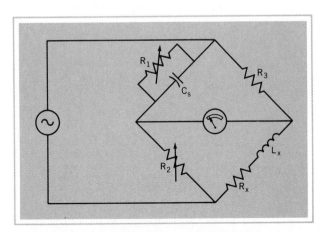

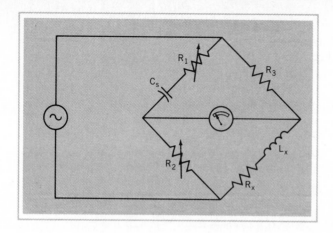

FIGURE 23-19
Hay Bridge.

order to obtain balance. For such coils, the Hay bridge of Figure 23-19 is preferred even though the equations turn out to be frequency dependent.

In this case,

$$\mathbf{Z}_1 = R_1 - j\frac{1}{\omega C_s} \qquad \mathbf{Z}_2 = R_2$$

$$\mathbf{Z}_3 = R_3 \qquad\qquad \mathbf{Z}_4 = R_x + j\omega L_x$$

Substituting in Equation (23-15),

$$\left(R_1 - j\frac{1}{\omega C_s}\right)(R_x + j\omega L_x) = R_2 R_3$$

$$R_1 R_x + j\omega L_x R_1 - \frac{jR_x}{\omega C_s} + \frac{L_x}{C_s} = R_2 R_3$$

Separating the real and j terms,

$$R_1 R_x + \frac{L_x}{C_s} = R_2 R_3 \tag{1}$$

$$\omega L_x R_1 = \frac{R_x}{\omega C_s} \tag{2}$$

From Equation (2), $$R_x = \omega^2 L_x C_s R_1 \tag{3}$$

Substituting in Equation (1) gives

$$\omega^2 L_x C_s R_1^2 + \frac{L_x}{C_s} = R_2 R_3$$

from which $$L_x = \frac{C_s R_2 R_3}{1 + \omega^2 C_s^2 R_1^2} \tag{23-20}$$

Substituting Equation (23-20) in Equation (3) gives

$$R_x = \frac{\omega^2 C_s^2 R_1 R_2 R_3}{1 + \omega^2 C_s^2 R_1^2} \tag{23-21}$$

EXAMPLE 23-6
The Maxwell bridge of Figure 23-18 balances when $C_s = 0.01$ μF, $R_1 = 470$ kΩ, $R_2 = 5.1$ kΩ, and $R_3 = 100$ kΩ. Determine the inductance and resistance of the inductor being measured.

SOLUTION

$$L_x = C_s R_2 R_3 = 0.01 \ \mu\text{F} \times 5.1 \ \text{k}\Omega \times 100 \ \text{k}\Omega = 5.1 \ \text{H}$$

$$R_x = \frac{R_2 R_3}{R_1} = \frac{5.1 \ \text{k}\Omega \times 100 \ \text{k}\Omega}{470 \ \text{k}\Omega} = 1085 \ \Omega$$

PROBLEMS

Problems 23-1 to 23-20 are drill problems in network solution. Solve odd-numbered problems by one method and the even-numbered problem for the same network by a different method.

23-1. Determine the source current drawn from each source in Figure 23-20(a).

23-2. Determine the voltage across the 50-Ω resistor in Figure 23-20(a).

23-3. Determine the capacitor current in Figure 23-20(b).

23-4. Determine the voltage drop across each of the 100-Ω resistors in Figure 23-20(b).

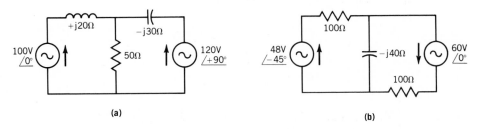

(a)   (b)

FIGURE 23-20

23-5. Determine the voltage drop across the $(10 - j10)$-Ω branch in Figure 23-21(a).

23-6. Determine both source currents in Figure 23-21(a).

23-7. Determine all three branch currents in Figure 23-21(b).

23-8. Reverse the polarity of the 20 V $\underline{/+90°}$ source and determine all three branch currents in Figure 23-21(b).

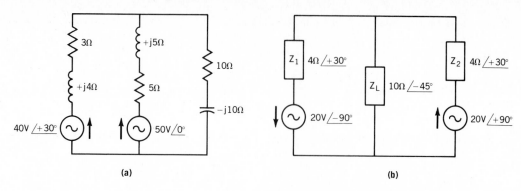

(a)   (b)

FIGURE   23-21

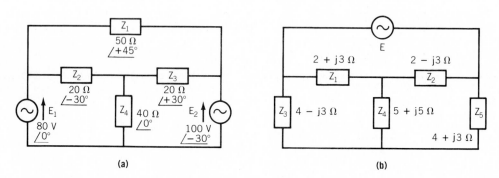

(a)   (b)

FIGURE   23-22

23-9. Determine the currents in Z_1 and Z_4 in Figure 23-22(a).

23-10. Determine the currents in Z_2 and Z_3 in Figure 23-22(a).

23-11. If the source voltage in Figure 23-22(b) is 10 V $\underline{/0°}$, determine the magnitude of the source current.

23-12. If the source voltage in Figure 23-22(b) is 25 V $\underline{/-45°}$, determine the magnitude of the voltage across Z_4.

23-13. Determine the magnitude of the current through the inductor in Figure 23-23(a).

23-14. Determine the magnitude of the voltage across the capacitor in Figure 23-23(a).

23-15. Determine the magnitude of the voltage across each constant-current source in Figure 23-23(b).

23-16. Determine the magnitude of the current through the capacitor in Figure 23-23(b).

23-17. Determine the load current in Figure 23-24(a).

23-18. Determine the load voltage in Figure 23-24(a) if the direction of the 100 V $\underline{/+30°}$ source is reversed.

23-19. Determine the voltage drop across the 5-kΩ resistor in Figure 23-24(b).

23-20. Determine the magnitude of the two capacitor currents in Figure 23-24(b) if the direction of the 200-V source is reversed.

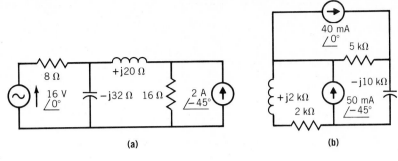

(a)   (b)

FIGURE 23-23

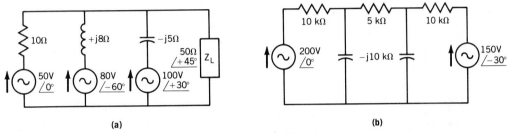

(a)   (b)

FIGURE 23-24

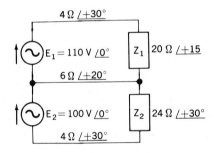

FIGURE 23-25

Use the method you think is most appropriate in solving the remaining applications networks.

23-21. Find the current in the common neutral lead of the three-wire, single-phase system shown in Figure 23-25.

23-22. The three-wire, single-phase system of Problem 23-21 can be converted into a two-phase system if we make $\mathbf{E}_2 = 110$ V $\underline{/90°}$. Find the current in $\mathbf{Z}_1$ under these circumstances.

23-23. Figure 23-26 shows one of the equivalent circuits for a transistor. It includes a current-controlled voltage source whose magnitude equals the mutual resistance r_m of the transistor $\times$ the current through r_e. If the signal source has an open-circuit voltage $E_g = 2$ V rms, and an internal resistance of $R_g = 200$ Ω, find the rms output voltage across a load resistor $R_L = 2$ kΩ.

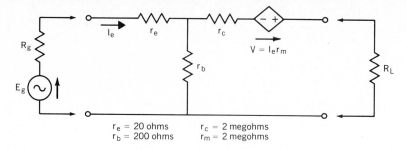

$r_e = 20$ ohms $r_c = 2$ megohms
$r_b = 200$ ohms $r_m = 2$ megohms

FIGURE 23-26
Transistor Equivalent Circuit.

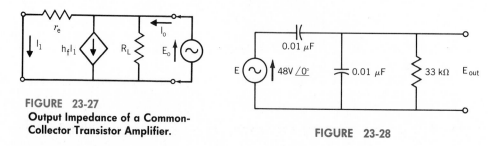

FIGURE 23-27
Output Impedance of a Common-Collector Transistor Amplifier.

FIGURE 23-28

23-24. Figure 23-27 shows an equivalent output circuit of a transistor amplifier. To determine its output impedance, we short circuit the input terminals and feed an ac signal E_o back into the output terminals and calculate the current I_o drawn from this source. From Ohm's law, the output impedance is simply $Z_o = E_o/I_o$. Set up an equation for the output impedance in terms of the circuit parameters given in Figure 23-27.

23-25. Find the Thévenin-equivalent source for the source shown in Figure 23-28 when **E** has a frequency of 400 Hz.

23-26. Find the Norton-equivalent source at 400 Hz for the source in Figure 23-28.

23-27. Using Thévenin's theorem, determine an equivalent source for the network of Figure 23-28 that will be applicable at *any* frequency. (*Hint:* Do not include the 33-kΩ resistor in the Thévenin transformation.)

23-28. Determine the parameters for a Norton-equivalent source that will be applicable at any frequency for the network of Figure 23-28. Compare with the solution for Problem 23-26.

23-29. A 26.5-μF capacitor is connected in series with a 100-Ω resistor to a 120-V 60-Hz source. What value of resistance must be connected in parallel with the capacitor to make the current in the capacitor have a magnitude of 0.5 A?

23-30. At what frequency will the output voltage of the network in Figure 23-29 be in phase with the input voltage?

23-31. A voltage amplifier and its *CR* output coupling network can be represented by the source shown in Figure 23-30. $\mathbf{E} = 10$ V $\underline{/180°}$; $r_a = 20$ kΩ; $R_b = 80$ kΩ; $C_c = 0.02$ μF; $R_g = 470$ kΩ. Using 1 kHz as a reference frequency, determine by how many degrees the phase of the output voltage will shift as the frequency decreases to 100 Hz.

23-32. An amplifier and its coupling network can be represented by the signal source shown in Figure 23-31. At what frequency will the output voltage drop to 70.7% of its magnitude at 1 kHz as a result of the increase in the reactance of the coupling capacitor as the frequency decreases?

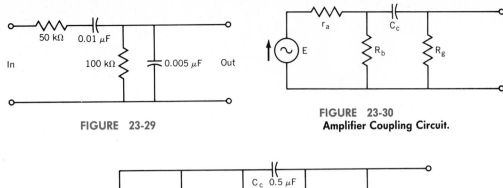

FIGURE 23-29

FIGURE 23-30
Amplifier Coupling Circuit.

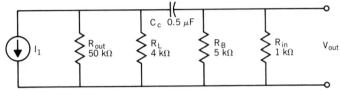

FIGURE 23-31
Equivalent Circuit of a Transistor Amplifier Coupling Circuit.

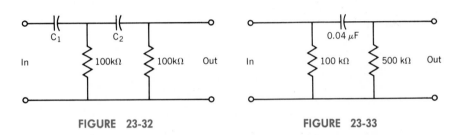

FIGURE 23-32

FIGURE 23-33

23-33. If the two capacitors in the circuit of Figure 23-32 are equal, what must their capacitance be for the open-circuit output voltage of this network to be 90° out of phase with the input voltage at 60 Hz?

23-34. Solve Example 23-2 by combining Thévenin's theorem with the superposition theorem; that is, for each of the circuit diagrams in Figure 23-5, find the *component* current through Z_L by Thévenin's theorem rather than by series-parallel impedance network techniques.

23-35. Find the equivalent T-network for the circuit of Figure 23-33 when the input frequency is 400 Hz.

23-36. Determine the equivalent π-network for the transistor equivalent circuit of Figure 23-26.

23-37. If the ammeter in Figure 23-34 has negligible internal resistance, solve for the ammeter reading, using the wye-delta transformation equations.

FIGURE 23-34

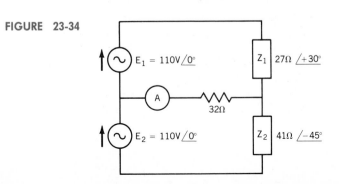

609

23-38. Find the source current in Problem 23-11 using wye-delta transformation.

23-39. The bridge circuit of Figure 23-16 balances when $Z_1 = 470\ \Omega\underline{/0°}$, Z_2 is a 0.01μF capacitor in series with an 860-kΩ resistor, $Z_3 = 5\ k\Omega\underline{/0°}$, and $f = 1$ kHz. Determine the components that constitute Z_4.

23-40. The bridge circuit of Figure 23-16 balances when $Z_2 = 10\ k\Omega\underline{/0°}$, $Z_3 = 5.6\ k\Omega\underline{/0°}$, Z_4 consists of a 0.01-μF capacitor in parallel with a 1.23-MΩ resistor, and $f = 1$ kHz. Determine the components that constitute Z_1.

23-41. The bridge circuit of Figure 23-16 balances when $Z_1 = 10\ k\Omega\underline{/0°}$, Z_3 is a 0.01-μF capacitor in series with a 27-Ω resistor, $Z_4 = 2700\ \Omega\underline{/0°}$, and $f = 1$ kHz. Determine the value of Z_2 in rectangular coordinates.

23-42. The bridge circuit of Figure 23-16 balances when $Z_1 = 475\ \Omega\underline{/0°}$, $Z_2 = 1780\ \Omega\underline{/+75°}$, $Z_3 = 1000\ \Omega\underline{/0°}$, and $f = 1$ kHz. Determine the components that constitute Z_4.

23-43. The bridge circuit of Figure 23-16 balances when $Z_1 = 200\ \Omega\underline{/0°}$, $Z_2 = 100\ \Omega\underline{/0°}$, $Z_3 = 20\ \Omega\underline{/0°}$, and Z_4 consists of a 200-pF capacitor in series with a coil whose inductance and resistance are unknown. $f = 1$ MHz. Determine L_x and R_x.

23-44. In the bridge circuit of Figure 23-16, $Z_1 = 200\ \Omega\underline{/-45°}$, $Z_2 = 1000\ \Omega\underline{/0°}$, $Z_3 = 10\ k\Omega\underline{/0°}$, and $Z_4 = 500\ \Omega\underline{/+60°}$. The null indicator has an impedance of 5 k$\Omega\underline{/0°}$. What is the total current drawn from a 50-V 1-kHz source?

REVIEW QUESTIONS

23-1. In writing loop equations for ac networks with more than one source, why is it necessary to identify which terminal of the source is which?

23-2. Compare the relative merits of the six solutions for Example 23-2 and select the method which you think is most appropriate. State your reasons for your choice.

23-3. What are the advantages and disadvantages of the superposition theorem as a general network theorem?

23-4. Thévenin's theorem is widely used in electronic circuit-analysis techniques. What is the main feature of Thévenin's theorem that would account for this popularity?

23-5. To be quite precise in stating Thévenin's theorem, we should have specified a network of fixed *linear, bilateral* impedances. In an ac system, what is the significance of (a) a *linear* impedance, and (b) a *bilateral* impedance?

23-6. We can solve Example 23-2 in one fell swoop by using the algebraic statement of Millman's theorem,

$$V_L = \frac{E_1 Y_1 + E_2 Y_2}{Y_1 + Y_L + Y_2} \tag{23-22}$$

Show that Equation (23-22) is valid for the network of Figure 23-2. (*Hint:* The solution for Example 23-2D is essentially a procedure based on Millman's theorem.)

23-7. One factor governing the choice between loop or nodal equations in solving an impedance network is the number of simultaneous equations required by each method. In some networks (such as Example 23-1), the same number of equations is required for each method. The choice then depends on whether we are asked to find one or more branch *voltages* or one or more branch *currents*. What would be your choice on this second basis?

610

23-8. Under what circumstances can we apply the equations that we developed for wye-delta transformations in *three*-terminal networks to *four*-terminal coupling networks?

23-9. Why do we not need to take the impedance of the null indicator into account in a balanced ac bridge circuit?

23-10. Why does a variation in the magnitude of the source voltage not affect the accuracy of a balanced bridge measurement?

23-11. How will the sensitivity of the null indicator affect the accuracy of an ac bridge measurement?

23-12. Why can we split the general bridge Equation (23-15) into two separate equations in deriving equations for the unknown components in the various ac bridge circuits?

23-13. After assembling a bridge in the form shown in Figure 23-35, we find that it cannot be balanced. Show why by deriving equations for L_x and R_x from the general bridge equation.

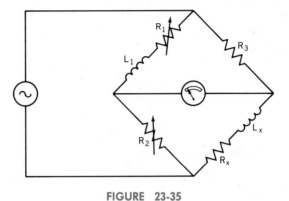

FIGURE 23-35

23-14. Why is the Hay bridge more suitable than the Maxwell bridge for measuring coils with a very low resistance?

23-15. Derive equations for C_x and R_x in the Schering bridge of Figure 23-36.

FIGURE 23-36
Schering Bridge.

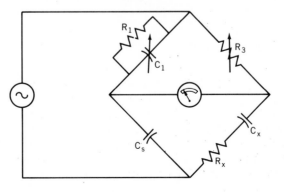

With electronic calculators, we no longer need to learn how to determine sines and tangents from trigonometry tables. Similarly, if personal computers and circuit-analysis software become as commonplace as calculators, we can solve complex electrical networks without learning the circuit-analysis techniques of this chapter. We don't need to know how a computer goes about circuit-analysis calculations. All we need to know is how to enter the requested network parameters.

Since the primary objective of this course is to understand why electric circuits must perform in a predictable manner, we have avoided circuit-analysis software in working with examples in this chapter. Now that we are familiar with circuit-analysis techniques, we are able to appreciate the computer's advantage of speed and accuracy.

*23-16. If you have access to a computer and circuit-analysis software, solve some of the problems in this chapter using the available circuit-analysis program.

*23-17. If you have access to a personal computer but not to circuit-analysis programs, produce your own programs for storage on diskette (or cassette) to perform the time-consuming and error-prone tasks of solving second- and third-order simultaneous equations in which both coefficients and unknowns are phasor quantities.

*23-18. Produce for storage on diskette (or cassette) programs to carry out delta-wye and wye-delta transformations involving impedance phasors.

24

RESONANCE

24-1 EFFECT OF VARYING FREQUENCY IN A SERIES *RLC* CIRCUIT

In most of the ac network examples we have considered, the frequency of the source has remained constant. This is the case in electric power systems. The standard power-line frequency in North America is 60 Hz, although there are a few areas where the frequency is 25 Hz. The standard frequency in Europe is 50 Hz. To reduce the weight of iron required for the magnetic circuits of motors and transformers, aircraft electrical systems use frequencies of 400 Hz or 1 kHz. In radio circuitry, however, frequency is an important variable. Therefore, we must now devote our attention to the behavior of electric circuits in which we can consider that the rms voltage remains constant but the frequency varies.

If the circuit connected to a variable-frequency source contains only resistance, the frequency of the source will have negligible effect on the magnitude or phase of the current drawn from the source, since $I = E/R$. But if we connect an inductor to a variable-frequency source, because $X_L = 2\pi f L$, its inductive reactance is directly proportional to the frequency. Similarly, the reactance of a capacitor *decreases* as the frequency increases.

In the circuit of Figure 24-1(a), resistance, inductance, and capacitance are all connected in series to a variable-frequency source. When the frequency is quite low, the inductive reactance is quite small and the capacitive reactance is quite large. Consequently, the net reactance is a large capacitive reactance, much greater in magnitude than the resistance, and thus the total impedance is practically equal to $-jX_C$. As we increase the frequency, not only does the capacitive reactance become

613

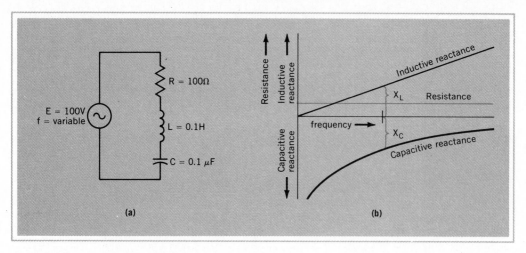

FIGURE 24-1
Series *RLC* Circuit Connected to a Variable-Frequency Source.

smaller, but the inductive reactance increases. We can show the effect of increasing frequency on the three components of the circuit of Figure 24-1(a) by plotting graphs of their resistance or reactance as a function of frequency, as shown in Figure 24-1(b). As we have noted, resistance is not frequency-sensitive. Hence, its graph is a horizontal line. And since X_L is *directly* proportional to frequency, its graph is a straight line through the origin with a slope proportional to the inductance of the coil. To show that we *subtract* X_C from X_L in determining the net reactance, we plot X_C along the negative Y-axis of the graph. Since X_C is *inversely* proportional to the frequency, its graph has the reciprocal-type curve as shown.

As we increase the frequency, the net reactance ($X_L - X_C$) decreases until we reach a frequency, marked by the brackets in Figure 24-1(b), at which $X_L = X_C$. The net reactance at this frequency is zero, and, therefore, the total impedance of the circuit is simply the resistance of R. As we increase the frequency past this point, the inductive reactance exceeds the capacitive reactance, and the total impedance becomes inductive, approaching the condition of being practically the same as the inductive reactance of the circuit when the frequency is quite high.

To examine the effect of varying frequency on the impedance of the series *RLC* circuit of Figure 24-1(a) in a more quantitative manner, we can program a computer to make repetitive calculations by increasing frequency in steps over a range of approximately 160 Hz to 16 kHz.[†] Table 24-1 contains selected calculations from such a computer printout. When the frequency of the source is 159 Hz, the capacitive reactance of the series circuit is 100 times as great as the inductive reactance, and the series *RLC* circuit appears to consist of a 100-ohm resistor in a series with a *capacitance* having a reactance of 9900 ohms.

As we noted in Figure 24-1(b), There is a frequency at which the inductive and capacitive reactances are numerically equal and cancel. This frequency is known as

---

[†]See the computer program we used in Section 6-7.

TABLE 24-1

EFFECT OF VARYING FREQUENCY IN A SERIES *RLC* CIRCUIT

f (Hz)	R (Ω)	X_L (Ω)	X_C (Ω)	$X_L - X_C$ (Ω)	$Z = \sqrt{R^2 + X^2}$	ϕ Arctan X/R	$I = E/Z$ (A)
159	100	100	10 K	$-j9900$	9900	$-89.4°$	0.0101
660	100	415	2415	$-j2000$	2002	$-87.2°$	.0499
985	100	618	1618	$-j1000$	1005	$-84.3°$	.0995
1245	100	781	1281	$-j500$	509	$-78.7°$	.1965
1440	100	905	1105	$-j200$	223	$-63.4°$	.448
1515	100	952	1052	$-j100$	141	$-45°$	.707
1590	100	1000	1000	0	100	$0°$	1.00
1675	100	1052	952	$+j100$	141	$+45°$	.707
1760	100	1105	905	$+j200$	223	$+63.4°$	.448
2040	100	1281	781	$+j500$	509	$+78.7°$	.1965
2575	100	1618	618	$+j1000$	1005	$+84.3°$	.0995
3840	100	2415	415	$+j2000$	2002	$+87.3°$	.0499
15 900	100	10 000	100	$+j9900$	9900	$+89.4°$	0.0101

the **resonant frequency** of the series *RLC* circuit and is identified by the letter symbol f_r. For the circuit of Figure 24-1(a), the resonant frequency is 1590 Hz. For a frequency of 15 900 Hz, Table 24-1 shows that X_L now greatly exceeds X_C and the series *RLC* circuit behaves as if it consisted of a 100-ohm resistor in series with an *inductance* having a reactance of 9900 ohms.

We can display the data of Table 24-1 graphically for each frequency by preparing a set of impedance diagrams similar to Figure 20-9. Figure 24-2 shows the impedance diagrams for the series *RLC* circuit of Figure 24-1(a) for the resonant frequency of 1590 Hz and for two specific frequencies: (a) when the frequency of the source is *lower* than the resonant frequency and (c) when the frequency is *higher* than the resonant frequency.

FIGURE 24-2

Impedance of a Series *RLC* Circuit when (a) $f < f_r$, (b) $f = f_r$, and (c) $f > f_r$.

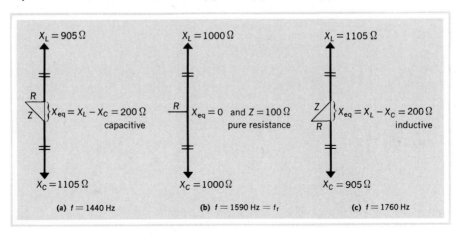

We would require thirteen separate impedance diagrams to show all the situations given in Table 24-1. Although these impedance diagrams show the impedance relationships for each specific frequency, they do not really show clearly the manner in which the impedance of the series RLC circuit varies as we vary the frequency of the source. Hence, we adopt the type of graph we used in Figure 24-1(b) where frequency is plotted as the independent variable on the X-axis of the graph. We can program the computer to display or print out such a graph directly from the calculations it made to produce Table 24-1. In plotting graphs of the last three columns of Table 24-1, we have used a *logarithmic* scale for the X-axis in Figure 24-3. Since $X_L - X_C$ is not a linear function of frequency, choosing a logarithmic frequency scale in both Table 24-1 and Figure 24-3 produces a graph with symmetrical "skirts."

If the source voltage remains at 100 volts rms in spite of the frequency variation, the magnitude of the current is inversely proportional to the impedance. Therefore, whereas the impedance becomes minimum at 1590 hertz, the current will be maximum at this frequency. As we have noted, at very low frequencies, the impedance of the series RLC circuit of Figure 24-1(a) is almost completely capacitive reactance, and thus the angle of the impedance is almost $-90°$. Consequently, at these low frequencies, the current *leads* the applied voltage by almost 90°. At 1590 hertz, the net reactance becomes zero and the total impedance is simply the resistance of the circuit with a 0° angle. Then, at very high frequencies, the impedance becomes almost completely inductive reactance, under which circumstances the current *lags* the ap-

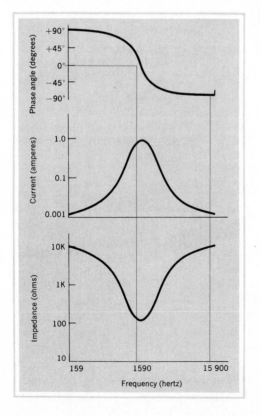

FIGURE 24-3

Effect of Varying Frequency on Impedance, Current, and Phase Angle of the Current in a Series *RLC* Circuit.

plied voltage by almost 90°. If we wish to show the effect of frequency on both the magnitude and phase of the current, we can prepare a circle diagram similar to Figure 22-20.

24-2 SERIES RESONANCE

In the series RLC circuit of Figure 24-1(a), when the frequency of the source is 1590 Hz:

1. The inductive reactance and the capacitive reactance of the circuit are equal.
2. Therefore, the impedance is minimum and is equal to the resistance of the circuit.
3. Hence, the current in the circuit is maximum and is equal to E/R.
4. The current through the circuit is in phase with the applied voltage.

This condition is known as **resonance.** The series RLC circuit of Figure 24-1(a) is called a **resonant circuit,** and the frequency at which resonance occurs is called the **resonant frequency** of the circuit.

The letter symbol for resonant frequency is f_r.

Since at resonance $X_L = X_C$, then $2\pi f_r L = 1/2\pi f_r C$, from which

$$f_r = \frac{1}{2\pi \sqrt{LC}} \tag{24-1}$$

where f_r is in hertz, L is in henrys, and C is in farads.

Although we have considered only the case of fixed values of L and C with the frequency varying to obtain resonance, rearranging Equation (24-1) will show that we can bring a given RLC circuit into resonance at a certain frequency by varying either the inductance or capacitance of the circuit. Inductance is usually varied by changing the reluctance of a coil's magnetic circuit. This is accomplished by moving a powdered-iron core on a threaded shaft in or out of the coil. Variable inductance is usually used for preset tuning adjustments. Continuously variable **tuning** is usually accomplished by rotating one set of plates of an air-dielectric capacitor between a set of fixed plates. More recently, electronic tuning varies the capacitance of a resonant circuit by varying the dc reverse-bias voltage applied to a specially-designed *pn*-junction called a **varactor.**

We note from Equation (24-1) that the resistance of the series-resonant circuit has no bearing on the resonant frequency. The resistance governs the minimum impedance of the circuit which occurs at resonance. As a result, the resistance governs the steepness of the "skirts" of the resonance curve. In the series-resonant circuit of Figure 24-4(a), the inductance and capacitance are the same as in the circuit of Figure 24-1(a). Therefore, its resonant frequency is still 1590 hertz. However, the resistance has been reduced from 100 ohms to 50 ohms. If we prepare a table similar to Table

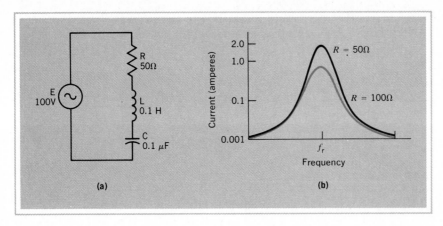

FIGURE 24-4
Effect of Resistance on the Shape of the Resonance Curve.

24-1 for the circuit of Figure 24-4(a), we can then plot the resonance curve shown in Figure 24-4(b). Off resonance, the total impedance has not been affected very much by the change in resistance. But at the resonant frequency, the impedance is only half that of the circuit of Figure 24-1(a), and thus the circuit current at resonance is twice as great.

We also note from Equation (24-1) that as long as the product of L and C is constant, the resonant frequency of a series-resonant circuit will remain constant. In the circuit of Figure 24-5(a) we have doubled the inductance and cut the capacitance in half. Therefore, the LC product and the resonant frequency are unchanged. However, we have increased the L/C *ratio* by a factor of 4 : 1. This means that at resonance, $X_L = X_C = 2000$ ohms instead of 1000 ohms, as in the circuit of Figure 24-1(a). Hence, at a frequency of 1515 Hz, $X_L - X_C = 200$ ohms instead of 100

FIGURE 24-5
Effect of *L/C* Ratio on the Shape of the Resonance Curve.

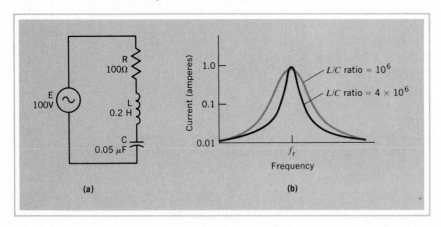

ohms, as in the circuit of Figure 24-1(a). This increase in the L/C ratio will, therefore, steepen the skirts of the resonance curve, as shown in Figure 24-5(b).

EXAMPLE 24-1

A coil in a tuned circuit in a radio receiver has an inductance of 300 μH and a resistance of 15 Ω. What value of capacitance must be connected in series with the coil for the circuit to be series resonant at 840 kHz?

SOLUTION

$$X_L = 2\pi fL = 2 \times \pi \times 840 \text{ kHz} \times 300 \text{ } \mu\text{H} = 1583 \text{ } \Omega$$

At resonance, $$X_C = X_L = 1583 \text{ } \Omega$$

$$\therefore C = \frac{1}{2 \times \pi \times 840 \text{ kHz} \times 1583 \text{ } \Omega} = 120 \text{ pF}$$

24-3 Q OF RESONANT CIRCUITS

Decreasing the resistance in a series resonant circuit and increasing the L/C ratio both have the effect of steepening the skirts of the resonance curve. Perhaps we can combine both these variables into a single factor relating to resonant circuits. As we noted when we considered power factor correction in Section 21-7, in an ac circuit containing both inductance and capacitance, the capacitance *stores* energy while the inductance *returns* energy stored in its magnetic field, and vice versa. Therefore, ac circuits containing both inductance and capacitance *exchange* reactive power between the inductance and capacitance. Hence, the source is required to supply only the difference between the reactive power of the inductance and the reactive power of the capacitance. This accounts for the net reactance of a series circuit being the *difference* between the inductive and capacitive reactance, and the reactive voltage of a series circuit being the *difference* between V_L and V_C.

At resonance, the reactive power of the inductance and capacitance are equal, and the source has to supply only the active power required by the resistance of the circuit.

The ratio between the reactive power of either the inductance or the capacitance at resonance and active power of a resonant circuit is called the Q factor of that resonant circuit.

At this point, standard letter symbols becomes rather confusing since the letter Q represents both reactive power and quality factor (and also electric charge). Therefore, in stating the definition of **quality factor** (more commonly called Q factor) in equation form, we shall revert to the secondary symbol for reactive power P_q.

$$\therefore Q = \frac{P_q}{P} \tag{24-2}$$

Since the resistance in a resonant circuit is usually the resistance associated with the coil, we often refer to the Q *of the coil*.

Because $\qquad$ $P_q = I^2 X_L$ and $P = I^2 R$, then $Q = \dfrac{I^2 X_L}{I^2 R}$.

But the current is common to all components in a series circuit.

$$\therefore Q = \frac{X_L}{R} = \frac{\omega L}{R} \qquad (24\text{-}3)$$

Equation (24-3) provides us with a simpler form for the calculation of Q than Equation (24-2). It is sometimes given as a definition of Q.

Since $\omega L = 2\pi f L$ and since $f_r = 1/2\pi\sqrt{LC}$,

$$\therefore Q = \frac{1}{R}\sqrt{\frac{L}{C}} \qquad (24\text{-}4)$$

The Q factor of a resonant circuit is, therefore, the single factor which takes into account the effect of both the resistance and the L/C ratio on the shape of a resonance curve. The higher the Q, the steeper the skirts of the resonance curve.

EXAMPLE 24-2

A series resonant circuit consists of a 50-μH inductance whose resistance is 5 Ω, and a 200-pF capacitor. What is its Q?

SOLUTION 1

$$Q = \frac{1}{R}\sqrt{\frac{L}{C}} = \frac{1}{5}\sqrt{\frac{50\ \mu\text{H}}{200\ \text{pF}}} = \frac{1}{5} \times 500 = \mathbf{100}$$

SOLUTION 2

Since it is quite likely that we shall need to know the resonant frequency of a circuit, this solution may be quicker in the long run.

$$f_r = \frac{1}{2\pi\sqrt{LC}} = \frac{1}{2 \times \pi \times \sqrt{50\ \mu\text{H} \times 200\ \text{pF}}} = 1.592\ \text{MHz}$$

$$X_L = 2\pi f L = 2 \times \pi \times 1.592\ \text{MHz} \times 50\ \mu\text{H} = 500\ \Omega$$

$$Q = \frac{X_L}{R} = \frac{500}{5} = \mathbf{100}$$

24-4 RESONANT RISE OF VOLTAGE

We have not as yet considered the effect on the voltage distribution in a series-resonant circuit of varying the frequency of the source. With respect to the circuit of Figure 24-1(a), we can prepare Table 24-2.

TABLE 24-2

RESONANT RISE OF VOLTAGE			
$f(Hz)$	$V_R = IR$	$V_L = IX_L$	$V_C = IX_C$
159	1.01 V	1.01 V	101 V
660	4.99	20.7	120
985	9.95	61.5	161
1245	19.65	153.5	252
1440	44.8	406	495
1515	70.7	675	747
1590	100	1000	1000
1675	70.7	747	675
1760	44.8	495	406
2040	19.65	252	153.5
2575	9.95	161	61.5
3840	4.99	120	20.7
15 900	1.01	101	1.01

Plotting the data from the right-hand column of Table 24-2, we obtain the graph of Figure 24-6, which shows the manner in which the voltage across the capacitor varies as we vary the frequency of the source. At frequencies well below the resonant frequency of the circuit, the voltage across the capacitor is practically the same as the applied voltage, since the total impedance of tne circuit at these frequencies is mainly the reactance of the capacitor. As the frequency approaches the resonant frequency, even though the reactance of the capacitor is decreasing, the increasing current due to the decreasing total impedance causes the voltage across the capacitance to exceed the applied voltage. Because V_L and V_C are equal but 180° out of phase at resonance, V_R is the same as the applied voltage, and V_C is much greater than the applied voltage. As we increase the frequency to many times the resonant frequency, the impedance of the circuit is practically the same as the inductive reactance of the circuit and the voltage across the capacitance becomes very small.

Since capacitive reactance decreases as frequency increases, the product IX_C reaches a maximum at a frequency just slightly *lower* than the resonant frequency of the circuit, as shown in Figure 24-6. This frequency difference is usually so small that we can neglect it. In Table 24-2, the frequency readings are not close enough to show the exact frequency at which V_C is maximum. We can plot a similar graph showing the

FIGURE 24-6
Effect of Varying Frequency on the Voltage across the Capacitance of a Series-Resonant Circuit.

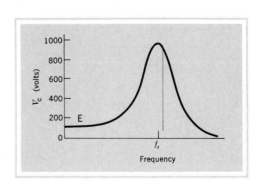

rise in the voltage across the inductance of the series-resonant circuit as we vary the frequency of the source through resonance. In this case, the exact frequency at which V_L is a maximum is slightly higher than the resonant frequency of the circuit. This increase in the voltage across the capacitance (and the inductance) of a series-resonant circuit at resonance, as shown in Figure 24-6, is called **resonant rise of voltage.**

$$V_C = IX_C$$

But $I = E/Z$ and at resonance, $Z = R$.

$$\therefore V_C = \frac{EX_C}{R}$$

But $X_C/R = Q$. Therefore, the resonant rise of voltage in a series-resonant circuit is

$$V_C = QE \tag{24-5}$$

where E is the source voltage at the resonant frequency. Because Q is usually much greater than unity, a voltage much greater than the source voltage appears across the capacitance and inductance of a series-resonant circuit at resonance. However, since the voltage drops across the inductance and capacitance are equal and opposite at resonance and the voltage drop across the resistance is equal to the applied voltage, the phasor sum of the voltage drops does equal the applied voltage, as required by Kirchhoff's voltage law. Equation (24-5) is sometimes given as a definition of Q.

Resonant rise of voltage can be put to practical use in radio receivers as a means of *increasing* the input voltage of a desired signal to which the circuit is tuned. Figure 24-7(a) shows a typical antenna-input circuit of a radio receiver. Antenna current in the primary coil induces a voltage in the tuned secondary coil. At first glance, it appears that the coil and tuning capacitor are in *parallel*. Actually, the induced voltage acts in *series* with the coil and capacitor, as shown by the equivalent circuit of Figure

FIGURE 24-7
Resonant Rise of Voltage in a Radio Receiver Input Circuit: (a) Actual Circuit; (b) Equivalent Circuit.

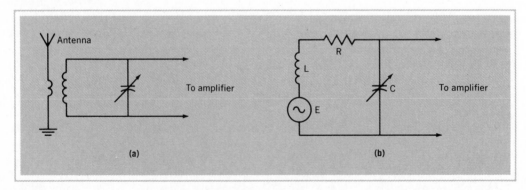

PART V   IMPEDANCE NETWORKS

24-7(b). At resonance, the voltage across the tuning capacitor (which becomes the input to the amplifier) is Q times the voltage induced into the coil by the antenna signal. As indicated by Equation (24-5), the higher the Q of the tuned circuit, the greater its **sensitivity.**

24-5 SELECTIVITY

Since the capacitance of the circuit of Figure 24-7 is variable, we can *select* the frequency at which the resonant rise of voltage provides us with a voltage gain of Q times the input voltage. The shape of the resonance curve of Figure 24-6 is such that frequencies close to the resonant frequency receive just about as great an increase in voltage as the resonant frequency itself. Therefore, we can say that a resonant circuit selects a certain *band* of frequencies. This is desirable in radio communications since, according to Hartley's bandwidth law, any signal carrying intelligence requires a band of frequencies whose bandwidth is directly proportional to the rate at which intelligence is transmitted.

However, the skirts of the resonance curve are not steep enough for us to be able to draw a definite boundary between those frequencies which will be selected and those frequencies which will be rejected. We must arbitrarily select a point on the resonance curve on either side of the resonant frequency, which we shall consider to be the practical boundary between selected and rejected frequencies.

Since the current is common to all components in a series circuit, we return to the graph of current versus frequency in a series-resonant circuit to determine the bandwidth of a tuned circuit. As the frequency of the source is decreased from the resonant frequency of a given tuned circuit, the impedance of the circuit will increase and the current will decrease, as shown in Figure 24-8(a). We consider the frequency f_1 at which the current is diminished to $1/\sqrt{2}$, or 0.707, of its value at resonance as the limit of the band of frequencies below the resonant frequency which the tuned circuit will accept. Similarly, the frequency f_2 in Figure 24-8(a), at which the current is again diminished to 0.707 of the maximum current at resonance, is considered to

FIGURE 24-8
Bandwidth of a Resonant Circuit.

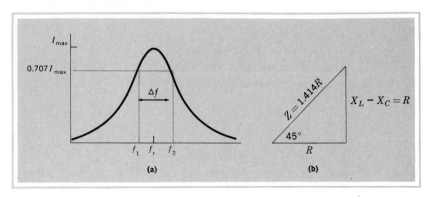

be the limit of the band of frequencies above the resonant frequency which the tuned circuit will accept. Therefore, the total bandwidth is

$$\Delta f = f_2 - f_1 \qquad (24\text{-}6)$$

Since $P = I^2R$, the active power input to the tuned circuit at frequencies f_1 and f_2 is

$$P = (0.707I_m)^2 R = 0.5P_m$$

Consequently, the frequencies f_1 and f_2 on a resonance curve are referred to as the **half-power points.**

For the current to drop to 0.707 of the current at resonance, the impedance of the tuned circuit must be $1/0.707$ or 1.414 times the impedance at resonance. At resonance $Z = R$; hence, the circuit impedance at the half-power frequencies must be $1.414R$. As indicated by Figure 24-8(b), the net reactance at the half-power points must be equal to the resistance of the tuned circuit for this to be the case. Since inductive reactance is directly proportional to frequency and capacitive reactance is inversely proportional to frequency, at frequency f_2, the value of X_L must have increased by an amount approximately equal to $0.5R$ and X_C must have decreased by an amount approximately equal to $0.5R$. From this, it follows that

$$2\pi f_2 L - 2\pi f_1 L = R, \quad \text{or} \quad 2\pi L(f_2 - f_1) = R$$

$$\therefore \Delta f = \frac{R}{2\pi L} \qquad (24\text{-}7)$$

Dividing both sides of Equation (24-7) by f_r,

$$\frac{\Delta f}{f_r} = \frac{R}{2\pi f_r L} = \frac{R}{X_L} = \frac{1}{Q}$$

$$\therefore \Delta f = \frac{f_r}{Q} \qquad (24\text{-}8)$$

Therefore, an increase in the Q of a resonant circuit not only increases the **sensitivity** by increasing the resonant rise of voltage, but it also increases the **selectivity** by decreasing the bandwidth of the tuned circuit.

EXAMPLE 24-3
(a) Determine the bandwidth of the resonant circuit of Example 24-2.
(b) If this circuit is connected to a 40-μV signal source, what is the voltage across the tuning capacitor at resonance?

SOLUTION

(a)
$$\Delta f = \frac{f_r}{Q} = \frac{1.592 \text{ MHz}}{100} = 15.92 \text{ kHz}$$

(b)
$$V_C = QE = 100 \times 40 \ \mu V = 4 \text{ mV}$$

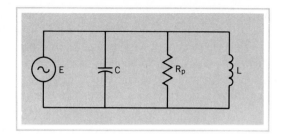

FIGURE 24-9
Theoretical Parallel-Resonant Circuit.

24-6 THEORETICAL PARALLEL-RESONANT CIRCUIT

The theoretical parallel-resonant circuit of Figure 24-9 consists of an ideal inductor, a capacitor, and a *high* resistance, all connected in parallel to a voltage source. Since the current in the inductive branch lags the applied voltage by 90° and the current in the capacitive branch leads the applied voltage by 90°,

$$I_\mathrm{T} = \sqrt{I_{R_\mathrm{p}}^2 + (I_C - I_L)^2}$$

Because $I_L = E/X_L$ and $I_C = E/X_C$, the current in the inductive branch will decrease, and the current in the capacitive branch will increase, as the frequency of the source increases. Therefore, there will be a particular frequency at which $I_L = I_C$. At this frequency, the current drawn from the source will be a minimum and will be equal to the current in the resistance branch of the theoretical parallel-resonant circuit.

For the parallel *RLC* circuit of Figure 24-9, we can plot a series of admittance diagrams similar to Figure 20-12 or a series of current phasor diagrams similar to Figure 20-11. Using current phasor diagrams, at the resonant frequency as shown in Figure 24-10(b), the currents in the inductance and capacitance branches cancel out

FIGURE   24-10
Total Current in a Parallel *RLC* Circuit when (a) $f < f_r$, (b) $f = f_r$, and (c) $f > f_r$.

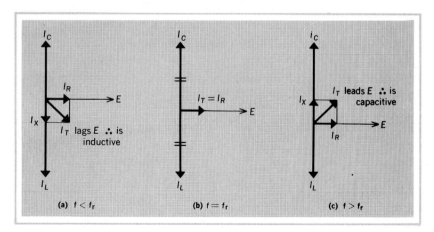

and the total current is minimum and is equal to the current in the R_p branch. At a frequency below the resonant frequency, X_L is less than X_C and the total current is greater than I_R and is *inductive,* as shown in Figure 24-10(a). Similarly, as shown in Figure 24-10(c), at a frequency above the resonant frequency, the total current in the theoretical parallel-resonant circuit is *capacitive.* Note that the parallel-resonant circuit is the exact opposite of the series-resonant circuit in this respect.

If we carefully plot a graph of total current versus frequency for the circuit of Figure 24-9, we note (lower graph of Figure 24-11) that we produce a resonance curve which has the same shape but is inverted in comparison with the current versus frequency graph for a series-resonant circuit. But, by now, we should be accustomed to the "similar but opposite" characteristics of parallel and series circuits. For a constant source voltage, total current is directly proportional to the total-admittance of the circuit. Hence, the total-admittance graph will have the same shape as the total-current resonance curve of Figure 24-11. And since the equivalent impedance of the theoretical parallel-resonant circuit is the reciprocal of the total admittance, it has the shape of the upper curve in Figure 24-11. Again, we should note that the symmetrical appearance of these resonance curves is due to using a *logarithmic* frequency scale, as in Figure 24-3. Because the resonance graphs for a theoretical parallel-resonant circuit are the reciprocal curves of the equivalent series-resonant circuit, parallel resonance is sometimes called **antiresonance.**

Since the current in the inductance branch and the current in the capacitance branch are equal at resonance, it follows that the resonant frequency of the theoretical parallel-resonant circuit is, as in the case of series resonance,

$$f_r = \frac{1}{2\pi\sqrt{LC}} \tag{24-1}$$

Although the *total* current in a parallel-resonant circuit is minimum at resonance,

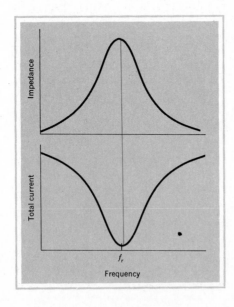

FIGURE 24-11
Parallel-Resonance Curves.

PART V   IMPEDANCE NETWORKS

if the reactances of the inductive and capacitive branches are considerably less than the resistance of the high-resistance branch, the current in the inductance and capacitance is many times as great as the line current. Remembering that the capacitance is charging when the magnetic field around the inductance is collapsing, and vice versa, we can consider this current as being the medium which carries the energy back and forth between the capacitance and the inductance. Therefore, at resonance, we can say that a considerable **tank current** flows in the closed loop consisting of the inductance and capacitance branches, while the resistance branch draws current from the source.

Again, we are concerned with the relationship between the resistance and reactance of the resonant circuit. We define the Q factor of the parallel-resonant circuit as the ratio between the reactive power of either the inductance or the capacitance at resonance and the active power of the resistance. Hence,

$$Q = \frac{P_q}{P} = \frac{E^2/X_L}{E^2/R_p}$$

$$Q = \frac{R_p}{X_L} \tag{24-9}$$

Note that this results in a ratio between resistance and reactance which is inverted as compared with the Q of a series-resonant circuit. But note also that the *parallel* resistance is quite high, 100 000 ohms or so, whereas the *series* resistance is only a few ohms.

At resonance, the tank current in the theoretical parallel-resonant circuit will be the same as I_L or I_C, and the total current will be the same as I_{R_p}. Since $E = I_R R_p = I_L X_L$,

$$\text{tank current} = Q \times \text{line current} \tag{24-10}$$

We can think of Equation (24-10) as representing a **resonant rise of tank current** similar to the resonant rise of voltage in a series-resonant circuit. However, when we examine the current graph of Figure 24-11, we find that there is a *decrease* in the current drawn from the source rather than a resonant rise in tank current. The problem with the circuit of Figure 24-9 is that we have neglected the internal resistance of the source. When we use a parallel-resonant circuit as the output load for a transistor amplifier, we find that the comparatively high internal resistance of the transistor makes it possible for us to translate the resonant rise of impedance of the parallel-resonant circuit into a resonant rise in output voltage.

In the equivalent circuit of Figure 24-12(a), from the voltage-divider principle.

$$\mathbf{V}_{\text{out}} = \mathbf{E} \times \frac{\mathbf{Z}_p}{R_{\text{int}} + \mathbf{Z}_p} \tag{7-1}$$

When the frequency of the signal source is considerably off resonance, the internal resistance of the source is much greater than the impedance of the parallel-resonant circuit. Consequently, most of the signal voltage generated by the constant-voltage source is lost across the internal resistance of the source. But, at the resonant fre-

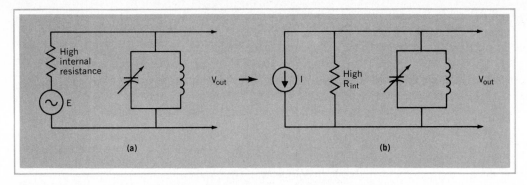

FIGURE 24-12
Resonant Rise in Output Voltage due to Resonant Rise of Impedance of Parallel-Resonant Circuit.

quency, the impedance of the parallel-resonant circuit is usually greater than the internal impedance of the signal source. Hence, the output voltage in Figure 24-12(a) is much greater at the resonant frequency than for frequencies off resonance.

In preparing equivalent circuits for transistor amplifiers, we usually find it more convenient to convert the constant-voltage source to its equivalent constant-current source of Figure 24-12(b). From Ohm's law,

$$\mathbf{V}_{out} = \mathbf{IZ}_T$$

And since $\mathbf{Z}_T$ consists of the internal resistance of the source and the impedance of the tuned circuit in parallel, V_{out} will be greatest at resonance when Z_p is maximum (Figure 24-11) and will approach the open-circuit voltage of the constant-voltage source of Figure 24-12. Off resonance, the output voltage will be considerably less than the open-circuit voltage, as shown in Figure 24-13.

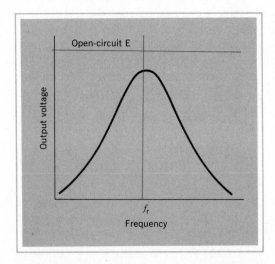

FIGURE 24-13
Output Voltage across a Parallel-Resonant Circuit when the Source Has Appreciable Internal Resistance.

24-7 PRACTICAL PARALLEL-RESONANT CIRCUITS

The losses in the dielectric of a practical capacitor are usually so small that we can neglect them. If we are required to take these losses into consideration, we can represent them as a high resistance in *parallel* with the capacitance. But, since capacitor losses are so small in comparison to the resistance of practical inductors, we can omit such parallel resistance from the equivalent circuit of Figure 24-14. However, we must represent the resistance of the coil as acting in *series* with its inductance, as shown in Figure 24-14.

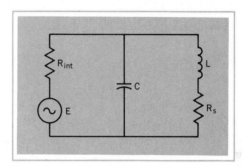

FIGURE 24-14
Practical Parallel-Resonant Circuit.

Since the practical parallel-resonant circuit is not a simple parallel circuit, we must be careful how we define its resonant frequency. We can define the resonant frequency of a practical parallel-resonant circuit by any of the following statements, all of which specify essentially the same conditions. **The resonant frequency of a practical parallel-resonant circuit is the frequency at which**

1. **The reactive power of the inductance and capacitance are equal.**
2. **The circuit has unity power factor.**
3. **The total impedance is completely resistive.**
4. **The line current is in phase with the source voltage.**

If we assume that the frequency of the voltage applied to the circuit of Figure 24-14 is such that $X_L = X_C$, the impedance of the inductive branch will be slightly greater than the reactance of the capacitive branch due to the resistance in series with the inductive reactance. Therefore, we must *decrease* the frequency of the source slightly to allow sufficient current in the inductive branch to make the reactive power of the inductive branch equal to the reactive power of the capacitance.

To assist us in solving for the actual resonant frequency of a practical parallel-resonant circuit, we can apply a technique we developed in Chapter 22. We can replace the series LR branch by an equivalent parallel circuit, as shown in Figure 24-15, where

$$R_p = \frac{Z^2}{R_s} = \frac{R_s^2 + X_s^2}{R_s} \tag{22-19}$$

$$X_p = \frac{Z^2}{X_s} = \frac{R_s^2 + X_s^2}{X_s} \tag{22-20}$$

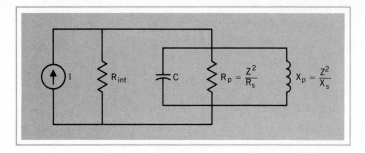

At resonance, the reactances of the inductive and capacitive branches of the *equivalent* parallel circuit must be equal.

$$\therefore \frac{1}{2\pi f_r C} = \frac{R_s^2 + (2\pi f_r L)^2}{2\pi f_r L} \tag{24-11}$$

$$\frac{L}{C} = R_s^2 + (2\pi f_r L)^2$$

from which

$$f_r^2 = \frac{L - CR_s^2}{4\pi^2 L^2 C}$$

$$\therefore f_r = \frac{1}{2\pi\sqrt{LC}} \times \sqrt{1 - \frac{CR_s^2}{L}} \tag{24-12}$$

Examining the factor under the square root sign in Equation (24-12), we can make several important observations about the resonant frequency of practical parallel-resonant circuits.

1. The magnitude of the factor under the square root sign is less than unity. Hence, the resonant frequency of a practical parallel-resonant circuit is slightly *lower* than the frequency given by Equation (24-1).
2. Unlike in series-resonant circuits or theoretical parallel-resonant circuits, the resonant frequency of a practical parallel-resonant circuit *is* dependent on the resistance of the circuit.
3. If the series resistance is large enough so that $CR_s^2/L > 1$, there is no parallel-resonant frequency for that particular circuit.

To evaluate the extent of the reduction in the resonant frequency of the practical parallel-resonant circuit of Figure 24-14, we first of all investigate the Q of a practical parallel-resonant circuit. We have already determined that the Q of a theoretical parallel-resonant circuit is

$$Q = \frac{R_p}{X_L} \tag{24-9}$$

Applying Equation (24-9) to the parallel equivalent of a practical parallel-resonant circuit shown in Figure 24-15,

$$\therefore Q = \frac{Z^2}{R_s} \times \frac{X_s}{Z^2} = \frac{X_s}{R_s}$$

Hence,

The Q of a practical parallel-resonant circuit is the same as the Q of the coil itself.

By substituting $2\pi f_r L/Q = R_s$ in Equation (24-11),

$$\frac{1}{2\pi f_r C} = \frac{(2\pi f_r L/Q)^2 + (2\pi f_r L)^2}{2\pi f_r L} = 2\pi f_r L \left(\frac{1 + Q^2}{Q^2}\right)$$

from which

$$f_r = \frac{1}{2\pi\sqrt{LC}} \times \sqrt{\frac{Q^2}{1 + Q^2}} \qquad (24\text{-}13)$$

From Equation (24-13), if the Q of the coil is reasonably high ($Q > 10$), the difference between the resonant frequency produced by Equation (24-13) and Equation (24-1) is small enough for us to consider the resonant frequency of most practical parallel-resonant circuits to be simply

$$f_r = \frac{1}{2\pi\sqrt{LC}} \qquad (24\text{-}1)$$

We have noted that the impedance of a parallel-resonant circuit is maximum at resonance and is

$$Z_P = R_p = \frac{R_s^2 + X_s^2}{R_s} = R_s + \frac{X_s^2}{R_s}$$

There are several useful forms of this equation. Factoring out R_s and substituting $Q = X_s/R_s$,

$$Z_p = R_s(1 + Q^2) \qquad (24\text{-}14)$$

Or simply substituting for X_s/R_s,

$$Z_p = R_s + QX_s \qquad (24\text{-}15)$$

Since R_s is usually very small compared to QX_s,

$$\therefore Z_p \approx QX_L \qquad (24\text{-}16)$$

Substituting for X_L from Equation (24-3),

$$Z_p \approx Q^2 R_s \qquad (24\text{-}17)$$

Or substituting for Q from Equation (24-3),

$$Z_p \approx \frac{X_L^2}{R_s} \tag{24-18}$$

and since

$$X_L^2 \approx X_L \times X_C = \frac{\omega L}{\omega C}$$

then

$$Z_p \approx \frac{L}{CR} \tag{24-19}$$

Again, if $Q > 10$, for practical purposes we can replace the "approximately-equals" signs in the above equations by "equals" signs.

EXAMPLE 24-4

A coil in a tuned circuit in a radio receiver has an inductance of 50 μH and a resistance of 25 Ω. It forms a parallel-resonant circuit with a 200-pF capacitor.
(a) At what frequency is the total impedance completely resistive?
(b) What is the magnitude of the parallel impedance at this frequency?

SOLUTION

(a)
$$f_r = \frac{1}{2\pi\sqrt{LC}} \times \sqrt{1 - \frac{CR_s^2}{L}}$$

$$= \frac{1}{2 \times \pi \times \sqrt{50 \ \mu H \times 200 \ pF}} \times \sqrt{1 - \frac{CR_s^2}{L}}$$

$$= 1.5915 \ \text{MHz} \times \sqrt{1 - \frac{200 \ pF \times (25 \ \Omega)^2}{50 \ \mu H}}$$

$$\therefore f_r = 1.5915 \ \text{MHz} \times 0.99875 = \mathbf{1.5895 \ MHz}$$

(b)
$$X_L = 2\pi f L = 2 \times \pi \times 1.5895 \ \text{MHz} \times 50 \ \mu H \approx 500 \ \Omega$$

$$Q = \frac{X_L}{R_s} \approx \frac{500 \ \Omega}{25 \ \Omega} \approx 20$$

$$\therefore Z_p = QX_L \approx 20 \times 500 \ \Omega \approx \mathbf{10 \ k\Omega}$$

24-8 SELECTIVITY OF PARALLEL-RESONANT CIRCUITS

As we noted in Section 24-5, the **selectivity** of a resonant circuit is a measure of its ability to differentiate between a desired frequency and undesired interference from other frequencies. We usually express selectivity in terms of the **bandwidth** Δf

between the **half-power points** (3 dB down points) either side of the resonant frequency. At these points, the voltage across the equivalent parallel-resonant circuit in Figure 24-15 will be 0.707 of its magnitude at resonance. With a constant-current source, this means that the equivalent impedance will have dropped to 0.707 of its value at resonance. Following the reasoning we used in Section 24-5, and since we are working with a *parallel* circuit, the *susceptance* of L and C must each change by an amount equal to $1/R_p$ between these two frequencies. Taking $B_C = \omega C$,

$$2\pi C(f_2 - f_1) = \frac{1}{R_p}$$

from which
$$\Delta f = \frac{1}{2\pi C R_p} = \frac{f_r}{2\pi f_r C R_p} = f_r \times \frac{X}{R_p}$$

And if $Q > 10$, this becomes

$$\Delta f = \frac{f_r}{Q} \qquad\qquad (24\text{-}8)^{\dagger}$$

As we discovered in Figure 24-12, the internal resistance of the signal source has a significant effect on the performance of a parallel-resonant circuit. The effective Q of a practical parallel-resonant circuit is seldom determined by R_p alone. In the equivalent circuit of Figure 24-15, the internal resistance of the source is directly in parallel with R_p. This reduces the parallel Q of the circuit and thus increases Δf, resulting in a reduction in selectivity.

EXAMPLE 24-5

The parallel-resonant circuit of Example 24-4 is connected across a signal source having an internal resistance of 40 kΩ. What is the bandwidth of the tuned circuit?

SOLUTION

The additional *parallel* resistance does not enter into the equation for the resonant frequency. Hence, the reactance of either branch at resonance is still 500 Ω. The parallel resistance due to the coil is

$$R_p = \frac{X^2}{R_s} = \frac{(500\ \Omega)^2}{25\ \Omega} = 10\ \text{k}\Omega$$

The equivalent parallel resistance becomes

$$R_{eq} = \frac{R_{int} \times R_p}{R_{int} + R_p} = \frac{40\ \text{k}\Omega \times 10\ \text{k}\Omega}{40\ \text{k}\Omega + 10\ \text{k}\Omega} = 8\ \text{k}\Omega$$

$$Q = \frac{R_{eq}}{X} = \frac{8\ \text{k}\Omega}{500\ \Omega} = 16$$

$$\therefore \Delta f = \frac{f_r}{Q} = \frac{1.5895\ \text{MHz}}{16} \approx 99\ \text{kHz}$$

---

†Note that this is the same equation as for a *series*-resonant circuit. This shouldn't surprise us when we remember that Q is defined as the ratio of reactive *power* to active *power*.

Returning to Equation (24-12), we noted that if R_s is large enough, there is no parallel-resonant frequency for a given coil and capacitor combination. Purposely increasing the series resistance (or reducing the Q) of a tuned circuit is called **damping.** When R_s is large enough for parallel resonance to be impossible, the result is **critical damping.** Just as we damp meter movements to keep the pointers from oscillating, we add resistance to peaking coils in television receivers to keep them from **ringing** at a frequency determined by their inductance and distributed capacitance.

24-9 RESONANT FILTER NETWORKS

Radio and television receivers are made up of transistor amplifier stages connected together by coupling networks. We can represent the coupling network by the box with two input and two output terminals in the block diagram of Figure 24-16. We have encountered many four-terminal coupling networks in various examples in earlier chapters. So far, we have not paid much attention to their *frequency* response.

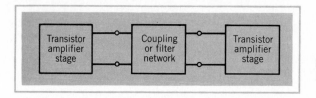

FIGURE 24-16
Block Diagram of an Interstage Filter Network.

Coupling networks between audio amplifier stages should have a uniform response over a range of at least 20 Hz to 20 kHz. However, in radio-frequency systems, we usually want to select or reject a specific narrow band of frequencies. Such coupling networks are called **filters.** The frequency response of low-frequency filters is often determined by CR or LR combinations. The filters have such descriptive names as **low-pass** and **high-pass filters.** By far the most common **band-pass** or **band-reject filters** are based on LC resonant or *tuned* circuits. We have already considered tuned circuits in terms of their ability to *select* a certain band of frequencies centered about the resonant frequency of the network and to *reject* all other frequencies. We can redraw the series-resonant circuit of Figure 24-7 and the parallel-resonant circuit of Figure 24-12 as the four-terminal band-pass filters of Figure 24-17(a) and (b), respectively.

We can also use resonant circuits to *reject* a specific frequency without affecting the response of the circuit to all other frequencies. Such an application is called a **wavetrap.** Figure 24-18 illustrates two wavetrap circuits. In both cases, the four-terminal network forms a voltage divider across the input. If we assume a very high-impedance load connected to the output terminals, then the output voltage of the circuit of Figure 24-18(a) will be

$$V_{out} = E_{in} \times \frac{Z_s}{R + Z_s} \qquad (24\text{-}20)$$

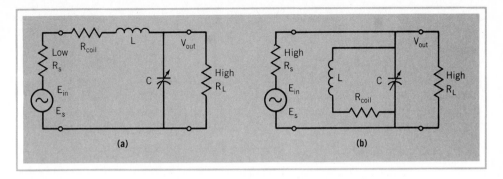

FIGURE 24-17
(a) Series-Resonant, and (b) Parallel-Resonant Bandpass Filters.

At resonance, Z_s will be equal to the resistance of the coil, which is quite small compared with the resistance R. With the values given in Figure 24-18(a), the output voltage will be only 1% of the input voltage at the resonant frequency. With a high-Q tuned circuit, at frequencies other than the resonant frequency, Z_s will be much greater than R and the fraction $Z_s/(R + Z_s)$ will approach 100%.

In the wavetrap circuit of Figure 24-18(b), the load voltage will be

$$V_{out} = E_{in} \times \frac{R_L}{R_L + Z_p} \qquad (24\text{-}21)$$

At resonance the impedance of the parallel-resonant circuit will be many times as great as the resistance of the load. This results in a very small voltage across the load at this frequency. At frequencies off resonance, Z_p becomes quite small, and, therefore, $R_L/(R_L + Z_p)$ approaches 100%.

FIGURE 24-18
Wavetraps.

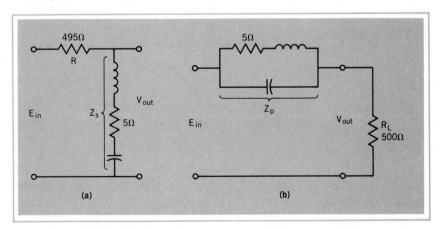

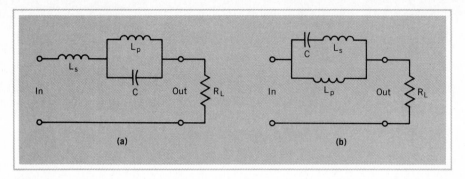

FIGURE 24-19
Double-Resonant Circuits.

In many communications circuits, it is desirable to include the selection of one frequency and the rejection of another frequency into a combined **double-resonant** circuit such as those shown in Figure 24-19.

EXAMPLE 24-6
If the capacitance in the circuit of Figure 24-19(a) is 200 pF, determine the inductances of L_p and L_s in order to reject a signal at 456 kHz and to accept a signal at 1200 kHz.

SOLUTION
Step I. L_p and C form a parallel-resonant circuit which will represent a high impedance in series with R_L at resonance. Hence, L_p and C form a wavetrap and must be tuned to the frequency which is to be rejected, 456 kHz.

$$X_C = \frac{1}{2\pi fC} = \frac{1}{2 \times \pi \times 456 \text{ kHz} \times 200 \text{ pF}} = 1745 \text{ } \Omega$$

$$L_p = \frac{X_L}{2\pi f} = \frac{1745 \text{ } \Omega}{2 \times \pi \times 456 \text{ kHz}} = 609 \text{ } \mu\text{H}$$

Step II. At 1200 kHz,

$$B_C = 2\pi fC = 2 \times \pi \times 1200 \text{ kHz} \times 200 \text{ pF} = 1.508 \text{ mS}$$

$$B_{L_p} = \frac{1}{2\pi fL} = \frac{1}{2 \times \pi \times 1200 \text{ kHz} \times 609 \text{ } \mu\text{H}} = 217.8 \text{ } \mu\text{S}$$

Therefore, the net susceptance of the parallel circuit at 1200 kHz is

$$B = 1.508 \text{ mS} - 217.8 \text{ } \mu\text{S} = 1.2902 \text{ mS capacitive}$$

For the whole circuit to be *series* resonant at 1200 kHz,

$$B_{L_s} = 1.2902 \text{ mS}$$

$$\therefore L_s = \frac{1}{2\pi f B_{L_s}} = \frac{1}{2 \times \pi \times 1200 \text{ kHz} \times 1.2902 \text{ mS}}$$

$$= 102.8 \ \mu\text{H}$$

Figure 24-19(b) illustrates an alternate circuit arrangement to achieve the results required in Example 24-6. Here L_s and C are selected first to form a series-resonant circuit at the desired frequency of 1200 kHz, thus comprising a low impedance in series with R_L at 1200 KHz. At 456 kHz, this series branch has a net capacitive reactance which can be tuned with L_p as a parallel-resonant circuit at 456 kHz. If the selected frequency is lower than the rejected frequency, L_s in the circuit of Figure 24-19(a) and L_p in the circuit of Figure 24-19(b) must be replaced with capacitors.

PROBLEMS

24-1. At what frequency is a circuit consisting of a 40-pF capacitor and a 90-μH coil with a Q of 80 series resonant?

24-2. What is the bandwidth of the circuit in Problem 24-1?

24-3. If the circuit of Problem 24-1 is connected to a 500-μV source, what voltage will appear across the capacitor at resonance?

24-4. What current will flow through the circuit at resonance?

24-5. What value of capacitance is required to tune a 500-μH coil to series resonance at 465 kHz?

24-6. What value of inductance in series with a 12-pF capacitor is resonant at 45 MHz?

24-7. A tuned circuit consisting of 40-μH inductance and 100-pF capacitance in series has a bandwidth of 25 kHz. What is its Q?

24-8. What is the resistance of the coil in the tuned circuit of Problem 24-7?

24-9. What is the resonant frequency of a tuned circuit consisting of the coil and capacitor of Problem 24-1 connected in parallel?

24-10. What is the impedance of the parallel-tuned circuit of Problem 24-9 at resonance?

24-11. If the resonant circuit of Problem 24-9 is connected to a 500-μV source, what is the tank current at resonance?

24-12. What will the parallel-resonant frequency be if loading the inductive branch reduces the Q of the tuned circuit of Problem 24-9 to 20?

24-13. A variable tuning capacitor in a radio receiver has a maximum capacitance of 365 pF and a minimum capacitance of 30 pF.
   (a) What inductance is required for the lowest frequency to which the circuit can tune to be 540 kHz?
   (b) What is the highest frequency to which this circuit can be tuned?

24-14. A coil with a Q of 90 and a capacitor form a parallel-resonant circuit tuned to 4.5 MHz. The total impedance of the circuit at resonance is 60 kΩ. Find
   (a) The inductance of the coil.
   (b) The capacitance of the capacitor.

24-15. The coil and capacitor of a tuned circuit have an L/C ratio of 10^5. The Q of the tuned circuit is 80 and its bandwidth is 5.8 kHz.
   (a) Calculate the upper and lower frequency limits of the bandpass.
   (b) Calculate the inductance and resistance of the coil.

24-16. A tuned circuit in a radio transmitter is composed of a tank coil with an inductance of 80 μH tuned by an 800-pF capacitor. The RF voltage across this tank circuit is 1200 V rms. The overall Q of the circuit is 30.
   (a) Calculate the tank current.
   (b) At what rate must RF energy be supplied to this circuit to maintain this amplitude?

24-17. A 4-H filter choke with a resistance of 20 Ω is connected in series with a pulsating dc source and a resistive load. What value of capacitance must be connected across the choke to provide the most effective filtering of a 120-Hz ripple component in the output of the pulsating source?

24-18. An audio filter for reducing the 10-kHz heterodyne between broadcast stations contains a parallel-resonant circuit whose capacitance branch is 0.02 μF. The inductance branch has a resistance of 100 Ω. What inductance should this branch possess?

24-19. Solve Example 24-6 with reference to the circuit of Figure 24-19(b).

24-20. Design a filter which can be used to pass the output of a 200-kHz powerline carrier transmitter whose bandwidth is 6 kHz but reject any third harmonic content in the transmitter output. This filter will contain an 0.5-mH coil.

REVIEW QUESTIONS

24-1. What is meant by the general term **resonance?**

24-2. What property of an electric circuit makes resonance possible?

24-3. Derive an equation to determine the capacitance required to tune a given inductance to series resonance at a given frequency.

24-4. Why does the resistance in a series-resonant circuit have no bearing on the resonant frequency?

24-5. What is the significance of the resistance in a series-resonant circuit?

24-6. What is the significance of the term **L/C ratio** in resonant circuits?

24-7. What is the significance of the Q factor of resonant circuits?

24-8. What is the significance of the term **resonant rise of voltage** in discussing resonant circuits?

24-9. What is the significance of the term **half-power points** in discussing resonant circuits?

24-10. How does a parallel-resonant circuit differ in circuit behavior from a series-resonant circuit?

24-11. Why are parallel-resonant circuits capacitive at frequencies above the resonant frequency, whereas series-resonant circuits are inductive at frequencies above the resonant frequency?

24-12. Why does *decreasing* the resistance of the coil of a parallel-resonant circuit *increase* the total impedance at resonance?

24-13. Why does a change in the resistance of a coil in a parallel-resonant circuit affect the resonant frequency?

24-14. If a resistor were placed in parallel with the capacitor of a series-resonant circuit, would the resonant frequency be affected? Explain.

24-15. Why does the resonant frequency of a parallel-resonant circuit tend to drift as the **loading** on the circuit is increased?

24-16. What is the significance of the term **tank current** in discussing resonant circuits?

24-17. What is the significance of the answer to Problem 24-18?

24-18. In the construction of the filter circuit of Figure 24-19(b), L_p and L_s are both made variable to allow the filter to be tuned. In which order must these inductances be adjusted? Why?

24-19. The filter network of Figure 24-19(a) performs equally well if the input and output terminals are reversed. Why does this not apply to the filter network of Figure 24-17(a)?

*24-20. Prepare (using either computer or calculator) a table similar to Table 24-1 for the circuit of Figure 24-4.

*24-21. Prepare a table similar to Table 24-1 for the circuit of Figure 24-5.

*24-22. Assuming that the input voltage remains constant for all frequencies, develop a computer program to display or print out a graph of the output voltage across $R_L = 100 \ \Omega$ in Example 24-6.

25

TRANSFORMERS

25-1 MUTUAL INDUCTION

We encountered mutual induction briefly in Chapter 15. Figure 15-2 showed a circuit in which *changing* the current in one winding (**primary**) induced a voltage in another winding (**secondary**), which was *linked* by the magnetic lines of force created by the current in the primary winding. The generation of a voltage in one winding by a *changing* current in another winding is called **mutual induction.** In Section 15-1, we considered mutual induction only to the extent that it illustrated Faraday's and Lenz's laws. Since mutual induction is based on *changing* current, we did not pursue its application to dc circuits. However, since the current in an ac circuit is continually changing, mutual induction plays a very important role in ac circuits.

25-2 TRANSFORMER ACTION

A practical device which depends on mutual induction for its operation is the **transformer.** In schematic diagrams, it is customary to connect the source to the *left*-hand coil (primary) and the load to the *right*-hand coil (secondary), as in Figure 25-1. For the moment, we shall assume that the secondary terminals are left open-circuit as shown.

According to Kirchhoff's voltage law, a voltage must appear across the primary coil such that it is exactly equal to the applied voltage at every instant in time. As we discovered in Section 18-1, if the resistance of the primary coil is negligible, this

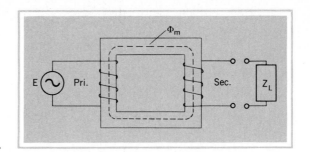

FIGURE 25-1
Transformer Action.

voltage must be induced into the coil by a changing flux in the core. Since the applied voltage is a sine wave, and since, according to Faraday's law, the induced voltage is proportional to the *rate of change* of flux, the flux in the transformer core must be a sine wave lagging the applied voltage by 90°.

Because the weber is defined as the magnetic flux which must cut (or link) an electric conductor in one second in order to induce an average voltage of one volt, and because *each* turn of a coil is linked by the same field, it follows that

$$E_{av} = \frac{N\Phi_m}{t} \tag{25-1}$$

where E_{av} is the average voltage induced into a coil, N is the number of turns in the coil, and t is the time (seconds) it takes the flux to rise from zero to Φ_m webers.

Since the flux in the transformer must be a sine wave, it must rise from zero to Φ_m in one-quarter of a cycle.

$$\therefore E_{av} = \frac{N\Phi_m}{1/4f} = 4fN\Phi_m$$

But the *average* value of a sine wave is 0.637 of the peak value, and the *rms* value of a sine wave is 0.707 of the peak value; hence.

$$E = 1.11E_{av} \tag{25-2}$$

$$\therefore E = 4.44fN\Phi_m \tag{25-3}$$

where f is the frequency of the applied voltage in hertz, N is the number of turns in the coil, and Φ_m is the peak value of the flux in webers. Equation (25-3) is called the **general transformer equation.**[†]

Applying the general transformer equation to the primary winding of the simple transformer of Figure 25-1, we obtain

$$E_p = 4.44fN_p\Phi_m \tag{25-4}$$

[†]See Appendix 2-9 for calculus derivation of the general transformer equation.

To create this sine wave of flux in the core, an alternating current must flow in the primary winding. And since

$$\Phi_p = \frac{F_m}{R_m} = \frac{N_p I_p}{R_m}$$ (13-2) and (13-1)

it follows that the peak value of the primary current is

$$I_{pm} = \Phi_m \frac{R_m}{N_p}$$

from which

$$I_p = \frac{\Phi_m}{\sqrt{2}} \times \frac{R_m}{N_p}$$ (25-5)

To apply Equations (25-4) and (25-5) to a practical transformer, we must know something about the *magnetic* circuit of the transformer core. First, for both the magnetic flux in the core and the primary current which induces it to be pure sine waves, the flux density in the core must remain in the linear region of the magnetization curve for the type of iron used. From the *BH* curve of Figure 14-4, B_m for sheet steel should not be greater than 1.1 teslas to stay below the upper knee (saturation) area of sheet steel. Maximum flux Φ_m will then depend on the cross-sectional area of the core. Once we decide on the maximum flux density for the core we plan to use, we can calculate the number of turns we need to use in our primary winding.

If we use Equation (25-5) to solve for the primary current, we must determine the reluctance of the magnetic circuit of the transformer core. If we use the magnetization curve of Figure 14-4, we can start with the maximum magnetizing force needed to create B_m. From Equation (13-6),

$$F_m = H_m l$$

And from Equation (13-1),

$$F_m = N I_m$$

Hence,

$$I_m = \frac{H_m l}{N}$$

and

$$I = \frac{H_m l}{\sqrt{2}\, N}$$ (25-6)

If, for the moment, we can neglect core losses, even though the secondary of the transformer is an open circuit, there must be a sine wave of primary current in phase with the flux and lagging the applied voltage by 90°. This primary current is called the **exciting** or **magnetizing** current of the transformer. In good transformers, this magnetizing current is usually less than 5% of the current which the primary winding can be expected to pass when full load is connected to the secondary.

EXAMPLE 25-1

The sheet-steel core of a transformer has a cross-sectional area of 5.0 cm^2 and an average path length of 25 cm. If the peak flux density in the core is to be 1.0 T (see the *BH* curves of Figure 14-4),

(a) How many turns are required on a primary winding which is to be connected to a 120-V 60-Hz source?

(b) What is the rms value of the exciting current in the primary winding?

SOLUTION

(a) $\Phi_m = 5.0$ cm$^2 \times 1.0$ T $= 0.0005$ Wb

$$N_p = \frac{E_p}{4.44\, f\Phi_m} = \frac{120\text{ V}}{4.44 \times 60\text{ Hz} \times 0.0005\text{ Wb}} = \textbf{901 turns}$$

(b) From the *BH* curve for sheet steel in Figure 14-4, when $B_m = 1.0$ T, $H_m = 425$ At/m.

$$\therefore I_p = \frac{H_m l}{\sqrt{2}\, N_p} = \frac{425\text{ At/m} \times 25\text{ cm}}{\sqrt{2} \times 901\text{ turns}} = \textbf{83.4 mA}$$

25-3 TRANSFORMATION RATIO

In the ideal transformer, all the flux created by primary current will link the secondary winding. Therefore, a voltage will be induced in the secondary winding,

$$E_s = 4.44\, fN_s\Phi_m \tag{25-7}$$

Dividing Equation (25-7) into Equation (25-4) gives

$$\frac{E_p}{E_s} = \frac{N_p}{N_s} = a \tag{25-8}$$

Consequently, when all the primary flux links the secondary winding, the ratio of the primary induced voltage to the secondary induced voltage is the same as the ratio of the number of primary turns to the number of secondary turns. These ratios are called the **transformation ratio,** which we represent by the letter symbol *a*.

Equation (25-8) shows us why transformers are so widely used in power systems. They provide a means of transforming a given source of alternating voltage into any desired voltage with minimum loss. If we arrange to produce a secondary voltage higher than the source voltage, we have a **step-up** transformer. Similarly, a **step-down** transformer is one in which the secondary voltage is less than the primary voltage.

EXAMPLE 25-2

The secondary of the transformer of Example 25-1 is to be wound to supply 6.3 V for the heater of a TV picture tube. How many turns must there be on the secondary winding?

SOLUTION

$$a = \frac{E_p}{E_s} = \frac{120}{6.3} = 19$$

$$N_s = \frac{N_p}{a} = \frac{901}{19} \approx \textbf{48 turns}$$

If we now connect the load impedance to the secondary winding in the simple transformer circuit of Figure 25-1, the voltage induced into the secondary winding by the primary flux will cause current to flow in the secondary circuit. This current flowing in the secondary winding will create a magnetomotive force of $N_s I_s$ ampere-turns, which, according to Lenz's law, tends to *reduce* the amplitude of the sine wave of flux in the core. But, in a lossless transformer, the amplitude of the flux sine wave must not change, since it is responsible for always inducing a voltage equal to the applied voltage into the primary winding. Therefore, when a load is connected to the secondary, *additional* primary current must flow in order to develop a primary mmf component which exactly cancels the secondary mmf. Hence,

$$N_p I'_p = N_s I_s \tag{25-9}$$

where I'_p is the additional or load component of the primary current. We can rearrange Equation (25-9) to give

$$\frac{I_s}{I'_p} = \frac{N_p}{N_s} = a \tag{25-10}$$

If the load connected to the secondary terminals of the transformer is a pure resistance, the secondary current will be in phase with the secondary voltage. Since E_p and E_s are both induced by the same changing flux, it follows that the additional I'_p is in phase with the applied voltage. Because the magnetizing current, I_{mag} in Figure 25-2, lags the applied voltage by almost 90°, the total primary current is the *phasor* sum of these two components.

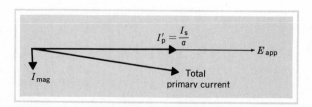

FIGURE 25-2
Total Primary Current of a Transformer with a Resistance Secondary Load.

25-4 IMPEDANCE TRANSFORMATION

When we connect a load to the secondary winding of the simple transformer of Figure 25-1, the primary current increases and shifts in phase to the position shown in the phasor diagram of Figure 25-2. But, as far as the source is concerned, we can treat the total primary current as being the phasor sum of two separate *parallel* branch currents. Hence, the right-hand circuit of Figure 25-3 is the equivalent circuit for the actual left-hand circuit. The imaginary parallel branch carrying I'_p appears to have an impedance of

$$Z_p = \frac{E_p}{I'_p}$$

From Equation (25-8),

$$E_p = \frac{N_p}{N_s} E_s = aE_s$$

and from Equation (25-10),

$$I'_p = \frac{N_s}{N_p} I_s = \frac{I_s}{a}$$

$$\therefore Z_p = \frac{E_s}{I_s} \left(\frac{N_p}{N_s}\right)^2$$

But E_s/I_s is dependent on the total impedance of the secondary circuit. In a good transformer, the impedance of the secondary winding will be small compared with the load impedance;

$$\therefore Z_p \approx \left(\frac{N_p}{N_s}\right)^2 Z_L \approx a^2 Z_L \tag{25-11}$$

where Z_p is called the **reflected impedance.** It is approximately equal to the *square* of the turns ratio times the load impedance.

Therefore, we can use transformers as a means of transforming a given load impedance into a different value with minimum loss. There are many circuit applications in which a transformer is used as an **impedance matching** device. One example is in coupling a given load to an electronic amplifier which requires a specific value of load impedance for maximum power output with minimum distortion, as illustrated in the following example.

FIGURE 25-3
Reflected Impedance.

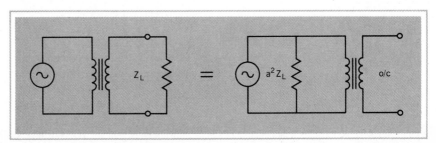

EXAMPLE 25-3

An audio transformer has 1200 turns in the primary winding. How many turns must be wound in the secondary winding to make a 4-Ω loudspeaker appear as a 5-kΩ load to the amplifier connected to the primary?

SOLUTION

$$a^2 = \frac{Z_p}{Z_L} = \frac{5000}{4} = 1250$$

$$\therefore a = 35.36$$

$$N_s = \frac{N_p}{a} = \frac{1200}{35.36} = \textbf{34 turns}$$

EXAMPLE 25-4

The audio output transformer of Figure 25-4 presents the correct load impedance to a transistor amplifier when *either* an 8-Ω load is connected to terminals A and B *or* a 16-Ω load is connected to terminals A and C. What load may be connected to terminals B and C alone to present the same reflected load?

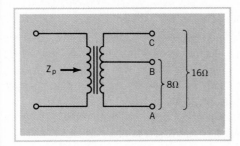

FIGURE 25-4
Tapped Audio Output Transformer.

SOLUTION

We can solve this problem either by assuming convenient values for the number of primary turns and reflected impedance or with algebraic symbols. Let N_1 represent the number of secondary turns between terminals A and B, and N_2 the total number of turns between terminals A and C.
   Then

$$Z_p = 8\left(\frac{N_p}{N_1}\right)^2 = 16\left(\frac{N_p}{N_2}\right)^2 = Z_x\left(\frac{N_p}{N_2 - N_1}\right)^2$$

from which $\qquad \dfrac{\sqrt{8}}{N_1} = \dfrac{\sqrt{16}}{N_2} = \dfrac{\sqrt{Z_x}}{N_2 - N_1}$

from which $\qquad \sqrt{Z_x} = \sqrt{16} - \sqrt{8}$, and $Z_x = (\sqrt{16} - \sqrt{8})^2 = \textbf{1.37 } \boldsymbol{\Omega}$

Note that the required impedance is *not* simply the difference between the two given loads.

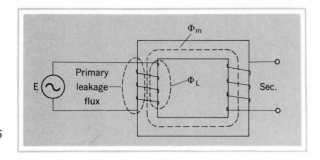

FIGURE 25-5
Leakage Flux.

25-5 LEAKAGE REACTANCE

In establishing the transformation ratio in Section 25-3, we assumed that *all* the flux created by the magnetizing current in the primary linked the secondary winding. However, since there is no magnetic insulator, some of the magnetic lines of force produced by the magnetizing current do not link the secondary, even in the best transformers. These lines of force are called the primary **leakage flux,** as shown in Figure 25-5. The remainder of the total primary flux does link the secondary and is called the **mutual flux.** Similarly, when load current flows in the secondary, some of the secondary magnetic lines of force do not link the primary. These lines of force are called the secondary leakage flux.

The effect of leakage flux is the same as if Φ_m/Φ_p of the total primary turns were developing lines of force that link the secondary winding and Φ_L/Φ_p of the primary turns act as a self-inductance in series with the primary. The reactance of these ineffective primary turns is called the primary **leakage reactance.** As far as the source is concerned, the transformer takes on the appearance shown in Figure 25-6.

In most iron-core transformers, the primary and secondary leakage reactances represent approximately the same percentage of their respective windings; thus, the transformation ratio of the perfect transformer portion of the equivalent circuit of

FIGURE 25-6

Effect of Leakage Reactance. R_{CL}, Core Losses; L_m, Primary Self-Inductance Governing Magnetizing Current; R_p, Primary Winding Resistance; X_p, Primary Winding Leakage Reactance; T, Ideal Transformer with Transformation Ratio a; X_s, Secondary Winding Leakage Reactance; R_s, Secondary Winding Resistance; Z_L, Secondary Load Impedance.

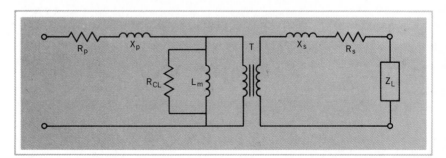

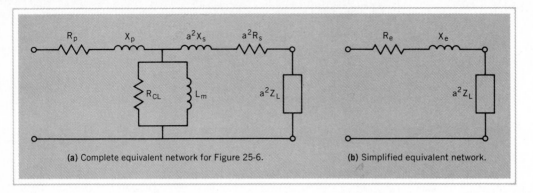

(a) Complete equivalent network for Figure 25-6. (b) Simplified equivalent network.

FIGURE 25-7
Equivalent Networks for an Iron-Core Transformer. $R_e = R_p + a^2R_s$ and
$X_e = X_p + a^2X_s$.

Figure 25-6 is still essentially N_p/N_s. Since this represents a perfect transformer, upon looking into its primary, the source "sees" a reflected load of

$$\mathbf{Z}_{ref} = a^2(R_s + jX_s + \mathbf{Z}_L) \qquad (25\text{-}12)$$

We can, therefore, reduce the transformer still further to the equivalent circuit of Figure 25-7(a). If the transformer is fully loaded, the magnetizing current in the branch of the equivalent circuit containing R_{CL} and L_m is so small compared with the load current that it creates negligible voltage drop across R_p and X_p. Consequently, when we wish to determine the effects of loading a transformer, we can omit the R_{CL} and L_m branch and simplify the equivalent circuit still further to the simple series circuit of Figure 25-7(b).

25-6 OPEN-CIRCUIT AND SHORT-CIRCUIT TESTS

We can determine experimentally the transformer constants for the equivalent networks of Figure 25-7 from two simple tests. If we leave the secondary terminals open-circuit, $\mathbf{Z}_L$ in Figure 25-7(a) is infinitely large, and current will flow only in the R_{CL} and $j\omega L_m$ branch of the equivalent network. If the transformer under test is designed with a 120-volt primary winding, we probably use the 120-volt winding as the primary winding for the test. However, if the transformer is designed to step a 550-volt source down to 110 volts, to protect the instruments, we use the 110-volt winding as the primary for the test, taking adequate steps to ensure that no one can come in contact with the open-circuit high-voltage winding in the setup of Figure 25-8(a).

We feed the winding we are using as the primary for the test with the exact rated voltage for the winding through a group of instruments consisting of wattmeter, voltmeter, and ammeter, connected as shown in Figure 25-8(a). Since the magnetizing current is only a very small fraction of the full-load primary current, we can assume that the I^2R loss due to magnetizing current alone in the primary winding is negligible. Therefore, the wattmeter shows the **core losses.** These core losses are due to eddy currents in the core and to the hysteresis loop of the core iron. The eddy currents produce a small counterflux in the core just as if we had connected a resistor across the secondary winding. The hysteresis loop of the iron causes the exciting current to lag the applied voltage by slightly less than 90°. The in-phase component of the magnetizing current contributes to the active power reading of the wattmeter.

From the voltmeter and ammeter readings, we can determine the total impedance of the magnetizing-current and core-loss branch of the equivalent network of Figure 25-7(a). As this diagram indicates, an increase in load current will slightly increase the voltage drop across R_p and X_p. This, in turn, will slightly reduce the current in the R_{CL} and L_m branch. Consequently, there is a slight reduction in magnetizing current as the transformer is fully loaded. For practical purposes, we can assume that the core losses remain constant as the load is increased from zero to full rated load.

For the short-circuit test, we short-circuit the low-voltage winding of the transformer, as shown in Figure 25-8(b). In the case of the step-down transformer, the low-voltage winding will be the normal secondary winding. The short circuit is equivalent to making $\mathbf{Z}_L$ zero in Figure 25-7(b). Since $R_e + jX_e$ is fairly small, we must greatly reduce the voltage applied to the high-voltage winding to keep the currents in the transformer windings from exceeding their rated values. With this

FIGURE 25-8
Meter Connections: (a) Open-Circuit Test; (b) Short-Circuit Test.

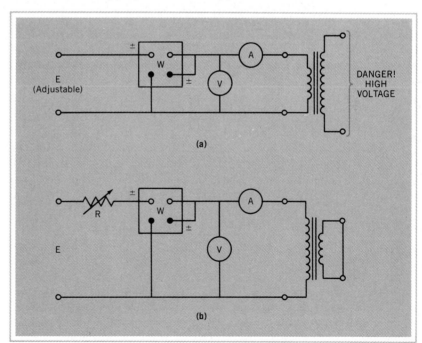

reduction in applied voltage, we obtain maximum protection for personnel and equipment by using the high-voltage winding as the primary, as shown in Figure 25-8(b). Rheostat R allows us to adjust the applied voltage to a value which will permit rated current to flow in the windings of the transformer. An applied voltage of less than 5% of the rated voltage for the winding being used as the primary is required to produce rated current in the windings. Since reducing the applied voltage reduces the exciting current to less than 5% of its normal value, the core losses are now less than 0.25% of their normal value, and we can assume that the wattmeter is showing only the copper losses from which we can determine the value of R_e.

EXAMPLE 25-5

When the secondary winding of a small step-down transformer is short-circuited, the applied primary voltage is 6 V, the primary current is 4 A, and the wattmeter reading is 12 W. Determine R_e and X_e.

SOLUTION

$$R_e = \frac{P}{I^2} = \frac{12 \text{ W}}{(4 \text{ A})^2} = 0.75 \ \Omega$$

$$Z_e = \frac{E}{I} = \frac{6 \text{ V}}{4 \text{ A}} = 1.5 \ \Omega$$

$$\therefore X_e = \sqrt{Z_e^2 - R_e^2} = \sqrt{1.5^2 - 0.75^2} = 1.3 \text{ ohms (inductive)}$$

25-7 EFFICIENCY

The core-loss data obtained from the open-circuit test are required in calculating the efficiency of a transformer under load. From Equation (5-6),

$$\eta = \frac{P_{out}}{P_{in}} = \frac{P_L}{P_L + \text{copper loss} + \text{core loss}} \tag{25-13}$$

The core loss remains constant and is equal to the wattmeter reading from the open-circuit test. But the copper loss is proportional to the *square* of current in the equivalent network of Figure 25-7(b). For practical purposes, the current, in turn, is proportional to the *apparent* power into the load. Hence, we can rewrite Equation (25-13),

$$\eta = \frac{S_L \cos \phi}{S_L \cos \phi + I^2 R_e + P_{CL}} \tag{25-14}$$

EXAMPLE 25-6

The transformer tested in Example 25-5 has a transformation ratio of 10 and develops a full-load secondary voltage of 12 V when the load current is 40 A at 0.866 lagging power factor. The open-circuit test indicates core losses of 5 W. Calculate the efficiency of the transformer with this load.

SOLUTION
    Step I.

$$S_L \cos \phi = 12 \text{ V} \times 40 \text{ A} \times 0.866 = 416 \text{ W}$$

Step II.   In the equivalent network of Figure 25-7(b),

$$I = \frac{I_s}{a} = \frac{40 \text{ A}}{10} = 4 \text{ A}$$

$$\text{Copper loss} = I^2 R_e = (4 \text{ A})^2 \times 0.75 \ \Omega = 12 \text{ W}$$

Step III.

$$\therefore \eta = \frac{416 \text{ W}}{416 \text{ W} + 12 \text{ W} + 5 \text{ W}} = 0.96 = 96\%$$

Since transformers are usually operated at somewhat less than full load, it is customary to design them so that *maximum* efficiency also occurs at somewhat less than full load. As we would expect from earlier consideration of the maximum power transfer theorem,

For maximum efficiency, copper loss = core loss.[†]

EXAMPLE 25-7

What secondary current must be drawn from the transformer of Example 25-6 for maximum efficiency?

SOLUTION
    Step I.

$$I^2 R_e = P_{CL} = 5 \text{ W}$$

from which
$$I = \sqrt{\frac{P}{R_e}} = \sqrt{\frac{5 \text{ W}}{0.75 \ \Omega}} = 2.58 \text{ A}$$

Step II.   The secondary current is

$$I_s = aI = 10 \times 2.58 \text{ A} = 25.8 \text{ A}$$

25-8 EFFECT OF LOADING A TRANSFORMER

One important effect of the leakage reactance and resistance of the windings in Figure 25-6 is that the load voltage V_L is not the same as the voltage induced into the secondary since, to satisfy Kirchhoff's voltage law,

[†]See Appendix 2-10.

$$\mathbf{E}_s = \mathbf{I}_s R_s + j\mathbf{I}_s X_s + \mathbf{V}_L$$

Similarly, the voltage induced into the primary by the mutual flux must be slightly less than the total applied voltage $\mathbf{V}_p$. Therefore, as the current drawn by the load is changed, the voltage across the load must also change, even though the applied voltage may remain constant.

We can show the effect of load current on the load voltage by preparing phasor diagrams for the equivalent network of Figure 25-7(b). The current flowing in this equivalent circuit is the primary load current, which is $1/a$ times the actual load current. Likewise, the voltage across $a^2\mathbf{Z}_L$ in Figure 25-7 is a times the actual load voltage. Because the current is common to all components, we use it as a reference phasor in the phasor diagrams of Figure 25-9. The total applied primary voltage will be the phasor sum of three voltages: $a\mathbf{V}_L$, which will have an angle with respect to the current depending on the power factor of the load (power-factor angle), $\mathbf{I}_p R_e$ (or $I_L/a \times R_e$), which will be in phase with the current, and $\mathbf{I}_p X_e$ (or $I_L/a \times X_e$), which will lead the current by 90°. Hence, when we reflect the secondary circuit elements back into the primary circuit to form the simple equivalent circuit of Figure 25-7(b),

$$\mathbf{V}_p = a\mathbf{V}_L + \mathbf{I}_p R_e + \mathbf{I}_p X_e \tag{25-15}$$

Although we could add these phasor voltages by the customary geometrical construction we learned in Chapter 19 (where all phasors are drawn from a common origin), our transformer phasor diagrams will be less confusing if we add each phasor, with its proper direction and magnitude, to the tip of the preceding phasor, as shown in Figure 25-9. Such phasors are called **funicular** (chain-like) phasors.

Since we are not including magnetizing current in our equivalent circuit, when we disconnect the load, $\mathbf{I}_p$ becomes zero. Hence, $\mathbf{I}_p R_e$ and $\mathbf{I}_p X_e$ in Figure 25-9 shrink to zero and $a\mathbf{V}_L$ and $\mathbf{V}_p$ are identical. But, as shown in Figure 25-9(a), when the load has a lagging power factor so that $\mathbf{V}_L$ leads the current, the load voltage will drop appreciably under load. However, with a leading power-factor load, as in Figure 25-9(b), the full-load voltage and no-load voltage are almost the same. The percentage change in secondary voltage from no load to full load with respect to the full-load voltage is called the percentage **voltage regulation** of the transformer.

FIGURE 25-9
Phasor Diagrams for a Transformer Under Load: (a) Lagging Power Factor Load; (b) Leading Power Factor Load.

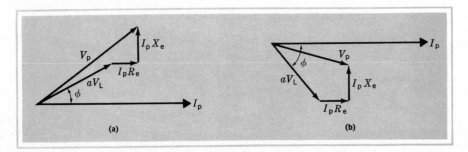

$$\text{Voltage regulation} = \frac{V_{NL} - V_{FL}}{V_{FL}} \times 100\% \qquad (25\text{-}16)$$

where V_{NL} is the no-load secondary voltage, and V_{FL} is the full-load secondary voltage.

EXAMPLE 25-8

Calculate the percentage voltage regulation for the transformer of Example 25-6.

SOLUTION

$$aV_L = 10 \times 12 \text{ V} = 120 \text{ V}$$

Since the power factor is 0.866 lagging, this voltage will lead the current by arccos 0.866 or 30°, as shown in Figure 25-10.

$$I_p R_e = \frac{40 \text{ A}}{10} \times 0.75 \text{ } \Omega = 3 \text{ V}$$

$$I_p X_e = \frac{40 \text{ A}}{10} \times 1.3 \text{ } \Omega = 5.2 \text{ V}$$

As shown in Figure 25-10, we must express the load voltage in rectangular coordinates in order to carry out the phasor addition.

$$aV_L = 120 \cos 30° + j120 \sin 30° = 104 + j60 \text{ V}$$
$$V_p = (104 + 3) + j(60 + 5.2) \text{ V}$$
$$\therefore V_p = \sqrt{107^2 + 65.2^2} = 125.3 \text{ V}$$
$$\text{No-load } V_L = \frac{V_p}{a} = \frac{125.3 \text{ V}}{10} = 12.53 \text{ V}$$
$$\text{Voltage regulation} = \frac{V_{NL} - V_{FL}}{V_{FL}} \times 100\%$$
$$= \frac{12.53 - 12.0}{12.0} \times 100\%$$
$$= 4.4\%$$

FIGURE 25-10
Phasor Diagram for Example 25-8.

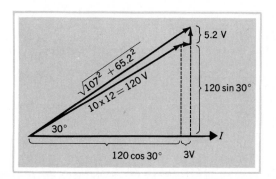

25-9 AUTOTRANSFORMERS

In many applications where we do not need the electrical isolation of separately insulated primary and secondary transformer windings, we can obtain the desired high-efficiency, voltage-changing ability of the transformer at reduced cost by using the **autotransformer** construction shown in Figure 25-11.

Instead of separate primary and secondary windings, the autotransformer of Figure 25-11 has one continuous tapped winding. Primary and secondary windings in a transformer must have separate load currents to produce equal and opposite mmfs. Since the lower 100 turns in Figure 25-11 are connected to the load, this becomes the secondary winding. Note that the primary consists of only the top 200 turns. Hence, the primary voltage is *not* the same as the source voltage. Since the source voltage is connected across all 300 turns, from Equation (25-8), the "primary" voltage is 80 V and the secondary voltage is 40 V.

The load in Figure 25-11 draws 15 A at 40 V for an apparent power of 600 VA. If we assume negligible core and copper losses, the apparent power drawn from the source is also 600 VA, from which $I_p' = 600/120 = 5$ A. From Kirchhoff's current law, the secondary current must be 10 A with a direction as shown in Figure 25-11. This agrees with Equation (25-10). The direction of I_s is *opposite* to the direction of I_p', which agrees with Lenz's law requirement that the two load components of the currents in the transformer windings must produce equal and opposite mmfs.

The apparent power delivered to the load by the secondary winding through transformer action is

$$S_s = V_s \times I_s = 40 \text{ V} \times 10 \text{ A} = 400 \text{ VA}$$

The remaining 200 VA reaches the load by direct conduction due to the primary current forming one third of the load current. Hence, an autotransformer used to reduce the voltage applied to a 600-VA load can be considerably smaller than a 600-VA separate-winding transformer.

We can connect a separate-winding transformer as an autotransformer as long as we arrange the two windings to act as a single continuous winding, as shown in Figure 25-12. We can also use an autotransformer in a step-up mode.

FIGURE 25-11
Autotransformer.

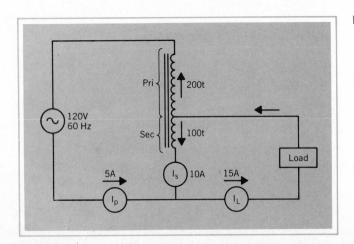

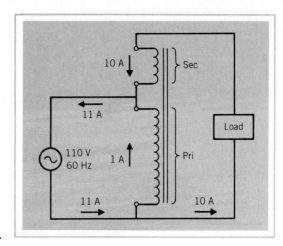

FIGURE 25-12
Separate-Winding Transformer Connected as a Step-up Autotransformer.

EXAMPLE 25-9

We want to raise the available 110-V 60-Hz power-line voltage to provide approximately 120 V to a 10-A load. Our junk box provides a 120-V to 12-V two-winding filament transformer rated at 120 VA. Show that this transformer can accomplish the task of feeding the 1200-VA load when it is connected as in Figure 25-12.

SOLUTION

From the transformer specifications,

$$a = \frac{V_p}{V_s} = \frac{120 \text{ V}}{12 \text{ V}} = 10$$

As connected in Figure 25-12, $V_p = E_{source}$.

$$\therefore V_s = \frac{110 \text{ V}}{10} = 11 \text{ V}$$

and $$V_L = 110 \text{ V} + 11 \text{ V} = 121 \text{ V}$$

Since I_L flows through the secondary winding,

$$I_p' = \frac{I_s}{a} = \frac{10 \text{ A}}{10} = 1 \text{ A}$$

Therefore, the current drawn from the source is

$$I_{source} = 1 \text{ A} + 10 \text{ A} = 11 \text{ A}$$

$$S_L = 121 \text{ V} \times 10 \text{ A} = 1210 \text{ VA}$$

But the apparent power delivered to the load by transformer action is only

$$S_p = 110 \text{ V} \times 1 \text{ A} = \mathbf{110 \text{ VA}} \quad \text{or} \quad S_s = 11 \text{ V} \times 10 \text{ A} = \mathbf{110 \text{ VA}}$$

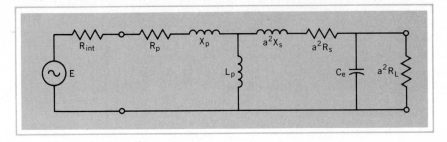

FIGURE 25-13
Equivalent Circuit of an Audio Transformer.

25-10 AUDIO TRANSFORMERS

We have already noted that transformers are very useful for **matching** a given load impedance to the output resistance of transistor amplifiers. In such applications, the frequency of the source can vary from 20 hertz to 20 kilohertz. To determine the effect of this range of input frequencies on the performance of an audio transformer, we must consider two additional circuit parameters in the equivalent circuit of Figure 25-13: the internal resistance of the source R_{int}, and, at the higher frequencies, the equivalent shunt capacitance C_e of the distributed capacitance between the turns.

As the frequency of the source decreases from a reference frequency of 1 kilohertz, the primary exciting current must increase in order to maintain the required mutual flux. Another way of saying the same thing is that, as the frequency of the source decreases, the reactance of the primary inductance decreases. This means that as the frequency decreases there will be more of the source voltage lost across R_{int} and less applied to the equivalent transformer circuit. We can consider the low-frequency limit of the transformer to be at the frequency at which the inductive reactance of the primary drops to the same value as the internal resistance of the source. Good low-frequency response, therefore, requires sources with low internal resistance and transformers with high primary inductance.

As the frequency increases, the voltage across the equivalent leakage reactance increases, which tends to reduce the output voltage, as shown by the shaded curve of Figure 25-14. But if the transformer is designed so that the equivalent capacitance and

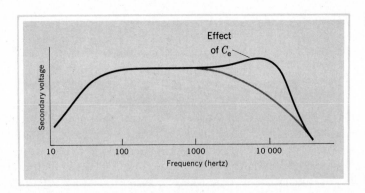

FIGURE 25-14
Frequency Response of a Typical Audio Transformer.

the leakage inductance are resonant at approximately 20 kilohertz, the resonant rise of voltage across this capacitance can be used to advantage to offset the natural roll-off at the high-frequency end of the audio spectrum. Excessive resonant rise would be undesirable, but this rise can be controlled in the transformer design by choosing the proper ratio between the reactance and resistance. At frequencies above the resonant frequency of X_e and C_e, the output voltage is very small.

PROBLEMS

25-1. What is the transformation ratio of a transformer having 40 turns on the secondary winding and 680 turns on the primary?

25-2. What is the transformation ratio of a transformer which has an open-circuit secondary voltage of 300 V rms when the primary is connected to a 120-V 60-Hz source?

25-3. What rms voltage must be applied to the primary of the transformer in Problem 25-1 to develop an open-circuit secondary voltage of 6.3 V?

25-4. What is the load component of the primary current when a 2-kΩ resistor is connected across the secondary winding of the transformer in Problem 25-2?

25-5. The secondary of the transformer in Problem 25-2 has 1400 turns. How many primary turns are required?

25-6. The primary of a certain transformer has 2.7 turns per volt of applied voltage. How many turns must there be on a 5-V secondary winding?

25-7. A 2300/230-V 15-kVA 60-Hz transformer is designed to have an induced voltage of 2.2 V per turn. Determine the number of turns and the full-load current of primary and secondary windings.

25-8. When an 0.8-Ω load is connected to the 24-V secondary of a step-down transformer, the primary current is measured as 8.0 A. What is the primary voltage of the transformer?

25-9. A small filament transformer has a laminated sheet steel core whose average path length is 20 cm and whose cross-sectional area is 5 cm^2. How many turns must there be on a primary winding to be connected to a 120-V 60-Hz source if the flux density must not exceed 1.1 T?

25-10. How many turns would there be on a $7\frac{1}{2}$-V secondary winding on the transformer of Problem 25-9?

25-11. Assuming the magnetizing current to be a pure sine wave, calculate the rms value of the primary current in the transformer of Problem 25-9 when the secondary is open-circuit. (Use the BH curves of Figure 14-4.)

25-12. The primary of the transformer of Problem 25-9 is connected to a 110-V 25-Hz source. What is the rms value of the exciting current?

25-13. What cross-sectional area must the core of the transformer in Problem 25-7 have if the flux density in it is not to exceed 1.1 T?

25-14. With the secondary open-circuit, the primary current of the transformer in Problem 25-7 is measured as 300 mA. What is the average path length of the magnetic circuit of the transformer (approximately)?

25-15. An audio output transformer has 1500 primary turns and 40 secondary turns. What is the reflected value of a 4-Ω load connected to its secondary?

25-16. An audio transformer is listed as 7-kΩ primary reflected impedance with a 4-Ω load at 1000 Hz. This transformer is to be used with an amplifier requiring a 4500-Ω load. What value of load must be connected to the transformer secondary?

25-17. An audio output transformer has 1200 primary turns and two secondary windings of 50 turns each. If the circuit feeding the transformer must see a reflected load of 5 kΩ, what load must be connected to the secondary windings when they are connected
(a) in series?
(b) in parallel?

25-18. An audio transformer has a 6-kΩ primary winding and an 8-Ω secondary with a tap at 4 Ω with respect to a common secondary terminal. Through error, a 4-Ω load is connected from the 4-Ω tap to the 8-Ω terminal. What is the reflected impedance?

25-19. A short-circuit test on a 2300/208-V 500-kVA 60-Hz distribution transformer yields the following data: V_p = 95 V, I_p = 218 A, and P_{in} = 8.2 kW. Calculate R_e and X_e for the equivalent circuit of Figure 25-7(b).

25-20. The open-circuit test on the transformer of Problem 25-19, with the high-voltage winding open and the low-voltage winding used as the primary, yields the following data: V = 208 V, I = 84 A, and P = 1.8 kW. Calculate the magnitude and phase angle of the exciting current when the transformer is used with the high-voltage winding as the primary.

Draw detailed and accurately-scaled phasor diagrams with the remaining problems.

25-21. Calculate the percentage voltage regulation for the transformer in Problem 25-19 with a 70.7% lagging power-factor load.

25-22. Calculate the percentage voltage regulation for the transformer in Problem 25-19 with an 86.6% leading power-factor load.

25-23. Given the following data on a 550/220-V 50-kVA 60-Hz transformer, calculate the percentage regulation with a 50% lagging power-factor load: R_p = 0.03 Ω, X_P = 0.05 Ω, R_s = 0.005 Ω, and X_s = 0.01 Ω.

25-24. The power input to the transformer in Problem 25-23 is 200 W when the secondary is open-circuit. Calculate the full-load efficiency with a unity power-factor load.

25-25. Calculate the efficiency and power output for maximum efficiency conditions with a 90% lagging power-factor load connected to the transformer of Problems 25-19 and 25-20.

REVIEW QUESTIONS

25-1. Why is mutual induction more significant in ac circuits than in dc circuits?

25-2. What is a **transformer?**

25-3. What is the purpose of an iron core in power and audio transformers?

25-4. What problems would be encountered in using iron cores in radio-frequency transformers?

25-5. What is meant by the **general transformer equation?** What is its application?

25-6. What is meant by **exciting current?**

25-7. What factors determine the magnitude of the exciting current?

25-8. The exciting current in a certain transformer lags the applied voltage by 87°. Account for this angle.

25-9. Account for the results obtained in Problem 25-12 and relate these results to a practical transformer.

25-10. What would be the effect of operating a 110-V 25-Hz transformer on a 120-V 60-Hz supply?

25-11. What is the relationship between the turns ratio of a transformer and its transformation ratio?

658

25-12. What is the relationship between the load current ratio and the turns ratio of a transformer?

25-13. By combining Equations (25-9) and (25-11), determine the power input to power output ratio of a transformer.

25-14. What is meant by **reflected impedance?**

25-15. Why is the impedance ratio of an audio transformer equal to the *square* of the turns ratio?

25-16. What is the significance of impedance ratio in practical transformers?

25-17. Complete the individual steps in reducing the initial equation in Example 25-4 to its final form.

25-18. What is meant by **mutual flux** and **leakage flux?**

25-19. Define **leakage reactance.**

25-20. Why is the total core loss of a transformer essentially independent of load current?

25-21. Why can we assume that the wattmeter reading on an open-circuit transformer test is essentially only the core loss?

25-22. Why is the core loss on a short-circuit transformer test less than $1/400$ of its normal value?

25-23. Explain why we can state that $R_e = R_p + a^2 R_s$ and $X_e = X_p + a^2 X_s$ in Figure 25-7(b).

25-24. Draw a phasor diagram showing the relationship between primary terminal voltage and secondary terminal voltage for a unity power-factor load.

25-25. What factor is primarily responsible for the difference between the no-load and full-load secondary voltages of a power transformer?

25-26. Why does a leading power-factor load give better voltage regulation than an equivalent lagging power-factor load?

25-27. Draw a circuit diagram for an autotransformer feeding a 208-V load from a 120-V source. Explain its operation with a numerical example.

25-28. By using the numerical value of load you selected in Question 25-27, show why an autotransformer can be appreciably smaller than the equivalent two-winding transformer.

25-29. What factors limit the lowest frequency that an audio transformer can handle satisfactorily?

25-30. What factors limit the highest frequency that an audio transformer can handle satisfactorily?

*25-31. Assuming a constant power factor of 0.866 lagging for the load connected to the transformer of Examples 25-5 and 25-6, and constant primary and secondary voltages, program a computer to prepare a table of percent efficiency (four significant digits) for load currents from 4 to 40 A in 2-A steps. In addition, display these data in graph form.

26

COUPLED CIRCUITS

26-1 DETERMINING COUPLING NETWORK PARAMETERS

There is an old saying about there being more than one way of skinning a cat. We have certainly discovered that this is true in solving electrical networks and in reducing them to simple equivalent circuits. We have already found several ways of solving the impedance network of Figure 26-1(a). Since this network is not too elaborate, we can view the network as a whole and write a set of Kirchhoff's voltage-law (loop) equations or Kirchhoff's current-law (nodal) equations from which we can solve for I_1 and I_2.

Thévenin's theorem provides us with a more powerful tool for working on this network. Thévenin's theorem views all networks as consisting of only two components, a two-terminal load and a two-terminal constant-voltage source. This technique is very useful for two reasons: (1) particularly in dealing with electric power systems, our basic notion of a simple electric circuit is that of connecting a load across a voltage source; and (2) Thévenin's theorem provides us with a laboratory method for determining the parameters of the Thévenin-equivalent source by laboratory measurements, even though we have no idea what the sealed two-terminal "black box" actually contains.

With the increasing significance of the signal-transmitting circuits of electronic communications and control systems, we need still another way of viewing the circuit of Figure 26-1(a). Suppose that we replace E_2 by a load impedance, as in Figure 26-1(b). The circuit now becomes a T-network, coupling the signal source E_1 to the load Z_L. Our experience with such networks tells us that I_2 will turn out to be a

660

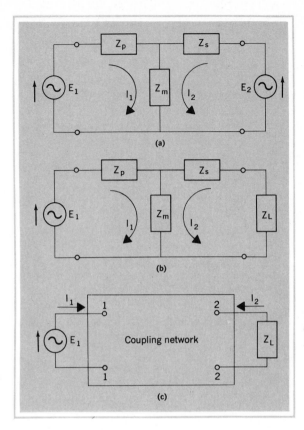

FIGURE 26-1
Coupling Networks: (a) Between Voltage Sources; (b) Between a Source and a Load; (c) Of Unknown Internal Composition.

negative quantity if we do not change its direction. However, to standardize our conventions, we think of the input current I_1 and the output current I_2 of the generalized coupling network of Figure 26-1(c) as having a direction *into* the network at both upper or "live" terminals of the network.

If we knew that the network of Figure 26-1(c) actually had the form of Figure 26-1(b), we could solve for I_2 (and I_1 if needed) by simple series-parallel impedance techniques. But if the coupling network is sealed up in a four-terminal "black box," as in Figure 26-1(c), we need some way of stating the network parameters in a form that we can obtain by the "black-box" measurement techniques we first used in Chapter 8. The simplest generalized coupling network will have one component or parameter which is common to both the input current I_1 and the output current I_2, one component through which only I_1 flows, and one through which only I_2 flows. We can, therefore, represent the coupling network, whatever its internal composition may be, by a simple T-network.

To determine the parameters of an unknown network, we connect the network in the test circuit shown in Figure 26-2, together with voltmeters whose internal resistances are high enough that we can neglect any voltmeter current. To find out what we "see" when looking back into the output terminals, we require E_2 rather than the actual load impedance.

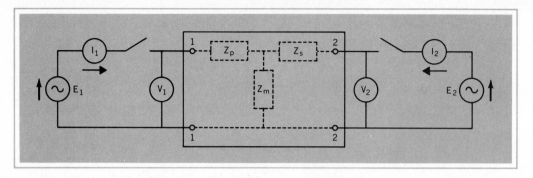

FIGURE 26-2
Determining the Open-Circuit Impedance Parameters of a Coupling Network.

26-2 OPEN-CIRCUIT IMPEDANCE PARAMETERS

To define the composition of a four-terminal, two-port network such as shown in Figure 26-1(c), we require four parameters. The test circuit of Figure 26-2 will give us one fairly common set of parameters called the **open-circuit impedance parameters**[†] of the network.

We start by opening the right-hand switch in Figure 26-2 so that $I_2 = 0$. Killing I_2 removes the possibility of the input to the network containing any component dependent on the output circuit. When we close the left-hand switch, the readings from voltmeter V_1 and ammeter I_1 give us a V/I ratio which represents only the open-circuit input *impedance* of the network. (Volts ÷ amperes = ohms.)

We now open the left-hand switch and close the right-hand switch. Since I_1 becomes zero, we won't be able to "see" the input impedance because there will be no voltage drop across it. However, it shouldn't surprise us to find a reading on voltmeter V_1 as a part of I_2 is coupled back to the input side of the network. We can create a parameter to express the amount of output-to-input coupling by recording the ratio of voltmeter V_1 reading to ammeter I_2 reading. Again we express a V/I ratio in ohms. Since it is the ratio between the voltage created at the *input* terminals (1,1) and the current flowing into the *output* terminals (2,2) to create it, we call this second parameter the **reverse-transfer impedance** of the coupling network. The same "measurement" technique provides similar output parameters. The complete set of open-circuit impedance parameters[‡] consists of

$$\mathbf{z}_{11} = \frac{\mathbf{E}_1}{\mathbf{I}_1} \qquad (\mathbf{I}_2 = 0) \quad \text{is the open-circuit input impedance}$$

(26-1)

[†]"Open-circuit" because we prevent the output circuit from affecting the measurement of input impedance by *opening* the right-hand switch in Figure 26-2.

[‡]We use lowercase letter symbols for impedance parameters because they are equivalent-circuit parameters, rather than actual circuit elements. Note carefully the subscript convention we observe.

$$\mathbf{z}_{12} = \frac{\mathbf{V}_1}{\mathbf{I}_2} \qquad (\mathbf{I}_1 = 0) \quad \text{is the } \textbf{reverse-transfer impedance} \qquad (26\text{-}2)$$

$$\mathbf{z}_{21} = \frac{\mathbf{V}_2}{\mathbf{I}_1} \qquad (\mathbf{I}_2 = 0) \quad \text{is the } \textbf{forward-transfer impedance} \qquad (26\text{-}3)$$

$$\mathbf{z}_{22} = \frac{\mathbf{E}_2}{\mathbf{I}_2} \qquad (\mathbf{I}_1 = 0) \quad \text{is the } \textbf{open-circuit output impedance} \qquad (26\text{-}4)$$

Having determined the four $\mathbf{z}$-parameters, we now close *both* switches in Figure 26-2 so that the network actually does couple the output and input circuits. Consequently, *both* $\mathbf{z}_{11}$ and $\mathbf{z}_{12}$ are now present in the input circuit and $\mathbf{V}_1$ consists of *two* voltages in series: the IZ drop due to $\mathbf{I}_1$ flowing in $\mathbf{z}_{11}$, plus a coupled voltage representing $\mathbf{I}_2$ flowing through the reverse-transfer impedance $\mathbf{z}_{12}$. We can write a Kirchhoff's voltage-law equation for the input terminals $(1,1)$ of the network as

$$\mathbf{z}_{11}\,\mathbf{I}_1 + \mathbf{z}_{12}\,\mathbf{I}_2 = \mathbf{E}_1 \qquad (26\text{-}5)$$

Similarly, for the output terminals $(2,2)$,

$$\mathbf{z}_{21}\,\mathbf{I}_1 + \mathbf{z}_{22}\,\mathbf{I}_2 = \mathbf{E}_2 \qquad (26\text{-}6)$$

The next step is to determine the equivalent circuit represented by the two Kirchhoff's voltage-law Equations (26-5) and (26-6). Using our "black-box" reasoning, we can investigate the two voltages we "see" looking into the input terminals of the coupling network. Obviously, the sum of two voltages represents a *series* circuit. With the left-hand switch open and the right-hand switch closed in Figure 26-2, we discovered a reading on voltmeter V_1 due to output current $\mathbf{I}_2$. Hence, we can argue that we are looking into a Thévenin-equivalent voltage source which is *dependent* on the current in some other part of the circuit. The open-circuit voltage of this current-controlled voltage source is $\mathbf{z}_{12}\,\mathbf{I}_2$. From the way we "measured" it, $\mathbf{z}_{11}$ appears to be the internal impedance of the dependent source. Looking into the output terminals, we see a similar Thévenin-equivalent dependent voltage source. Hence, Figure 26-3 becomes the equivalent circuit for the open-circuit impedance parameters of a four-terminal coupling network.

FIGURE 26-3
z-Parameter Equivalent Circuit.

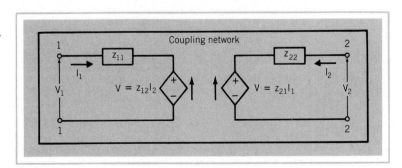

We can now perform the same "measurements" on paper with the dotted T-network in position in Figure 26-2. With the left-hand switch closed and the right-hand switch open,

$$\mathbf{I}_1 = \frac{\mathbf{E}_1}{\mathbf{Z}_p + \mathbf{Z}_m}$$

from which
$$\mathbf{Z}_p + \mathbf{Z}_m = \frac{\mathbf{E}_1}{\mathbf{I}_1} = \mathbf{z}_{11} \qquad (26\text{-}7)$$

While the left-hand switch is still closed and the right-hand switch open, voltmeter V_2 reads

$$\mathbf{V}_2 = \mathbf{Z}_m \mathbf{I}_1$$

Hence,
$$\mathbf{Z}_m = \frac{\mathbf{V}_2}{\mathbf{I}_1} = \mathbf{z}_{21} \qquad (26\text{-}8)$$

Similarly, with the left-hand switch in Figure 26-2 open and the right-hand switch closed,

$$\mathbf{Z}_s + \mathbf{Z}_m = \frac{\mathbf{E}_2}{\mathbf{I}_2} = \mathbf{z}_{22} \qquad (26\text{-}9)$$

and
$$\mathbf{Z}_m = \frac{\mathbf{V}_1}{\mathbf{I}_2} = \mathbf{z}_{12} \qquad (26\text{-}10)$$

We can also rearrange Equations (26-7) to (26-10) to express the three T-network components in terms of $\mathbf{z}$-parameters.

$$\mathbf{Z}_m = \mathbf{z}_{12} = \mathbf{z}_{21} \qquad (26\text{-}11)$$

$$\mathbf{Z}_p = \mathbf{z}_{11} - \mathbf{z}_{12} \qquad (26\text{-}12)$$

and
$$\mathbf{Z}_s = \mathbf{z}_{22} - \mathbf{z}_{21} \qquad (26\text{-}13)$$

If we know the $\mathbf{z}$-parameters for a passive coupling network, we can readily establish the equivalent T-network, or vice versa. To use a T-network, note that both forward- and reverse-transfer impedances must be the same. This is characteristic of any *passive* network containing only resistance and reactance elements.[†] Consequently, we call $\mathbf{Z}_m$ the **mutual impedance** of the coupling network. In *active* networks containing transistors, $\mathbf{z}_{12}$ and $\mathbf{z}_{21}$ must be different for amplification to take place. The $\mathbf{z}$-parameter equivalent circuit of Figure 26-3 can represent both passive and active coupling networks.

[†]The reciprocity theorem of Section 9-8 depends on this equal forward- and reverse-transfer impedance characteristic of coupling networks.

EXAMPLE 26-1

Check Equations (26-5) and (26-6) by writing Kirchhoff's voltage-law equations for the T-network in Figure 26-2.

SOLUTION

$$\mathbf{I}_1 \mathbf{Z}_p + (\mathbf{I}_1 + \mathbf{I}_2)\mathbf{Z}_m = (\mathbf{Z}_p + \mathbf{Z}_m)\mathbf{I}_1 + \mathbf{Z}_m \mathbf{I}_2 = \mathbf{E}_1$$

$$\therefore (\mathbf{z}_{11} - \mathbf{z}_{12} + \mathbf{z}_{12})\mathbf{I}_1 + \mathbf{z}_{12}\mathbf{I}_2 = \mathbf{E}_1 \qquad (26\text{-}5)$$

$$\mathbf{I}_2 \mathbf{Z}_s + (\mathbf{I}_1 + \mathbf{I}_2)\mathbf{Z}_m = (\mathbf{Z}_s + \mathbf{Z}_m)\mathbf{I}_2 + \mathbf{Z}_m \mathbf{I}_1 = \mathbf{E}_2$$

$$\therefore (\mathbf{z}_{22} - \mathbf{z}_{21} + \mathbf{z}_{21})\mathbf{I}_2 + \mathbf{z}_{21}\mathbf{I}_1 = \mathbf{E}_2 \qquad (26\text{-}6)$$

EXAMPLE 26-2

Write Equations (26-5) and (26-6) as they apply to the network in Figure 26-4.

SOLUTION

In this example, $\mathbf{Z}_L$ becomes the internal impedance of a source producing zero voltage. As Figure 26-4 shows, this internal impedance acts in series with $\mathbf{Z}_s$. There is no change in the equation for $\mathbf{E}_1$:

$$\mathbf{z}_{11}\mathbf{I}_1 + \mathbf{z}_{12}\mathbf{I}_2 = \mathbf{E}_1 \qquad (26\text{-}5)$$

But the second equation is

$$(\mathbf{Z}_L + \mathbf{Z}_s + \mathbf{Z}_m)\mathbf{I}_2 + \mathbf{Z}_m \mathbf{I}_1 = 0$$

from which
$$\mathbf{z}_{21}\mathbf{I}_1 + (\mathbf{z}_{22} + \mathbf{Z}_L)\mathbf{I}_2 = 0 \qquad (26\text{-}14)$$

Equations (26-5) and (26-14) provide us with a means of developing a simpler expression for the total input impedance of the T-coupling network of Figure 26-4 (and any generalized coupling network) than an equation for $\mathbf{Z}_{in}$ in terms of series and parallel impedances. From Equation (26-14),

$$\mathbf{I}_2 = \frac{-\mathbf{z}_{21}\mathbf{I}_1}{\mathbf{z}_{22} + \mathbf{Z}_L} \qquad (26\text{-}15)$$

FIGURE 26-4
T-Coupling Network.

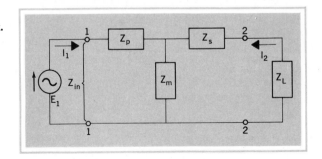

Substituting in Equation (26-5),

$$\mathbf{z}_{11}\mathbf{I}_1 - \frac{\mathbf{z}_{12}\mathbf{z}_{21}\mathbf{I}_1}{\mathbf{z}_{22} + \mathbf{Z}_L} = \mathbf{E}_1$$

Hence,

$$\mathbf{Z}_{in} = \frac{\mathbf{E}_1}{\mathbf{I}_1} = \mathbf{z}_{11} - \frac{\mathbf{z}_{12}\mathbf{z}_{21}}{\mathbf{z}_{22} + \mathbf{Z}_L}$$

But from Equation (26-11),

$$\mathbf{z}_{12}\mathbf{z}_{21} = \mathbf{Z}_m^2$$

Therefore,

$$\mathbf{Z}_{in} = \mathbf{z}_{11} - \frac{\mathbf{Z}_m^2}{\mathbf{z}_{22} + \mathbf{Z}_L} \tag{26-16}$$

The second term in Equation (26-16) is a significant parameter in coupled-circuit techniques; hence, we give it a special name:

$$\textbf{Coupled impedance} = \frac{\mathbf{Z}_m^2}{\mathbf{z}_{22} + \mathbf{Z}_L} \tag{26-17}$$

Equation (26-16) represents an equivalent circuit consisting of two impedances connected in *series*. The negative sign with the coupled impedance means that the coupled impedance tends to *reduce* the input impedance of the network so that energy may be coupled to the load. Thus, our coupled-circuit technique gives us a *series* **coupled impedance** in contrast to the *parallel* **reflected impedance** which we developed with the transformation-ratio technique for dealing with iron-core transformers in Chapter 25. However, as far as the source is concerned, both techniques produce the same results. In Chapter 25, disconnecting the load made the reflected impedance in *parallel* with the primary impedance infinitely large so that $\mathbf{Z}_{in}$ was simply the open-circuit input impedance of the transformer. With **z**-parameters, disconnecting the load so that $\mathbf{Z}_L$ in Equation (26-17) goes to infinity results in the coupled impedance in *series* with the open-circuit input impedance $\mathbf{z}_{11}$ becoming zero. Again, $\mathbf{Z}_{in}$ is simply the open-circuit input impedance.

EXAMPLE 26-3

What is the input impedance of the H-attenuator in Figure 26-5 when a 600-Ω resistive load is connected to its output terminals? (Check the results by series-parallel resistance calculations.)

SOLUTION

We may regard the H-attenuator as a balanced form of a T-network. The equivalent T-network has $\mathbf{Z}_p = \mathbf{Z}_s = 200 \ \Omega$. Hence,

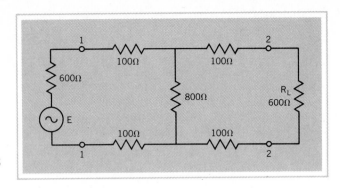

FIGURE 26-5
H-Attenuator.

$$\mathbf{z}_{11} = \mathbf{z}_{22} = 200 + 800 = 1000 \ \Omega$$

and $\qquad \mathbf{Z}_m = 800 \ \Omega \qquad$ (all $\underline{/0°}$)

$$\mathbf{Z}_{in} = \mathbf{z}_{11} - \frac{\mathbf{Z}_m^2}{\mathbf{z}_{22} + \mathbf{Z}_L} = 1000 - \frac{640\,000}{1000 + 600}$$

$$\therefore Z_{in} = \mathbf{600 \ \Omega}$$

EXAMPLE 26-4

Calculate the power into a 75-Ω load connected to the high-pass filter circuit in Figure 26-6.

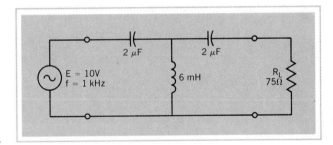

FIGURE 26-6
High-Pass Filter Network.

SOLUTION

$$\mathbf{z}_{11} = \mathbf{z}_{22} = -j\frac{1}{\omega C} + j\omega L$$

$$= -j\frac{1}{2 \times \pi \times 1 \text{ kHz} \times 2 \ \mu F} + j(2 \times \pi \times 1 \text{ kHz} \times 6 \text{ mH})$$

$$= -j79.58 + j37.70 = -j41.88 \ \Omega$$

$$\mathbf{Z}_m^2 = (+j37.7)^2 = -1421^†$$

†Remember that $(+j)^2 = (-j)^2 = -1$.

Hence,
$$\mathbf{Z}_{in} = \mathbf{z}_{11} - \frac{\mathbf{Z}_m^2}{\mathbf{z}_{22} + \mathbf{Z}_L} \qquad (26\text{-}16)$$

$$= -j41.88 + \frac{1421}{75 - j41.88} = 14.44 - j33.82$$

and
$$Z_{in} = 36.77 \ \Omega$$

(We do not need to record the angle of $\mathbf{Z}_{in}$ to determine the magnitude of the load current.)

$$I_{in} = \frac{E}{Z_{in}} = \frac{10 \text{ V}}{36.77 \ \Omega} = 272 \text{ mA}$$

$$\mathbf{I}_2 = \frac{-\mathbf{z}_{21}\mathbf{I}_1}{\mathbf{z}_{22} + \mathbf{Z}_L} \qquad (26\text{-}15)$$

$$\therefore I_2 = \frac{37.7}{85.9} \times 272 \text{ mA} = 119.4 \text{ mA}$$

Since the load contains only resistance,

$$P_L = I_2^2 R_L = (119.4 \text{ mA})^2 \times 75 \ \Omega = \mathbf{1.07 \ W}$$

26-3 SHORT-CIRCUIT ADMITTANCE PARAMETERS

In Figure 26-7, we want to represent the generalized coupling network by the π-network shown in dotted lines. If we follow the rules associated with Equations (26-1) to (26-4) carefully, we can still derive equations relating the three π-network impedances to $\mathbf{z}$-parameters for the network.[†] But by now we should be sufficiently familiar with the *dual* form of network equations to suspect that it might be simpler to work with *admittances* when we encounter a coupling network in the form of a π-network (which is a dual for a T-network). Although the resulting **short-circuit admittance parameters** are not often used in modern equivalent-circuit techniques, we shall derive the **y**-parameters of a generalized four-terminal coupling network as an introduction to the hybrid parameters of Section 26-4.

In the test circuit of Figure 26-7, to ensure that the source $\mathbf{E}_2$ does not interfere with the input meter readings when we are "measuring" the input admittance, we place the right-hand switch in the position which short-circuits the output terminals (2,2) through ammeter I_2. Similarly, when we apply $\mathbf{E}_2$ to the output terminals, the left-hand switch must short-circuit the input terminals (1,1) through ammeter I_1.

Starting with $\mathbf{E}_1$ applied to the input terminals, we write the input meter readings as an I/V ratio. Hence, we are stating the input *admittance* of the network with the output terminals short-circuit. Again we should not be surprised to find a reading on ammeter I_2 short-circuiting the output terminals as a portion of the input current is

[†]See Review Question 26-15.

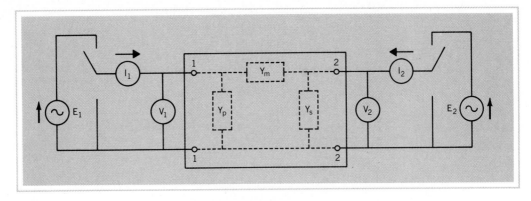

FIGURE 26-7
Determining the Short-Circuit Admittance Parameters of a Coupling Network.

coupled through to the output. Looking back into the short-circuited output terminals, we see a Norton-equivalent source containing a *dependent* current source controlled by the voltage applied to the input terminals. To find the constant that relates I_2 to E_1, we simply divide ammeter I_2 reading by voltmeter V_1 reading. And since amperes ÷ volts = siemens, we call this constant the **forward-transfer admittance** of the network. Reversing the position of the switches allows us to "measure" a **short-circuit output admittance** and a **reverse-transfer admittance.** Hence, the four admittance parameters of any four-terminal, two-port network are

$$y_{11} = \frac{I_1}{E_1} \qquad (E_2 = 0) \quad \text{is the } \textbf{short-circuit input admittance} \qquad (26\text{-}18)$$

$$y_{12} = \frac{I_1}{E_2} \qquad (E_1 = 0) \quad \text{is the } \textbf{reverse-transfer admittance} \qquad (26\text{-}19)$$

$$y_{21} = \frac{I_2}{E_1} \qquad (E_2 = 0) \quad \text{is the } \textbf{forward-transfer admittance} \qquad (26\text{-}20)$$

$$y_{22} = \frac{I_2}{E_2} \qquad (E_1 = 0) \quad \text{is the } \textbf{short-circuit output admittance} \qquad (26\text{-}21)$$

Since a Norton-equivalent current source has to be in *parallel* with the input admittance, when we look into the input terminals of the network, we see *two* paths for I_1 to follow when it enters the network. Writing the Kirchhoff's current-law equation for I_1 entering the input terminals $(1,1)$,

$$y_{11}\,E_1 + y_{12}\,E_2 = I_1 \qquad (26\text{-}22)$$

Similarly, for the output terminals $(2,2)$,

$$y_{21}\,E_1 + y_{22}\,E_2 = I_2 \qquad (26\text{-}23)$$

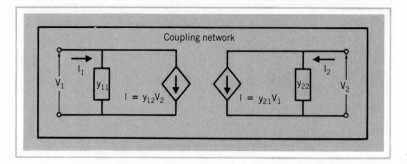

FIGURE 26-8
y-Parameter Equivalent Circuit.

Drawing the equivalent circuit which satisfies Equations (26-22) and (26-23) gives us the equivalent circuit for the short-circuit admittance parameters of a four-terminal coupling network, as shown in Figure 26-8.

If we now assume that the coupling network consists of the π-network shown in dotted lines in Figure 26-7, we can repeat the "measurements" of Equations (26-18) to (26-21) using values from the π-network. Starting with ammeter I_2 placing a short circuit across the output terminals,

$$\mathbf{I}_1 = \mathbf{E}_1(\mathbf{Y}_p + \mathbf{Y}_m)$$

$$\therefore \mathbf{Y}_p + \mathbf{Y}_m = \frac{\mathbf{I}_1}{\mathbf{E}_1} = \mathbf{y}_{11} \qquad (26\text{-}24)$$

Similarly,
$$\mathbf{Y}_s + \mathbf{Y}_m = \frac{\mathbf{I}_2}{\mathbf{E}_2} = \mathbf{y}_{22} \qquad (26\text{-}25)$$

With the left-hand switch in the short-circuit position, there can be no voltage drop across $\mathbf{Y}_p$—hence, no current through $\mathbf{Y}_p$. All the current through $\mathbf{Y}_m$ must flow through ammeter I_1. As a result, $\mathbf{I}_1$ will have the same *magnitude* as the current through $\mathbf{Y}_m$. But the current through $\mathbf{Y}_m$ is caused by $\mathbf{E}_2$ and, as we can see from Figure 26-7, has the *opposite* direction to $\mathbf{I}_1$. We would find that ammeter I_1 reads *backwards* in the short-circuit position of the switch. Hence,

$$\mathbf{I}_1 = -\mathbf{I}_{Y_m} = -\mathbf{Y}_m\mathbf{E}_2$$

Therefore,
$$\mathbf{Y}_m = \frac{-\mathbf{I}_1}{\mathbf{E}_2} = -\mathbf{y}_{12} \qquad (26\text{-}26)$$

Similarly
$$\mathbf{Y}_m = \frac{-\mathbf{I}_2}{\mathbf{E}_1} = -\mathbf{y}_{21} \qquad (26\text{-}27)$$

Rearranging Equations (26-24) to (26-27), we can now state the equivalent π-network in terms of the y-parameters of the coupling network as

$$\mathbf{Y}_m = -\mathbf{y}_{12} = -\mathbf{y}_{21} \qquad (26\text{-}28)$$

$$\mathbf{Y}_p = \mathbf{y}_{11} + \mathbf{y}_{12} \qquad (26\text{-}29)$$

and
$$\mathbf{Y}_s = \mathbf{y}_{22} + \mathbf{y}_{21} \qquad (26\text{-}30)$$

EXAMPLE 26-5

Determine the short-circuit admittance parameters for the low-pass filter circuit in Figure 26-9.

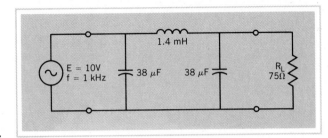

FIGURE 26-9
Low-Pass Filter Network.

SOLUTION

$$\mathbf{Y}_p = \mathbf{Y}_s = +j\omega C = +j(2 \times \pi \times 1 \text{ kHz} \times 38 \ \mu\text{F}) = +j238.8 \text{ mS}$$

$$\mathbf{Y}_m = -j\frac{1}{\omega L} = -j\frac{1}{2 \times \pi \times 1 \text{ kHz} \times 1.4 \text{ mH}} = -j113.7 \text{ mS}$$

The **y**-parameters are

$$\mathbf{y}_{11} = \mathbf{y}_{22} = \mathbf{Y}_p + \mathbf{Y}_m = +j238.8 \text{ mS} - j113.7 \text{ mS} = +\mathbf{j125.1 \text{ mS}}$$

$$\mathbf{y}_{12} = \mathbf{y}_{21} = -\mathbf{Y}_m = -(-j113.7 \text{ mS}) = +\mathbf{j113.7 \text{ mS}}$$

26.4 HYBRID PARAMETERS

Instead of being made up of all impedances or all admittances, as the name indicates, hybrid parameters use a mixture of units. In analyzing circuits containing active devices such as transistors, we find it more convenient to think of the input terminals of a four-terminal coupling network as a Thévenin-equivalent voltage source and the output terminals as a Norton-equivalent current source. We can still use the same lab measurement technique to determine the four hybrid parameters for any four-terminal coupling network.

To find the open-circuit voltage of the Thévenin-equivalent source at input terminals (1,1) in Figure 26-10(a), we feed $\mathbf{V}_2$ into the output terminals (2,2). In this case, we consider the Thévenin-equivalent source to be a *voltage*-controlled voltage source. Hence the parameter which represents the fraction of the output voltage appearing at the input terminals is $\mathbf{h}_{12} = \mathbf{V}_1/\mathbf{V}_2$. Since volts ÷ volts = a ratio without units, $\mathbf{h}_{12}$ is the **reverse-voltage** *ratio* parameter.

Since we are treating the dependent source as a *voltage*-controlled voltage source, to kill $\mathbf{V}_2$ while we measure the input voltage and current, we short circuit the output terminals, as in Figure 26-10(b). By recording the ratio $\mathbf{V}_1/\mathbf{I}_1$, the parameter $\mathbf{h}_{11}$ is expressed in ohms and represents the **short-circuit input impedance** of the

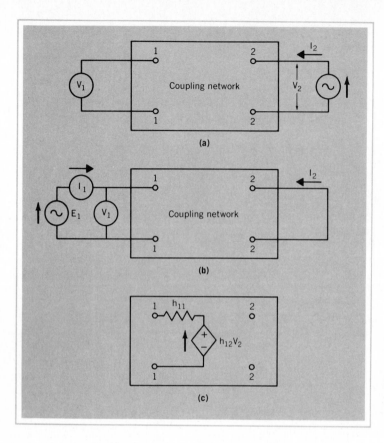

FIGURE 26-10
Finding the Thévenin-Equivalent Input Circuit of a Four-Terminal Network: (a) Open-Circuit Reverse Voltage; (b) Internal Input Resistance; (c) Network Input Parameters.

network. Since $h_{12}V_2$ is a *voltage* source, the equivalent input circuit for the coupling network shows the dependent voltage source and input impedance in series, as in Figure 26-10(c).

To determine the short-circuit current of the Norton-equivalent source at the output terminals (2,2) in Figure 26-11(a), we feed I_1 into the input terminals and short-circuit the output terminals through ammeter I_2. Because the network is linear, I_2 will be a constant fraction of the input current I_1. In this case, amperes ÷ amperes = a ratio without units and h_{21} becomes the **forward-current ratio** parameter.

Since the output impedance of a Norton-equivalent source is in *parallel* with the current source, it is more convenient to think in terms of output admittance. And since we are treating this dependent source as a *current*-controlled current source, to kill I_1 while we measure I_2 and V_2 in Figure 26-11(b), we leave the input terminals of the network open-circuit. By recording the ratio I_2/V_2, the parameter h_{22} is expressed in siemens and represents the **open-circuit output admittance.** Summarizing the four sets of measurements to determine the hybrid parameters of a four-terminal coupling network,

$$h_{11} = \frac{V_1}{I_1} \qquad (V_2 = 0) \quad \text{is the } \textbf{short-circuit input impedance} \qquad (26\text{-}31)$$

$$h_{12} = \frac{V_1}{V_2} \qquad (I_1 = 0) \quad \text{is the reverse-voltage ratio} \tag{26-32}$$

$$h_{21} = \frac{I_2}{I_1} \qquad (V_2 = 0) \quad \text{is the forward-current ratio} \tag{26-33}$$

$$h_{22} = \frac{I_2}{V_2} \qquad (I_1 = 0) \quad \text{is the open-circuit output admittance} \tag{26-34}$$

For the Thévenin-equivalent source we see looking into the input terminals (1,1) of the network, we can write a Kirchhoff's voltage-law equation, as we did for **z**-parameters. For the Norton-equivalent source looking back into the output terminals (2,2), we write a Kirchhoff's current-law equation, as we did for **y**-parameters. The two unknowns in this case are I_1 and V_2.

$$h_{11}I_1 + h_{12}V_2 = E_1 \tag{26-35}$$

$$h_{21}I_1 + h_{22}V_2 = I_2 \tag{26-36}$$

FIGURE 26-11
Finding the Norton-Equivalent Output Circuit of a Four-Terminal Network: (a) Short-Circuit Forward Current; (b) Output Conductance; (c) Complete Hybrid Parameters.

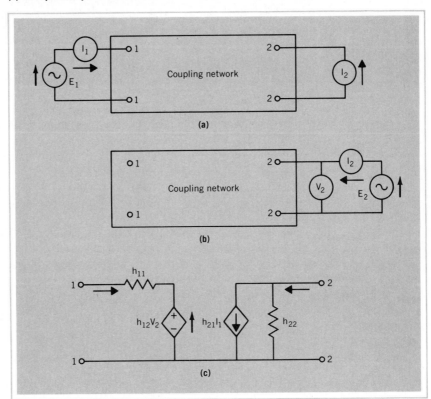

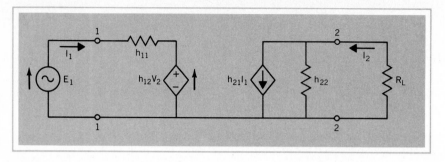

FIGURE 26-12
Hybrid Parameters of a Simple Transistor Amplifier.

The transistor amplifier equivalent circuit of Figure 26-12 is a typical example of hybrid parameters.

EXAMPLE 26-6

Given the following parameters for the transistor amplifier equivalent circuit of Figure 26-12, calculate the voltage gain when a 2.2-kΩ resistance load is connected across the output terminals.

$$\mathbf{h}_{11} = 50 \ \Omega \qquad \mathbf{h}_{12} = 3 \times 10^{-4}$$

$$\mathbf{h}_{21} = -0.95 \qquad \mathbf{h}_{22} = 2 \times 10^{-5} \ \text{S}$$

SOLUTION

We define voltage gain as V_2/E_1. One solution is to derive a formula for voltage gain by rearranging Equations (26-35) and (26-36) and then substituting for the parameters. We shall substitute directly and solve for V_2. Since there are no reactance elements in this equivalent circuit, we can solve Equations (26-35) and (26-36) by linear algebra. There are four variables in Equations (26-35) and (26-36). With only two equations, we must make two of them independent. We can pick some convenient value for E_1, say 1 V rms. And from Ohm's law,

$$I_2 = \frac{-V_2}{R_L}$$

The negative sign, as we have noted, is because the $\mathbf{I}_2$ direction in Equations (26-35) and (26-36) was established by a generator in place of R_L. For the same reason, $\mathbf{h}_{21}$ ($\mathbf{I}_2/\mathbf{I}_1$) in the above data is also a negative quantity. Substituting these values in Equations (26-35) and (26-36),

$$50 \ I_1 + 3 \times 10^{-4} \ V_2 = 1 \tag{1}$$

$$-0.95 \ I_1 + 2 \times 10^{-5} \ V_2 = \frac{-V_2}{2.2 \ \text{k}\Omega} \tag{2}$$

From (1),
$$I_1 = \frac{1 - 3 \times 10^{-4} \ V_2}{50}$$

674

Substituting in (2),

$$-0.95\left(\frac{1 - 3 \times 10^{-4}V_2}{50}\right) + 2 \times 10^{-5}V_2 = \frac{-V_2}{2.2 \text{ k}\Omega}$$

from which
$$V_2 = 39.9 \text{ V}$$

Hence, the voltage gain of the amplifier is

$$\frac{V_2}{E_1} = 40$$

In Section 9-5, we discovered that we cannot use the method stated in Thévenin's theorem to obtain the Thévenin-equivalent internal resistance of a dependent source when the controlling element is included in the transformation. Since this is the case when we want to determine input and output impedances of coupling networks, we must "measure" the V/I ratio. We determined $\mathbf{h}_{11}$ with a short circuit across the output terminals of the network. There will be a slightly different value for $\mathbf{Z}_{in}$ in the circuit of Figure 26-12 where reverse coupling occurs.

EXAMPLE 26-7

The source $\mathbf{E}_1$ in the equivalent circuit of Figure 26-12 has an internal resistance of 50 Ω. Determine the input and output resistances for the transistor amplifier circuit of Example 26-6.

SOLUTION

For this example, we shall derive formulas and then substitute the given parameters.

(a) In Example 26-6, we noted that

$$I_2 = \frac{-V_2}{R_L}$$

Substituting this expression for $\mathbf{I}_2$ in Equation (26-36) and collecting terms gives

$$V_2 = \frac{-\mathbf{h}_{21}\,\mathbf{Z}_L\,\mathbf{I}_1}{1 + \mathbf{h}_{22}\,\mathbf{Z}_L}$$

Substituting for $\mathbf{V}_2$ in Equation (26-35) and collecting terms,[†]

$$V_1 = \mathbf{h}_{11}\,\mathbf{I}_1 - \frac{\mathbf{h}_{12}\mathbf{h}_{21}\,\mathbf{Z}_L\,\mathbf{I}_1}{1 + \mathbf{h}_{22}\,\mathbf{Z}_L}$$

$$\therefore \mathbf{Z}_{in} = \frac{\mathbf{V}_1}{\mathbf{I}_1} = \mathbf{h}_{11} - \frac{\mathbf{h}_{12}\mathbf{h}_{21}\,\mathbf{Z}_L}{1 + \mathbf{h}_{22}\,\mathbf{Z}_L} \qquad (26\text{-}37)$$

[†]We use $\mathbf{V}_1$ rather than $\mathbf{E}_1$ because $\mathbf{V}_1$ will not be the same as $\mathbf{E}_1$ if the source has any internal resistance. The source resistance is *not* included in the input resistance of the network.

Substituting for the parameters in Equation (26-37),

$$R_{in} = \mathbf{Z}_{in} = 50 - \frac{-0.95 \times 3 \times 10^{-4} \times 2.2 \times 10^3}{1 + 2 \times 10^{-5} \times 2.2 \times 10^3}$$

$$= 50 + 0.601 = \mathbf{50.6\ \Omega}$$

(b) To obtain $\mathbf{h}_{22}$, we killed $\mathbf{I}_1$ by leaving the input terminals $(1, 1)$ in Figure 26-11(b) open-circuit. To "measure" the output impedance, we kill source $\mathbf{E}_1$ but leave its internal resistance $\mathbf{Z}_s$ connected across terminals $(1, 1)$. The Thévenin-equivalent source in Figure 26-11(c) creates a current through $\mathbf{h}_{11}$ and $\mathbf{Z}_s$ in the *opposite* direction to $\mathbf{I}_1$. Hence,

$$\mathbf{I}_1 = \frac{-\mathbf{h}_{12}\mathbf{V}_2}{\mathbf{h}_{11} + \mathbf{Z}_s}$$

Substituting for $\mathbf{I}_1$ in Equation (26-36),

$$\mathbf{I}_2 = \frac{-\mathbf{h}_{12}\mathbf{h}_{21}\mathbf{V}_2}{\mathbf{h}_{11} + \mathbf{Z}_s} + \mathbf{h}_{22}\mathbf{V}_2$$

$$= \mathbf{V}_2 \frac{\mathbf{h}_{22}(\mathbf{h}_{11} + \mathbf{Z}_s) - \mathbf{h}_{12}\mathbf{h}_{21}}{\mathbf{h}_{11} + \mathbf{Z}_s}$$

from which

$$\frac{\mathbf{V}_2}{\mathbf{I}_2} = \frac{\mathbf{h}_{11} + \mathbf{Z}_s}{\mathbf{h}_{22}(\mathbf{h}_{11} + \mathbf{Z}_s) - \mathbf{h}_{12}\mathbf{h}_{21}}$$

Dividing both numerator and denominator by $(\mathbf{h}_{11} + \mathbf{Z}_s)$,

$$\mathbf{Z}_{out} = \frac{1}{\mathbf{h}_{22} - \dfrac{\mathbf{h}_{12}\mathbf{h}_{21}}{\mathbf{h}_{11} + \mathbf{Z}_s}} \tag{26-38}$$

Substituting for the parameters in Equation (26-38),

$$\mathbf{Z}_{out} = \frac{1}{2 \times 10^{-5} - \dfrac{-0.95 \times 3 \times 10^{-4}}{50 + 50}} = \mathbf{43.76\ k\Omega}$$

26-5 AIR-CORE TRANSFORMERS

We encounter a good example of the coupled-circuit technique of circuit analysis in dealing with the magnetically coupled circuits used in radio circuitry. The equivalent circuit we developed in Chapter 25 for iron-core power and audio transformers based on the **transformation ratio** is valid only because the leakage flux is very small compared to the mutual flux. We cannot use the transformation-ratio technique with

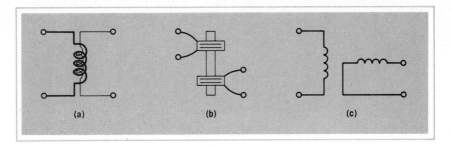

FIGURE 26-13
Variations in Magnetic Coupling: (a) Tight Coupling; (b) Loose Coupling; (c) Minimum Coupling.

loosely coupled air-core transformers, since the mutual flux in these transformers is only a fraction of the total primary flux. This fraction is called the **coefficient of coupling**. Therefore, by definition,

$$k = \frac{\Phi_m}{\Phi_p} \qquad (26\text{-}39)$$

where k is the coefficient of coupling.

The value of k can range from unity when all the primary flux links the secondary, as in the ideal transformer of Figure 25-1, down to zero when the secondary winding is completely outside the magnetic field of the primary or is purposely positioned at right angles to the primary flux [Figure 26-13(c)]. The physical arrangement of Figure 26-13(c) is used in radio circuits where magnetic coupling between tuned circuits is to be kept to a minimum and the physical dimensions are rather restricted. In order to obtain *tight* coupling with k close to unity in radio-frequency coils where iron cores are not practical, the two strands of wire are interwound on the coil form as a **bifilar** winding [Figure 26-13(a)]. Magnetic coupling of *tuned* coils requires a very *loose* coupling with a coefficient of coupling of approximately 0.01, as shown by the physical placement of the two coils on the coil form of Figure 26-13(b).

26-6 MUTUAL INDUCTANCE

To develop an equivalent coupling network for an air-core transformer, we require one parameter which is common to both primary and secondary circuits. This parameter is the **mutual inductance** of the pair of coils. In Chapter 15, we found that a coil has a *self*-inductance of one henry when current in that coil *changing* at a rate of one ampere per second induces a voltage of one volt in the coil. Applying the same line of thought to the process of mutual induction, we can say that

A pair of magnetically coupled coils has the mutual inductance of one henry when current changing at the rate of one ampere per second in one coil induces an average voltage of one volt in the other coil.

The letter symbol for mutual inductance is M.

Therefore, by definition,

$$E_{s_{av}} = \frac{MI_{p_m}}{t} \tag{26-40}$$

Since the instantaneous current must rise from zero to maximum in one-quarter of a cycle of a sine wave,

$$E_{s_{av}} = \frac{MI_{p_m}}{1/4f} = 4fMI_{p_m}$$

But

$$E_{av} = \frac{2}{\pi} E_m \tag{17-17}$$

Thus,

$$\frac{2}{\pi} E_{s_m} = 4fMI_{p_m}$$

from which

$$E_s = 2\pi fMI_p \tag{26-41}$$

Rearranging Equation (26-41) gives

$$\frac{E_s}{I_p} = 2\pi fM = X_M \tag{26-42}$$

where X_M is called the **mutual reactance** of the magnetically coupled windings.

EXAMPLE 26-8

120 V, 60 Hz is applied to the primary of a transformer whose primary inductance is 5 H. The open-circuit secondary voltage is 40 V. Neglecting losses, determine the mutual inductance between the two windings.

SOLUTION

$$E_s = \omega MI_p$$

But

$$I_p = \frac{E_p}{\omega L_p}$$

$$\therefore E_s = \frac{M}{L_p} E_p$$

and

$$M = L_p \frac{E_s}{E_p} = 5 \times \frac{40}{120} = \mathbf{1.67\ H}$$

Combining Equations (25-1) and (26-40)

$$\frac{N\Phi_m}{t} = \frac{MI_{p_m}}{t}$$

But from Equation (26-39)
$$\Phi_m = k\Phi_p$$
$$\therefore MI_{P_m} = N_s k \Phi_p \tag{1}$$

If we now reverse the windings and use the original secondary winding as the primary, and vice versa, it follows that

$$MI_{s_m} = N_p k \Phi_s \tag{2}$$

Multiplying Equations (1) and (2),

$$M^2 = k^2 \left(\frac{N_p \Phi_p}{I_{P_m}}\right)\left(\frac{N_s \Phi_s}{I_{s_m}}\right) \tag{3}$$

But from the definition of self-inductance,

$$E_{av} = \frac{LI_m}{t}$$

and from the definition of the weber,

$$E_{av} = \frac{N\Phi}{t}$$

$$\therefore L = \frac{N\Phi}{I_m}$$

Substituting in Equation (3) gives

$$M = k\sqrt{L_p L_s} \tag{26-43}$$

EXAMPLE 26-9

If the secondary winding of the transformer of Example 26-8 has a self-inductance of 0.8 H, what is the coefficient of coupling between the windings?

SOLUTION

$$k = \frac{M}{\sqrt{L_p L_s}} = \frac{1.67}{\sqrt{5 \times 0.8}} = 0.835$$

One method for experimentally determining mutual inductance is shown in Figure 26-14. We can measure the total inductance of the series-connected coils on an inductance bridge. If we connect the windings as shown and check by our hand rule, the flux produced by the current in one winding is in the same direction around the core as the flux produced by the current in the other winding. This increases the total flux, which increases the voltage induced by a given alternating current and, in turn, increases the total inductance. All the induced voltages will be in phase. There

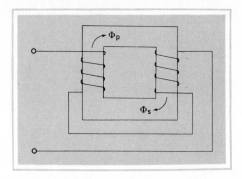

FIGURE 26-14
Determining Mutual Inductance.

are four induced voltages: the self-induced voltage in the primary, the voltage mutually induced in the primary by current in the secondary, the voltage mutually induced in the secondary by current in the primary, and the self-induced voltage in the secondary:

$$\therefore E = I(\omega L_p + \omega M + \omega M + \omega L_s)$$

from which $L_T = L_p + L_s + 2M$.

If, however, we reverse the leads to the secondary, the mutually induced voltages in each coil are 180° out of phase with the self-induced voltages, resulting in

$$L_T = L_p + L_s - 2M$$

Therefore, we can extend our original Equation (15-6) for two inductances in series to include magnetic coupling between them;

$$\therefore L_T = L_p + L_s \pm 2M \tag{26-44}$$

The mutual inductance of the magnetically-coupled coils in Figure 26-14 will be one-quarter of the difference between the total inductance readings with the coils connected series-aiding and then series-opposing.

26-7 COUPLED IMPEDANCE

Before we set up the equivalent coupling network for a loosely coupled transformer, we should note a useful practical advantage that transformer coupling has over other forms of coupling networks. For a given direction of $\mathbf{E}_1$ in Figure 26-15, we can reverse the phase of the output voltage by reversing the direction of one of the windings, or simply by reversing the leads to one of the windings. Where it is necessary to keep track of this phase relationship in a circuit diagram, we mark one end of each winding with a dot, as shown in Figure 26-15(a) and (b). If, at a certain moment, changing mutual flux induces an instantaneous voltage into the primary winding with a polarity such that the dotted end of the winding is positive with respect to the undotted end, then the same mutual flux change must induce an instantaneous voltage into the secondary winding with a polarity such that the dotted end of the secondary winding is positive with respect to the undotted end.

Although reversing the secondary polarity will reverse the current through the external circuit, note that I_2 in Figure 26-15(b) still must flow into the dot end of the secondary winding, just as it does in Figure 26-15(a). Hence, the primary circuit is not affected by any phase reversal we obtain by reversing the secondary leads. In selecting the direction for these current arrows, we must remember that the *same* mutual flux (which is produced by these currents) induces voltages into both primary and secondary windings. So if we draw I_1 pointing into the dot end of the primary winding, we must also draw I_2 pointing into the dot end of the secondary winding.

When we write the Kirchhoff's voltage-law (loop) equation for the primary loop, there are three voltages which we add as phasors to equal E_1: an *IR* drop across the resistance of the primary circuit, a *self*-induced voltage due to I_1 flowing through the primary winding, and a *mutually*-induced voltage due to I_2 flowing in the secondary circuit. Figure 26-15(c) shows one form of equivalent circuit for this transformer, in which the self-induced voltage is represented by an inductance symbol and the mutually induced voltage by a CCVS symbol. To satisfy Lenz's law, the polarity of the CCVS must be such that it opposes changes in I_1 due to the applied voltage E_1. Hence, changing the polarity of the CCVS in the secondary circuit will *not* change the polarity of the voltage I_s $(+j\omega M)$ in the primary circuit. Consequently, Kirchhoff's voltage-law equation for the primary loop becomes

$$I_p(R_p + j\omega L_p) + I_s(+j\omega M) = E_p$$

or

$$I_p Z_p + I_s(+j\omega M) = E_p \qquad (26\text{-}45)$$

FIGURE 26-15
Equivalent Circuit of an Air-Core Transformer.

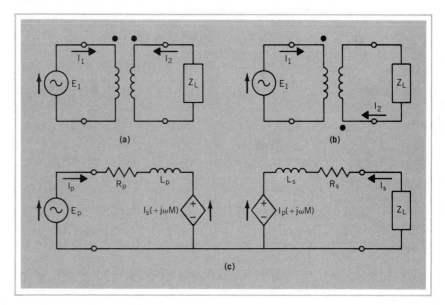

where $\mathbf{Z}_p$ is the open-circuit impedance of the primary circuit by itself, and ωM is the mutual reactance of the two windings. Note that Lenz's law gives us the same direction for $\mathbf{I}_2$ that we chose for our generalized coupling network. [Compare Equation (26-45) with Equation (26-5).] Hence, we write the loop equation for the secondary circuit of Figure 26-15(c) as

$$\mathbf{I}_s(\mathbf{Z}_L + R_s + j\omega L_s) + \mathbf{I}_p(+j\omega M) = 0$$

or
$$\mathbf{I}_s(\mathbf{Z}_L + \mathbf{Z}_s) + \mathbf{I}_p(+j\omega M) = 0 \tag{26-46}$$

where $\mathbf{Z}_s$ is the open-circuit impedance of the secondary winding by itself. [Compare Equations (26-46) and (26-14).] From Equation (26-46),

$$\mathbf{I}_s = \frac{-\mathbf{I}_p(+j\omega M)}{\mathbf{Z}_s + \mathbf{Z}_L} \tag{26-47}$$

Substituting in Equation (26-45),

$$\mathbf{I}_p\mathbf{Z}_p - \mathbf{I}_p\frac{(+j\omega M)^2}{\mathbf{Z}_s + \mathbf{Z}_L} = \mathbf{E}_p \tag{26-48}$$

Equation (26-48) is the equivalent of Equation (26-16). In this case, we can go one step further by substituting -1 for j^2. Thus,

$$\mathbf{E}_p = \mathbf{I}_p\mathbf{Z}_p + \mathbf{I}_p\frac{(\omega M)^2}{\mathbf{Z}_s + \mathbf{Z}_L} \tag{26-49}$$

and, dividing through by I_p,

$$\mathbf{Z}_{in} = \mathbf{Z}_p + \frac{(\omega M)^2}{\mathbf{Z}_s + \mathbf{Z}_L} \tag{26-50}$$

where $(\omega M)^2/(\mathbf{Z}_s + \mathbf{Z}_L)$ is the **coupled impedance** *for transformer coupling.* Note that when we substitute -1 for j^2, $(\omega M)^2$ has now become an impedance with a $0°$ angle.

We can check Equation (26-50) by considering the effect of loading the secondary winding of a pair of magnetically-coupled coils. If the secondary is left open-circuit, $\mathbf{Z}_L$ is infinitely large and, therefore, the coupled impedance in Equation (26-50) becomes zero. As a result, the total impedance is simply the primary impedance alone as we would expect. If we connect a resistance across the secondary winding, $\mathbf{Z}_L$ will have a $0°$ angle. Since $\mathbf{Z}_s$ is largely the inductive reactance of the secondary winding, the total secondary-circuit impedance is inductive. Consequently, when we carry out the phasor division, dividing an impedance with a $+$ angle into $(\omega M)^2$ with its $0°$ angle results in a *capacitive* coupled impedance. Because $\mathbf{Z}_p$ is largely *inductive*, the *capacitive* impedance coupled into the primary circuit via the mutual inductance adds the coupled resistance component in series with the primary resistance component but *reduces* the total primary reactance. As a result, the total impedance becomes smaller as the secondary is loaded with a resistance, thus allowing more primary current to flow in order to transfer energy to the secondary circuit. This then checks with our first approach to transformer action in which we discovered that primary current must increase when a transformer is loaded in order to keep the amplitude of the mutual-flux sine wave constant.

EXAMPLE 26-10

If the resistance of the windings can be neglected, determine the total input impedance to the transformer of Examples 26-8 and 26-9, (a) with the secondary open-circuit, and (b) with a 50-Ω resistance connected to the secondary.

SOLUTION

(a)
$$\mathbf{Z}_{in} = \mathbf{Z}_p = \omega L_p = 377 \times 5 = +j1885 \ \Omega$$

(b)
$$\text{Coupled } \mathbf{Z} = \frac{(\omega M)^2}{\mathbf{Z}_s + \mathbf{Z}_L} = \frac{(377 \times 1.667)^2}{+j(377 \times 0.8) + 50}$$

$$= \frac{395\,000\,\underline{/0°}}{305.7\,\underline{/80.59°}} = 1292 \ \Omega\,\underline{/-80.59°} = 211.2 - j1275 \ \Omega$$

$$\therefore \mathbf{Z}_{in} = (+j1885) + (211.2 - j1275) = 211.2 + j610 \ \Omega$$

$$= 645.5 \ \Omega\,\underline{/+70.9°}$$

26-8 TUNED TRANSFORMERS

The most important application of loose coupling is the use of tuned transformers in radio circuitry. If we consider the source to be a constant-voltage source, as in Figure 26-16(a), the primary winding is designed to form a series-resonant circuit with the

FIGURE 26-16
Tuned Transformer: (a) Fed from a Low-Impedance Constant-Voltage Source; (b) Fed from a High-Impedance Constant-Current Source.

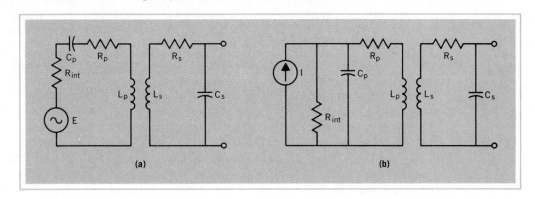

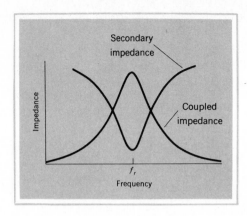

FIGURE 26-17
Effect of a Tuned Secondary Circuit.

capacitive reactance of C_p equal to the inductive reactance of the primary winding at the desired resonant frequency. This will result in maximum current in the primary winding at resonance (neglecting for the moment any coupled impedance due to secondary current). Since the transistor amplifiers we use to feed the primary of a tuned transformer have a relatively high internal resistance, they behave as constant-current sources. Accordingly, C_p must be connected in *parallel* with the primary winding so that the resonant rise of impedance will increase the voltage across (and the tank current in) the primary at resonance, as in Figure 26-16(b).

When we consider the effect of tuning the secondary circuit, the total secondary-circuit impedance [$\mathbf{Z}_s + \mathbf{Z}_L$ in Equation (26-50)] will be minimum and equal to R_s at resonance. Therefore, the secondary circuit behaves like a series-resonant circuit. Since $(\omega M)^2$ is essentially constant over the small range of frequencies near resonance, when $\mathbf{Z}_s + \mathbf{Z}_L$ becomes a *minimum,* the coupled impedance becomes a *maximum,* as shown in Figure 26-17. When the source frequency is slightly lower than the resonant frequency, X_{C_s} is greater than X_{L_s} and the secondary impedance becomes capacitive. But dividing a *capacitive* impedance into $(\omega M)^2$ with its 0° angle results in an *inductive* coupled impedance. Thus, as far as the signal source is concerned, the secondary behaves as if it were a *parallel*-resonant circuit in *series* with the primary winding.

The primary impedance by itself in the circuit of Figure 26-16(a) is attempting to become a minimum at resonance in order to allow maximum primary current to flow. But the coupled impedance is attempting to raise the primary impedance at resonance, thus limiting the maximum primary current. The extent to which the coupled secondary-tuned circuit affects the primary resonance curve depends on the degree of coupling between the coils. When the coupling is very loose, the mutual inductance is very small and, even at resonance, the coupled impedance is *smaller* than the resistance of the primary circuit. Under these circumstances, the only effect that the coupled secondary has on the shape of the primary-current resonance curve is to limit its peak value slightly. The secondary current tends to have the usual series resonance curve. But, with loose coupling, since the primary current rises to a maximum at resonance and since the secondary induced voltage depends on the primary current, the secondary resonance curve is much sharper than that of a single-tuned circuit having the same Q.[†]

We recall from earlier studies that maximum transfer of energy occurs when the load resistance is equal to the resistance of the source. If we apply this principle to magnetically coupled tuned circuits, maximum energy transfer will take place when the coupled resistance is equal to the primary-circuit resistance. Since the coupled resistance depends on the mutual inductance, which in turn depends on the coefficient of coupling, there is a **critical coupling** for a given pair of tuned circuits at which maximum energy transfer from primary to secondary takes place.

At resonance,
$$\mathbf{Z}_p = R_p$$

and the coupled
$$\mathbf{Z} = \frac{(\omega M)^2}{R_s}$$

Therefore, for critical coupling,

$$M^2 = R_p R_s / \omega^2$$

But
$$M^2 = k^2 L_p L_s \qquad (26\text{-}43)$$

Hence,
$$k^2 = \frac{R_p}{\omega L_p} \times \frac{R_s}{\omega L_s}$$

from which
$$k_c = \frac{1}{\sqrt{Q_p Q_s}} \qquad (26\text{-}51)$$

With critical coupling, the secondary current attains its greatest value. But, with the coupled impedance rising at resonance, the primary current is no longer maximum at resonance. As we have already noted, at a frequency slightly below resonance, the coupled impedance is *inductive,* whereas the impedance of the series-tuned primary is *capacitive*. Therefore, there are two frequencies, one slightly above the resonant frequency and one slightly below where the coupled reactance tunes the primary reactance to give minimum total primary impedance and thus maximum primary current. This double hump in the primary current tends to flatten the peak of the secondary current response curve because the voltage induced into the secondary at the resonant frequency is not quite as great as at the two hump frequencies either side of resonance.

If the tuned circuits are overcoupled, the increase in coupled impedance moves the primary-current humps further apart. This, in turn, causes such a decrease in primary current at resonance that the secondary current also starts to show a double-humped curve, as shown in Figure 26-18. In practice, a coefficient of coupling of 1.5 times the critical coefficient of coupling produces such a slight dip in the secondary current at resonance that the resulting secondary-current resonance curve has a very desirable flat top with steep skirts.

[†]The bandwidth is approximately $\Delta f \approx k f_r$.

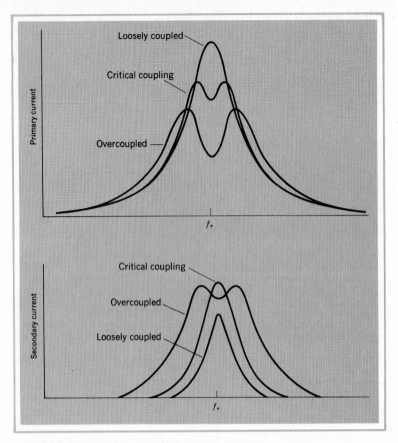

FIGURE 26-18
Effect of Coefficient of Coupling on the Resonance Curves of a Tuned Transformer.

PROBLEMS

*In Problems 26-1 to 26-4, calculate the open-circuit **z**-parameters for each T-network. Use the subscript notation of Figure 26-4.*

26-1. $\mathbf{Z}_p = R_p = 356\ \Omega$, $\mathbf{Z}_s = R_s = 36\ \Omega$, $\mathbf{Z}_m = R_m = 286\ \Omega$.

26-2. $\mathbf{Z}_p = R_p = 2500\ \Omega$, $\mathbf{Z}_s = -j500\ \Omega$, $\mathbf{Z}_m = R_m = 10\ k\Omega$.

26-3. $\mathbf{Z}_p = \mathbf{Z}_s = 20 + j800\ \Omega$, $\mathbf{Z}_m = -j800\ \Omega$.

26-4. $\mathbf{Z}_p = \mathbf{Z}_s = $ a 50-μH inductance in series with a 200-pF capacitance, $\mathbf{Z}_m = $ a 200-μH inductance, $f = 1$ MHz.

*In Problems 26-5 to 26-8, find the equivalent T-networks for the "black boxes" having the given **z**-parameters.*

26-5. $\mathbf{z}_{11} = 600\ \Omega\underline{/0°}$, $\mathbf{z}_{12} = \mathbf{z}_{21} = 400\ \Omega\underline{/0°}$, $\mathbf{z}_{22} = 1200\ \Omega\underline{/0°}$.

26-6. $\mathbf{z}_{11} = 500\ k\Omega - j5\ k\Omega$, $\mathbf{z}_{12} = \mathbf{z}_{21} = \mathbf{z}_{22} = 500\ k\Omega + j0\ \Omega$.

686

26-7. $z_{11} = z_{22} = 50 + j0 \ \Omega$, $z_{12} = z_{21} = -j1200 \ \Omega$.

26-8. $z_{11} = 100 + j1500 \ \Omega$, $z_{12} = z_{21} = 40 + j500 \ \Omega$, $z_{22} = 80 + j1200 \ \Omega$.

26-9. What is the coupled impedance when a 1500-Ω resistive load is connected to the output terminals of the network in Problem 26-5?

26-10. What is the coupled impedance when a capacitive reactance of 1240 Ω is connected to the output terminals of the network in Problem 26-8?

26-11. What is Z_{in} when a 250-Ω resistive load is connected to the output terminals of the T-network in Problem 26-1?

26-12. What is Z_{in} when a 1-kΩ resistive load is connected to the output terminals of the T-network in Problem 26-4?

26-13. In the T-attenuator circuit of Problem 26-11, determine the ratio between the total power input to terminals (1,1) and the power into the load at terminals (2,2) and express the attenuation in decibels.

26-14. With the techniques used in deriving Equations (26-7) to (26-10), and noting that the internal resistance of the internal source symbol is zero, find the z-parameters for the transistor in Figure 23-26.

*In Problems 26-15 to 26-18, find the open-circuit **z**-parameters for each π-network. Use the subscript notation of Figure 26-19.*

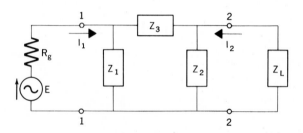

FIGURE 26-19
Pi-Coupling Network.

26-15. $\mathbf{Z}_1 = R_1 = 3470 \ \Omega$, $\mathbf{Z}_2 = R_2 = 350 \ \Omega$, $\mathbf{Z}_3 = R_3 = 440 \ \Omega$.

26-16. $\mathbf{Z}_1 = 80 \ k\Omega \underline{/0°}$, $\mathbf{Z}_2 = 470 \ k\Omega \underline{/0°}$, $\mathbf{Z}_3$ is a 0.02-μF capacitor, and $f = 100$ Hz.

26-17. $\mathbf{Z}_1 = \mathbf{Z}_2 = -j3600 \ \Omega$, $\mathbf{Z}_3 = +j3600 \ \Omega$.

26-18. $\mathbf{Z}_1 = \mathbf{Z}_2 = 150 \ \mu$H, $\mathbf{Z}_3 = 600 \ \Omega$, $f = 500$ kHz.

26-19. Find Z_{in} when a 300-Ω resistive load is connected to the output terminals of the π-network in Problem 26-15.

26-20. Find Z_{in} when a load of $300 - j400 \ \Omega$ is connected to the output terminals of the π-network in Problem 26-18.

*In Problems 26-21 and 26-22, find the short-circuit **y**-parameters for each π-network. Use the subscript notation of Figure 26-7.*

26-21. $\mathbf{Y}_p = G_p = 2$ mS, $\mathbf{Y}_s = G_s = 4$ mS, $\mathbf{Y}_m = G_m = 5$ mS.

26-22. $\mathbf{Y}_p = \mathbf{Y}_s = +j400 \ \mu$S, $\mathbf{Y}_m = -j400 \ \mu$S.

*In Problems 26-23 and 26-24, find the equivalent π-networks for the "black boxes" having the given **y**-parameters.*

26-23. $\mathbf{y}_{11} = 400\ \mu S\ \underline{/0°}$, $\mathbf{y}_{12} = \mathbf{y}_{21} = 200\ \mu S\ \underline{/180°}$, $\mathbf{y}_{22} = 600\ \mu S\ \underline{/0°}$.

26-24. $\mathbf{y}_{11} = \mathbf{y}_{22} = 10\ mS\ \underline{/-45°}$, $\mathbf{y}_{12} = \mathbf{y}_{21} = 10\ mS\ \underline{/-135°}$.

26-25. Find the **y**-parameters for the network of Problem 26-18.

26-26. Find the **z**-parameters for the network of Problem 26-23.

26-27. Hybrid parameters are not used for simple resistance coupling networks. However, they may readily be derived. What would the **h**-parameters for the network of Problem 26-1 look like?

26-28. Find the hybrid parameters for the transistor in Figure 23-26.

26-29. For a common-base configuration, a certain transistor has the following parameters:

$$\mathbf{h}_{11} = 40\ \Omega \qquad \mathbf{h}_{12} = 4 \times 10^{-4}$$

$$\mathbf{h}_{21} = -0.96 \qquad \mathbf{h}_{22} = 10^{-5}\ S$$

The transistor feeds into a 4.7-kΩ resistive load and is fed from a 1-kHz signal source having an open-circuit voltage of 10 mV and an internal resistance of 250 Ω. Calculate the voltage gain (the ratio of output to input voltage).

26-30. For a common-emitter configuration, the same transistor as used in Problem 26-29 has the following parameters:

$$\mathbf{h}_{11} = 1000\ \Omega \qquad \mathbf{h}_{12} = 9.6 \times 10^{-3}$$

$$\mathbf{h}_{21} = 24 \qquad \mathbf{h}_{22} = 2.5 \times 10^{-4}\ S$$

For the same signal source and load as in Problem 26-29, calculate the voltage gain for this circuit configuration.

26-31. An air-core transformer with adjustable coupling has a primary self-inductance of 200 mH and a secondary self-inductance of 240 mH. If a 500-μA current at $f = 1$ kHz is flowing in the primary, what coefficient of coupling is required to induce a secondary voltage of 50 mV?

26-32. Two coils wound on a common core have self-inductances of 1.6 and 3.2 H, respectively. When connected in series, their total inductance is 8 H.
   (a) What is the coefficient of coupling?
   (b) What is the total inductance when the leads of one of the coils are reversed?

26-33. An air-core transformer has a primary inductance of 200 mH, a secondary inductance of 50 mH, and a coefficient of coupling of 0.1. If the resistance of the windings can be neglected, what is the coupled impedance at 100 kHz when a 10-kΩ resistor is connected across the secondary?

26-34. If the voltage applied to the primary winding of the transformer in Problem 26-33 has an rms value of 20 mV, what voltage will appear across the 10-kΩ resistor in the secondary circuit?

26-35. The primary and secondary windings of an intermediate-frequency transformer in a radio receiver each have a self-inductance of 360 μH. The total equivalent primary-circuit series resistance is 30 Ω, and the total equivalent secondary-circuit series resistance is 24 Ω.
   (a) What value of tuning capacitors are required to tune each winding to 456 kHz?
   (b) What is the critical coupling coefficient?
   (c) What is the approximate bandwidth of the IF transformer if its coils are critically coupled?

26-36. An IF transformer in a television receiver has an untuned primary winding and a tuned secondary. The primary self-inductance is 2.4 μH, the secondary inductance is 3.0 μH, and the coefficient of coupling is 0.15. The secondary circuit has an effective resistance of 20 Ω. If the primary current contains a 42-MHz component with an rms value of 50 μA, what voltage will appear across the tuning capacitance in the secondary when the secondary is tuned to resonance?

REVIEW QUESTIONS

26-1. Why will I_2 in the circuit of Figure 26-1(b) turn out to be a negative quantity?

26-2. What is meant by the "black-box" technique of determining the equivalent circuit of a network?

26-3. Why do we require at least three components in the equivalent circuit of a generalized coupling network?

26-4. Give a practical example (with a circuit diagram) in which one of these components has an impedance of 0 Ω.

26-5. Why are z_{12} and z_{21} of the open-circuit impedance parameters called **transfer impedances?**

26-6. With reference to Equations (26-5) and (26-6), discuss the merit of our choice of direction for I_1 and I_2.

26-7. What is the **mutual impedance** of a coupling network?

26-8. Under what circumstances will a coupling network have no mutual impedance?

26-9. Figure 23-26 shows the (older) R-parameters of a transistor equivalent circuit. What is the reason for the internal voltage source symbol in this circuit?

26-10. Account for the $-$ sign in Equation (26-15).

26-11. Account for the $-$ sign in Equation (26-16).

26-12. At a given frequency, it would make no difference to Z_{in} in Figure 26-4 if Z_m were a pure capacitive reactance of X ohms or a pure inductive reactance of X ohms. Explain.

26-13. Why is the H-attenuator network of Figure 26-5 called a "balanced" network, whereas the equivalent T-network is not?

26-14. Under what circumstances must we use the short-circuit parameters of a coupling network, rather than the open-circuit parameters?

26-15. By carefully following the rules for Equations (26-1) to (26-4), determine the open-circuit **z**-parameters for the π-network of Figure 26-19.

26-16. From the **z**-parameters derived in Question 26-15, find the equivalent T-network.

26-17. Derive an equation for Y_{in} when a load admittance replaces E_2 in Figure 26-7 which will be the dual of Equation (26-16).

26-18. Why is the term **hybrid** used to describe the parameters of Figure 26-12?

26-19. State the units in which each of the four hybrid parameters is expressed.

26-20. Is it possible to apply the Thévenin-equivalent circuit technique used in developing the hybrid parameters to a *passive* coupling network? Explain with numerical examples.

26-21. Define **coefficient of coupling.**

26-22. What is meant by **loose** coupling and **tight** coupling?

26-23. Express coefficient of coupling in terms of mutual flux and leakage flux.

26-24. What effect has the coefficient of coupling on the mutual inductance of a pair of magnetically-coupled coils?

26-25. What is meant by **mutual reactance?**

26-26. How does the coupled impedance of loose-coupled coils differ from the reflected impedance of an audio transformer?

26-27. How is the coupled impedance related to the coefficient of coupling?

26-28. Why does an inductive secondary circuit appear as a capacitive coupled impedance?

26-29. Define **critical coupling.**

26-30. Why is the secondary current greater at frequencies slightly off resonance than at the resonant frequency in an overcoupled tuned transformer?

26-31. Why does increasing the coupling between two tuned circuits increase their bandwidth?

26-32. Does resonant rise of voltage take place in the secondary circuit of a tuned transformer? Explain.

*26-33. Write a computer program which will print out the voltage gain of a transistor amplifier when you enter the four hybrid parameters of the amplifier and R_L.

27

THREE-PHASE SYSTEMS

27-1 ADVANTAGES OF POLYPHASE SYSTEMS

In most of the ac networks we have considered up to this point, we have encountered only one source of alternating voltage. Such circuits are referred to as **single-phase** circuits. In Chapter 17, we were introduced to the sinusoidal nature of alternating voltages by considering the instantaneous voltage induced in an electric conductor rotating at a constant angular velocity in a uniform magnetic field. We also noted that most of our electric energy is generated at the expense of mechanical energy by **alternators** based on the principle of a loop rotating in a magnetic field. However, when we consider practical systems for the generation and distribution of large amounts of electric energy, the single-phase system has several disadvantages.

If we add a second loop to the simple alternator of Chapter 17, as shown in Figure 27-1(a), two separate sine-wave voltages will be induced in the one machine. Since both loops are mounted on the same rotor assembly, both rotate at the same angular velocity. Therefore, both voltages have the same frequency. If both loops have the same number of turns, both voltages will have the same rms value. But what is most important is that loop A is mounted on the rotor 90° ahead (in the direction of rotation) of loop B. Consequently, loop A always cuts a certain magnetic line of force 90° ahead (in time) of loop B. Therefore, although we can use the two voltages separately if we wish, as in Figure 27-2(a), the voltage in loop A is always 90° out of phase with the voltage in loop B, as shown in Figure 27-1(b). This simple alternator is referred to as a **two-phase** alternator.

In Figure 27-2, we have connected identical loads to the two windings of the simple two-phase alternator. Note the customary procedure of using a coil to represent

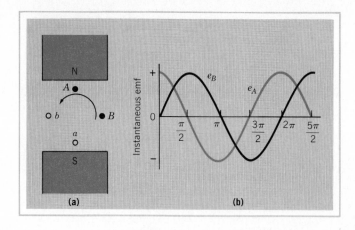

FIGURE 27-1
Simple Two-Phase Generator.

the alternator winding rather than the general ac generator symbol. This procedure enables us to draw the coils at right angles in a circuit diagram to indicate that their voltages are 90° out of phase. Even though we can connect each winding independently to its load, as in Figure 27-2(a), it is customary to use a common or **neutral** lead for the two circuits, as in Figure 27-2(b). This neutral conductor reduces the number of conductors required from four to three. Although, as the circuit diagram of Figure 27-2(b) indicates, the neutral current is the sum of the two load currents, this is a *phasor* sum. Therefore, if the two loads are identical, the neutral current is not twice the current in each of the other two conductors, but, as shown in the phasor diagram of Figure 27-3, it is only $\sqrt{2}$, or 1.414, times the current in each of the other conductors. Therefore, the total copper cross section to feed the two 120-ohm loads at 120 volts in the two-phase system is only the copper cross section required to handle one ampere, 1.414 amperes, and one ampere, a total of 3.414 amperes. The equivalent

FIGURE 27-2
Simple Two-Phase System.

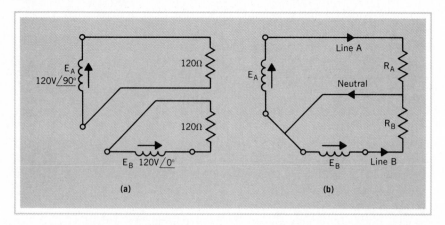

PART V   IMPEDANCE NETWORKS

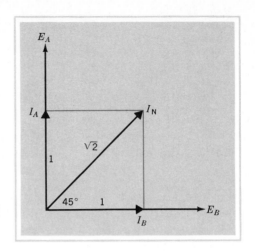

FIGURE 27-3
Phasor Diagram for the Neutral Current in a Simple Two-Phase System.

single-phase system to supply the two 120-ohm loads in parallel at 120 volts would require two conductors, each carrying two amperes, or a total copper cross section to handle four amperes. This then is the first advantage of any polyphase system. **Less copper is required to supply a given load power at a given voltage with a polyphase system than with a single-phase system.**

We can illustrate the second advantage of a polyphase system by plotting on a common graph, as in Figure 27-4, the graphs of the instantaneous power to the two identical loads in the circuits of Figure 27-2. The total instantaneous power supplied by the alternator at any instant is the sum of the instantaneous power to the two loads at that instant. Since the two voltages are 90° out of phase, the instantaneous power to one load becomes maximum as the other becomes zero. Even though the instantaneous power in each load is the customary pulsating sine-squared wave, if we check carefully, we find that at every instant in time the sum of the two instantaneous powers is constant. Therefore, **if the load on each phase of a polyphase source is identical, the instantaneous power output of the alternator is constant.** This is an important advantage in large machines since it allows a steady conversion of mechanical energy into electric energy.

FIGURE 27-4
Instantaneous Power in a Balanced Two-Phase System.

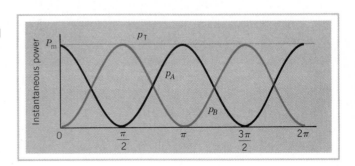

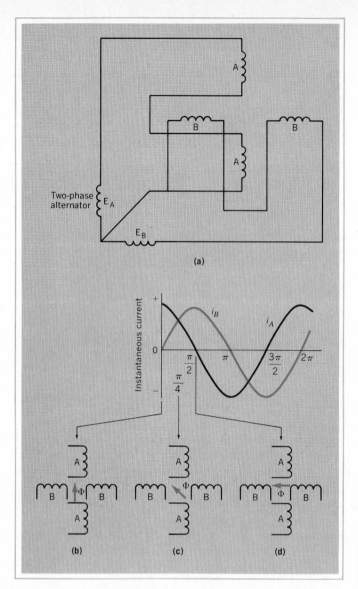

FIGURE 27-5
Producing a Rotating Magnetic Field in a Two-Phase System.

The third advantage can be demonstrated by connecting the two-phase alternator to a set of coils placed at right angles, as shown in Figure 27-5(a). Each coil will pass a sine wave of current, but the current in coil B will be 90° out of phase with the current in coil A. The magnetic flux produced by the two currents at any instant depends on the magnitude at that instant of the two instantaneous currents. At the 0° position, the current in coil B is zero and the current in coil A is maximum in the positive direction. This produces a magnetic field in which the direction of the magnetic lines of force is that indicated by the arrow in Figure 27-5(b). At 45°, both coils pass $0.707I_m$, resulting in a net flux which is the *phasor* sum of these two components. This results in a total flux with the *same* total flux density but with the direction of the magnetic lines of force being that shown in Figure 27-5(c). At the 90° point in the cycle, coil

694

A current is zero and coil B current is maximum positive, resulting in a magnetic field which is now at right angles to its original direction at $0°$ in the cycle.

If we continue the diagrams commenced in Figure 27-5, we find that the coil arrangement of Figure 27-5(a) connected to a two-phase source develops **a magnetic field which has a constant flux density and rotates at the frequency of the applied sine wave.** If we place a compass needle in the center of the coils in Figure 27-5(a), it will rotate with the magnetic field at a **synchronous speed,** which, in this case, is the same as the frequency of the two-phase sine wave. This rotating magnetic field, characteristic of all polyphase systems, greatly simplifies ac motor construction. A single-phase system produces only a magnetic field with increases and decreases in flux density and *reverses* its direction each $180°$ but does *not* rotate.

27-2 GENERATION OF THREE-PHASE VOLTAGES

The simple two-phase system that we used to show the advantages of polyphase electric power generation has been used to a limited extent in the distribution of electric energy. However, its main application in present-day equipment is in servomechanisms. The auto pilot of an aircraft is an example of a servomechanism. If the aircraft deviates from its course, a gyrocompass senses this error and develops voltages which operate a servomotor, which, in turn, operates the rudder to bring the aircraft back on course. Two-phase motors are ideal for servosystem applications because their rotation depends on the presence of an error voltage and their direction of rotation depends on whether this sine-wave error voltage leads or lags a reference voltage by $90°$.

In search of an improvement on the two-phase system for the distribution of electric energy, let us suppose that we place two coils $180°$ apart on the rotor of a simple ac generator and connect them to two identical loads with a common neutral lead, as shown in Figure 27-6. The neutral current is now the phasor sum of two equal currents which are $180°$ out of phase. Therefore, with identical loads, the neutral current in this system is zero. Consequently, this system requires less copper than the two-phase system of Figure 27-2(b) to supply each load with 115 V at a given power.

FIGURE 27-6
Edison Three-Wire Single-Phase System.

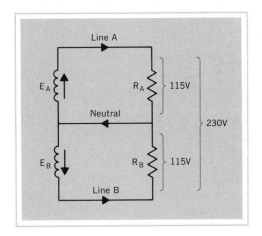

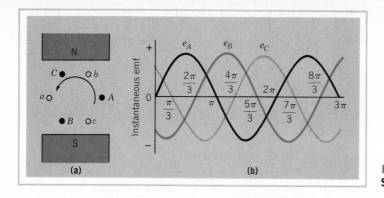

FIGURE 27-7
Simple Three-Phase Generator.

However, the total voltage from line A to line B is greater than in the two-phase system. And the instantaneous power in each coil of the generator in Figure 27-6 reaches its peak at the *same* instant, and thus the total instantaneous power *is* pulsating in nature. Also, with two currents 180° out of phase, we cannot produce a rotating magnetic field. Therefore, this system is *not* a two-phase system. It is called the Edison three-wire, single-phase system and is widely used for distributing electric energy to residences because of the reduction in line current (and the resulting reduction in wire size required) for a given load power and voltage, as compared with the simple two-wire, single-phase system.

The system which is universally used for electric power distribution is a *three*-phase system, based on the simple generator of Figure 27-7(a), which has three coils mounted on the rotor at 120° intervals. This alternator generates three sine-wave voltages which are 120° out of phase with one another, as shown in Figure 27-7(b). Figure 27-8(a) shows one method of connecting these coils to a load. With a **balanced** load, the currents in each arm of the load will be equal in magnitude and 120° out of phase with one another, as shown in the phasor diagram of Figure 27-8(b).[†] The neutral current is the phasor sum of these three load currents. As the phasor diagram of Figure 27-8(b) indicates, this neutral current will be zero with a balanced load. As a result, with a balanced load, even less copper is required to convey energy at a given rate and given voltage in the three-phase system than in an equivalent two-phase system.

With identical load resistors in Figure 27-8(a),

$$p_T = \frac{e_A^2 + e_B^2 + e_C^2}{R}$$

Careful inspection of Figure 27-7(b) shows us that $e_A^2 + e_B^2 + e_C^2$ is always constant. Therefore, the instantaneous power in a balanced three-phase system is a constant. We also find that we can use the voltages generated by a three-phase alternator to produce a rotating magnetic field of constant flux density, just as we did with the two-phase system. Consequently, the three-phase system has all the advantages of the two-phase system plus others that we shall discover as we investigate the behavior of three-phase circuits.

[†]A balanced load is one in which the impedance of each arm of the load is identical, both in magnitude and phase (power factor) angle. A balanced load need not be only resistance.

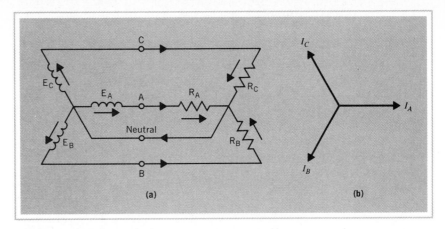

FIGURE 27-8
Simple Three-Phase System.

27-3 DOUBLE-SUBSCRIPT NOTATION

The three-phase (3ϕ) circuit of Figure 27-8 represents an impedance network with three voltage sources acting simutaneously. Reversing the leads to one of the coils in an ac generator has the same effect on an ac circuit as reversing the leads to a battery in a dc circuit containing several dc voltage sources. Hence, we cannot take the six leads from the coils in Figure 27-8(a), connect them in a random manner, and then assume that we obtain the current relationships shown by the phasor diagram of Figure 27-8(b). When we consider that we can reverse the leads to each coil of the simple 3ϕ generator, thus changing the angle of the source voltage by 180°, we can obtain quite a few possible combinations.

In the circuit of Figure 27-8(a), we indicate the manner in which we connected the leads to each winding by drawing an arrow alongside the symbol representing the alternator winding. Using the angle given with the source voltage, we must consider the tracing direction around the loop to be in the direction indicated by the arrow. However, the arrow system by itself does not allow us to show the connection of the leads to each coil when we write a Kirchhoff's voltage-law equation for the loop. Neither does the letter symbol $\mathbf{E}_A$ give us any indication of which lead is which.

This problem can be solved by using a **double-subscript notation.** In one of the common systems of double-subscript notation, one end of each coil is indicated by a lowercase letter and the other by an uppercase letter, as in Figure 27-9(a). In any complete electric circuit, these coils will be included in tracing loops around the circuit. If, as we follow one of these tracing loops, it enters coil A by the terminal marked with the lowercase letter and leaves by the terminal marked with the uppercase letter, we would represent the induced voltage of that coil as $\mathbf{E}_{aA}$. **The first of the two subscripts always represents the end of a circuit component at which the tracing loop enters the component, and the second subscript represents the end of the component from which the tracing loop leaves the component.**

When we use double-subscript notation to identify voltages, we must also be careful to distinguish between a **voltage rise** *inside* the source and a **voltage drop** in the *external* circuit. In the simplified diagram of Figure 27-10, we have connected a

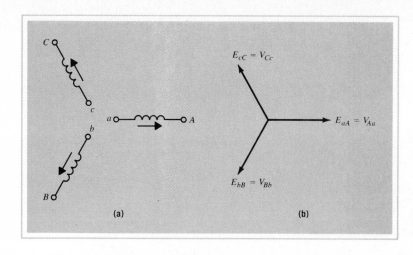

(a) (b)

FIGURE 27-9
Double-Subscript Notation for Three-Phase Voltages.

load directly across an alternator winding whose terminals are marked a and A. The direction of the induced voltage for this winding, as shown by the arrow in Figure 27-10, is from a to A *inside* the alternator. Therefore, we identify the voltage rise inside the alternator as $\mathbf{E}_{aA}$. To be consistent with the tracing loop, the voltage drop across the load $\mathbf{V}_{Aa}$ must reverse the order of the subscripts. From Kirchhoff's voltage law,

$$\mathbf{E}_{aA} = \mathbf{V}_{Aa} \tag{27-1}$$

Hence, the same phasors in Figure 27-9(b) can be used to represent either the voltage rises *inside* the source or the voltage drops in the external loads.

When we write loop equations based on Kirchhoff's voltage law, we include both $\mathbf{E}$ and $\mathbf{V}$ in the same equations [as in Equation (27-1)]. Hence, proper designation of letter symbols and subscripts is quite important. But when we write Kirchhoff's current-law equations for nodal analysis, we use only the external voltage drop between the two nodes and encounter a further convention for subscripts. **In the circuit external to the source, the *second* subscript with the voltage drop symbol V is the *reference* node.** Hence, in Figure 27-10, terminal a is the reference node and the voltmeter is measuring the voltage from terminal A with respect to terminal a.

In the following sections, when we are concerned with the manner in which we connect alternator windings, we use the internal voltage rise symbol $\mathbf{E}$. If we reverse the leads to coil A so that the tracing loop in a given circuit enters the coil at end A

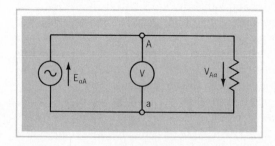

FIGURE 27-10
Double-Subscript Notation for Voltage Rise and Voltage Drop.

and leaves at end a, we would represent the source voltage in Kirchhoff's voltage-law equations as $\mathbf{E}_{Aa}$, where $\mathbf{E}_{Aa}$ has the same rms magnitude as $\mathbf{E}_{aA}$ but its angle is displaced by an additional 180°. From our study of subtraction of phasors in Chapter 19, since $\mathbf{E}_{Aa}$ is 180° out of phase with $\mathbf{E}_{aA}$,

$$\therefore \mathbf{E}_{Aa} = -\mathbf{E}_{aA} \tag{27-2}$$

When we make calculations in the *external* circuit, we usually use the appropriate voltage drop symbol $\mathbf{V}$ with conventional double subscripts.

27-4 FOUR-WIRE WYE-CONNECTED SYSTEM

We can assume that, in winding the coils on the simple 3ϕ generator of Figure 27-7, we were methodical enough that if we take the tracing direction from the lowercase letter to the uppercase letter in each coil, the three voltages will be 120° out of phase. We assume that these three generated voltages for the alternator of Figure 27-7 are

$$\mathbf{E}_{aA} = 120 \text{ V } \underline{/0°}$$
$$\mathbf{E}_{bB} = 120 \text{ V } \underline{/-120°}$$
$$\mathbf{E}_{cC} = 120 \text{ V } \underline{/+120°}$$

Since one of the reasons for using a three-phase system is to cut down on the conductors required to carry electric energy to a load, we can connect one end of each generator coil to a common neutral. **It is conventional to consider the tracing direction as going out from the generator to the load along the individual lines and returning along the common neutral lead;** consequently, we connect the lower case letter ends of the three windings to the neutral, as shown in Figure 27-11. To show that points a, b, and c are all connected to the neutral, we can replace these lower case letters with the letter o. Hence, the three source voltages become $\mathbf{E}_{oA} =$

FIGURE 27-11
Wye-Connected Alternator Windings.

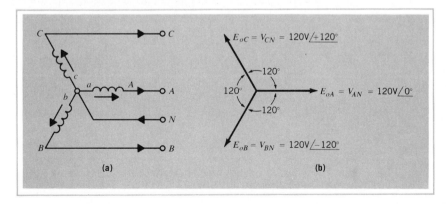

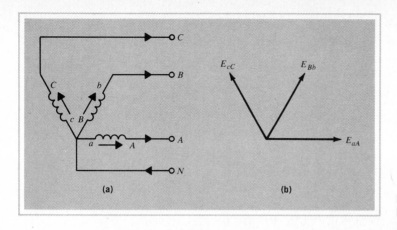

FIGURE 27-12
Incorrect Wye Connection.

$\mathbf{E}_{aA}$, $\mathbf{E}_{oB} = \mathbf{E}_{bB}$, and $\mathbf{E}_{oC} = \mathbf{E}_{cC}$. As shown in Figure 27-11(b), we designate the equivalent voltage drops from line to neutral as $\mathbf{V}_{AN}$, $\mathbf{V}_{BN}$, and $\mathbf{V}_{CN}$. Connecting the alternator coils in this manner produces a **wye-** (Y)-connected three-phase source, so called because of the appearance of the circuit diagram and the phasor diagram of Figure 27-11.[†] The voltage measured from line to neutral with a wye-connected source is called the **phase voltage** of the source.

We could also form a wye-connected 3ϕ source by connecting all the uppercase letter ends of the coils to the common neutral. However, we must avoid the situation shown in Figure 27-12. Coils A and C have their lowercase letter ends connected to the neutral, but coil B leads have been reversed. Since the tracing direction must be out along the line and back along the neutral, we must reverse the subscripts and represent the voltage induced in coil B as $\mathbf{E}_{Bb}$ in Figure 27-12(b), which is 180° displaced from $\mathbf{E}_{bB}$ in Figure 27-11(b). If we connect such a source to three equal resistors, as the phasor diagram of Figure 27-12(b) suggests, there will be considerable neutral current.

Figure 27-13 shows three impedances connected as a four-wire wye load. If we trace the wiring carefully, we note that $\mathbf{Z}_A$ is connected directly across coil A of the generator. We also find that the current in $\mathbf{Z}_A$, the current in line A, and the current in coil A must all be the same since they are all part of the same series loop. Because it is conventional to consider the tracing direction as outward from the alternator along the individual lines and back toward the alternator along the neutral, we do not need double subscripts to represent the current in a four-wire wye system. Therefore, $\mathbf{I}_A$ represents the current in $\mathbf{Z}_A$, line A, and generator coil A. Hence,

$$\mathbf{I}_A = \frac{\mathbf{E}_{oA}}{\mathbf{Z}_A} = \frac{\mathbf{V}_{AN}}{\mathbf{Z}_A}, \; \mathbf{I}_B = \frac{\mathbf{E}_{oB}}{\mathbf{Z}_B} = \frac{\mathbf{V}_{BN}}{\mathbf{Z}_B}, \quad \text{and} \quad \mathbf{I}_C = \frac{\mathbf{E}_{oC}}{\mathbf{Z}_C} = \frac{\mathbf{V}_{CN}}{\mathbf{Z}_C} \qquad (27\text{-}3)$$

and
$$\mathbf{I}_N = \mathbf{I}_A + \mathbf{I}_B + \mathbf{I}_C \qquad (27\text{-}4)$$

where all quantities are phasors.

[†]A system in which one terminal of each phase is connected to a common point (star point) can also be called a **star**-connected system.

700

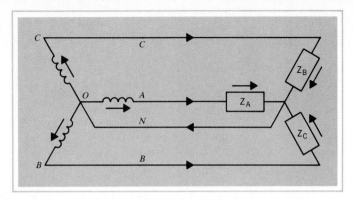

FIGURE 27-13
Four-Wire Wye-Connected Load.

The three load impedances of the four-wire wye load may be independent single-phase loads. For example, $\mathbf{Z}_A$ may be a bank of lamps, $\mathbf{Z}_B$ may be a heater, and $\mathbf{Z}_C$ a single-phase motor. It is customary in dividing the total single-phase load among the three phases to make the line currents as nearly equal in magnitude as possible. Also, $\mathbf{Z}_A$, $\mathbf{Z}_B$, and $\mathbf{Z}_C$ may represent the windings of a 3ϕ motor, in which case $\mathbf{Z}_A = \mathbf{Z}_B = \mathbf{Z}_C$ and the load is a balanced load. In such cases, the three line currents will be equal in magnitude, and the neutral current will be zero. For a four-wire balanced wye load,

$$I_{\mathrm{L}} = \frac{V_{\mathrm{P}}}{Z}$$

(27-5)

where I_{L} is the magnitude of the current in each line, V_{P} is the magnitude of the phase voltage, and Z is the impedance from each line to neutral. The load may also be made up of a combination of single-phase and three-phase loads.

EXAMPLE 27-1
In the circuit of Figure 27-13, $\mathbf{Z}_A = \mathbf{Z}_B = \mathbf{Z}_C = 30\ \Omega\underline{/+30°}$. The phase voltages are as shown in Figure 27-11(b).
(a) Find the current in each line.
(b) Find the neutral current.
(c) Draw a phasor diagram showing phase voltages and line currents.

SOLUTION

(a)
$$\mathbf{I}_A = \frac{\mathbf{V}_{AN}}{\mathbf{Z}_A} = \frac{120\ \mathrm{V}\underline{/0°}}{30\ \Omega\underline{/+30°}} = 4\ \mathrm{A}\underline{/-30°}$$

$$\mathbf{I}_B = \frac{\mathbf{V}_{BN}}{\mathbf{Z}_B} = \frac{120\ \mathrm{V}\underline{/-120°}}{30\ \Omega\underline{/+30°}} = 4\ \mathrm{A}\underline{/-150°}$$

$$\mathbf{I}_C = \frac{\mathbf{V}_{CN}}{\mathbf{Z}_C} = \frac{120\ \mathrm{V}\underline{/+120°}}{30\ \Omega\underline{/+30°}} = 4\ \mathrm{A}\underline{/+90°}$$

(b)
$$\mathbf{I}_N = \mathbf{I}_A + \mathbf{I}_B + \mathbf{I}_C = 4\ \mathrm{A}\underline{/-30°} + 4\ \mathrm{A}\underline{/-150°} + 4\ \mathrm{A}\underline{/+90°}$$

$$= (3.464 - j2) + (-3.464 - j2) + (0 + j4) = 0\ \mathrm{A}$$

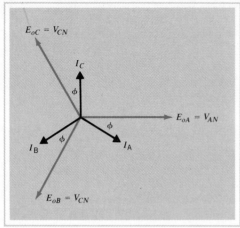

FIGURE 27-14
Phasor Diagram for Example 27-1.

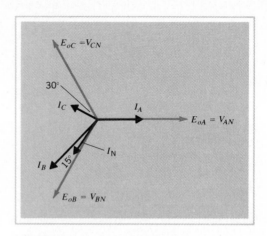

FIGURE 27-15
Phasor Diagram for Example 27-2.

(c) Figure 27-14 shows the phasor diagram for the phase voltage and line currents. Once we realize that the neutral current with a balanced wye load is zero, it is not necessary to go through the calculations of step (b). Note that each load current lags the voltage applied to that load by 30°. This is called the **power-factor angle, ϕ.**

EXAMPLE 27-2

In the circuit of Figure 27-13, $\mathbf{Z}_A$ is a 360-W lamp, $\mathbf{Z}_B$ passes a current of 4 A at a 0.966 lagging power factor, and $\mathbf{Z}_C = 60\ \Omega\,\underline{/-30°}$. Find the neutral current.

SOLUTION

Because the lamp possesses only resistance, $I_A = P_A/V_{AN} = 360\ \text{W}/120\ \text{V} = 3\ \text{A}$ in phase with $\mathbf{V}_{AN}$ (Figure 27-15).

$$\therefore \mathbf{I}_A = 3\ \text{A}\,\underline{/0°}$$

Since $\mathbf{Z}_B$ has a 0.966 lagging power factor, the power-factor angle is arccos $0.966 = 15°$. Therefore, $\mathbf{I}_B$ lags $\mathbf{V}_{BN}$ by 15° and

$$\therefore \mathbf{I}_B = 4\ \text{A}\,\underline{/-135°}$$

$$\mathbf{I}_C = \frac{\mathbf{V}_{CN}}{\mathbf{Z}_C} = \frac{120\ \text{V}\,\underline{/+120°}}{60\ \Omega\,\underline{/-30°}} = 2\ \text{A}\,\underline{/+150°}$$

$$\mathbf{I}_N = \mathbf{I}_A + \mathbf{I}_B + \mathbf{I}_C$$

$$= 3\ \text{A}\,\underline{/0°} + 4\ \text{A}\,\underline{/-135°} + 2\ \text{A}\,\underline{/+150°}$$

$$= (3 + j0) + (-2.828 - j2.828) + (-1.732 + j1)$$

$$= -1.56 - j1.828\ \text{A} = \mathbf{2.4\ \text{A}\,\underline{/-130.5°}}$$

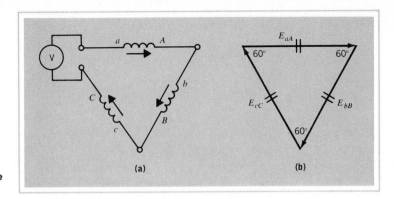

FIGURE 27-16
Delta Connection of a Three-Phase
Source.

27-5 DELTA-CONNECTED SYSTEM

The phasor diagram of Figure 27-9(b) suggests that, when properly connected, the **phasor sum of the voltages generated in the three coils of a three-phase source is zero.**[†] To add voltages by circuit connection, we connect the coils in series (like adding batteries in series). We start by connecting end b of coil B to end A of coil A and end c of coil C to end B of coil B, as shown in Figure 27-16(a). Finally, we connect a voltmeter between the remaining two terminals to read the phasor total voltage.

When we are required to add several phasors diagrammatically, we can cut down on the geometrical construction and also obtain a less cluttered phasor diagram by switching to funicular phasors (Section 25-8). Since the head end of the phasor for $\mathbf{E}_{aA}$ represents the uppercase letter end of the coil [Figure 27-16(a)], we can duplicate the circuit connection of Figure 27-16(a) by drawing the phasor $\mathbf{E}_{bB}$ with its proper magnitude and direction from the tip of the phasor $\mathbf{E}_{aA}$. Similarly, we construct the phasor for $\mathbf{E}_{cC}$ from the tip of the phasor $\mathbf{E}_{bB}$. This returns the tip of the phasor $\mathbf{E}_{cC}$ back to the origin, forming an equilateral triangle. If we have actually connected the coils as we have specified, the voltmeter in Figure 27-16(a) will read zero. Therefore, it is safe to connect terminal C to terminal a without any short-circuit current flowing around the loop. In practice, we would always check with a voltmeter before making this final connection, because reversing the leads to any one of the coils (coil C in Figure 27-17) results in a voltmeter reading of twice the phase voltage. This would cause a very heavy current, limited only by the impedance of the three coils, to flow around the loop.

Since we can connect the windings of a three-phase source into a delta without any short-circuit current flowing, we require only three wires to connect the three sets of voltages to a load which can also be connected in delta fashion, as in Figure 27-18.[‡] In the **delta** (Δ)-connected system, each arm of the load is connected directly across one of the generator coils. We can label the three lines according to the *uppercase* or

---

[†]The three-phase source need not be a rotating machine. The secondary windings of three transformers fed from a three-phase system can be considered as representing the coils of a three-phase source.

[‡]So called because of the similarity to the Greek letter Δ.

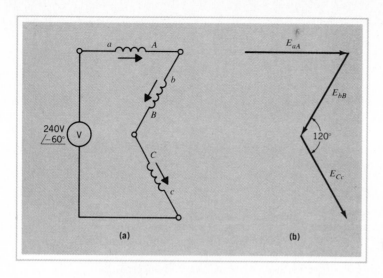

FIGURE 27-17
Incorrect Delta Connection.

second subscript letters of the various phase voltages connected to each line. Thus, $\mathbf{E}_{aA}$ becomes $\mathbf{E}_{CA}$, $\mathbf{E}_{bB}$ becomes $\mathbf{E}_{AB}$, and $\mathbf{E}_{cC}$ becomes $\mathbf{E}_{BC}$. The equivalent voltage drops become $\mathbf{V}_{AC}$, $\mathbf{V}_{BA}$, and $\mathbf{V}_{CB}$.

In the wye-connected load where each arm of the load is connected from line to neutral, we found that a single subscript for each impedance was quite satisfactory. In the delta load, each arm is connected from line to line. Consequently, to identify each impedance properly, we use two subscripts, these being the two lines between which the arm in question is connected. The order in which we write the subscripts is taken from the tracing direction around each loop. Let us examine carefully the loop between lines A and B. The tracing direction through generator coil B is from b to B. Therefore, the voltage generated in that coil is written $\mathbf{E}_{AB}$. This loop then passes through the load impedance from line B to line A. Hence, the impedance is identified as $\mathbf{Z}_{BA}$. The subscripts are in the same order as the subscripts for the voltage drop $\mathbf{V}_{BA}$. Hence, when working with delta-connected loads, it will be less confusing if we work with $\mathbf{V}$ rather than $\mathbf{E}$, since this allows us to write the voltage, current, and impedance subscripts at the load all with the same order.

FIGURE 27-18
Delta-Connected Load.

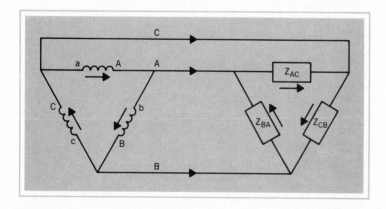

704

The phase current of the source is defined as the current flowing in a particular coil of the source, and the phase current of a load is the current flowing in a particular arm of the load. In the wye-connected system, the source phase current, load phase current, and line current are all one and the same. But careful inspection of the delta system of Figure 27-18 shows us that each line has to carry current for *two* arms of the load. Closer inspection shows us that the tracing direction for one of these currents is *away* from the source and the other is *toward* the source. Therefore, the line current to a delta load must be the phasor *difference* between the two load phase currents flowing in that line. (However, if two phasor quantities have an angle between them of greater than 90°, their phasor difference is *greater* than either one by itself.)

Hence, the load phase currents in Figure 27-18 are

$$\mathbf{I}_{AC} = \frac{\mathbf{V}_{AC}}{\mathbf{Z}_{AC}} = \frac{\mathbf{E}_{CA}}{\mathbf{Z}_{AC}} \qquad (27\text{-}6A)$$

$$\mathbf{I}_{CB} = \frac{\mathbf{V}_{CB}}{\mathbf{Z}_{CB}} = \frac{\mathbf{E}_{BC}}{\mathbf{Z}_{CB}} \qquad (27\text{-}6B)$$

$$\mathbf{I}_{BA} = \frac{\mathbf{V}_{BA}}{\mathbf{Z}_{BA}} = \frac{\mathbf{E}_{AB}}{\mathbf{Z}_{BA}} \qquad (27\text{-}6C)$$

Even in a three-wire system, it is conventional to consider the direction of the line current as being *outward* along all *three* lines *away* from the source, as shown in Figure 27-18. Therefore, with reference to Figure 27-18,

$$\mathbf{I}_A = \mathbf{I}_{AC} - \mathbf{I}_{BA} \qquad (27\text{-}7A)$$

$$\mathbf{I}_B = \mathbf{I}_{BA} - \mathbf{I}_{CB} \qquad (27\text{-}7B)$$

$$\mathbf{I}_C = \mathbf{I}_{CB} - \mathbf{I}_{AC} \qquad (27\text{-}7C)$$

In a 3ϕ, three-wire system, there is no neutral for a return tracing direction from the load back to the generator. Consequently, even with an unbalanced load, in any 3ϕ, three-wire system, the phasor sum of the three line currents must be zero. We can show this by adding the three Equations (27-7).

$$\mathbf{I}_A + \mathbf{I}_B + \mathbf{I}_C = 0 \qquad (27\text{-}8)$$

This relationship is useful in checking the solution of 3ϕ, three-wire ac circuit problems.

EXAMPLE 27-3
The impedances of the branches of the delta load in Figure 27-18 are $\mathbf{Z}_{AC} = 60\ \Omega\underline{/0°}$, $\mathbf{Z}_{CB} = 30\ \Omega\underline{/-30°}$, and $\mathbf{Z}_{BA} = 30\ \Omega\underline{/+45°}$. Determine the three line currents. Use the alternator voltages given in Figure 27-11(b).

SOLUTION
Step I. Start to prepare a phasor diagram, adding phasors as we determine their data. The complete phasor diagram is shown in Figure 27-19.

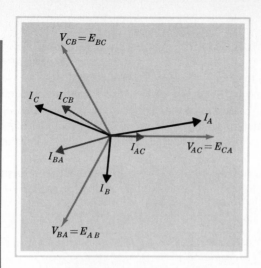

FIGURE 27-19
Phasor Diagram for Example 27-3.

Step II. Determine the phase currents.

$$\mathbf{I}_{AC} = \frac{\mathbf{V}_{AC}}{\mathbf{Z}_{AC}} = \frac{120\ V\ \underline{/0°}}{60\ \Omega\ \underline{/0°}} = 2\ A\ \underline{/0°}$$

$$\mathbf{I}_{CB} = \frac{\mathbf{V}_{CB}}{\mathbf{Z}_{CB}} = \frac{120\ V\ \underline{/+120°}}{30\ \Omega\ \underline{/-30°}} = 4\ A\ \underline{/+150°}$$

$$\mathbf{I}_{BA} = \frac{\mathbf{V}_{BA}}{\mathbf{Z}_{BA}} = \frac{120\ V\ \underline{/-120°}}{30\ \Omega\ \underline{/+45°}} = 4\ A\ \underline{/-165°}$$

Step III. Determine the line currents.

$$\mathbf{I}_A = \mathbf{I}_{AC} - \mathbf{I}_{BA} = 2\ A\ \underline{/0°} - 4\ A\ \underline{/-165°}$$
$$= (2 + j0) - (-3.864 - j1.035)$$
$$= 5.864 + j1.035\ A = \mathbf{5.96\ A\ \underline{/+10°}}$$

$$\mathbf{I}_B = \mathbf{I}_{BA} - \mathbf{I}_{CB} = 4\ A\ \underline{/-165°} - 4\ A\ \underline{/+150°}$$
$$= (-3.864 - j1.035) - (-3.464 + j2)$$
$$= -0.4 - j3.035\ A = \mathbf{3.06\ A\ \underline{/-97.5°}}$$

$$\mathbf{I}_C = \mathbf{I}_{CB} - \mathbf{I}_{AC} = 4\ A\ \underline{/+150°} - 2\ A\ \underline{/0°}$$
$$= (-3.464 + j2) - (2 + j0)$$
$$= -5.464 + j2\ A = \mathbf{5.82\ A\ \underline{/+159.9°}}$$

Check.

$$\mathbf{I}_A + \mathbf{I}_B + \mathbf{I}_C = (5.864 + j1.035) + (-0.4 - j3.035) + (-5.464 + j2)$$
$$= 0 + j0\ A$$

706

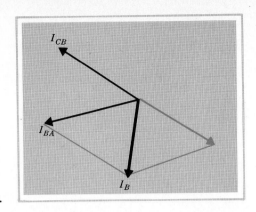

FIGURE 27-20
Phasor Subtraction of Phase Currents.

Perhaps we can understand the relationship between line current and phase current in a delta system better if we carry out the phasor subtraction by geometrical construction. From the solution for Example 27-3,

$$\mathbf{I}_B = \mathbf{I}_{BA} - \mathbf{I}_{CB} = 4\underline{/-165°} - 4\underline{/+150°}$$

As we noted in Chapter 19, we can draw a phasor for $-\mathbf{I}_{CB}$ in the phasor diagram of Figure 27-20 by drawing a phasor equal in length to the $\mathbf{I}_{CB}$ phasor but in the opposite direction from the origin. Having constructed the $-\mathbf{I}_{CB}$ phasor, we can now add the $\mathbf{I}_{BA}$ phasor and the $-\mathbf{I}_{CB}$ phasor by the usual geometrical construction to complete the parallelogram.

In a balanced delta load, the impedances of all three arms are equal in magnitude and angle. Since the phase voltages are equal and 120° out of phase, it follows that the phase currents will also be equal and 120° out of phase, as shown in Figure 27-21. Therefore, $\mathbf{I}_{CB}$ and $-\mathbf{I}_{AC}$ are equal in magnitude and have a 60° angle between them.

We can label the parallelogram by which we construct the line-current phasor as shown in Figure 27-21. Since $\mathbf{I}_{CB}$ and $-\mathbf{I}_{AC}$ are equal in magnitude and have a 60° angle between them, OWY is an equilateral triangle which is bisected by the diagonal OX which bisects WY at right angles at point Z

$$\therefore OW = 2WZ$$

From Pythagoras' theorem,

$$OZ = \sqrt{OW^2 - WZ^2} = WZ\sqrt{2^2 - 1^2} = \sqrt{3}\,WZ = \frac{\sqrt{3}}{2}OW$$

FIGURE 27-21
Showing the Relationship between Line Current and Phase Current with a Balanced Delta Load.

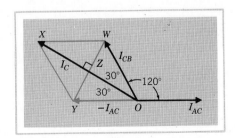

Because the diagonals of a parallelogram bisect each other,

$$OX = 2OZ = \sqrt{3}\,OW$$

Hence,

The line current to a balanced delta load has a magnitude of $\sqrt{3}$ times the phase current in each arm of the load and is displaced 30° from the phase current.

27-6 WYE-DELTA SYSTEM

In discussing the four-wire wye-connected three-phase system, we considered only the source voltages as measured from line to neutral. Since the coils of the wye-connected source are connected from line to neutral, this voltage is the phase voltage of the wye-connected source. However, in working with three-phase systems, it is customary to work in terms of the voltage measured between lines and called the **line voltage.** In fact, unless it is specified otherwise, the voltages given for a three-phase system are **line** voltages. When the source is delta-connected, this distinction poses no problem. Because the coils of the source are connected from line to line, the phase voltage and line voltage of a delta-connected source are one and the same thing.

In the wye-connected source of Figure 27-22, each voltmeter reads a line voltage which depends on the voltages generated in *two* coils. In writing the subscripts for the three line voltages, we must decide whether to read clockwise or counterclockwise around the phasor diagram for the source. As long as we follow the same direction for all three line voltages, they will have the same relative angular positions. In Figure 27-22, we have drawn the tracing arrows for the source in the direction which will give us the same subscripts for the line voltages in this wye-connected source as we had for the delta-connected source of Figure 27-18.

Following the tracing arrow from line A to line B through the wye-connected source in Figure 27-22 shows us that

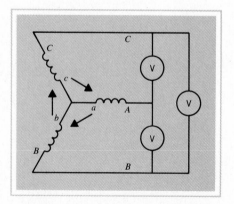

FIGURE 27-22
Line Voltages of a Wye-Connected Three-Phase Source.

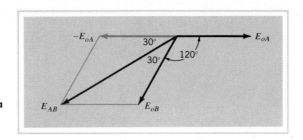

FIGURE 27-23
Determining the Line Voltages of a Wye-Connected Source by Geometrical Construction.

$$\mathbf{E}_{AB} = \mathbf{E}_{Aa} + \mathbf{E}_{bB} = -\mathbf{E}_{aA} + \mathbf{E}_{bB} \qquad (27\text{-}9A)$$

Therefore, the line voltage in a wye-connected source is the phasor *difference* between two phase voltages. Since the angle between the phase voltages is greater than 90°, this phasor difference is greater than either phase voltage by itself. Using the values given with Figure 27-11, we get

$$\mathbf{E}_{AB} = -120 \text{ V}\underline{/0°} + 120 \text{ V}\underline{/-120°}$$
$$= -(120 + j0) + (-60 - j104)$$
$$= -180 - j104 \text{ V} = 208 \text{ V}\underline{/-150°}$$

We can also solve for $\mathbf{E}_{AB}$ by geometrical construction, as shown in Figure 27-23. First, we construct phasors for the given phase voltages $\mathbf{E}_{oA}$ and $\mathbf{E}_{oB}$. Then we construct the phasor for $-\mathbf{E}_{oA}$ by reversing the direction of the $\mathbf{E}_{oA}$ phasor. $\mathbf{E}_{AB}$ then becomes the phasor sum of $-\mathbf{E}_{oA}$ and $\mathbf{E}_{oB}$.

If the coils of the wye-connected source are properly connected, the phase voltages are always equal in magnitude and 120° out of phase. Consequently, just as in the case of the line current to a balanced delta load, the line voltage from a wye must be

$$V_{\text{L}} = \sqrt{3}\ V_{\text{P}} = 1.732\ V_{\text{P}} \qquad (27\text{-}10)$$

For the phase voltages given with the wye-connected source of Figure 27-11, the three line voltages are

$$\mathbf{V}_{BA} = \mathbf{E}_{AB} = \mathbf{E}_{Aa} + \mathbf{E}_{bB} = \mathbf{E}_{bB} - \mathbf{E}_{aA} = 208 \text{ V}\underline{/-150°} \qquad (27\text{-}9A)$$
$$\mathbf{V}_{CB} = \mathbf{E}_{BC} = \mathbf{E}_{Bb} + \mathbf{E}_{cC} = \mathbf{E}_{cC} - \mathbf{E}_{bB} = 208 \text{ V}\underline{/+90°} \qquad (27\text{-}9B)$$
$$\mathbf{V}_{AC} = \mathbf{E}_{CA} = \mathbf{E}_{Cc} + \mathbf{E}_{aA} = \mathbf{E}_{aA} - \mathbf{E}_{cC} = 208 \text{ V}\underline{/-30°} \qquad (27\text{-}9C)$$

The complete phasor diagram for both line and phase voltages of a wye-connected three-phase source is shown in Figure 27-24(a). The geometrical construction lines have been omitted in order to keep the diagram as simple as possible. Note that the line voltages are always displaced 30° from the phase voltages. Since these

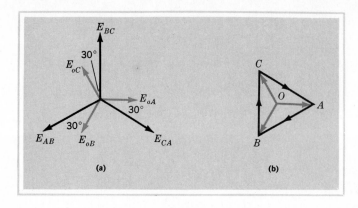

(a)   (b)

FIGURE 27-24
Phasor Diagrams for a
Wye-Connected Three-Phase Source.

are generated voltages, changes in the load connected to a wye-connected source cannot alter the phasors of Figure 27-24.

Because the opposite sides of a parallelogram are equal in length and parallel, a line joining the tips of the $\mathbf{E}_{oA}$ and $\mathbf{E}_{oB}$ phasors in Figure 27-24 would have the same magnitude and direction as $\mathbf{E}_{AB}$. Therefore, we can check rapidly for the magnitude and direction of the line-voltage phasors with respect to the phase-voltage phasors for a wye-connected three-phase source without special geometrical construction by joining the tips of the phase-voltage phasors, as in Figure 27-24(b).

An important feature of the wye-connected three-phase source is that two different values of three-phase source voltage are available in the one system. This is why a 3ϕ source consisting of wye-connected transformer secondary windings rated at 120/208 volts is probably the most widely-used distribution system in North America. If we wish to apply 120 volts, 3ϕ to the load, we connect $\mathbf{Z}_A$, $\mathbf{Z}_B$ and $\mathbf{Z}_C$ as a four-wire wye load, as in Figure 27-13. But if we wish to operate the same load from a 208-V 3ϕ source, we can connect the three arms of the load in delta fashion, as shown in Figure 27-25. Since the various load impedances are connected from line to line, we use the same double-subscript identification that we used with the simple delta-connected system.

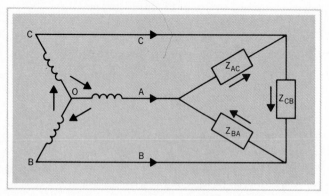

FIGURE 27-25
Wye-Connected Source Feeding a
Delta-Connected Load.

EXAMPLE 27-4

A balanced load, each arm of which has an impedance of 52 $\Omega\underline{/+45°}$, is delta-connected to a 120/208-V 3ϕ wye-connected source, as in Figure 27-25. What is the magnitude of the current in each coil of the source?

SOLUTION

Since each arm of the load is connected from line to line, the load voltage is the line voltage.

$$\therefore I_{BA} = I_{CB} = I_{AC} = \frac{208 \text{ V}}{52 \text{ }\Omega} = 4 \text{ A}$$

The load is a balanced load; hence,

$$I_L = 1.732 I_P = 1.732 \times 4 \text{ A} = 6.93 \text{ A}$$

As Figure 27-25 indicates, the current in each winding of a wye-connected source is the same as the current in the line connected to that particular coil. Therefore, the current in each winding of the source is **6.93 A.**

When a delta load is connected to a wye-connected source, as in Figure 27-25, the phase voltages of the *source* are not used. Thus, once we have determined the line voltages, we can disregard the phase voltages completely. Hence, the procedure for solving an unbalanced wye-delta circuit is exactly the same as that given in Example 27-3.

27-7 POWER IN A BALANCED THREE-PHASE SYSTEM

In a balanced three-phase load, the power in each of the three arms will be identical. Consequently the total power is three times the power in each phase. Since, as far as the load is concerned, *phase* voltage is the voltage across each arm of the load and *phase* current is the current through each arm of the load,

$$P_T = 3V_P I_P \cos \phi \qquad\qquad (27\text{-}11A)$$

Because the junction points on three-phase loads such as three-phase motors are not always accessible, it is more convenient to express the total power in terms of *line* voltage and *line* current, which we can readily measure. In the wye-connected system of Figure 27-13,

$$V_L = \sqrt{3} \; V_P \qquad \text{and} \qquad I_L = I_P$$

In the balanced delta-connected load of Figure 27-18,

$$V_L = V_P \qquad \text{and} \qquad I_L = \sqrt{3} \; I_P$$

In both cases, substitution in Equation (27-11A) gives

$$P_T = \sqrt{3}\ V_L I_L \cos \phi \qquad\qquad (27\text{-}11\text{B})$$

EXAMPLE 27-5

What is the total power input to the load in Example 27-4?

SOLUTION

Since each impedance has a power-factor angle of $+45°$,

$$\cos \phi = \cos 45° = 0.707 \text{ lagging}$$

$$\therefore P_T = 1.732 \times V_L I_L \cos \phi = 1.732 \times 208 \text{ V} \times 6.93 \text{ A} \times 0.707 = \mathbf{1765\ W}$$

EXAMPLE 27-6

A 550-V 3ϕ motor delivers 15 hp to a mechanical load with an 80% efficiency. If the power factor of the motor is 0.9 lagging, what is the line current?

SOLUTION

$$P_T = \frac{\text{hp} \times 746 \text{ W}}{\eta} = \frac{15 \times 746 \text{ W}}{0.8} = 14 \text{ kW}$$

From Equation (27-11),

$$I_L = \frac{P_T}{1.732 V_L \cos \phi} = \frac{14 \text{ kW}}{1.732 \times 550 \text{ V} \times 0.9} = \mathbf{16.32\ A}$$

27-8 MEASURING POWER IN A THREE-PHASE SYSTEM

One method of measuring the total power in a three-phase load is to connect three wattmeters so that each meter measures the voltage across and the current through one of the arms of the load, as in Figure 27-26. The total power is, therefore, the sum of the three individual phase powers. This system applies to either balanced or unbalanced loads. However, if we know that the load is balanced (all wattmeters read the same), we can leave out two of the meters and take the total power as three times the reading of the remaining wattmeter.

As we noted in Section 21-8, reversing the leads to one of the coils of a wattmeter results in a negative torque, which drives the pointer off scale to the left. To assist us in connecting the wattmeter correctly in the first instance, one terminal of each wattmeter coil is marked with a $\pm$ identification. As shown in Figure 27-26, we connect the identified terminal of the *current* coil to the source and the identified terminal of the *voltage* coil to the line containing the current coil. Where we use a separate wattmeter for each arm of the load, as in Figure 27-26, a negative wattmeter reading indicates an incorrect connection which we can readily correct. However, with

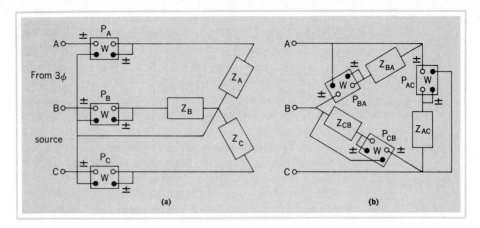

FIGURE 27-26
Measuring Individual Phase Powers.

the two-wattmeter method of measuring three-phase power, which we shall consider in Section 27-9, it is important that we start out with the wattmeters properly connected.

With many three-phase loads, the junction points in both wye- and delta-connected loads are not accessible for measuring phase currents and phase voltages independently, as in Figure 27-36. We can, however, switch to the wattmeter connections shown in Figure 27-27 to measure the total power to either a wye- or a delta-connected load. By connecting the identical voltage coils of the three wattmeters as a wye load, each wattmeter will have a voltage equivalent to the phase voltage of

FIGURE 27-27
Three-Wattmeter Measurement of Three-Phase Power.

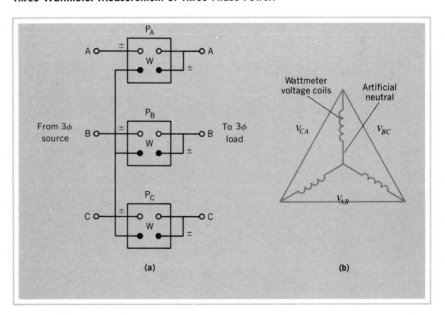

a wye-connected system $(V_L/\sqrt{3})$, as shown in Figure 27-27(b). On this basis, regardless of load connections, the line current will appear to the wattmeters as if it were the same as the phase current of a wye-connected load. Therefore, the total power is the sum three wattmeter readings.

Again, if the load is balanced, all wattmeters will read alike. But, in this case, we cannot remove two wattmeters without disturbing the voltage to the voltage coil of the remaining wattmeter. Consequently, when we remove two wattmeters, we must provide some means of restoring the proper voltage to the remaining wattmeter voltage coil. One method is to complete the wye connection with two impedance coils whose impedances are exactly equal to the impedance of the voltage coil of the remaining meter. A second method is to create an artificial neutral to which to connect the voltage coil, by connecting three equal resistors as a wye load on the three-phase system.

A third method of using a single wattmeter to indicate total power in a balanced three-phase load is shown in Figure 27-28. The unidentified terminal of the voltage coil of the wattmeter is returned to a center-tapped resistor across the two lines not containing the current coil. With this connection, the voltage coil receives a voltage of

$$\mathbf{V}_M = \mathbf{V}_{AB} + 0.5\mathbf{V}_{BC}$$

Constructing the resultant phasor in Figure 27-28(b), we find that the voltage coil receives a voltage having the same phase position as if it were connected to a neutral, that is, the same angle as a phase voltage in an equivalent wye system. But, unlike the phase voltage of the equivalent wye, which has a magnitude of $V_L/\sqrt{3}$, the wattmeter voltage, as Figure 27-28(b) shows, is $V_L \cos 30°$, which is $(\sqrt{3}/2)V_L$. Therefore, since $P_T = \sqrt{3}\, V_L I_L \cos \phi$, the wattmeter reads *one-half* of the total power to a balanced three-phase load.

FIGURE 27-28
Single-Wattmeter Measurement of Three-Phase Power in a Balanced Load.

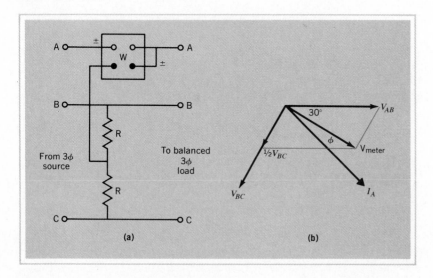

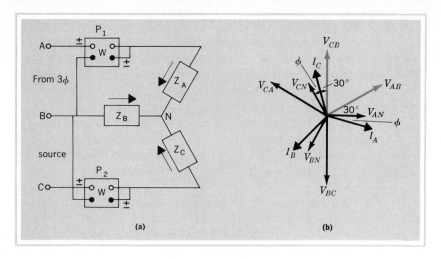

FIGURE 27-29
Two-Wattmeter Power Measurement with a Balanced Wye Load.

27-9 TWO-WATTMETER MEASUREMENT OF THREE-PHASE POWER

The most widely used method of measuring three-phase power is based on the two-wattmeter connections of Figures 27-29 and 27-30. Its advantages are: (1) the same connections apply to both wye- and delta-connected loads; (2) it shows the total power in both balanced and unbalanced loads with one less meter than the three-wattmeter systems; (3) with balanced loads, we can determine load power factor from the two wattmeter readings. To determine what each wattmeter reads, we shall consider first the lagging power-factor balanced wye load of Figure 27-29.

FIGURE 27-30
Two-Wattmeter Power Measurement with a Balanced Delta Load.

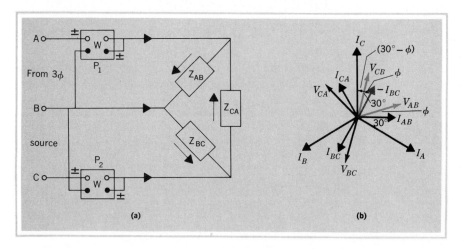

Examining the connections to wattmeter P_1, we find that the current coil carries the current $\mathbf{I}_A$, which lags the voltage $\mathbf{V}_{AN}$ across $\mathbf{Z}_A$ by the power-factor angle ϕ. However, the voltage coil is connected across the *line* voltage $\mathbf{V}_{AB}$. As we have already noted, the line voltage in a wye-connected three-phase system is displaced $30°$ from the phase voltage. Locating $\mathbf{V}_{AB}$ on the phasor diagram (from $\mathbf{V}_{AB} = \mathbf{V}_{AN} + \mathbf{V}_{NB} = \mathbf{V}_{AN} - \mathbf{V}_{BN}$), we find that the total angle between $\mathbf{I}_A$ and $\mathbf{V}_{AB}$ is $(30° + \phi)$. Thus, wattmeter P_1 reads

$$P_1 = V_{AB} I_A \cos(30° + \phi)$$

The current coil of wattmeter P_2 carries the current $\mathbf{I}_C$. But note that the voltage coil is connected to read $\mathbf{V}_{CB}$ rather than $\mathbf{V}_{BC}$ (the systematically arranged line voltage). Therefore, locating $\mathbf{V}_{CB}$ on the phasor diagram of Figure 27-29(b) (colored line), we find that the angle between $\mathbf{I}_C$ and $\mathbf{V}_{BC}$ is $(30° - \phi)$. Hence, in general terms,

$$P_1 = V_L I_L \cos(30° + \phi) \qquad (27\text{-}12)$$

$$P_2 = V_L I_L \cos(30° - \phi) \qquad (27\text{-}13)$$

In the balanced delta-connected load of Figure 27-30, each phase current lags the line voltage by the power-factor angle ϕ. As we have already noted, with a *balanced* load, the line current is displaced $30°$ from the phase current in each arm of the load. Therefore, locating $\mathbf{I}_A$ on the phasor diagram (from $\mathbf{I}_A = \mathbf{I}_{AB} - \mathbf{I}_{CA}$), we find that the angle between $\mathbf{I}_A$ and $\mathbf{V}_{AB}$ is $(30° + \phi)$. To find the angle between $\mathbf{I}_C$ and the colored phasor $\mathbf{V}_{CB}$, we can draw in the $-\mathbf{I}_{BC}$ phasor. With a balanced load, there is a $30°$ angle between $-\mathbf{I}_{BC}$ and $\mathbf{I}_C$. The angle between $\mathbf{V}_{CB}$ and $-\mathbf{I}_{BC}$ is the same as the angle between $\mathbf{V}_{BC}$ and $\mathbf{I}_{BC}$, the power-factor angle ϕ. Consequently, the angle between $\mathbf{I}_C$ and $\mathbf{V}_{CB}$ is $(30° - \phi)$, and again

$$P_1 = V_L I_L \cos(30° + \phi) \qquad (27\text{-}12)$$

$$P_2 = V_L I_L \cos(30° - \phi) \qquad (27\text{-}13)$$

Adding Equations (27-12) and (27-13) gives

$$P_1 + P_2 = V_L I_L [\cos(30 + \phi) + \cos(30 - \phi)]$$

To expand the terms in brackets, we substitute the trigonometric expansion for $\cos(A + B)$ and $\cos(A - B)$. Therefore,

$$P_1 + P_2 = V_L I_L (\cos 30 \cos \phi - \sin 30 \sin \phi + \cos 30 \cos \phi + \sin 30 \sin \phi)$$

$$= V_L I_L \, 2 \cos 30 \cos \phi$$

But $2 \cos 30° = \sqrt{3}$.

$$\therefore \; P_1 + P_2 = \sqrt{3} \, V_L I_L \cos \phi$$

which is the total power in a balanced 3ϕ system.

We can also show that the sum of the two wattmeter readings is the total power for an unbalanced load by returning to the circuit of Figure 27-30(a) and considering *instantaneous* power values. In the delta-connected load,

$$p_t = (v_{AB}i_{AB}) + (v_{BC}i_{BC}) + (v_{CA}i_{CA}) \tag{1}$$

But, as we noted when we delta-connected a 3ϕ source,

$$v_{AB} + v_{BC} + v_{CA} = 0 \tag{2}$$

from which
$$v_{CA} = -v_{AB} - v_{BC}$$

Substituting in Equation (1) gives

$$p_t = (v_{AB}i_{AB}) + (v_{BC}i_{BC}) - (v_{AB}i_{CA}) - (v_{BC}i_{CA})$$

Collecting the voltage terms and substituting $-v_{CB}$ for v_{BC},

$$p_t = v_{AB}(i_{AB} - i_{CA}) + v_{CB}(i_{CA} - i_{BC}) = (v_{AB}i_A) + (v_{CB}i_C)$$

Inspection of the circuit of Figure 27-30(a) shows us that this expression indicates precisely what the two wattmeters are connected to read. By a similar algebraic procedure, we can show that the total instantaneous power in *any* wye-connected load can also be reduced to the sum of the two wattmeter readings. Therefore, with the two wattmeters connected as in Figures 27-29 and 27-30, **total power input to any three-phase load, balanced or unbalanced, is the algebraic sum of the two wattmeter readings.**

The third advantage we claimed for the two-wattmeter system of three-phase power measurement was that with balanced loads we can determine the load power factor from the two wattmeter readings. First, let us consider the effect of various load power factors on the meter readings. With unity power factor, Equations (27-12) and (27-13) both become

$$P_1 = P_2 = V_L I_L \cos 30° = \frac{\sqrt{3}}{2} V_L I_L$$

Hence, **each wattmeter reads half of the total power.**

If the power-factor angle is $+30°$,

$$P_1 = V_L I_L \cos 60° = 0.5 \, V_L I_L$$
$$P_2 = V_L I_L \cos 0° = V_L I_L$$

One wattmeter reads one-third of the total power and the other reads two-thirds. A power-factor angle of $-30°$ simply reverses the wattmeter readings.

If the power-factor angle is $60°$,

$$P_1 = V_L I_L \cos 90° = 0$$

$$P_2 = V_L I_L \cos - 30° = \frac{\sqrt{3}}{2} V_L I_L$$

One wattmeter reads zero, and the other then registers the total power. Since this represents a power factor of 0.5, note that the total power is half that at unity power factor for the same line voltages and currents.

If the power factor drops below 50%, the angle $(30° + \phi)$ exceeds 90°, and the cosine becomes a negative quantity. Therefore, in the wattmeter movement, the counterclockwise torque exceeds the clockwise torque, and the meter tries to read off scale to the left. To read the total active power to this low power-factor load, we must reverse the wattmeter voltage *or* current coil leads to bring the reading onto the scale and then *subtract* this reading from that of the other meter.[†]

When we know the power factor of a balanced load, we can express the ratio between the two wattmeter readings by dividing Equation (27-13) into Equation (27-12). This gives

$$\frac{P_1}{P_2} = \frac{\cos (30° + \phi)}{\cos (30° - \phi)} \tag{27-14}$$

To use the two wattmeter readings to determine the load power factor, we start by subtracting Equation (27-12) from Equation (27-13).

$$\therefore P_2 - P_1 = V_L I_L \cos (30° - \phi) - V_L I_L \cos (30° + \phi)$$

When we substitute the trigonometric expansions for $\cos (30° - \phi)$ and $\cos (30° + \phi)$ and simplify,

$$P_2 - P_1 = V_L I_L \, 2 \sin 30° \sin \phi$$

But $\sin 30° = 0.5$,

$$\therefore P_2 - P_1 = V_L I_L \sin \phi$$

Dividing this expression by the sum of Equations (27-12) and (27-13),

$$\frac{P_2 - P_1}{P_2 + P_1} = \frac{V_L I_L \sin \phi}{\sqrt{3} \, V_L I_L \cos \phi} = \frac{\tan \phi}{\sqrt{3}}$$

$$\therefore \tan \phi = \sqrt{3} \, \frac{P_2 - P_1}{P_2 + P_1} \tag{27-15}$$

[†]If we encounter a negative wattmeter reading when conducting a laboratory assignment on power input to a three-phase load, it can mean either a load power factor of less than 50%, or incorrect connection of the wattmeter coils in the first instance. Conversely, if one of the wattmeters shows a small positive power reading, we do not know for sure that we have not already reversed the wattmeter and are really reading a power which should be subtracted due to a low power-factor load. There are several ways of checking if we are not sure that the original meter connections were correct. Perhaps the simplest test is to "single-phase" the system by momentarily opening the line containing the current coil of the *other* wattmeter. If the wattmeter in question reads negative, it was incorrectly connected in the first place. However, if it shows a positive power reading under single-phase conditions, any negative reading when three-phase source voltages are restored to the load is due to low power factor. We must then reverse its voltage coil to obtain a reading which we then *subtract* from the other wattmeter reading.

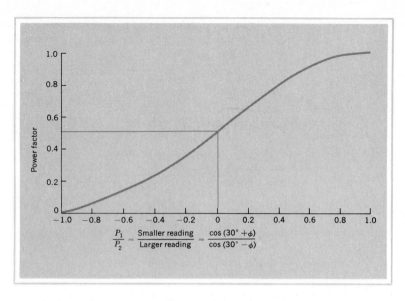

FIGURE 27-31
Power Factor from Ratio of Two-Wattmeter Readings.

$$\frac{P_1}{P_2} = \frac{\text{Smaller reading}}{\text{Larger reading}} = \frac{\cos (30° + \phi)}{\cos (30° - \phi)}$$

Knowing the two wattmeter readings, we can solve for the power-factor angle ϕ and then determine the power factor of the load from cos ϕ.

EXAMPLE 27-7

The total power input to the load in Example 27-4 is measured by the two-wattmeter method. What are the two wattmeter readings?

SOLUTION

$$P_1 = V_L I_L \cos (30° + 45°) = 208 \text{ V} \times 6.93 \text{ A} \times \cos 75° = 373 \text{ W}$$

$$P_2 = V_L I_L \cos (30° - 45°) = 208 \text{ V} \times 6.93 \text{ A} \times \cos - 15° = 1392 \text{ W}$$

For laboratory tests on variable power-factor three-phase loads such as induction motors, we can avoid repeated calculation of power factor from Equation (27-15) by plotting an accurate graph based on Equation (27-14). This power-factor graph has the shape shown in Figure 27-31.

27-10 PHASE SEQUENCE

In the three-phase circuit problems we have encountered so far, we have been given data from which we could determine the exact angle (with respect to the reference axis) of all the phase and line voltages. When a three-phase service is supplied by an electrical utility company, we know that all three line voltages have the same magnitude and are displaced from one another by a 120° angle. We can measure the *magnitude* of the voltages quite readily. However, this does not tell us whether $\mathbf{V}_{BN}$ *lags* $\mathbf{V}_{AN}$ by 120° or *leads* by 120°. This information is called the **phase sequence** (or phase rotation) of a three-phase system. It is important information, since it governs the direction of rotation of three-phase motors and also governs the division of the

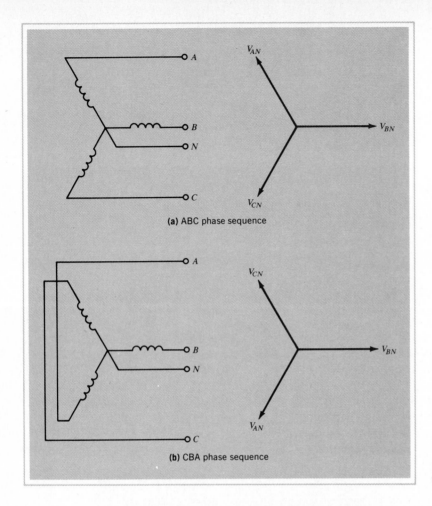

(a) ABC phase sequence

(b) CBA phase sequence

FIGURE 27-32
Phase Sequence of a
Three-Phase Source.

current among the three lines feeding an unbalanced load. It is also important that the electrical utility maintain the same phase sequence.

In the 3ϕ source of Figure 27-32(a), we are given information that shows that $\mathbf{V}_{BN}$ lags $\mathbf{V}_{AN}$ by 120° and $\mathbf{V}_{CN}$ lags $\mathbf{V}_{BN}$ by 120°. Therefore, the instantaneous voltage to line A passes through its positive peak value first; then the instantaneous voltage to line B passes through its positive peak value, followed by the instantaneous voltage to line C. As a result, we say that the system in Figure 27-32(a) has an ABC phase sequence. If we reverse the leads to two of the generator coils, as in Figure 27-32(b), $\mathbf{V}_{BN}$ now *leads* $\mathbf{V}_{AN}$ by 120° and $\mathbf{V}_{CN}$ *leads* $\mathbf{V}_{BN}$ by 120°. Thus the phase sequence has been reversed and is stated as a CBA phase sequence, because the instantaneous voltage to line C passes a certain point in its cycle 120° before the instantaneous voltage to line B, which in turn is 120° ahead of the instantaneous voltage to line A.

Once we have prepared a phasor diagram for the voltages in a 3ϕ system, we can read the phase sequence from the phasor diagram quite readily. Since the direction of rotation of a phasor is counterclockwise, the phase sequence is the order in which the voltage phasors would pass the reference axis if they were rotated in a counter-clockwise direction. If we prefer to leave the phasors stationary, we can read the phase sequence from a phasor diagram by reading *clockwise* around the diagram.

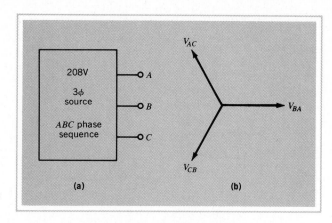

FIGURE 27-33
Determining Phase Sequence from a Phasor Diagram.

When we are given the line voltage and phase sequence of a three-phase source, as in Figure 27-33(a), we can prepare a phasor diagram by placing one phasor (usually $\mathbf{V}_{BA}$) along the reference axis. To identify the other two voltage phasors, as we read *clockwise* around the phasor diagram of Figure 27-33(b), *either* all first subscript letters *or* all second subscript letters must agree with the specified phase sequence.

We can show the effect of phase sequence on the line currents to an unbalanced load by considering the following example.

EXAMPLE 27-8
$\mathbf{Z}_{AB} = 52\ \Omega\,\underline{/-30°}$, $\mathbf{Z}_{BC} = 52\ \Omega\,\underline{/+45°}$, and $\mathbf{Z}_{CA} = 104\ \Omega\,\underline{/0°}$, are connected as a delta load to a 208-V 3ϕ source. Find the magnitude of the current in each line if the phase sequence of the source is (a) *ABC* and (b) *CBA*.

SOLUTION
(a) With *ABC* phase sequence.

Step I.

$$\mathbf{I}_{AB} = \frac{\mathbf{V}_{AB}}{\mathbf{Z}_{AB}} = \frac{208\ \mathrm{V}\underline{/0°}}{52\ \Omega\,\underline{/-30°}} = 4\ \mathrm{A}\underline{/+30°}$$

$$\mathbf{I}_{BC} = \frac{\mathbf{V}_{BC}}{\mathbf{Z}_{BC}} = \frac{208\ \mathrm{V}\underline{/-120°}}{52\ \Omega\,\underline{/+45°}} = 4\ \mathrm{A}\underline{/-165°}$$

$$\mathbf{I}_{CA} = \frac{\mathbf{V}_{CA}}{\mathbf{Z}_{CA}} = \frac{208\ \mathrm{V}\underline{/+120°}}{104\ \Omega\,\underline{/0°}} = 2\ \mathrm{A}\underline{/+120°}$$

Step II. From the circuit diagram of Figure 27-34(a),

$$\mathbf{I}_A = \mathbf{I}_{AB} - \mathbf{I}_{CA} = 4\ \mathrm{A}\underline{/+30°} - 2\ \mathrm{A}\underline{/+120°}$$

$$= (3.464 + \mathrm{j}2) - (-1 + \mathrm{j}1.732) = 4.464 + \mathrm{j}0.268\ \mathrm{A}$$

$$\therefore I_A = \mathbf{4.47\ A}$$

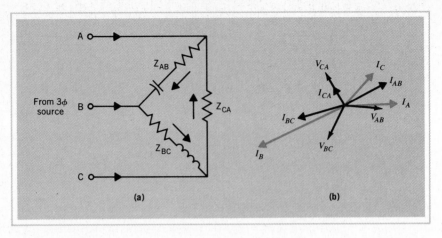

FIGURE 27-34
Phasor Diagram for Example 27-8(a).

$$\mathbf{I}_B = \mathbf{I}_{BC} - \mathbf{I}_{AB} = 4\ A\underline{/-165°} - 4\ A\underline{/+30°}$$
$$= (-3.863 - j1.035) - (3.464 + j2) = -7.327 - j3.035\ A$$
$$\therefore I_B = 7.93\ A$$

$$\mathbf{I}_C = \mathbf{I}_{CA} - \mathbf{I}_{BC} = 2\ A\underline{/+120°} - 4\ A\underline{/-165°}$$
$$= (-1 + j1.732) - (3.863 - j1.035) = 2.863 + j2.767\ A$$
$$\therefore I_C = 3.98\ A$$

(b) With *CBA* phase sequence (as shown in Figure 27-35).

Step I.

$$\mathbf{I}_{AB} = \frac{\mathbf{V}_{AB}}{\mathbf{Z}_{AB}} = \frac{208\ V\underline{/0°}}{52\ \Omega\underline{/-30°}} = 4\ A\underline{/+30°}$$

$$\mathbf{I}_{BC} = \frac{\mathbf{V}_{BC}}{\mathbf{Z}_{BC}} = \frac{208\ V\underline{/+120°}}{52\ \Omega\underline{/+45°}} = 4\ A\underline{/+75°}$$

$$\mathbf{I}_{CA} = \frac{\mathbf{V}_{CA}}{\mathbf{Z}_{CA}} = \frac{208\ V\underline{/-120°}}{104\ \Omega\underline{/0°}} = 2\ A\underline{/-120°}$$

Step II.

$$\mathbf{I}_A = \mathbf{I}_{AB} - \mathbf{I}_{CA} = 4\ A\underline{/+30°} - 2\ A\underline{/-120°}$$
$$= (3.464 + j2) - (-1 - j1.732) = 4.464 + j3.732\ A$$
$$\therefore I_A = 5.82\ A$$

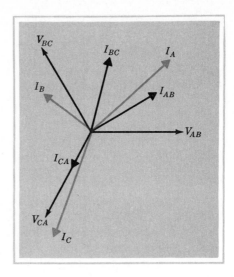

FIGURE 27-35
Phasor Diagram for Example 27-8(b).

$$\mathbf{I}_B = \mathbf{I}_{BC} - \mathbf{I}_{AB} = 4\ \text{A}\underline{/+75°} - 4\ \text{A}\underline{/+30°}$$

$$= (1.035 + j3.863) - (3.464 + j2) = -2.429 + j1.863\ \text{A}$$

$$\therefore I_B = \mathbf{3.06\ A}$$

$$\mathbf{I}_C = \mathbf{I}_{CA} - \mathbf{I}_{BC} = 2\ \text{A}\underline{/-120°} - 4\ \text{A}\underline{/+75°}$$

$$= (-1 - j1.732) - (1.035 + j3.863) = -2.035 - j5.595\ \text{A}$$

$$\therefore I_C = \mathbf{5.95\ A}$$

27-11 UNBALANCED THREE-WIRE WYE LOADS

We have been able to solve unbalanced four-wire wye loads and delta loads with no particular difficulty since, in both cases, we know the phase voltage across each arm of the load. But with the unbalanced three-wire wye load of Figure 27-36, we do not know what the phase voltages will be, since the wye junction point is *floating*. There

FIGURE 27-36
Three-Wire Wye Load.

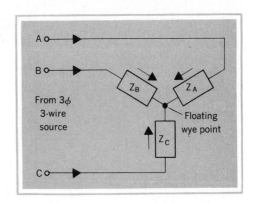

are two fairly straightforward procedures based on the ac network solutions we studied in Chapter 23. The first is based on writing Kirchhoff's-law equations for the wye load of Figure 27-36.

Following the conventional tracing arrows through the wye load in Figure 27-36, we can write three loop equations,

$$\mathbf{I}_A \mathbf{Z}_A - \mathbf{I}_B \mathbf{Z}_B = \mathbf{E}_{BA} = \mathbf{V}_{AB} \tag{1}$$

$$\mathbf{I}_B \mathbf{Z}_B - \mathbf{I}_C \mathbf{Z}_C = \mathbf{E}_{CB} = \mathbf{V}_{BC} \tag{2}$$

$$\mathbf{I}_C \mathbf{Z}_C - \mathbf{I}_A \mathbf{Z}_A = \mathbf{E}_{AC} = \mathbf{V}_{CA} \tag{3}$$

Although we have three equations for three unknown line currents, if we try to solve the determinant, we discover that the three equations are not independent, due to the symmetrical relationship of the left-hand sides. Hence, we require an additional equation. Because there is no neutral lead to this wye load, from Kirchhoff's current law,

$$\mathbf{I}_A + \mathbf{I}_B + \mathbf{I}_C = 0 \tag{4}$$

Omitting Equation (2) which does not contain $\mathbf{I}_A$, we can rearrange Equations (1), (3), and (4) in the form

$$\mathbf{Z}_A \mathbf{I}_A - \mathbf{Z}_B \mathbf{I}_B \qquad = \mathbf{V}_{AB} \tag{1}$$

$$-\mathbf{Z}_A \mathbf{I}_A \qquad + \mathbf{Z}_C \mathbf{I}_C = \mathbf{V}_{CA} \tag{3}$$

$$\mathbf{I}_A + \quad \mathbf{I}_B + \quad \mathbf{I}_C = 0 \tag{4}$$

from which,

$$\mathbf{I}_A = \frac{\begin{vmatrix} \mathbf{V}_{AB} & -\mathbf{Z}_B & \\ \mathbf{V}_{CA} & & \mathbf{Z}_C \\ 1 & 1 \end{vmatrix}}{\begin{vmatrix} \mathbf{Z}_A & -\mathbf{Z}_B & \\ -\mathbf{Z}_A & & \mathbf{Z}_C \\ 1 & 1 & 1 \end{vmatrix}} = \frac{0 + (-0) + 0 - 0 - \mathbf{V}_{AB}\mathbf{Z}_C - (-\mathbf{V}_{CA}\mathbf{Z}_B)}{0 + (-\mathbf{Z}_B\mathbf{Z}_C) + 0 - 0 - \mathbf{Z}_A\mathbf{Z}_C - \mathbf{Z}_A\mathbf{Z}_B}$$

$$\therefore \mathbf{I}_A = \frac{\mathbf{V}_{AB}\mathbf{Z}_C - \mathbf{V}_{CA}\mathbf{Z}_B}{\mathbf{Z}_A\mathbf{Z}_B + \mathbf{Z}_B\mathbf{Z}_C + \mathbf{Z}_C\mathbf{Z}_A} \tag{27-16A}$$

Similarly,
$$\mathbf{I}_B = \frac{\mathbf{V}_{BC}\mathbf{Z}_A - \mathbf{V}_{AB}\mathbf{Z}_C}{\mathbf{Z}_A\mathbf{Z}_B + \mathbf{Z}_B\mathbf{Z}_C + \mathbf{Z}_C\mathbf{Z}_A} \tag{27-16B}$$

and
$$\mathbf{I}_C = \frac{\mathbf{V}_{CA}\mathbf{Z}_B - \mathbf{V}_{BC}\mathbf{Z}_A}{\mathbf{Z}_A\mathbf{Z}_B + \mathbf{Z}_B\mathbf{Z}_C + \mathbf{Z}_C\mathbf{Z}_A} \tag{27-16C}$$

EXAMPLE 27-9

The phase sequence tester shown in Figure 27-37 is connected to a 208-V 3ϕ 60-Hz source whose phase sequence is ABC. The choke has an inductance of 0.5 H with negligible resistance, and the lamps each have a resistance of 100 Ω. (We shall assume that the resistance remains constant even though the current through the lamp changes). Using Equations (27-16), determine the magnitude of the current in each lamp.

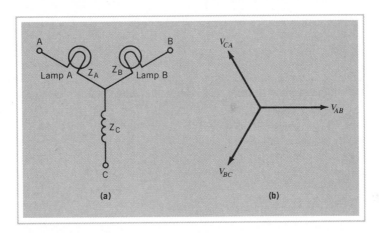

FIGURE 27-37
Phase-Sequence Tester.

SOLUTION

$$\mathbf{Z}_C = +j\omega L = +j377 \times 0.5\ \text{H} = 188.5\ \Omega\underline{/+90°}$$

$$\mathbf{I}_A = \frac{\mathbf{V}_{AB}\,\mathbf{Z}_C - \mathbf{V}_{CA}\,\mathbf{Z}_B}{\mathbf{Z}_A\,\mathbf{Z}_B + \mathbf{Z}_B\,\mathbf{Z}_C + \mathbf{Z}_C\,\mathbf{Z}_A}$$

$$= \frac{(208\underline{/0°} \times 188.5\underline{/+90°}) - (208\underline{/+120°} \times 100\underline{/0°})}{(100\underline{/0°} \times 100\underline{/0°}) + (100\underline{/0°} \times 188.5\underline{/90°}) + (188.5\underline{/90°} \times 100\underline{/0°})}$$

$$= \frac{39\,208\underline{/+90°} - 20\,800\underline{/120°}}{10\,000\underline{/0°} + 18\,850\underline{/+90°} + 18\,850\underline{/+90°}}$$

$$= \frac{(0 + j39\,208) - (-10\,400 + j18\,013)}{10\,000 + j37\,700} = \frac{10\,400 + j21\,195}{10\,000 + j37\,700}$$

$$\therefore I_A = \frac{23\,609}{39\,003} = \mathbf{0.605\ A}$$

Similarly,

$$\mathbf{I}_B = \frac{(208\,\underline{/-120^\circ} \times 100\,\underline{/0^\circ}) - (208\,\underline{/0^\circ} \times 188.5\,\underline{/+90^\circ})}{39\,003}$$

$$= \frac{20\,800\,\underline{/-120^\circ} - 39\,208\,\underline{/+90^\circ}}{39\,003}$$

$$= \frac{(-10\,400 - j18\,013) - (0 + j39\,208)}{39\,003} = \frac{-10\,400 - j57\,221}{39\,003}$$

and

$$I_B = \frac{58\,158}{39\,003} = \mathbf{1.49\ A}$$

Therefore, with an ABC phase sequence, lamp B is brighter than lamp A.

The appearance of Equations (27-16A), (27-16B), and (27-16C) may suggest to us the second method for solving three-phase circuits with an unbalanced three-wire wye load. In Chapter 23, we set down equations which would allow us to substitute an equivalent delta circuit for the original wye-connected load. Once we know the impedance of the delta arms, we can solve for the line currents in the usual manner. Therefore, in the circuit of Figure 27-38,

$$\mathbf{Z}_{BC} = \frac{\mathbf{Z}_A\,\mathbf{Z}_B + \mathbf{Z}_B\,\mathbf{Z}_C + \mathbf{Z}_C\,\mathbf{Z}_A}{\mathbf{Z}_A} \qquad (23\text{-}5)$$

$$\mathbf{Z}_{CA} = \frac{\mathbf{Z}_A\,\mathbf{Z}_B + \mathbf{Z}_B\,\mathbf{Z}_C + \mathbf{Z}_C\,\mathbf{Z}_A}{\mathbf{Z}_B} \qquad (23\text{-}6)$$

$$\mathbf{Z}_{AB} = \frac{\mathbf{Z}_A\,\mathbf{Z}_B + \mathbf{Z}_B\,\mathbf{Z}_C + \mathbf{Z}_C\,\mathbf{Z}_A}{\mathbf{Z}_C} \qquad (23\text{-}7)$$

FIGURE 27-38
Using a Wye-Delta Transformation to Solve Unbalanced Three-Wire Wye Loads.

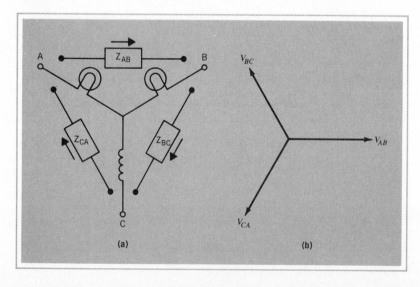

(a)

(b)

EXAMPLE 27-10

If the phase sequence of the 208-V 3ϕ 60-Hz source in Example 27-9 is changed to *CBA*, determine the magnitude of the current in each lamp, using the wye-delta transformation equations.

SOLUTION

Since we have already solved the numerator of the wye-delta transformation equations in solving Example 27-9,

Step I.

$$\mathbf{Z}_{AB} = \frac{39\,003\,\underline{/+75.14°}}{188.5\,\underline{/+90°}} = 206.9\ \Omega\,\underline{/-14.86°}$$

$$\mathbf{Z}_{BC} = \frac{39\,003\,\underline{/+75.14°}}{100\,\underline{/0°}} = 390\ \Omega\,\underline{/+75.14°}$$

$$\mathbf{Z}_{CA} = \frac{39\,003\,\underline{/+75.14°}}{100\,\underline{/0°}} = 390\ \Omega\,\underline{/+75.14°}$$

Step II.

$$\mathbf{I}_{AB} = \frac{\mathbf{V}_{AB}}{\mathbf{Z}_{AB}} = \frac{208\ \text{V}\,\underline{/0°}}{206.9\ \Omega\,\underline{/-14.86°}} = 1.005\ \text{A}\,\underline{/+14.86°}$$

$$\mathbf{I}_{BC} = \frac{\mathbf{V}_{BC}}{\mathbf{Z}_{BC}} = \frac{208\ \text{V}\,\underline{/+120°}}{390\ \Omega\,\underline{/+75.14°}} = 0.533\ \text{A}\,\underline{/+44.86°}$$

$$\mathbf{I}_{CA} = \frac{\mathbf{V}_{CA}}{\mathbf{Z}_{CA}} = \frac{208\ \text{V}\,\underline{/-120°}}{390\ \Omega\,\underline{/+75.14°}} = 0.533\ \text{A}\,\underline{/+164.86°}$$

Step III. From Figure 27-38,

$$\mathbf{I}_A = \mathbf{I}_{AB} - \mathbf{I}_{CA} = 1.005\ \text{A}\,\underline{/+14.86°} - 0.533\ \text{A}\,\underline{/+164.86°}$$

$$= (0.9714 + j0.2577) - (-0.5145 + j0.1392)$$

$$= 1.486 + j0.1185\ \text{A}$$

$$\therefore I_A = \mathbf{1.49\ A}$$

and

$$\mathbf{I}_B = \mathbf{I}_{BC} - \mathbf{I}_{AB} = 0.533\ \text{A}\,\underline{/+44.86°} - 1.005\ \text{A}\,\underline{/+14.86°}$$

$$= (0.3778 + j0.3760) - (0.9714 + j0.2577)$$

$$= -0.5936 + j0.1183\ \text{A}$$

$$\therefore I_B = \mathbf{0.605\ A}$$

Hence, with a CBA phase sequence, lamp A is brighter than lamp B.

Once we have solved for line currents, we can return to the original wye-connected circuit and solve for the various phase voltages on the basis that $\mathbf{V}_{AN} = \mathbf{I}_A \mathbf{Z}_A$, etc.

PROBLEMS

27-1. Given a 60-Hz 3ϕ four-wire wye-connected source consisting of $\mathbf{E}_{oA} = 120 \text{ V}\underline{/0°}$, $\mathbf{E}_{oB} = 120 \text{ V}\underline{/+120°}$, and $\mathbf{E}_{oC} = 120 \text{ V}\underline{/-120°}$. A balanced four-wire wye-connected load consists of three 10-Ω resistors connected to this source.
(a) Draw a fully labeled schematic diagram showing tracing arrows.
(b) Determine algebraically the magnitude and phase of the current in each line.
(c) Draw a phasor diagram of the phase voltages and currents.
(d) Determine algebraically the neutral current.

27-2. The load in Problem 27-1 is replaced with a balanced four-wire wye-connected load each arm of which is rated at 200 W with an 80% lagging power factor. Repeat all four sections of Problem 27-1. What will the neutral current be if line A becomes open-circuit?

27-3. The load in Problem 27-1 is replaced with a four-wire wye-connected load consisting of $\mathbf{Z}_A = 30 \; \Omega\underline{/+30°}$, $\mathbf{Z}_B = 40 \; \Omega\underline{/-30°}$ and $\mathbf{Z}_C = 50 \; \Omega\underline{/-90°}$. Repeat all four sections of Problem 27-1.

27-4. The load in Problem 27-1 is replaced with an unbalanced four-wire wye-connected load in which $\mathbf{Z}_A$ is a 30-Ω heater, $\mathbf{Z}_B$ is a 150-W lamp, and $\mathbf{Z}_C$ is an 0.1-H inductance whose resistance is 5 Ω. Repeat all four sections of Problem 27-1.

27-5. Given a 60-Hz 3ϕ delta-connected source in which $\mathbf{E}_{CA} = 550 \text{ V}\underline{/0°}$, $\mathbf{E}_{AB} = 550 \text{ V}\underline{/+120°}$, and $\mathbf{E}_{BC} = 550 \text{ V}\underline{/-120°}$. A balanced delta-connected load consisting of three 11-Ω resistors is connected to this source.
(a) Draw a fully labeled schematic diagram showing tracing arrows.
(b) Determine the phase currents algebraically.
(c) Draw a phasor diagram of the phase voltages and currents.
(d) Determine the line currents by geometrical construction on a phasor diagram.
(e) Determine the line currents by phasor algebra.

27-6. The load in Problem 27-5 is replaced by a balanced delta load which has a total power rating of 15 kW at 96.6% leading power factor. Repeat all five sections of Problem 27-5.

27-7. The load in Problem 27-5 is replaced by a delta load consisting of $\mathbf{Z}_{AC} = 5 \; \Omega\underline{/+45°}$, $\mathbf{Z}_{CB} = 10 \; \Omega\underline{/-36.9°}$, and $\mathbf{Z}_{BA} = 11 \; \Omega\underline{/0°}$. Repeat all five sections of Problem 27-5.

27-8. The load in Problem 27-5 is replaced by a delta-connected load in which $\mathbf{Z}_{AC} = 10 \text{ k}\Omega\underline{/0°}$, $\mathbf{Z}_{CB}$ consists of a 10-H choke with a resistance of 50 Ω, and $\mathbf{Z}_{BA}$ is a 0.5-μF capacitor. Repeat all five sections of Problem 27-5.

27-9. The load of Problem 27-5 is connected to the source of Problem 27-1.
(a) Draw a fully labeled schematic diagram showing tracing arrows.
(b) Determine the line voltages by phasor construction.
(c) Determine the line voltages by phasor algebra.
(d) Determine the phase currents.
(e) Determine the line currents by phasor algebra.
(f) Draw a complete phasor diagram for the circuit.

27-10. The load of Problem 27-6 is connected to the source of Problem 27-1. Repeat all six sections of Problem 27-9.

728

27-11. The load of Problem 27-7 is connected to the source of Problem 27-1. Repeat all sections of Problem 27-9.

27-12. The load of Problem 27-8 is connected to the source of Problem 27-1. Repeat all sections of Problem 27-9.

27-13. What is the total power input to the load in Problem 27-1?

27-14. What is the total power input to the load in Problem 27-5?

27-15. What is the total power input to the load in Problem 27-3?

27-16. What is the total power input to the load in Problem 27-7?

27-17. What is the phase sequence of the source in Problem 27-1?

27-18. What is the phase sequence of the source in Problem 27-5?

27-19. A wattmeter is connected with its current coil in line B and its voltage coil from line A to line B in the circuit of Problem 27-1. What is its reading?

27-20. A wattmeter is connected with its current coil in line C and its voltage coil from line A to line B in the circuit of Problem 27-7. What is its reading?

27-21. If the phase sequence of Problem 27-19 is reversed, what is the wattmeter reading?

27-22. If the phase sequence of Problem 27-20 is reversed, what is the wattmeter reading?

27-23. The total power input to the load in Problem 27-5 is checked by the two-wattmeter method. What is the reading on each wattmeter?

27-24. The total power input to the load in Problem 27-2 is checked by the two-wattmeter method. What is the reading on each wattmeter?

27-25. A balanced delta load connected to a 208-V 3ϕ source is checked by the two wattmeter method whose readings are 2 kW and 1 kW, respectively.
(a) What is the total three-phase power?
(b) What is the power factor of the load?
(c) What is the magnitude of the line current?
(d) What is the impedance (magnitude and angle) of each arm of the load?
(e) What is the angle between the line voltage and the line current?

27-26. Repeat the five sections of Problem 27-25 when one wattmeter reads 3 kW and the other reads zero.

27-27. The load of Problem 27-1 is connected to the source of Problem 27-5. Repeat the four sections of Problem 27-1.

27-28. The neutral lead in Problem 27-3 becomes open circuit. Determine the current in each line.

27-29. Determine the three line currents in Problem 27-28 if the phase sequence of the source is reversed.

27-30. The load of Problem 27-7 is connected to a 208-V 3ϕ source with a CBA phase sequence through three conductors; each of which has an impedance of $2 + j1$ Ω. Determine the magnitude of the current in each line.

REVIEW QUESTIONS

27-1. What are the three main advantages of a polyphase system compared to a single-phase power distribution system?

27-2. Why is it desirable to arrange that the total instantaneous power input to a polyphase system be free from the customary sine-squared fluctuations?

27-3. Draw a graph of the total instantaneous power if one of the resistors in Figure 27-2 is changed to 180 Ω.

27-4. Continue the vector representation of the magnetic flux in Figure 27-5 at 45° intervals for the remainder of the cycle.

27-5. Why does the flux density of the rotating magnetic field in Figure 27-5 remain constant?

27-6. What is the significance of the term **synchronous speed?**

27-7. How could we reverse the direction of rotation of the magnetic field in Figure 27-5?

27-8. Why can we not produce a rotating magnetic field with the Edison three-wire system of Figure 27-6?

27-9. What is the advantage of using the system shown in Figure 27-6 for electric power distribution?

27-10. What voltage would appear across a load connected between line A and line B in the two-phase system of Figure 27-2(b)?

27-11. What is meant by a **balanced load** in a polyphase system?

27-12. Why does a load not have to have unity power factor to be classed as a balanced load?

27-13. Draw a graph showing that the total instantaneous power input to a balanced three-phase load is constant.

27-14. Why can we state that, in a balanced three-phase load, the currents in the three arms are equal in magnitude and displaced 120° in phase from one another regardless of the power factor of the load?

27-15. Show (with diagrams) how a rotating magnetic field can be produced when a three-phase source is available.

27-16. What is meant by the terms **phase voltage** and **line voltage** in a three-phase system? How do these terms apply to: (a) a wye-connected source? (b) a delta-connected source?

27-17. What is meant by the terms **phase current** and **line current** in a three-phase system? How do these terms apply to: (a) a wye-connected load? (b) a delta-connected load?

27-18. Draw a phasor diagram for a wye-connected source in which the upper-case letter ends of the windings in Figure 27-11 are connected to the neutral.

27-19. What is the function of the voltmeter when delta-connecting the coils of a three-phase source?

27-20. How would you test for incorrect wye connection of a three-phase source with an ac voltmeter?

27-21. Three transformers whose primary windings are fed from a three-phase source have the following secondary voltages: $\mathbf{E}_{aA} = 240$ V$\underline{/0°}$, $\mathbf{E}_{bB} = 240$ V$\underline{/-120°}$, and $\mathbf{E}_{cC} = 240$ V$\underline{/+120°}$. Each secondary is center-tapped, and these center taps are connected to a common neutral, as shown in Figure 27-39. Draw a fully labeled phasor diagram for the six phase source available from these **star**-connected secondary windings.

27-22. Show with a phasor diagram why the line current to a balanced delta load is greater than the phase current even though we *subtract* one phase current from the other to obtain the line current.

27-23. Why must the phasor sum of the three line currents in any three-phase, three-wire system be zero?

27-24. Show with a phasor diagram that the line voltage from a wye-connected source is $\sqrt{3}$ times the voltage generated in each winding and displaced 30° from the voltages in the windings.

27-25. $I_L = \sqrt{3}\, I_p$ in a delta-connected load depends on the load being a balanced load. However, $V_L = \sqrt{3}\, V_p$ in a wye-connected source is not dependent on a balanced load. Explain.

27-26. What is the advantage of expressing power input to a balanced three-phase load in terms of *line* voltage and current rather than *phase* voltage and current?

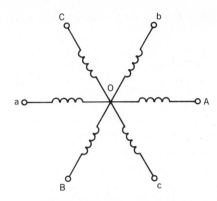

FIGURE 27-39
Star Connection of Three-Phase
Transformer Secondary Windings.

27-27. Why do the wattmeters in Figure 27-26 read the *active* power of each arm of the load?

27-28. What advantage does the three-wattmeter system of Figure 27-27 have over that of Figure 27-26?

27-29. With reference to Figure 27-27, what is meant by an **artificial neutral?**

27-30. What advantage does the two-wattmeter system of three-phase power measurement have over a single wattmeter?

27-31. With the two-wattmeter system of three-phase power measurement, why can the two wattmeters have different readings even though a balanced load is used?

27-32. Show that the total instantaneous power in *any* **wye**-connected load is equal to the sum of the two wattmeter readings in Figure 27-29.

27-33. The horizontal axis scale of the power factor graph of Figure 27-31 ranges from -1.0 through 0 to $+1.0$. Explain the reason for these particular limits.

27-34. Why is it necessary for an electrical utility company to maintain the same phase sequence?

27-35. What effect will reversing the tracing arrows when writing the voltage subscripts in the wye-connected source of Figure 27-22 have on the phase sequence of the source? Explain.

27-36. Why are we instructed to read *clockwise* around a phasor diagram when determining the phase sequence?

27-37. Explain the principle of operation of the phase sequence tester of Figure 27-37(a).

27-38. Account for the changes in load phase voltages when the neutral lead becomes open-circuit in Problem 27-28.

27-39. What is meant by a **floating wye point** in a three-phase system?

27-40. Derive Equation (27-16B) from the basic Kirchhoff's-law equations.

27-41. We can solve three-phase networks with the superposition theorem. The individual calculations are simpler, but there are many repetitive component-current calculations. However, we must use *phase* voltages for the source rather than *line* voltages. Using diagrams, explain this restriction.

*27-42. Write a program to print out the magnitude of the three line currents I_A, I_B, and I_C when any three-wire wye load is connected to a 208-V 3-ϕ source. The program should ask the operator to select phase sequence *ABC* or *CBA* for the source and enter the magnitude and angle for $\mathbf{Z}_A$, $\mathbf{Z}_B$, and $\mathbf{Z}_C$.

28

HARMONICS

28-1 NONSINUSOIDAL WAVES

In Chapter 17, we discovered that the source voltage in an alternating-current circuit is continuously changing according to a definite pattern which repeats itself periodically. We also noted that the fundamental pattern for periodic waves is the **sine wave.** In developing the behavior of ac circuits, we have taken for granted that the sources we have employed do produce sine-wave voltages. For the distribution of electric power, sine waves of voltage and current are desirable, and, in designing ac machinery, we strive to obtain such waveforms. In practice, it is not always possible to produce an alternating source voltage which has a perfectly sinusoidal form. For example, in the simple alternator of Chapter 17, a sine wave of induced voltage depends on a *uniform* magnetic field. If the flux density is not uniform, the induced voltage will not be a true sine wave.

We can also generate special nonsinusoidal waveforms such as square waves, sawtooth waves, and short-duration pulses for radar, television, and computer systems. Cycle after cycle, the instantaneous value of these waves has exactly the same shape as the preceding cycle. Hence, these geometric waveforms are also *periodic* waves.

In audio reproducing equipment, the waveforms are seldom sinusoidal. The musical note created by feeding a sine wave from an audio oscillator into a loudspeaker is very uninteresting to listen to. If we play a single note on a musical instrument and examine the waveform of the instantaneous voltage generated by a microphone picking up this sound wave, we discover a very complex nonsinusoidal alternating voltage, as shown by the black waveform in Figure 28-1. Since this

732

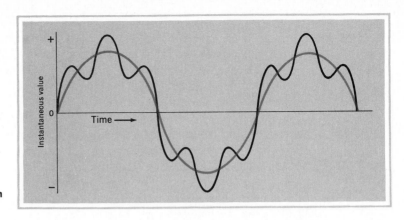

FIGURE 28-1
Nonsinusoidal Waveform from a Musical Instrument.

waveform also repeats exactly the same sequence of instantaneous values cycle after cycle, it too is a *periodic* wave.

Examining the complex wave of Figure 28-1 closely, we notice a component, somewhat sinusoidal in shape, which seems to be going through five complete cycles during one cycle of the overall waveform. This becomes more noticeable when we draw in the colored line which cuts across the complex wave halfway between the peaks of the minor excursions of the instantaneous voltage. It would seem then that there are at least two distinct frequencies present in this nonsinusoidal waveform. One is the frequency at which the overall complex waveform repeats itself. This is called the **fundamental frequency** of the periodic wave. The fundamental frequency is the lowest frequency present in the complex wave. The other predominant frequency appears to be exactly five times the fundamental frequency. This frequency is called the **fifth harmonic** of the fundamental frequency. With further investigation, we find that **any nonsinusoidal periodic wave which maintains the same complex waveform cycle after cycle is made up of a series of harmonically related pure sine waves.** As shown in Figure 28-2, a harmonic must be an integral multiple of the fundamental frequency.

FIGURE 28-2
Harmonically Related Sine Waves.

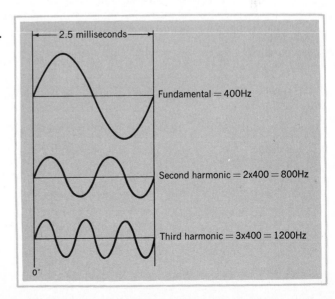

28-2 FOURIER SERIES

To determine the behavior of ac circuits in the presence of nonsinusoidal waveforms, we must be able to determine the fundamental frequency, which harmonics are present, their relative amplitudes, and their phase relationship with respect to the fundamental sine wave. One method for determining the harmonic content of a nonsinusoidal wave is based on a mathematical analysis originally suggested over 160 years ago by the French mathematician Jean-Baptiste Joseph Fourier.

Fourier demonstrated that any *periodic* complex waveform is composed of a fixed term, plus an infinite series of harmonically related cosine terms, plus an infinite series of harmonically related sine terms. If we represent the instantaneous magnitude of the nonsinusoidal waveform as a function of time by the mathematical expression $f(t)$, the general equation for the Fourier series is

$$f(t) = A_0 + A_1 \cos \omega t + A_2 \cos 2\omega t + A_3 \cos 3\omega t + \cdots + A_n \cos n\omega t$$
$$+ B_1 \sin \omega t + B_2 \sin 2\omega t + B_3 \sin 3\omega t + \cdots + B_n \sin n\omega t \qquad (28\text{-}1)$$

where A_0 is a steady-state term representing the average value or dc component of the complex wave. The coefficient of each cosine or sine term can have any magnitude. However, for most nonsinusoidal waveforms, the magnitudes of the coefficients diminish rapidly as the order of the harmonic increases. Often it is only necessary to include the first few sine and cosine terms of the infinite series to determine the shape of the complex wave.

Since evaluation of the coefficients of the terms of a Fourier series requires calculus, we shall find an alternate method of analyzing nonsinusoidal waveforms. Before we leave the topic, we can use the Fourier series to illustrate several characteristics of nonsinusoidal waves. As in Figure 28-1, we can represent $f(t)$ by a linear graph with the Y-axis representing *instantaneous magnitude* and the X-axis representing *time*. Figure 28-3(a) shows the graph of a sawtooth waveform in which the area under the instantaneous value above the X-axis equals the area under the instantaneous value below the X-axis. This means that A_0 in the Fourier series is zero and the sawtooth waveform has no dc component—it is an ac waveform. The same sawtooth waveform in Figure 28-3(b) has all positive instantaneous values, hence it is a **pul-**

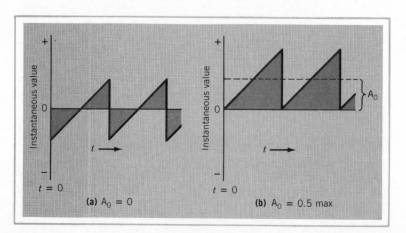

FIGURE 28-3
Average or DC Component of a Nonsinusoidal Wave.

(a) $A_0 = 0$

(b) $A_0 = 0.5$ max

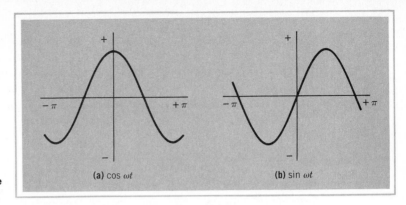

FIGURE 28-4
Symmetry with Respect to the Vertical Axis.

(a) cos ωt

(b) sin ωt

sating dc waveform. To achieve this condition, A_0 in the Fourier series must equal $0.5 \times$ the peak instantaneous dc value of the sawtooth wave.

When we plot a graph of cos ωt from $-\pi/2$ to $+\pi/2$ radians, as in Figure 28-4(a), we note that the graph is horizontally symmetrical about the $t = 0$ vertical axis. The graph of sin ωt, on the other hand, is *not* symmetrical. Sin ωt is positive for values of ωt from 0 to $+\pi/2$ radians; but it is negative from $-\pi/2$ to 0 radians. Consequently, if the graph of a nonsinusoidal wave has such vertical-axis symmetry, as in Figure 28-5(a), we know that all coefficients of the series of sine terms in Equation (28-1) will be zero. By shifting the phase of the square wave $\pi/2$ radians, as in Figure 28-5(b), it now consists of the series of sine terms with zero coefficients for the cosine terms. No matter how we shift the phase of the sawtooth wave in Figure 28-3(a), we cannot get rid of the sine terms to produce vertical-axis symmetry.

28-3 ADDITION OF HARMONICALLY RELATED SINE WAVES

A second method for determining the harmonic content of a nonsinusoidal wave is based on the principle of **reciprocity.** *If we can exactly duplicate a given nonsinusoidal wave by adding together certain harmonically related sine waves with proper magnitudes and phase relationships, it follows that the complex wave must contain this same series of harmonically related sine waves.* Since this latter method

FIGURE 28-5
Effect of Cosine and Sine Fourier Series on Vertical-Axis Symmetry.

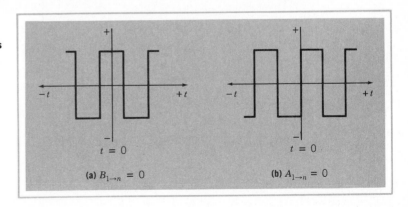

(a) $B_{1 \to n} = 0$

(b) $A_{1 \to n} = 0$

is more in keeping with our undertaking to *understand* what goes on in electric circuits, we shall use it for the remainder of this chapter. It is interesting to note that some electronic music synthesizers use this principle to duplicate the tone quality of various musical instruments.

We can add harmonics to a fundamental sine wave in our pencil-and-paper complex-waveform synthesis by adding instantaneous values on a linear graph, as we did in Chapter 19. On the graph of Figure 28-6, we have drawn a fundamental sine wave and a fifth harmonic with an amplitude of about 30% of that of the fundamental. We say that this harmonic is shown in phase with the fundamental since both sets of instantaneous values are zero at the origin of the graph and both start out in a positive direction. If we wish to show the harmonic π rad out of phase with the fundamental, we simply invert the harmonic waveform on the graph. It would start with an instantaneous value of zero at the origin but would proceed first to a *negative* peak, whereas the fundamental is proceeding first to a positive peak. If we wish to phase the harmonic so that it lags the fundamental by $\pi/2$ rad, we must draw it on the graph so that at the same instant that the instantaneous value of the fundamental is passing through zero and proceeding in a positive direction, the instantaneous value of the harmonic is at its negative peak. Note that we consider the π-rad and $\pi/2$-rad designations on the basis of one cycle of the *harmonic* consisting of 2π radians.

By adding the instantaneous values of these two sine waves in Figure 28-6 from instant to instant along the horizontal axis of the graph, we obtain the resultant waveform shown by the colored line. The resultant no longer is a sine wave as it is when we add sine waves of the same frequency; it resembles the complex wave of Figure 28-1. Since the fifth harmonic is an exact multiple of the fundamental frequency, the succeeding cycles will have exactly the same shape as the one cycle we have plotted.

When we are discussing the *shape* of an ac waveform, we are thinking in terms of the instantaneous values of voltage or current. In Chapter 17, we described the shape of a sine wave alternating voltage by writing the general equation for the sine wave voltage thus:

$$e = E_\mathrm{m} \sin \omega t \tag{17-4}$$

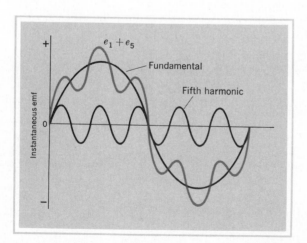

FIGURE 28-6
Duplicating the Complex Wave of Figure 28-1 by Adding a Fifth Harmonic to the Fundamental Sine Wave.

736

By substituting suitable values for E_m and ω, we could determine the amplitude and frequency of the alternating voltage as well as its shape. We can, therefore, follow a similar procedure and describe the shape of a complex wave by writing its equation. Since we formed the complex wave in Figure 28-6 by adding the instantaneous values of two sine waves, the equation for the complex wave becomes the sum of the equations for the two sine-wave components.

$$\therefore \; e_T = e_1 + e_5 = E_m(\sin \omega t + 0.3 \sin 5\omega t) \tag{28-2}$$

If we had phased the harmonic π rad out of phase with the fundamental, it would simply reverse the $+$ and $-$ signs for all instantaneous values of the harmonic sine wave. The complex wave would have a different appearance and would then be described by the equation

$$e_T = E_m(\sin \omega t - 0.3 \sin 5\omega t) \tag{28-3}$$

Examining the tables of trigonometrical functions will show us that the cosine of an angle starts at $+1$ at $t = 0$ and decreases to zero at $\pi/2$ rad, whereas the sine of an angle is zero at $t = 0$ and increases to $+1$ at $\pi/2$ rad. Therefore, a *cosine* wave has the same shape as a sine wave but *leads* the sine wave by $\pi/2$ rad (since the cosine reached $+1$ $\pi/2$ rad ahead of the sine). This then gives us a convenient way of expressing a $\pi/2$ rad phase difference between a harmonic and the fundamental when we write the equation of a complex wave. If we wished to draw the fifth harmonic in the graph of Figure 28-6 with its *negative* peak coinciding with the instant when the instantaneous value of the fundamental is passing through zero and proceeding in a positive direction, we would write the equation describing the complex wave thus:

$$e_T = E_m(\sin \omega t - 0.3 \cos 5\omega t) \tag{28-4}$$

In practice, most nonsinusoidal waves do not consist of a fundamental and a single harmonic, as in Figure 28-6. Most complex waves consist of a *series* of harmonically related sine waves with the amplitude of each harmonic diminishing as the multiple of the fundamental frequency becomes higher. However, we can get a fair approximation of the shape of the complex waveform by adding together the instantaneous values of only the first few harmonics in the series. On analyzing the harmonic content of some of the more common nonsinusoidal waveforms, we shall discover that there are two series of harmonics which govern the general appearance of nonsinusoidal waves. Complex waves such as that of Figure 28-1, in which the positive and negative half-cycles are *symmetrical*, are composed mainly of the *odd*-order harmonics (third, fifth, seventh, ninth, etc.). The *even*-order series of harmonics produces a symmetrical complex waveform if all the even harmonics are exactly in phase or π rad out of phase with the fundamental. For a harmonic to be in phase with the fundamental, when the instantaneous value of the fundamental is passing through zero and going in a positive direction, the instantaneous value of the harmonic must be passing through zero, also going in a positive direction, as shown in Figures 28-1 and 28-6. When a complex wave is composed of an even-order series of harmonics, one or more of which is *not* exactly in phase or π rad out of phase with the fundamental, the positive and negative half-cycles of the nonsinusoidal wave are quite different in appearance, as shown in Figure 28-8(b).

EXAMPLE 28-1

Determine the shape of the complex wave whose equation is

$$e = 60 \sin 2512t + 20 \cos 7536t$$

SOLUTION

$$\omega = 2\pi f \tag{17-5}$$

fundamental $\quad f = \dfrac{\omega}{2\pi} = \dfrac{2512}{6.28} = 400 \text{ Hz}$

and harmonic $\quad f = \dfrac{7536}{6.28} = 1200 \text{ Hz} = \text{third harmonic}$

Draw as accurately as possible on a sheet of graph paper a sine wave representing the fundamental. Select the number of degrees per horizontal scale division in such a manner that the third harmonic can be accurately located in phase.

Draw a third-harmonic sine wave with $\frac{20}{60}$ or $\frac{1}{3}$ the amplitude of the fundamental. Since this is a positive cosine component, start with the positive peak value phased to match the zero instantaneous value of the fundamental.

Add instantaneous values every few degrees by counting the sum of the vertical scale divisions on the graph. Mark these plots on the graph and join them to show the waveform of the complex wave, as in Figure 28-7.

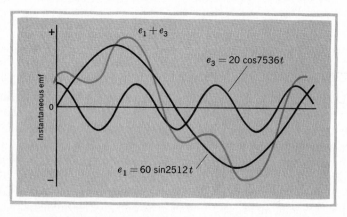

FIGURE 28-7
Complex Waveform for Example 28-1.

28-4 GENERATION OF HARMONICS

In stating the ac network theorems in Chapter 23, we were assuming that the impedance of every circuit component remained constant throughout the cycle of the sine-wave voltage and current waveforms. Such an impedance is called a **linear** imped-

738

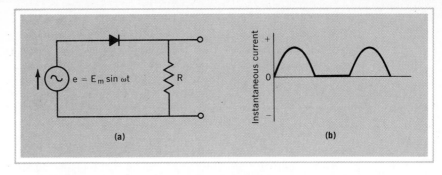

$e = E_m \sin \omega t$ R

(a)   (b)

FIGURE 28-8
Complex Waveform Created by a Half-Wave Rectifier.

ance. However, the circuit of Figure 28-8(a) contains a **nonlinear** circuit element in the form of a rectifier. The ideal rectifier has no resistance to current in one direction through it but infinite resistance to current in the opposite direction. As a result, the instantaneous current in the resistor in Figure 28-8(a) during the positive half-cycle follows the customary sine-wave variation. But during the negative half-cycle, there is no current in the resistor. Therefore, although a sine wave is applied to the circuit of Figure 28-8(a), the output waveform is the complex wave shown in Figure 28-8(b). This particular nonsinusoidal wave is called a **half-wave-rectified sine wave.** If our method of analyzing complex waves is to remain valid, we should be able to duplicate the half-wave rectified output by adding a series of harmonics to the input sine wave. Since the waveform of Figure 28-8(b) is quite unsymmetrical, the series of harmonics required will be the even-order series. Their phasing will have to be such that the peaks of these even harmonics coincide with the peak of the fundamental. To achieve this phase relationship, all the harmonics must lag the fundamental by $\pi/2$ rad, and thus all the harmonics will be minus cosine terms.

Examining the complex waveform of Figure 28-8(b), we also note that all the instantaneous values are positive quantities. Consequently, when we average the *complete* cycle, instead of an average value of zero, for this particular waveform we obtain an average value of 0.318 of the peak instantaneous value. Whenever the average of a complete cycle of a waveform is not zero, it is an indication that there is a dc component present in addition to the harmonically related sine waves. Its presence in this waveform is due to the characteristic of the rectifier that allows current through R in one direction but not in the other. This is usually the one component of this particular complex waveform that we are interested in when we construct the circuit of Figure 28-8(a) in practical equipment. When we remove the parentheses in Equation (28-5), the first term gives us the magnitude of this dc component.

When we write an equation combining all these specifications, the Fourier series for the half-wave-rectified sine wave developed by an ideal rectifier becomes

$$e = \frac{E_m}{\pi}\left(1 + \frac{\pi}{2}\sin \omega t - \frac{2}{3}\cos 2\omega t - \frac{2}{15}\cos 4\omega t \right.$$
$$\left. - \frac{2}{35}\cos 6\omega t - \frac{2\cos n\omega t}{n^2 - 1}\right) \quad (28\text{-}5)$$

where n is an even number.

CHAPTER 28   HARMONICS   **739**

The significance of Equation (28-5), insofar as this chapter is concerned, is that **harmonics not contained in the input waveform are generated whenever an ac waveform is applied to a nonlinear circuit element.**

28-5 GENERATION OF HARMONICS IN AN ELECTRONIC AMPLIFIER

In an ideal transistor amplifier, varying the signal input voltage 1.0 volt either side of its quiescent (no-signal) value should swing the output current by the same number of milliamperes either side of its quiescent value. This symmetry would be indicated by using a straight line for the dynamic transfer characteristic on the graph of Figure 28-9. But many electronic amplifiers have slightly curved transfer characteristics, as shown by Figure 28-9. As a result, a 1.0-volt swing of the input voltage in one direction produces a 2.7-milliampere change in output current, whereas a 1.0-volt swing of the input voltage in the other direction produces only a 2.3-milliampere change in output current.

We note that, with respect to the quiescent value of the output current, the positive and negative half-cycles of the waveform are unsymmetrical. Therefore, we can assume that the transistor has added a series of even-order harmonics (neither in phase nor π rad out of phase) to the input sine wave. Most significant of this series is the second harmonic, and we can approximately duplicate the output waveform by

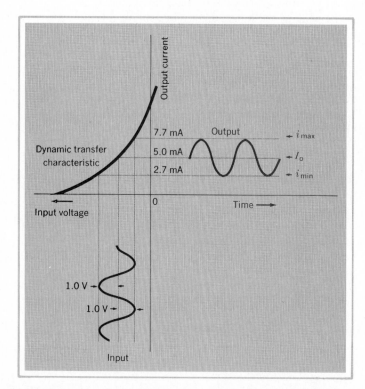

FIGURE 28-9
Nonlinear Transfer Characteristic of an Electronic Amplifier.

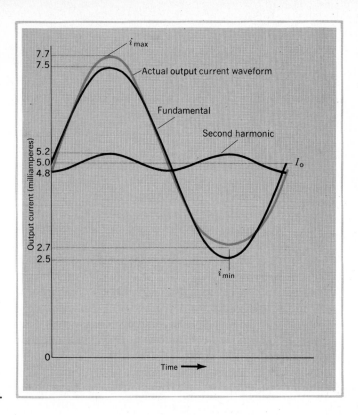

FIGURE 28-10
Second-Harmonic Content in the
Output of an Electronic Amplifier.

considering that it consists of a sine-wave fundamental with a peak amplitude of 2.5 milliamperes and a second harmonic with a peak value of 0.2 milliampere superimposed on the quiescent dc component, as shown in Figure 28-10.

Careful examination of Figure 28-10 will show us how we can determine the magnitude of this second-harmonic distortion from the measurement of the peak deviations in the actual output current (colored curve) from the quiescent (no-signal) output current I_0. Since the positive peaks of the second harmonic coincide with the peaks of the fundamental, the *peak-to-peak* amplitude of the fundamental is the same as the peak-to-peak deviation in output current. The rms value of a sine wave is $1/\sqrt{2}$ of the peak value; therefore, the rms value of the sine-wave fundamental is $1/2\sqrt{2}$ of the peak-to-peak value of the sine wave.

$$\therefore I_{\text{fund}} = \frac{i_{\max} - i_{\min}}{2\sqrt{2}} \qquad (28\text{-}6)$$

Again, from Figure 28-10, the peak-to-peak amplitude of the second harmonic is the difference in the deviations of the instantaneous output current from its quiescent value. Hence,

$$\text{peak-to-peak second harmonic} = (i_{\max} - I_0) - (I_0 - i_{\min})$$

and
$$I_{\text{2nd harm}} = \frac{i_{\max} + i_{\min} - 2I_0}{2\sqrt{2}} \qquad (28\text{-}7)$$

28-6 GENERATION OF HARMONICS IN AN IRON-CORE TRANSFORMER

In Chapter 25, we noted that when we connect the primary of a transformer across a sine-wave source, in order for the primary to develop a sine-wave self-induced voltage, the flux in the core must also be a sine wave. If the permeability of the core were constant, this would require a sine wave of exciting current in the primary winding. However, with an iron core, because of saturation effect, the permeability decreases as the flux density approaches its maximum value. As a result, a greater than normal increase in the instantaneous primary current is required in order to maintain the sine-wave shape for the total flux waveform.

We can use the *BH* curve of Figure 28-11 as a transfer curve to determine the nature of the exciting current waveform.[†] We note that the peaks of the exciting-current waveform are narrower and greater in amplitude than those of a sine wave. Because both half-cycles of the exciting-current waveform are identical, we can expect that odd-order harmonics will be present. Since the third harmonic is the predominant harmonic of this series, we can approximate the waveform of the exciting current in an iron-core transformer possessing negligible hysteresis by adding a third harmonic to the fundamental, as shown in Figure 28-12.

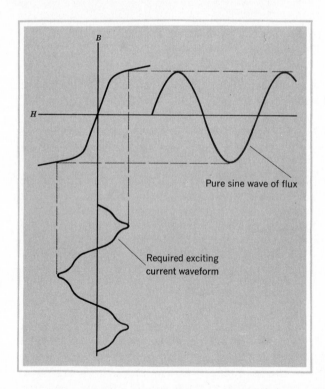

Pure sine wave of flux

Required exciting current waveform

FIGURE 28-11
Producing a Sine Wave of Flux in an Iron Core Having Negligible Hysteresis.

[†]In a practical iron-core transformer, we must use the hysteresis loop of the iron as a transfer characteristic rather than the simple *BH* curve of Figure 28-11. The exciting current in a practical transformer will have a slightly different appearance from that shown in Figure 28-11. This is due to a shift in the phase relaitonships among the component sine waves.

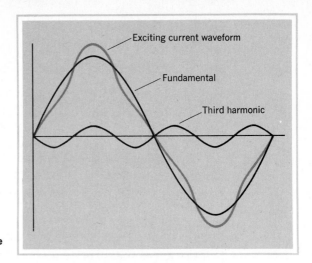

FIGURE 28-12
Third-Harmonic Content in the Exciting Current of an Iron-Core Transformer Possessing Negligible Hysteresis.

Therefore, the primary current of an iron-core transformer contains appreciable third harmonic content. If we reduce the amount of iron in the core of a given transformer, we have to operate the core further around the knee of the saturation curve in order to obtain the required sine wave of flux. Consequently, the amplitude of the third harmonic must increase appreciably. In a single-phase system, this third harmonic (180 Hz) must flow in the lines connecting the transformer to the source and can cause trouble by inducing audible 180 Hz voltages into telephone lines.

This leads to an additional advantage for three-phase systems of electric power distribution. If we connect the primary windings of a three-phase transformer bank as a delta, the line current is the difference between the phase currents. With a balanced load, the fundamental frequency components of the phase currents are $2\pi/3$ rad (120°) out of phase, so the fundamental current in the line is $\sqrt{3}\,I_p$. However, as we can see from examining Figure 28-2, a $2\pi/3$ rad phase shift at the fundamental frequency is the same as three times $2\pi/3$ rad, or a 2π rad phase shift for the third harmonic. Therefore, subtracting the third harmonic components of the transformer primary phase currents results in complete cancellation of any third-harmonic current in the line. The third-harmonic current necessary to operate the iron-core transformer properly flows around the delta and not in the transmission lines. However, the phase difference between the fifth-harmonic components of the primary phase currents is five times $2\pi/3$ rad, which is the same as $-2\pi/3$ rad. Thus, there will be a fifth-harmonic component in the line current with a magnitude which is the customary $\sqrt{3}$ times the fifth-harmonic content of the phase current.

28-7 RMS VALUE OF A NONSINUSOIDAL WAVE

In Chapter 17, we found that in order to obtain the rms value of a sine wave of voltage or current, we had to determine the average power. The rms voltage then became $E = \sqrt{P_{av}R}$. We have already noted that a regularly-recurring nonsinusoidal wave consists of a fundamental sine wave plus a series of harmonic sine waves. Applying the principle of the superposition theorem, we can think of a nonsinusoidal source as consisting of the harmonically-related sine-wave sources all acting independently to contribute to the total power.

$$\therefore P_T = \frac{E_1^2}{R} + \frac{E_2^2}{R} + \frac{E_3^2}{R} + \text{etc.}$$

Since each of these terms is based on a sine wave of voltage, we can apply the definition for rms value to the total power. Hence, the rms value of a nonsinusoidal wave becomes

$$E = \sqrt{P_T R} = \sqrt{E_1^2 + E_2^2 + E_3^2 + \text{etc.}} \qquad (28\text{-}8)$$

where E_1, E_2, and E_3 are the rms values of the fundamental, second, and third harmonics, respectively.

EXAMPLE 28-2

The terminal voltage of a three-phase wye-connected alternator is known to possess a certain amount of third and fifth harmonics. No other harmonic content of any significance is present. The phase voltage of the alternator is measured as 122 V and the line voltage is 208 V. What is the rms value of the third harmonic voltage in each alternator coil?

SOLUTION

Since the fundamental phase voltages are $2\pi/3$ rad out of phase, the line voltage contains a fundamental component which is $\sqrt{3}$ times the fundamental component of the phase voltage.

Since the fifth-harmonic phase voltages are $10\pi/3$ rad (600°), or $-2\pi/3$ rad out of phase, the line voltage contains a fifth-harmonic component which is $\sqrt{3}$ times the fifth-harmonic component of the phase voltage.

But the third-harmonic phase voltages are 2π rad out of phase and, therefore, there is no third harmonic in the line voltage. Hence,

$$208 \text{ V} = 1.73\sqrt{E_1^2 + E_5^2} \qquad \text{and} \qquad 122 \text{ V} = \sqrt{E_1^2 + E_3^2 + E_5^2}$$

Squaring both equations, subtracting, and taking the square root of both sides gives

$$E_3 = \sqrt{122^2 - 120^2} = \mathbf{22 \text{ V}}$$

28-8 SQUARE WAVES AND SAWTOOTH WAVES

From our discussion of the generation of harmonics in electronic amplifiers and iron-core transformers, it would seem that nonlinear circuit elements should be avoided. However, there are many applications in which we purposely distort an input sine wave in order to obtain a special wave shape. One such example is the **square wave** of Figure 28-13(a). Although our method of determining the harmonic content of a nonsinusoidal wave has been to synthesize the required waveform by adding together the appropriate harmonics, we must note that in practical equipment these waveforms are usually produced by passing a sine wave through a nonlinear impedance network. Thus, we can produce a square wave by passing a large-amplitude sine wave through a nonlinear network which limits the peak value of the output voltage

744

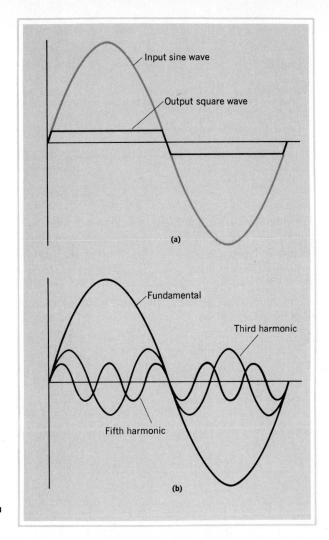

FIGURE 28-13
Production and Synthesis of a Square Wave.

Labels within figure: Input sine wave; Output square wave; (a); Fundamental; Third harmonic; Fifth harmonic; (b)

to $\pm$ a few volts. As a result, we can say that the square wave of Figure 28-13(a) has been produced by *clipping* the peaks of the input sine wave.

Although the square wave has been produced by clipping a sine wave, the nonlinear impedance network producing this effect has added harmonics to the input sine wave. We can produce square waves in which the positive pulse is either wider or narrower than the negative pulse. But the square wave of Figure 28-13(a) is completely symmetrical. Consequently, as we would expect, it is made up of a fundamental sine wave plus a series of odd-order harmonics. As the simple geometrical shape of the square wave suggests, this series of harmonics has a simple harmonic relationship between the amplitude and the frequency of each harmonic; the amplitude of each harmonic is inversely proportional to its frequency. And in order to contribute to the flat top and steep sides of the square wave, all harmonics are in phase with the fundamental. Therefore, the Fourier series for the instantaneous value of a symmetrical square wave with no dc component becomes

$$e = E_m \left(\sin \omega t + \tfrac{1}{3} \sin 3\omega t + \tfrac{1}{5} \sin 5\omega t + \tfrac{1}{7} \sin 7\omega t + \text{etc.} \right) \qquad (28\text{-}9)$$

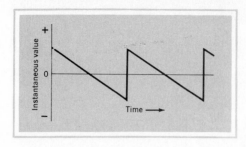

FIGURE 28-14
Sawtooth Waveform.

The first three terms of this equation are shown superimposed on the linear graph in Figure 28-13(b) as we would prepare to determine the shape of the waveform represented by Equation (28-9) by graphical addition of instantaneous values.

Another common nonsinusoidal wave is the **sawtooth wave,** shown in Figure 28-14. This waveform contains the same odd-order series of harmonics as the square wave, but it also includes the even-order series of harmonics as well. Hence, the Fourier series for the waveform of Figure 28-14 becomes

$$e = E_m(\sin \omega t + \tfrac{1}{2} \sin 2\omega t + \tfrac{1}{3} \sin 3\omega t + \tfrac{1}{4} \sin 4\omega t$$
$$+ \tfrac{1}{5} \sin 5\omega t + \tfrac{1}{6} \sin 6\omega t + \text{etc.}) \qquad (28\text{-}10)$$

28-9 NONSINUSOIDAL WAVES IN LINEAR IMPEDANCE NETWORKS

We have investigated the generation of harmonics which combine to form nonsinusoidal waveforms by applying sine waves to networks containing nonlinear circuit elements. We must now consider how we can handle a circuit problem in which a nonsinusoidal wave is applied to a network of linear impedances.

We have noted that we can think of a nonsinusoidal wave as consisting of a sine wave at the fundamental frequency plus a series of harmonically-related sine waves. In determining the rms value of a nonsinusoidal wave, we referred to the superposition theorem. According to the superposition theorem, we can determine the net effect of several sources in a network on any branch of that network by determining the effect of each source acting independently, with the other sources switched off. This provides us with the basic line of action for determining the effect of nonsinusoidal waves in linear impedance networks.

We can, therefore, think of the nonsinusoidal source of Figure 28-15(a) as being equivalent to the harmonically related sine-wave generators connected in series, as in Figure 28-15(b). Applying the superposition theorem to this circuit, we would determine the current in the load resistor independently for each frequency. If we wish to know the rms value of the total current, we simply apply Equation (28-8) to the individual harmonic currents in the load resistor. If we wish to know the shape of the output waveform, we can plot the various harmonic components of the output on a linear graph and add their instantaneous values.

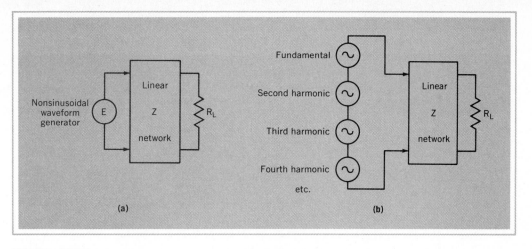

(a)   (b)

FIGURE 28-15
Applying the Superposition Theorem to Linear Impedance Networks Fed from Nonsinusoidal Sources.

EXAMPLE 28-3
The nonsinusoidal waveform described as

$$e = 45 \sin 6280t - 15 \sin 18\,840t$$

is applied to the circuit of Figure 28-16.
(a) Determine the waveform of the input wave.
(b) Determine the waveform of the output wave.
(c) Determine the overall power factor of the circuit.

SOLUTION
(a) The fundamental has a frequency of

$$f_{\mathrm{f}} = \frac{\omega}{2\pi} = \frac{6280}{6.28} = 1 \text{ kHz}$$

and the harmonic in the nonsinusoidal wave has a frequency of

$$f_{\mathrm{h}} = \frac{18\,840}{6.28} = 3 \text{ kHz}$$

which is the third harmonic.

As the equation indicates, the harmonic and the fundamental are 180° out of phase, and the amplitude of the third harmonic is one-third of the amplitude of the

FIGURE 28-16
Schematic Diagram for Example 28-3.

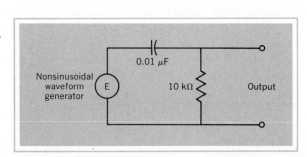

747

fundamental. Therefore, the input voltage has the waveform shown in Figure 28-17.

(b) $v_{out} = iR$, and since resistance is a constant, the output voltage waveform will have the same shape as the total instantaneous current waveform. At the fundamental frequency,

$$X_C = \frac{1}{\omega C} = \frac{1}{6280 \times 0.01 \ \mu F} = 15.92 \ k\Omega$$

$$\phi = \arctan \frac{-15.92 \ k\Omega}{10 \ k\Omega} = -57.87°$$

$$\mathbf{Z} = \frac{10 \ k\Omega}{\cos - 57.87°} = 18.8 \ k\Omega \underline{/-57.87°}$$

$$E_f = 0.707E_m = 0.707 \times 45 \ V = 31.8 \ V$$

$$\therefore \mathbf{I}_f = \frac{\mathbf{E}}{\mathbf{Z}} = \frac{31.8 \ V\underline{/0°}}{18.8 \ k\Omega \underline{/-57.87°}} = 1.69 \ mA \underline{/+57.87°}$$

At the third harmonic,

$$X_C = \frac{1}{18\,840 \times 0.01 \ \mu F} = 5.31 \ k\Omega$$

$$\phi = \arctan \frac{-5.31 \ k\Omega}{10 \ k\Omega} = -27.97°$$

$$\mathbf{Z} = \frac{10 \ k\Omega}{\cos - 27.97°} = 11.32 \ k\Omega \underline{/-27.97°}$$

$$E_3 = 0.707 \times 15 \ V = 10.6 \ V$$

$$\therefore \mathbf{I}_3 = \frac{\mathbf{E}_3}{\mathbf{Z}_3} = \frac{10.6 \ V\underline{/+180°}}{11.32 \ k\Omega \underline{/-27.97°}} = 0.937 \ mA \underline{/+207.97°}$$

In drawing the fundamental and third-harmonic current sine waves on a linear graph, we must note, first of all, that the amplitude of the harmonic is now 55% of the amplitude of the fundamental, since the reactance of the capacitor is less at the third-harmonic frequency than at the fundamental frequency. In the input wave,

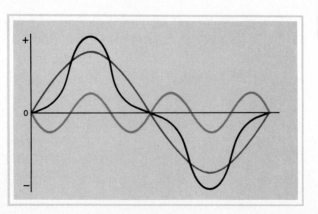

FIGURE 28-17
Input-Voltage Waveform.

748

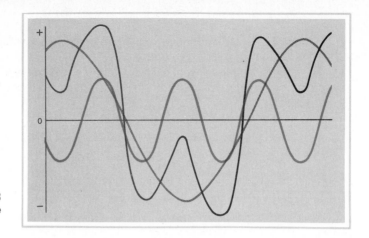

**FIGURE 28-18
Output-Voltage
Waveform.**

the 180° point of the third harmonic coincides with the 0° point of the fundamental at $t = 0$ on the horizontal axis. But, in the current waveform, the fundamental has been advanced 57.8° (360° to a fundamental cycle) and the third harmonic has been advanced only 27.9° (360° to a cycle of third harmonic). Therefore, in drawing the output voltage waveform, the 57.8° point on the fundamental and the 207.9° point on the third harmonic must coincide as shown in Figure 28-18.

(c) The total average power is

$$P_T = I_f^2 R + I_3^2 R = [(1.69 \text{ mA})^2 + (0.937 \text{ mA})^2] \times 10 \text{ k}\Omega$$
$$= 37.34 \text{ mW}$$

The rms value of the nonsinusoidal input voltage is

$$E_T = \sqrt{E_f^2 + E_3^2} = \sqrt{31.8^2 + 10.6^2} = 33.5 \text{ V}$$
$$I_T = \sqrt{(1.69 \text{ mA})^2 + (0.937 \text{ mA})^2} = 1.93 \text{ mA}$$

Therefore, the apparent power in the network is

$$S = E_T I_T = 33.5 \text{ V} \times 1.93 \text{ mA} = 64.7 \text{ mVA}$$

And, by definition, the power factor is

$$\cos \phi = \frac{P}{S} = \frac{37.34 \text{ mW}}{64.7 \text{ mVA}} = \textbf{57.7\% leading}$$

PROBLEMS

28-1. Determine graphically the shape of the complex wave described by Equation (28-3).

28-2. Determine graphically the shape of the complex wave described by Equation (28-4).

28-3. Determine graphically the shape of the complex wave based on only the fundamental and the first two harmonic terms of the equation for the square wave, as shown in Figure 28-13(b).

28-4. Determine graphically the shape of the complex wave described by Equation (28-5), using all terms up to and including the sixth harmonic.

28-5. Determine graphically the shape of the nonsinusoidal wave described by the equation

$$e = 120 \sin 1570t - 60 \sin 3140t + 40 \sin 4710t - 30 \sin 6280t + 24 \sin 7850t$$

Compare this with the waveform of Figure 28-14.

28-6. Determine graphically the shape of the nonsinusoidal wave described by the equation

$$e = 100(\sin 1256t - \tfrac{1}{9} \sin 3768t + \tfrac{1}{25} \sin 6280t - \tfrac{1}{49} \sin 8792t)$$

Describe the shape of the resulting complex wave.

28-7. What is the rms value of the complex voltage in Problem 28-5?

28-8. A wye-connected three-phase source has a phase voltage of 110 V and a line voltage of 188 V. The induced voltage is known to contain odd-order harmonics. If the amplitudes of the harmonics above the seventh are small enough that they can be neglected, what is the rms value of the third-harmonic content of the phase voltage?

28-9. When a sine wave is applied to a transistor amplifier, the collector current swings 8 mA above its quiescent value and 7 mA below. Determine the percentage second-harmonic distortion by expressing the second-harmonic current and the fundamental current as a ratio.

28-10. (a) What is the rms value of the current in a series circuit consisting of 50-mH inductance and 50-Ω resistance when the applied voltage has the nonsinusoidal waveform of Problem 28-5?
(b) What is the average power in this circuit?
(c) What is the overall power factor of this circuit?

28-11. Determine the output voltage waveform if the positions of the capacitor and resistor in Example 28-3 are reversed.

28-12. Determine the equation for the total current when a voltage described by the equation

$$e = 180 \sin 377t - 60 \sin 1131t + 36 \sin 1885t$$

is applied to a circuit consisting of two parallel branches. Branch I has an impedance of 90 Ω $\underline{/-60°}$ at the fundamental frequency, and Branch II has an impedance of 40 + j30 Ω at the fundamental frequency.

REVIEW QUESTIONS

28-1. Suggest several reasons why it is desirable to obtain a *sine*-wave source for electric power distribution systems.

28-2. Define the terms **fundamental** and **harmonic.**

28-3. Why is it possible to describe the shape of a nonsinusoidal wave by an equation such as Equation (28-2)?

28-4. Show that Equation (28-5) for a half-wave rectified sine wave is consistent with the general form of the Fourier series in Equation (28-1).

28-5. Under what conditions will the coefficient A_0 in the Fourier series be zero?

28-6. Is it necessary for a waveform to be a pulsating dc waveform for A_0 in the Fourier series to be other than zero? Explain.

28-7. What is the significance of a negative value of t in the graphs of Figure 28-5?

28-8. What is the significance of a $-$ sign in front of a term in the equation for a nonsinusoidal wave?

28-9. What is the significance of a cosine term in the equation for a nonsinusoidal wave?

28-10. How would you describe the phase relationship among the fundamental and a the harmonics shown in Figure 28-5?

28-11. Why is it usually possible to obtain a fairly good duplication of a nonsinusoidal wave by considering only the first five or six terms of the harmonic series?

28-12. What is the distinction between the odd- and even-order of harmonics in governing the shape of nonsinusoidal waves?

28-13. What is meant by a **linear bilateral** circuit element? Give several examples.

28-14. What is meant by a **unilateral** circuit element?

28-15. Give an example of a nonlinear bilateral circuit element.

28-16. When a transistor introduces harmonic distortion into a circuit, where do these harmonics which are not present in the input wave come from?

28-17. If a transistor amplifier is developing 8% second-harmonic distortion and 6% third-harmonic distortion, what is the percentage total harmonic distortion?

28-18. The waveforms shown for the transformer in Figure 28-11 are based on a sine-wave *voltage* source. Draw the waveforms which apply when the transformer is fed from a sine-wave *current* source. What order of harmonics is introduced?

28-19. The *BH*-curve transfer characteristic in Figure 28-11 is based on only alternating current in the primary. Draw a similar sketch to show the appearance of the flux waveform when the transformer primary is connected in series with the collector of a transistor in which the collector current varies in a sinusoidal manner about its quiescent value. What order of harmonics is introduced by such an output transformer?

28-20. Why does none of the third-harmonic content of the exciting current in the primary windings of a three-phase delta-connected transformer bank appear in the line current?

28-21. Why does the seventh-harmonic content of the line current to a delta-connected transformer bank have the customary $\sqrt{3}$ relationship to the seventh-harmonic current in the transformer primary windings?

28-22. Suggest why the harmonic content of an alternator terminal voltage contains odd-order harmonics rather than even-order harmonics.

28-23. What percentage of the ninth harmonic induced into an alternator winding will appear in the line voltage of a wye-connected three-phase source?

28-24. A symmetrical square wave of voltage has a peak value of 100 V. What is its rms value?

28-25. What effect does reversing the sign in front of the even-order harmonic terms of Equation (28-10) have on the shape of the complex wave?

28-26. A certain amplifier advances the phase of a 500-Hz sine wave by 2°. In order to pass a complex wave without changing its shape, this amplifier must have a 4° phase advance at 1 kHz, a 6° phase advance at 1.5 kHz, and so on. Explain.

28-27. Explain the significance of the superposition theorem in solving linear ac networks which are fed from nonsinusoidal sources.

28-28. The output waveform in Example 28-3 is taken across a resistor whose resistance is not dependent on frequency. Why then does the output waveform not have the same shape as the input waveform?

*28-29. In working with time-domain equations, such as the Fourier series Equation (28-1), personal computers have the ability to perform repetitive calculations for small increments in the independent variable rapidly and accurately. For Example 28-3, program a computer to
(a) Plot a graph of the input waveform.
(b) Determine the equation for the output waveform.
(c) Plot a graph of the output waveform.

APPENDICES

APPENDIX 1 DETERMINANTS

In solving simultaneous equations, we require as many equations as there are unknowns. For the network examples in this text, we can limit our discussion to second- and third-order simultaneous equations and write them in the general form shown below, arranging the order of the terms so that the *same* unknown appears in a vertical column.

Second-Order Equations

$$a_1 x + b_1 y = k_1 \quad (1)$$
$$a_2 x + b_2 y = k_2 \quad (2)$$

Third-Order Equations

$$a_1 x + b_1 y + c_1 z = k_1 \quad (A)$$
$$a_2 x + b_2 y + c_2 z = k_2 \quad (B)$$
$$a_3 x + b_3 y + c_3 z = k_3 \quad (C)$$

where x, y, and z are the unknowns and a, b, c, and k are numerical coefficients which can be either $+$ or $-$ quantities or even zero in some cases.

In our search for a systematic procedure, we can solve the generalized second-order simultaneous equations above by the more familiar elimination method.

$$b_2 \times (1) \text{ gives} \quad a_1 b_2 x + b_1 b_2 y = k_1 b_2$$
$$b_1 \times (2) \text{ gives} \quad a_2 b_1 x + b_1 b_2 y = k_2 b_1$$

Subtracting,

$$x(a_1 b_2 - a_2 b_1) = k_1 b_2 - k_2 b_1$$

from which

$$x = \frac{k_1 b_2 - k_2 b_1}{a_1 b_2 - a_2 b_1} \tag{3}$$

Similarly,

$$y = \frac{a_1 k_2 - a_2 k_1}{a_1 b_2 - a_2 b_1} \tag{4}$$

We can, therefore, solve *any* second-order simultaneous equations by *memorizing* Equations (3) and (4). The numerators and denominators of these generalized equations are called **determinants.** Note that the *same* determinant appears in the denominators of both Equation (3) and Equation (4). Hence, we can rewrite Equations (3) and (4) as

$$x = \frac{D_x}{D} \quad \text{and} \quad y = \frac{D_y}{D}$$

These equations are called **Cramer's rule.**

We now need a simple way of memorizing the formulas for D, D_x, and D_y. We can achieve this by writing the numerical coefficients which make up the determinants in matrix form. We first write the simultaneous equations in the systematic form of Equations (1) and (2). Directly below, we write the matrices for the determinants. Each determinant matrix contains as many vertical columns and horizontal rows as there are unknowns. The denominator matrix D simply reproduces the numerical coefficients of the unknowns in the *same* matrix position as in the original equations.

$$
\begin{array}{ccc}
\textit{Column} & \textit{Column} & \textit{Substitution} \\
\textit{1} & \textit{2} & \textit{column}
\end{array}
$$

$$a_1 x \; + \; b_1 y \; = \; k_1 \tag{1}$$

$$a_2 x \; + \; b_2 y \; = \; k_2 \tag{2}$$

$$x = \frac{\begin{vmatrix} k_1 & b_1 \\ k_2 & b_2 \end{vmatrix}}{\begin{vmatrix} a_1 & b_1 \\ a_2 & b_2 \end{vmatrix}} \tag{5}$$

$$y = \frac{\begin{vmatrix} a_1 & k_1 \\ a_2 & k_2 \end{vmatrix}}{\begin{vmatrix} a_1 & b_1 \\ a_2 & b_2 \end{vmatrix}} \tag{6}$$

The numerator matrix for the determinant D_x replaces the coefficients of x (a_1 and a_2) with the numerical constants k_1 and k_2 from the right-hand side of Equations (1) and (2). Similarly, the numerator matrix for the determinant D_y replaces the coefficients of y (b_1 and b_2) with the same numerical constants k_1 and k_2.

To evaluate the three determinants so as to arrive at the same result as Equations (3) and (4), we use the following procedure. Using the denominator determinant D as an example, we require the product $a_1 b_2$ *minus* the product $a_2 b_1$. This can be accomplished in the D matrix by multiplying coefficients along *diagonals*, always moving diagonally to the right. To obtain the required solution ($a_1 b_2 - a_2 b_1$) we always *subtract* the *upward* diagonal from the *downward* diagonal.

Second operation—subtractive diagonal
Reverse sign of the product

$$\begin{vmatrix} a_1 & b_1 \\ a_2 & b_2 \end{vmatrix} = a_1 b_2 - a_2 b_1$$

First operation—additive diagonal

Checking against the numerator determinants in Equations (3) and (4), we find that the same rule applies to the evaluation of the matrices for D_x and D_y. The best way to memorize the procedure is to work numerical examples.

$$\begin{vmatrix} 4 & 2 \\ 3 & 5 \end{vmatrix} = (4 \times 5) - (3 \times 2) = 14$$

$$\begin{vmatrix} 3 & 5 \\ 4 & 2 \end{vmatrix} = 6 - 20 = -14$$

$$\begin{vmatrix} -2 & -4 \\ -3 & 5 \end{vmatrix} = (-2 \times 5) - (-3 \times -4) = -10 - 12 = -22$$

Using the simultaneous equations from Step V of Example 8-1,

$$60I_1 + 10I_2 = 80$$

$$10I_1 + 35I_2 = 80$$

$$I_1 = \frac{\begin{vmatrix} 80 & 10 \\ 80 & 35 \end{vmatrix}}{\begin{vmatrix} 60 & 10 \\ 10 & 35 \end{vmatrix}} = \frac{2800 - 800}{2100 - 100} = \frac{2000}{2000} = 1 \text{ A}$$

$$I_2 = \frac{\begin{vmatrix} 60 & 80 \\ 10 & 80 \end{vmatrix}}{\begin{vmatrix} 60 & 10 \\ 10 & 35 \end{vmatrix}} = \frac{4800 - 800}{2000} = \frac{4000}{2000} = 2 \text{ A}$$

We can check our solution by substituting the answers into the original equations:

$$60 \times 1 + 10 \times 2 = 80$$

$$10 \times 1 + 35 \times 2 = 80$$

Adding a third unknown greatly increases the amount of numerical computation required in the solution of the third-order simultaneous equations. However, the procedure for evaluating third-order determinants is exactly the same as the one we have developed for evaluating second-order determinants. If we take the time to solve the three generalized Equations (A), (B), and (C) for x, we obtain

$$x = \frac{k_1 b_2 c_3 + b_1 c_2 k_3 + c_1 k_2 b_3 - k_3 b_2 c_1 - b_3 c_2 k_1 - c_3 k_2 b_1}{a_1 b_2 c_3 + b_1 c_2 a_3 + c_1 a_2 b_3 - a_3 b_2 c_1 - b_3 c_2 a_1 - c_3 a_2 b_1}$$

The equations for y and z are similar, all with the same denominator determinant. Our third order determinant matrices must provide us with three downward additive diagonals and three upward subtractive diagonals, each diagonal being the product of three coefficients. First, we set up the required matrices from Equations (A), (B), and (C).

	Column 1		Column 2		Column 3		Substitution column	
	$a_1 x$	$+$	$b_1 y$	$+$	$c_1 z$	$=$	k_1	(A)
	$a_2 x$	$+$	$b_2 y$	$+$	$c_2 z$	$=$	k_2	(B)
	$a_3 x$	$+$	$b_3 y$	$+$	$c_3 z$	$=$	k_3	(C)

$$x = \frac{\begin{vmatrix} k_1 & b_1 & c_1 \\ k_2 & b_2 & c_2 \\ k_3 & b_3 & c_3 \end{vmatrix}}{\begin{vmatrix} a_1 & b_1 & c_1 \\ a_2 & b_2 & c_2 \\ a_3 & b_3 & c_3 \end{vmatrix}}$$

$$y = \frac{\begin{vmatrix} a_1 & k_1 & c_1 \\ a_2 & k_2 & c_2 \\ a_3 & k_3 & c_3 \end{vmatrix}}{D}$$

$$z = \frac{\begin{vmatrix} a_1 & b_1 & k_1 \\ a_2 & b_2 & k_2 \\ a_3 & b_3 & k_3 \end{vmatrix}}{D}$$

To obtain the required diagonals from the 3×3 matrices, we can think of them as being printed on a vertical cylinder, so that there are three rows to each diagonal, no matter where we start. To achieve the same effect on flat paper, we repeat the first two columns in sequence on the right-hand side of the original matrix.

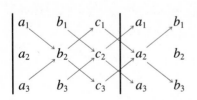

Second operation—three subtractive diagonals
Reverse sign of these three products

First operation—three additive diagonals

$$\therefore D = a_1 b_2 c_3 + b_1 c_2 a_3 + c_1 a_2 b_3 - a_3 b_2 c_1 - b_3 c_2 a_1 - c_3 a_2 b_1$$

EXAMPLE

$$D = \begin{vmatrix} 4 & 3 & -2 \\ 5 & -2 & 4 \\ 6 & 0 & 1 \end{vmatrix}$$

SOLUTION

$$= -8 + 72 + 0 - 24 - 0 - 15 = +25$$

APPENDIX 2   CALCULUS SOLUTIONS

2-1 Maximum Power-Transfer Theorem

Given a voltage source with a fixed internal resistance R_{int}, we wish to determine the value of load resistance R_L in the circuit of Fig. A2-1 which will permit maximum power in the load resistance.

First, we set up an equation for power output in terms of the variable load resistance:

$$P_L = I^2 R_L$$

but $I = E/(R_{int} + R_L)$; hence,

$$P_L = \frac{E^2 R_L}{(R_{int} + R_L)^2} \tag{7}$$

Maximum power transfer will occur when the value of R_L is such that the derivative of P_L with respect to R_L is zero:

$$\frac{dP_L}{dR_L} = \frac{E^2[(R_{int} + R_L)^2 - R_L(2R_{int} + 2R_L)]}{(R_{int} + R_L)^4} = 0$$

Since $E \neq 0$, the factor in square brackets must equal zero, from which

$$R_{int} + R_L = 2R_L$$

Therefore, maximum power output occurs when $R_L = R_{int}$. $\hspace{2cm}$ (8)

2-2 Instantaneous PD in a CR Circuit

When we throw the switch to position 1 in the circuit of Fig. A2-2,

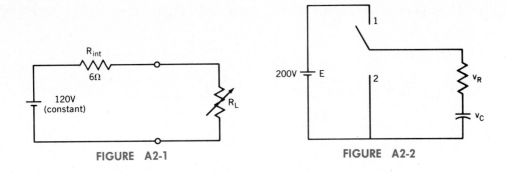

FIGURE A2-1   FIGURE A2-2

$$E = iR + v_C \tag{9}$$

Since $i = dq/dt$ and since $q = Cv_C$,

$$i = C\frac{dv_C}{dt}$$

Substituting in Equation (9),

$$E = CR\left(\frac{dv_C}{dt}\right) + v_C \tag{10}$$

To solve this differential equation, we rearrange it in the form

$$\frac{dv_C}{dt} = \frac{E - v_C}{CR}$$

$$\frac{dv_C}{E - v_C} = \frac{dt}{CR}$$

Multiplying both sides by -1,

$$\frac{dv_C}{v_C - E} = -\frac{dt}{CR}$$

Integrating gives

$$\ln{(v_C - E)}\Big]_0^{v_C} = \frac{-t}{CR}\Big]_0^t$$

$$\ln{(v_C - E)} - \ln{(0 - E)} = \frac{-t}{RC} + 0$$

which becomes

$$\ln\left(\frac{v_C - E}{-E}\right) = \frac{-t}{CR}$$

Switching from the logarithmic form to the exponential form,

$$\frac{v_C - E}{-E} = e^{-t/CR}$$

Rearranging this to solve for v_C,

$$v_C = E - Ee^{-t/CR}$$

Thus,

$$v_C = E(1 - e^{-x}) \tag{11}$$

where $e = 2.718$ (*the base of natural logarithms*) *and* $x = t/CR$.

Once the capacitor has charged, if we throw the switch in the circuit of Figure A2-2 to position 2, the applied voltage in the loop consisting of the capacitor and the resistor is zero, and

$$0 = iR + v_C = CR\frac{dv_C}{dt} + v_C$$

If we rewrite the equation as $\dfrac{dv_C}{v_C} = \dfrac{-dt}{CR}$ and integrate,

$$\ln v_C \Big]_0^{v_C} = \frac{-t}{CR}\Big]_0^t$$

$$\ln\left(\frac{v_C}{V_0}\right) = \frac{-t}{CR} \quad \text{or} \quad \frac{v_C}{V_0} = e^{-t/CR}$$

Thus,

$$v_C = V_0 e^{-x} \tag{12}$$

where V_0 is the initial voltage across the capacitor, $e = 2.718$, and $x = t/CR$.

2-3 Energy Stored by a Capacitor

To determine the energy stored by a capacitor, we calculate the area under the instantaneous power graph of Figure 12-19 by establishing the integral

$$w = \int_0^t p \, dt \tag{13}$$

But

$$p = vi \quad \text{and} \quad i = C\frac{dv}{dt}$$

Therefore,
$$w = \int_0^t Cv\frac{dv}{dt}\,dt$$

$$= \int_0^v Cv\,dv$$

and
$$w = \tfrac{1}{2}Cv^2$$

Hence, no matter how a capacitor charges, the energy stored by it at *any* instant in time is proportional to its capacitance and the square of the potential difference between its plates.

2-4 Instantaneous Current in an LR Circuit

When we throw the switch to position 1 in the circuit of Fig. A2-3,

$$E = v_R + v_L = iR + L\left(\frac{di}{dt}\right) \tag{15}$$

To solve this differential equation, we rearrange it in the form

$$\frac{di}{i - \dfrac{E}{R}} = -\frac{R}{L}dt$$

Integrating gives

$$\ln\left(i - \frac{E}{R}\right)\Bigg]_0^i = -\frac{R}{L}t\Bigg]_0^t$$

$$\ln\left(i - \frac{E}{R}\right) - \ln\left(0 - \frac{E}{R}\right) = -\frac{R}{L}t + 0$$

FIGURE A2-3

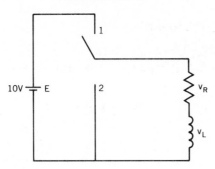

which becomes

$$\ln\left(\frac{i - \dfrac{E}{R}}{-\dfrac{E}{R}}\right) = -\frac{R}{L}t$$

Switching from the logarithmic form to the exponential form,

$$\frac{i - \dfrac{E}{R}}{-\dfrac{E}{R}} = e^{-(R/L)t}$$

Rearranging this to solve for i,

$$i = \frac{E}{R} - \frac{E}{R}(e^{-(R/L)t})$$

Thus,

$$i = \frac{E}{R}(1 - e^{-x}) \tag{16}$$

where $e = 2.718$ (the base of natural logarithms) and $x = tR/L$.

Once the current has reached its steady-state value of $I = E/R$, if we throw the switch in the circuit of Figure A2-3 to position 2, the applied voltage in the loop consisting of the inductance and the resistor is zero, and

$$0 = v_R + v_L = iR + L\left(\frac{di}{dt}\right)$$

If we rewrite the equation as $di/i = -R\,dt/L$ and integrate,

$$\ln i\Big]_{I_0}^{i} = \frac{-Rt}{L}\Big]_0^t$$

$$\ln\left(\frac{i}{I_0}\right) = \frac{-Rt}{L} \quad \text{or} \quad \frac{i}{I_0} = e^{-Rt/L}$$

Thus,

$$i = I_0 e^{-x} \tag{17}$$

where I_0 is the initial current through the inductance, $e = 2.718$, and $x = tR/L$.

2-5 Energy Stored by an Inductor

To determine the energy stored in the magnetic field resulting from current through an inductor, we calculate the area under the instantaneous power graph of Figure 16-12 by establishing the integral

$$w = \int_0^t p\, dt \tag{18}$$

But

$$p = iv_L = iL\frac{di}{dt}$$

Hence the integral becomes

$$w = \int_0^i Li\, di$$

and

$$w = \tfrac{1}{2}Li^2 \tag{19}$$

Hence, the energy stored by *any* inductor at *any* instant in time is proportional to its inductance and the square of the current through it at that instant in time.

2-6 Rms and Average Values of a Sine Wave

By using integration techniques to average instantaneous power and instantaneous current waveforms, we can derive expressions for rms and average values for any ac waveform. To find the rms value of an alternating current, we start with the definition that

> *The rms value of an alternating current is equal to the direct current which will produce the same average heating effect in a given resistance.*

To find the average heating effect of an alternating current, we can integrate the instantaneous power over a complete cycle (2π radians) and divide by the number of radians included in the integration. Hence,

$$P_{av} = \frac{1}{2\pi}\int_0^{2\pi} p\, d\phi = \frac{1}{2\pi}\int_0^{2\pi} i^2R\, d\phi$$

From the foregoing definition,

$$P_{av} = I^2R$$

where I is the rms value of the alternating current.

Hence,
$$I^2R = \frac{R}{2\pi}\int_0^{2\pi} i^2\, d\phi$$

and
$$I = \sqrt{\frac{1}{2\pi}\int_0^{2\pi} i^2\, d\phi} \tag{20}$$

Equation (20) gives the rms value of an alternating current of *any* waveshape.[†] For a sine wave, we substitute $i = I_m \sin \phi$.

$$I^2 = \frac{1}{2\pi} \int_0^{2\pi} I_m^2 \sin^2 \phi \, d\phi$$

From the relationship, $\sin^2 \phi = \frac{1}{2}(1 - \cos 2\phi)$[‡]

$$I^2 = \frac{I_m^2}{4\pi} \int_0^{2\pi} (1 - \cos 2\phi) \, d\phi$$

$$= \frac{I_m^2}{4\pi} [\phi - \tfrac{1}{2} \sin 2\phi]_0^{2\pi}$$

Since $\sin 0$ and $\sin 4\pi$ are both zero,

$$I^2 = \frac{2\pi I_m^2}{4\pi}$$

and
$$I = \frac{I_m}{\sqrt{2}} \tag{21}$$

We find the average value of a symmetrical alternating-current waveform from our decision to average the instantaneous current for one half-cycle only. Therefore,

$$I_{av} = \frac{1}{\pi} \int_0^{\pi} i \, d\phi \tag{22}$$

Applying Equation (22) to a sine wave,

$$I_{av} = \frac{1}{\pi} \int_0^{\pi} I_m \sin \phi \, d\phi$$

$$= \frac{I_m}{\pi} [-\cos \phi]_0^{\pi}$$

$$\therefore I_{av} = \frac{2}{\pi} I_m \tag{23}$$

2-7 Inductive Reactance

When an ideal inductor is connected across a sine-wave voltage source, the instantaneous voltage drop across the inductance must equal the instantaneous applied voltage. Hence,

---

[†]Equation (20) shows why we can call this value of an alternating current its **root-mean-square** or **rms** value.

[‡]See footnote on page 440.

$$L\frac{di}{dt} = E_m \sin \omega t$$

from which
$$\frac{di}{dt} = \frac{E_m}{L} \sin \omega t$$

Proceeding with the indefinite integration, we can solve for the instantaneous current in an inductance.

$$i = \frac{E_m}{L} \int \sin \omega t \, dt$$

$$= \frac{E_m}{\omega L} \times -\cos \omega t + C$$

Except when dealing with transients caused by opening or closing a switch in the circuit, the constant of integration C is zero. And if we check the values of sines and cosines for a complete revolution, we find that a *negative* cosine wave is the same as a sine wave *lagging* by 90° or $\pi/2$ radians. Hence,

$$i_L = \frac{E_m}{\omega L} \sin \left(\omega t - \frac{\pi}{2} \right) \tag{24}$$

The instantaneous current i in Equation (24) will attain its maximum value when $\sin (\omega t - \pi/2) = 1$. Hence,

$$I_m = \frac{E_m}{\omega L}$$

Since I_L and V_L are both 0.707 of their peak values, from the definition of inductive reactance,

$$X_L = \frac{V_L}{I_L} = \frac{E_m}{I_m} = \omega L$$

$$\therefore \quad X_L = \omega L \tag{25}$$

where $\omega = 2\pi f$.

2-8 Capacitive Reactance

In Equation (18-7) for the instantaneous current in a capacitor,

$$i = C\frac{dv}{dt} \tag{18-7}$$

we substitute $E_m \sin \omega t$ for the instantaneous voltage across the capacitor.

$$i = C\frac{d}{dt}(E_m \sin \omega t)$$

$$= \omega C E_m \cos \omega t$$

Since $\cos \phi$ is the same as the sine of an angle *leading* ϕ by 90° or $\pi/2$ radians

$$i_C = \omega C E_m \sin\left(\omega t + \frac{\pi}{2}\right) \tag{26}$$

By noting that $i = I_m$ when $\sin(\omega t + \pi/2) = 1$ in Equation (26), we obtain

$$I_m = \omega C E_m$$

from which
$$X_C = \frac{1}{\omega C} \tag{27}$$

2-9 General Transformer Equation

We can arrive at the general transformer equation (25-3) by writing Equation (15-1) as it applies to the *instantaneous* voltage induced into a transformer winding. Therefore,

$$e = N\frac{d\phi}{dt} \tag{15-1}$$

Since the flux in the core of the transformer must be a sine wave,

$$\phi = \Phi_m \sin \omega t \qquad \text{and} \qquad \frac{d\phi}{dt} = \omega \Phi_m \cos \omega t$$

Hence,
$$e = \omega N \Phi_m \cos \omega t \tag{28}$$

Because the instantaneous voltage in Equation (28) will be at its peak value when $\cos \omega t = 1$,

$$E_m = 2\pi f N \Phi_m$$

and
$$E = \frac{E_m}{\sqrt{2}} = 4.44 f N \Phi_m \tag{25-3}$$

2-10 Maximum Transformer Efficiency

The efficiency of a transformer is

$$\eta = \frac{S_L \cos \phi}{S_L \cos \phi + I^2 R_e + P_{CL}} \tag{25-14}$$

Substituting $S = IV$ in Equation (25-14) gives

$$\eta = \frac{IV \cos \phi}{IV \cos \phi + I^2 R_e + P_{CL}}$$

If we now consider that the current is the only variable, maximum efficiency will occur when the derivative of η with respect to I is zero.

$$\frac{d\eta}{dI} = \frac{(IV \cos \phi + I^2 R_e + P_{CL})V \cos \phi - IV \cos \phi(V \cos \phi + 2IR_e)}{(IV \cos \phi + I^2 R_e + P_{CL})^2}$$

For the derivative to equal zero, the numerator must equal zero, from which

$$V \cos \phi(I^2 R_e + P_{CL} - 2I^2 R_e) = 0$$

and

$$P_{CL} = I^2 R_e \tag{29}$$

APPENDIX **3**   AMERICAN WIRE GAUGE TABLE

Although we have adopted SI units for our pencil and paper analysis of electric circuits, practical electric conductors in North America will still be manufactured for some time using the foot as the unit of length and the mil (one-thousandth of an inch) as the unit of diameter. Before we can use Equation (4-4) to calculate the resistance of a practical conductor of a given AWG size, we must determine its cross-sectional area in square meters using the conversion factor

$$1 \text{ mil} = 0.0254 \text{ mm}$$

In the case of copper conductors, most wire tables spare us the tedious calculations by including a column giving the resistance in ohms per thousand feet at 20°C for each AWG conductor size.

It is worthwhile noting that in setting up the AWG wire table (Table A3), the cross-sectional dimensions were chosen so that each *decrease* of one gauge number represents a 25% *increase* in cross-sectional *area*. On this basis, a decrease of *three* gauge numbers represents an increase in cross-sectional area of $1.25 \times 1.25 \times 1.25$ or approximately a 2 : 1 increase. Similarly, a change of 10 wire gauge numbers represents a 10 : 1 change in cross-sectional area. Also, since doubling the cross-sectional area cuts the resistance in half, a decrease of three wire gauge numbers cuts the resistance of the conductor of a given length in half. Therefore, if we memorize the dimensions for any one wire gauge number, we can approximate the dimensions of any other wire gauge number. For this purpose, AWG No. 10 wire is quite convenient.

AWG gauge number   10   diameter (mils) 100 (approx)
ohms/thousand feet   1
   (copper at 20°C)

TABLE A3

AMERICAN WIRE GAUGE CONDUCTOR SIZES

AWG Number	Diameter (mils)*	Ohms per 1000 ft†	AWG Number	Diameter (mils)*	Ohms per 1000 ft†
0000	460.0	0.049 01	19	35.89	8.051
000	409.6	0.061 80	20	31.96	10.15
00	364.8	0.077 93	21	28.46	12.80
0	324.9	0.098 27	22	25.35	16.14
1	289.3	0.1239	23	22.57	20.36
2	257.6	0.1563	24	20.10	25.67
3	229.4	0.1970	25	17.90	32.37
4	204.3	0.2485	26	15.94	40.81
5	181.9	0.3133	27	14.20	51.47
6	162.0	0.3951	28	12.64	64.90
7	144.3	0.4982	29	11.26	81.83
8	128.5	0.6282	30	10.03	103.2
9	114.4	0.7921	31	8.928	130.1
10	101.9	0.9989	32	7.950	164.1
11	90.74	1.260	33	7.080	206.9
12	80.81	1.588	34	6.305	260.9
13	71.96	2.003	35	5.615	329.0
14	64.08	2.525	36	5.000	414.8
15	57.07	3.184	37	4.453	523.1
16	50.82	4.016	38	3.965	659.6
17	45.26	5.064	39	3.531	831.8
18	40.30	6.385	40	3.145	1049.0

*1 mil–0.001 inch.
†Copper at 20°C.

EXAMPLE

If the resistance of AWG No. 10 copper wire is 1 Ω per thousand feet at 20°C, what is the resistance at 20°C of 200 ft of AWG No. 14 copper wire?

SOLUTION

Since the cross-sectional area of AWG No. 14 wire is $1/(1.25)^4$ that of AWG No. 10 wire,

$$R = 2.5 \times \frac{200}{1000} \times 1 \ \Omega \approx 0.5 \ \Omega$$

APPENDIX 4 RESISTOR COLOR CODE

Most carbon-composition resistors are too small to print the resistance values on them clearly. Hence, this information is provided by a series of colored bands using a standard color code. Note that the bands are near one end of the resistor so that there is no mistaking the order in which they are read from the end toward the center. The resistance value consists of two significant digits multiplied by a power of ten (the number of zeros after the two digits). A tolerance band indicates either ±5% or ±10%

tolerance. If this band is missing, the resistance value is within $\pm 20\%$ of the stated value and the resistor costs less. Hence, the resistance of a 470-kΩ resistor without a tolerance band can be anywhere between 376 kΩ and 564 kΩ.

Color	Significant Digit	Decimal Multiplier	Tolerance	
Blank	—	—	20%	1st Significant digit
Silver	—	10^{-2}	10%	2nd Significant digit
Gold	—	10^{-1}	5%	Decimal multiplier
Black	0	1	—	Tolerance
Brown	1	10	—	
Red	2	10^2	—	
Orange	3	10^3	—	
Yellow	4	10^4	—	
Green	5	10^5	—	
Blue	6	10^6	—	
Violet	7	—	—	
Gray	8	—	—	
White	9	—	—	

EXAMPLES

Yellow, violet, yellow $\quad = 47 \times 10^4 \ \Omega = 470 \ \text{k}\Omega \quad \pm 20\%$

Green, blue, black, silver $= 56 \times 1 \ \Omega = 56 \ \Omega \qquad \pm 10\%$

Red, yellow, gold, gold $\quad = 24 \times 0.1 \ \Omega = 2.4 \ \Omega \qquad \pm 5\%$

Not all possible combinations of colors will be found in practice, since resistors are manufactured according to the standard values given in the following table. Only the two significant digits are shown. Any permissible decimal multiplier can be used with these selected digits.

Tolerance			
5% 10% 20%	5% only	5% and 10%	5% only
Significant Digits			
10	11	12	13
15	16	18	20
22	24	27	30
33	36	39	43
47	51	56	62
68	75	82	91

ANSWERS TO PROBLEMS

NOTE As a check on numerical computation, most answers are given to four significant calculator digits. Final answers should be rounded off to reflect the accuracy of the given data.

CHAPTER 1

1-1. (a) 4.55×10^2; (b) 5.9×10^4;
    (c) 10^4; (d) 7.65×10^{-2};
    (e) 3.7×10^{-4};
    (f) 5.47×10^{-1}
1-3. (a) 47 kΩ; (b) 45 MHz;
    (c) 1.5 kW; (d) 50.5 ms;
    (e) 0.5 mV or 500 μV;
    (f) 390 pF
1-5. (a) -723; (b) 230.21;
    (c) 6450; (d) 6.25×10^{-14};
    (e) 786.48474; (f) 195
1-7. 2.5×10^4 m
1-9. 8.999×10^{-28} g
1-11. 1.65 kg
1-13. 16.67 ms
1-15. 22 g
1-17. 1257 mm^2
1-19. 2144 Ω
1-21. 1 ft^2 = 0.0929 m^2
1-23. 22.35 m/s
1-25. 2.0806 mm^2

CHAPTER 2

2-1. 5 A
2-3. 1.6×10^{-4} A
2-5. 18 C
2-7. 32 s
2-9. 0.25 V
2-11. 0.4167 C
2-13. 0.3 J
2-15. 16.5 J
2-17. 9.028 V
2-19. 150 mA

CHAPTER 4

4-1. 48 Ω
4-3. 22 Ω
4-5. 5 A
4-7. 8.167 A
4-9. 45 V
4-11. 89.6 V
4-13. 2 kΩ
4-15. 583.3 Ω
4-17. 25 mA
4-19. 218 μA
4-21. 4.4 V
4-23. 3.96 V
4-25. 0.86 Ω
4-27. 1.719 Ω
4-29. 0.86 Ω
4-31. 2.252 Ω
4-33. 47.9 cm
4-35. 0.25 mm
4-37. 3.0×10^{-5} Ω.m
4-39. 2.82 Ω
4-41. 25.5 Ω
4-43. 2.535 Ω
4-45. 0.002
4-47. 8.135 Ω
4-49. tungsten

CHAPTER 5

5-1. 2156 J
5-3. 5.859×10^5 J
5-5. 600 W
5-7. 3.78 W
5-9. 468.75 W
5-11. 40 mW
5-13. 1-W rating (0.7258 W)
5-15. 22 mA
5-17. 200 V
5-19. 120 V
5-21. 16.58 A
5-23. 320 W
5-25. 88.2%
5-27. 82.89%
5-29. 2.79 min
5-31. 15.81 A
5-33. 30 Ω
5-35. 7.95 MJ
5-37. 18 kWh
5-39. $1.85

CHAPTER 6

6-1. 30 Ω
6-3. 20 V; 40 V; 60 V
6-5. 0.923 Ω; 39 W
6-7. 90 Ω; 100 Ω
6-9. 3889 V
6-11. 119.1 Ω
6-13. (a) 4.8 V; (b) 59.51 W;
    (c) 75%; (d) 180 W
6-15. (a) 13.8 kV (b) 13.2 kV;
    (c) 7.5 mA
6-17. 0.5 Ω
6-19. 411.4 W
6-21. 44 A
6-23. 1.333 Ω
6-25. 2.5 kΩ; 909.1 Ω
6-27. 30 kΩ
6-29. 9.229 mA; 6.153 mA; 4.615 mA

CHAPTER 7

7-1. (a) 36 Ω; (b) 1800 Ω;
    (c) 1395 Ω
7-3. (a) 250 mA; (b) 981.5 mA;
    (c) 55.94 μA
7-5. 173.3 Ω; 93.33 Ω; 373.3 Ω;
    293.3 Ω; 240 Ω; 360 Ω
7-7. (a) 361.4 Ω; (b) 354.3 Ω
7-9. (a) 75 Ω; (b) 77.38 Ω
7-17. 10.44 V (or 0.9566 V)
7-19. 30 kΩ
7-21. R_1 = 320 Ω; R_2 = 533 Ω
7-23. R_1 = 1.307 MΩ; R_F = 996 kΩ
7-25. Point A = +20 V w.r.t. ground;
    Point B = -4 V w.r.t. ground
7-27. R_s = 1.2 Ω, 7.5 W;
    R_1 = 26 Ω, 6.5 W;
    R_2 = 31.58 Ω, 4.56 W;
    R_3 = 4.858 Ω, 29.64 W
7-29. Left = 10 mA; right = 5 mA
7-31. 250 Ω

CHAPTER 8

8-1. 27 V and 5 Ω
8-3. (a) 0.37 μA and 68 kΩ;
    (b) 333 mA and 225 Ω;
    (c) 2 mA and 40 kΩ
8-5. 6.667 V + at bottom

8-7. 4.25 A
8-9. 2.434 A
8-11. $I_A = 363$ mA, $I_B = 116.5$ mA
8-13. Left = 85.955 mA,
right = 79.092 mA
8-15. Left to right, 147 mA, 117.6 mA,
29.4 mA, 51.46 mA, 22.06 mA
8-17. 5.333 mA
8-19. Mesh currents, -10.33 mA,
-11.5 mA, 6mA; 6 mA; arm currents, 1.167 mA, zero, 16.33 mA,
-17.5 mA
8-21. Left to right, -12.38 mA,
2.14 mA, 10.24 mA
8-23. 4 V
8-25. 2.943 mA
8-27. $V_{10} = 80$ V, $V_{20} = 20$ V,
$V_{30} = 270$ V
8-29. (a) 8.8 V;   (b) 11.51 V
8-31. Left 2.53 mA, right 3.84 mA
8-33. (a) $+79.49$ V;   (b) $+19.6$ V
8-35. 573 V

CHAPTER 9

9-1. 160 V and 180 Ω
9-3. 80.02 mA
9-5. 888.9 mA and 180 Ω
9-7. 620 Ω
9-9. 887.8 kΩ
9-11. 35.56 W
9-13. 875 mA
9-15. 120 V
9-17. 237.08 V and 36.01 kΩ
9-19. 1.5 A
9-21. 10.06 V
9-23. 20.97 V and 1.0306 kΩ
9-25. 33.86 V
9-27. 18 V and 500 Ω
9-29. 13.33 V
9-31. 200 Ω, 500 Ω, 200 Ω
9-33. 117.6 mA
9-35. 2.944 mA
9-37. 48 Ω
9-39. $+264$ mV; $+15.84$ V

CHAPTER 10

10-1. 0.6 N
10-3. 0.050 25 Ω
10-5. 0.450 225 Ω; 0.045 022 5 Ω;
0.005 002 5 Ω
10-7. 199 kΩ; 800 kΩ; 1 MΩ; 8 MΩ
10-9. 10.91 V
10-11. 100 kΩ
10-13. 0.4948 V
10-15. 30 kΩ
10-17. 120 kΩ

10-19. 53.33 MΩ
10-21. 210 kΩ
10-23. 40 Ω
10-25. 9.09 km

CHAPTER 11

11-1. 2.877×10^{-10} N
11-3. 8.015×10^{-15} N
11-5. 17.81 pF
11-7. 13.9 pF
11-9. 885 pF
11-11. (a) 50 μF;   (b) 4 mC, 16 mC
11-13. (a) 5.714 μF;   (b) 2.286 mC;
(c) 228.6 V, 114.3 V, 57.15 V
11-15. 27.8 pF
11-17. 12 μF; 192 V; 48 V; 240 V
11-19. 680 V

CHAPTER 12

12-1. (a) 1.25 mA;   (b) 0 V;
(c) 62.5 V/s;   (d) 4 s;
(e) 20 s
12-3. (a) 175 V;   (b) 0.75 mA;
(c) 6.5 s;   (d) 2.8 s
12-5. (a) 94 s;   (b) 1.06 mA
12-7. (a) 43 s;   (b) 290 V;
(c) 0.394 mA;   (d) 26.2 s
12-9. (a) 178.4 V;   (b) 0.7582 mA;
(c) 6.438 s;   (d) 2.772 s
12-11. (a) 40 μA, 0.8 V/s;   (b) 173.3 s;
(c) 297.4 μA, 5.949 V/s;
(d) 145.36 V
12-13. 172.5 V
12-15. (a) -33.58 V;   (b) $+12.93$ V
12-17. (a) 0.625 J;   (b) 1.28 μJ
12-19. 25.6 mJ

CHAPTER 13

13-5. 5×10^5 At/Wb
13-7. 10^{-3} H/m
13-9. 800 turns
13-11. 4×10^{-4} H/m
13-13. 1.61 T (approx.)
13-15. 5.085×10^{-4} H/m (approx.)
13-17. 2.5×10^{-4} H/m (approx.)

CHAPTER 14

14-1. 3.701×10^{-7} Wb
14-3. 811 turns
14-5. 0.475 A (approx.)
14-7. 387 mA (approx.)
14-9. 1.361 A (approx.)
14-11. 144.3 N

CHAPTER 15

15-1. 0.25 H
15-3. 333.3 V
15-5. 7.896 H
15-7. 2.573 H
15-9. 270 turns
15-11. (a) 120 mH;   (b) 12.77 mH

CHAPTER 16

16-1. 1.5 s
16-3. 0.8 V
16-5. (a) 0 A;   (b) 1.2 A/s;
(c) 1.5 A;   (d) 6.25 s
16-7. (a) 0.3 s;   (b) 0.21 s;
(c) 2.465 A
16-9. (a) 1.373 s;   (b) 1.197 A
16-11. (a) 37.18 V;   (b) 1.88 μs
16-13. (a) 4.685 V
16-15. (a) 22.5 J;   (b) 36 W
16-17. (a) 66.67 Ω;   (b) 0.375 s;
(c) 22.5 J
16-19. 1.557 A

CHAPTER 17

17-1. 106.07 V
17-3. 951 V
17-5. 126.2 V
17-7. -88.17 V
17-9. 8.507 A
17-11. 200 sin 2513t V
17-13. 325 V
17-15. 15.9 V
17-17. 40 mA
17-19. (a) 469 mA;   (b) 166 V
17-21. 41.67 min
17-23. 4.8 V per side = 9.6 V

CHAPTER 18

18-1. 29.95 Ω
18-3. 16.67 kΩ
18-5. 1508 Ω
18-7. 7.958 kΩ
18-9. 636.6 Hz
18-11. 469.1 μH
18-13. 124.3 Hz
18.15. 7.958 μF
18-17. 306.6 mH
18-19. 215 μH

CHAPTER 19

19-1. 280 V
19-3. $i_T = 110.5$ sin $(377t - 0.325)$ A
19-5. 0 V

19-7. 25.9 V $\underline{/-5°}$
19-9. (a) $+19.99 + j15.01$;
   (b) $-27.58 + j23.14$;
   (c) $+89 - j154.15$;
   (d) $-30 - j51.96$;
   (e) $-5.909 + j12.91$;
   (f) $+5.501 + j3.997$
19-11. $26.4\underline{/+24.62°}$
19-13. $198.12\underline{/-133.74°}$
19-15. $134.39\underline{/+77.28°}$
19-17. $0.8256\underline{/-16.39°}$
19-19. (a) $53.3\underline{/-51°}$;
   (b) $0.3154\underline{/-77°}$
19-21. $10.75\ \Omega\underline{/-30°}$

CHAPTER 20

20-1. (a) 22.36 V;   (b) 12.5 mV
20-3. $96\ V\underline{/+36.87°}$
20-5. (a) $133.3\ \Omega\underline{/+72°}$;
   (b) $41.2 + j126.8\ \Omega$
20-7. (a) $8\ \Omega\underline{/+20°}$;
   (b) $7.518 + j2.636\ \Omega$
20-9. $842.9\ \Omega\underline{/+53.62°}$
20-11. $1312\ \Omega\underline{/-23.85°}$
20-13. (a) $409.53\ \Omega$;   (b) $+56.9°$
20-15. 2.078 kΩ and 0.9549 H
20-17. $150\ \Omega\underline{/+48.18°}$
20-19. $3.063\ \mu F$
20-21. 291.7 mH; 48 Ω
20-23. $6.531\ \mu F$ or $0.9619\ \mu F$
20-25. $2.327\ \mu F$
20-27. 42.1 Ω; 983.8 mH
20-29. (a) $170\ mA\underline{/-28.07°}$;
   (b) $1.327\ A\underline{/+90°}$
20-31. (a) $2.062\ A\underline{/+75.96°}$;
   (b) $17.18\ mS\underline{/+75.96°}$
   (c) $58.2\ \Omega\underline{/-75.96°}$
20-33. $44.72\ mS\underline{/+63.43°}$;
   $22.36\ \Omega\underline{/-63.43°}$
20-35. $13.12\ mS\underline{/-23.85°}$
20-37. $G = 433\ mS;\ B = 250\ mS$
   (inductive)
20-39. $R_p = 2.772\ k\Omega;\ L_p = 3.82\ H$
20-41. $R = 2.943\ \Omega;\ L = 31.22\ mH$
20-43. $4.745\ \mu F$
20-45. (a) $2.277\ mS\underline{/-49.77°}$;
   (b) $1.48\ mS\underline{/-6.498°}$;
   (c) $2.188\ mS\underline{/+47.76°}$
20-47. $78.102\ mS\underline{/+2.936°}$
20-49. $2.507\ A\underline{/+4.447°}$

CHAPTER 21

21-1. (a) 660 W;   (b) 1320 W
21-3. 150 W
21-5. (a) 0 W;   (b) 7.54 mvars;
   (c) 7.54 mW

21-7. (a) 75.78 VA;   (b) 9.571 W
   (c) 12.63% lagging
21-9. 30.07 kW
21-11. (a) 10.368 Ω;   (b) 36.67 mH
21-13. (a) 24.24 A;   (b) 2667 W
21-15. 98%
21-17. 147.58 V
21-19. (a) $45.9\ \mu F$;
   (b) 6.0 A, 3.6 A, 4.8 A
21-21. 99.8% leading
21-23. 680.79 W
21-25. 1.59 MHz
21-27. 469.8 Ω
21-29. (a) 12 Ω, 156 mH;
   (b) 15.43 Ω, 1.768 H
21-31. +34 dB
21-33. 792.4 mW
21-35. 7 dB

CHAPTER 22

22-1. $1119.17\ \Omega\underline{/+45.36°}$
22-3. $3.5\ mS\underline{/-29.64°}$
22-5. (a) $1.071\ A\ \underline{/-47.98°}$;
   (b) $228.53\ V\underline{/+14.07°}$
22-7. 519.62 Ω and $0.5305\ \mu F$
22-9. 416.1 Ω and 0.7026 H
22-11. $63.95\ \Omega\underline{/+15.95°}$
22-13. $287.4\ mA\ \underline{/+69.83°}$;
   $230\ mA\ \underline{/+106.7°}$
22-15. $265.6\ mA\ \underline{/+54.9°}$
22-17. $51.88\ mV\underline{/+3.62°}$
22-19. $533\ mV\underline{/-80°}$
22-21. (a) 26.332 kΩ and $0.0532\ \mu F$;
   (b) 14.78 kΩ and $0.04218\ \mu F$
22-23. $315 + j107\ \Omega$
22-25. $148.242 - j86.843\ \Omega$
22-27. $240\ \mu W$
22-29. 43.3 Ω and $0.0478\ \mu H$
22-33. $50\ \Omega\underline{/-30°}$; $50\ \Omega\underline{/-30°}$; $Z_1 = Z_2$
   with a 120° difference in angle

CHAPTER 23

23-1. $13.183\ A\ \underline{/-21.14°}$ and
   $12.198\ A\ \underline{/-179.23°}$
23-3. $334\ mA\ \underline{/-88.8°}$
23-5. $43.27\ V\underline{/+4.49°}$
23-7. $935\ mA\ \underline{/+124.59°}$;
   $935\ mA\ \underline{/+124.59°}$;
   $1.87\ A\ \underline{/+124.59°}$
23-9. $1.009\ A\ \underline{/+52.52°}$;
   $1.434\ A\ \underline{/-21.34°}$
23-11. 2.845 A
23-13. 684 Ma
23-15. 153.25 V; 173.1 V
23-17. $2.84\ A\ \underline{/+64.88°}$
23-19. $20.12\ V\underline{/+35.66°}$

23-21. $930\ mA\ \underline{/+18.22°}$
23-23. 18.15 V
23-25. $20.55\ V\underline{/+31.08°}$ and
   $17.037\ k\Omega\underline{/-58.92°}$
23-27. $24\ V\underline{/0°}$, $0.02\ \mu F$, and 33 kΩ
23-29. 84.62 Ω
23-31. 8.36°
23-33. $0.0265\ \mu F$
23-35. $83.32\ k\Omega\underline{/+0.95°}$,
   $1.658\ k\Omega\underline{/-89.05°}$,
   $8.288\ k\Omega\underline{/-89.05°}$
23-37. 1.625 A
23-39. 9.149 MΩ and 940 pF
23-41. $2.878 + j1696.5\ \Omega$
23-43. $126.7\ \mu H$ and 10 Ω

CHAPTER 24

24-1. 2.653 MHz
24-3. 40 mV
24-5. 234.3 pF
24-7. 100.66
24-9. 2.652 MHz
24-11. $0.3332\ \mu A$
24-13. (a) $23\ \mu H$;   (b) 1.8835 MHz
24-15. (a) 466.9 kHz and 461.1 kHz;
   (b) 3.953 Ω and $108.47\ \mu H$
24-17. $0.44\ \mu F$
24-19. $87.95\ \mu H$ and $521.1\ \mu H$

CHAPTER 25

25-1. 17
25-3. 107.1 V
25-5. 560 turns
25-7. $N_p = 1045$ turns,
   $N_s = 104.5$ turns.
   $I_p = 6.522\ A,\ I_s = 65.22\ A$
25-9. 819 turns
25-11. 91 mA (approx.)
25-13. 75.1 cm²
25-15. 5625 Ω
25-17. (a) 34.72 Ω;   (b) 8.68 Ω
25-19. $R_e = 0.1725\ \Omega,\ X_e = 0.4002\ \Omega$
25-21. 3.846%
25-23. 2.118%
25-25. 98.3%, 211.45 kW

CHAPTER 26

26-1. $\mathbf{z}_{11} = 642\ \Omega,\ \mathbf{z}_{12} = \mathbf{z}_{21} = 286\ \Omega$,
   $\mathbf{z}_{22} = 322\ \Omega$
26-3. $\mathbf{z}_{11} = \mathbf{z}_{22} = 20 + j0\ \Omega$,
   $\mathbf{z}_{12} = \mathbf{z}_{21} = -j800\ \Omega$
26-5. $\mathbf{Z}_m = 400\ \Omega\underline{/0°},\ \mathbf{Z}_p = 200\ \Omega\underline{/0°}$,
   $\mathbf{Z}_s = 800\ \Omega\underline{/0°}$

26-7. $\mathbf{Z}_m = -j1200\ \Omega$,
$\mathbf{Z}_p = \mathbf{Z}_s = 50 + j1200\ \Omega$

26-9. $59.26\ \Omega\underline{/0°}$

26-11. $499\ \Omega\underline{/0°}$

26-13. 9 dB

26-15. $\mathbf{z}_{11} = 643.5\ \Omega$,
$\mathbf{z}_{22} = 321.24\ \Omega$,
$\mathbf{z}_{12} = \mathbf{z}_{21} = 285.09\ \Omega$

26-17. $\mathbf{z}_{11} = \mathbf{z}_{22} = 0$,
$\mathbf{z}_{12} = \mathbf{z}_{21} = -j3600\ \Omega$

26-19. $512.7\ \Omega\underline{/0°}$

26-21. $\mathbf{y}_{11} = 7\ \text{mS}\underline{/0°}$,
$\mathbf{y}_{22} = 9\ \text{mS}\underline{/0°}$,
$\mathbf{y}_{12} = \mathbf{y}_{21} \doteq 5\ \text{mS}\underline{/180°}$

26-23. $\mathbf{Y}_m = 200\ \mu\text{S}\underline{/0°}$,
$\mathbf{Y}_p = 200\ \mu\text{S}\underline{/0°}$,
$\mathbf{Y}_s = 400\ \mu\text{S}\underline{/0°}$

26-25. $\mathbf{y}_{11} = \mathbf{y}_{22} = 2.698\ \text{mS}\underline{/-51.85°}$,
$\mathbf{y}_{12} = \mathbf{y}_{21} = 1.667\ \text{mS}\underline{/180°}$

26-27. $\mathbf{h}_{11} = 387.98\ \Omega$, $\mathbf{h}_{12} = 0.888$,
$\mathbf{h}_{21} = -0.888$, $\mathbf{h}_{22} = 3.106\ \text{mS}$

26-29. V.G. of transistor = 103.284,
overall V.G. = 14.775

26-31. 0.07264

26-33. $1197.43\ \Omega\underline{/-72.34°}$

26-35. (a) 338.38 pF; (b) 0.026;
(c) 11.86 kHz

27-1. (b) $12\ \text{A}\ \underline{/0°}$, $12\ \text{A}\ \underline{/+120°}$,
$12\ \text{A}\ \underline{/-120°}$; (d) 0 A

27-3. (b) $4\ \text{A}\ \underline{/-30°}$, $3\ \text{A}\ \underline{/+150°}$,
$2.4\ \text{A}\ \underline{/-30°}$;
(d) $3.3995\ \text{A}\ \underline{/-30°}$

27-5. (b) $50\ \text{A}\ \underline{/0°}$, $50\ \text{A}\ \underline{/+120°}$,
$50\ \text{A}\ \underline{/-120°}$;
(e) $86.6\ \text{A}\ \underline{/-30°}$,
$86.6\ \text{A}\ \underline{/+90°}$,
$86.6\ \text{A}\ \underline{/-150°}$

27-7. (b) $110\ \text{A}\ \underline{/-45°}$, $50\ \text{A}\ \underline{/+120°}$,
$55\ \text{A}\ \underline{/-83.1°}$;
(e) $158.82\ \text{A}\ \underline{/-49.67°}$,
$102.88\ \text{A}\ \underline{/+107.88°}$,
$74.85\ \text{A}\ \underline{/+161.96°}$

27-9. (c) $208\ \text{V}\underline{/+30°}$, $208\ \text{V}\underline{/+150°}$,
$208\ \text{V}\underline{/-90°}$
(d) $18.909\ \text{A}\ \underline{/+30°}$
$18.909\ \text{A}\ \underline{/+150°}$,
$18.909\ \text{A}\ \underline{/-90°}$
(e) $32.752\ \text{A}\ \underline{/0°}$,
$32.751\ \text{A}\ \underline{/+120°}$;
$32.751\ \text{A}\ \underline{/-120°}$

27-11. (c) $208\ \text{V}\underline{/+30°}$, $208\ \text{V}\underline{/+150°}$,
$208\ \text{V}\underline{/-90°}$

(d) $41.6\ \text{A}\ \underline{/-15°}$,
$18.909\ \text{A}\ \underline{/+150°}$,
$20.8\ \text{A}\ \underline{/-53.1°}$
(e) $60.07\ \text{A}\ \underline{/-19.67°}$,
$38.91\ \text{A}\ \underline{/+137.89°}$,
$28.31\ \text{A}\ \underline{/-168.04°}$

27-13. 4.32 kW

27-15. 727.44 W

27-17. CBA

27-19. 2.162 kW

27-21. 2.162 kW

27-23. 41.25 kW

27-25. (a) 3 kW; (b) 0.866;
(c) 9.616 A;
(d) $37.47\ \Omega\underline{/+30°}$; (e) 60°

27-27. (b) $31.754\ \text{A}\ \underline{/-30°}$,
$31.754\ \text{A}\ \underline{/+90°}$,
$31.754\ \text{A}\ \underline{/-150°}$;
(d) 0 A

27-29. $5.893\ \text{A}\ \underline{/+12.6°}$,
$1.31\ \text{A}\ \underline{/-167.4°}$,
$4.584\ \text{A}\ \underline{/-167.4°}$

28-7. 102.65 V

28-9. 6.667%

INDEX

A

Acceptor atom, 56
ac resistance, 535
Accuracy, numerical, 7
Active networks, 114, 664
Active power, 513, 619
Addition:
    by geometrical construction, 468, 707, 709
    of instantaneous values, 464, 737
    of nonsinusoidal waves, 735
    phasor, 468, 470, 477
    of sine waves, 462, 477, 735
Admittance, 500, 503, 549, 668, 673
Admittance diagram, 502
Air-core coils, 364
Air-core transformers, 676, 681
Air gaps, 374, 379
Algebra:
    matrix, 169
    phasor, 462–82
Alternating current:
    average value, 262, 442, 762
    effective value, 443
    instantaneous value, 443, 437, 439, 736
    nonsinusoidal, 441, 732–49
    rms value, 443, 743, 761
    sine-wave, 429–46, 450–59, 511–18
Alternating-current bridges, 601
Alternating-current circuits:
    impedance networks, 573–604
    parallel, 498, 548
    power in, 511–37
    series, 488, 545
    simple, 427, 437
    three-phase, 695–727
Alternating voltage:
    average value, 442, 762
    effective value, 443
    generation of, 426
    instantaneous value, 433, 437, 736
    nature of, 427
    nonsinusoidal, 732–49
    peak value, 432
    rms value, 443, 743, 761
    sine-wave, 426–46, 451–57, 511–18, 732–38

Alternator, 426, 691, 695
American wire gauge, 44, 765
Ammeter:
    ac, 263
    dc, 246
    multirange, 248
Ammeter shunts, 151
Ampacity, 39
Amperage, 39
Ampere, definition, 30, 343
Ampere-turn, 345
Amplitude of a sine wave, 453, 741
Analysis:
    dimensional, 12
    nodal, 188, 593
Angle:
    impedance, 488
    phase, 431, 487, 524
    power-factor, 524, 652, 702
    in radians, 435
Angular velocity, 436
Anode, 49
Antiresonance, 626
Apparent power, 518, 650
Applied voltage, 38
Atom:
    acceptor, 56
    Bohr, 23, 350
    combination of, 25
    donor, 55
    Rutherford, 22
    structure of, 22
Atomic model, 23, 350
Attenuator, 537, 667
Audio transformer, 646, 656
Autotransformer, 654
Average permeability, 357
Average value:
    current, 262, 442, 761
    power, 444, 516, 761
    voltage, 442, 761
Ayrton shunt, 248

B

B, symbol for flux density, 347
B, symbol for susceptance, 501
Balanced three-phase load, 696, 708, 711

Band:
    conduction, 48
    energy, 46
    valence, 48
Bandwidth:
    resonant circuit, 623, 632
    tuned transformer, 685
Barrier potential, 57
BASIC computer language, 14, 317, 431, 479
Battery:
    dry-cell, 59
    lead-acid storage, 61
    manganese-alkaline, 60
    mercury, 60
    nickel-cadmium, 63
    primary, 59
    secondary, 61
    solar, 66
    Weston standard, 255
    wet-cell, 34, 58
    zinc-carbon, 60
BH curves, 354, 368
Bias, *pn*-junction, 58
Bifilar winding, 677
Bleeder resistor, 145
Bonding:
    covalent, 25
    ionic, 25
    metallic, 25
Bridge, general equation, 261, 602
Bridged-T network, 201, 239
Bridge networks, 159, 171, 601
Bridge rectifier, 262
Bridges:
    ac, 601
    capacitance-comparison, 602
    Hay, 604
    Maxwell, 603
    Schering, 611
    slide-wire, 268
    Varley loop, 268
    Wheatstone, 159, 171, 212, 232, 260

C

C, symbol for capacitance, 285
Calculator, electronic, 6, 313, 406, 417
Calculus solutions:
    capacitive reactance, 763
    energy stored by a capacitor, 758
    energy stored by an inductor, 760
    general transformer equation, 764
    inductive reactance, 762
    instantaneous current in an *LR* circuit, 759

    instantaneous PD in a *CR* circuit, 756
    maximum power transfer, 756
    maximum transformer efficiency, 764
    rms and average values of a sine wave, 761
Calibrating resistor, meter, 246
Capacitance, 274–93
    in ac circuits, 494, 495, 498
    in dc circuits, 298–327
    definition, 286
    factors governing, 286
    farad, 285
    stray, 326
Capacitance-comparison bridge, 602
Capacitive reactance:
    definition, 456
    factors governing, 457, 763
Capacitors:
    charging, 283, 298, 301
    construction, 282
    definition, 284
    dielectric, 280
    discharging, 306
    electrolytic, 282
    energy stored, 322, 758
    filter, 324
    graphic symbols, 4
    leaky, 326
    parallel, 290
    power-factor improving, 529
    reservoir, 325
    semiconductor, 283
    series, 291
    smoothing, 325
    varactor, 283
    working voltage, 283
Carrier, charge, 27, 39, 49, 52, 57
Cathode, 49, 51
Cell:
    electrolytic, 49
    fuel, 64
    manganese-alkaline, 60
    mercury, 60
    photovoltaic, 65
    primary, 59
    secondary, 61
    voltaic, 34
    Weston standard, 255
    zinc-carbon, 60
Charge, electric:
    capacitors, 284
    carriers, 27, 39, 49, 52, 57
    coulomb, 29
    definition, 20
    positive and negative, 20
Charge-separating device, 35
Charging current, 63, 298

Chemical energy, 2, 34, 58, 64
Choke, filter, 420
Circle diagram, 565
Circuit:
  ac, 427, 437, 511, 544–604
  capacitive dc, 324
  coupled, 660–85
  CR, 302, 756
  dc, 107–235
  double-resonant, 636
  electric, 2, 29, 68
  equivalent, 112, 136, 160, 206, 557,
      663, 670
  integrated, 13, 44, 85
  LR, 401, 415, 759
  magnetic, 343, 363–80
  magnetically-coupled, 677
  parallel, 124, 129, 503, 548
  resonant, 617
  series, 110, 116, 486, 495, 545
  series-parallel, 135, 555
  three-phase, 695–727
Circuit diagram, 2, 69
Coefficient:
  of coupling, 677, 685
  temperature, 79
Coercive force, 357
Coils, air-core, 364
Color code, resistor, 766
Complex algebra, 462
Complex waves, 441, 733
Computers, 13, 82, 121, 170, 317
Conductance, 126, 188, 500
Conduction:
  electrolytic, 49
  in gases, 51
  metallic, 26
  in vacuum, 51
Conduction band, 48, 54
Conductivity, 129
Conductors:
  electric, 2, 26, 43, 243
  nonmetallic, 49
  semi-, 52
Conjugate quantity, 482, 565
Conservation of energy, 20, 95
Constant-current source, 162, 215, 323,
    563
Constant-voltage source, 160, 563
Controlled source, 120, 223, 663, 681
Conventional current direction, 39, 51,
    114
Conversion, source, 164, 562
Conversion factors, 13, 101, 436
Conversion of units, 12
Coordinates:
  polar, 472

  rectangular, 473, 477, 558
Coulomb, definition, 29
Coulomb's law, 21, 275
Counter emf, 389
Coupled circuits, 660–85
Coupled impedance, 666, 680
Coupling:
  close, loose, tight, 677
  coefficient of, 677, 685
  critical, 685
  mutual, 393, 680
  over-, 685
Coupling networks, 660–85
  four-terminal, 598, 660–85
  hybrid parameters, 671
  mutual impedance, 664
  open-circuit impedance parameters, 662
  short-circuit admittance parameters, 669
Covalent bonding, 25
Cramer's rule, 752
Critical coupling, 685
Critical damping, 634
CR time constant, 303, 308
CR waveshaping, 309
Curie temperature, 352
Current:
  alternating, 437, 450, 454
  ampere, 5, 30
  branch, 125
  capacitive, 299, 306, 454
  charging, 63, 299, 301
  collapse, 412
  conventional, 39, 51, 114
  definition, 26, 29
  dependent variable, 90
  direction, 39, 51, 114
  discharging, 63, 306, 417
  displacement, 280, 457
  eddy, 359, 535
  electric, 26, 29
  electron flow, 26, 40
  exciting, 642
  inductive, 396
  inrush, 85
  instantaneous, 299, 398, 401, 406, 433,
      437, 450, 454, 759
  leakage, 57
  line, 701, 705, 708
  loop, 167, 574, 704
  magnetizing, 642
  neutral, 700
  nonsinusoidal, 746
  phase, 705
  pulsating, 309
  rate of change, 390, 397, 759
  reactive, 499
  resonant rise, 627

current: (*cont.*)
  rise, 398
  short-circuit, 161, 163
  sine-wave, 429, 438
  steady-state, 299, 401
  tank, 627
  total, 124, 499
Current-divider principle, 149, 553
Cycle, 429, 433, 442

D

D, symbol for electric flux density, 279
Damping, resonant circuits, 634
D'Arsonval movement, 245
Decay, current, 412
Decibel, 536
Degrees, electrical and mechanical, 434
Delta-connected three-phase system, 703, 708
Delta-wye transformation, 230, 598, 726
Dependent source, 223, 663, 681
Depletion layer, 57
Depolarizer, batteries, 60
Determinants, 169, 752
Diagram:
  admittance, 502
  circle, 565
  circuit, 2, 69
  energy-band, 46, 53
  impedance, 489
  phasor, 466, 693–727
  power-factor, 520
  symbols, 4
Diamagnetic materials, 349
Dielectric absorption, 281
Dielectric constant, 289
Dielectric hysteresis, 536
Dielectrics, 280
Dielectric strength, 281
Difference in potential, 33
Differential permeability, 356
Digital meters, 266
Dimensional analysis, 12
Diode:
  semiconductor, 57, 89, 262
  vacuum, 52, 89
Direct current, 35
Discharge:
  battery, 63
  capacitor, 306
Discharge resistor, 408, 414
Displacement current, 280, 457
Dissociation, 50
Distortion, 740

Divider:
  current, 149
  voltage, 144
Division, phasor, 481
Domain:
  frequency, 463
  magnetic, 351
  time, 463
Donor atoms, 55
Double-resonant circuit, 636
Double-subscript notation, 111, 697
Drift, electron, 27
Drop of potential, 34, 90
Dry cell, 60
Duality, 124, 148, 163, 232, 598
Dynamo, 32, 64

E

E, symbol for applied voltage, 36, 38
E, symbol for electric field intensity, 276
Eddy current, 359, 535
Edison three-wire system, 178, 695
Effective resistance, 534
Effective value of a sine wave (*see* rms value)
Efficiency, 100, 650, 764
Elastance, 293
Electrical degrees, 434
Electric charge, 20, 29, 284
Electric energy, 2, 36, 96
Electric field, 275
Electric field intensity, 276, 349
Electric field strength, 276
Electric flux, 279
Electric flux density, 279, 349
Electricity:
  definition, 2
  nature of, 19, 333
  static, 274
Electric lines of force, 275
Electric motor equation, 245
Electric potential, 33
Electric quantity, 29
Electrode, 49
Electrodynamometer, 264, 532
Electrolysis, 49
Electrolyte, 49, 59
Electrolytic cell, 49
Electrolytic conduction, 49
Electromagnet, 341, 380
Electromagnetic induction, 344, 383
Electromagnetism, 339
Electromotive force:
  counter, 389

definition, 32, 37
induced, 344
source of, 34
volt, 38
Electronic calculator, 6, 313
Electronic voltmeter, 254
Electrons:
  charge, 22
  drift, 27
  energy levels, 23
  flow, 26, 29, 40
  free, 26, 51
  mass, 22
  shells, 23
  theory, 21
  valence, 24, 47, 52
  velocity, 30
Electroplating, 50
Electrostatic field, 275
Electrostatic force, 275
Electrostatic induction, 279
Electrostatics, 275
Energy:
  chemical, 2, 34, 58, 64
  conversion of, 20, 95
  definition, 20, 95
  electric, 2, 36, 96
  forms of, 2, 95
  heat, 2, 65, 97
  joule, 36, 96
  kinetic, 52
  levels in an atom, 23, 47
  mechanical, 2, 19, 66, 95, 101
  potential, 23, 32
  storage, by capacitors, 322, 758
  storage, by inductors, 409, 760
  work, 20, 36, 95
Energy-band diagrams, 46, 53
Engineering notation, 10
Equations:
  CR transient, 319
  delta-wye transformation, 231
  electric motor, 245
  general bridge, 261, 602
  general transformer, 641, 764
  Kirchhoff's law, 140, 159, 166, 185, 574, 593
  loop, 166, 174, 574
  LR transient, 418
  mesh, 181, 580
  nodal, 188, 593
Equipotential surface, 278
Equivalent circuit, 112
  circuit simplification, 136, 160, 206, 557
  delta-wye, 230, 598

Norton, 216, 219, 591
parallel, 504, 557
series, 504, 557
Thévenin, 209, 586
transformer, 647, 656, 681
transistor, 208, 608, 674
y-parameter, 670
z-parameter, 663
Equivalent dc value (see rms value)
Equivalent impedance, 502
Equivalent reactance, 496
Equivalent resistance, 112, 127
Equivalent susceptance, 501
Exciting current, 642
Exponential curves, 305, 402

F

f, symbol for frequency, 7, 433, 617
F, symbol for mechanical force, 36, 275, 380
F_m, symbol for magnetomotive force, 345
Factor:
  conversion, 13, 101, 436
  form, 262, 446
  leakage, 371
  power, 523
  reactive, 526
Farad, definition, 285
Faraday's law, 50, 385, 641
Ferrites, 353
Ferroelectric materials, 351
Field:
  electric, 275
  magnetic, 242, 334, 339, 695
Field intensity:
  electric, 276, 349
  magnetic, 348
Field strength:
  electric, 276, 349
  magnetic, 348
Filter capacitors, 324
Filter chokes, 420
Filter networks, 634, 667, 671
Flux, electric, 279
Flux, magnetic:
  definition, 343
  density, 347
  fringing, 375
  leakage, 370, 647
  lines of force, 335, 343
  linkages, 385, 640
  mutual, 647
  weber, 344, 641
Flux density:
  electric, 279, 349

Flux density: (*cont.*)
   magnetic, 347
Forbidden energy levels, 48
Force:
   coercive, 357
   electromotive, 32, 37, 344, 389
   field, 19, 333
   gravitational, 19, 32
   magnetizing, 348
   magnetomotive, 345
   mechanical, 19, 36, 243, 379
   tractive, 379
Form factor, 262, 446
Forward-current ratio, 671
Fourier series, 734
Four-terminal networks, 598, 634, 660–85
Four-wire three-phase system, 699
Free electrons, 26, 51
Frequency:
   definition, 433
   fundamental, 733
   harmonic, 733
   resonant, 615, 617, 629
Frequency domain, 463
Fringing, flux, 375
Fuel cell, 64
Fundamental, 733
Funicular phasors, 652, 703
Fuse, 4

G

G, symbol for conductance, 126, 501
Gain, power, 536
Galvanometer, 245, 260, 284, 344
General bridge equation, 261, 602
General transformer equation, 641, 764
Generation of harmonics:
   in electronic amplifiers, 740
   in iron-core transformers, 742
   in nonlinear impedances, 738
Generator:
   ac, 426, 691
   definition, 32, 64
   four-pole, 434
   graphic symbol, 4
   three-phase, 695
   two-phase, 691
   two-pole, 434
Geometric construction, phasor diagrams, 468, 707, 709
Gradient, voltage, 278
Graphic symbols, 4, 87, 224
Ground, 4, 146, 278

H

H, symbol for magnetic field intensity, 348
Half-cycle average, 443
Half-power points, 624, 633
Half-wave rectification, 442, 739
Hand-rules, magnetic field direction, 340, 342
Harmonics, 732-49
   definition, 733
   generation, 738
Hartley's bandwidth law, 623
Hay bridge, 604
Heat energy, 2, 65, 97
Henry:
   definition, 390, 677
   unit of permeance, 346
Hertz, 433
Hole flow, 56
Hooke's law, 244
Horsepower, 101
Hybrid parameters, 671
Hydraulic analogy, 32
Hysteresis:
   dielectric, 536
   magnetic, 357
Hysteresis loop, 357, 649, 742
Hysteresis loss, 264, 358, 535, 649

I

I, symbol for electric current, 6, 30
Imaginary component, 462, 473
Impedance, 486–505
   conjugate, 565
   coupled, 666, 680
   definition, 488, 534
   input, 662, 671
   linear, 738, 746
   mutual, 664
   nonlinear, 739
   parallel, 498, 548
   reflected, 645, 666
   series, 488, 545
   series-parallel, 544–65
   total, 546
   transfer, 663
Impedance angle, 488
Impedance diagram, 489
Impedance matching, 645, 656
Impedance networks, 573–604
Impedance parameters, 662
   forward-transfer, 663
   open-circuit input, 662

open-circuit output, 663
reverse-transfer, 662
Impedance transformation, 645
Incremental permeability, 355
Independent source, 223
Independent variable, 90
Inductance, 334, 383–94
  in ac circuits, 486, 495, 498
  in dc circuits, 396–420
  factors governing, 391
  henry, 389, 677
  mutual, 677
  self-, 389
Induction:
  electromagnetic, 344, 383
  electrostatic, 279
  magnetic, 339
  mutual, 385, 640
  self-, 388
Inductive reactance:
  definition, 452
  factors governing, 453, 762
Inductors:
  definition, 391
  energy stored, 409, 760
  filter, 420
  graphic symbol, 4
  ideal, 396, 450
  parallel, 393
  practical, 398, 491, 503, 620, 629
  series, 392, 680
  smoothing, 420
Initial permeability, 356
Input impedance, 662, 671
Inrush current, 85
Instantaneous value:
  addition, 464
  current, 299, 307, 398, 401, 406, 433,
    437, 450, 454, 759
  power, 322, 410, 439, 512, 513, 693,
    696
  symbols, 299, 433
  voltage, 299, 307, 433, 756
Instrument, measuring, 242–66
Insulation breakdown, 48
Insulators, 45
Integrated circuits, 13, 44, 85
Intensity:
  electric field, 276, 349
  magnetic field, 348, 349
Internal resistance, 118, 160, 162, 164
International system of units, 5
Ionic bonding, 25
Ionization potential, 47
Ions, 25, 34, 50, 52, 59
IR drop, 90

Iron, magnetic properties, 351
Iron-core coils, 366, 393
Iron-core transformers, 641–56, 742
Iron losses, 535, 649

J

j operator, 473
Joule, definition, 36, 96
Junctions:
  node, 115
  pn-, 57, 262, 283

K

Kilowatthour, definition, 101
Kilowatthour meter, 102
Kinetic energy, 52
Kirchhoff's-law equations, 140, 159, 166,
  185, 574, 593
Kirchhoff's laws:
  in ac circuits, 464, 545
  current law, 125, 185, 186, 188, 548,
    593
  in dc circuits, 115, 140, 159
  voltage law, 115, 166, 396, 545, 651

L

L, symbol for inductance, 8, 389
Ladder network, 138, 151
Lamination, 360
Lattice network, 201, 239
Laws:
  Coulomb's, 21, 275
  Faraday's, 50, 385, 641
  Hartley's bandwidth, 623
  Hooke's, 244
  Kirchhoff's, 115, 125, 140, 159, 166,
    185, 186, 188, 396, 545, 548
  Lenz's, 386, 640
  Ohm's, 68, 110, 438, 449
LC ratio, 618, 620
Leakage current, 57
Leakage factor, 371
Leakage flux, 370, 647
Leakage reactance, 647
Lenz's law, 386, 640
Levels, energy, 23, 47
Linear impedance network, 738, 746
Linear magnetic circuit, 367
Linear resistor, 83, 88
Line current, 701, 705, 708

Lines of force:
    electric, 275
    magnetic, 335, 336, 343
Line voltage, 708
Linkages, flux, 385, 640
Load:
    balanced, 696, 708, 711
    delta-connected, 704, 708
    electric, 3, 35, 519
    sink, 3
    unbalanced, 702, 705, 723
    wye-connected, 701, 723
Loading, transformer, 651
Loading effect, 252
Load resistance, 117, 426, 519
Loop, tracing, 167, 574
Loop current, 167, 574, 704
Loop equations, 166, 174, 574
Loss:
    copper, 649
    core, 649
    eddy-current, 535
    hysteresis, 264, 358, 535, 649
    iron, 535
    power, 535, 536
    radiation, 97, 535
LR time constant, 401, 414

M

M, symbol for mutual inductance, 677
Maghet:
    bar, 335
    electro-, 341, 380
    permanent, 242, 336, 353
    temporary, 339, 353
Magnetic circuit, 343, 363–80
    linear, 367
    nonlinear, 367, 368
    parallel, 377
    series, 371
Magnetic domain, 351
Magnetic field, 242, 334, 339, 695
Magnetic field intensity, 348
Magnetic field strength, 348
Magnetic flux, 243, 343
Magnetic flux density, 243, 347
Magnetic induction, 339
Magnetic lines of force, 242, 335, 336,
    343
Magnetic materials:
    diamagnetic, 349
    ferrite, 353
    ferromagnetic, 351
    nonmagnetic, 349
    paramagnetic, 349

Magnetic moment, 350
"Magnetic Ohm's law," 346
Magnetic poles, 243, 335
Magnetic retentivity, 353, 357
Magnetic saturation, 352, 354
Magnetic shielding, 360
Magnetism, 333–60
    residual, 353
Magnetization, 351
Magnetizing current, 642
Magnetization curves, 353, 368
Magnetizing force, 348
Magnetomotive force, 345
Magneton, 350
Majority carrier, 57
Maximum power transfer, 120, 564, 756
Maxwell bridge, 603
Mechanical energy, 2, 19, 66, 95, 101
Mechanical force, 19, 36, 242, 379
Mesh, definition, 181
Mesh equations:
    ac networks, 580
    dc networks, 181–85
Metallic bonding, 26
Meter calibrating resistor, 246
Meter movement:
    D'Arsonval, 245
    electrodynamometer, 264, 532
    moving-coil, 245, 261
    moving-iron, 263
    Weston, 245
Meter multiplier resistor, 251
Meters:
    ammeter, 4, 246
    digital, 266
    electronic, 254
    galvanometer, 245, 261, 284, 344
    multimeter, 263
    ohmmeter, 256
    varmeter, 534
    voltmeter, 4, 250
    wattmeter, 265, 533, 712, 715
Meter shunt resistor, 247
Meter swamping resistor, 247
Millman's theorem, 218, 610
mmf gradient, 348
Molecule, 25
Motor action, 245
Moving-coil movement, 245, 261
Moving-iron movement, 263
Multimeter, 263
Multiplication, phasor, 481
Multiplier, voltmeter, 251
Multirange meters, 248, 251
Mutual coupling, 393, 680
Mutual flux, 647
Mutual impedance, 664

Mutual inductance, 677
Mutual induction, 385, 640
Mutual reactance, 678

N

Negative charge, 22, 24
Negative resistance, 600
Network theorems:
    maximum power transfer, 120, 564, 756
    Millman's, 218, 610
    Norton's, 215, 590
    reciprocity, 200, 235
    substitution, 234
    superposition, 194, 582, 747
    Thévenin's, 207, 218, 585
Networks:
    active, 114, 664
    attenuator, 537, 667
    bridge, 159, 171
    bridged-T, 201, 239, 574
    coupling, 660–85
    delta, 230, 598
    Edison three-wire, 178, 695
    filter, 634, 667, 671
    four-terminal, 598, 634, 660–85
    impedance, 573–604
    ladder, 138, 151
    lattice, 201, 239
    multisource, 173, 214, 218
    passive, 114, 168, 664
    pi, 236, 239, 598, 687
    planar, 181
    resistance, 159–98
    tee, 210, 236, 239, 598, 665
    three-terminal, 230, 598
    two-terminal, 209, 215, 585, 590
    Wheatstone bridge, 159, 171, 212, 232,
        260
    wye, 230, 598, 697, 699, 723
Neutral, 692, 714
Neutron, 22
Newton, 34, 36, 244, 380
Nodal equations, 188, 594
Node, 115, 185, 575, 594, 698
Nonlinear impedance, 739
Nonlinear magnetic circuit, 367, 368
Nonlinear resistors, 85
Nonmagnetic materials, 349
Nonsinusoidal waves, 732–49
    addition, 735
    definition, 733
    generation, 738, 740, 742
    half-wave rectified, 442, 739
    in linear impedance networks, 746
    pulse, 326, 441, 734

rms value, 743
    sawtooth, 441, 734, 746
    square, 309, 441, 744
Normal permeability, 355
Norton's theorem, 215, 590
Notation:
    double-subscript, 114, 697
    engineering, 10
    scientific, 7
n-type semiconductor, 55
Nucleus, 21
Numerical accuracy, 7

O

Ohm, definition, 70, 453, 457, 488
Ohmmeter, 256, 271
Ohm meter, definition, 74
Ohm's law, 68, 110, 438, 449
Open-circuit impedance parameters, 662
Open-circuit terminal voltage, 118, 161,
    208
Open-circuit transformer test, 648
Operator j, 473
Orbits, electron, 22
Oscillator, 434
Oscilloscope, 442
Output admittance, 673
Overload, 123

P

P, symbol for power, 97, 513
P_m, symbol for permeance, 346
Parallel capacitors, 290
Parallel circuits:
    electric, 124, 129, 499
    magnetic, 377
Parallel impedances, 548
Parallel inductors, 393
Parallel resistors, 124
Parallel resonance, 625, 629
Paramagnetic materials, 349
Parameters:
    hybrid, 671
    open-circuit z, 662
    short-circuit y, 669
Passive networks, 114, 168, 664
Peak-to-peak value, 442, 741
Peak value of a sine wave, 432
Period, 429, 433, 441, 454
Periodic wave, 441, 733
Permanent magnet, 242, 336, 353
Permeability:
    average, 357
    definition, 346

Permeability: (cont.)
  differential, 356
  free space, 349
  incremental, 355
  initial, 356
  normal, 355
Permeance, 346
Permittivity, 287, 349
  absolute, 288
  free space, 289
  relative, 289
Personal computers, 13, 82, 121, 170, 479
Phase angle, 431, 487, 524
Phase current, 705
Phase rotation, 719
Phase sequence, 719, 725
Phase shifter, 571, 572
Phase voltage, 700, 709
Phasor, 429, 462–82
  funicular, 652, 703
  reference, 469
Phasor diagram, 466, 693–727
Phasor power, 521
Phasor quantity, 466, 472, 479, 481
Photoresistor, 87
Photovoltaic cell, 65
Piezoelectric effect, 66
Pi-network, 236, 239, 556, 598, 687
Planar network, 181
pn-junction, 57, 262, 283
Polar coordinates, 472
Polarity of voltage drop, 112, 168
Polarization:
  atom, 281, 290
  battery, 59
Poles, magnetic, 243, 335
Polymers, 45
Polyphase systems, 691–727
Positive charge, 20, 24
Potential:
  barrier, 57
  depletion layer, 57
  ionization, 47
Potential difference, 33, 36, 37
  instantaneous, 299, 307
Potential drop, 34
Potential energy, 32
Potential rise, 34, 303
Potentiometer, 143, 254
Power:
  in ac circuits, 511–37
  active, 513, 619
  average, 444, 514, 516
  decibel, 536
  definition, 97, 513
  factor, 523, 719
  gain, 536

instantaneous, 322, 410, 439, 512, 513, 693, 696
  loss, 536
  maximum transfer of, 120, 564, 756
  measurement, 265, 532, 712, 715
  peak, 440, 512
  phasor, 521
  quadrature, 521
  ratio, 536
  reactive, 515, 516, 619
  real, 513, 521
  three-phase, 711, 712
  true, 513
  var, 515
  voltampere, 519
  watt, 98, 515
Power factor:
  angle, 524, 652, 702
  correction, 526
  definition, 523
  lagging, 524
  leading, 524
  three-phase, 719
  unity, 529
Power triangle, 519
Prefixes:
  SI unit, 9
  table of, 9
Primary, transformer, 384, 640
Primary cell, 59
Printed-circuit board, 44
Proton, 22
p-type semiconductor, 56
Pulsating current, 309, 420, 734
Pythagoras' theorem, 470, 492, 520, 707

Q

Q, symbol for electric charge, 29
Q, symbol for reactive power, 515
Q factor, resonant circuits, 619, 630
Quadrature coordinate, 473
Quadrature power, 521
Quantity of electric charge, 29, 275, 277, 285

R

R, symbol for resistance, 69
R_m, symbol for magnetic reluctance, 346
Radian, 435
Radiation loss, 97, 535
Radio frequencies, 535, 616, 677, 683
Rate of change, 300, 396
Reactance, 449–60
  capacitive, 456, 459, 763

inductive, 452, 459, 762
    leakage, 647
    mutual, 678
    net, 496
Reactive factor, 526
Reactive power, 515, 516, 619
Reactive voltage, 496
Real component, 462, 473
Real power, 513, 521
Reciprocal farad, 293
Reciprocal henry, 346, 390
Reciprocity, 735
    theorem, 200, 235
Rectangular coordinates, 473, 477, 558
Rectifier, 52, 58, 262, 739
Reference axis, 429, 466
Reference node, 186, 575, 698
Reference phasor, 469
Reflected impedance, 645, 666
Regulation, voltage, 120, 145, 652
Relative permittivity, 289
Reluctance, 345, 365, 371
Remanence, 353
Reservoir capacitor, 325
Residual magnetism, 353
Resistance:
    ac, 535
    definition, 69, 286, 535
    effective, 534
    equivalent, 112, 125
    factors governing, 72
    internal, 118, 160, 162, 164
    load, 117, 426, 519
    measurement of, 256
    nature of, 71, 459
    negative, 600
    ohm, 70
    ohmic, 534
    specific, 74
    temperature coefficient, 71, 79
    total, 111
Resistance networks, 159–98
Resistivity, 73
Resistors:
    bleeder, 145
    color code, 766
    construction, 83
    definition, 71
    discharge, 408, 414
    graphic symbol, 4, 87
    linear, 83, 88
    meter calibrating, 246
    multiplier, 251
    nonlinear, 85
    parallel, 124
    photo-, 87
    series, 110

series-dropping, 144
series-parallel, 135
shunt, meter, 247
swamping, meter, 247
thermistor, 86
varistor (thyrite), 87
Resonance, 416, 613–37
    definition, 617
    parallel, 625, 629
    series, 617
Resonant circuit, 617, 619, 625, 629
Resonant filter networks, 634
Resonant frequency, 615, 617, 629
Resonant rise of current, 627
Resonant rise of voltage, 620, 628
Retentivity, 353, 357
Reverse-voltage ratio, 673
Rheostat, 117, 384
Right-hand rule, flux direction, 340, 342
Rise, current:
    ideal inductor, 396
    LR circuit, 398
Rise, potential, 34, 303
rms value, nonsinusoidal wave, 743
rms value, sine wave, 443, 761
Rotating magnetic field, 695
Rotation, phase, 719

S

S, symbol for apparent power, 518
S, symbol for elastance, 293
Saturation, magnetic, 352, 354
Sawtooth wave, 441, 734, 746
Scalar quantities, 433, 467
Schering bridge, 611
Scientific notation, 7
Secondary, transformer, 384, 640
Secondary cell, 61
Selectivity, resonant circuit, 623, 632
Self-inductance:
    definition, 389
    factors governing, 391
    graphic symbol, 4
Self-induction, 388
Semiconductors, 52, 89, 208
Sensitivity:
    resonant circuit, 623
    voltmeter, 251, 252
Sequence, phase, 719, 725
Series capacitors, 291
Series circuit:
    electric, 110, 116, 486, 494, 495, 545
    magnetic, 371
Series-dropping resistor, 144
Series impedances, 545
Series inductors, 392, 680

Series-parallel circuits, 135–52, 555
Series resistors, 110, 144
Series resonance, 617
Shells, electron, 23
Shielding, magnetic, 360
Short-circuit admittance parameters, 669
Short-circuit current, 161, 163
Short-circuit transformer test, 648
Shunt:
    ammeter, 151, 247
    Ayrton, 248
Siemens, definition, 126, 501
Sine wave:
    addition, 462, 477, 735
    amplitude, 453, 741
    average value, 262, 442, 762
    cycle, 429, 433, 442
    definition, 429, 432, 732
    effective value, (see rms value)
    half-wave rectified, 441, 739
    instantaneous value, 433, 438
    peak-to-peak value, 442
    peak value, 432
    period, 429, 442, 454
    rms value, 443, 761
Sink, 3
SI unit prefixes, 9
SI units, 5
Skin effect, 535
Slide-wire bridge, 268
Smoothing capacitor, 325
Smoothing choke, 420
Solar battery, 66
Solenoid, 341
Source:
    constant-current, 162, 215
    constant-voltage, 160
    controlled, 120, 223, 663, 687
    dependent, 223
    energy, 3
    independent, 223
    three-phase, 693, 703, 708
    voltage, 34, 38, 58, 64, 118, 160, 173,
        426
Source conversion, 164, 562
Specific resistance, 74
Square wave, 309, 441, 744
Standard cell, 255
Star-connected system, 700, 731
Static electricity, 274
Steady-state values, 299, 319, 401, 418
Storage batteries, 61
Substitution theorem, 234
Subtraction, phasor, 479, 707
Superconductivity, 77
Superposition theorem, 194, 582, 747
Susceptance, 500

Swamping resistor, meter, 247
Switch, electric, 3–4
Symbols, graphic, 4, 87, 224
Symmetrical waves, 735, 745
Synchronous speed, 695

T

t, symbol for time, 6, 299, 385, 397, 435
T, symbol for temperature, 76
Tables:
    American wire gauge, 766
    dielectric constants, 290
    dielectric strengths, 281
    graphic symbols, 4, 87, 224
    interrelationship of basic electrical units,
        103
    maximum power transfer, 122
    resistivity, 75
    resistor color code, 767
    series and parallel ac circuit characteris-
        tics, 505
    SI unit prefixes, 9
    SI units, 6
    temperature coefficient of resistance,
        77, 80
Tank current, 627
Temperature, 79
    Curie, 352
Temperature coefficient of resistance, 77,
    79
Temperature rise, 78
Temporary magnet, 339, 353
Terminal voltage, 118, 161, 163, 208
Tesla, definition, 347
Theorems:
    maximum power transfer, 120, 564, 756
    Millman's, 218, 610
    Norton's, 215, 590
    Pythagoras, 470, 492, 520, 707
    reciprocity, 235
    substitution, 234
    superposition, 194, 582, 747
    Thévenin's, 207, 218, 585
Thermistor, 86
Thermocouple, 65
Thévenin's theorem, 207, 218, 585
Three-phase systems, 695–727
    definition, 696
    delta, 703
    four-wire wye, 699
    power, 711, 712, 715
    three-wire wye, 708, 723
    wye-delta, 708
Three-terminal networks, 230, 598

Three-wire distribution system, 178, 695
Time constant:
    CR networks, 303, 308
    LR networks, 401, 414
Time domain, 463
T-network, 210, 236, 239, 598, 665
Toroid, 367
Torque, 244, 261
Tracing loops, 167, 574
Tractive force, 379
Transfer impedance, 663
Transformation ratio, 643, 676
Transformations:
    delta-wye, 230, 598, 726
    impedance, 645
    Thévenin, 228, 586
Transformer action, 640
Transformer equation, 641, 764
Transformers, 640–56, 681
    air-core, 676, 681
    audio, 646, 656
    auto-, 654
    efficiency, 650, 764
    equivalent circuit, 647, 656, 681
    iron-core, 641–56, 742
    loading, 651
    step-up and step-down, 643, 654
    tuned, 683
Transformer tests, 648
Transient response:
    CR networks, 318
    LR networks, 417
Transistor, 208, 237, 254, 608, 674, 740
True power, (see Real power)
Tuned circuits, 617
Tuned transformers, 683
Two-phase systems, 691
Two-terminal networks, 209, 215, 585, 590
Two-wattmeter power measurement, 715

U

Unbalanced three-phase loads, 702, 705, 717, 723
Units:
    conversion, 12
    international system, 5
    interrelationship, 103
    MKS, 5
    prefixes, 9
    SI, 5
Universal exponential graphs:
    CR circuits, 305
    LR circuits, 402

V

V, symbol for voltage drop, 36, 38
Vacuum tube, 51, 89
Valence band, 48
Valence electrons, 24, 47, 52
Var, 515
Varactor, 283, 617
Varistor, 86
Varley loop, 268
Varmeter, 534
Vectors, 277, 467
Volt, definition, 36
Voltage:
    definition, 36–37
    induced, 384, 389
    line, 708
    open-circuit, 118, 161, 208
    phase, 700, 709
    rate of change, 300, 455, 757
    reactive, 496
    resonant rise of, 620
    terminal, 118, 161, 163, 208
    working, 283
Voltage divider, 144
Voltage-divider principle, 142, 547
Voltage drop, 38, 90, 113, 396, 697
Voltage gradient, 278
Voltage regulation, 120, 145, 652
Voltage source, 34, 38, 58, 64, 118, 160, 173, 426, 563
Voltaic cell, 34, 58
Voltampere, 519
Volt-ampere characteristics, 88
Voltmeter, 250
    electronic, 254
Voltmeter loading effect, 252
Voltmeter sensitivity, 251–52

W

W, symbol for work and energy, 36
Watt, definition, 98, 515
Wattmeter, 265, 533, 712, 715
Wave:
    nonsinusoidal, 732–49
    periodic, 441, 733
    pulse, 326, 441
    sawtooth, 441, 734, 746
    sine, 429–45, 449–60, 511–18, 732–49
    square, 309, 441, 744
Waveshaping circuits, 309, 424
Wavetraps, 634
Weber, definition, 344, 641
Weston meter movement, 245
Weston standard cell, 255
Wet-cell battery, 34, 59

Wheatstone bridge, 159, 171, 212, 232, 260
Wire table, 766
Work, 20, 36, 95
Wye-connected three-phase system, 699, 708
Wye-delta three-phase system, 708
Wye-delta transformation, 231, 598, 726

X

X, symbol for reactance, 8, 452, 457, 496, 678

Y

Y, symbol for admittance, 501
y-parameters, 669

Z

Z, symbol for impedance, 488
z-parameters, 663